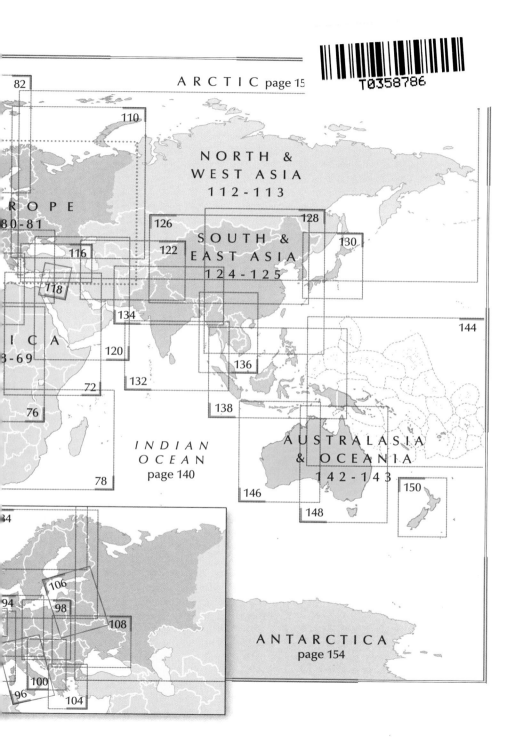

ARCTIC page 15

82

110

NORTH &
WEST ASIA
112-113

ROPE
80-81

126

128

SOUTH &
EAST ASIA
124-125

116

122

130

118

134

ICA
3-69

120

136

144

132

72

138

76

INDIAN
OCEAN
page 140

AUSTRALASIA
& OCEANIA
142-143

78

146

150

148

84

106

94

98

108

ANTARCTICA
page 154

100

96

104

ESSENTIAL WORLD ATLAS

ELEVENTH EDITION

Senior Cartographic Editor Simon Mumford
Managing Editor Gareth Jones
Managing Art Editor Lee Griffiths
Production Editor Robert Dunn
Production Controller Nancy-Jane Maun
Cover Design Development Manager Sophia MTT

FIRST EDITION

Project Cartography and Design Julia Lunn, Julie Turner, Katy Wall
Cartographers James Anderson, Roger Bullen, Martin Darlison,
Simon Mumford, John Plumer, Peter Winfield
Index-Gazetteer Natalie Clarkson, Ruth Duxbury,
Margaret Hynes, Margaret Stevenson
Art Direction Chez Picthall
Editorial Direction Andrew Heritage

This American Edition, 2024
First American Edition, 1997
Published in the United States by DK Publishing
a division of Penguin Random House LLC
1745 Broadway, 20th Floor, New York, NY 10019

A catalog record for this book is available from the Library of Congress.
ISBN 978-0-5938-4413-7

DK books are available at special discounts when purchased in bulk for
sales promotions, premiums, fund-raising, or educational use. For details, contact:
DK Publishing Special Markets, 1745 Broadway, 20th Floor, New York, NY 10019

SpecialSales@dk.com

Printed and bound in the UAE

www.dk.com

MIX
Paper | Supporting
responsible forestry
FSC™ C018179

This book was made with Forest
Stewardship Council™ certified
paper – one small step in DK's
commitment to a sustainable future.
Learn more at **www.dk.com/uk/
information/sustainability**

Key to map symbols

Physical features

Elevation

19,686ft/6000m
13,124ft/4000m
9843ft/3000m
6562ft/2000m
3281ft/1000m
1640ft/500m
820ft/250m
0
Below sea level

△ Mountain

▽ Depression

⌂ Volcano

)(Pass/tunnel

Sandy desert

Drainage features

Major perennial river

Minor perennial river

- - - Seasonal river

Canal

| Waterfall

Perennial lake

Seasonal lake

Wetland

Ice features

Permanent ice cap/ice shelf

Winter limit of pack ice

Summer limit of pack ice

Borders

Full international border

- - - - Disputed de facto border

. . . . . Territorial claim border

x x x Cease-fire line

- - - - Undefined boundary

Internal administrative boundary

Communications

Major road

Minor road

Railroad

✈ International airport

Settlements

⊡ Above 500,000

◉ 100,000 to 500,000

○ 50,000 to 100,000

○ Below 50,000

● National capital

◉ Internal administrative capital

Miscellaneous features

+ Site of interest

ᴜᴜᴜᴜ Ancient wall

Graticule features

Line of latitude/longitude/Equator

- - - Tropic/Polar circle

25° Degrees of latitude/longitude

Names

Physical features

Andes
Sahara   Landscape features
Ardennes

Land's End   Headland

Mont Blanc
4,807m   Elevation/volcano/pass

Blue Nile   River/canal/waterfall

Ross Ice Shelf   Ice feature

PACIFIC
OCEAN
Sulu Sea   Sea features
Palk Strait

Chile Rise   Undersea feature

Regions

FRANCE   Country

BERMUDA   Dependent territory
(to UK)

KANSAS   Administrative region

Dordogne   Cultural region

Settlements

PARIS   Capital city

SAN JUAN   Dependent territory capital city

Chicago
Kettering   Other settlements
Burke

Inset map symbols

Urban area

City

Park

▪ Place of interest

□ Suburb/district

Contents

The World Today

Flags of the World 6-7
The Political World 8-9
The Physical World 10-11
Standard Time Zones 12-13
Geology and Structure 14-15
World Climate 16-17
Ocean Currents 18-19
Life Zones 20-21
Population 22-23
Languages 24-25
Religion 26-27
The Global Economy 28-29
The Gobal Internet 30-31

The World's Regions

North & Central America

North & Central America 34-35
Western Canada & Alaska 36-37
Eastern Canada 38-39
USA: The Northeast 40-41
USA: The Southeast 42-43
   Bermuda
USA: Central States 44-45
USA: The West 46-47
   Los Angeles & Hawaii

USA: The Southwest 48-49
Mexico 50-51
Central America 52-53
The Caribbean 54-55
   Jamaica, St. Lucia, & Barbados

South America

South America 56-57
Northern South America 58-59
Western South America 60-61
   Galápagos Islands
Brazil 62-63
Southern South America 64-65
The Atlantic Ocean 66-67

Africa

Africa 68-69
Northwest Africa 70-71
Northeast Africa 72-73
West Africa 74-75
Central Africa 76-77
   Sao Tome & Principe
Southern Africa 78-79

Europe

Europe 80-81
The North Atlantic 82-83
Scandinavia & Finland 84-85
The Low Countries 86-87
United Kingdom & Ireland 88-89
   London

continued....

Europe *continued*

France, Andorra, & Monaco...............90-91
  Paris, Andorra, & Monaco
Spain & Portugal............................92-93
  Azores & Gibraltar
Germany & the Alpine States.............94-95
  Liechtenstein
Italy...96-97
  San Marino & Vatican City
Central Europe..............................98-99
Southeast Europe.......................100-101
  Bosnia & Herzegovina
The Mediterranean.....................102-103
  Malta & Cyprus
Bulgaria & Greece......................104-105
The Baltic States & Belarus...........106-107
Ukraine, Moldova,
  & Romania.............................108-109
European Russia........................110-111

North & West Asia

North & West Asia.......................112-113
Russia & Kazakhstan...................114-115
Turkey & the Caucasus.................116-117
Eastern Mediterranean.................118-119
  West Bank
Southwest Asia..........................120-121
Central Asia.............................122-123

South & East Asia

South & East Asia.......................124-125
Western China & Mongolia............126-127
Eastern China & Korea.................128-129
  Hong Kong
Japan.....................................130-131
  Tokyo & Nansei-Shoto
South India & Sri Lanka................132-133
North India, Pakistan,
  & Bangladesh.........................134-135
Mainland Southeast Asia..............136-137
Maritime Southeast Asia...............138-139
  Singapore
The Indian Ocean......................140-141

Australasia & Oceania

Australasia & Oceania..................142-143
The Southwest Pacific..................144-145
Western Australia.......................146-147
Eastern Australia.......................148-149
  Sydney
New Zealand............................150-151
The Pacific Ocean......................152-153
Antarctica...................................154
The Arctic Ocean..........................155

Index– Gazetteer

Overseas Territories
  & Dependencies......................156-159
Glossary of Geographical
  Terms..................................160-161
Continental Factfiles...................162-163
Geographical Comparisons...........164-166
Country Profiles........................167-179
Index-Gazetteer........................180-256

Flags of the World

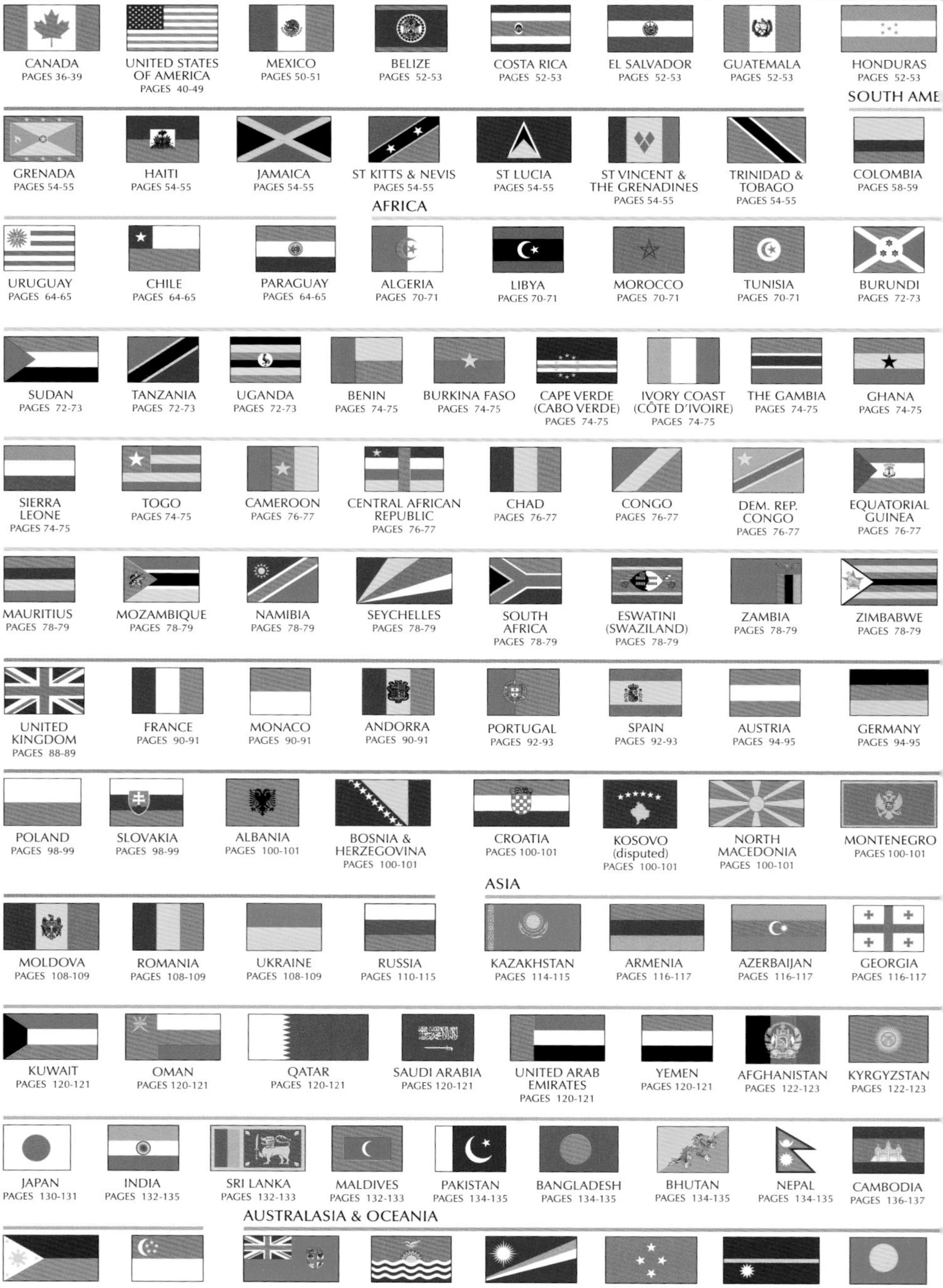

NORTH & CENTRAL AMERICA

CANADA PAGES 36-39
UNITED STATES OF AMERICA PAGES 40-49
MEXICO PAGES 50-51
BELIZE PAGES 52-53
COSTA RICA PAGES 52-53
EL SALVADOR PAGES 52-53
GUATEMALA PAGES 52-53
HONDURAS PAGES 52-53
GRENADA PAGES 54-55
HAITI PAGES 54-55
JAMAICA PAGES 54-55
ST KITTS & NEVIS PAGES 54-55
ST LUCIA PAGES 54-55
ST VINCENT & THE GRENADINES PAGES 54-55
TRINIDAD & TOBAGO PAGES 54-55

SOUTH AME

COLOMBIA PAGES 58-59
URUGUAY PAGES 64-65
CHILE PAGES 64-65
PARAGUAY PAGES 64-65

AFRICA

ALGERIA PAGES 70-71
LIBYA PAGES 70-71
MOROCCO PAGES 70-71
TUNISIA PAGES 70-71
BURUNDI PAGES 72-73
SUDAN PAGES 72-73
TANZANIA PAGES 72-73
UGANDA PAGES 72-73
BENIN PAGES 74-75
BURKINA FASO PAGES 74-75
CAPE VERDE (CABO VERDE) PAGES 74-75
IVORY COAST (CÔTE D'IVOIRE) PAGES 74-75
THE GAMBIA PAGES 74-75
GHANA PAGES 74-75
SIERRA LEONE PAGES 74-75
TOGO PAGES 74-75
CAMEROON PAGES 76-77
CENTRAL AFRICAN REPUBLIC PAGES 76-77
CHAD PAGES 76-77
CONGO PAGES 76-77
DEM. REP. CONGO PAGES 76-77
EQUATORIAL GUINEA PAGES 76-77
MAURITIUS PAGES 78-79
MOZAMBIQUE PAGES 78-79
NAMIBIA PAGES 78-79
SEYCHELLES PAGES 78-79
SOUTH AFRICA PAGES 78-79
ESWATINI (SWAZILAND) PAGES 78-79
ZAMBIA PAGES 78-79
ZIMBABWE PAGES 78-79

UNITED KINGDOM PAGES 88-89
FRANCE PAGES 90-91
MONACO PAGES 90-91
ANDORRA PAGES 90-91
PORTUGAL PAGES 92-93
SPAIN PAGES 92-93
AUSTRIA PAGES 94-95
GERMANY PAGES 94-95
POLAND PAGES 98-99
SLOVAKIA PAGES 98-99
ALBANIA PAGES 100-101
BOSNIA & HERZEGOVINA PAGES 100-101
CROATIA PAGES 100-101
KOSOVO (disputed) PAGES 100-101
NORTH MACEDONIA PAGES 100-101
MONTENEGRO PAGES 100-101
MOLDOVA PAGES 108-109
ROMANIA PAGES 108-109
UKRAINE PAGES 108-109
RUSSIA PAGES 110-115

ASIA

KAZAKHSTAN PAGES 114-115
ARMENIA PAGES 116-117
AZERBAIJAN PAGES 116-117
GEORGIA PAGES 116-117
KUWAIT PAGES 120-121
OMAN PAGES 120-121
QATAR PAGES 120-121
SAUDI ARABIA PAGES 120-121
UNITED ARAB EMIRATES PAGES 120-121
YEMEN PAGES 120-121
AFGHANISTAN PAGES 122-123
KYRGYZSTAN PAGES 122-123
JAPAN PAGES 130-131
INDIA PAGES 132-135
SRI LANKA PAGES 132-133
MALDIVES PAGES 132-133
PAKISTAN PAGES 134-135
BANGLADESH PAGES 134-135
BHUTAN PAGES 134-135
NEPAL PAGES 134-135
CAMBODIA PAGES 136-137
PHILIPPINES PAGES 138-139
SINGAPORE PAGES 138-139

AUSTRALASIA & OCEANIA

FIJI PAGES 144-145
KIRIBATI PAGES 144-145
MARSHALL ISLANDS PAGES 144-145
FED. STATES OF MICRONESIA PAGES 144-145
NAURU PAGES 144-145
PALAU PAGES 144-145

FLAGS OF THE WORLD

NICARAGUA
PAGES 52-53

PANAMA
PAGES 52-53

ANTIGUA &
BARBUDA
PAGES 54-55

THE BAHAMAS
PAGES 54-55

BARBADOS
PAGES 54-55

CUBA
PAGES 54-55

DOMINICA
PAGES 54-55

DOMINICAN
REPUBLIC
PAGES 54-55

GUYANA
PAGES 58-59

SURINAME
PAGES 58-59

VENEZUELA
PAGES 58-59

BOLIVIA
PAGES 60-61

ECUADOR
PAGES 60-61

PERU
PAGES 60-61

BRAZIL
PAGES 62-63

ARGENTINA
PAGES 64-65

DJIBOUTI
PAGES 72-73

EGYPT
PAGES 72-73

ERITREA
PAGES 72-73

ETHIOPIA
PAGES 72-73

KENYA
PAGES 72-73

RWANDA
PAGES 72-73

SOMALIA
PAGES 72-73

SOUTH
SUDAN
PAGES 72-73

GUINEA
PAGES 74-75

GUINEA–BISSAU
PAGES 74-75

LIBERIA
PAGES 74-75

MALI
PAGES 74-75

MAURITANIA
PAGES 74-75

NIGER
PAGES 74-75

NIGERIA
PAGES 74-75

SENEGAL
PAGES 74-75

GABON
PAGES 76-77

SAO TOME &
PRINCIPE
PAGES 76-77

ANGOLA
PAGES 78-79

BOTSWANA
PAGES 78-79

COMOROS
PAGES 78-79

LESOTHO
PAGES 78-79

MADAGASCAR
PAGES 78-79

MALAWI
PAGES 78-79

EUROPE

ICELAND
PAGES 82-83

DENMARK
PAGES 84-85

FINLAND
PAGES 84-85

NORWAY
PAGES 84-85

SWEDEN
PAGES 84-85

BELGIUM
PAGES 86-87

LUXEMBOURG
PAGES 86-87

NETHERLANDS
PAGES 86-87

IRELAND
PAGES 88-89

LIECHTENSTEIN
PAGES 94-95

SLOVENIA
PAGES 94-95

SWITZERLAND
PAGES 94-95

ITALY
PAGES 96-97

MALTA
PAGES 96-97

SAN MARINO
PAGES 96-97

VATICAN
CITY
PAGES 96-97

CZECHIA
(CZECH REPUBLIC)
PAGES 98-99

HUNGARY
PAGES 98-99

SERBIA
PAGES 100-101

CYPRUS
PAGES 102-103

BULGARIA
PAGES 104-105

GREECE
PAGES 104-105

BELARUS
PAGES 106-107

ESTONIA
PAGES 106-107

LATVIA
PAGES 106-107

LITHUANIA
PAGES 106-107

TURKEY
(TÜRKIYE)
PAGES 116-117

ISRAEL
PAGES 118-119

JORDAN
PAGES 118-119

LEBANON
PAGES 118-119

SYRIA
PAGES 118-119

BAHRAIN
PAGES 120-121

IRAN
PAGES 120-121

IRAQ
PAGES 120-121

TAJIKISTAN
PAGES 122-123

TURKMENISTAN
PAGES 122-123

UZBEKISTAN
PAGES 122-123

CHINA
PAGES 126-129

MONGOLIA
PAGES 126-127

NORTH KOREA
PAGES 128-129

SOUTH KOREA
PAGES 128-129

TAIWAN
PAGES 128-129

LAOS
PAGES 136-137

MYANMAR
(BURMA)
PAGES 136-137

THAILAND
PAGES 136-137

VIETNAM
PAGES 136-137

BRUNEI
PAGES 138-139

EAST TIMOR
(TIMOR-LESTE)
PAGES 138-139

INDONESIA
PAGES 138-139

MALAYSIA
PAGES 138-139

PAPUA NEW
GUINEA
PAGES 144-145

SAMOA
PAGES 144-145

SOLOMON
ISLANDS
PAGES 144-145

TONGA
PAGES 144-145

TUVALU
PAGES 144-145

VANUATU
PAGES 144-145

AUSTRALIA
PAGES 146-149

NEW ZEALAND
(AOTEAROA)
PAGES 150-151

The Political World

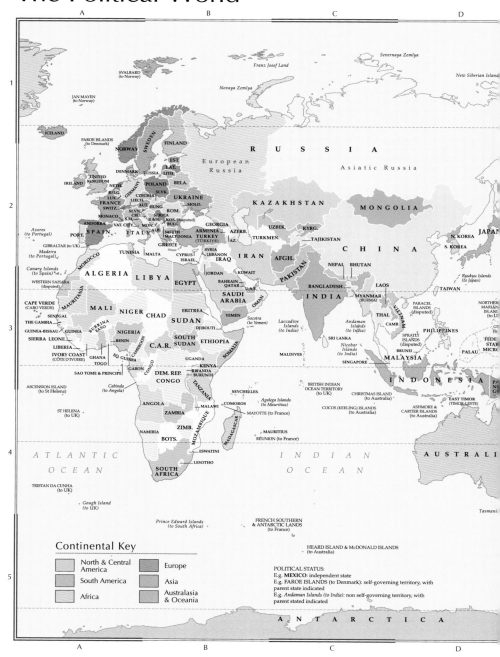

Continental Key

- North & Central America
- South America
- Africa
- Europe
- Asia
- Australasia & Oceania

POLITICAL STATUS:
E.g. **MEXICO**: independent state
E.g. FAROE ISLANDS (to Denmark): self-governing territory, with parent state indicated
E.g. *Andaman Islands (to India)*: non self-governing territory, with parent stated indicated

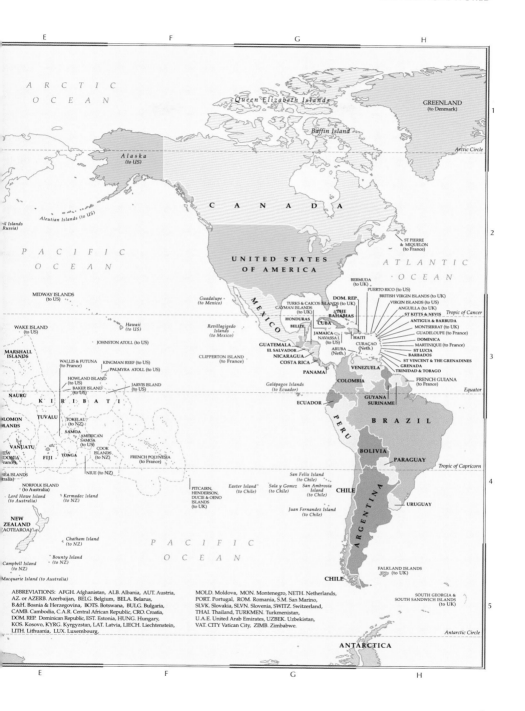

The Physical World

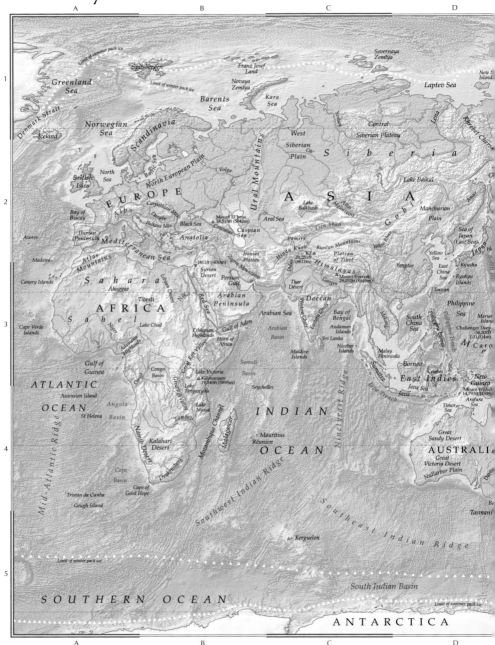

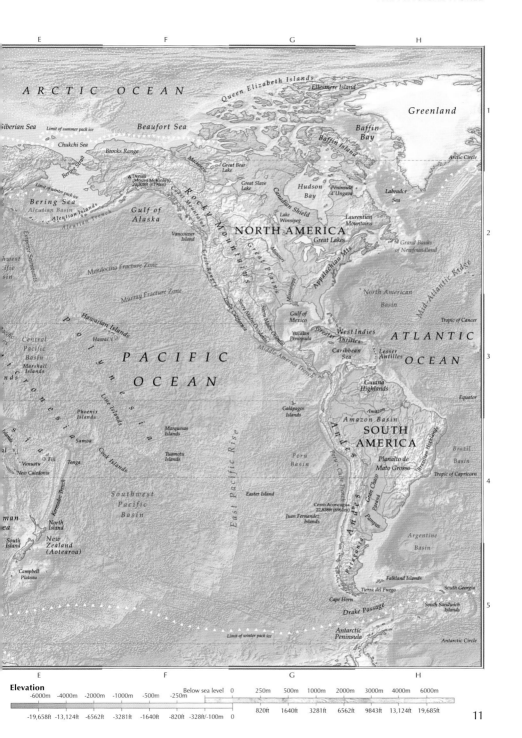

E F G H

1

ARCTIC OCEAN
Queen Elizabeth Islands
Ellesmere Island
Greenland

Siberian Sea
Limit of summer pack ice
Beaufort Sea
Baffin
Bay
Arctic Circle
Chukchi Sea
Brooks Range
Baffin Island

Mackenzie
Great Bear
Lake

△ Denali
(Mount McKinley)
20,308ft (6190m)
Great Slave
Lake
Hudson
Bay
Peninsula
d'Ungava
Labrador
Sea

Bering Sea
Limit of winter pack ice
Aleutian Basin
Aleutian Islands
Aleutian Trench
Gulf of
Alaska
Rocky Mountains
Canadian Shield
Lake
Winnipeg
Laurentian
Mountains

Emperor Seamounts
Vancouver
Island
NORTH AMERICA
Great Lakes
Grand Banks
of Newfoundland

2

west
ific
sin
Mendocino Fracture Zone
Coast Ranges
Great Plains
Missouri
Appalachian Mts
North American
Basin
Mid-Atlantic Ridge

Murray Fracture Zone
Sierra Madre Occidental
Mississippi
Tropic of Cancer

Central
Pacific
Basin
Hawaiian Islands
Hawai'i
Sierra Madre Oriental
Gulf of
Mexico
Yucatán
Peninsula
Greater
Antilles
West Indies
Lesser
Antilles
ATLANTIC

**PACIFIC
OCEAN**
Polynesia
Middle America Trench
Caribbean
Sea
OCEAN

3

nds
Marshall
Islands
Line Islands
Guiana
Highlands
Equator

Phoenix
Islands
Marquesas
Islands
Galápagos
Islands
Amazon
Amazon Basin
**SOUTH
AMERICA**
Brazil
Basin

al
Samoa
Cook Islands
Tuamotu
Islands
Perú
Basin
Planalto de
Mato Grosso
Brazilian Highlands
Tropic of Capricorn

Vanuatu
Fiji
Tonga
East Pacific Rise
Peru-Chile Trench
Andes
Gran Chaco
Paraná

4

New Caledonia
Southwest
Pacific
Basin
Easter Island
Cerro Aconcagua
22,838ft (6960m)
Juan Fernández
Islands
Pampas
Argentine
Basin

man
ea
Kermadec Trench
North
Island
Patagonia
Andes

South
Island
New
Zealand
(Aotearoa)
Falkland Islands
South Georgia

Campbell
Plateau
Tierra del Fuego
South Sandwich
Islands
Cape Horn
Drake Passage

5

Limit of winter pack ice
Antarctic
Peninsula
Antarctic Circle

E F G H

Elevation

| Below sea level | | | | | | | 0 | 250m | 500m | 1000m | 2000m | 3000m | 4000m | 6000m |

-6000m -4000m -2000m -1000m -500m -250m

-19,658ft -13,124ft -6562ft -3281ft -1640ft -820ft -328ft/-100m 0

820ft 1640ft 3281ft 6562ft 9843ft 13,124ft 19,685ft

Standard Time Zones

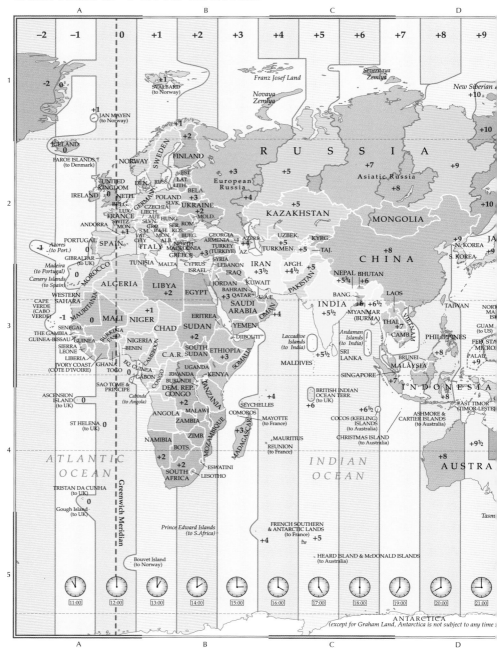

The numbers represented thus; +2/-2, indicate the number of hours each time zone is ahead or behind UCT (Coordinated Universal Time)

The clocks and 24-hour times given at the bottom of the map show time in each time zone when it is 12.00 hours noon UCT

Geology & Structure

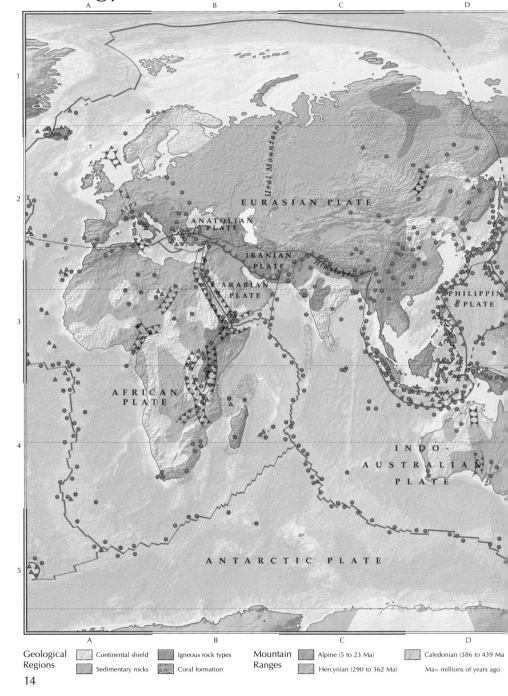

EURASIAN PLATE

ANATOLIAN PLATE

IRANIAN PLATE

ARABIAN PLATE

AFRICAN PLATE

PHILIPPINE PLATE

INDO-AUSTRALIAN PLATE

ANTARCTIC PLATE

Ural Mountains

Alps

Himalayas

Geological Regions

Continental shield

Sedimentary rocks

Igneous rock types

Coral formation

Mountain Ranges

Alpine (5 to 23 Ma)

Hercynian (290 to 362 Ma)

Caledonian (386 to 439 Ma)

Ma= millions of years ago

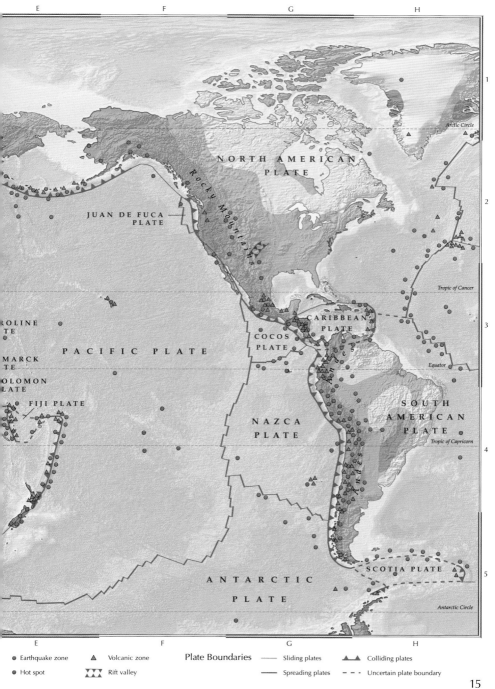

E   F   G   H

Arctic Circle

NORTH AMERICAN PLATE

Rocky Mountains

JUAN DE FUCA PLATE

Tropic of Cancer

CAROLINE PLATE

BISMARCK PLATE

SOLOMON PLATE

PACIFIC PLATE

FIJI PLATE

COCOS PLATE

CARIBBEAN PLATE

Andes

Equator

SOUTH AMERICAN PLATE

Tropic of Capricorn

NAZCA PLATE

Andes

ANTARCTIC PLATE

SCOTIA PLATE

Antarctic Circle

E   F   G   H

● Earthquake zone   ▲ Volcanic zone   **Plate Boundaries**   —— Sliding plates   ▲▲ Colliding plates

● Hot spot   ⋀⋀⋀ Rift valley   —— Spreading plates   --- Uncertain plate boundary

World Climate

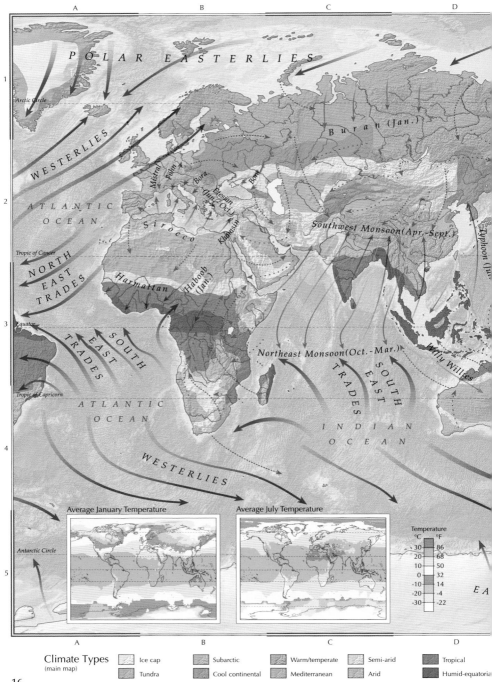

Average January Temperature

Average July Temperature

Temperature	
°C	°F
30	86
20	68
10	50
0	32
-10	14
-20	-4
-30	-22

Climate Types
(main map)

- Ice cap
- Tundra
- Subarctic
- Cool continental
- Warm/temperate
- Mediterranean
- Semi-arid
- Arid
- Tropical
- Humid-equatorial

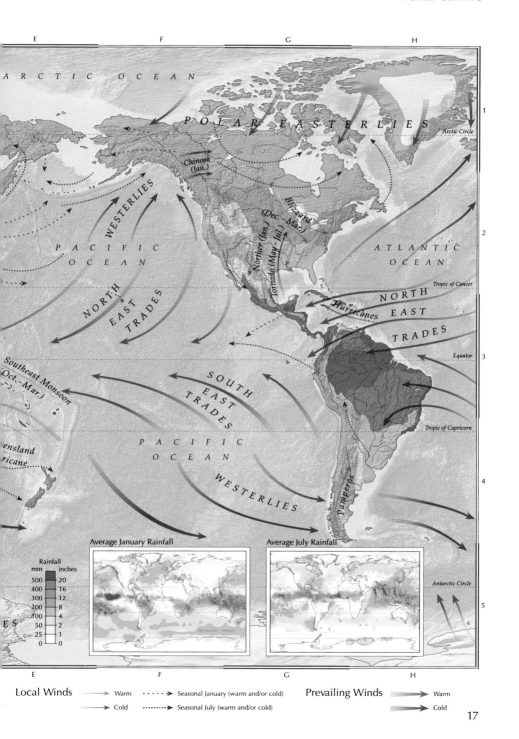

ARCTIC OCEAN

POLAR EASTERLIES

Arctic Circle

Chinook
(Jan.)

WESTERLIES

Blizzard
(Dec. - Mar.)

Norther (Jan.)

Tornado (May - Jul.)

PACIFIC
OCEAN

ATLANTIC
OCEAN

Tropic of Cancer

NORTH

EAST
TRADES

NORTH

EAST

TRADES

Hurricanes

Equator

Southeast Monsoon
(Oct. - Mar.)

SOUTH
EAST
TRADES

Pampas

Tropic of Capricorn

ensland
ricane

PACIFIC
OCEAN

WESTERLIES

Pampero

Antarctic Circle

Average January Rainfall

Average July Rainfall

Rainfall
mm | inches
500 — 20
400 — 16
300 — 12
200 — 8
100 — 4
50 — 2
25 — 1
0 — 0

ES

Local Winds

→ Warm
→ Cold
••••• → Seasonal January (warm and/or cold)
•••••• → Seasonal July (warm and/or cold)

Prevailing Winds

→ Warm
→ Cold

17

Ocean Currents

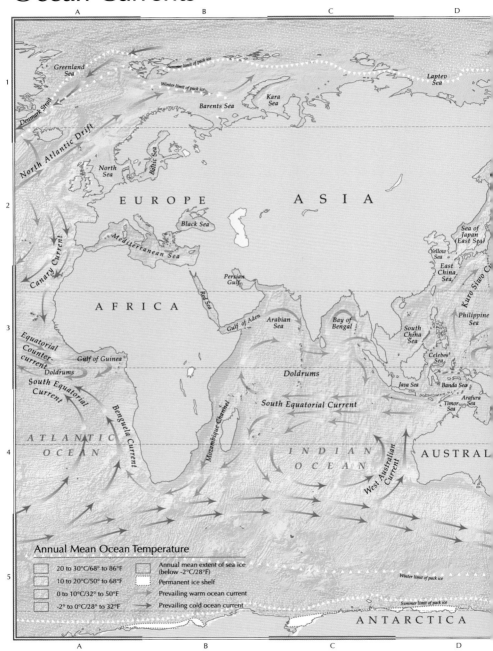

Annual Mean Ocean Temperature

- 20 to 30°C/68° to 86°F
- 10 to 20°C/50° to 68°F
- 0 to 10°C/32° to 50°F
- -2° to 0°C/28° to 32°F
- Annual mean extent of sea ice (below -2°C/28°F)
- Permanent ice shelf
- Prevailing warm ocean current
- Prevailing cold ocean current

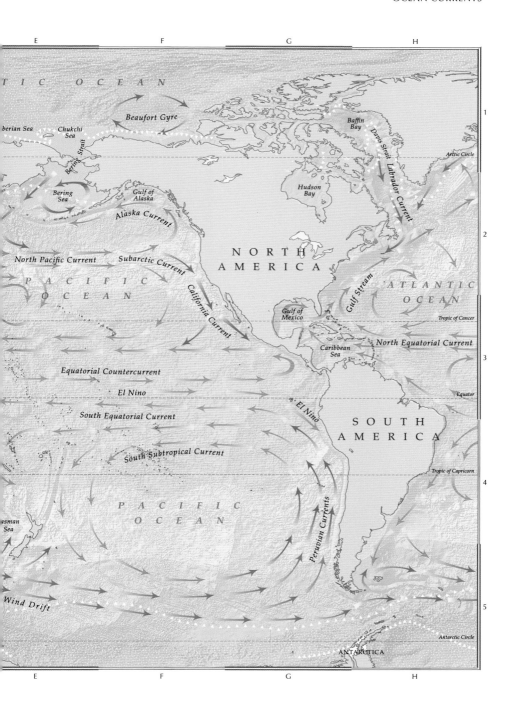

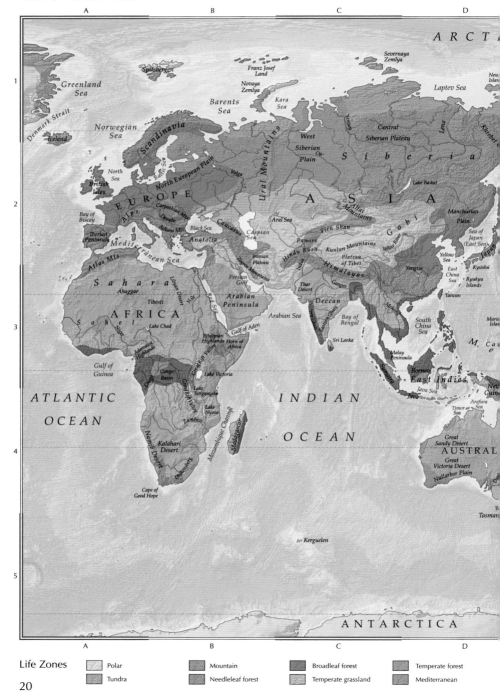

Life Zones

Polar	Mountain
Tundra	Needleleaf forest

Broadleaf forest	Temperate forest
Temperate grassland	Mediterranean

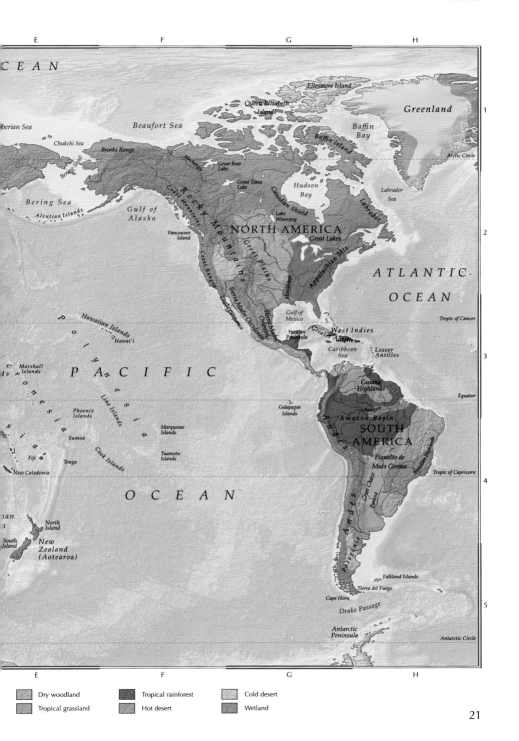

| E | F | G | H |

1

Ellesmere Island

*Queen Elizabeth
Islands*

Greenland

berian Sea

Beaufort Sea

*Baffin
Bay*

Chukchi Sea

Baffin Island

Brooks Range

Mackenzie

Arctic Circle

*Great Bear
Lake*

Bering Strait

*Great Slave
Lake*

*Hudson
Bay*

*Labrador
Sea*

Bering Sea

Aleutian Islands

*Gulf of
Alaska*

Canadian Shield

Labrador

2

*Vancouver
Island*

NORTH AMERICA

*Lake
Winnipeg*

Great Lakes

Coast Mountains

Rocky Mountains

Great Plains

Appalachian Mts

ATLANTIC-

OCEAN

Coast Ranges

*Sierra Madre
Occidental*

Baja California

Mississippi

*Gulf of
Mexico*

Tropic of Cancer

Hawaiian Islands

Hawai'i

*Yucatán
Peninsula*

*Sierra Madre
Oriental*

Greater Antilles

West Indies

*Lesser
Antilles*

3

*Marshall
Islands*

ds

PACIFIC

*Caribbean
Sea*

*Phoenix
Islands*

Line Islands

*Galápagos
Islands*

*Guiana
Highlands*

Equator

Amazon

Samoa

*Marquesas
Islands*

Amazon Basin

SOUTH
AMERICA

Andes

Fiji

Tonga

Cook Islands

*Tuamotu
Islands*

*Planalto de
Mato Grosso*

Brazilian Highlands

New Caledonia

Tropic of Capricorn

4

OCEAN

Gran Chaco

Pampas

ian

*North
Island*

*South
Island*

New
Zealand
(Aotearoa)

Andes

Patagonia

Falkland Islands

5

Tierra del Fuego

Cape Horn

Drake Passage

*Antarctic
Peninsula*

Antarctic Circle

| E | F | G | H |

| | Dry woodland | | Tropical rainforest | | Cold desert |
| | Tropical grassland | | Hot desert | | Wetland |

Population

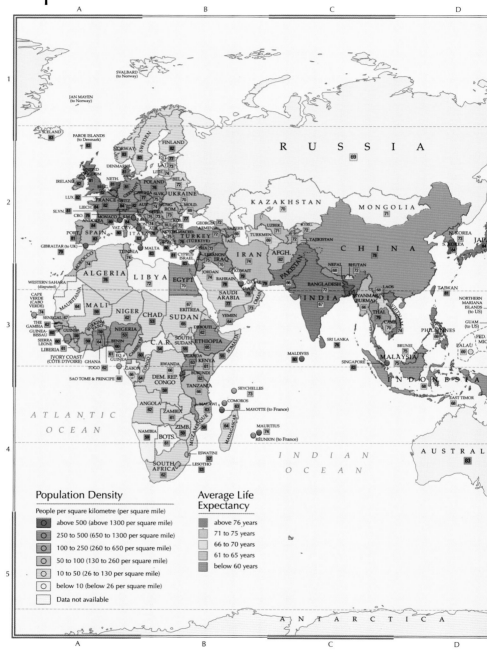

Population Density

People per square kilometre (per square mile)

- above 500 (above 1300 per square mile)
- 250 to 500 (650 to 1300 per square mile)
- 100 to 250 (260 to 650 per square mile)
- 50 to 100 (130 to 260 per square mile)
- 10 to 50 (26 to 130 per square mile)
- below 10 (below 26 per square mile)
- Data not available

Average Life Expectancy

- above 76 years
- 71 to 75 years
- 66 to 70 years
- 61 to 65 years
- below 60 years

Languages

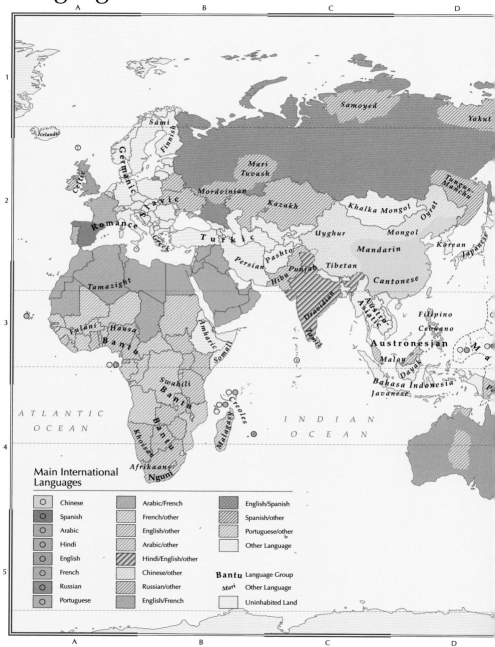

Main International Languages

○	Chinese		Arabic/French			English/Spanish
○	Spanish		French/other			Spanish/other
○	Arabic		English/other			Portuguese/other
○	Hindi		Arabic/other			Other Language
○	English		Hindi/English/other			
○	French		Chinese/other	**Bantu**	Language Group
○	Russian		Russian/other	*Mari*	Other Language
○	Portuguese		English/French		Uninhabited Land

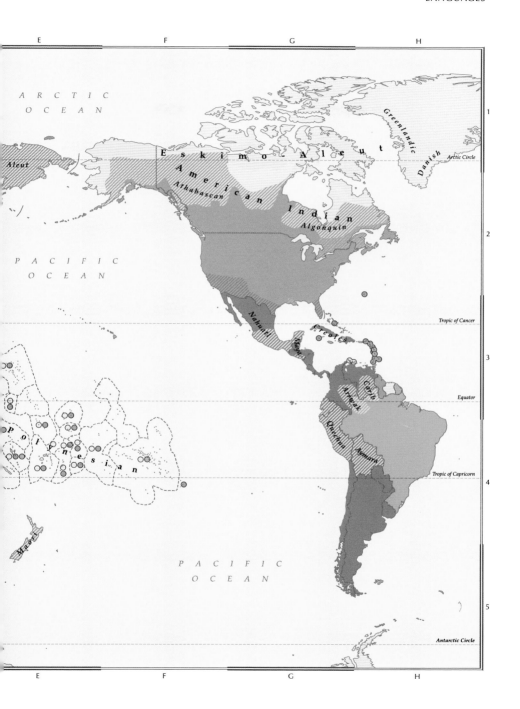

Religion

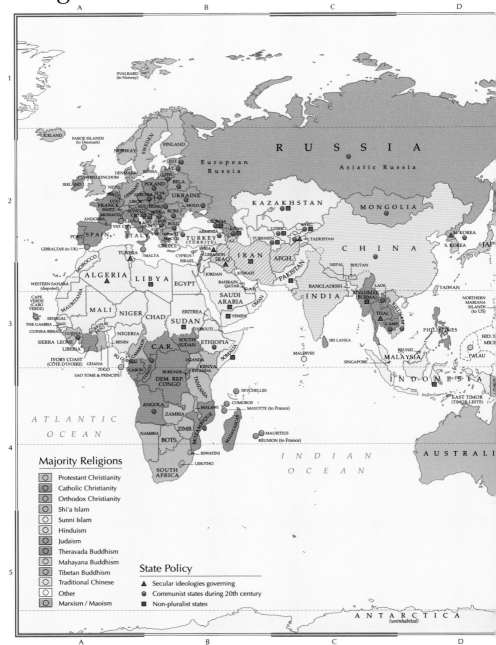

Majority Religions

- Protestant Christianity
- Catholic Christianity
- Orthodox Christianity
- Shi'a Islam
- Sunni Islam
- Hinduism
- Judaism
- Theravada Buddhism
- Mahayana Buddhism
- Tibetan Buddhism
- Traditional Chinese
- Other
- Marxism / Maoism

State Policy

- ▲ Secular ideologies governing
- ● Communist states during 20th century
- ■ Non-pluralist states

The Global Economy

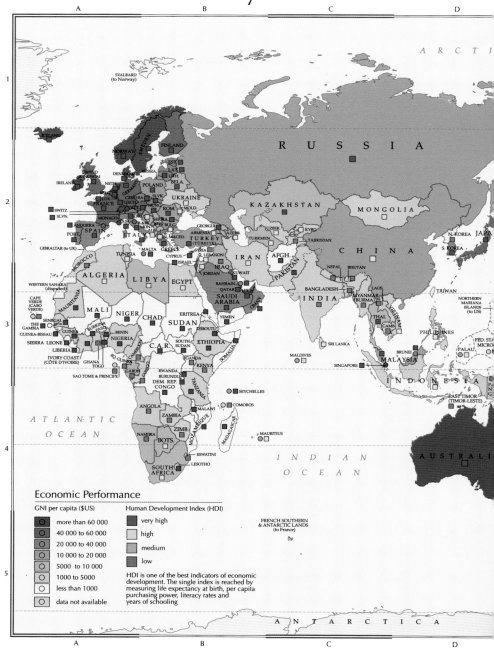

Economic Performance

GNI per capita ($US)
- more than 60 000
- 40 000 to 60 000
- 20 000 to 40 000
- 10 000 to 20 000
- 5000 to 10 000
- 1000 to 5000
- less than 1000
- data not available

Human Development Index (HDI)
- very high
- high
- medium
- low

HDI is one of the best indicators of economic development. The single index is reached by measuring life expectancy at birth, per capita purchasing power, literacy rates and years of schooling

The Global Internet

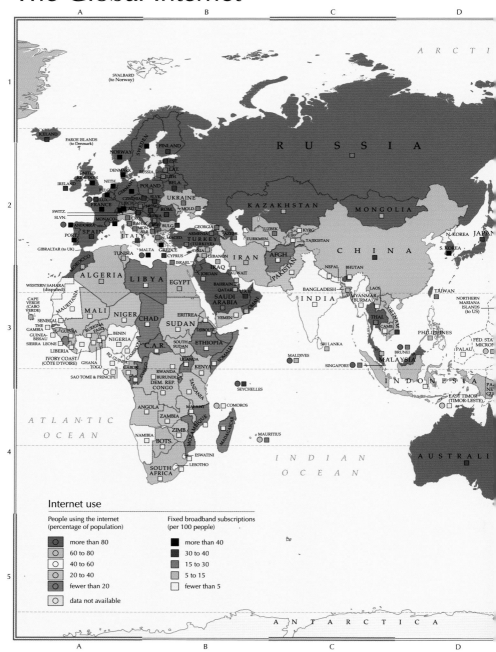

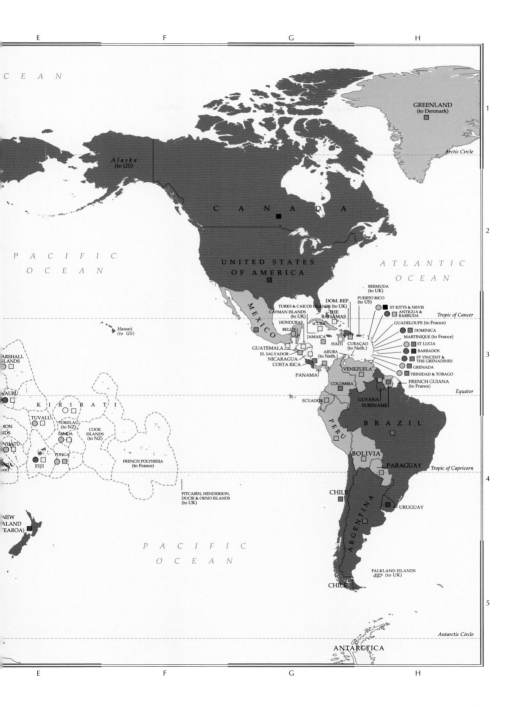

E   F   G   H

GREENLAND
(to Denmark)

Arctic Circle

1

Alaska
(to US)

C   A   N   A   D   A

2

P A C I F I C
O C E A N

UNITED STATES
OF AMERICA

A T L A N T I C
O C E A N

BERMUDA
(to UK)

Hawaii
(to US)

M E X I C O

TURKS & CAICOS ISLANDS (to UK)
CAYMAN ISLANDS
(to UK)
HONDURAS
BELIZE

PUERTO RICO
(to US)

THE
BAHAMAS
CUBA

JAMAICA

ST KITTS & NEVIS
ANTIGUA &
BARBUDA

Tropic of Cancer

GUADELOUPE (to France)
DOMINICA

GUATEMALA
EL SALVADOR
NICARAGUA
COSTA RICA

PANAMA

HAITI
DOM. REP.
CURAÇAO
(to Neth.)
ARUBA
(to Neth.)

VENEZUELA

COLOMBIA

MARTINIQUE (to France)
ST LUCIA
BARBADOS
ST VINCENT &
THE GRENADINES
GRENADA
TRINIDAD & TOBAGO
FRENCH GUIANA
(to France)

3

MARSHALL
ISLANDS

NAURU

K I R I B A T I

TUVALU

TOKELAU
(to NZ)
SAMOA

MON
DS

NUATU

TONGA

FIJI

COOK
ISLANDS
(to NZ)

FRENCH POLYNESIA
(to France)

ECUADOR

P E R U

GUYANA
SURINAME

B R A Z I L

BOLIVIA

PARAGUAY

Equator

Tropic of Capricorn

4

NEW
ALAND
EAROA)

PITCAIRN, HENDERSON,
DUCIE & OENO ISLANDS
(to UK)

CHILE

A R G E N T I N A

URUGUAY

P A C I F I C
O C E A N

FALKLAND ISLANDS
(to UK)

CHILE

5

Antarctic Circle

ANTARCTICA

E   F   G   H

31

The
WORLD'S
REGIONS

North & Central America

Population

- National capital
- o below 50,000
- o 50,000 to 100,000
- ◉ 100,000 to 500,000
- ▣ above 500,000

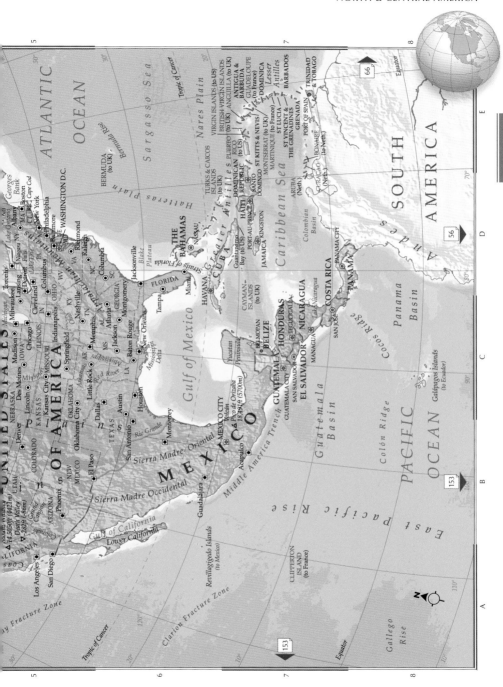

Western Canada & Alaska

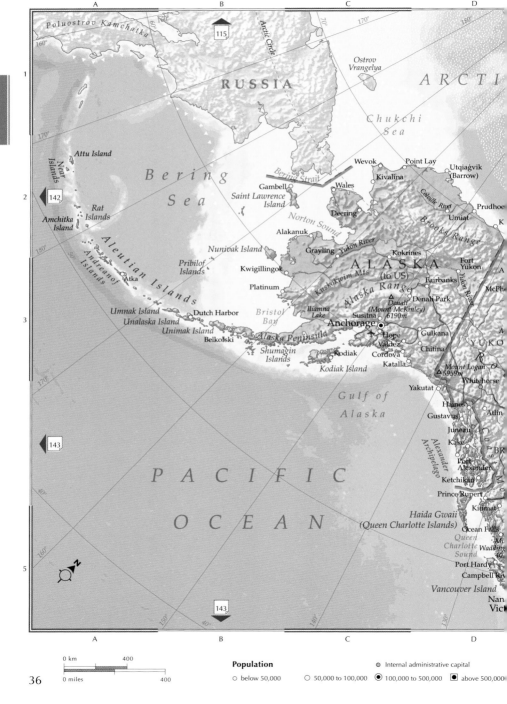

0 km 400

0 miles 400

Population

○ below 50,000   ○ 50,000 to 100,000   ◉ 100,000 to 500,000   ■ above 500,000

◉ Internal administrative capital

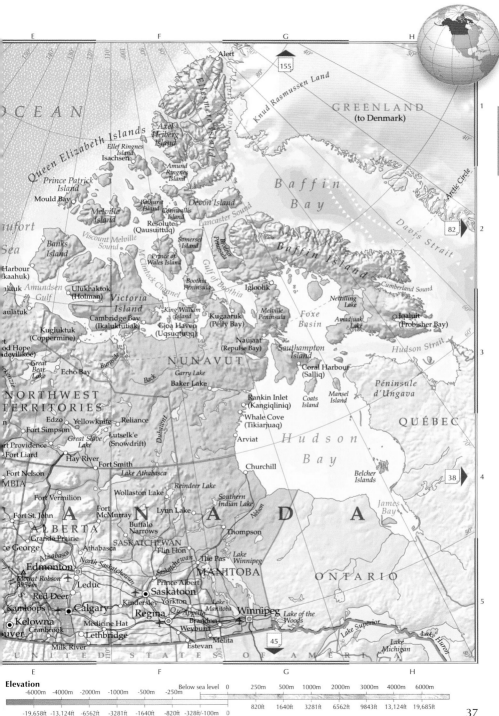

OCEAN

Queen Elizabeth Islands

Alert

155

Knud Rasmussen Land

GREENLAND
(to Denmark)

Axel
Heiberg
Island

Ellef Ringnes
Island
Isachsen

Amund
Ringnes
Island

Prince Patrick
Island

Mould Bay

*Baffin
Bay*

Melville
Island

Bathurst
Island

Cornwallis
Island

Devon Island

Resolute
(Qausuittuq)

Lancaster Sound

Davis Strait

82

Banks
Island

Viscount Melville
Sound

Prince of
Wales Island

Somerset
Island

Baffin Island

Harbour
(kaahuk)

*Amundsen
Gulf*

Ulukhaktok
(Holman)

Victoria
Island

King William
Island

Gjoa Haven
(Uqsuqtuuq)

Kugaaruk
(Pelly Bay)

Melville
Peninsula

Boothia
Peninsula

Igloolik

Cumberland Sound

aulatuk

Cambridge Bay
(Ikaluktutiak)

Gulf of Boothia

Nettilling
Lake

Foxe
Basin

Amadjuak
Lake

Iqaluit
(Frobisher Bay)

Kugluktuk
(Coppermine)

Naujaat
(Repulse Bay)

*Southampton
Island*

Hudson Strait

od Hope
deyilikóé)

Great
Bear
Lake

Echo Bay

Burnside

NUNAVUT

Garry Lake

Baker Lake

Coral Harbour
(Salliq)

*Péninsule
d'Ungava*

NORTHWEST

Back

Rankin Inlet
(Kangiqliniq)

Coats
Island

Mansel
Island

TERRITORIES

Edzo

Yellowknife

Reliance

Whale Cove
(Tikiarjuaq)

QUÉBEC

Fort Simpson

*Great Slave
Lake*

Lutselk'e
(Snowdrift)

Dubawnt

Arviat

Hudson

rt Providence

Fort Liard

Hay River

Fort Smith

Lake Athabasca

Churchill

Bay

Belcher
Islands

38

Fort Nelson

MBIA

Fort Vermilion

Wollaston Lake

Reindeer Lake

*Southern
Indian Lake*

Nelson

*James
Bay*

Fort St. John

A

Fort
McMurray

Lynn Lake

Thompson

D

A

ALBERTA

Grande Prairie

SASKATCHEWAN

Buffalo
Narrows

ONTARIO

ce George

Athabasca

Athabasca

Flin Flon

The Pas

*Lake
Winnipeg*

Edmonton

North Saskatchewan

Saskatchewan

Prince Albert

MANITOBA

Mount Robson
3954m

Leduc

Saskatoon

*Lake
Manitoba*

Red Deer

Kindersley

Yorkton

Kamloops

Calgary

Regina

Qu'Appelle

Winnipeg

*Lake of the
Woods*

Kelowna

Medicine Hat

Brandon

Lake Superior

*Lake
Huron*

uver

Cranbrook

Lethbridge

Weyburn

Milk River

Estevan

Melita

45

*Lake
Michigan*

UNITED   STATES   OF   AMERICA

Elevation

| -6000m | -4000m | -2000m | -1000m | -500m | -250m | Below sea level 0 | 250m | 500m | 1000m | 2000m | 3000m | 4000m | 6000m |

| -19,658ft | -13,124ft | -6562ft | -3281ft | -1640ft | -820ft | -328ft/-100m | 0 | 820ft | 1640ft | 3281ft | 6562ft | 9843ft | 13,124ft | 19,685ft |

Eastern Canada

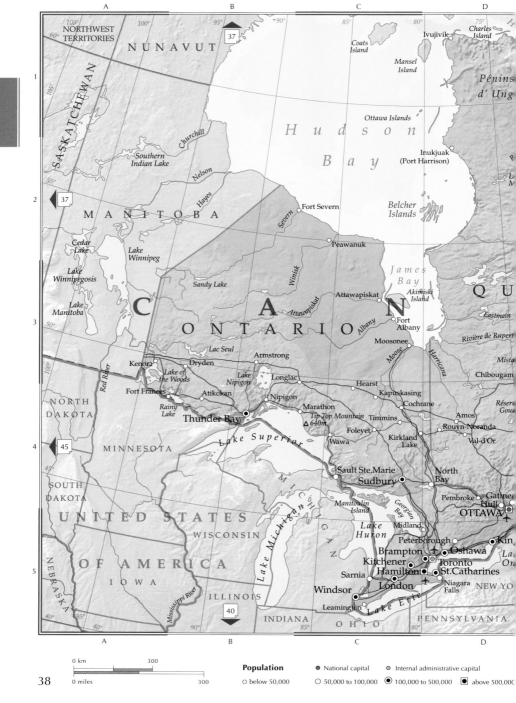

NORTHWEST TERRITORIES

NUNAVUT

SASKATCHEWAN

Churchill

Southern Indian Lake

Nelson

Hayes

MANITOBA

Cedar Lake

Lake Winnipeg

Lake Winnipegosis

Lake Manitoba

Sandy Lake

C A N

O N T A R I O

Red River

Kenora

Lac Seul

Dryden

Lake of the Woods

Fort Frances

Rainy Lake

Atikokan

Armstrong

Lake Nipigon

Longlac

Nipigon

Marathon

Tip Top Mountain
△ 640m.

Thunder Bay

Lake Superior

NORTH DAKOTA

MINNESOTA

SOUTH DAKOTA

UNITED STATES

WISCONSIN

OF AMERICA

IOWA

NEBRASKA

ILLINOIS

INDIANA

Mississippi River

Lake Michigan

MICHIGAN

Manitoulin Island

Georgian Bay

Lake Huron

Sault Ste.Marie

Sudbury

North Bay

Pembroke

Midland

Peterborough

Brampton

Kitchener

Hamilton

Sarnia

London

Windsor

Leamington

Lake Erie

Niagara Falls

St.Catharines

Toronto

Oshawa

OTTAWA

Gatineau
Hull

Kin

La

On

NEW YO

PENNSYLVANIA

OHIO

Coats Island

Mansel Island

Ivujivik

Charles Island

Pénins d' Ung

Hudson

Bay

Ottawa Islands

Inukjuak
(Port Harrison)

Fort Severn

Belcher Islands

Peawanuk

Severn

Winisk

Attawapiskat

James Bay

Attawapiskat

Akimiski Island

QU

Albany

Fort Albany

Moosonee

Moose

Harricana

Eastmain

Rivière de Rupert

Mista

Chibougam

Réser
Gou

Hearst

Kapuskasing

Cochrane

Amos

Rouyn-Noranda

Val-d'Or

Timmins

Foleyet

Kirkland Lake

Wawa

0 km   300

0 miles   300

Population

● National capital   ○ Internal administrative capital

○ below 50,000   ○ 50,000 to 100,000   ◉ 100,000 to 500,000   ◼ above 500,000

Labrador Sea

Baffin Island
Resolution Island
Button Islands
Akpatok Island
Ungava Bay
íjuaq
Rivière à la Baleine
Caniapiscau
Nain
Hopedale
Makkovik
Cape Harrison
Cartwright
Schefferville
NEWFOUNDLAND
Smallwood Reservoir
Lake Melville
Churchill
& LABRADOR
Strait of Belle Isle
St.Anthony
rvoir de iapiscau
E C
D
A
Gagnon
Réservoir anticouagan
Havre-St-Pierre
Corner Brook
Gander
Grand Falls
St.John's
Sept-Îles
Île d'Anticosti
Newfoundland
Cape Race
Baie-Comeau
Gaspé
Gulf of St. Lawrence
Chicoutimi
St.Lawrence
Péninsule de Gaspé
Channel-Port aux Basques
ST PIERRE & MIQUELON (to France)
Matane
Rimouski
Cabot Strait
Bathurst
Îles de la Madeleine
Rivière-du-Loup
Edmundston
PRINCE EDWARD ISLAND
Sydney
Glace Bay
Charlesbourg
Québec
NEW BRUNSWICK
Charlottetown
Cape Breton Island
uque
Drummondville
St-Georges
Moncton
Amherst
New Glasgow
Fredericton
Oromocto
Truro
NOVA SCOTIA
réal
MAINE
Saint John
Dartmouth
Halifax
Sherbrooke
Bay of Fundy
Liverpool
NEW HAMPSHIRE
Yarmouth
Sable Island
SACHUSETTS
Cape Cod
ATLANTIC
NNECTICUT
RHODE ISLAND
OCEAN

Elevation

-6000m	-4000m	-2000m	-1000m	-500m	-250m	Below sea level 0	250m	500m	1000m	2000m	3000m	4000m	6000m
-19,658ft	-13,124ft	-6562ft	-3281ft	-1640ft	-820ft	-328ft/-100m	0	820ft	1640ft	3281ft	6562ft	9843ft	13,124ft	19,685ft

39

USA: The Northeast

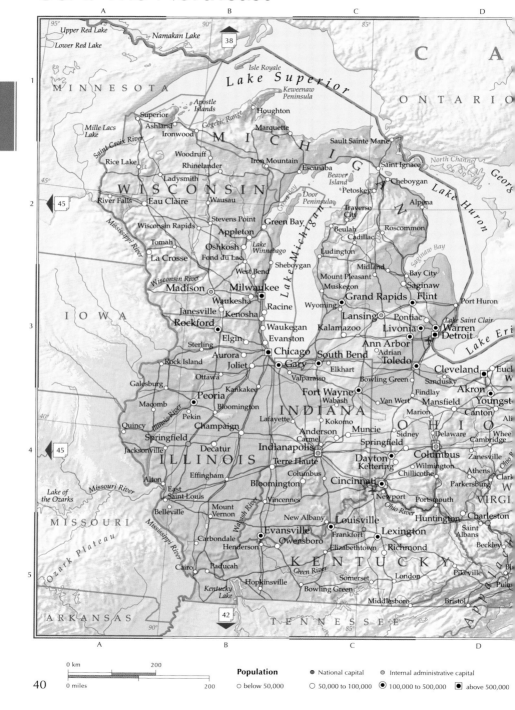

0 km 200

0 miles 200

Population

- National capital
- Internal administrative capital
- ○ below 50,000
- ○ 50,000 to 100,000
- ◉ 100,000 to 500,000
- ■ above 500,000

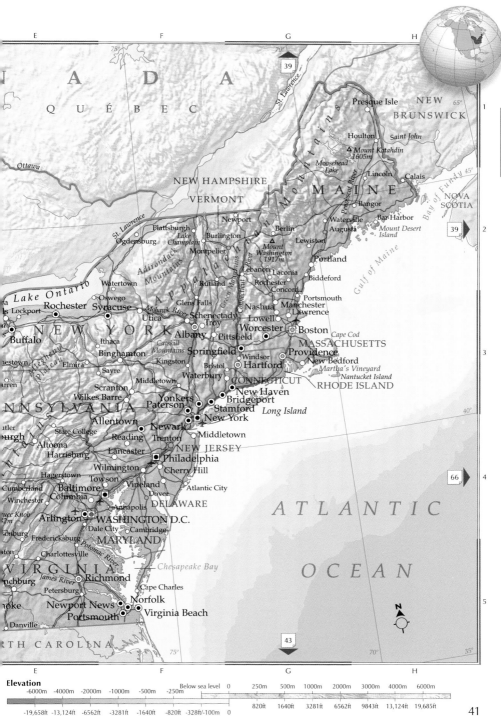

C A N A D A

Q U É B E C

Ottawa

NEW BRUNSWICK

Presque Isle

Houlton   *Saint John*

△ Mount Katahdin
   1605m

Moosehead
Lake

Lincoln   Calais

NOVA
SCOTIA

Bangor

NEW HAMPSHIRE

VERMONT

M A I N E

St. Lawrence

Plattsburgh   Newport   Berlin

Ogdensburg   Lake
              Champlain   Burlington

Montpelier

Mount
Washington
1917m

Lebanon   Laconia

Rochester   Concord

Biddeford

Waterville   Bar Harbor
Augusta   Mount Desert
           Island

Lewiston

Portland

Portsmouth

Gulf of Maine

Watertown

Lake Ontario

Oswego

Rochester   Syracuse   Utica

Lockport

Buffalo

NEW YORK

Ithaca

Binghamton

Elmira

Sayre

Glens Falls
Schenectady
Troy
Albany

Mohawk River

Catskill
Mountains

Kingston

Pittsfield

Nashua
Lowell

Worcester

Springfield

Windsor

Bristol

Hartford

Manchester
Lawrence
Boston

Cape Cod

MASSACHUSETTS

Providence
New Bedford

Martha's Vineyard
Nantucket Island

RHODE ISLAND

Scranton

Wilkes Barre

Middletown

Waterbury

CONNECTICUT

New Haven

PENNSYLVANIA

Yonkers
Paterson

Bridgeport
Stamford

New York

Long Island

Allentown

State College

Altoona

Harrisburg

Reading

Lancaster

Newark

Trenton

Middletown

NEW JERSEY

Philadelphia

Cherry Hill

Hagerstown

Cumberland

Winchester

Wilmington

Towson

Vineland

Dover

Atlantic City

Baltimore
Columbia

Annapolis

DELAWARE

Arlington   WASHINGTON D.C.

Dale City   Cambridge

Fredericksburg

MARYLAND

Charlottesville

Potomac River

V I R G I N I A

James River   Richmond

Chesapeake Bay

Petersburg

Cape Charles

Newport News   Norfolk

Portsmouth   Virginia Beach

Danville

NORTH CAROLINA

ATLANTIC

OCEAN

N

39

39

66

43

Elevation

					Below sea level	0	250m	500m	1000m	2000m	3000m	4000m	6000m

-6000m   -4000m   -2000m   -1000m   -500m   -250m

-19,658ft   -13,124ft   -6562ft   -3281ft   -1640ft   -820ft   -328ft/-100m   0   820ft   1640ft   3281ft   6562ft   9843ft   13,124ft   19,685ft

USA: The Southeast

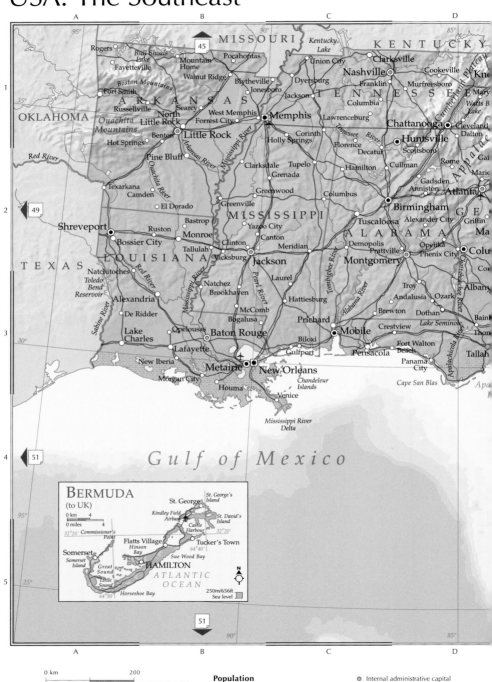

Population

○ below 50,000 ○ 50,000 to 100,000 ◉ 100,000 to 500,000 ■ above 500,000

◉ Internal administrative capital

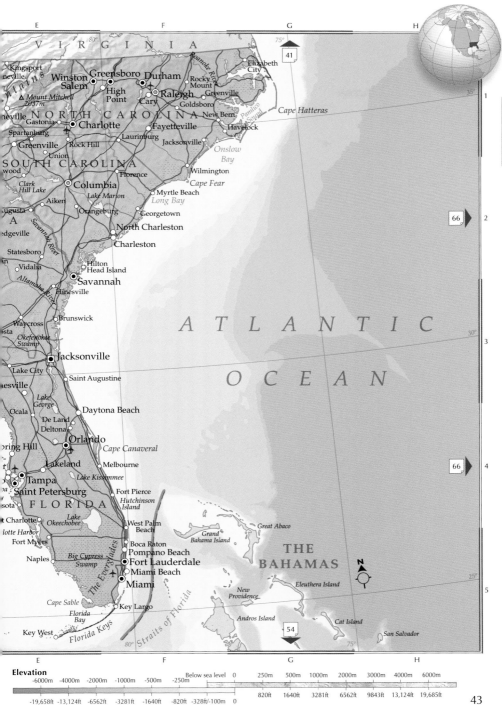

E   F   G   H

75°

41

35°

VIRGINIA
80°

Kingsport
neville
Kingsport
Winston
Salem
Greensboro
Durham
High
Point
Raleigh
Cary
Mount Mitchell
2037m
NORTH CAROLINA
Gastonia
Charlotte
neville
Spartanburg
Rock Hill
Greenville
Union
SOUTH CAROLINA
wood
Clark
Hill Lake
Columbia
Aiken
ugusta
A
Orangeburg
edgeville
North Charleston
Statesboro
n
Vidalia
Hilton
Head Island
Savannah
Hinesville
Waycross
sta
Okefenokee
Swamp
Brunswick
Jacksonville
Lake City
esville
Lake
George
Ocala
Daytona Beach
De Land
Deltona
Orlando
ring Hill
Lakeland
Melbourne
Tampa
Lake Kissimmee
Saint Petersburg
Fort Pierce
sota
FLORIDA
Hutchinson
Island
t Charlotte
Lake
Okeechobee
lotte Harbor
West Palm
Beach
Fort Myers
Big Cypress
Swamp
Boca Raton
Naples
Pompano Beach
Fort Lauderdale
Miami Beach
Miami
Cape Sable
Key Largo
Florida
Bay
Key West
Florida Keys
Straits of Florida
80°

Rocky
Mount
Greensboro
High Point
Raleigh
Cary
Goldsboro
New Bern
Fayetteville
Havelock
Laurinburg
Jacksonville
Florence
Wilmington
Myrtle Beach
Cape Fear
Long Bay
Georgetown
Charleston

Roanoke River
Elizabeth
City
Pamlico
Sound
Cape Hatteras

Onslow
Bay

Savannah River
Altamaha River

ATLANTIC

OCEAN

Cape Canaveral

The Everglades

Grand
Bahama Island
Great Abaco

THE
BAHAMAS

New
Providence
Eleuthera Island

Andros Island
Cat Island

San Salvador

75°

66

30°

66

25°

N

54

E   F   G   H

1

2

3

4

5

Elevation

-6000m	-4000m	-2000m	-1000m	-500m		Below sea level	0	250m	500m	1000m	2000m	3000m	4000m	6000m
					-250m									

-19,658ft -13,124ft -6562ft -3281ft -1640ft -820ft -328ft/-100m 0   820ft 1640ft 3281ft 6562ft 9843ft 13,124ft 19,685ft

43

USA: Central States

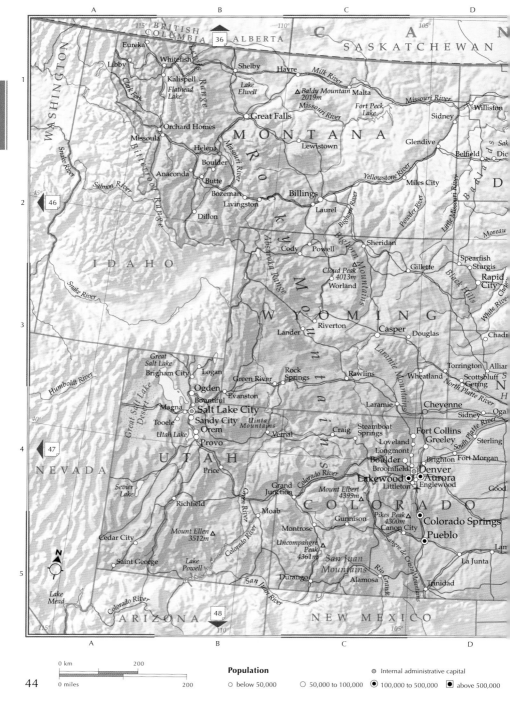

Population		
○ below 50,000	⊙ Internal administrative capital	
○ 50,000 to 100,000	◉ 100,000 to 500,000	■ above 500,000

0 km   200
0 miles   200

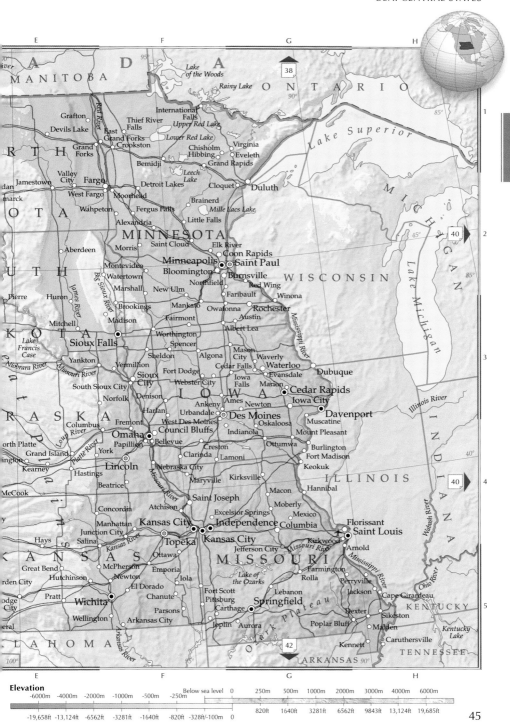

Elevation

-6000m	-4000m	-2000m	-1000m	-500m	Below sea level	0	250m	500m	1000m	2000m	3000m	4000m	6000m
-19,658ft	-13,124ft	-6562ft	-3281ft	-1640ft	-820ft -328ft/-100m	0	820ft	1640ft	3281ft	6562ft	9843ft	13,124ft	19,685ft

USA: The West

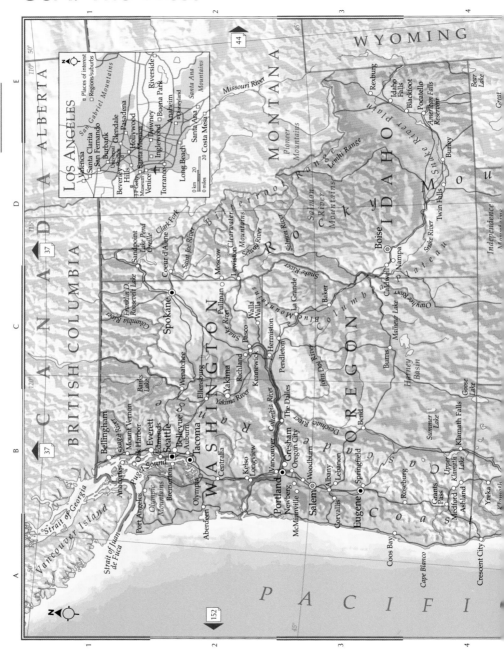

LOS ANGELES

- Places of interest
- Regions/suburbs

San Gabriel Mountains
Valencia
Santa Clarita
San Fernando
Burbank
Universal Glendale
Studios
Pasadena
Beverley Hollywood
Hills Santa Monica
Getty Venice Downey
Museum Inglewood Disneyland
Torrance Buena Park
Anaheim
Long Beach Santa Ana
Costa Mesa

Riverside
Santa Ana
Mountains

0 km 20
0 miles 20

CANADA

ALBERTA

BRITISH COLUMBIA

Vancouver Island

Strait of Georgia

Strait of Juan de Fuca

WASHINGTON

Bellingham
Anacortes
Mount Vernon
Oak Harbor
Everett
Edmonds
Port Angeles
Olympic Mountains
Bremerton
Seattle
Bellevue
Auburn
Tacoma
Olympia
Aberdeen
Centralia
Puget Sound
Wenatchee
Ellensburg
Yakima
Richland
Pasco
Kennewick
Kelso
Longview
Vancouver
Portland
Gresham
Oregon City
Newberg
McMinnville
Woodburn
Salem
Albany
Lebanon
Springfield
Corvallis
Eugene

Skagit River
Yakima River
Columbia River
Snake River
Deschutes River

OREGON

The Dalles
Bend
Roseburg
Grants Pass
Medford
Ashland
Klamath Falls
Klamath Lake
Upper Klamath Lake
Summer Lake
Goose Lake
Yreka

John Day River
Burns
Harney Basin
Malheur Lake

Coos Bay
Cape Blanco
Crescent City

WYOMING

MONTANA
Pioneer Mountains

Missouri River
Bitterroot Mountains
Clearwater Mountains
Salmon River Mountains
Lemhi Range
Clark Fork
St Joe River
Moscow
Lewiston
Pullman
Walla Walla
Hermiston
Pendleton
La Grande
Baker

Sandpoint
Lake Pend Oreille
Coeur d'Alene
Franklin D. Roosevelt Lake
Spokane
Columbia River

IDAHO
ROCKY MOUNTAINS
Salmon River
Snake River
Boise
Nampa
Caldwell
Owyhee River
Burns
Independence Mountains

Rexburg
Idaho Falls
Blackfoot
Pocatello
American Falls Reservoir
Burley
Twin Falls
Snake River Plain
Bear Lake
Great

PACIFIC

0 km 200
0 miles 200

Population
○ below 50,000
○ 50,000 to 100,000
◉ 100,000 to 500,000
■ above 500,000

⊙ Internal administrative capital

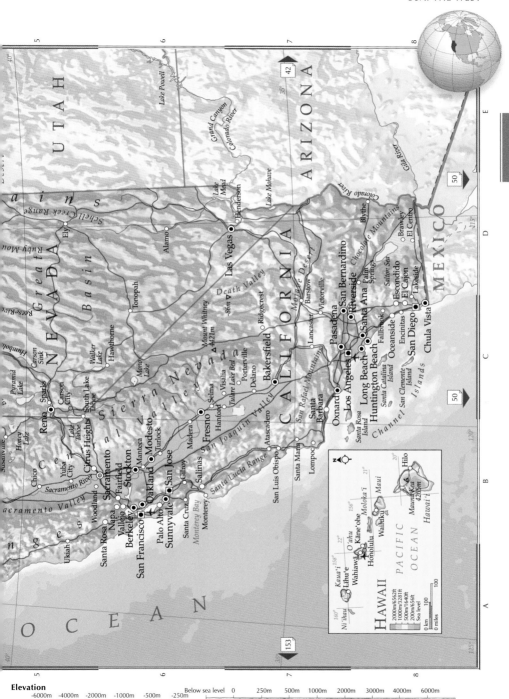

Elevation

-6000m	-4000m	-2000m	-1000m	-500m	-250m	Below sea level	0	250m	500m	1000m	2000m	3000m	4000m	6000m
-19,658ft	-13,124ft	-6562ft	-3281ft	-1640ft	-820ft	-328ft/-100m	0	820ft	1640ft	3281ft	6562ft	9843ft	13,124ft	19,685ft

USA: The Southwest

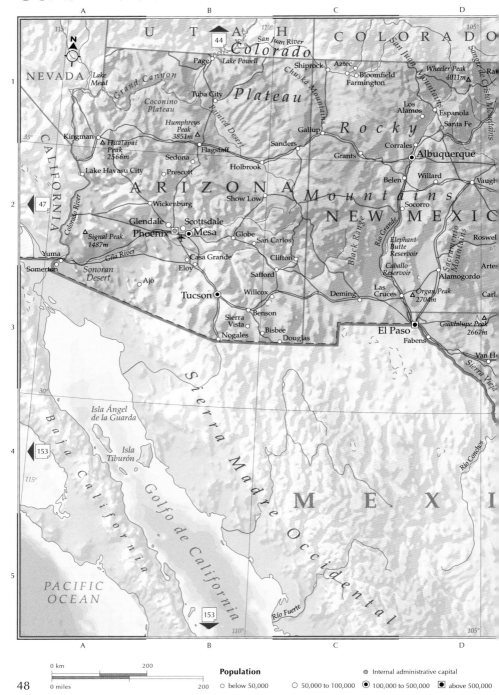

Population
- ○ below 50,000
- ○ 50,000 to 100,000
- ◉ 100,000 to 500,000
- ◉ Internal administrative capital
- ◼ above 500,000

0 km   200

0 miles   200

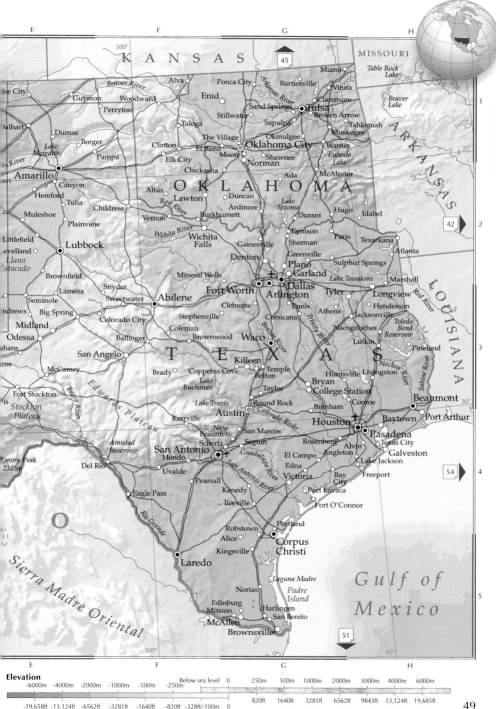

100°
95°

MISSOURI

K A N S A S

45

Beaver River
Alva
Ponca City
Bartlesville
Miami
Table Rock
Lake
Beaver
Lake

ise City
Guymon
Woodward
Enid
Sand Springs
Vinita
Claremore

Perryton
Stillwater
Sapulpa
Tulsa
Broken Arrow

alhart
Taloga
The Village
El Reno
Tahlequah

Dumas
Borger
Clinton
Oklahoma City
Moore
Warner
Muskogee

Lake
Meredith
Pampa
Elk City
Norman
Eufaula
Lake

n River
Amarillo
Chickasha
Ada
McAlester

ari
Canyon
O K L A H O M A

Hereford
Altus
Lawton
Duncan
Lake
Texoma
Hugo
Idabel

Tulia
Childress
Vernon
Burkburnett
Durant

Muleshoe
Plainview
Wichita River
Wichita
Falls
Denison
Paris
Texarkana
Atlanta

Littlefield
Lubbock
Mineral Wells
Gainesville
Sherman
Sulphur Springs
Marshall

evelland
Llano
Estacado
Denton
Greenville
Lake Tawakoni

Brownfield
Plano
Garland

Lamesa
Snyder
Fort Worth
Dallas
Tyler
Longview

Seminole
Sweetwater
Abilene
Arlington
Ennis
Henderson

ndrews
Big Spring
Colorado City
Cleburne
Corsicana
Athens
Jacksonville
Toledo
Bend
Reservoir

Midland
Stephenville
Nacogdoches

Odessa
Ballinger
Coleman
Brownwood
Waco
Lufkin
Pineland

hans
San Angelo
T E X A S
Neches River
Sabine River

cos
McCamey
Brady
Copperas Cove
Temple
Killeen
Huntsville
Livingston

Fort Stockton
Stockton
Plateau
Edwards Plateau
Lake
Buchanan
Belton
Taylor
Bryan
College Station
Conroe
Beaumont

Amistad
Reservoir
Lake Travis
Round Rock
Brenham
Houston
Baytown
Port Arthur

Emory Peak
2385ft
Del Rio
Kerrville
New
Braunfels
San Marcos
Austin
Colorado River
Pasadena
Texas City

O
Uvalde
San Antonio
Hondo
Schertz
Segun
Rosenberg
Alvin
Galveston

Eagle Pass
Pearsall
El Campo
Angleton
Lake Jackson

Kenedy
Edna
Victoria
Bay
City
Freeport

Beeville
Port Lavaca

Rio Grande
Robstown
Portland
Port O'Connor

Alice
Corpus
Christi

Sierra Madre Oriental
Kingsville
Laredo

Laguna Madre
Padre
Island
Gulf of

Norias
Mexico

Edinburg
Mission
Harlingen

McAllen
San Benito

Brownsville

51

42
54
51
45

Pecos River
Guadalupe River
San Antonio River
Brazos River
Trinity River
Red River
Arkansas River

A R K A N S A S
L O U I S I A N A

35°
30°

Elevation

| -6000m | -4000m | -2000m | -1000m | -500m | -250m | Below sea level 0 | 250m | 500m | 1000m | 2000m | 3000m | 4000m | 6000m |

| -19,658ft | -13,124ft | -6562ft | -3281ft | -1640ft | -820ft | -328ft/-100m | 0 | 820ft | 1640ft | 3281ft | 6562ft | 9843ft | 13,124ft | 19,685ft |

Mexico

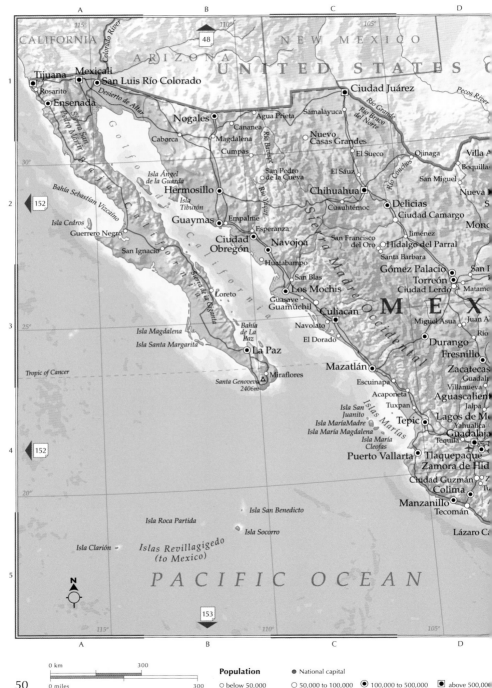

CALIFORNIA

48

NEW MEXICO

ARIZONA

UNITED STATES

Colorado River

115°

110°

105°

Tijuana
Mexicali
San Luis Río Colorado
Rosarito
Ensenada
Desierto de Altar
Ciudad Juárez
Pecos River

Nogales
Agua Prieta
Samalayuca
Río Grande
del Norte

Cananea
Caborca
Magdalena
Cumpas
Nuevo
Casas Grandes
Río Bravo
del Norte

Sierra San Pedro Mártir

Bahía Sebastián Vizcaíno

30°

El Sueco
Ojinaga
Villa A

Isla Ángel
de la Guarda
San Pedro
de la Cueva
El Sáuz
Boquillas

Hermosillo
Chihuahua
San Miguel
Nueva

Isla
Tiburón
Cuauhtémoc
Delicias
S

152

Guaymas
Empalme
Ciudad Camargo
Mon

Isla Cedros
Guerrero Negro
Esperanza
San Francisco
del Oro
Jiménez
Hidalgo del Parral

San Ignacio
Ciudad
Obregón
Navojoa
Santa Barbara

Huatabampo
Gómez Palacio
San

San Blas
Torreón
Matam

Loreto
Los Mochis
Ciudad Lerdo

Guasave
Guamúchil
Culiacán
M E X

25°
Isla Magdalena
Navolato
Miguel Asua
Juan A

Isla Santa Margarita
Bahía
de La
Paz
El Dorado
Durango
Río

La Paz
Fresnillo

Tropic of Cancer
Mazatlán
Zacatecas

Santa Genoveva
2406m
Miraflores
Escuinapa
Guadal

Acaponeta
Villanueva
Aguascalien

Isla San
Juanito
Tuxpan
Jalpa

Isla MaríaMadre
Tepic
Lagos de M

Isla María Magdalena
Xahualica

Isla María
Cleofas
Tequila
Guadalaj

152

Puerto Vallarta
Tlaquepaque
Zamora de Hid

Ciudad Guzmán
Z

20°
Colima
Tu

Manzanillo
Tecomán

Isla San Benedicto

Isla Roca Partida
Lázaro Cá

Isla Socorro

Isla Clarión
Islas Revillagigedo
(to Mexico)

PACIFIC OCEAN

153

115°
110°
105°

0 km   300

0 miles   300

Population

● National capital

○ below 50,000   ○ 50,000 to 100,000   ◉ 100,000 to 500,000   ■ above 500,00

E   F   G   H

95   90   85°

Brazos River

Red River

MISSISSIPPI

ALABAMA

42

FLORIDA

Sabine River

LOUISIANA

30°

MERICA

S

Colorado River

Mississippi River

Mississippi River
Delta

as Negras

Nuevo Laredo

Padre Island

25°

85°

66

2

oinas
dalgo
Ciudad
Miguel Alemán

Reynosa   Río
Bravo   Matamoros

Monterrey

Montemorelos

Linares

Tropic of Cancer

Laguna Madre

Gulf of

Mexico

Ciudad Victoria

Yucatan Channel

3

his

Ciudad
Mante

Rio Lagartos   Tizimín   Cancún

Progreso   Isla
Cozumel

Ciudad Madero

Mérida   Motul

Valladolid

Pánuco   Tampico

Umán

20°

erde

Ciudad Valles

Ticul   Peto

Oolores
Hidalgo

Tamazunchale

Laguna de Tamiahua

Oxkutzcab
Tekax

Felipe Carrillo
Puerto

anajuato

Tuxpán

Bahía de Campeche

Campeche   *Yucatan*

Querétaro   Poza Rica
Papantla

Champotón   *Peninsula*

Chetumal

52

4

uato Pachuca

Tulancingo

Teziutlán

*Laguna de
Términos*

DAD DE MÉXICO
(MEXICO CITY)

Perote   Xalapa

Frontera   Francisco Escárcega

Carmen

Toluca

Tlaxcala

Veracruz

Comalcalco

an   Cuernavaca

Puebla   Córdoba

Alvarado

Coatzacoalcos

Villahermosa

BELIZE

el   Zacatepec
illa   Popocatépetl
5452m

Tehuacán   San
Andrés

Minatitlán

Macuspana
Teapa

Taxco   Cuautla

Tuxtepec

Tuxtla

San Cristóbal
de Las Casas

Iguala

*Istmo de
Tehuantepec*

Tuxtla

Sierra   Chilpancingo

Huajuapan

Oaxaca

Ocozocuautla

Chiapa de
Corzo   Comitán

Madre del Sur

Ixtepec   Matías Romero

Tecpan   Tehuantepec   Juchitán   Arriaga   Presa de la
Angostura

Acapulco

Pinotepa
Nacional

Miahuatlán   Salina Cruz

Pijijiapán

15°

HONDURAS

Puerto
Escondido

Puerto
Angel

*Golfo de
Tehuantepec*

Escuintla

Huixtla

GUATEMALA

Tapachula   Ciudad Hidalgo

EL SALVADOR

100°   95°   90°

153

E   F   G   H

Elevation

-6000m   -4000m   -2000m   -1000m   -500m

Below sea level   0   250m   500m   1000m   2000m   3000m   4000m   6000m

-250m

-19,658ft -13,124ft -6562ft -3281ft -1640ft -820ft -328ft/-100m   0   820ft   1640ft   3281ft   6562ft   9843ft 13,124ft 19,685ft

51

Central America

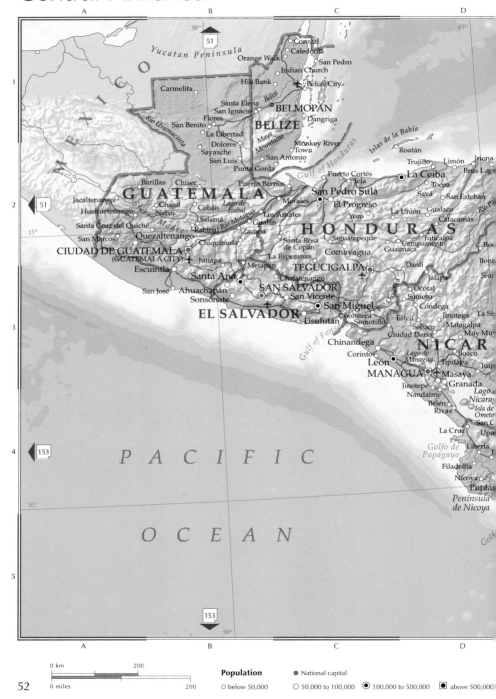

Yucatan Peninsula

51

90°

M E X I C O

Carmelita

Corozal
Caledonia
Orange Walk
San Pedro
Indian Church
Hill Bank
Belize City
Santa Elena
San Ignacio
Flores
San Benito
La Libertad
Dolores
Sayaxché
San Luis
Punta Gorda

BELMOPAN
Belize
BELIZE
Dangriga

Maya Mountains

Monkey River Town
San Antonio

Islas de la Bahía

Rio Usumacinta

Barillas Chisec
GUATEMALA
Jacaltenango
Huehuetenango
Santa Cruz del Quiché
San Marcos Quezaltenango

Puerto Barrios
Gulf of Honduras
Puerto Cortés
Tela
La Ceiba
Tocoa
Savá
San Esteban
Trujillo Limón Iriona
Brus Lag

Chajul
Nebaj Cobán Lago de Izabal
Salamá Río Motagua
Rabinal Zacapa Gualán
Chiquimula
Morales
Los Amates

San Pedro Sula
El Progreso
Yoro
La Unión Gualaco
Cátacamas

Sierra Madre

CIUDAD DE GUATEMALA
(GUATEMALA CITY)
Escuintla
Jutiapa
Metapán

HONDURAS
Santa Rosa de Copán
Comayagua
Siguatepeque
Guaimaca
Campamento
Juticalpa
Boc
Bon

San José
Ahuachapán
Sonsonate

Santa Ana
SAN SALVADOR
San Vicente

Chalatenango
La Esperanza
TEGUCIGALPA
Danlí
Jalapa
Siu

EL SALVADOR
Usulután Choluteca
Somotillo Somoto
Ocotal
Condega
Estelí Jinotega La Si
Sébaco Matagalpa
Ciudad Darío Muy Mu

San Miguel

Chinandega
Corinto
León
MANAGUA

Gulf of Fonseca

NICAR
Boaco
Lago de Managua
Tipitapa Juig
Masaya
Granada

Jinotepe
Nandaime
Belén
Rivas

Lago
Nicarag
Isla de Omete
San C
Upa

La Cruz

Golfo de
Papagayo

Liberia

Filadelfia
Nicoya
Punta
Península de Nicoya

P A C I F I C

153

O C E A N

15°

10°

153

90°

85°

Population

○ below 50,000
○ 50,000 to 100,000
◉ 100,000 to 500,000
■ above 500,000

● National capital

0 km   200

0 miles   200

N

Santanilla
(Honduras)

Bajo Nuevo
(to Colombia)

Cayo de Serranilla
(to Colombia)

de Caratasca
Puerto Lempira

Cayo de Serrana
(to Colombia)

C a r i b b e a n

am
Cayos Miskitos
Tuapi
is
Puerto Cabezas

S e a

Prinzapolka

Isla de Providencia
(to Colombia)

Barra de Río Grande

Laguna de Perlas

Isla de San Andrés
(to Colombia)

ma
Islas del Maíz

Bluefields

Punta Gorda

San Juan del Norte

Juan

COSTA RICA
ada
Siquirres
Heredia
SAN JOSÉ Limón
Cartago

Istmo de Panamá

El Porvenir

Gulf of

Portobelo
Colón
Cristóbal
Aligandí
Darien

Guabito
Almirante
Golfo de los
Mosquitos
Panama Canal
Lago Gatún
Cordillera de San Blas
Lago Bayano

Chiripó
Grande
3819m
Laguna
de Chiriquí
Balboa
San Miguelito
Chimán
Puerto Obaldía

Buenos Aires
Cortés
Volcán Barú 3475m
Capira
CIUDA DE
PANAMÁ
La Palma

Palmar Sur
Boquete
Cordillera Central
Aguadulce
Penonomé
(PANAMA CITY)
Archipiélago
de las Perlas
Isla
del Rey
Yaviza

Bahía
oronado
La Concepción
David
P A N A M A
El Real
Garachiné

ula de Osa
Santiago
Chitré

Golfo Dí
Golfo
de Chiriquí
Guarumal
Ocú
Las Tablas
Golfo
Jaqué

Isla de Coiba
Isla
Cébaco
Península de
Azuero
de   P a n a m á

COLOMBIA

Elevation

						Below sea level	0		250m	500m	1000m	2000m	3000m	4000m	6000m
-6000m	-4000m	-2000m	-1000m	-500m	-250m										
-19,658ft	-13,124ft	-6562ft	-3281ft	-1640ft	-820ft	-328ft/-100m	0		820ft	1640ft	3281ft	6562ft	9843ft	13,124ft	19,685ft

The Caribbean

UNITED STATES OF AMERICA

Gulf of Mexico

Tropic of Cancer

Florida Keys

Straits of Florida

The Everglades

Grand Bahama Island

Freeport

Marsh Harbour

Great Abaco

Bimini Islands

Berry Islands

Northeast Providence Channel

Nicholls Town

NASSAU

Eleuthera Island

New Providence

Rock Sound

Cat Island

Andros Town

Exuma Cays

Exuma Sound

San Salvador

Anguilla Cays

Andros Island

THE BAHAMAS

Cay Sal

George Town

Rum Cay

Great Exuma Island

Long Island

Clarence Town

Crooked Island

Crooked Island Passage

Acklins Island

Mayaguana Passage

Caicos Pa

May

Ragged Island Range

Archipiélago de Camagüey

Little Inagua

Lake Rosa

Matthew Town

Great

Windward Passage

Yucatan Channel

LA HABANA
(HAVANA)

Guanabacoa

Cárdenas

Artemisa

Consolación del Sur

Matanzas

Sagua la Grande

Santa Clara

Placetas

Morón

Ciego de Ávila

Pinar del Río

La Fé

Nueva Gerona

Cienfuegos

Sancti Spíritus

CUBA

Camagüey

Nuevitas

Holguín

Isla de la Juventud

Cayo Largo

Bahía de Cochinos

Archipiélago de los Canarreos

Archipiélago de los Jardines de la Reina

Las Tunas

Manzanillo

Bayamo

Guantánamo

Palma Soriano

Santiago de Cuba

Guantánamo Bay (to US)

Gonaïves

Little Cayman

Cayman Brac

GEORGE TOWN

Grand Cayman

CAYMAN ISLANDS
(to UK)

Montego Bay

Spanish Town

Portmore

KINGSTON

JAMAICA

Pedro Cays

Greater

Jamaica Channel

NAVASSA ISLAND (to US)

Île de la Gonâve

Jérémie

Cayes

PORT-AU-PRINC

Hai

H

Jac

Caribbean

HONDURAS

NICARAGUA

COSTA RICA

Caribbean Sea

JAMAICA

Montego Bay

Lucea

Falmouth

Discovery Bay

St Ann's Bay

Caribbean Sea

The Cockpit Country

Ocho Rios

Annotto Bay

Buff Bay

Port Antonio

Cambridge

Christiana

Ewarton

Savanna-La-Mar

Mandeville

Spanish Town

Blue Mountain Peak

▲2258m

Black River

May Pen

Old Harbour

KINGSTON

Portmore

Morant Bay

Portland Bight

Caribbean Sea

0 km 20
0 miles 20

2000m/6562ft
1000m/3281ft
500m/1640ft
200m/656ft
Sea level

COLOMB

0 km 200
0 miles 200

Population

○ below 50,000
○ 50,000 to 100,000
◉ 100,000 to 500,000
■ above 500,000

● National capital

ST LUCIA

N

CASTRIES

Caribbean Sea

Gros Islet

14°00′

Anse La Raye

Dennery

Soufrière

Mount Gimie 950m

Micoud

500m/1640ft
200m/656ft
Sea level

0 km   10

0 miles   10

61°00′

Vieux Fort

BARBADOS

N

ATLANTIC OCEAN

Speightstown

200m/656ft
Sea level

Mt Hillaby 340m

Bathsheba

Holetown

Welchman Hall

13°10′

BRIDGETOWN

The Crane

0 km   10

0 miles   10

59°30′

Oistins

DOMINICAN REPUBLIC

Puerto Plata

Santiago

San Francisco de Macorís

La Vega

La Romana

SANTO DOMINGO

Isla Saona

Isla Mona

Mona Passage

SAN JUAN

Caguas

Ponce

Mayagüez

PUERTO RICO
(to US)

Si Croix

A T L A N T I C   O C E A N

Leeward Islands

BRITISH VIRGIN ISLANDS
(to UK)

VIRGIN ISLANDS
(to US)

ROAD TOWN

CHARLOTTE AMALIE

ANGUILLA
(to UK)

THE VALLEY

ST MARTIN (to France)

ST BARTHÉLEMY (to France)

SINT MAARTEN
(Netherlands)

BASSETERRE

SAINT KITTS & NEVIS

Barbuda

ST JOHN'S

Antigua

ANTIGUA & BARBUDA

BRADES

MONTSERRAT
(to UK)

BASSE-TERRE

Grande Terre

Pointe-à-Pitre

Basse-Terre

Marie-Galante

GUADELOUPE
(to France)

DOMINICA

ROSEAU

Martinique Passage

MARTINIQUE
(to France)

FORT-DE-FRANCE

St Lucia Channel

ST LUCIA

CASTRIES

Vieux Fort

BARBADOS

BRIDGETOWN

Saint Vincent Passage

Saint Vincent

SAINT VINCENT & THE GRENADINES

Kingstown

The Grenadines

GRENADA

ST GEORGE'S

A n t i l l e s

L e s s e r   A n t i l l e s

S e a

L e s s e r   A n t i l l e s

W i n d w a r d   I s l a n d s

ARUBA
(Netherlands)

ORANJESTAD

CURAÇAO
(Netherlands)

BONAIRE
(to Neth.)

KRALENDIJK

WILLEMSTAD

Islas Los Roques

Isla La Orchila

Isla Blanquilla

Islas Los Testigos

Isla de Margarita

Isla La Tortuga

de Venezuela

V E N E Z U E L A

Tobago

TRINIDAD & TOBAGO

PORT OF SPAIN

Trinidad

San Fernando

Gulf of Paria

TURKS & CAICOS ISLANDS

BURN TOWN

Tropic of Cancer

66

66

66

59

Elevation

-6000m	-4000m	-2000m	-1000m	-500m	-250m	Below sea level	0	250m	500m	1000m	2000m	3000m	4000m	6000m
-19,658ft	-13,124ft	-6562ft	-3281ft	-1640ft	-820ft	-328ft/-100m	0	820ft	1640ft	3281ft	6562ft	9843ft	13,124ft	19,685ft

South America

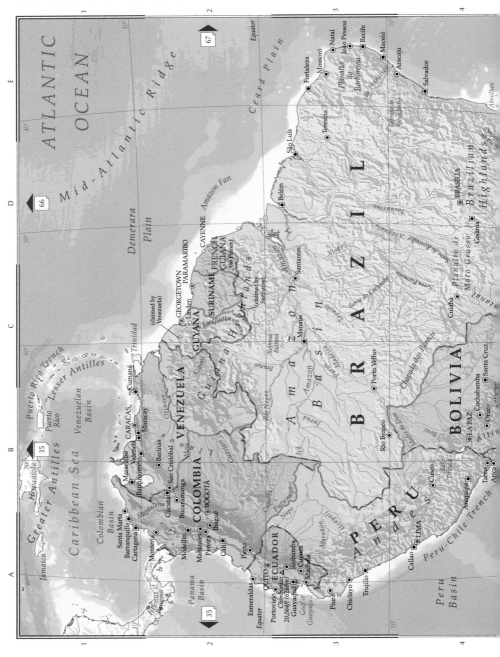

Population ● National capital

○ below 50,000   ○ 50,000 to 100,000   ◉ 100,000 to 500,000   ▣ above 500,000

0 km   500

0 miles   500

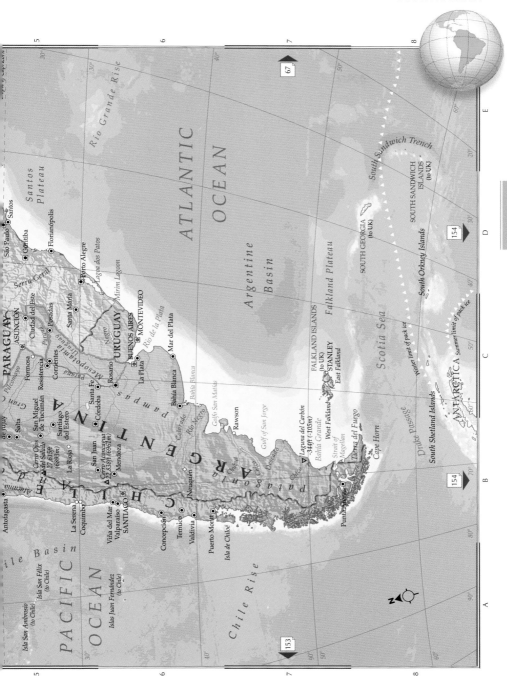

ATLANTIC OCEAN

PACIFIC OCEAN

Chile Basin

Chile Rise

Rio Grande Rise

Santos Plateau

Argentine Basin

Falkland Plateau

Scotia Sea

South Sandwich Trench

PARAGUAY

ASUNCIÓN

URUGUAY

MONTEVIDEO

BUENOS AIRES

ARGENTINA

CHILE

SANTIAGO

Patagonia

Pampas

Gran Chaco

Sierras de Córdoba

Mesopotamia

Serra Geral

São Paulo
Santos
Curitiba
Florianópolis
Pôrto Alegre
Lagoa dos Patos
Mirim Lagoon
Santa Maria
Ciudad del Este
Posadas
Formosa
Resistencia
Corrientes
Santa Fe
Rosario
Córdoba
La Plata
Mar del Plata
Bahía Blanca
Rawson
Santiago del Estero
San Miguel de Tucumán
Salta
La Rioja
San Juan
Mendoza
Neuquén
Antofagasta
La Serena
Coquimbo
Viña del Mar
Valparaíso
Concepción
Temuco
Valdivia
Puerto Montt
Isla de Chiloé
Punta Arenas

Río de la Plata

Negro

Colorado

Río Negro

Golfo San Matías

Gulf of San Jorge

Desado

Bahía Grande

Laguna del Carbón
-344ft (-105m)

Strait of Magellan

Tierra del Fuego

Cape Horn

Drake Passage

South Shetland Islands

South Orkney Islands
(to UK)

SOUTH GEORGIA
(to UK)

SOUTH SANDWICH ISLANDS
(to UK)

FALKLAND ISLANDS
(to UK)
STANLEY
East Falkland
West Falkland

ANTARCTICA

Summer limit of pack ice

Winter limit of pack ice

Isla San Ambrosio
(to Chile)
Isla San Félix
(to Chile)
Islas Juan Fernández
(to Chile)

Cerro Ojos del Salado
27,615ft
(6893m)
Cerro Aconcagua
22,838ft (6962m)

Pilcomayo

Bermejo

Paraná

Uruguay

Cerro Aconcagua

Río Bravo

Pampas

Río Negro

Z

67

153

154

154

Northern South America

Caribbean Sea

Península de la Guajira

Lesser An

ARUBA (Netherlands)
CURAÇAO (Neth.)
BONAIRE (to Neth.)

Puerto López
Punto Fijo
Coro
Puerto Camarebo
Islas Los Roques

Ríohacha
Maicao
Golfo de Venezuela

Santa Marta
Barranquilla
Ciénaga
Dabajuro
Sabaneta

Pico Cristobal Colón 5775m
Maracaibo
Puerto Cabello
CARA

Soledad
Sabanalarga
La Concepción
Cabimas
San Felipe
Maraca

Cartagena
Valledupar
Machiques
Ciudad Ojeda
Carora
Barquisimeto
Valencia
San Jua de los

El Carmen de Bolívar
Sincelejo
Magangué
San Carlos del Zulia
Lago de Maracaibo
Valera
Acarigua

Gulf of Darien
Montería
Cereté
Planeta Rica
Aguachica
Ocaña
El Vigía
Mérida
Guanare
Calabozo
Vall la Pa

PANAMA
Caucasia
Cúcuta
Pico Bolívar 5007m
Barinas
Río Guanare
San Fern

Golfo de Panamá
Dabeiba
San Cristóbal
Río Apure
L
V E N

Yarumal
Pamplona
Arauca
Río Arauca

Bello
Bucaramanga
Barrancabermeja
Río Meta
Puerto Ca

Nuquí
Itagüí
Medellín
Puerto Berrío
Sogamoso
Río Meta
Puerto Ay

PACIFIC OCEAN
Quibdó
Manizales
Tunja
Yopal
Orinoquía

Pereira
Zipaquira
Río Meta
Río Guaviare
Puerto Inírida

Armenia
BOGOTÁ
Girardot
Villavicencio

Buenaventura
Ibagué
Espinal
C O L O M B I A

Tuluá
Buga
Palmira
Neiva

Cali
Popayán
Garzón
San José del Guaviare

Tumaco
Pitalito
Río Vaupés
Mitú

Nevado de Cumbal 4764m
Pasto
Mocoa
Florencia
A m a z o n i a
Río Apaporis

Ipiales
Orito

Equator

E C U A D O R
Río Putumayo
Río Napo
Río Caquetá
Río Japurá

P E R U
Río Içá
Amazon
A

0 km 200
0 miles 200

Population ● National capital
○ below 50,000 ○ 50,000 to 100,000 ◉ 100,000 to 500,000 ■ above 500,000

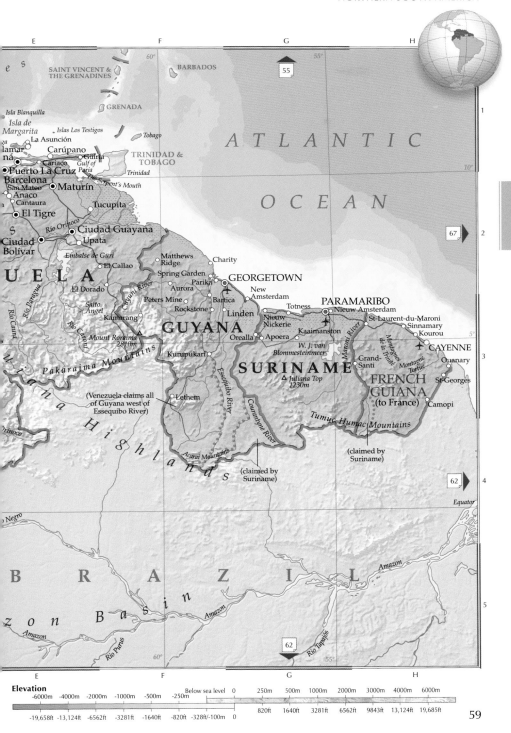

ATLANTIC

OCEAN

SAINT VINCENT & THE GRENADINES

BARBADOS

GRENADA

Isla Blanquilla
Isla de Margarita
Islas Los Testigos
Tobago
La Asunción
lamar-
ná
Carúpano
Güiria
Cariaco
Gulf of Paria
Puerto La Cruz
Trinidad
TRINIDAD & TOBAGO
Barcelona
San Mateo
The Serpent's Mouth
Anaco
Maturín
Cantaura
El Tigre
Tucupita
Río Orinoco
Ciudad Guayana
Upata
Ciudad
Bolívar
Embalse de Guri
Matthews Ridge
Charity
UELA
El Callao
Spring Garden
New Amsterdam
GEORGETOWN
El Dorado
Parika
PARAMARIBO
Río Paragua
Cuyuni River
Aurora
Peters Mine
Bartica
Nieuw Amsterdam
Totness
Kamarang
Rockstone
Nieuw-
St-Laurent-du-Maroni
Salto
Angel
Linden
Nickerie
Sinnamary
Río Caroní
Mount Roraima
2810m
GUYANA
Orealla
Apoera
Kaaimanston
Kourou
Río Caura
Pakaraima Mountains
Kurupukari
W. J. van
Blommesteinmeer
Maroni River
Grand-
Santi
CAYENNE
Montagne de la Trinité
Ouanary
SURINAME
Juliana Top
1230m
FRENCH
GUIANA
(to France)
Montagne Tortue
St-Georges
Venezuela claims all
of Guyana west of
Essequibo River)
Lethem
Essequibo River
Courantyne River
Camopi
Tumuc-Humac Mountains
Acarai Mountains
(claimed by Suriname)
Orinoco
(claimed by
Suriname)
Equator
Negro
Amazon
B R A Z I L
Amazon
z o n
B a s i n
Amazon
Río Purús
Río Tapajós
Amazon

Elevation

-6000m	-4000m	-2000m	-1000m	-500m	Below sea level	0	250m	500m	1000m	2000m	3000m	4000m	6000m
-19,658ft	-13,124ft	-6562ft	-3281ft	-1640ft	-820ft	-328ft/-100m	0	820ft	1640ft	3281ft	6562ft	9843ft	13,124ft	19,685ft

Western South America

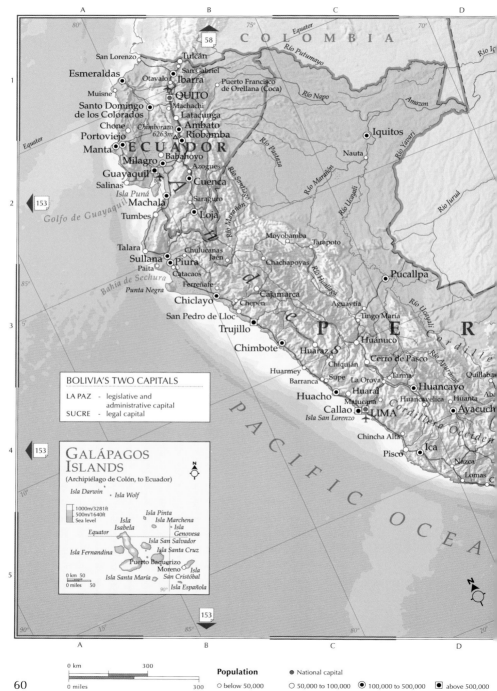

BOLIVIA'S TWO CAPITALS

LA PAZ - legislative and
administrative capital

SUCRE - legal capital

GALÁPAGOS ISLANDS
(Archipiélago de Colón, to Ecuador)

Isla Darwin · Isla Wolf

1000m/3281ft
500m/1640ft
Sea level

Isla Pinta
Isla   Isla Marchena
Isabela   · Isla
Equator   Genovesa
              Isla San Salvador
Isla Fernandina   · Isla Santa Cruz
              Puerto Baquerizo
                 Moreno ○ Isla
0 km 50   San Cristóbal
0 miles 50   Isla Santa María ○
                Isla Española

Population

○ below 50,000

○ 50,000 to 100,000

◉ 100,000 to 500,000

◼ above 500,000

● National capital

0 km 300

0 miles 300

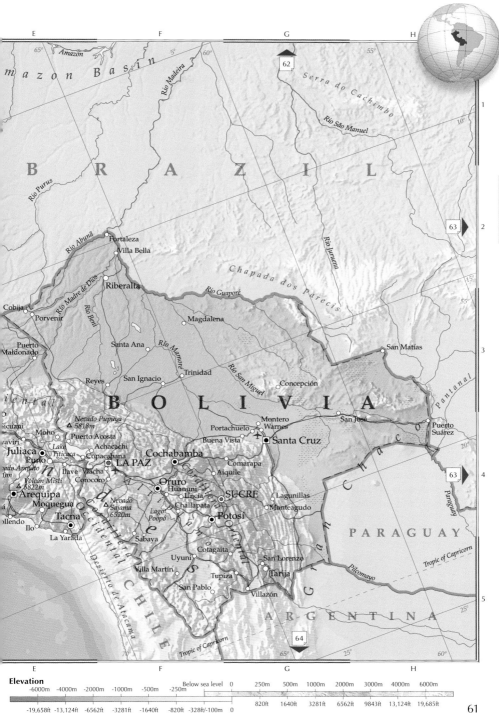

E F G H

65° *Amazon* 5° 60° 55° 10°

Amazon Basin

Serra do Cachimbo

Rio Madeira

Rio São Manuel

62

1

B R A Z I L

Rio Purus

Rio Abuñá

Fortaleza
Villa Bella

63 ▶ 2

Chapada dos Parecis

Rio Madre de Dios

Riberalta

Rio Guaporé

Rio Juruena

55°

15°

Cobija

Porvenir

Rio Beni

Magdalena

San Matías

3

Puerto
Maldonado

Santa Ana

Rio Mamoré

Trinidad

Rio San Miguel

Pantanal

Reyes

San Ignacio

Concepción

iental

B O L I V I A

Nevado Pupuya
△ 5818m

Montero
Warnes

San José

Puerto
Suárez

cuani

Moho

Puerto Acosta
Achacachi

Portachuelo
Buena Vista

Santa Cruz

G

20°

avivi

Lake
Titicaca
Copacabana

LA PAZ

Cochabamba

Comarapa

r

a

63 ▶ 4

do Apupilo

Ilave
Viacha

Aiquile

n

Juliaca

Puno

Corocoro

Oruro

C

Paraguay

Volcán Misti
△ 5822m

Nevado
Sajama
△ 6520m

Huanuni
Uncía

Lagunillas

h

Arequipa

Moquegua

Lago
Poopó

Challapata

SUCRE

Monteagudo

a

Tacna

Potosí

c

P A R A G U A Y

ollendo

Ilo

La Yarada

Sabaya

Cotagaita

C
H
I
L
E

Uyuni

San Lorenzo

o

Villa Martín

Tarija

Pilcomayo

Tropic of Capricorn

Desierto de Atacama

San Pablo

Tupiza

Villazón

25°

Occidental

70° Tropic of Capricorn 65° 25° 60°

A R G E N T I N A

64 ▼

5

E F G H

Elevation

-6000m	-4000m	-2000m	-1000m	-500m	Below sea level	-250m	0	250m	500m	1000m	2000m	3000m	4000m	6000m

| -19,658ft | -13,124ft | -6562ft | -3281ft | -1640ft | -820ft | -328ft/-100m | 0 | 820ft | 1640ft | 3281ft | 6562ft | 9843ft | 13,124ft | 19,685ft |

Brazil

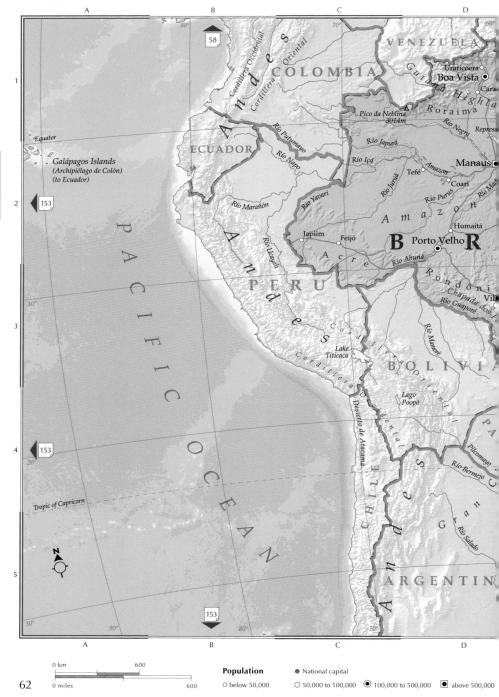

58

153

VENEZUELA

COLOMBIA

Uraricoera
Boa Vista

Cara

Guiana Highla

Roraima

Equator

ECUADOR

*Galápagos Islands
(Archipiélago de Colón)
(to Ecuador)*

Pico da Neblina
3014m

Río Negro

Represa

Río Putumayo

Río Napo

Río Içá

Río Japurá

Tefé

Amazon

Manaus

Coari

Río Juruá

Río Purus

Río Maď

Río Marañón

Río Yavarí

A m a z o n

Japiim

Feijó

Humaitá

B Porto Velho **R**

Río Ucayali

Acre

Río Abunã

Rondôni

Vil

Chapada dos

Río Guaporé

P E R U

n

d

e

C

o

r

d

i

l

l

e

r

a

Lake
Titicaca

BOLIVIA

O

r

i

e

n

t

a

l

Río Mamoré

PA

P A C I F I C

Lago
Poopó

Desierto de Atacama

Pilcomayo

Río Bermejo

C

O C E A N

Tropic of Capricorn

O

c

c

i

d

e

n

t

a

l

C H I L E

A n d e s

G

r

a

n

Río Salado

N

153

A R G E N T I N

Population

○ below 50,000   ○ 50,000 to 100,000   ◉ 100,000 to 500,000   ■ above 500,000

● National capital

0 km   600

0 miles   600

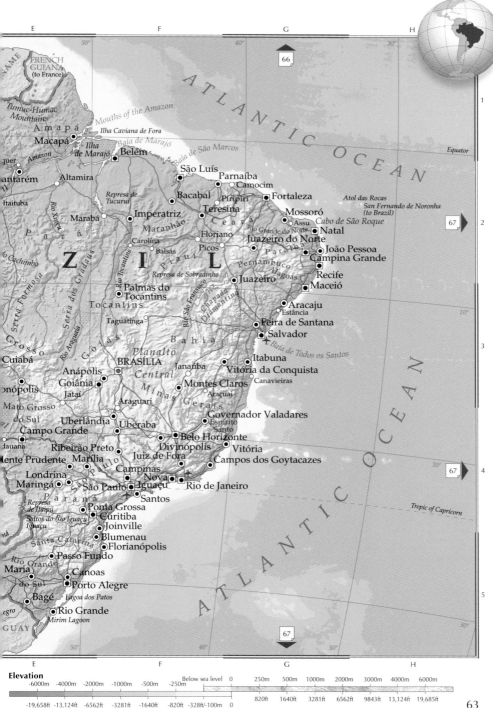

ATLANTIC OCEAN

ATLANTIC OCEAN

Elevation

-6000m	-4000m	-2000m	-1000m	-500m	-250m	Below sea level 0	250m	500m	1000m	2000m	3000m	4000m	6000m
-19,658ft	-13,124ft	-6562ft	-3281ft	-1640ft	-820ft	-328ft/-100m 0	820ft	1640ft	3281ft	6562ft	9843ft	13,124ft	19,685ft

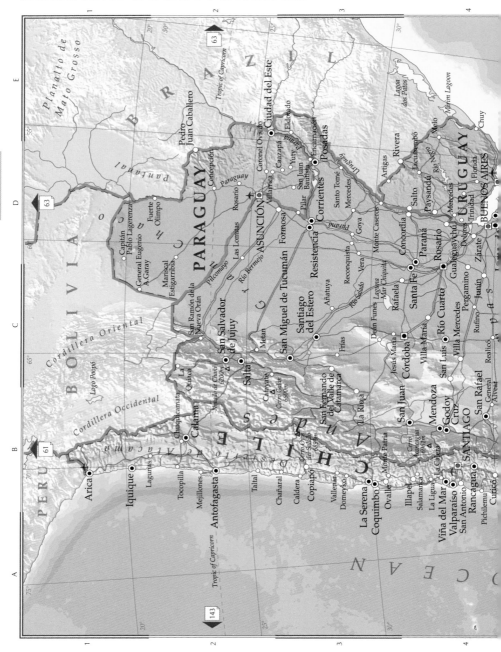

Planalto de Mato Grosso

BRAZIL

BOLIVIA

PERU

PARAGUAY

Pantanal

Tropic of Capricorn

Pedro Juan Caballero

Ciudad del Este

Coronel Oviedo

Concepción

Capitán Pablo Lagerenza

General Eugenio A.Garay

Mariscal Estigarribia

Fuerte Olimpo

San Ramón de la Nueva Orán

San Salvador de Jujuy

Chuquicamata

Calama

Nevado de Chañi 6200m

La Quiaca

Cafayate

Cerro Galán 6600m

Salta

Metán

San Miguel de Tucumán

Santiago del Estero

Frías

Añatuya

Río Salado

Vera

Reconquista

Resistencia

Formosa

Pilar

Corrientes

Goya

Monte Caseros

Santo Tomé

Mercedes

Paraná

Encarnación

Posadas

Paraguay

Villarrica

San Juan Bautista

Caazapá

Yuty

Eldorado

Rosario

Las Lomitas

Las Lomas

ASUNCIÓN

Pilcomayo

Río Bermejo

Cordillera Oriental

Cordillera Occidental

Lago Poopó

Arica

Iquique

Lagunas

Tocopilla

Mejillones

Antofagasta

Taltal

Chañaral

Caldera

Copiapó

Vallenar

Domeyko

La Serena

Coquimbo

Ovalle

Illapel

Salamanca

La Ligua

Viña del Mar

Valparaíso

San Antonio

Pichilemu

Curicó

San Rafael

Rancagua

SANTIAGO

Godoy Cruz

Mendoza

San Luis

Villa Mercedes

San Juan

Córdoba

Villa María

Jesús María

Rafaela

Doán Funes

Laguna Mar Chiquita

Santa Fe

Río Cuarto

Pergamino

Junín

Rufino

Realicó

General Alvear

Monte Patria

Cerro Aconcagua 6961m

Cerro Ojos del Salado 6893m

San Fernando del Valle de Catamarca

La Rioja

CHILE

ATACAMA

Atacama

Tropic of Capricorn

OCEAN

PERU

Paraná

Uruguay

Rivera

Artigas

Salto

Paysandú

Concordia

Gualeguaychú

Dolores

Zárate

BUENOS AIRES

Trinidad

Mercedes

URUGUAY

Melo

Jacuarembó

Río Negro

Florida

Chuy

Mirim Lagoon

Lagoa dos Patos

63

63

61

143

0 km   200

0 miles   200

Population   ● National capital

○ below 50,000   ○ 50,000 to 100,000   ● 100,000 to 500,000   ▣ above 500,000

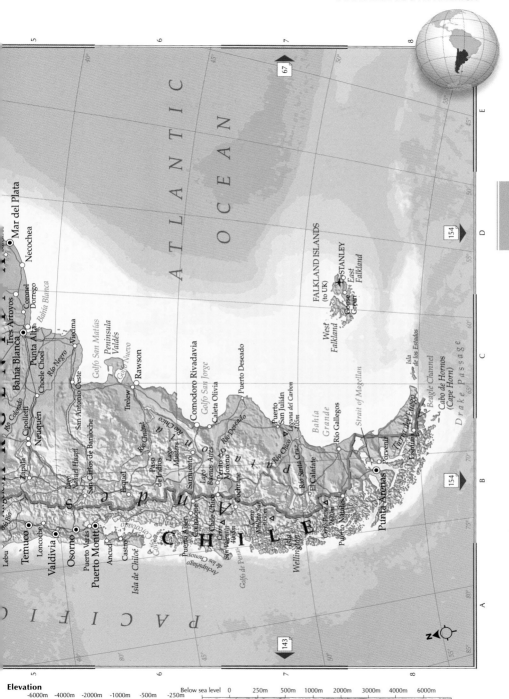

67

154

154

143

143

Elevation

-6000m	-4000m	-2000m	-1000m	-500m	-250m	Below sea level 0	250m	500m	1000m	2000m	3000m	4000m	6000m

| -19,658ft | -13,124ft | -6562ft | -3281ft | -1640ft | -820ft | -328ft/-100m | 0 | 820ft | 1640ft | 3281ft | 6562ft | 9843ft | 13,124ft | 19,685ft |

The Atlantic Ocean

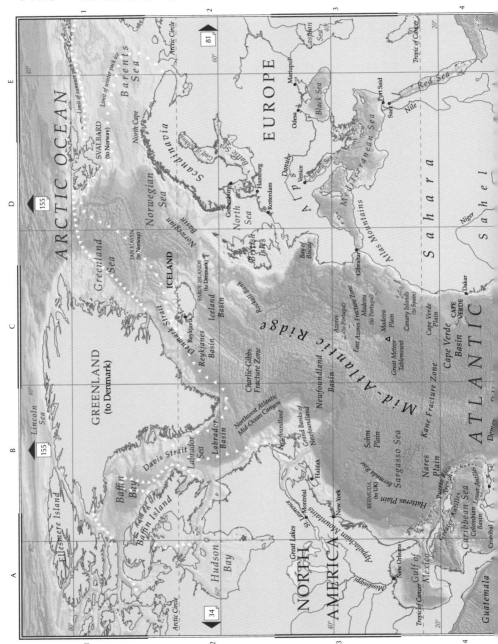

Major port

0 km 1000
0 miles 1000

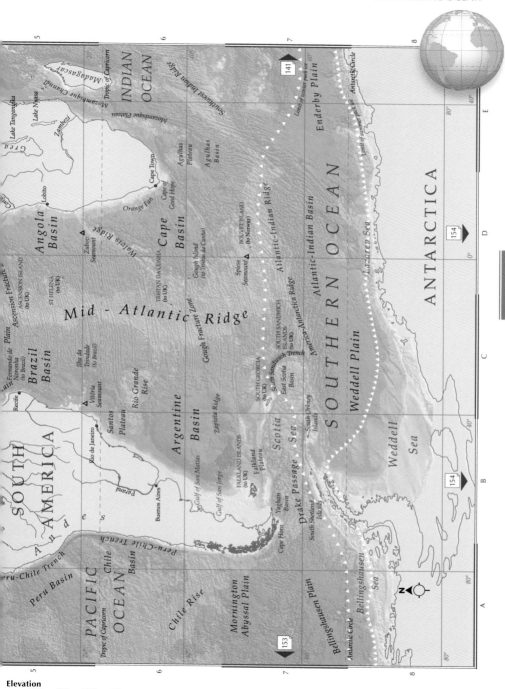

Elevation

-6000m	-4000m	-2000m	-1000m	-250m	0
-19,658ft	-13,124ft	-6562ft	-3281ft	-820ft	0

Africa

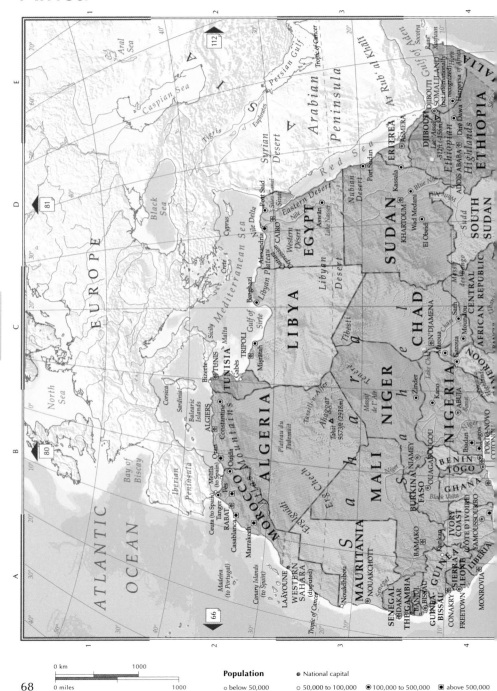

Population

- o below 50,000
- National capital
- o 50,000 to 100,000
- ⊙ 100,000 to 500,000
- ▪ above 500,000

0 km   1000

0 miles   1000

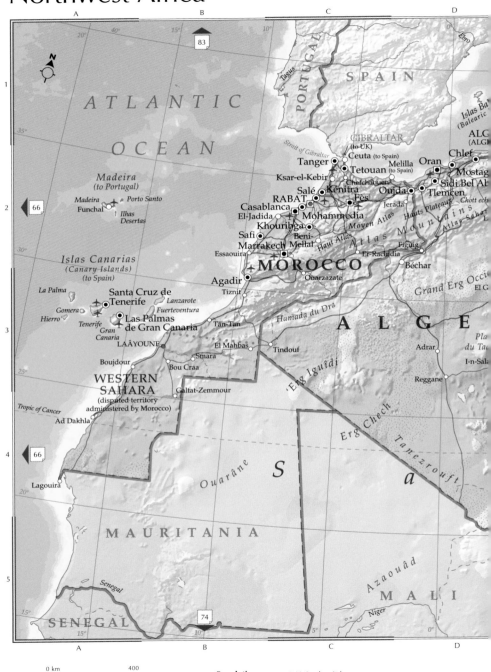

ATLANTIC

OCEAN

SPAIN

PORTUGAL

GIBRALTAR (to UK)
Ceuta (to Spain)
Tanger
Tetouan
Melilla (to Spain)
Oran
Chlef
Mostag
Ksar-el-Kebir
Chefchaouen
Sidi Bel Al
Salé
Kénitra
Oujda
Tlemcen
RABAT
Fès
Jerada
Casablanca
El-Jadida
Mohammedia
Khouribga
Beni-
Safi
Marrakech
Mellal
MOROCCO
Essaouira
Er-Rachidia
Béchar
Agadir
Ouarzazate
Tiznit
Grand Erg Occi
El G
Santa Cruz de Tenerife
Tan-Tan
Hamada du Dra
ALGE
El Mahbas
Tindouf
Adrar
Pla
du Ta
I-n-Sala
LAÂYOUNE
Smara
'Erg Iguîdi
Boujdour
Bou Craa
Reggane
WESTERN SAHARA
(disputed territory administered by Morocco)
Galtat-Zemmour
Erg Chech
Tanezrouft
Tropic of Cancer
Ad Dakhla
Lagouira
Ouarâne
S
a
MAURITANIA
Azaouâd
MALI
Senegal
Niger
SENEGAL

Madeira (to Portugal)
Madeira
Porto Santo
Funchal
Ilhas Desertas

Islas Canarias (Canary Islands) (to Spain)
La Palma
Gomera
Hierro
Tenerife
Gran Canaria
Las Palmas de Gran Canaria
Lanzarote
Fuerteventura

Islas Bal (Balearic
ALG (ALG

Strait of Gibraltar

Tagus

Ebro

Moyen Atlas
Haut Atlas
Atlas
Mountains
Hauts Plateaux
Chott ech
Atlas Sahar

83
66
66
74

0 km   400
0 miles   400

Population
● National capital
○ below 50,000
○ 50,000 to 100,000
◉ 100,000 to 500,000
▣ above 500,000

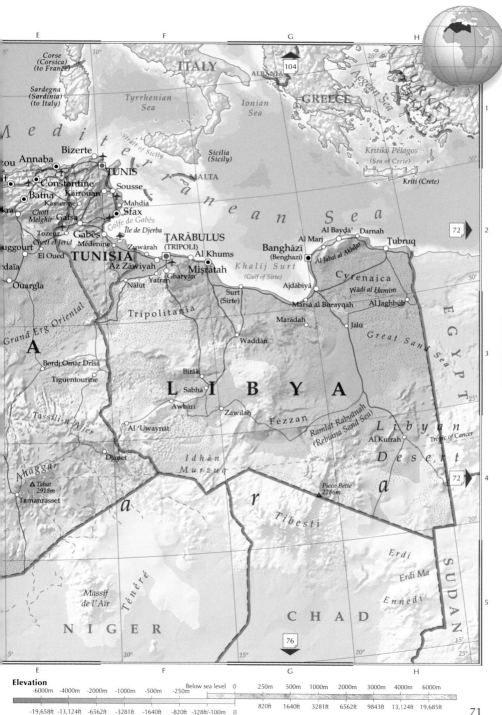

Elevation

-6000m	-4000m	-2000m	-1000m	-500m	-250m	Below sea level	0	250m	500m	1000m	2000m	3000m	4000m	6000m
-19,658ft	-13,124ft	-6562ft	-3281ft	-1640ft	-820ft	-328ft/-100m	0	820ft	1640ft	3281ft	6562ft	9843ft	13,124ft	19,685ft

Northeast Africa

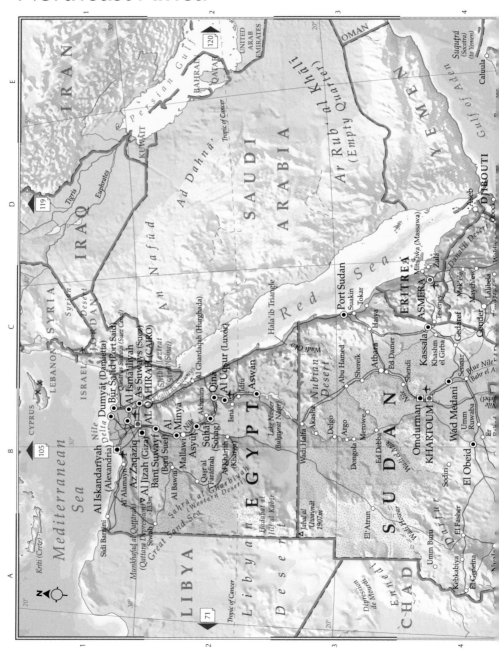

Population

● National capital

○ below 50,000 ○ 50,000 to 100,000 ◉ 100,000 to 500,000 ◼ above 500,000

0 km 400

0 miles 400

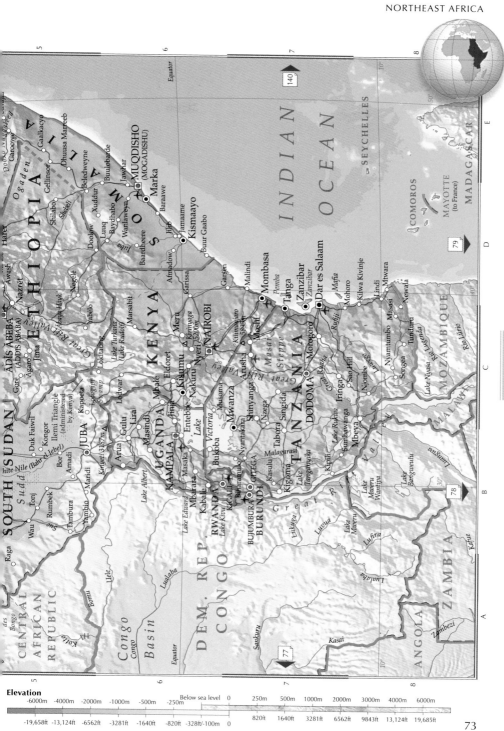

Elevation

	Below sea level									
-6000m	-4000m	-2000m	-1000m	-500m	-250m	0	250m	500m	1000m	2000m	3000m	4000m	6000m
-19,658ft	-13,124ft	-6562ft	-3281ft	-1640ft	-820ft	-328ft/-100m	0	820ft	1640ft	3281ft	6562ft	9843ft	13,124ft	19,685ft

73

West Africa

A   **B**   **C**   **D**

66

N

66

Tropic of Cancer

WESTERN SAHARA
(disputed territory
administered by Morocco)

Aïn Ben Tili

Bir Mogreïn

Kâghet

'Erg Igu

El Hamk

Fdérik   Zouérat

Touâjîl

S

Nouâdhibou

Choûm

Atâr   Chinguetti

Akjoujt   Oujeft

El Mreyyé

Ouarâne

CAPE VERDE
(CABO VERDE)

Ilhas de Barlavento

Santo Antão

Mindelo

São   Pedra Lume

Vicente   São   Sal

Nicolau   Boa Vista

Santiago

Fogo   Maio

PRAIA

Ilhas de Sotavento

NOUAKCHOTT   Idîni

M A U R I T A N I A

Tidjikja   Tichît

Boutilimit   Magta Lahjar

Boûmdeïd   Oualâta

Aoukâr

Rkîz

Rosso   Aleg   Kaédi   Kiffa   Tâmchekket   Ayoûn el 'Atroûs   Néma

Richard Toll   Dagana   Kobenni   Amourj

Saint Louis   Louga   Matam   Sélibabi   Timbedgha   Bassi

Mékhé   **SENEGAL**   Kobenni

DAKAR   Thiès   Mbaké   Nioro   S

Mbour   Diourbel

Kaolack   Kayes   Ténén

Sokone   Toukoto   Kolokani   Nio

BANJUL   **THE GAMBIA**   Tambacounda   Kita   Ségou

Bignona   Kolda   Gambia   Koulikoro

Ziguinchor   Sédhiou   Bafatá   **BAMAKO**

BISSAU   Gaoual   Koi

GUINEA-   Boké   Labé   Dinguiraye   Siguiri   Bougouni

BISSAU   Pita   Mamou   Kankan   Tengréla

Kindia   Faranah   Odienné   Ferkess

CONAKRY   Tokounou   Boundiali   Ko

Makeni   Kissidougou   Katio

SIERRA   Beyla   **IVORY**

FREETOWN   **LEONE**   **COAST**

Bo   Kenema   Nzérékoré   **CÔTE D'IVOI**

Gbanga   Danane   Dive

Tubmanburg   **YAMOUSSOUKRO**

MONROVIA   Harbel   Gagnoa

Buchanan   Zwedru   **LIBERIA**

Harper   Sass
San-Pédr

A   T   L   A   N   T   I   C   O   C   E   A   N

0 km   400
0 miles   400

Population   ● National capital

○ below 50,000   ○ 50,000 to 100,000   ◉ 100,000 to 500,000   ▣ above 500,000

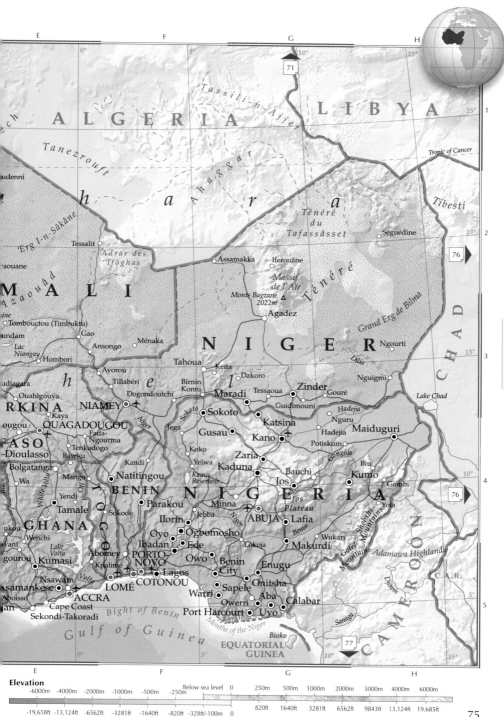

ALGERIA

LIBYA

Tassili-n-Ajjer

71

Tanezrouft

Tropic of Cancer

Ahaggar

Tibesti

S *a* *h* *a* *r* *a*

Ténéré du Tafassâsset

Séguédine

76

udenni

Erg I-n-Sâkâne

Tessalit

Assamakka

Iferouâne

Adrar des Ifôghas

aouane

A z a o u â d

MALI

Tombouctou (Timbuktu)

Gao

Ménaka

Massif de l'Aïr

Monts Bagzane △ 2022m

Agadez

Ténéré

Grand Erg de Bilma

CHAD

ine

Ansongo

Lac Niangay

Hombori

A z a o u â d

NIGER

Ngourti

20°

Dilia

Nguigmi

Lake Chad

udiagara

h

Ayorou

Tahoua

Keita

Dakoro

e

Tillabéri

Birnin Konni

Maradi

Tessaoua

Zinder

Gouré

l

Ouahigouya

Dogondoutchi

Guidimouni

Hadejia

RKINA

NIAMEY

Jega

Sokoto

Katsina

Nguru

Maiduguri

Kaya

Gusau

Kano

Hadejia

ougou

QUAGADOUGOU

Fada-Ngourma

Koko

Potiskum

ASO

Tenkodogo

Bawku

Yelwa

Zaria

Gongola

Biu

Kumo

-Dioulasso

Bolgatanga

Mango

Kandi

Kainji Reservoir

Kaduna

Bauchi

Gombi

Wa

Yendi

Natitingou

Jos

Yola

Tamale

Sokodé

BENIN

Minna

Jos Plateau

Lafia

GHANA

Wenchi

Parakou

Jebba

ABUJA

Benue

Wukari

Adamawa Highlands

Lake Volta

Ilorin

Tokoja

Makurdi

Gotel Mountains

ukou

Oyo

Ogbomosho

NIGERIA

C.A.R.

yani

Kumasi

Abomey

Ibadan

Ede

Owo

Benin City

Enugu

gourou

Kpalimé

PORTO-NOVO

Onitsha

Nsawam

LOMÉ

COTONOU

Lagos

Sapele

Aba

Calabar

CAMEROON

amankese

ACCRA

Warri

Owerri

Uyo

Aboisso

Cape Coast

Port Harcourt

Sanaga

jan

Sekondi-Takoradi

Bight of Benin

Mouths of the Niger

Bioko

77

Gulf of Guinea

EQUATORIAL GUINEA

Elevation

-6000m   -4000m   -2000m   -1000m   -500m   -250m   Below sea level   0   250m   500m   1000m   2000m   3000m   4000m   6000m

-19,658ft   -13,124ft   -6562ft   -3281ft   -1640ft   -820ft   -328ft/-100m   0   820ft   1640ft   3281ft   6562ft   9843ft   13,124ft   19,685ft

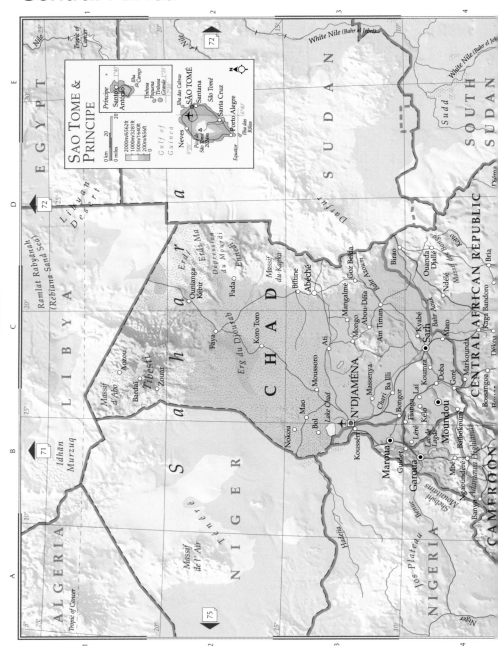

Population

● National capital

○ below 50,000   ○ 50,000 to 100,000   ◉ 100,000 to 500,000   ◼ above 500,000

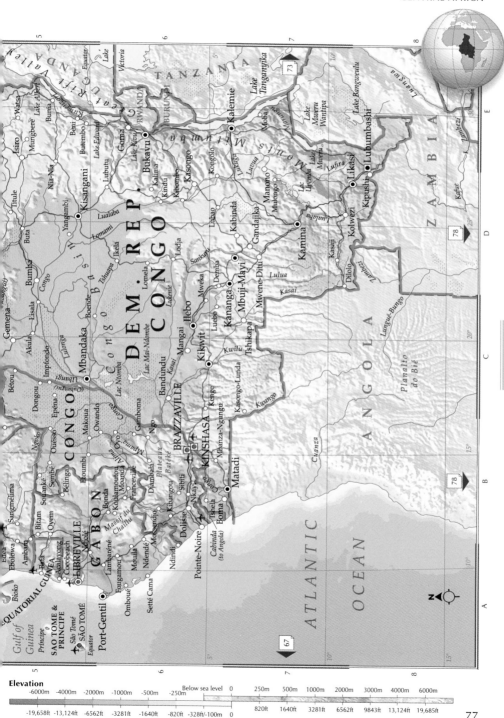

Elevation

-6000m	-4000m	-2000m	-1000m	-500m	-250m	Below sea level	0	250m	500m	1000m	2000m	3000m	4000m	6000m
-19,658ft	-13,124ft	-6562ft	-3281ft	-1640ft	-820ft	-328ft/-100m	0	820ft	1640ft	3281ft	6562ft	9843ft	13,124ft	19,685ft

Southern Africa

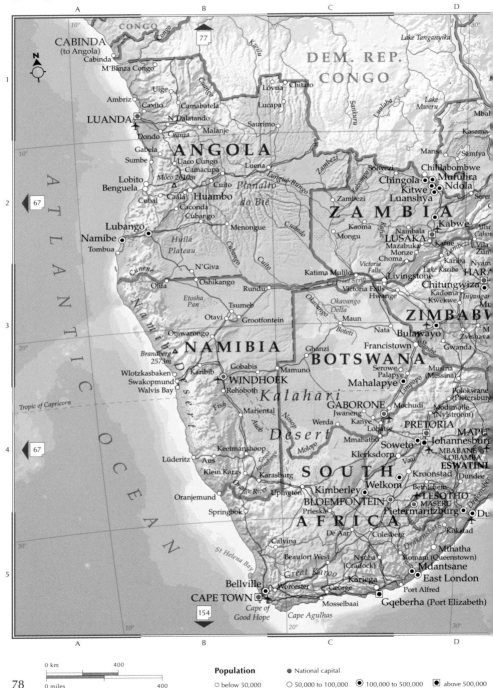

| A | B | C | D |

Population   ● National capital

○ below 50,000   ○ 50,000 to 100,000   ◉ 100,000 to 500,000   ◼ above 500,000

0 km   400

0 miles   400

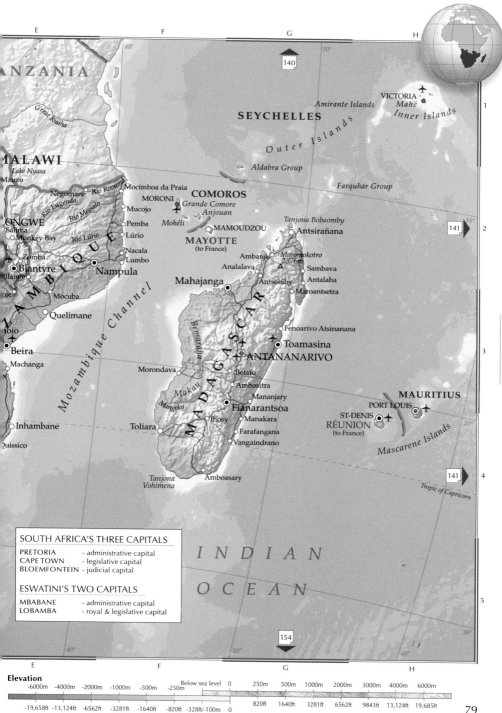

ANZANIA

Great Ruaha

MALAWI
Lake Nyasa
Mzuzu

140

Amirante Islands

VICTORIA
Mahé

SEYCHELLES

Inner Islands

1

Outer Islands

Aldabra Group

Farquhar Group

Negomane
Rio Rovuma
Rio Lugenda
Rio Messalo

Mocímboa da Praia

COMOROS

MORONI
*Grande Comore
Anjouan*

Tanjona Bobaomby

141

2

ONGWE
Salima
Monkey Bay
Zomba

Mucojo

Pemba
Rio Lúrio
Lúrio
Nacala
Lumbo

Mohéli

MAMOUDZOU

MAYOTTE
(to France)

Antsirañana

*Maromokotro
2876m*

Ambanja
Analalava
Antsohihy

Sambava
Antalaha

Blantyre

ZAMBIQUE

Nampula

Mahajanga

Bemaraha

Maroantsetra

nje
Mocuba

Quelimane

Fenoarivo Atsinanana

3

bio

Beira

M
o
z
a
m
b
i
q
u
e

C
h
a
n
n
e
l

Machanga

Morondava

Betafo

Ambositra

Mananjary

ANTANANARIVO

Toamasina

MADAGASCAR

Makay

Mangoky

Manakara

Fianarantsoa

Ihosy

MAURITIUS

PORT LOUIS

ST-DENIS
RÉUNION
(to France)

Mascarene Islands

Inhambane

Toliara

Farafangana

Vangaindrano

20°

Quissico

*Tanjona
Vohimena*

Amboasary

141

4

Tropic of Capricorn

SOUTH AFRICA'S THREE CAPITALS

PRETORIA — administrative capital
CAPE TOWN — legislative capital
BLOEMFONTEIN — judicial capital

ESWATINI'S TWO CAPITALS

MBABANE — administrative capital
LOBAMBA — royal & legislative capital

I N D I A N

O C E A N

5

154

Elevation

-6000m	-4000m	-2000m	-1000m	-500m	-250m	Below sea level	0	250m	500m	1000m	2000m	3000m	4000m	6000m
-19,658ft	-13,124ft	-6562ft	-3281ft	-1640ft	-820ft	-328ft/-100m	0	820ft	1640ft	3281ft	6562ft	9843ft	13,124ft	19,685ft

79

Europe

155

66

66

40° 60° 30° 20° 70° 10° 0°

Limit of winter pack ice

REYKJAVÍK
ICELAND
Vatnajökull

Arctic Circle

Reykjanes Ridge

Iceland Basin

Norwegian Basin

Faroe-Iceland Ridge

FAROE ISLANDS
(to Denmark)

Norwegian Sea

Hatton Ridge

Trondheim

Faroe-Shetland Trough

Shetland Islands

Rockall Bank

Rockall Trough

Outer Hebrides

Bergen

Stavanger

OSLO

NORW

Mid - Atlantic Ridge

Charlie - Gibbs Fracture Zone

British Isles

Orkney Islands

Glasgow
Edinburgh

North Sea

Gothenburg
Aalborg
Jön

Porcupine Plain

Ireland
Belfast

UNITED

Jutland

DENMARK
COPENH
M

Ireland
IRELAND
DUBLIN

Isle of Man

KINGDOM

Odense

Liverpool
Manchester
Birmingham

Britain

Celtic Sea

Cardiff

LONDON

NETHERLANDS
THE
HAGUE
AMSTERDAM
Rotterdam

Hamburg

Hanover

BERLIN

N

Celtic Shelf

English Channel

Channel Islands
le Havre

BELGIUM
BRUSSELS
Liège

Düsseldorf
Bonn

GERMANY

We

A T L A N T I C

Biscay Plain

Rennes

LUXEMBOURG
LUXEMBOURG

PARIS

Frankfurt
am Main

Stuttgart

CZECH
PRAG

(CZECH REP)

O C E A N

Azores-Biscay Rise

Charcot Seamounts

Iberian Plain

Nantes
Orléans

Strasbourg

Munich

FRANCE

Zurich

BR
SALZBUR

Bay of Biscay

A Coruña

Galicia Bank

Bordeaux

Lyon

BERN
SWITZERLAND
Innsbruck

VIENN

LIECH.

AUSTR

Porto

Duero

Cordillera Cantábrica

Bilbao

Mont Blanc
15,774ft
(4808m)

Milan

LJUBLJANA

SLOVEN

Iberian Plain

PORTUGAL

Iberian Peninsula

Zaragoza

Duero

Toulouse

Massif Central

ANDORRA

Nice

Turin

Venice

TRIESTE

Bologna

Adriatic

Tagus Plain

LISBON

MADRID

Tagus

Marseille

MONACO

Pisa

SAN MARINO

Horseshoe Seamounts

SPAIN

Seville

Guadalquivir

Barcelona

Corsica

VATICAN CITY
ROME

I

Madeira
(to Portugal)

Málaga

Valencia

Palma

Balearic Islands

Sardinia

Naples

Ba

GIBRALTAR
(to UK)

Ceuta
(to Spain)

Strait of Gibraltar

Algerian Basin

Cagliari

Tyrrhenian Sea

Cosenza

Canary Islands
(to Spain)

Melilla
(to Spain)

M e d i t e r r

Palermo

Mount Etna
10,922ft
(3329m)

Catar

Sicily

A t l a s   M o u n t a i n s

68

MALTA
VALLETTA

AFRICA

a n e

10° 30° 0° 70°

80

0 km   500

0 miles   500

Population   ● National capital

○ below 50,000   ○ 50,000 to 100,000   ◉ 100,000 to 500,000   ■ above 500,000

Barents Sea

North Cape

Ostrov Kolguyev

Murmansk
Kola
Peninsula

White
Sea

Archangel

FINLAND

Tampere

Turku
HELSINKI

TALLINN
IOLM

ESTONIA

LATVIA

RIGA

THUANIA
Kaunas

VILNIUS

LININGRAD
(to Russia)

Babruysk/
Bobruysk

BELARUS

Brest

ND

RSAW

Lviv

Chernivtsi

UKRAINE

MOLDOVA

CHIŞINĂU

Cluj-Napoca

ROMANIA

Braşov

GRADE

BUCHAREST

Danube

BULGARIA
Balkan Mountains

SOFIA

Burgas

KOPJE
NORTH
MACEDONIA

GREECE
ATHENS

Piraeus
ponnese

Irákleio

Crete

Arctic Circle

Ural Mountains

R U S S I A

Oh'

Irtysh

Perm'

Vologda

Saint Petersburg

Lake Onega

Lake Ladoga

Yaroslavl'

Nizhniy
Novgorod

MOSCOW

Kazan'

Ul'yanovsk

Samara

Ufa

Orenburg

Volga Uplands

Vitsyebsk/
Vitebsk

Central
Russian
Upland

Homyel'/
Gomel'

Voronezh

Pripyat
Marshes

Dnieper Lowland

Don

KYIV

Kharkiv

Dnipro

Donetsk

Rostov-na-Donu

Volgograd

Astrakhan'

Volga Delta
-98ft (-28m)

Syr Darya

Aral Sea

Amu Darya

Ural

Stavropol'

Sea of
Azov

Odesa

Crimea

Simferopol

Caucasus

El'brus 18,510ft
(5642m)

Black
Sea

Constanţa

Varna

Caspian Sea

A

S

I

A

Anatolia

TURKEY
(TÜRKİYE)

Cyprus

Zagros Mountains

Tigris

Euphrates

20° 30° 40° 50° 60° 70° 80° 70° 80° 60° 50° 70°

E F G H

155

112

112

118

The North Atlantic

Population

- ○ below 50,000
- ○ 50,000 to 100,000
- ◉ 100,000 to 500,000
- ◼ above 500,00[0]

● National capital

ARCTIC OCEAN

Lincoln Sea

Kap Morris Jesup

Wandel Sea

Independence Fjord

Nord

Zemlya Frantsa-Iosifa

Kvitøya

SVALBARD
(to Norway)

Nordaustlandet

Kong Karls Land

Novaya Zemlya

155

Kong Frederik VIII Land

Spitsbergen

Barentsøya

Edgeøya

Barents Sea

110

LONGYEARBYEN
Barentsburg

Storfjorden

Limit of winter pack ice

Greenland Sea

Christian X Land

Bjørnøya
(to Norway)

Nordkapp
(North Cape)

Limit of summer pack ice

Daneborg

FINLAND

Petermann Bjerg
2940m

Mohns Ridge

Kong Oscar Fjord

Kangertittivaq

Ittoqqortoormiit

Kangikajik

JAN MAYEN
(to Norway)

Norwegian Sea

Norwegian Basin

Vestfjorden

Arctic Circle

84

SWEDEN

a i t

ICELAND

ngarvík
Siglufjörður
Raufarhöfn

Húsavík

Akureyri

Seyðisfjörður

Stykkishólmur
Neskaupstaður

REYKJAVÍK

Selfoss
Vatnajökull
Djúpivogur

ákshöfn
Hvannadalshnúkur
2119m

Vestmannaeyjar

NORWAY

Gulf of Bothnia

FAROE ISLANDS
(to Denmark)

TÓRSHAVN

85

Shetland Islands

N

1
2
3
4
5

Elevation

-6000m	-4000m	-2000m	-1000m	-500m	-250m	Below sea level 0	250m	500m	1000m	2000m	3000m	4000m	6000m
-19,658ft	-13,124ft	-6562ft	-3281ft	-1640ft	-820ft	-328ft/-100m 0	820ft	1640ft	3281ft	6562ft	9843ft	13,124ft	19,685ft

Scandinavia & Finland

RUSSIA

Barents Sea

Nordkapp
(North Cape)

ARCTIC OCEAN

Norwegian Sea

FINLAND

Arctic Circle

Arctic Circle

0 km   200

0 miles   200

Population

○ below 50,000

● National capital

○ 50,000 to 100,000   ◉ 100,000 to 500,000   ■ above 500,000

RUSSIA

BELARUS

Ladozhskoye
Ozero

Saimaa
Hauklvesl

Varkaus
Savitaipale

Lappeenranta
(Willmanstrand)

Imatra
Joutseno

Jyväskylä
Kouvola
Kotka

Kuusankoski
(Kouvola)

Lahti (Lahtis)

Hämeenlinna
(Tavastehus)

Tampere (Tammerfors)

Nokia

Riihimäki

Kerava
(Kervo)

Porvoo
(Borgå)

HELSINKI/HELSINGFORS

Vantaa
(Vanda)

Espoo
(Esbo)

Salo

Turku
(Åbo)

Hanko
(Hangö)

Pori
(Björneborg)

Rauma

Kankaanpää

Seinäjoki

Närpiö
(Närpes)

Nästviken

Kaurru

Gulf of Finland

Lake Peipus

ESTONIA

Gulf of
Riga

Hiiumaa

Saaremaa

Western Dvina

LATVIA

LITHUANIA

Neman

Courland Lagoon

Gulf of
Gdańsk

KALININGRAD
(to Russia)

Wisła

POLAND

Odra

GERMANY

Elbe

Weser

Ems

Åland

Åland Hav

STOCKHOLM

Norrtälje

Uppsala

Täby

Tierp

Sala

Märsta

Solna

Enköping

Södertälje

Eskilstuna

Hallsberg

Askersund

Örebro

Karlstad

Arvika

Säffle

Åmål

Grums

Mellerud

Lidköping

Vänern

Trollhättan

Uddevalla

Vänersborg

Göteborg
(Gothenburg)

Kungsbacka

Varberg

Mölndal

Borås

Jönköping

Vättern

Vänersborg

Motala

Linköping

Norrköping

Nyköping

Gotland

Visby

Borgholm

Öland

Kalmar

Oskarshamn

Växjö

Ljungby

Karlskrona

Kristianstad

Karlstad

Hanöbukten

Baltic Sea

Ronne

Bornholm

SWEDEN

Gävle

Söderhamn

Hudiksvall

Sundsvall

Härnösand

Ånge

Timrå

Ljusdal

Bollnäs

Leksand

Rättvik

Falun

Borlänge

Ludvika

Mora

Nalung

Orsa

Sveg

Idre

Klarälven

NORWAY

OSLO

Ski

Moss

Sarpsborg

Fredrikstad

Strömstad

Halden

Lillestrøm

Drammen

Sandvika

Hamar

Elverum

Lillehammer

Gjøvik

Mjøsa

Gol

Hønefoss

Ringerike

Kongsvinger

Glåma

Dombås

Oppdal

Dovrefjell

Glittertind
2472m

Jotunheimen

Galdhøpiggen
2469m

Andalsnes

Ålesund

Bergen

Voss

Odda

Haugesund

Stavanger

Sandnes

Egersund

Eigersund

Flekkefjord

Mandal

Lyngdal

Farsund

Kristiansand

Arendal

Grimstad

Risør

Porsgrunn

Skien

Horten

Tønsberg

Sandefjord

Larvik

Kongsberg

Notodden

Rjukan

Setesdal

Skagerrak

North
Sea

Hirtshals

Hjørring

Frederikshavn

Aalborg

Læsø

Randers

Viborg

Holstebro

Ringkøbing Fjord

Varde

Esbjerg

Ribe

Rømø

Kolding

Vejle

Fredericia

Jylland

Lille Bælt

Odense

DENMARK

KØBENHAVN
(Copenhagen)

Sjælland

Roskilde

Helsingør

Lund

Malmö

Helsingborg

Landskrona

Ystad

Store Bælt

Nyborg

Korsør

Slagelse

Næstved

Møn

Falster

Lolland

Nykøbing

Aarhus

Skanderborg

Horsens

Silkeborg

KALININGRAD

Elevation

Below sea level											
-6000m	-4000m	-2000m	-1000m	-500m	-250m	0	250m	500m	1000m	2000m	3000m 4000m 6000m
-19,658ft	-13,124ft	-6562ft	-3281ft	-1640ft	-820ft	-328ft/-100m	0	820ft	1640ft	3281ft	6562ft 9843ft 13,124ft 19,685ft

The Low Countries

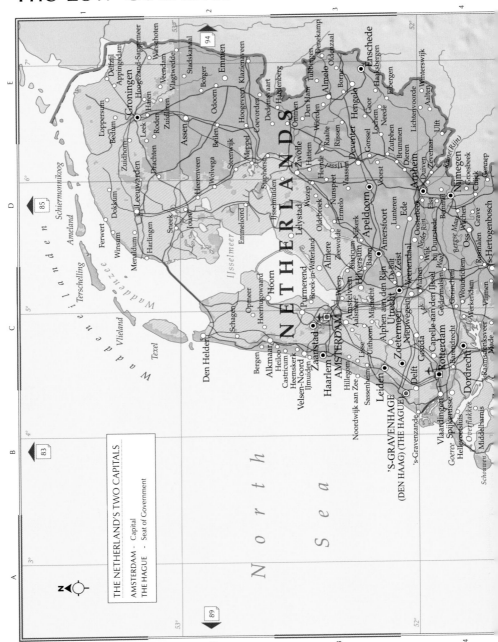

THE NETHERLAND'S TWO CAPITALS

AMSTERDAM - Capital
THE HAGUE - Seat of Government

Population

- ○ below 50,000
- ○ 50,000 to 100,000
- ◉ 100,000 to 500,000
- ■ above 500,000
- ● National capital

0 km   50
0 miles   50

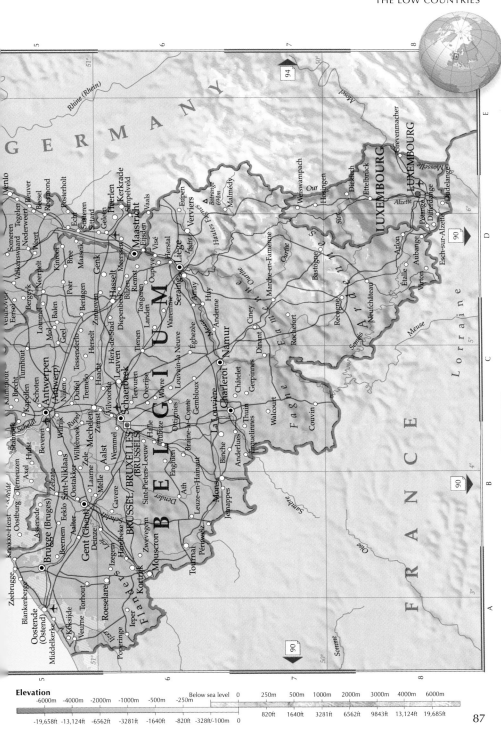

Elevation

Below sea level							
-6000m	-4000m	-2000m	-1000m	-500m -250m	0	250m 500m 1000m 2000m 3000m 4000m 6000m	
-19,658ft	-13,124ft	-6562ft	-3281ft	-1640ft -820ft	-328ft/-100m 0	820ft 1640ft 3281ft 6562ft 9843ft 13,124ft 19,685ft	

United Kingdom & Ireland

Population

○ below 50,000

● National capital

○ 50,000 to 100,000

◉ 100,000 to 500,000

◎ Internal administrative capital

■ above 500,000

Elevation

					Below sea level	0	250m	500m	1000m	2000m	3000m	4000m	6000m
-6000m	-4000m	-2000m	-1000m	-500m	-250m								
-19,658ft	-13,124ft	-6562ft	-3281ft	-1640ft	-820ft	-328ft/-100m	0	820ft	1640ft	3281ft	6562ft	9843ft	13,124ft	19,685ft

France, Andorra & Monaco

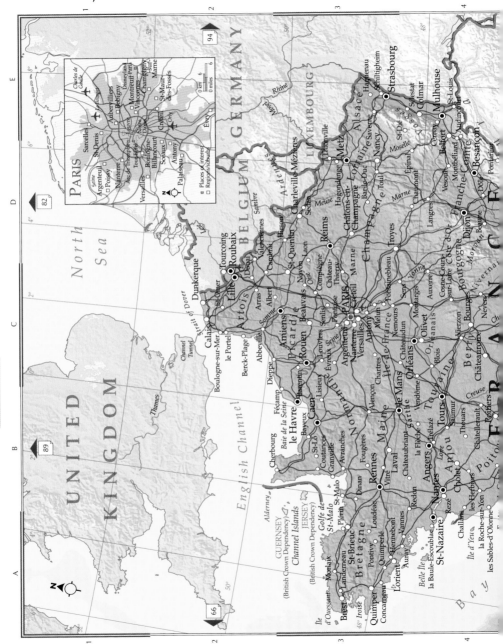

Population   ● National capital

○ below 50,000   ○ 50,000 to 100,000   ◉ 100,000 to 500,000   ◼ above 500,000

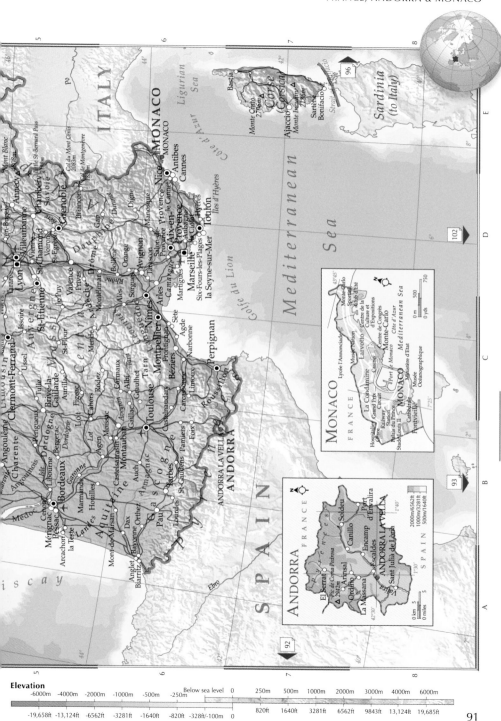

Elevation

					Below sea level	0	250m	500m	1000m	2000m	3000m	4000m	6000m
-6000m	-4000m	-2000m	-1000m	-500m	-250m								
-19,658ft	-13,124ft	-6562ft	-3281ft	-1640ft	-820ft	-328ft/-100m	0	820ft	1640ft	3281ft	6562ft	9843ft	13,124ft	19,685ft

Spain & Portugal

A Coruña
Ferrol
Luarca Avilés
Gijón/ Costa Verd
Xixón Santan
Betanzos
Vilalba
Pravia
Tineo
Villaviciosa Llanes
Santa Cataliña de Armada
Cabo Fisterra
Galicia
Asturias
Oviedo/Uviéu
Mieres del Camín Torre
Outes
La Pola
Cabanaquinta Cant
Muros
Lugo
Cordillera Cantábrica
Reinosa
Santiago de Compostela
Chantada
Ponferrada
León
Bu
Santa Uxía de Ribeira
Lalín
Monforte
O Carballiño de Lemos
Astorga
Pontevedra
Marín
Castilla y León
Vigo
Ourense
Benavente
Palencia
Ponteareas
Xinzo de Limia
Valladolid
Viana do Castelo
Bragança Embalse de
Ricobayo
Zamora
Ar
de D
Ponte da Barca
Chaves
Toro
Póvoa de Varzim
Braga
Guimarães
Medina del Campo
Vila do Conde
Vila Real
Salamanca
Matosinhos
Porto
Douro Lamego
Embalse S
de Almendra
Segovia
Vila Nova de Gaia
São João da Madeira
Ovar Albergaria-a-Velha
Viseu
Avila
P
Aveiro
Guarda
Ciudad-Rodrigo
Béjar Sistema Cent
MADRI
Ílhavo
Alto da Torre
1993▲
Coimbra
Sierra de Gredos Get
Figueira da Foz
Serra da Estrela
Covilhã
Plasencia
Talavera
de la Reina
Ar
PORTUGAL
Coria Toledo
Leiria
Castelo Branco
Tagus Embalse de
Alcántara Cáceres
Embalse de
Valdecañas
Entroncamento
Tomar Abrantes
Trujillo
Peniche
Caldas da Rainha
Santarém
Portalegre
Extremadura Herrera
del Duque
Torres Vedras
Coruche
D
Sintra
Estremoz Elvas
Mérida Villanueva de la Serena Ciudad Rea
LISBOA (LISBON)
Barreiro
Serra d'Ossa
Badajoz
Don Benito
Puertollano
Cascais
Almada
Setúbal
Évora
Almendralejo
Castuera
Villafranca de los Barros
Pozoblanco
Alcácer do Sal
Barragem
do Alqueva
Zafra
Azuaga La Ca
Baía de Setúbal
Jerez de los Caballeros
Morena
Montoro
Sines
Beja
Córdoba
Bujalance
Cortegana Sierra Martos Alca
Ourique
Nerva
Guadalquivir Palma del Río
Andalu
Valverde del Camino La Algaba
Carmona Écija Lucena Sist
Algarve
Ayamonte Lepe
Sevilla
(Seville)
Osuna Grar
Portimão
Faro Tavira Isla
Cristina
Huelva
Dos
Hermanas
Antequera Archid
Lagos Olhão Las Cabezas de San Juan
Olvera Álora
Cabo de
São Vicente
Golfo de Cádiz
Lebrija
Ubrique Ronda
Con Mála
Sanlúcar de Barrameda Fuengirol
El Puerto de Santa María
Cádiz
Jerez de la Frontera
Marbella
San Fernando Estepona Costa
Vejer de la Frontera
Barbate de Franco
GIBRALTAR
(to UK)
Algeciras Ceuta (to Spain)
Strait of Gibraltar
MOROCCO

ATLANTIC

OCEAN

66

66

66

70

AZORES (to Portugal)

Corvo
São Graciosa
Flores Jorge
Faial Terceira
Pico São Miguel
Ponta Delgada
Santa Maria

0 km 100

0 miles 100

200m/656ft
Sea level

0 km 100

0 miles 100

Population

○ below 50,000

● National capital

○ 50,000 to 100,000

◉ 100,000 to 500,000

■ above 500,000

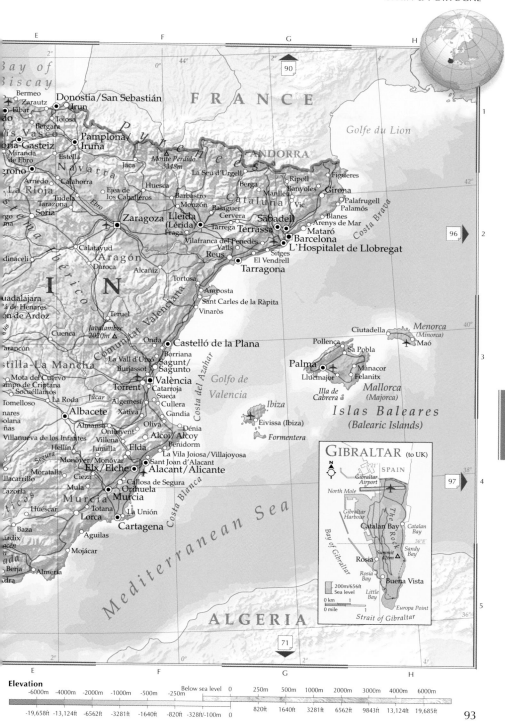

Bay of
Biscay

Bermeo
Zarautz
Eibar
Tolosa
Donostia/San Sebastián
Irun
ís Vasco
Bergara
Pamplona/
Iruña
ria-Gasteiz
Miranda
de Ebro
Estella
Calahorra
La Rioja
Arnedo
groño
Soria
Tudela
Tarazona
ma
dinaceli
F R A N C E
P y r é n é e s
Monte Perdido
3348m
La Seu d'Urgell
ANDORRA
Navarra
Jaca
Huesca
Ejea de
los Caballeros
Barbastro
Monzón
Catalunya
Berga
Ripoll
Manlleu
Vic
Banyoles
Figueres
Girona
Palafrugell
Palamós
Costa Brava
Blanes
Arenys de Mar
Golfe du Lion

Calatayud
Aragón
Daroca
Alcañiz
N
I
Zaragoza
Lleida
(Lérida)
Fraga
Tàrrega
Cervera
Balaguer
Vilafranca del Penedès
Valls
Reus
Terrassa
Sabadell
Barcelona
L'Hospitalet de Llobregat
Sitges
El Vendrell
Tarragona

uadalajara
á de Henares
ón de Ardoz
rico
Teruel
Javalambre
2020m
Tortosa
Amposta
Sant Carles de la Ràpita
Vinaròs

rancón
Cuenca
Onda
Comunitat Valenciana
Castelló de la Plana
Borriana
Sagunt/
Sagunto
La Vall d'Uixó
Burjassot
stilla-La Mancha
Mota del Cuervo
ampo de Criptana
Socuéllamos
Tomelloso
La Roda
nares
olana
ñas
Villanueva de los Infantes
Albacete
Almansa
Hellín
Jumilla
Moratalla
Cieza
Mula
Huéscar
Baza
acén
ada
da
Berja
Almería
Mojácar
 ticos
Torrent
Catarroja
Sueca
Cullera
Gandia
València
Algemesí
Xàtiva
Oliva
Dénia
Ontinyent
Villena
Elda
Alcoi/Alcoy
Benidorm
La Vila Joiosa/Villajoyosa
Monòver/Monóvar
Sant Joan d'Alacant
Elx/Elche
Alacant/Alicante
Callosa de Segura
Orihuela
Murcia
Totana
Lorca
La Unión
Cartagena
Aguilas
Costa del Azahar
Golfo de
Valencia
Ibiza
Eivissa (Ibiza)
Formentera
Costa Blanca

Mediterranean Sea

A L G E R I A

Ciutadella
Menorca
(Minorca)
Maó
Pollença
Sa Pobla
Palma
Manacor
Felanitx
Lluemajor
Illa de
Cabrera
Mallorca
(Majorca)
Islas Baleares
(Balearic Islands)

GIBRALTAR (to UK)
N
SPAIN
Gibraltar
Airport
North Mole
Gibraltar
Harbour
Bay of Gibraltar
Catalan Bay
Catalan
Bay
The Rock
Sandy
Bay
Rosia
Summit
699m
Rosia
Bay
Little
Bay
Buena Vista
Europa Point
Strait of Gibraltar
200m/656ft
Sea level
0 km   1
0 mile   1

Elevation

-6000m	-4000m	-2000m	-1000m	-500m	-250m	Below sea level	0	250m	500m	1000m	2000m	3000m	4000m	6000m
-19,658ft	-13,124ft	-6562ft	-3281ft	-1640ft	-820ft	-328ft/-100m	0	820ft	1640ft	3281ft	6562ft	9843ft	13,124ft	19,685ft

Germany & the Alpine States

LIECHTENSTEIN

AUSTRIA

SWITZERLAND

Ringgell
Bendern
Mauren
Schaan
Planken

VADUZ
Triesenberg
Triesen
Balzers

Samínatal

	2000m/6562ft
	1000m/3281ft
	500m/1640ft
	250m/820ft

0 km   4
0 miles   4

98

85

86

85

POLAND

SWEDEN

DENMARK

NETHERLANDS

GERMANY

Baltic Sea

North Sea

Bornholm
(to Denmark)

Rügen
Sassnitz
Bergen
Stralsund
Greifswald Bay
Greifswalder Bucht
Wolgast
Oderhaff

Pomeranian Bay

Frankfurt an der Oder
Oder
Noteć

Eisenhüttenstadt
Guben
Cottbus
Finsterwalde
Senftenberg
Hoyerswerda
Bautzen
Görlitz
Döbeln
Riesa
Torgau
Lübben
Luckenwalde
Lübbenau
Spree

Eberswalde-Finow
Bad Freienwalde
Angermünde
Bernau
BERLIN
Potsdam
Ludwigsfelde
Dessau
Magdeburg
Bernburg
Halberstadt
Halle-Neustadt
Leipzig
Eisleben
Nordhausen
Göttingen
Northeim
Seesen
Schönebeck
Brandenburg
Neuruppin
Oranienburg
Wittenberge
Perleberg
Pritzwalk
Neustrelitz
Prenzlau
Pasewalk
Anklam
Demmin
Teterow
Waren
Malchin
Güstrow
Schwerin
Parchim
Ludwigslust
Müritz
Stendal
Salzwedel

Rostock
Warnemünde
Wismar
Grevesmühlen
Neubrandenburg

Kiel
Neumünster
Lübeck
Norderstedt
Hamburg
Heide
Rendsburg
Schleswig
Flensburg
Kappeln
Husum
Oldenburg
Eutin
Fehmarn
Fehmarnbelt
Grömitz
Plattgarden
Kieler Bucht
Mecklenburger Bucht

Salzgitter
Braunschweig
Wolfsburg
Peine
Gifhorn
Celle
Uelzen
Soltau
Lüneburg
Dannenberg
Boizenburg
Wittenberge
Elbe

Hannover
(Hanover)
Hildesheim
Minden
Herford
Bielefeld
Gütersloh
Paderborn
Warburg
Kassel
Marsberg

Bremen
Bremerhaven
Wilhelmshaven
Verden
Bassum
Diepholz
Osnabrück
Rheine
Nordhorn
Lingen
Cloppenburg
Delmenhorst
Oldenburg
Emden
Leer
Weener
Meppen
Dülmen
Ahlen
Hamm
Münster
Bocholt
Borken
Recklinghausen
Dortmund
Bochum
Duisburg
Essen
Krefeld
Wuppertal
Düsseldorf

Cuxhaven
Stade
Rosengarten
Buchholz
Scheeßel
Elmshorn
Itzehoe
Wesel

North Frisian Islands
(Nordfriesische Inseln)
Ostfriesische Inseln
Helgoländer Bucht
Helgoland

Jylland
Sjælland
Fyn
Falster
Lolland
Westerland

Weser
Ems
Rhine
Ijsselmeer

52°
54°
56°

6°
8°
10°
12°
14°
16°
18°

0 km   100
0 miles   100

Population

○ below 50,000
○ 50,000 to 100,000
◉ 100,000 to 500,000
◼ above 500,000
● National capital

Elevation

-6000m	-4000m	-2000m	-1000m	-500m	-250m	Below sea level	0	250m	500m	1000m	2000m	3000m	4000m	6000m
-19,658ft	-13,124ft	-6562ft	-3281ft	-1640ft	-820ft	-328ft/-100m	0	820ft	1640ft	3281ft	6562ft	9843ft	13,124ft	19,685ft

Population

● National capital

○ below 50,000   ○ 50,000 to 100,000   ◉ 100,000 to 500,000   ■ above 500,000

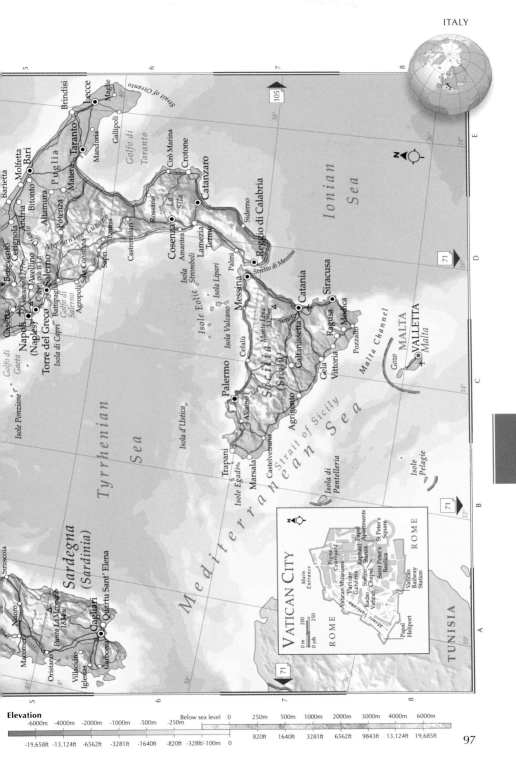

Strait of Otranto

Golfo di Taranto

Strait of Sicily

Ionian Sea

Tyrrhenian Sea

Mediterranean Sea

Malta Channel

Lecce
Maglie
Brindisi
Gallipoli
Manduria
Taranto
Bari
Molfetta
Barletta
Andria
Bitonto
Altamura
Cerignola
Matera
Benevento
Avellino
Potenza
Vesuvio 1277m
Napoli (Naples)
Torre del Greco
Caserta
Battipaglia
Salerno
Agropoli
Sapri
San Consilina
Castrovillari
Rossano
La Sila
Cosenza
Cirò Marina
Crotone
Catanzaro
Siderno
Amantea
Lamezia Terme
Palmi
Reggio di Calabria
Stretto di Messina
Messina
Catania
Siracusa
Modica
Pozzallo
Ragusa
Vittoria
Gela
Caltanissetta
Agrigento
Cefalù
Palermo
Alcamo
Castelvetrano
Marsala
Trapani

Isole Eolie
Isola Stromboli
Isola Lipari
Isola Vulcano
Monte Etna 3329m
Simeto

Sicilia (Sicily)

Isole Egadi
Isola di Pantelleria

Isole Pelagie

VALLETTA
MALTA
Malta
Gozo

Sardegna (Sardinia)
Sassari
Nuoro
Macomer
Oristano
Villacidro
Iglesias
Carbonia
Quartu Sant'Elena
Cagliari
Punta La Marmora 1834m

Golfo di Gaeta
Isole Ponziane
Isola di Capri
Golfo di Salerno

Isola d'Ustica

Campania
Puglia
Apennino Lucano

TUNISIA

105
71
71
71

VATICAN CITY

ROME

Main Entrance
Pigna Courtyard
Vatican Museums
Vatican Gardens
Radio Vatican
Raphael Stanza
Sistine Chapel
Saint Peter's Basilica
Raphael Apartments
St Peter's Square
Monte Vaticano
Vatican Railway Station
Papal Heliport

0 m   200
0 yds   250

Elevation

					Below sea level	0		250m	500m	1000m	2000m	3000m	4000m	6000m
-6000m	-4000m	-2000m	-1000m	-500m	-250m									
-19,658ft	-13,124ft	-6562ft	-3281ft	-1640ft	-820ft	-328ft/-100m	0		820ft	1640ft	3281ft	6562ft	9843ft	13,124ft	19,685ft

Central Europe

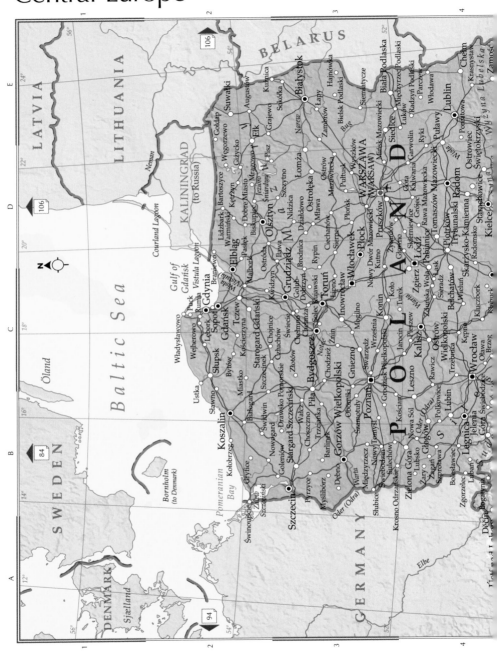

LATVIA

LITHUANIA

BELARUS

KALININGRAD
(to Russia)

Baltic Sea

SWEDEN

DENMARK

GERMANY

POLAND

WARSZAWA
(WARSAW)

Gulf of Gdańsk

Courland Lagoon

Vistula Lagoon

Pomeranian Bay

Bornholm (to Denmark)

Öland

Sjælland

Neman

Elbe

Oder (Odra)

Suwałki · Gołdap · Węgorzewo · Giżycko · Ełk · Grajewo · Szczuczyn · Augustów · Sokółka · Kuźnica · Białystok · Łapy · Hajnówka · Bielsk Podlaski · Siemiatycze · Biała Podlaska · Międzyrzec Podlaski · Parczew · Radzyń Podlaski · Włodawa · Chełm · Krasnystaw · Zamość

Lidzbark Warmiński · Bartoszyce · Kętrzyn · Mrągowo · Szczytno · Pisz · Kolno · Łomża · Ostrów Mazowiecka · Zambrów · Wysokie Mazowieckie · Wyszków · Sokołów Podlaski · Siedlce · Łuków · Radzyń · Lublin

Elbląg · Braniewo · Pasłęk · Orneta · Olsztyn · Dobre Miasto · Biskupiec · Nidzica · Działdowo · Mława · Ciechanów · Pułtusk · Nowy Dwór Mazowiecki · Legionowo · Mińsk Mazowiecki · Garwolin · Kozienice · Puławy · Ryki · Opole Lubelskie

Gdynia · Sopot · Gdańsk · Tczew · Pruszcz Gdański · Malbork · Kwidzyn · Iława · Ostróda · Grudziądz · Brodnica · Rypin · Sierpc · Płońsk · Pruszków · Radom · Kielce · Starachowice · Ostrowiec Świętokrzyski · Skarżysko-Kamienna · Końskie · Świętokrzyskie

Władysławowo · Wejherowo · Puck · Rumia · Lębork · Kościerzyna · Starogard Gdański · Chojnice · Czersk · Świecie · Chełmno · Chełmża · Toruń · Dobrzyń · Lipno · Włocławek · Płock · Gostynin · Kutno · Łowicz · Skierniewice · Rawa Mazowiecka · Tomaszów Mazowiecki · Piotrków Trybunalski · Radomsko · Opoczno

Słupsk · Bytów · Miastko · Szczecinek · Złotów · Więcbork · Koronowo · Bydgoszcz · Inowrocław · Gniezno · Znin · Mogilno · Koło · Turek · Łódź · Zgierz · Pabianice · Zduńska Wola · Sieradz · Łask · Bełchatów

Ustka · Sławno · Koszalin · Kołobrzeg · Białogard · Szczecinek · Drawsko Pomorskie · Czaplinek · Wałcz · Piła · Chodzież · Wągrowiec · Oborniki · Poznań · Września · Słupca · Konin · Kalisz · Ostrów Wielkopolski · Kępno · Wieluń

Świnoujście · Wolin · Goleniów · Nowogard · Świdwin · Choszczno · Trzcianka · Czarnków · Szamotuły · Środa Wielkopolska · Jarocin · Pleszew · Krotoszyn · Rawicz · Trzebnica · Oleśnica · Wrocław · Oława · Brzeg

Szczecin · Stargard Szczeciński · Pyrzyce · Myślibórz · Barlinek · Gorzów Wielkopolski · Międzyrzecz · Nowy Tomyśl · Grodzisk Wielkopolski · Kościan · Leszno · Gostyń · Góra · Rawicz · Wołów

Dębno · Warta · Sulęcin · Świebodzin · Wolsztyn · Nowa Sól · Głogów · Lubin · Polkowice · Jawor · Legnica · Złotoryja

Gryfice · Stubice · Krosno Odrzańskie · Zielona Góra · Żary · Żagań · Szprotawa · Bolesławiec · Lubań · Zgorzelec · Bogatynia · Jelenia Góra · Świeradów · Świebodzice

Stubice · Lubsko · Zary · Sulechów · Zielona Góra · Zawidów · Góra Świętej Anny

Děčín

0 km   100

0 miles   100

Population   ● National capital

○ below 50,000   ○ 50,000 to 100,000   ◉ 100,000 to 500,000   ◼ above 500,000

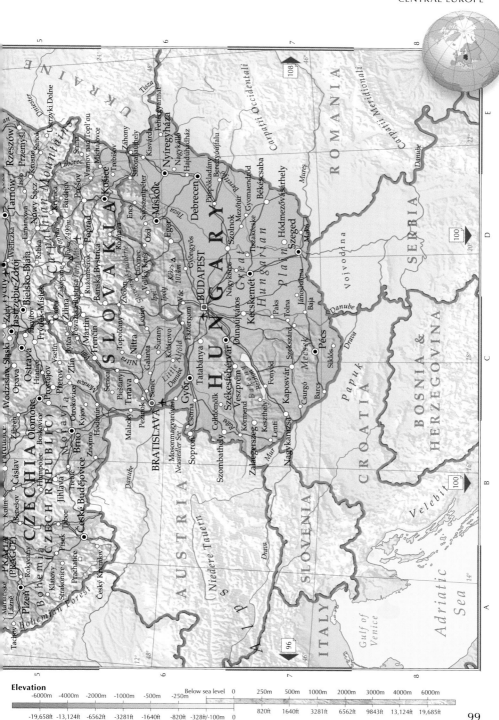

Elevation

Below sea level													
-6000m	-4000m	-2000m	-1000m	-500m	-250m	0	250m	500m	1000m	2000m	3000m	4000m	6000m
-19,658ft	-13,124ft	-6562ft	-3281ft	-1640ft	-820ft	-328ft/-100m	0	820ft	1640ft	3281ft	6562ft	9843ft	13,124ft	19,685ft

Southeast Europe

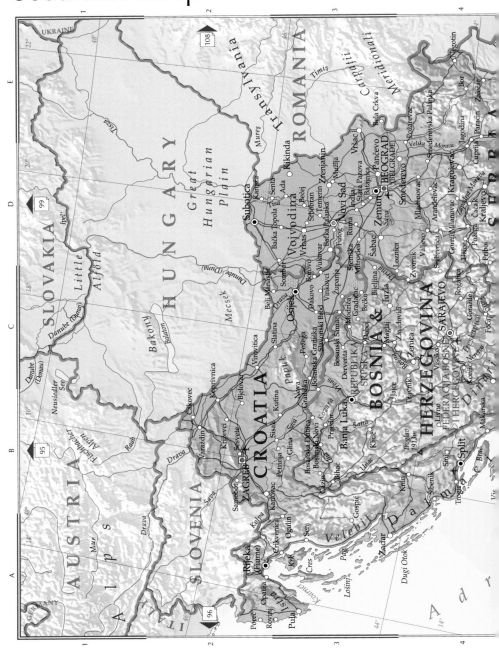

Population

- National capital
- Internal administrative capital
- below 50,000
- 50,000 to 100,000
- 100,000 to 500,000
- above 500,000

0 km   100

0 miles   100

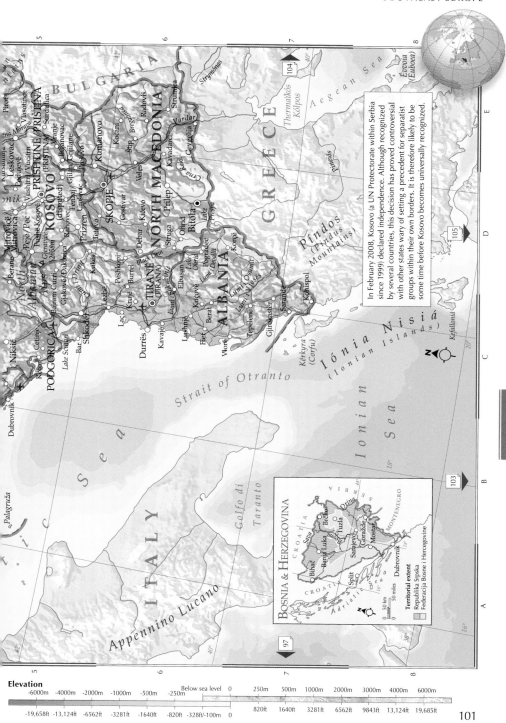

In February 2008, Kosovo (a UN Protectorate within Serbia since 1999) declared independence. Although recognized by several countries, this decision has proved controversial with other states wary of setting a precedent for separatist groups within their own borders. It is therefore likely to be some time before Kosovo becomes universally recognized.

BULGARIA

GREECE

NORTH MACEDONIA

KOSOVO
PRISHTINË / PRISTINA

SKOPJE

ALBANIA

TIRANË (TIRANA)

PODGORICA

Pindos
(Pindus
Mountains)

Aegean Sea

Strait of Otranto

Ionian Sea

Iónia Nisiá
(Ionian Islands)

ITALY

Appennino Lucano

Golfo di Taranto

Adriatic Sea

BOSNIA & HERZEGOVINA

CROATIA

SERBIA

MONTENEGRO

Banja Luka
Brčko
Tuzla
Sarajevo
Goražde
Mostar
Dubrovnik
Bihać
Split

Territorial extent
Republika Srpska
Federacija Bosne i Hercegovine

Elevation

Below sea level											
-6000m	-4000m	-2000m	-1000m	-500m	-250m	0	250m	500m	1000m	2000m	3000m	4000m	6000m
-19,658ft	-13,124ft	-6562ft	-3281ft	-1640ft	-820ft	-328ft/-100m	0	820ft	1640ft	3281ft	6562ft	9843ft	13,124ft	19,685ft

The Mediterranean

Population

- o below 50,000
- ● National capital
- o 50,000 to 100,000
- ◉ 100,000 to 500,000
- ■ above 500,000

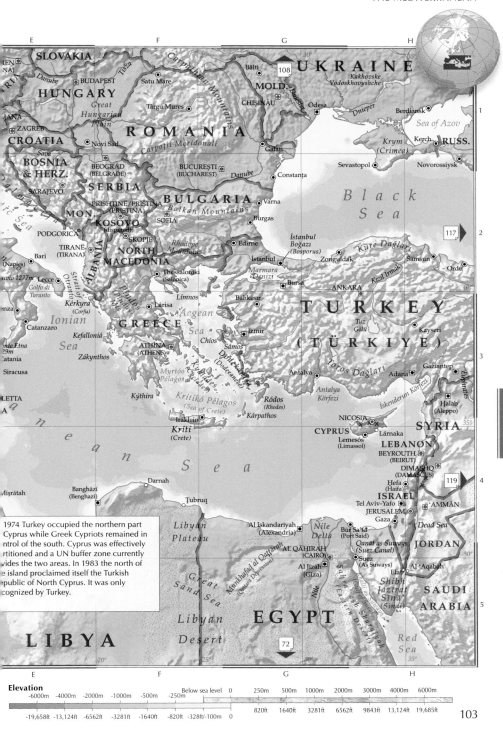

In 1974 Turkey occupied the northern part of Cyprus while Greek Cypriots remained in control of the south. Cyprus was effectively partitioned and a UN buffer zone currently divides the two areas. In 1983 the north of the island proclaimed itself the Turkish Republic of North Cyprus. It was only recognized by Turkey.

Elevation

Below sea level								250m	500m	1000m	2000m	3000m	4000m	6000m
-6000m	-4000m	-2000m	-1000m	-500m	-250m		0							
-19,658ft	-13,124ft	-6562ft	-3281ft	-1640ft	-820ft	-328ft/-100m	0	820ft	1640ft	3281ft	6562ft	9843ft	13,124ft	19,685ft

Bulgaria & Greece

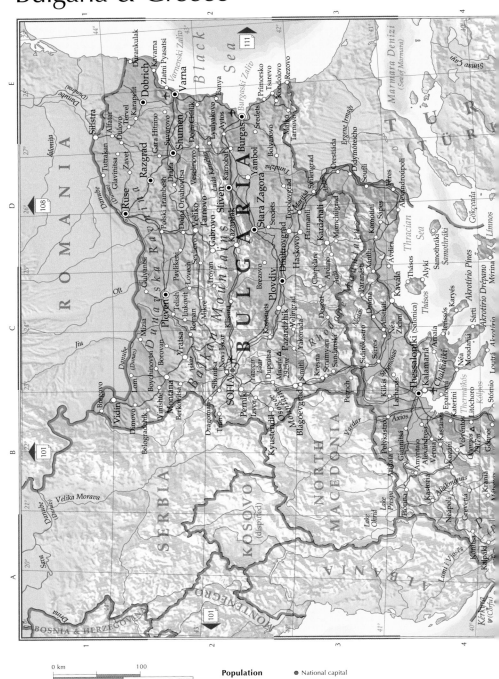

Population ● National capital

○ below 50,000 ○ 50,000 to 100,000 ◉ 100,000 to 500,000 ▣ above 500,000

0 km   100

0 miles   100

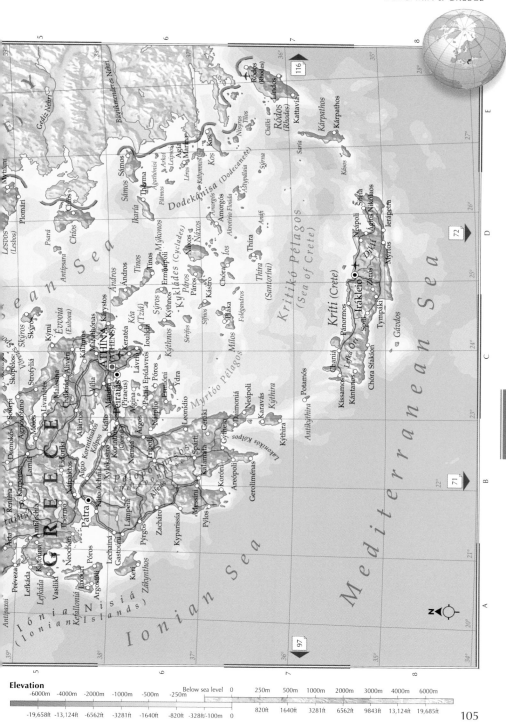

Ródos
(Rhodes)
Lindos
Kattaviá
Kárpathos
Kárpathos
Saría
Kásos

116

E

Ródos
(Rhodes)
Chálki
Agía
Nísyros
Marína
Kos
Kálymnos
Astypálaia
Sýrna
Sími

72

Léros
Lévitha
Arkoí
Agathónisi
Pátmos
Amorgós
Akrotírio Floúda

D

Sámos
Sámos
Ikaría
Tírma
Léros
Dodekánisa (Dodecanese)
Anáfi
Thíra

Kritikó Pélagos
(Sea of Crete)

Chíos
Psará
Antípsara
Mýkonos
Náxos
Íos
Thíra
(Santoríni)

Neápoli
Sitía
Ágios Nikólaos
Ierápetra

Lésmos
(Lesbos)
Mytilíni

Myrtóo Pélagos

Kalámata

Gýnteni

Messíni

Pýlos
Zacháro

Kyparissía

Gastoúni

Lechaná

Kefí

Argostóli
Ithouri

Zákynthos

Iónia Nisiá
(Ionian Islands)

Ionian Sea

Mediterranean Sea

Elevation

-6000m	-4000m	-2000m	-1000m	-500m	-250m	Below sea level 0	250m	500m	1000m	2000m	3000m	4000m	6000m
-19,658ft	-13,124ft	-6562ft	-3281ft	-1640ft	-820ft	-328ft/-100m 0	820ft	1640ft	3281ft	6562ft	9843ft	13,124ft	19,685ft

The Baltic States & Belarus

0 km 100
0 miles 100

Population ● National capital

○ below 50,000 ○ 50,000 to 100,000 ◉ 100,000 to 500,000 ◼ above 500,000

RUSSIA

UKRAINE

POLAND

B E L A R U S

Dnieper Lowland

Dnieper (Dnyapro/Dnepr)

Mazury

Wyżyna Lubelska

Pripyat Marshes

Minskaya Wzvyshsha

Byelaruskaya Hrada

Navapolatsk/Novopolotsk
Polatsk/Polotsk
Vitsyebsk/Vitebsk
Haradok/Gorodok
Mahilyow/Mogilëv
Orsha
Babruysk/Bobruysk
Homyel'/Gomel
Mazyr/Mozyr'
MINSK
Barysaw/Borisov
Hrodna/Grodno
Brest
Pinsk
Slutsk
Salihorsk/Soligorsk
Baranavichy/Baranovichi
Lida
Slonim
Navahrudak/Novogrudok
Zhlobin
Svetlogorsk
Svyetlahorsk
Kalinkavichy

Kastsyukovichy
Klimavichy
Krychaw/Krichëv
Cherykaw
Chausy
Chavusy
Kholdasy
Slawharad
Rahachow/Rogachëv
Dobrush
Tsyerakhowka
Byval'ki
Loyew
Khoyniki
Narowlya
Yel'sk
Lyel'chytsy
Dabryn'/Dobryn
Milashavichy
Tonyezh
Pyetrykaw
Kaptsevichy
Zhytkavichy
Mikashevichy
Luninyets/Luninets
Bastyn'
Starobin
Semezhava
Lyusina
Lyakhavichy
Hantsavichy
Gantsevichi
Abrova
Ivatsevichy
Ruzhany
Zhabinka
Damachava
Makrany
Haradzyets/Gorodets
Drahichyn/Drogichin
Ivanava
Iyanava
Kobryn/Kobrin
Pruzhany
Zhabinka

Vezyaryshcha
Surazh
Lyozna/Lëzno
Western Dvina
Harany/Goryany
Obal'
Bacheykava
Shumlina
Bahushewsk
Chashniki
Kruhlaye
Shklow
Horki/Gorki
Talachyn/Tolochin
Sava
Khodasy
Byalynichy
Belynichi
Dashkawka
Chachevichy
Abidavichy
Obidovichi
Aktsyabrski
Shchadryn
Brozha
Ptsich
Yasyel'da
Styr
Horyn'
Sluch

Elevation

-6000m -4000m -2000m -1000m -500m -250m   Below sea level 0 250m 500m 1000m 2000m 3000m 4000m 6000m

-19,658ft -13,124ft -6562ft -3281ft -1640ft -820ft -328ft/-100m 0   820ft 1640ft 3281ft 6562ft 9843ft 13,124ft 19,685ft

Ukraine, Moldova & Romania

Population

○ below 50,000

○ 50,000 to 100,000

● National capital

◉ 100,000 to 500,000

◼ above 500,000

0 km ___ 100

0 miles ___ 100

110

Following the annexation of the Ukrainian territory of Crimea by Russia in 2014, in February 2022 Russia launched a full-scale invasion of Ukraine, which has caused devastation to cities and infrastructure, and displacement of its people. Resistance from Ukraine has been bolstered by international political and military support for Ukraine and sanctions against Russia. The conflict is ongoing.

110

110

RUSSIA

110

Sea of Azov

Black Sea

116

Elevation

						Below sea level	0	250m	500m	1000m	2000m	3000m	4000m	6000m
-6000m	-4000m	-2000m	-1000m	-500m	-250m									

| -19,658ft | -13,124ft | -6562ft | -3281ft | -1640ft | -820ft | -328ft/-100m | 0 | | 820ft | 1640ft | 3281ft | 6562ft | 9843ft | 13,124ft | 19,685ft |

European Russia

0 km   300

0 miles   300

Population   ● National capital

○ below 50,000   ○ 50,000 to 100,000   ◉ 100,000 to 500,000   ◼ above 500,00

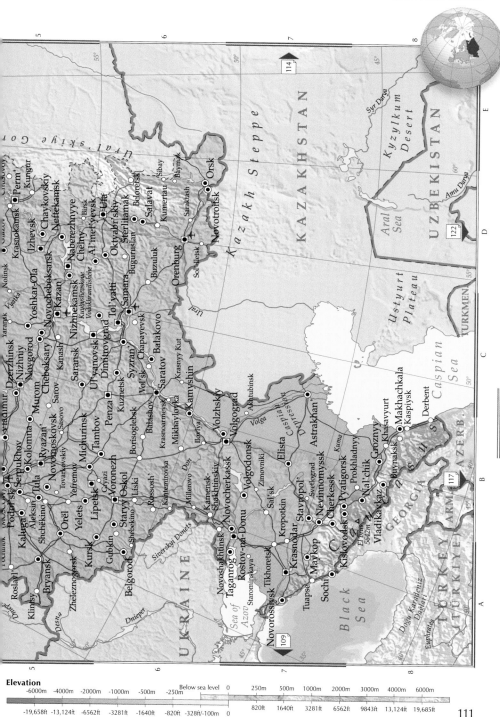

Elevation

					Below sea level							
-6000m	-4000m	-2000m	-1000m	-500m	-250m	0	250m	500m	1000m	2000m	3000m	4000m	6000m

| -19,658ft | -13,124ft | -6562ft | -3281ft | -1640ft | -820ft | -328ft/-100m | 0 | 820ft | 1640ft | 3281ft | 6562ft | 9843ft | 13,124ft | 19,685ft |

North & West Asia

155

Franz Josef Land

ARCTI

Severnaya Z

Ostrov Komsomolets

Ostrov Oktyabr'skoy Revolyutsii
Ostrov Bol'shevik

Summer limit of pack ice

Winter limit of pack ice

Novaya Zemlya

East Novaya Zemlya Trench

Kara Sea

Poluostrov Taymy

Poluostrov Yamal

North Sibe

Kheta

Central

Siberia

Platea

Noril'sk

Norwegian
Sea   North Cape

Barents
Sea

Ostrov
Kolguyev

Arctic Circle

81

Murmansk

Kola
Peninsula

White Sea

Archangel

Lake
Onega

Lake Ladoga

Northern
Dvina

Ob'

West Siberian
Plain

R   U   S   S
i

Pechora

Reka Nizhnaya

Stony Tunguska

Angara

Saint Petersburg

Yaroslavl'

Vologda

Nizhniy
Novgorod

Perm'

Yekaterinburg

Irtysh

Ob'

Tomsk

Chulym

Krasnoyarsk

Kaliningrad

MOSCOW

Central
Russian
Upland

Ul'yanovsk

Kazan'

Ufa

Chelyabinsk

Omsk

Novosibirsk

Novokuznetsk

KALININGRAD
(to Russia)

Voronezh

Saratov

Samara

Orenburg

Oral/Ural'sk

ASTANA

Qaraghandy/
Karagandy

Sayanskiy Khrebet

Semey

A

S

Volga

Ural Mountains

Baltic Sea

Gulf of Bothnia

E   U   R   O   P   E

Volgograd

Rostov-na-Donu

Don

Danube

Black Sea

Ural

Aral/Aral'sk

Kazakh
Steppe

Kazakh Uplands

Zhosaly Kolti

Ozero Zaysan

Altai Mountains

KAZAKHSTAN

Istanbul

Stavropol'

Astrakhan'

El'brus
18,510ft
(5642m)

Caucasus

Aktau

Ustyurt
Plateau

Dasoguz

Aral
Sea

Syr Darya

Kyzyl
Kum

Qyzylorda/Kyzylorda

Taraz

Lake
Balkhash

Ili

Almaty

Jengish Chokusu/Tömür Feng
24,406ft (7439m)

GEORGIA

ARMENIA

TBILISI

Küre Dağları

AZERB.

BAKU

Caspian Sea

UZBEKISTAN

Garagum

TASHKENT

BISHKEK

Tien Shan

KYRGYZSTAN

Istanbul

ANKARA

TURKEY
(TÜRKIYE)

YEREVAN

Tabriz

TURKMENISTAN

ASGABAT

DUSHANBE

TAJIKISTAN

Adana

Gaziantep

Mosul

Qom

THRAN

KABUL

Hindu Kush

Kunlun Mountains

CYPRUS

Aleppo

SYRIA

IRAQ

Isfahan

IRAN

Herat

AFGHANISTAN

Jalalabad

Khyber Pass

Mediterranean Sea

103

BEIRUT

LEBANON

ISRAEL

DAMASCUS

BAGHDAD

AMMAN

Basra

Iranian
Plateau

Zagros Mountains

Thar Desert

Himalayas

JERUSALEM

JORDAN

Dead Sea
-1411ft
(-430m)

KUWAIT

KUWAIT
CITY

An Nafud

Euphrates

Persian Gulf

Shiraz

Zahedan

Bandar-e Abbas

Dubai

Zagros Mountains

Indus Fan

Ganges

SAUDI
ARABIA

MANAMA

BAHRAIN

RIYADH

QATAR

DOHA

U.A.E.

ABU
DHABI

MUSCAT

Sur

Gulf of Oman

Murray Ridge

Ganges Fan

Tropic of Cancer

Jeddah

At Ta'if

Red Sea

Arabian
Peninsula

Ar Rub' al Khali

OMAN

Bay of
Bengal

Nile

AFRICA

SANAA

Ta'izz

Aden

YEMEN

Gulf of Aden

Socotra
(to Yemen)

Arabian
Sea

N

69

0 km   800

0 miles   800

Population ● National capital

○ below 50,000   ○ 50,000 to 100,000   ◉ 100,000 to 500,000   ▣ above 500,000

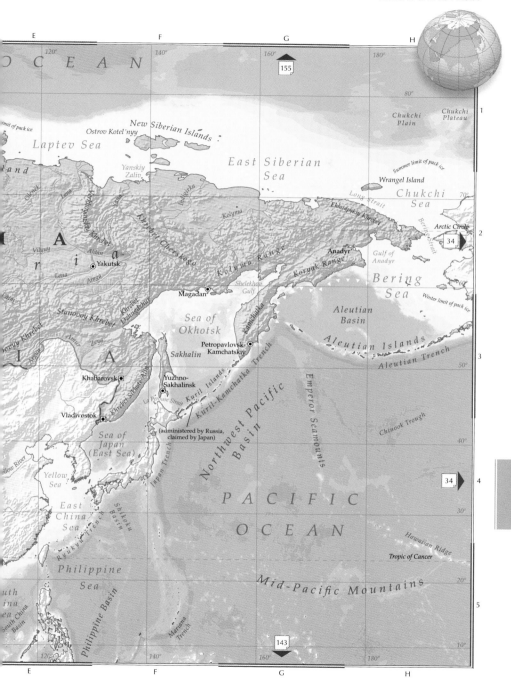

E 120° F 140° G 160°

155

80°

O C E A N

limit of pack ice

Chukchi
Plain

Chukchi
Plateau

1

Laptev Sea

Ostrov Kotel'nyy

New Siberian Islands

Yanskiy
Zaliv

East Siberian
Sea

Summer limit of pack ice

Wrangel Island

70°

Chukchi
Sea

Long Strait

Ekhabinskiy Khrebet

Arctic Circle

Bering Strait

34

2

Olenek

Lena

Verkhoyanskiy Khrebet

Aldan

Indigirka

Khrebet Cherskogo

Kolyma

Kolyma Range

Koryak Range

Anadyr

Gulf of
Anadyr

60°

i

a

Vilyuy

Yakutsk

Lena

Amga

Shelekhov
Gulf

Bering
Sea

Winter limit of pack ice

r

Magadan

Kamchatka

Aleutian
Basin

Stanovoy Khrebet

Khrebet
Dzhugdzhur

Sea of
Okhotsk

Aleutian Islands

50°

3

orvyy Khrebet

Amur

Leya

Petropavlovsk-
Kamchatskiy

Aleutian Trench

Khabarovsk

Yuzhno-
Sakhalinsk

Sakhalin

Sikhote Alin

Chinook Trough

Vladivostok

La Pérouse Strait

Kuril Islands

Kuril-Kamchatka Trench

Northwest Pacific
Basin

Emperor Seamounts

40°

4

Sea of
Japan
(East Sea)

(administered by Russia,
claimed by Japan)

34

ellow River

Yellow
Sea

Japan Trench

P A C I F I C

30°

East
China
Sea

Shikoku
Basin

Shikoku

O C E A N

Hawaiian Ridge

Tropic of Cancer

20°

Ryukyu Trench

Philippine
Sea

Mid-Pacific Mountains

uth
ina
ea
South China
Basin

Philippine Basin

Mariana Trench

143

10°

120° 140° 160° 180°

E F G H

Russia & Kazakhstan

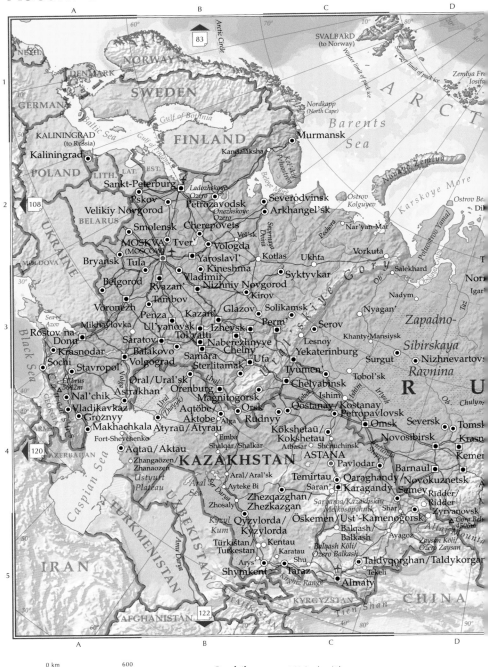

	A	B	C	D

NORWAY
SWEDEN
83
Arctic Circle
60°
70°
80°
10°
SVALBARD
(to Norway)
ARCTIC
Summer limit of pack-ice
Zemlya Fr
Iosifa

NETH.
DENMARK
70°
GERMANY
50°
KALININGRAD
(to Russia)
Kaliningrad
POLAND
LITH.
LAT.
EST.
Baltic Sea
Gulf of Bothnia
FINLAND
Gulf of Finland
Nordkapp
(North Cape)
Murmansk
Kandalaksha
Beloye More
Kol'skiy
Poluostrov
Barents Sea
Winter limit of pack-ice
Nordkapp
Novaya Zemlya
Karskoye More
Ostrov
Kolguyev
Nar'yan-Mar
Poluostrov Yamal
Ostrov Be
Di

BELARUS
Sankt-Peterburg
Ladozhskoye
Ozero
Pskov
Velikiy Novgorod
Petrozavodsk
Onezhskoye
Ozero
Vel'sk
Severodvinsk
Arkhangel'sk
Severnaya
Dvina
Pechora

UKRAINE
MOLDOVA
30°
Smolensk
Cherepovets
Vologda
MOSKVA
(MOSCOW)
Tver
Bryansk
Tula
Yaroslavl'
Kineshma
Vladimir
Nizhniy Novgorod
Kirov
Kotlas
Syktyvkar
Ukhta
Vorkuta
Salekhard
Nadym
Nor
Igar'
Ob'
Taz
Zapadno-

Belgorod
Ryazan'
Tambov
Voronezh
Penza
Kazan'
Glazov
Solikamsk
Perm'
Serov
Khanty-Mansiysk
Nyagan'
Sibirskaya

Mikhaylovka
Ul'yanovsk
Izhevsk
Lesnoy
Yekaterinburg
Surgut
Nizhnevartovsk
Ravnina

Sea of Azov
Rostov-na-Donu
Saratov
Tol'yatti
Naberezhnyye
Chelny
Ufa
Tyumen'
Chelyabinsk
Tobol'sk
Ob'
Chulym
R **U**

Krasnodar
Balakovo
Volgograd
Samara
Sterlitamak
Ural
Ishim
Irtysh

Sochi
Stavropol'
Elbrus
5642m
Nal'chik
Astrakhan'
Oral/Ural'sk
Orenburg
Magnitogorsk
Aqtöbe/
Aktobe
Alga
Orsk
Rudnyy
Qostanay/Kostanay
Petropavlovsk
Omsk
Novosibirsk
Tomsk
Krasn

GEORGIA
Vladikavkaz
Groznyy
Makhachkala
Fort-Shevchenko
Shalkar/Shalqar
Atyraū/Atyrau
Emba
Kökshetaū/
Kokshetau
Atbasar
Shchuchinsk
ASTANA
Pavlodar
Barnaul
Novokuznetsk
Kemer
A

AZERBAIJAN
Aqtaū/Aktau
Zhangaözen/
Zhanaozen
Ustyurt
Plateau
KAZAKHSTAN
Temirtau
Qaraghandy/
Karagandy
Semey
Ridder
Zyryanovsk

Caspian Sea
Aral/Aral'sk
Ayteke Bi
Saran'
Sarı arqa/Kazakhskiy
Melkosopochnik
Shar
Gora Belukha
4506m
Altai
Mountai

TURKMENISTAN
Aral Sea
Syr Darya
Zhosaly
Zhezqazghan/
Zhezkazgan
Qyzylorda/
Kyzylorda
Öskemen/
Ust'-Kamenogorsk
Balqash/
Balkash
Ayagoz
Zaysan Köli/
Ozero Zaysan
UZBEKISTAN
Kyzyl
Kum
Balqash Köli/
Ozero Balkash

IRAN
Amu Darya
Türkistan/
Turkestan
Kentau
Karatau
Shu
Taldyqorghan/Taldykorgan
Tekeli
CHINA

TAJIKISTAN
AFGHANISTAN
122
Arys
Shymkent
Taraz
Kyrgyz Range
Almaty
KYRGYZSTAN
Tien Shan
30°
60°
70°
80°
90°
40°

	A	B	C	D

0 km 600
0 miles 600

Population
○ below 50,000
○ 50,000 to 100,000
◉ 100,000 to 500,000
◼ above 500,00
● National capital

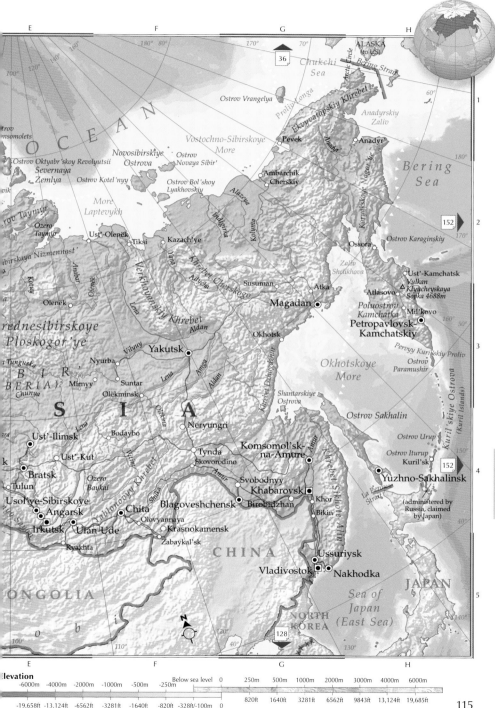

ALASKA
(to US)

Chukchi
Sea

36

Arctic Circle

Bering Strait

Ostrov Vrangelya

Anadyrskiy
Zaliv

O C E A N

trov
nsomolets

Vostochno-Sibirskoye
More

Pevek

Anabar

Anadyr'

Bering
Sea

Novosibirskiye
Ostrova

Ostrov
Novaya Sibir'

Ostrov Oktyabr'skoy Revolyutsii
Severnaya
Zemlya

Ostrov Kotel'nyy

Ostrov Bol'shoy
Lyakhovskiy

Ambarchik
Cherskiy

Ekvyvatapskiy Khrebet

Koryakskoye Nagor'ye

152

170°

More
Laptevykh

trov Taymyr

Alazeya

Ossora

Ostrov Karaginskiy

Ozero
Taymyr

Ust'-Olenëk

Tiksi

Kazach'ye

Indigirka

Kolyma

Zaliv
Shelikhova

Ust'-Kamchatsk
Vulkan
Klyuchevskaya
Sopka 4688m

vik

irskaya Nizmennost'

Anabar

Olenëk

Yana

Khrebet Cherskogo

Susuman

Atka

Atlasovo

Poluostrov
Kamchatka

Mil'kovo

Konyу

Olenëk

Lena

Adycha

Verkhoyanskiy

Magadan

Okhotsk

Petropavlovsk-
Kamchatskiy

rednesibirskoye
Ploskogor'ye

Nyurba

Vilyuy

Yakutsk

Khrebet

Aldan

Pervyy Kuril'skiy Proliv

B I R '
BERIA)
S I A

Tunguska

Chunya

Mirnyy

Suntar

Olëkminsk

Amga

Aldan

Khrebet Dzhugdzhur

Okhotskoye
More

Ostrov
Paramushir

Lena

Olëkma

Neryungri

Shantarskiye
Ostrova

Ostrov Sakhalin

Ust'-Ilimsk

Bodaybo

Ostrov Urup

Ust'-Kut

Tynda

Komsomol'sk-
na-Amure

Ostrov Iturup
Kuril'sk

152

Bratsk

Ozero
Baykal

Skovorodino

Amur

Svobodnyy

Amur

Kuril'skiye Ostrova
(Kuril Islands)

Tulun

Vitim

Yablonovyy Khrebet

Khabarovsk

Yuzhno-Sakhalinsk

Usol'ye-Sibirskoye
Angarsk

Chita

Blagoveshchensk

Birobidzhan

Khor

La Perouse
Strait

Irkutsk

Ulan-Ude

Olovyannaya

Shilka

Krasnokamensk

Bikin

(administered by
Russia, claimed
by Japan)

Ryakhta

Zabaykal'sk

C H I N A

Khrebet Sikhote Alin'

ONGOLIA

o   b   i

Ussuriysk

Vladivostok

Nakhodka

JAPAN

Sea of
Japan
(East Sea)

N

NORTH
KOREA

128

Elevation

| | | | | | Below sea level | 0 | 250m | 500m | 1000m | 2000m | 3000m | 4000m | 6000m |
| -6000m | -4000m | -2000m | -1000m | -500m | -250m | | | | | | | | |

-19,658ft -13,124ft -6562ft -3281ft -1640ft -820ft -328ft/-100m 0   820ft 1640ft 3281ft 6562ft 9843ft 13,124ft 19,685ft

Turkey & the Caucasus

ROMANIA

Lacul Sinoie

108

N

UKRAINE

Krym (Crimea)

Danube

BULGARIA

Varnenski Zaliv

B l a c k   S e a

Maritsa

Burgaski Zaliv

104

Kırklareli

Edirne

Cide

İnebolu

Sinop

Gerze

Bartın

K ü r e   D a ğ l a r ı

Kastamonu

Bafra

Ergene Çayı

Çorlu

Zonguldak

Karabük

Kargı

Samsun

Tekirdağ

Marmara Denizi (Sea of Marmara)

İstanbul

Devrek

Çerkeş

Canık Dağları

İzmit

Adapazarı

Gerede

Çankırı

Kızıl Irmak

Merzifon

Bandırma

Yalova

İznik Gölü

Bolu

Çorum

Tokat

Çanakkale

Bursa

Bilecik

Kalecik

Alaca

Yıldıze

Çanakkale Boğazı (Dardanelles)

Balıkesir

Bozüyük

Eskişehir

ANKARA

Kırıkkale

Sorgun

Edremit

Ayvalık

Kütahya

T U R K E Y

Boğazlıyan

Şarkışla

Lésvos

Akhisar

Simav

Gediz

Polatlı

Hırfanlı Barajı

Bünyan

Gürün

Chíos

Menemen

Manisa

Uşak

Afyon

Kulu

Tuz Gölü

T Ü R K İ Y E

İncesu

Kayseri

İzmir

Ödemiş

Alaşehir

Cihanbeyli

Nevşehir

Sámos

Aydın

Nazilli

Büyükmenderes Nehri

Dinar

Akşehir

Aksaray

Göksun

Söke

Denizli

Beyşehir Gölü

A n a t o l i a

Niğde

Burdur Gölü

Konya

Kahraman Gazi

Milas

Muğla

Tavas

İsparta

Burdur

Suğla Gölü

Ereğli

Karaman

T o r o s   D a ğ l a r ı

Ceyhan

Osmani

Bodrum

105

Marmaris

Dalaman

Antalya

Manavgat

Mut

Tarsus

Adana

İskenderun

Kilis

Dodekánisa (Dodecanese)

Fethiye

Kaş

Finike

Antalya Körfezi

Alanya

Silifke

Antakya

Kırıkha

Ródos (Rhodes)

Anamur

Kárpathos

CYPRUS

TURKISH REPUBLIC OF NORTHERN CYPRUS (recognized only by Turkey)

Orontes

M e d i t e r r a n e a n   S e a

72

LEBANON

---

0 km   200

0 miles   200

Population

○ below 50,000

○ 50,000 to 100,000

● 100,000 to 500,000

■ above 500,000

● National capital

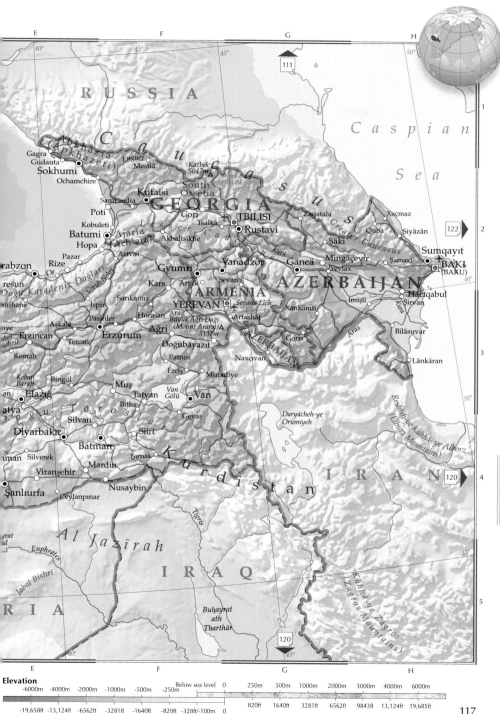

Elevation

-6000m	-4000m	-2000m	-1000m	-500m	-250m	Below sea level	0	250m	500m	1000m	2000m	3000m	4000m	6000m
-19,658ft	-13,124ft	-6562ft	-3281ft	-1640ft	-820ft	-328ft/-100m	0	820ft	1640ft	3281ft	6562ft	9843ft	13,124ft	19,685ft

Eastern Mediterranean

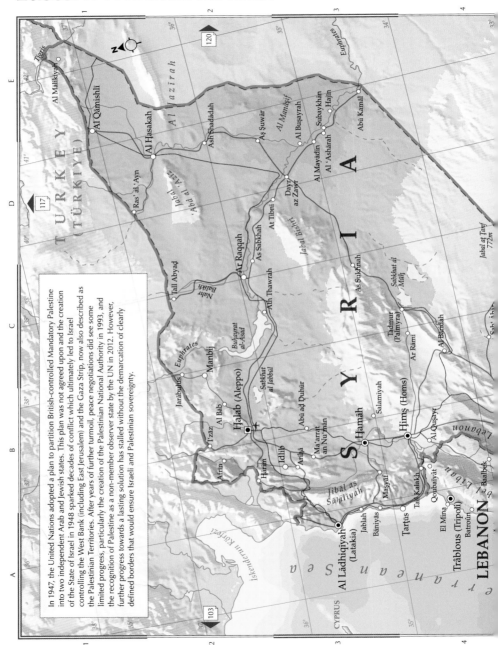

In 1947, the United Nations adopted a plan to partition British-controlled Mandatory Palestine into two independent Arab and Jewish states. This plan was not agreed upon and the creation of the State of Israel in 1948 sparked decades of conflict which ultimately led to Israel controlling the West Bank (including East Jerusalem) and the Gaza Strip, now also described as the Palestinian Territories. After years of further turmoil, peace negotiations did see some limited progress, particularly the creation of the Palestinian National Authority in 1993, and the recognition of Palestine as a non-member observer state by the UN in 2012. However, further progress towards a lasting solution has stalled without the demarcation of clearly defined borders that would ensure Israeli and Palestinian sovereignty.

TURKEY
(TÜRKİYE)

Al Māliķīyah
Al Qāmishlī
Al Ḩasakah
Ras' al 'Ayn

Al Jazīrah
'Abd al 'Azīz

Ash Shadādah
As Şuwār
Al Manājif
Al Buşayrah
Subaykhān
Ḩajīn
Abū Kamāl

Al Mayādīn
Al 'Asharah

Dayr
az Zawr
At Tibnī

Jabal Bishrī

Tall Abyaḑ
Nahr
Balīkh

Ar Raqqah
As Sabkhah
Ath Thawrah

As Sukhnah

Sabkhat al
Mūḩ

S Y R I A

Jabal aţ Ţanf
772m

Tadmur
(Palmyra)
Ar Rami

Al Baridah

Euphrates
Jarābulus
Manbij
Buḩayrat
al-Asad

Sabkhat
al Jabbūl

Ḩalab (Aleppo)
Al Bāb
A'zāz
Afrīn
Ḩārim
Idlib
Arīḩā
Ma'arrat
an Nu'mān

Abū aḑ Ḑuhūr

Salamīyah

Ḩamāh
Ḩimş (Homs)
Al Quşayr

Masyāf

Jibāl as
Sāḩilīyah

İskenderun Körfezi

Al Lādhiqīyah
(Latakia)
Jablah
Bāniyās
Tarţūs
Tall Kalakh
Qouḩaiyāt

El Mina
Ţrāblous (Tripoli)
Baţroûn

Baalbek

Lebanon

Jebel Liban

LEBANON

Mediterranean Sea

CYPRUS

0 km   100

0 miles   100

Population   ● National capital

○ below 50,000   ○ 50,000 to 100,000   ◉ 100,000 to 500,000   ◼ above 500,000

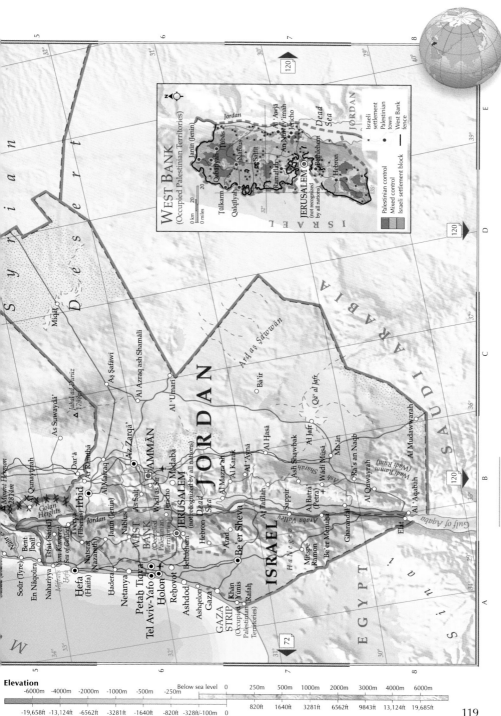

WEST BANK
(Occupied Palestinian Territories)

Janīn (Jenin)
Qalqīlyah
Ţūbās
Nāblus
Tūlkarm
Qalqīlyah
Salfīt
Al 'Awjā
An Nuway'imah
Jericho
Ramallah
Ar Ram
JERUSALEM
(not recognized
by all nations)
Bethlehem
Hebron

Jordan

Dead Sea

JORDAN

ISRAEL

Israeli settlement
Palestinian town
West Bank fence

Palestinian control
Mixed control
Israeli settlement block

0 km 20
0 miles 20

120

120

120

72

Elevation

					Below sea level								
-6000m	-4000m	-2000m	-1000m	-500m	-250m	0	250m	500m	1000m	2000m	3000m	4000m	6000m
-19,658ft	-13,124ft	-6562ft	-3281ft	-1640ft	-820ft -328ft/-100m	0	820ft	1640ft	3281ft	6562ft	9843ft	13,124ft	19,685ft

Southwest Asia

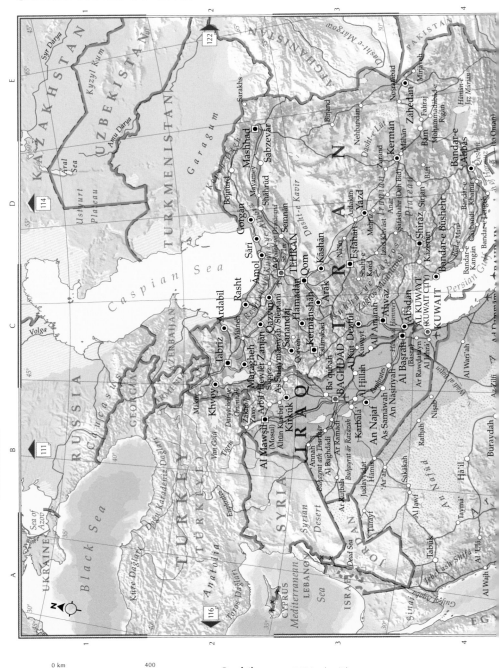

Population

● National capital

○ below 50,000   ○ 50,000 to 100,000   ◉ 100,000 to 500,000   ▣ above 500,000

0 km   400

0 miles   400

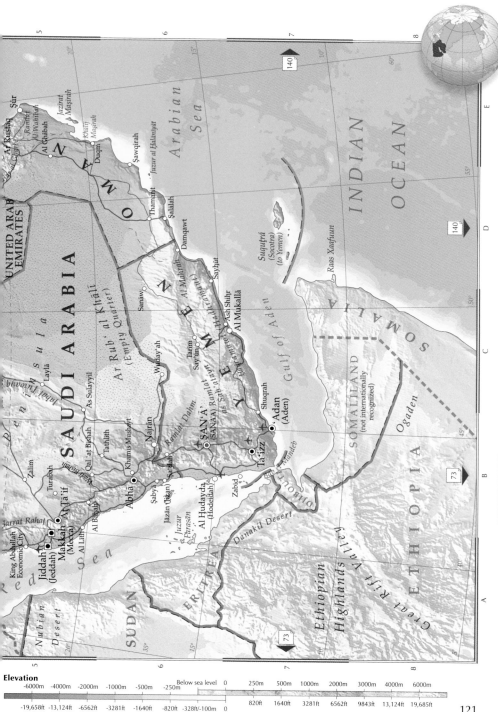

Şūr
Afrāsiāq
Ramlat
Al Wahibah
Jazīrat
Masīrah
Al Ghabah
Khalīj
Masīrah
Duqm
Şawqirah

Arabian Sea

O M A N

UNITED ARAB
EMIRATES

Thamarīt
Jazīrat al Ḩalāniyāt
Salalah

Damqawt

Y E M E N
Al Mahrah

Arabian

Peninsula

Sanāw

Suquţrá
(Socotra)
(to Yemen)

Raas Xaafuun

INDIAN

OCEAN

Ar Rub' al Khālī
(Empty Quarter)

S A U D I   A R A B I A

As Sulayyil

Wadīyah
Tarīm
Sayhūt
Say'ūn
Ash Shiḩr
Al Mukallā

Hadhramawt

Hadhramaut

Layla

Gulf of Aden

S O M A L I A

Zalim

Najrān
Ramlat Dahm
Ramlat Sab'atayn
SAN'Ā'
(SANAA)
Shuqrah
Adan
(Aden)

SOMALILAND
(not internationally
recognized)

Ogaden

Turabah
Tathlīth
Qal' at Bishah
Khamīs Mushayt
Sardah
Ta'izz

Abhā
Şabyā

Al Bāḩah

Zabīd

Danakil Desert

Ethiopian

E T H I O P I A

Jarrat Rahat
Makkah
(Mecca)

At Ţā'if
Jāzān (Īzān)
Juzur
Farasān
Al Lith
Al Hudaydah
(Hodeidah)

Bab el Mandeb

King Abdullah
Economic City
Jiddah
(Jeddah)

E R I T R E A

Highlands

Great Rift Valley

Red   Sea

Nubian
Desert

S U D A N

Elevation

-6000m	-4000m	-2000m	-1000m	-500m	-250m	Below sea level 0	250m	500m	1000m	2000m	3000m	4000m	6000m
-19,658ft	-13,124ft	-6562ft	-3281ft	-1640ft	-820ft/-328ft/-100m	0	820ft	1640ft	3281ft	6562ft	9843ft	13,124ft	19,685ft

Central Asia

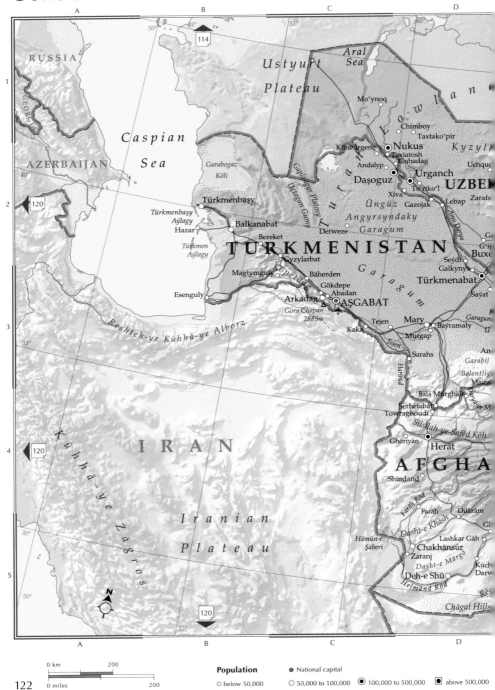

RUSSIA

GEORGIA

AZERBAIJAN

Caspian Sea

Ustyurt Plateau

Aral Sea

Mo'ynoq

Chimboy

Taxtako'pir

Köneürgenç

Nukus

Taxiatosh

Andalyp

Cubadag

Uchquc

Garabogaz Köli

Daşoguz

Urganch

UZBEK

Xiva

To'rtko'l

Zarafs

Türkmenbaşy

Uçtagan Gumy

Ungüz

Gazojak

Lebap

Türkmenbaşy Aýlagy

Balkanabat

Bereket

Derweze

Angyrsyndaky

Garagum

G'ij

Hazar

Gö

Buxo

Türkmen Aýlagy

TURKMENISTAN

Seýdi

Galkynyş

Türkmenabat

Magtymguly

Gyzylarbat

Bäherden

Gökdepe

Garagum

Esenguly

Arkadag

Abadan

AŞGABAT

Gora Chapan 2889m

Kaka

Tejen

Mary

Bayramaly

Sayat

Garagum

Murgap

An

Reshteh-ye Kūhhā-ye Alborz

Tejen

Sarahs

Garabil

Belentlig

Main

Harirūd

Bālā Murghāb

ye M

Serhetabat

Towraghoudi

Silsilah-ye Safēd Kōh

Ghōriyān

Herāt

I R A N

AFGHA

Shindand

Kūhhā-ye Zagros

Iranian Plateau

Farah

Farāh Rōd

Dilārām

Gi

Dasht-e Khāsh

Lashkar Gāh

Hāmūn-e Şāberi

Chakhānsūr

Küch

Zaranj

Dasht-e Margo

Darw

Deh-e Shū

Helmand Rōd

Rēs

Chāgai Hills

0 km 200
0 miles 200

Population

● National capital

○ below 50,000

○ 50,000 to 100,000

◉ 100,000 to 500,000

◼ above 500,000

KAZAKHSTAN

Balqash Köli/
Ozero Balkash

Peski
Moyynkum

Peski
Saryyesik-Atyrau

Peski Taukum

Borohoro Shan

Sur Darya

Ili

BISHKEK
Kara-Balta   Tokmok
Talas   Kemin   Tüp   Jyrgalang
Gora Manas   Bakay-Ata   Ysyk-Köl   Karakol
4482m   Balykchy   Kyzyl-Suu
TOSHKENT   Jengish Chokusu/
(TASHKENT)   KYRGYZSTAN   Kaji-Say   Kara-Say   Tömür Feng
Yangiyo'l   7439m
Chirchiq   Tash-Kömür   Naryn   Karakol
Angren   Namangan   Jalal-Abad
Olmaliq   Chatyr-Tash
Guliston   Bekobod   Qo'qon   Andijon   Kök-Art
Khujand   Farg'ona   Osh
Kattaqo'rg'on   Sary-Tash
Samarqand   Istaravshan   Sülüktü   Aydarken   Daroot-Korgon   Qarokül
Kitob   Urgut   Zeravshan
Qarshi   DUSHANBE   Qullai Ismoili Somoni
Denov   Norak   Qalaikhumb   Ghüdara   7495m
Boysun   Danghara   Murghob
Bokhtar   Kŭlob   Maskav
Termiz   Jarqo'rg'on   Dŭsti   Farkhor   Khorugh
Balkh   Kunduz   Talugan   Faizabad   Jelondi   Qizilrabot
Mazar-e   Khänäbäd   Ishkoshim
Sharif   Baghlän
Pul-e Khumri
Bari Köl
Charikär   Mahmüd-e Räqi
KÄBUL   Asadäbäd
Maidän Shahr   Mehtar Läm
Jaläläbäd
Ghazni   Khyber Pass
Gardez   1080m
Khöst   (A 'line of control
was agreed between
Zarghün   India and Pakistan
Shahr   in 1972)
Qalät
dahär
Boldak
Toba Käkar Range

XINJIANG
UYGUR
ZIZHIQU

Taklimakan
Shamo

CHINA

(claimed by India)

AKSAI CHIN
(administered by China,
claimed by India)

Aksai
Chin

(administered
by Pakistan,
claimed by India)

Karakoram Range

DÊMQOG/
DEMCHOK
(administered by China,
claimed by India)

XIZANG
ZIZHIQU
(Tibet)

(administered by India,
claimed by Pakistan)

(administered by China,
claimed by India)

Himalayas

PAKISTAN   INDIA

NEPAL

126
126
134

Elevation

-6000m	-4000m	-2000m	-1000m	-500m	-250m	Below sea level	0	250m	500m	1000m	2000m	3000m	4000m	6000m
-19,658ft	-13,124ft	-6562ft	-3281ft	-1640ft	-820ft	-328ft/-100m	0	820ft	1640ft	3281ft	6562ft	9843ft	13,124ft	19,685ft

South & East Asia

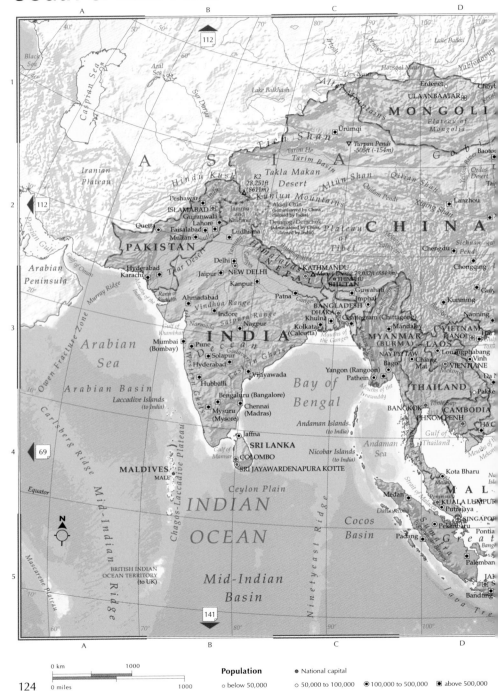

Population   ● National capital

○ below 50,000   ○ 50,000 to 100,000   ◉ 100,000 to 500,000   ◼ above 500,000

0 km   1000

0 miles   1000

E   F   G   H

113

Sakhalin

Kuril Islands

Qiqihar

Manchuria
Plain •Harbin *Lake Khanka* Sapporo Hokkaido

Changchun

Lino He

JAPAN

Northwest
Pacific
Basin

Shatskiy Rise

Mapmaker Seamounts

henyang •

NORTH
KOREA
PYONGYANG Sendai

Sea of
Japan
(East Sea)

anjin • •Dalian

Bo Hai

SEOUL **SOUTH**
KOREA

Sejong

152

nan •

•Qingdao

Nagoya Yokohama
Kyoto •TOKYO
Osaka *Fuji-san*
Hiroshima *12,389ft (3776m)*

Shikoku
Kitakyushu
Kyushu

Yellow
Sea

ou •

ing •

East China •Shanghai
Sea

Shikoku Basin

Mid-Pacific Mountains

Hangzhou •

Ryukyu Trench

Ryukyu Islands

K y u s h u

P A C I F I C

Nanchang •

ha

Fuzhou • •TAIPEI

TAIWAN

gzhou • •Kaohsiung

Tokara Trench

P h i l i p p i n e   S e a

West
Mariana
Basin

O C E A N

East
Mariana
Basin

ng Kong •

Luzon Strait

EL
OS
ed)

Baguio •

Luzon

B a s i n

Mariana
Trench

Melanesian
Basin

•Mindoro

MANILA/MAYNILA

PHILIPPINES

Samar

M i c r o n e

s i

h China
Sea

ILY ISLANDS
disputed)

Panay

•Bacolod •Cebu

Negros

Palawan

Sulu
Sea

•Davao

•Zamboanga Mindanao

Philippine Basin

P a l a u   R i d g e

Yap Trench

Eauripik Rise

Ontong
Java
Rise

a

Equator

152

BANDAR
SERI BEGAWAN

A

Celebes
Sea

Manado • Halmahera

M e l a n e

s i a

neo

kpapan

Moluccas

Jayapura •

Bismarck Archipelago

Solomon
Islands

n d a

I s l a n d s

Serui •
Ambon •

Pegunungan Maoke

Solomon
Sea

D O N E S I A

•Banjarmasin

Buru

•Makassar

Celebes *Banda Sea*

New Guinea

baya •

Flores
Sea

Lesser Sunda Islands

Flores DILI •

Sumba *Timor*

EAST TIMOR
(TIMOR-LESTE)

Arafura
Sea

Coral
Sea

ng • *Bali*

Timor Trough

Timor
Sea

AUSTRALIA

142

E   F   G   H

Western China & Mongolia

R U S S

Kulunda
Steppe

Yenisey

Zapadnyy Sayan

Hövsgöl
Nuur

K A Z A K H S T A N

Uvs Nuur

Ulaangom

Olgiy

Hyargas
Nuur

Har Us Nuur

Har Nuur

Hangayn Nuruu

Tsetser

Saryarqa/
Kazakhskiy Melkosopchnik

Zaysan Köli/
Ozero Zaysan

Altay

Hovd

M O N

123

Balqash Köli
(Ozero Balkash)

Ulungur
Hu

Karamay

Gurbantünggüt
Shamo

Altay

Bayanhongo

Borohoro Shan

Kuytun

Shihezi

Fukang

Jimsar

Aj Bogd Uul
3802m

Yining

Ürümqi

Qitai

Atas Bogd
2695m

Ysyk-Köl

Tien Shan

Jengish Chokusu/Tömür Feng
7439m

Turpan

Hami

G

K Y R G Y Z S T A N

Bosten Hu

Turpan
Pendi

Dalain H

Korla

Kuruktag

Xingxingxia

GANSU

Kashi

Tarim He

Tarim Basin

Lop Nur

Qilian Shan

Yengisar

Shache

X I N J I A N G   U Y G U R

Z I Z H I Q U

Ruoqiang

Danghe Nanshan

Qinghai

Yecheng
(claimed
by India)

Pishan

Taklimakan

Shamo

Altun Shan

Qaidam Pendi

Moyu

K2
8611m

Hotan

Qira

Kunlun Shan

Golmud

Burhan Budai Shan

Dulan

AKSAI
CHIN

(AKSAICHIN
administered by
China, claimed
by India)

Qingzang Gaoyuan
(Plateau of Tibet)

C H

Q I N G H A I

134

JAMMU
AND
KASHMIR

Rutog

Tongtian He

Bayan Har Shan

Yushu

Mekong

DEMQOG/DEMCHOK
(administered by China,
claimed by India)

Gar Xincun

Zanda

X I Z A N G

Tanggula Shan

Amdo

Qamdo

Z I Z H I Q U

Gozhê

Siling Co

Nagqu

Tangra
Yumco

Gyaring
Co

Nam Co

Damxung

(Tibet)

Ngangzê
Co

Nyainqêntanglha Shan

Brahmaputra

Lhazê

Xigazê

Maizhokunggar

ARUNACHAL
PRADESH
(claimed by China)

Lhasa

Yamuna

N E P A L

Mount Everest
8849m

Gyangzê

Gonggar

Ganges

H

a

l

a

y

a

s

I N D I A

I N D I A

MYANMAR
(BURMA)

135

BHUTAN

0 km   400

0 miles   400

Population

○ below 50,000

○ 50,000 to 100,000

● National capital

◉ Internal administrative capital

◉ 100,000 to 500,000

■ above 500,000

RUSSIA

HEILONGJIANG

Baykal

Shilka

Argun (Ergun He)

Ergun

Jagdaqi

Amur (Heilong Jiang)

Lake
Khanka

Onon

Hulun Buir
(Hailar)

Manzhouli

Hulun
Nur

Sühbaatar

Selenga

Darhan
net

Choybalsan

Onon Gol

A

JILIN

Da Hinggan Ling

Menengiyn
Tal

ULAANBAATAR

Dzuunmod

Ondörhaan

Kerulen

Baruun-Urt

Holin Gol

Tongliao

128

Sea of
Japan
(East Sea)

L I A

Saynshand

Erenhot

Xilinhot

Chifeng
(Ulanhad)

Liao He

135°

Dalandzadgad

Nuriu

Mongolia (Inner Mongol) ZIZHIQU

Ulan Qab (Jining)

NORTH
KOREA

SOUTH
KOREA

i

Lang Shan

Hohhot

Baotou

Huang He
(Yellow River)

Linxi

Korea
Bay

Bo Hai

Liaodong Wan

TIANJIN

Japan

129

EI MONGOL

Wuhai
(Haibowan)

Mu Us
Shadi

HEBEI

SHANDONG

Yellow
Sea

gger
hamo

Great Wall of China

NINGXIA

SHANXI

Huang He (Yellow River)

JIANGSU

East

g

N

A

GANSU

SHAANXI

Han Shui

HENAN

ANHUI

SHANGHAI SHI

China

129

HUAN

HUBEI

ZHEJIANG

Sea

Chang Jiang (Yangtze)

CHONGQING

JIANGXI

Nansei-shotō
(to Japan)

HUNAN

FUJIAN

YUNNAN

GUIZHOU

Tropic of Cancer

TAIWAN

129

Elevation

| -6000m | -4000m | -2000m | -1000m | -500m | -250m | Below sea level | 0 | 250m | 500m | 1000m | 2000m | 3000m | 4000m | 6000m |

| -19,658ft | -13,124ft | -6562ft | -3281ft | -1640ft | -820ft | -328ft/-100m | 0 | 820ft | 1640ft | 3281ft | 6562ft | 9843ft | 13,124ft | 19,685ft |

127

Eastern China & Korea

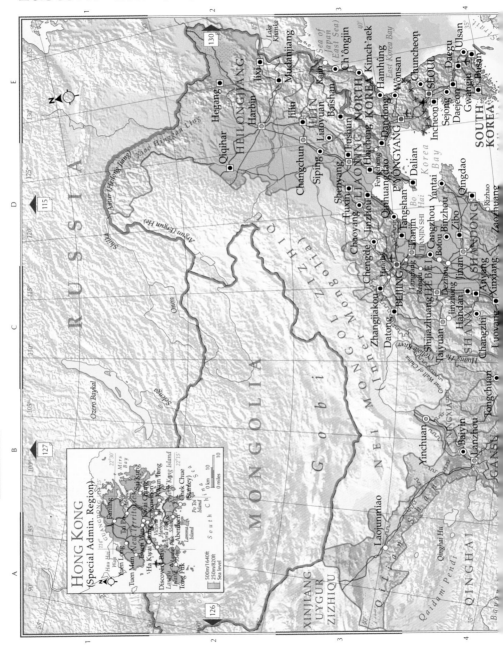

Population

- National capital
- Internal administrative capital
- ○ below 50,000
- ○ 50,000 to 100,000
- ◉ 100,000 to 500,000
- ◼ above 500,000

0 km 400
0 miles 400

PACIFIC

OCEAN

PHILIPPINES

East China
Sea

Okinawa

Nansei-shotō (Ryukyu Islands)

Tropic of Cancer

Keelung

TAIPEI

TAIWAN

Taichung

Chiayi

Tainan

Kaohsiung

Shanghai

Ningbo

Wenzhou

Suzhou

Wuxi

Yaxing

Hangzhou

Jinhua

Shangrao

Fuzhou

Quanzhou

Xiamen

Shantou

Dongguan

Hong Kong
(S.A.R.)

Macao
(Special Administrative Region)

HEBEI

Hefei

ANHUI

Wuhu

Anqing

Wuhan

HUBEI

Xichang

Jingdezhen

ZHEJIANG

Nanchang

JIANGXI FUJIAN

Jiujiang

Yichang

Huangshi

Yueyang

Changsha

Loudi

Xiangtan

Nanping

Longyan

Yong'an

Fuzhou

Zhangzhou

GUANGDONG

Shaoguan

Ganzhou

GUANGXI

Liuzhou

Zhaoqing

Jiangmen

Maoming

Zhanjiang

Haikou

Hainan Dao

HAINAN

Danzhou

Dongfang

Beihai

Xuwen

Qinzhou

Yulin

Nanning

Guilin

Quanzhou

Hengyang

Yongzhou

Chenzhou

Guiyang

GUIZHOU

Anshun

Zunyi

Huaihua

HUNAN

Shaoyang

Dongting Hu

CHONGQING

Chongqing

Neijiang

SICHUAN

Zigong

Leshan

Ya'an

Chengdu

Deyang

Panzhihua

Xichang

Hengduan Shan

YUNNAN

Kunming

Dali

Baoshan

Wuliang Shan

Lincang

Salween

Mekong

Chang Jiang

Min Shan

Sichuan Pendi

Wanzhou

Luzhou

Yibin

Tiger Leaping Gorge

Red River

MYANMAR
(BURMA)

INDIA

LAOS

THAILAND

CAMBODIA

VIETNAM

Gulf of Tonkin

Gulf of Thailand

Mekong

South China

Sea

Luzon Strait

Tropic of Cancer

PARACEL
ISLANDS
(disputed by China,
Taiwan and Vietnam)

Amphitrite Group

Crescent Group

Triton Island

Thitu
Island

Flat Island

Nanshan Island

Loaita Island

Namyit Island

Lansdowne Reef

Spratly Island

SPRATLY ISLANDS
(disputed by Brunei, China,
Malaysia, Philippines,
Taiwan and Vietnam)

139

136

136

Elevation

					Below sea level								
-6000m	-4000m	-2000m	-1000m	-500m	-250m	0	250m	500m	1000m	2000m	3000m	4000m	6000m
-19,658ft	-13,124ft	-6562ft	-3281ft	-1640ft	-820ft	-328ft/-100m	0	820ft	1640ft	3281ft	6562ft	9843ft	13,124ft	19,685ft

Japan

Kuril Islands
(administered by Russia,
claimed by Japan)

Kuril'sk

Ostrov
Iturup

Ostrov
Shikotan

Ostrov
Kunashir

Nemuro

Akkeshi

Kushiro

Shari

Kitami

Abashiri

Asahi-dake
2290m

Obihiro

Hiroshiri-dake
2052m

Mombetsu

Ebetsu

Chitose

Tomakomai

Noboribetsu

Muroran

Hakodate

Nayoro

Shibetsu

Takikawa

Asahikawa

Otaru

Sapporo

Iwanai

Wakkanai

Rebun-tō

Rishiri-tō

Sea of
Okhotsk

La Pérouse Strait

Ostrov Sakhalin
(to Russia)

Hokkaidō

Ishikari-wan

Uchiura-wan

Oshiri-tō

Tsugaru-kaikyō

Mutsu-wan

Hachinohe

Kuji

Iwate

Morioka

Miyako

Kesennuma

Shizugawa

Ishinomaki

Aomori

Odate

Yokote

Shinjō

Funakawa

Goshogawara

Hirosaki

Noshiro

Gojome

Honjō

Akita

Sakata

Tsuruoka

Sea of

R U S S I A

Amur

C H I N A

TŌKYŌ

Sumitomo
Building

Tōkyō University

National Museum

Tōkyō Stock Exchange

Imperial Palace

Tōkyō Tower

World Trade
Center

Kawasaki

Yokohama

Yokohama
Bay Bridge

Haneda

Tōkyō Bay

Chiba

z

0 km   10

0 miles   10

■ Places of interest

□ Regions/suburbs

NANSEI-
SHOTŌ

Kyūshū

Ōsumi-shotō

Satsunan-shotō

Amami-guntō

Amami-
ō-shima

Naze

Okinawa

Naha

Okinoerabu

Tokunoshima

Ishigaki-jima

Iriomote-jima

Senkaku-
shotō

Sakishima-shotō

N a n s e i - s h o t ō
(Ryūkyū Islands)

0 km   100

0 miles   100

z

500m/1640ft

Sea level

115

152

115

128

Population

○ below 50,000

○ 50,000 to 100,000

◉ 100,000 to 500,000

National capital

■ above 500,00

0 km   200

0 miles   200

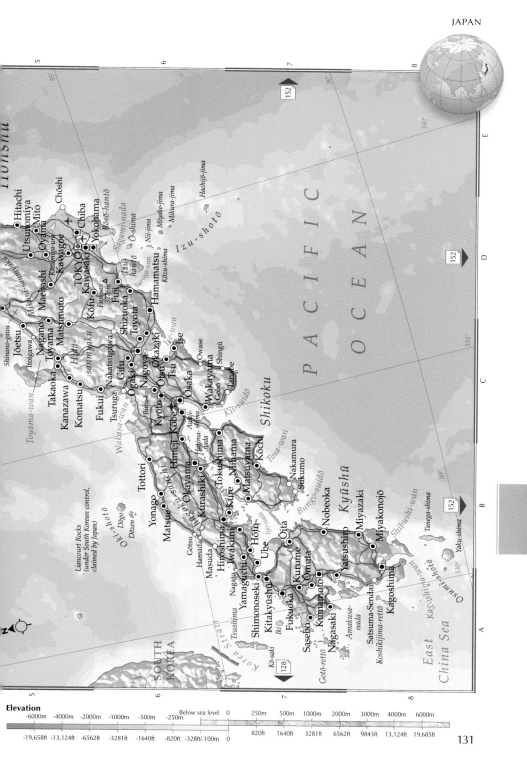

Southern India & Sri Lanka

SRI LANKA'S TWO CAPITALS

COLOMBO — Capital
SRI JAYEWARDENAPURA KOTTE — Administrative capital

Population

○ below 50,000
○ 50,000 to 100,000
◉ 100,000 to 500,000
◼ above 500,000

● National capital

0 km 300
0 miles 300

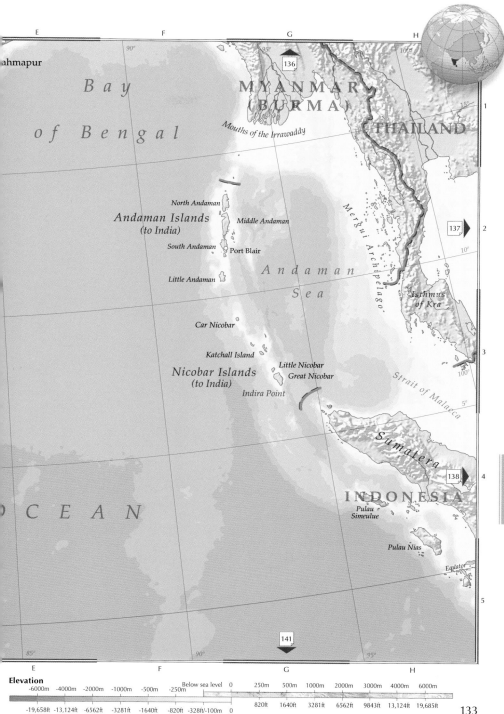

ahmapur

Bay

of Bengal

MYANMAR
(BURMA)

THAILAND

Mouths of the Irrawaddy

North Andaman

Andaman Islands
(to India)

Middle Andaman

South Andaman

Port Blair

Little Andaman

Andaman

Sea

Mergui Archipelago

Isthmus
of Kra

Car Nicobar

Katchall Island

Nicobar Islands
(to India)

Little Nicobar
Great Nicobar

Indira Point

Strait of Malacca

Sumatera

INDONESIA

Pulau
Simeulue

Pulau Nias

Equator

OCEAN

136

137

138

141

Elevation

Below sea level							0	250m	500m	1000m	2000m	3000m	4000m	6000m
-6000m	-4000m	-2000m	-1000m	-500m	-250m									
-19,658ft	-13,124ft	-6562ft	-3281ft	-1640ft	-820ft	-328ft/-100m	0		820ft	1640ft	3281ft	6562ft	9843ft	13,124ft	19,685ft

Northern India, Pakistan & Bangladesh

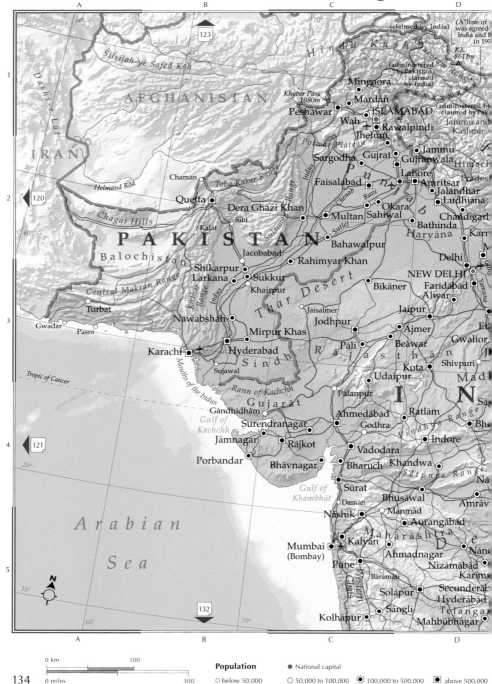

123

Silsilah-ye Safēd Kōh

Dasht-e Lūt

Hindu Kush

(claimed by India)

(A"line of was agreed India and in 197

K2 8611m

AFGHANISTAN

Khyber Pass 1080m

Mingaora

Mardan

(administered by Pakistan, claimed by India)

(administered by claimed by Pak

IRAN

Peshawar

Wāh

ISLAMABAD

Jammu and Kashmīr

Rawalpindi

Jhelum

Jammu

Chaman

Toba Kakar Range

Pothwar Plateau

Sargodha

Gujrāt

Gujranwala

Himach

Helmand Rōd

Indus

Lahore

Prades

120

Chagai Hills

Quetta

Sulaiman Range

Faisalabad

Amritsar

Jalandhar

Ludhiāna

Dera Ghazi Khan

Kalat

Sibi

Chenab

Rāvi

Okara

Sahiwal

Chandigarl

Multan

Bathinda

Karr

P A K I S T A N

Sutlej

Bahawalpur

Haryāna

Balochistan

Jacobabad

Rahimyar Khan

Delhi

Shikarpur

Larkana

Sukkur

Bīkāner

NEW DELHI

Faridābād

Alwar

Khairpur

Central Makran Range

Kirthar Range

Indus

Thar Desert

Jaisalmer

Jaipur

Turbat

Nawabshah

Jodhpur

Ajmer

Et

Gwalior

Gwadar

Pasni

Mirpur Khas

Pāli

Beāwar

Karachi

Hyderabad

Rā

Shivpuri

Mā

Sindh

j

a

s

t

h

ā

n

Sujawal

Kota

Tropic of Cancer

Mouths of the Indus

Rann of Kachchh

Udaipur

I

N

Pālanpur

Gujarāt

Gāndhīdhām

Ahmedābād

Ratlām

Bh

Gulf of Kachchh

Surendranagar

Godhra

Indore

Jāmnagar

Rājkot

Vadodara

Porbandar

Bhāvnagar

Bharūch

Khandwa

Na

Sātpura Range

Sūrat

Bhusāwal

Amrāv

Gulf of Khambhāt

Daman

Manmād

Nāshik

Aurangabad

Arabian

Maharāshtra

D

Mumbai (Bombay)

Kalyān

Ahmadnagar

Nan

Sea

Pune

Nizāmābād

Karim

Baramati

Secunderāb

Hyderābād

N

Solāpur

Telangāna

Sangli

Mahbūbnagar

Kolhapur

132

0 km   300

0 miles   300

Population

⦿ National capital

○ below 50,000   ○ 50,000 to 100,000   ◉ 100,000 to 500,000   ■ above 500,000

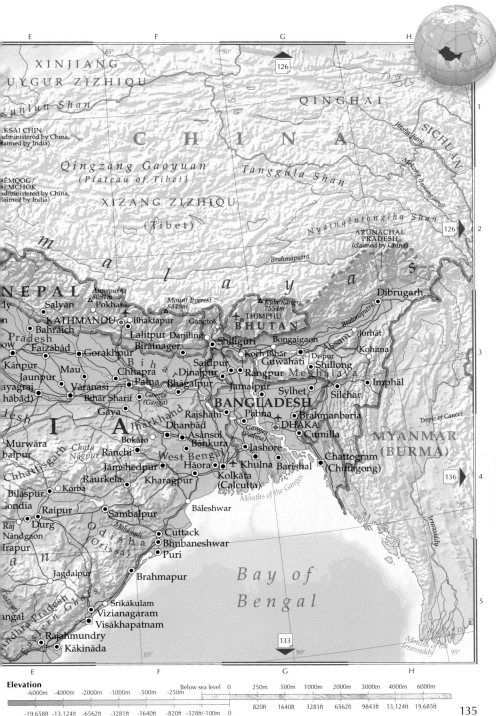

E   F   G   H

XINJIANG
UYGUR ZIZHIQU

Kunlun Shan

AKSAI CHIN
administered by China,
claimed by India)

QINGHAI

C H I N A

SICHUAN

DÊMQOG/
DEMCHOK
administered by China,
claimed by India)

Qingzang Gaoyuan
(Plateau of Tibet)

Tanggula Shan

XIZANG ZIZHIQU
(Tibet)

Nyainqêntanglha Shan

126

ARUNACHAL
PRADESH
(claimed by China)

Brahmaputra

m   a   l   a   y   a   s

NEPAL

Annapurna
8091m △

• Salyan   • Pokhara

△ *Mount Everest*
8849m

△ *Kula Kangri*
7554m

Dibrugarh •

• KATHMANDU   • Bhaktapur
Gangtok •

• Bahraich   • Lalitpur Darjiling •
Pradesh   • Birātnagar

THIMPHU +
BHUTAN
• Shiliguri   Bongaigaon •

Jorhat •
Assam
Kohima •

ow   Faizabad • Gorakhpur •
B   i   h   a   r

Koch Bihar •   Guwahati Dispur • Shillong •
Meghalaya

Kānpur • Mau
Jaunpur •   Chhapra • Dinajpur • Rangpur •
Saidpur •

Imphal •

ayagraj • Vārānasi   Patna • Bhagalpur •
hābād)   Jamalpur •

Sylhet •
Silchar •

• Bihar Sharif   *Ganges*
(Ganga)

BANGLADESH

• Gaya   Rajshahi •   Pabna •   Brahmanbaria •

desh   Jharkhand   Dhanbad •
• Asansol • DHAKA
Ganges   Cumilla •
I   A   Bokaro • Bankura •   *(Padma)*

MYANMAR
(BURMA)

• Murwāra
balpur

Chota
Nāgpur Ranchi •

Jashore •

Tropic of Cancer

Chhattisgarh

• Jamshedpur   Hāora •   Khulna • Barishal •
Raurkela •   Kharagpur • Kolkāta
(Calcutta)

Chattogram
(Chittagong) •

Bilāspur •
Gondia   • Korba

Rāj
Nāndgaon • Durg

Irapur •

Bāleshwar •

Mouths of the Ganges

a   n

136

• Raipur   Sambalpur •

O   *Mahānadi*
d   Cuttack •
i   • Bhubaneshwar
s
h   • Puri
(Orissa) a

• Jagdalpur   Brahmapur •

Bay of
Bengal

ngal   • Srikākulam
Pradesh   • Vizianagaram
• Visākhapatnam

△ • Rajahmundry
△ • Kākināda

133

Mouths of the
Irrawaddy

E   F   G   H

Elevation

					Below sea level	0	250m	500m	1000m	2000m	3000m	4000m	6000m
-6000m	-4000m	-2000m	-1000m	-500m	-250m									
-19,658ft	-13,124ft	-6562ft	-3281ft	-1640ft	-820ft	-328ft/-100m	0		820ft	1640ft	3281ft	6562ft	9843ft	13,124ft 19,685ft

Mainland Southeast Asia

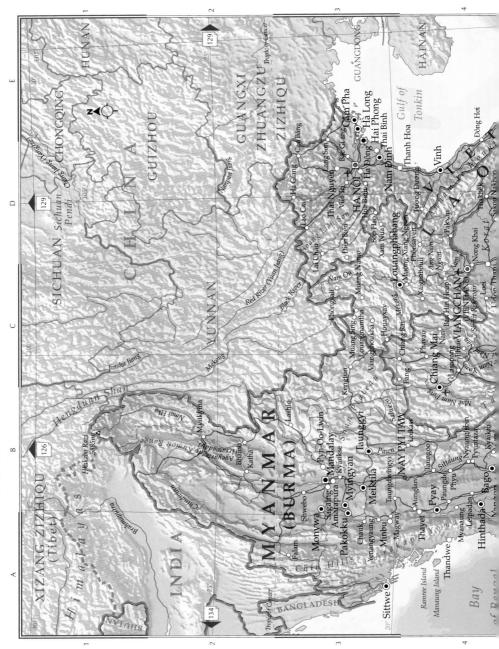

0 km   200

0 miles   200

Population   ● National capital

○ below 50,000   ○ 50,000 to 100,000   ◉ 100,000 to 500,000   ◼ above 500,000

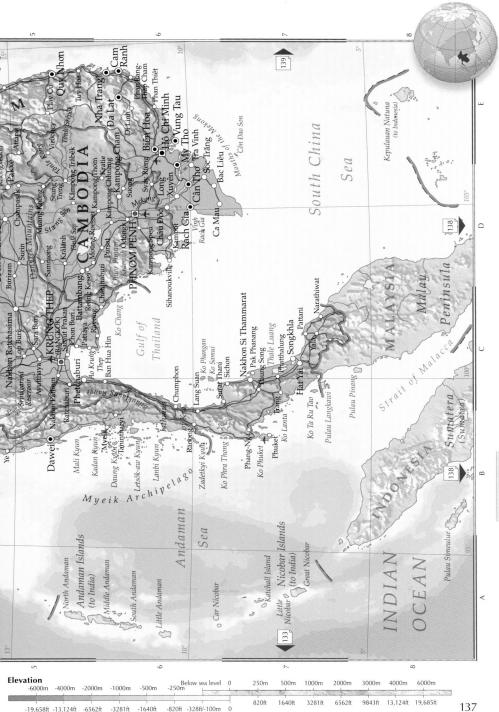

Elevation

-6000m	-4000m	-2000m	-1000m	-500m	-250m	Below sea level	0	250m	500m	1000m	2000m	3000m	4000m	6000m

-19,658ft	-13,124ft	-6562ft	-3281ft	-1640ft	-820ft	-328ft/-100m	0		820ft	1640ft	3281ft	6562ft	9843ft	13,124ft	19,685ft

Maritime Southeast Asia

SINGAPORE

MALAYSIA

Johore Strait

Causeway
Lim Chu
Kang
Choa Chu
Kang
Jurong
Industrial
Estate
Selat Pandan
Pulau Sudong
Pulau Pawai

Bukit Panjang Newtown
Queenstown
Teluk Blangah
Sentosa

Hougang
City
Bedok
New Town
Changi

Pulau Ubin
Pulau Tekong

Strait of Singapore

Urban areas
Open areas
Nature reserves

MYANMAR
(BURMA)

LAOS

VIETNAM

Gulf of
Tonkin

Hainan Dao
(to China)

THAILAND

CAMBODIA

South Chi

Sea

PARACEL ISLANDS
(disputed by China, Taiwan
and Vietnam)

SPRATLY ISLANDS
(disputed by Brunei, China, Malaysi
Philippines, Taiwan and Vietnam)

Andaman
Sea

Nicobar Islands
(to India)

Gulf of
Thailand

Isthmus of Kra

Mouths of
the Mekong

Banda Aceh
Sigli
Meulaboh
Langsa

George
Town
Butterworth
Taiping
Ipoh

Kota Bharu

Kuala Terengganu

Dungun

Pulau
Pinang

Strait of Malacca

Cukai

Kuantan

Kepulauan
Natuna

Balabac
Gunung Kin

Kota Kinabalu
BANDAR SERI
BEGAWAN
BRUNEI
Miri

Medan
Tebingtinggi
Pematangsiantar

Pulau Simeulue

Kepulauan
Banyak

Danau
Toba
Sibolga

Pulau Nias

Klang
Putrajaya

KUALA LUMPUR

MALAYSIA

Bintulu

Bituni

Sibu
Batang Rajang
Sri Aman

Sungai K

Sungai Mahe

Melaka
Muar
Batu Pahat

Keluang
Johor Bahru

SINGAPORE

Kuching
Sarawak

Selat Serasan

Kuching

Borneo

Pegunungan

Equator

Padang

Pulau Siberut

Kepulauan
Mentawai

Singkawang
Pekanbaru

Solok
Rengat

Batang Hari

Jambi

Sungaipenuh

Pangkalpinang

Sidas

Pontianak

Sungai Kapuas

Kalimantan

Sampit

Samarind
Balikpapan

Sungai Barito

Amun
Kanda

Kepulauan
Lingga

Kualatungkal

Bangka

Selat Karimata

Bengkulu
Lahat

Palembang

INDONESIA

Pulau
Belitung

Banjarmasi

Pulau
Laut

Sumatera
(Sumatra)

Kotabumi

Bandar Lampung

Cirebon

Tegal

Java Sea

Serang
Sukabumi

JAKARTA
Bogor
Bandung

Pekalongan
Semarang
Kudus

Ma

INDIAN

OCEAN

Selat Sunda

Tasikmalaya
Cilacap
Magelang
Yogyakarta
Surakarta

Jawa
(Java)

Pulau
Madura

Surabaya
Probolinggo
Jember Ma

Malang
Kediri
Madiun

Bali
Denp
Pulau
Lombo

0 km   200

0 miles   200

Population

● National capital

○ below 50,000   ○ 50,000 to 100,000   ◉ 100,000 to 500,000   ■ above 500,000

Elevation

						Below sea level	0		250m	500m	1000m	2000m	3000m	4000m	6000m
-6000m	-4000m	-2000m	-1000m	-500m	-250m										
-19,658ft	-13,124ft	-6562ft	-3281ft	-1640ft	-820ft	-328ft/-100m	0		820ft	1640ft	3281ft	6562ft	9843ft	13,124ft	19,685ft

The Indian Ocean

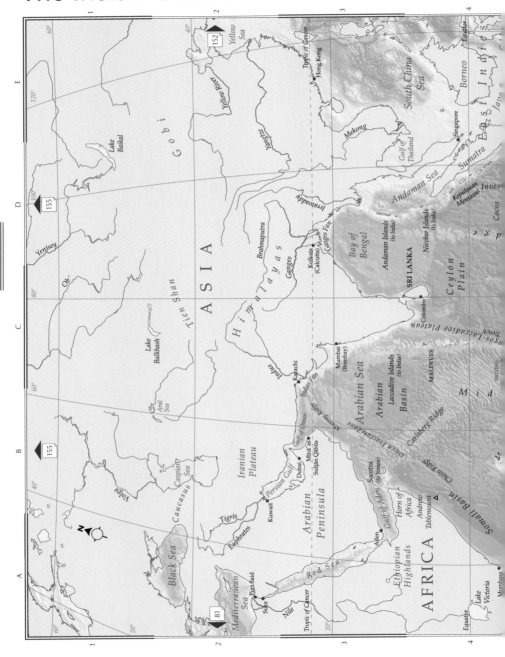

0 km   1500

0 miles   1500

• Major port

Elevation

-6000m	-4000m	-2000m	-1000m	-250m	0
-19,658ft	-13,124ft	-6562ft	-3281ft	-820ft	0

Australasia & Oceania

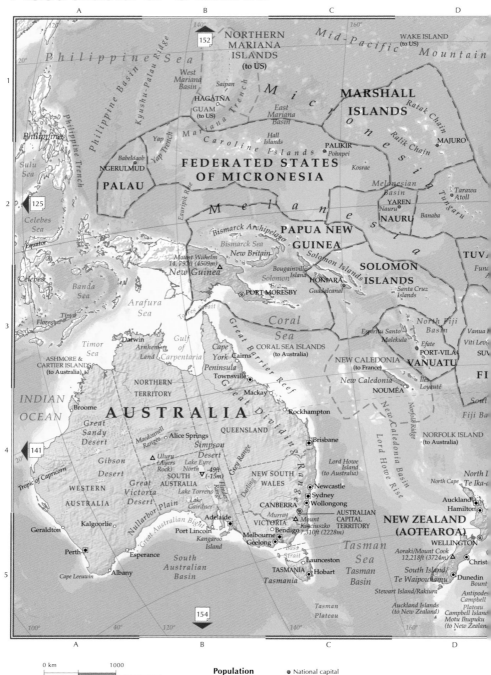

NORTHERN MARIANA ISLANDS (to US)

152

Mid-Pacific Mountain

WAKE ISLAND (to US)

Philippine Sea

Philippine Basin Ridge

Kyushu-Palau Ridge

West Mariana Basin

Saipan

HAGÅTÑA
GUAM (to US)

Mariana Trench

East Mariana Basin

M i c r o n e s i a

MARSHALL ISLANDS

Ratak Chain

Ralik Chain

MAJURO

Philippine Trench

Yap

Hall Islands

PALIKIR
Pohnpei

Kosrae

Caroline Islands

Eauripik Rise

FEDERATED STATES OF MICRONESIA

Melanesian Basin

Tarawa Atoll

YAREN
Nauru

NAURU

Banaba

Tungaru

PALAU

Babeldaob
NGERULMUD

Yap Trench

125

Celebes Sea

Equator

Celebes

Banda Sea

Arafura Sea

M e l a n e s i a

Bismarck Archipelago

Bismarck Sea

New Britain

PAPUA NEW GUINEA

Mount Wilhelm 14,793ft (4509m) △

New Guinea

Solomon Islands

Bougainville

Solomon Sea

HONIARA

Guadalcanal

SOLOMON ISLANDS

Santa Cruz Islands

TUV
Fun

Timor

Timor Sea

Flores

Darwin

Arnhem Land

Gulf of Carpentaria

Cape York

Cairns

PORT MORESBY

Torres Strait

Coral Sea

CORAL SEA ISLANDS (to Australia)

North Fiji Basin

Espíritu Santo
Malekula

Efate
PORT-VILA

Vanua

Viti Levi

SU

VANUATU

FI

ASHMORE & CARTIER ISLANDS (to Australia)

NORTHERN TERRITORY

Peninsula

Townsville

Mackay

Great Barrier Reef

NEW CALEDONIA (to France)

New Caledonia

NOUMÉA

Iles Loyauté

Sout

Fiji Ba

INDIAN OCEAN

Broome

AUSTRALIA

Great Sandy Desert

Macdonnell Ranges

Alice Springs

Simpson Desert

QUEENSLAND

Rockhampton

■ Brisbane

New Caledonia Ridge

New Caledonia Basin

NORFOLK ISLAND (to Australia)

141

Gibson Desert

Tropic of Capricorn

WESTERN AUSTRALIA

Uluru (Ayers Rock) △
SOUTH AUSTRALIA

Great Victoria Desert

Lake Eyre North -49ft (-15m) ▽

Lake Torrens

NEW SOUTH WALES

Grey Range

Darling

Great Dividing Range

Lord Howe Island (to Australia)

Lord Howe Rise

Norfolk Ridge

North I
Te Ika-a

North Cape

Auckland
Hamilton

Kalgoorlie

Nullarbor Plain

Lake Gairdner

Flinders Ranges

Adelaide

Murray

● Newcastle
● Sydney
● Wollongong

CANBERRA

AUSTRALIAN CAPITAL TERRITORY

NEW ZEALAND (AOTEAROA)

Geraldton

Great Australian Bight

Port Lincoln

Kangaroo Island

VICTORIA
Bendigo
Melbourne

Mount Kosciuszko 7,310ft (2228m) △

WELLINGTON

Tasman Sea

Aoraki/Mount Cook 12,218ft (3724m) △

Christ

Perth ■

Esperance

South Australian Basin

Geelong

Bass Strait

Launceston

TASMANIA Hobart

Tasman Sea

South Island/ Te Waipounamu

Dunedin
Bount

Cape Leeuwin

Albany

Tasmania

Tasman Basin

Stewart Island/Rakiura

Antipodes
Campbell
Plateau

Tasman Plateau

154

Auckland Islands (to New Zealand)

Campbell Islan
Motu Ihupuku (to New Zealan

0 km 1000

0 miles 1000

Population

● National capital

○ below 50,000

○ 50,000 to 100,000

◉ 100,000 to 500,000

■ above 500,000

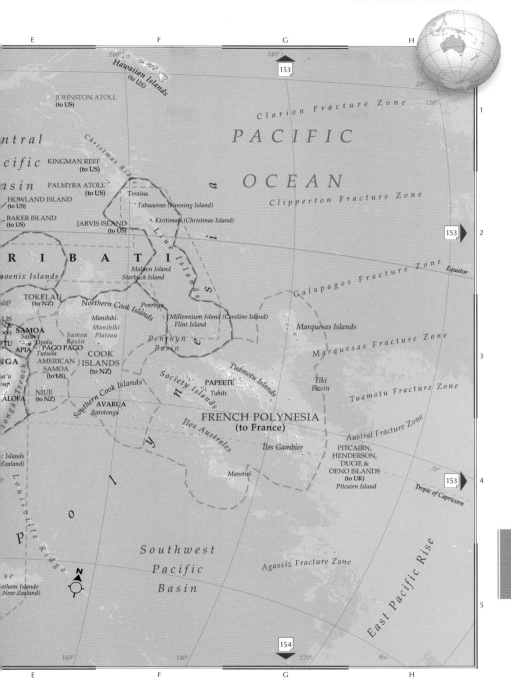

E   F   G   H

153

Hawaiian Islands
(to US)

JOHNSTON ATOLL
(to US)

PACIFIC

Clarion Fracture Zone

ntral

cific   KINGMAN REEF
(to US)

asin   PALMYRA ATOLL
(to US)

OCEAN

Clipperton Fracture Zone

HOWLAND ISLAND
(to US)

Teraina

BAKER ISLAND
(to US)

Tabuaeran (Fanning Island)

JARVIS ISLAND
(to US)

Kiritimati (Christmas Island)

153

R   I   B   A   T   I

Malden Island

Equator

Galapagos Fracture Zone

oenix Islands

Starbuck Island

TOKELAU
(to NZ)

Northern Cook Islands

Penrhyn

Marquesas Islands

LIS   *Manihiki*
Manihiki
NA   *Plateau*
ce)

Samoa
Basin

SAMOA

Savai'i

Upolu

Penrhyn
Basin

Millennium Island (Caroline Island)
Flint Island

Marquesas Fracture Zone

TU   APIA   *Tutuila*   PAGO PAGO

COOK
ISLANDS
(to NZ)

AMERICAN
SAMOA
(to US)

Society Islands

Tuamotu Islands

Tiki
Basin

Tuamotu Fracture Zone

GA

a'u
up

NIUE
(to NZ)

Southern Cook Islands

PAPEETE
Tahiti

ALOFA   AVARUA
Rarotonga

FRENCH POLYNESIA
(to France)

Îles Australes

Austral Fracture Zone

Îles Gambier

PITCAIRN,
HENDERSON,
DUCIE &
OENO ISLANDS
(to UK)
Pitcairn Island

Marotiri

Tropic of Capricorn

153

Islands
Zealand)

Southwest
Pacific
Basin

Agassiz Fracture Zone

N

tham Islands
New Zealand)

East Pacific Rise

154

E   F   G   H

143

The Southwest Pacific

A B C D

152

MARSHALL
ISLANDS

Saipan
Tinian NORTHERN
Rota MARIANA
GUAM ISLANDS (to US)
(to US) HAGÅTÑA

Enewetak Bikini Atoll Rongelap
Atoll Atoll
Ailuk

Yap Ujelang Atoll Wotje
Kwajalein Ma
Hall Islands Atoll Atoll Atc
Caroline Islands Namu Atoll Majuro
Babeldaob Chuuk Ailinglaplap Atoll
NGERULMUD Islands Pohnpei Jaluit Atoll Mili
PALIKIR

PALAU FEDERATED STATES Kosrae Ebon Atoll
OF MICRONESIA

139

Te
A
Equator Abe

YAREN
NAURU Banaba

Admiralty
Islands St.Matthias Group
New Guinea Bismarck Archipelago New Ireland
Bismarck Sea
INDONESIA Madang PAPUA NEW GUINEA Bougainville
Central Range Island
Mount Wilhelm New Choiseul
4509m Lae Britain Santa Isabel
Arafura Sea Solomon Sea New Georgia Malaita SOLOMON
Gulf of Islands HONIARA
Port Moresby Papua D'Entrecasteaux Guadalcanal ISLANDS
PORT MORESBY Islands San Cristobal Santa Cruz
Torres Strait Louisiade Rennell Islands
Archipelago

Arnhem Groote Banks Islands
Land Eylandt Coral Sea
Gulf of Cape Espiritu Santo Maéwo
146 Carpentaria York CORAL SEA ISLANDS Pentecost
Barkly Tableland Peninsula (to Australia) Malekula Ambrym
Efate Epi
PORT-VILA
Great NEW VANUATU
CALEDONIA Erromango
NORTHERN Barrier (to France) Tanna
Great Reef Ouvéa Aneityum
TERRITORY Dividing New Lifou
Tropic of Capricorn Range Caledonia Maré
Macdonnell QUEENSLAND Îles Loyauté
Ranges NOUMÉA
AUSTRALIA

149

A B C D

0 km 750

0 miles 750

Population ● National capital
○ below 50,000 ○ 50,000 to 100,000 ◉ 100,000 to 500,000 ■ above 500,000

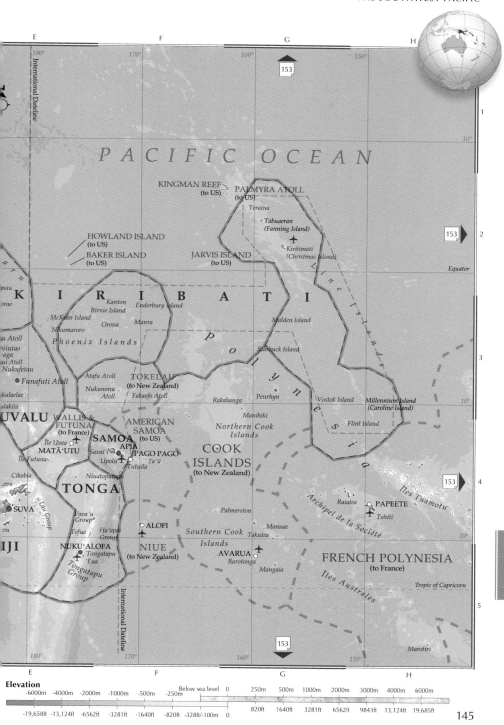

PACIFIC OCEAN

KINGMAN REEF
(to US)

PALMYRA ATOLL
(to US)

Teraina

Tabuaeran
(Fanning Island)

HOWLAND ISLAND
(to US)

Kiritimati
(Christmas Island)

BAKER ISLAND
(to US)

JARVIS ISLAND
(to US)

K I R I B A T I

Kanton
Birnie Island   Enderbury Island

Malden Island

McKean Island
Nikumaroro   Orona   Manra

Phoenix Islands   Starbuck Island

a Atoll
Niutao
aga
ui Atoll
Nukufetau

Funafuti Atoll

Atafu Atoll   TOKELAU
Nukunonu   (to New Zealand)
Atoll   Fakaofo Atoll

Rakahanga   Penrhyn   Vostok Island   Millennium Island
(Caroline Island)

kulaelae
evu
alakita

UVALU   WALLIS &
FUTUNA
(to France)

AMERICAN
SAMOA
(to US)

Manihiki   Flint Island

Île Uvea   SAMOA
MATĀ'UTU   APIA
Savai'i

Northern Cook
Islands

Île Futuna   'Upolu   Ta'ū   COOK
Tutuila   ISLANDS
(to New Zealand)

PAGO PAGO

Cikobia   Niuatoputapu
evu   TONGA

Raiatea   PAPEETE
Tahiti

SUVA   Lau Group
vu   Vava'u
Group

Palmerston   Manuae
Takutea

Archipel de la Société

Îles Tuamotu

Tofua   Ha'apai
Group

ALOFI

Southern Cook
Islands

IJI   NUKU'ALOFA   NIUE
Tongatapu   (to New Zealand)
'Eua
Tongatapu
Group

AVARUA
Rarotonga
Mangaia

FRENCH POLYNESIA
(to France)

Îles Australes

Tropic of Capricorn

Marotiri

Elevation

-6000m	-4000m	-2000m	-1000m	-500m	Below sea level	-250m	0	250m	500m	1000m	2000m	3000m	4000m	6000m
-19,658ft	-13,124ft	-6562ft	-3281ft	-1640ft	-820ft	-328ft/-100m	0	820ft	1640ft	3281ft	6562ft	9843ft	13,124ft	19,685ft

Western Australia

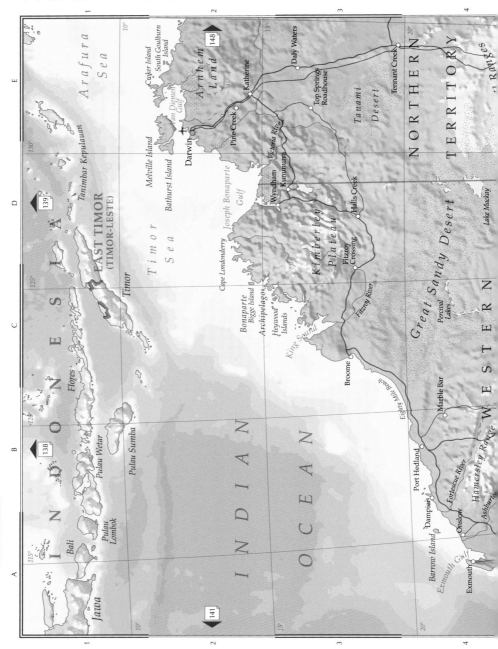

Population

○ below 50,000 ○ 50,000 to 100,000 ◉ 100,000 to 500,000 ■ above 500,000

◉ Internal administrative capital

0 km 300

0 miles 300

WESTERN AUSTRALIA

SOUTH AUSTRALIA

AUSTRALIA

Uluru (Ayers Rock) 867m

Musgrave Ranges

Coober Pedy

Great Victoria Desert

Tarcoola
Lake Everard
Penong
Lake Gairdner
Ceduna
Elliston
Port Lincoln

Nullarbor Plain

Great Australian Bight

Reid

Eucla

AUSTRALIA

Lake Carnegie
Lake Wells

Robinson Range

Meekatharra

Mount Magnet

Lake Carey

Lake Rebecca

Zanthus

Lake Cowan

Balladonia

Kalgoorlie
Coolgardie
Southern Cross
Merredin

Norseman

Esperance

Lake Barlee

Lake Moore

Northam
Brookton
Narrogin
Wagin
Katanning
Kojonup
Collie
Manjimup

Albany

INDIAN OCEAN

Murchison River

Carnarvon
Denham
Shark Bay
Dorre Island
Dirk Hartog Island

Kalbarri

Geraldton

Moora
Gingin
Perth
Fremantle
Rockingham
Mandurah
Bunbury
Busselton
Augusta

Elevation

-6000m	-4000m	-2000m	-1000m	-500m	-250m	Below sea level	0	250m	500m	1000m	2000m	3000m	4000m	6000m
-19,658ft	-13,124ft	-6562ft	-3281ft	-1640ft	-820ft	-328ft/-100m	0	820ft	1640ft	3281ft	6562ft	9843ft	13,124ft	19,685ft

Eastern Australia

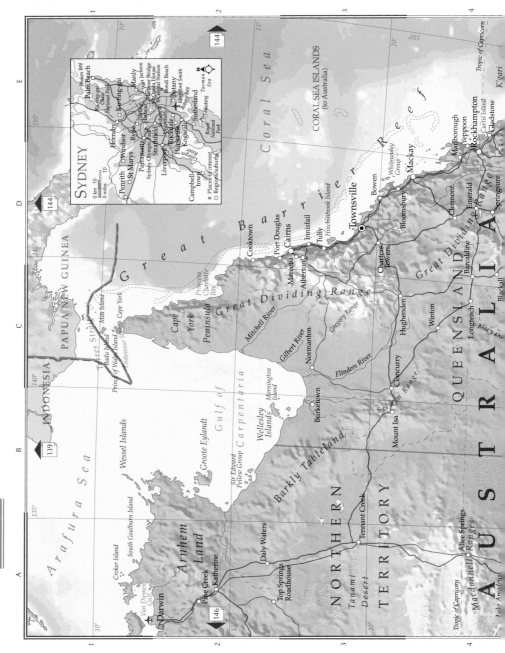

SYDNEY

■ Places of interest
□ Regions/suburbs

Population

● National capital ○ below 50,000 ○ 50,000 to 100,000 ◉ 100,000 to 500,000 ■ above 500,000

◎ Internal administrative capital

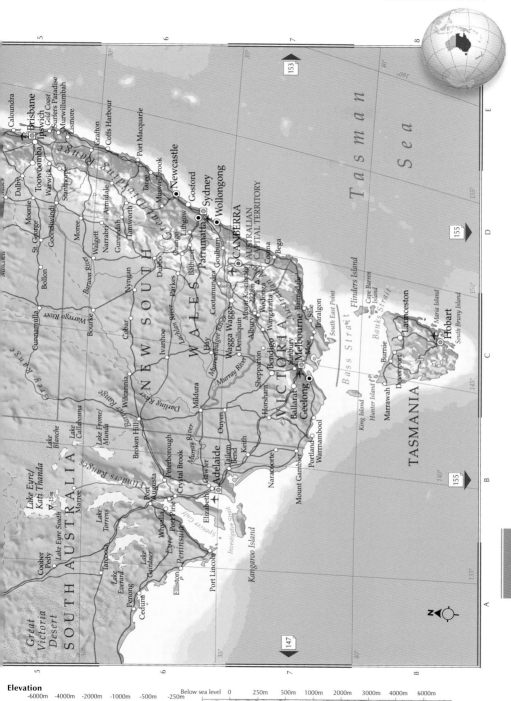

New Zealand (Aotearoa)

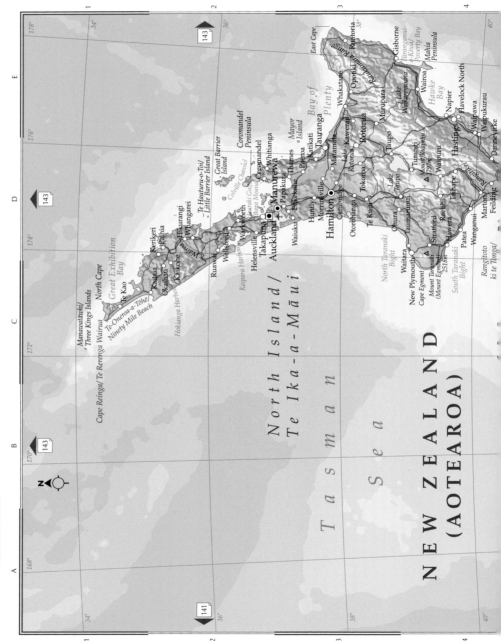

Population

- ○ below 50,000
- ○ 50,000 to 100,000
- ◉ 100,000 to 500,000
- ▣ above 500,000
- ● National capital

0 km   100
0 miles   100

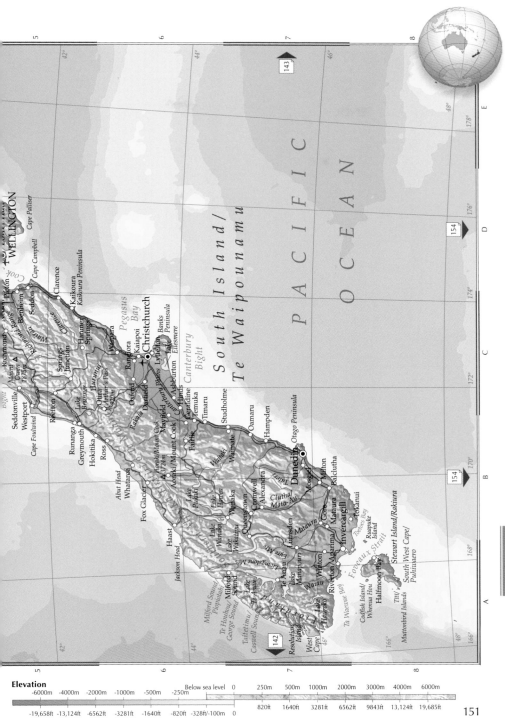

Elevation

-6000m	-4000m	-2000m	-1000m	-500m	-250m	Below sea level	0	250m	500m	1000m	2000m	3000m	4000m	6000m
-19,658ft	-13,124ft	-6562ft	-3281ft	-1640ft	-820ft	-328ft/-100m	0	820ft	1640ft	3281ft	6562ft	9843ft	13,124ft	19,685ft

The Pacific Ocean

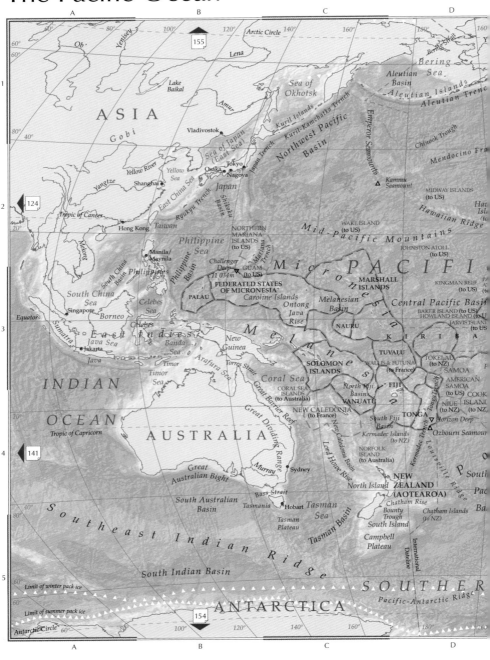

60° 80° 100° 120° Arctic Circle 140° 160° 180° 160°

155

60°

Ob' Yenisey Lena Bering Strait

Bering Sea

1 Lake Baikal Sea of Okhotsk Aleutian Basin Aleutian Islands Aleutian Trench

A S I A Amur Kuril Islands Kuril-Kamchatka Trench Northwest Pacific Basin Chinook Trough

Gobi 80° 40° Vladivostok Kuril Trench Emperor Seamounts Mendocino Fra

Yellow River Sea of Japan (East Sea) Japan Trench Kammu Seamount MIDWAY ISLANDS (to US)

Yangtze Yellow Sea Osaka Tokyo Nagoya Japan Shikoku Basin Hawaiian Ridge

2 124 Shanghai Ryukyu Trench WAKE ISLAND (to US) Mid-Pacific Mountains

Tropic of Cancer 20° Hong Kong Taiwan East China Sea NORTHERN MARIANA ISLANDS (to US) JOHNSTON ATOLL (to US)

Philippine Sea Mariana Trench M i c r P A C I F I

South China Sea Manila/ Philippine Basin Challenger Deep GUAM (to US) o MARSHALL ISLANDS KINGMAN REEF (to US)

Basilan Philippines 11 034m FEDERATED STATES OF MICRONESIA n Melanesian Basin Central Pacific Basin

South China Sea Celebes Sea PALAU Caroline Islands e BAKER ISLAND (to US) HOWLAND ISLAND (to US)

Equator Singapore Borneo Ontong Java Rise s NAURU JARVIS ISLAND (to US)

3 E a s t I n d i e s Celebes i K I R I B

Sumatra Java Sea Banda Sea New Guinea M a TUVALU TOKELAU (to NZ)

Jakarta Timor Arafura Sea Torres Strait l SOLOMON ISLANDS WALLIS & FUTUNA (to France) SAMOA

Java Timor Sea Great Barrier Reef Coral Sea a n North Fiji Basin FIJI AMERICAN SAMOA (to US) COOK ISLANDS

INDIAN CORAL SEA ISLANDS (to Australia) e VANUATU NIUE (to NZ) (to NZ,

NEW CALEDONIA (to France) s TONGA Horizon Deep

OCEAN 20° New Caledonia Basin South Fiji Basin Ozbourn Seamour

Tropic of Capricorn A U S T R A L I A Great Dividing Range Kermadec Islands (to NZ)

4 141 NORFOLK ISLAND (to Australia) P South

Murray Sydney Lord Howe Rise NEW ZEALAND (AOTEAROA) Pac

Great Australian Bight North Island Chatham Rise

South Australian Basin Bass Strait Tasman Sea Bounty Trough Chatham Islands (to NZ) Ba

40° 80° Tasmania Hobart Tasman Plateau South Island

S o u t h e a s t I n d i a n R i d g e Campbell Plateau

5 60° Limit of winter pack ice South Indian Basin S O U T H E R

60° Limit of summer pack ice Pacific-Antarctic Ridge

Antarctic Circle 60° 80° A N T A R C T I C A 100° 120° 140° 160° 180° 60°

154

A B C D

0 km 2000 • Major port

0 miles 2000

E   F   G   H

120°   Arctic Circle   100°   80°   60°   40°   20°   60°

155

Rocky Mountains

Hudson Bay

Labrador Sea

NORTH AMERICA

Vancouver

Cascadia Basin

Great Lakes

San Francisco

Colorado

ATLANTIC

Appalachian Mountains

OCEAN

ay Fracture Zone   Long Beach

kai Fracture Zone

Gulf of California

Gulf of Mexico

Greater Antilles

Tropic of Cancer

66

Lesser Antilles

Clarion Fracture Zone

Caribbean Sea

C E A N

CLIPPERTON ISLAND (to France)

Middle America Trench

perton Fracture Zone

Guatemala Basin

Cocos Ridge

Panama City

N

Galapagos Fracture Zone   Gallego Rise

Galápagos Islands (to Ecuador)

Equator

Amazon

Marquesas Islands   Marquesas Fracture Zone

Bauer Basin

Galapagos Rise

Peru Basin

Callao

SOUTH AMERICA

Tiki Basin

FRENCH POLYNESIA (to France)

Austral Fracture Zone

Mendaña Fracture Zone

Peru-Chile Trench

Nazca Ridge

Îles Gambier

PITCAIRN, HENDERSON, DUCIE & OENO ISLANDS (to UK)

Sala y Gomez (to Chile)

Sala y Gomez Ridge

Chile Basin

Andes

Easter Fracture Zone

Paraná

Tropic of Capricorn

Australes

Easter Island (to Chile)

Isla San Félix (to Chile)

Isla San Ambrosio (to Chile)

67

Islas Juan Fernández (to Chile)

Valparaíso

Challenger Fracture Zone

East Pacific Rise

Agassiz Fracture Zone

Chile Rise

ATLANTIC

Eltanin Fracture Zone

Mornington Abyssal Plain

OCEAN

Cape Horn

C E A N   Southeast Pacific Basin   Bellingshausen Plain   Drake Passage

5

Amundsen Plain

PETER I ØY (to Norway)

154

120°   100°   80°   60°   40°   20°   0°

Antarctic Circle

E   F   G   H

Elevation

-6000m	-4000m	-2000m	-1000m	-250m	0
-19,658ft	-13,124ft	-6562ft	-3281ft	-820ft	0

153

Antarctica

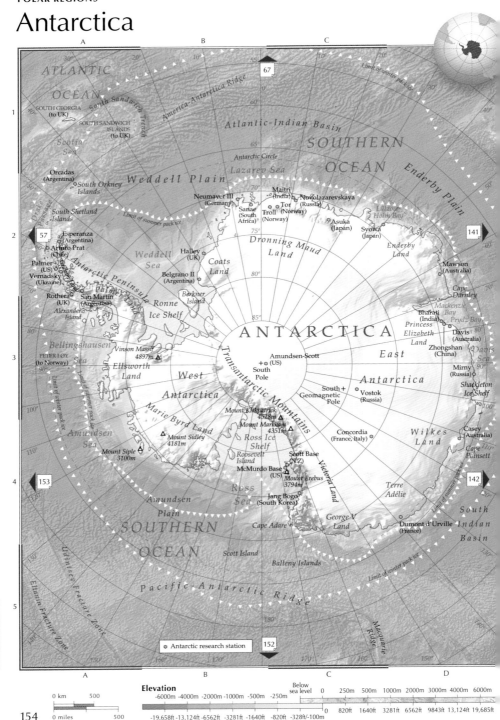

Elevation

-6000m	-4000m	-2000m	-1000m	-500m	-250m	Below sea level	0	250m	500m	1000m	2000m	3000m	4000m	6000m
-19,658ft	-13,124ft	-6562ft	-3281ft	-1640ft	-820ft	-328ft/-100m	0	820ft	1640ft	3281ft	6562ft	9843ft	13,124ft	19,685ft

0 km 500
0 miles 500

Arctic Ocean

NORTH AMERICA

ALASKA (to US)

ASIA

RUSSIA

Saint Lawrence Island

Providensiya

Bering Sea

Arctic Circle

152

Chukchi Sea

Ostrov Vrangelya

East Siberian Sea

113

Tuktoyaktuk

Limit of summer pack ice

Northwind Plain

Chukchi Plain

Novosibirskiye Ostrova

Limit of permanent ice cap

36

Beaufort Sea

Chukchi Plateau

Amundsen Gulf

Canada Basin

Mendeleyev Ridge

Wrangel Plain

Laptev Sea

Victoria Island

ARCTIC

Makarov Basin

Severnaya Zemlya

CANADA

Queen Elizabeth Islands

Alpha Cordillera

Lomonosov Ridge

North Pole

Fram Basin

Gakkel Cordillera

Dikson

Baffin Island

Ellesmere Island

North Geomagnetic Pole

Nares Strait

Lincoln Sea

Nansen Basin

Svyataya Anna Trough

Kara Sea

Ostrov Belyy

OCEAN

Franz Josef Land

East Novaya Zemlya Trough

Baffin Bay

Knud Rasmussen Land

Kap Morris Jesup

Wandel Sea

Novaya Zemlya

112

38

Kong Frederik VIII Land

SVALBARD (to Norway)

Spitsbergen

Longyearbyen

Ostrov Kotel'nyy

Chëshskaya Gulf

GREENLAND (to Denmark)

Greenland Sea

Limit of winter pack ice

Bjørnøya (to Norway)

Barents Sea

North Cape

Murmansk

Kola Peninsula

Archangel

Limit of summer pack ice

JAN MAYEN (to Norway)

Mohns Ridge

NORWAY

White Sea

FINLAND

Limit of winter pack ice

Denmark Strait

Iceland Plateau

Norwegian Sea

SWEDEN

EUROPE

66

Elevation

| | -6000m | -4000m | -2000m | -1000m | -250m | 0 |

-19,658ft -13,124ft -6562ft -3281ft -820ft 0

0 km   500

0 miles   500

• Major port

Overseas territories & dependencies

Despite the rapid process of global decolonization since the Second World War, around 8 million people in more than 50 territories around the world continue to live under the protection of France, Australia, the Netherlands, Denmark, Norway, New Zealand, the UK, or the USA. These remnants of former colonial empires may have persisted for economic, strategic or political reasons and are administered in a variety of ways.

AUSTRALIA

Australia's overseas territories have not been an issue since Papua New Guinea became independent in 1975. Consequently there is no overriding policy toward them. Norfolk Island is inhabited by descendants of the H.M.S Bounty mutineers and more recent Australian migrants.

Ashmore & Cartier Islands
Indian Ocean
Status: External territory
Claimed: 1931
Capital: Not applicable
Population: None
Area: 2 sq miles
(5.2 sq km)

Christmas Island
Indian Ocean
Status: External territory
Claimed: 1958
Capital: The Settlement
Population: 1692
Area: 52 sq miles
(135 sq km)

Cocos (Keeling) Islands
Indian Ocean
Status: External territory
Claimed: 1955
Capital: West Island
Population: 544
Area: 5.5 sq miles (14 sq km)

Coral Sea Islands
South Pacific
Status: External territory
Claimed: 1969
Capital: None
Population: below 10 (scientists)
Area: Less than 1.2 sq miles
(3 sq km)

Heard I. & McDonald Is.
Indian Ocean
Status: External territory
Claimed: 1947
Capital: Not applicable
Population: None
Area: 159 sq miles
(412 sq km)

Norfolk Island
South Pacific
Status: External territory
Claimed: 1914
Capital: Kingston
Population: 2188
Area: 13 sq miles (34 sq km)

DENMARK

The Faroe Islands have been under Danish administration since Queen Margreth I of Denmark inherited Norway in 1380. The Home Rule Act of 1948 gave the Faroese control over all their internal affairs. Greenland first came under Danish rule in 1380. Today, Denmark is responsible for the island's foreign affairs and defense.

Faroe Islands
North Atlantic
Status: Self-governing territory of Denmark
Claimed: 1380
Capital: Tórshavn
Population: 52,269
Area: 538 sq miles (1393 sq km)

Greenland
North Atlantic
Status: Self-governing territory of Denmark
Claimed: 1380
Capital: Nuuk
Population: 57,792
Area: 822,700 sq miles
(2,130,783 sq km)

FRANCE

France has developed economic ties with its *Territoires d'Outre-Mer,* thereby stressing interdependence over independence. Overseas *départements,* officially part of France, have their own governments. Territorial *collectivités* and overseas *territoires* have varying degrees of autonomy.

Clipperton Island
East Pacific
Status: Administered directly from France
Claimed: 1935
Capital: Not applicable
Population: None
Area: 2 sq miles (6 sq km)

French Guiana
South America
Status: Overseas department
Claimed: 1946
Capital: Cayenne
Population: 304,641
Area: 32,252 sq miles
(83,534 sq km)

French Polynesia
South Pacific
Status: Overseas collectivity
Claimed: 1842
Capital: Papeete
Population: 299,356
Area: 1609 sq miles
(4167 sq km)

French Southern & Antarctic Lands
Indian Ocean
Status: Overseas territory
Claimed: 1924
Capital: Port-aux-Français
Population: 150
Area: 2991 sq miles (7747 sq km)

Guadeloupe
West Indies
Status: Overseas department
Claimed: 1635
Capital: Basse-Terre
Population: 395,752
Area: 629 sq miles (
1628 sq km)

Martinique
West Indies
Status: Overseas department
Claimed: 1635
Capital: Fort-de-France
Population: 367,507
Area: 436 sq miles (1128 sq km)

Mayotte
Indian Ocean
Status: Overseas department
Claimed: 1843
Capital: Mamoudzou
Population: 326,206
Area: 144 sq miles (374 sq km)

New Caledonia
South Pacific
Status: Self-governing territory
Claimed: 1853
Capital: Nouméa
Population: 297,160
Area: 7172 sq miles
(18,575 sq km)

Réunion
Indian Ocean
Status: Overseas department
Claimed: 1638
Capital: Saint-Denis
Population: 974,157
Area: 970 sq miles
(2511 sq km)

St. Barthélemy
West Indies
Status: Overseas collectivity
Claimed: 1878
Capital: Gustavia
Population: 10,000
Area: 10 sq miles (25 sq km)

St Martin
West Indies
Status: Overseas collectivity
Claimed: 1648
Capital: Marigot
Population: 32,500
Area: 20 sq miles (53 sq km)

St. Pierre & Miquelon
North America
Status: Overseas collectivity
Claimed: 1604
Capital: Saint-Pierre
Population: 5257
Area: 93 sq miles
(242 sq km)

Wallis & Futuna
South Pacific
Status: Overseas collectivity
Claimed: 1842
Capital: Matá'Utu
Population: 15,891
Area: 55 sq miles
(142 sq km)

NETHERLANDS

The country's remaining overseas
territories were formerly part of the
Dutch West Indies. The Netherlands
Antilles dissolved in 2010 leaving
the constituent islands with varying
degrees of autonomy, but the
Netherlands remains responsible for
their security.

Aruba
West Indies
Status: Self-governing country
of the Netherlands
Claimed: 1643
Capital: Oranjestad
Population: 122,320
Area: 69 sq miles
(180 sq km)

Bonaire
West Indies
Status: Special municipality
of the Netherlands
Claimed: 1816
Capital: Kralendijk
Population: 22,573
Area: 111 sq miles
(288 sq km)

Curaçao
West Indies
Status: Self-governing country
of the Netherlands
Claimed: 1815
Capital: Willemstad
Population: 152,379
Area: 171 sq miles
(444 sq km)

Saba
West Indies
Status: Special municipality
of the Netherlands
Claimed: 1816
Capital: The Bottom
Population: 1911
Area: 5 sq miles (13 sq km)

Sint-Eustatius
West Indies
Status: Special municipality
of the Netherlands
Claimed: 1784
Capital: Oranjestad
Population: 3242
Area: 8 sq miles (21 sq km)

Sint-Maarten
West Indies
Status: Constituent country
of the Netherlands
Claimed: 1648
Capital: Philipsburg
Population: 45,126
Area: 13 sq miles (34 sq km)

NEW ZEALAND

New Zealand's government
has no desire to retain any overseas
territories. However, the economic
weakness of its dependent territory
Tokelau and its freely associated
states, Niue and the Cook Islands,
has forced New Zealand to
remain responsible for their
foreign policy and defense.

Cook Islands
South Pacific
Status: Associated territory
Claimed: 1901
Capital: Avarua
Population: 8128
Area: 91 sq miles
(235 sq km)

Niue
South Pacific
Status: Associated territory
Claimed: 1901
Capital: Alofi
Population: 2000
Area: 100 sq miles
(260 sq km)

Continued on page158

Overseas territories & dependencies

Tokelau
South Pacific
Status: Dependent territory
Claimed: 1925
Capital: Not applicable
Population: 1871
Area: 5 sq miles (12 sq km)

NORWAY

In 1920, 41 nations signed the Spits-bergen Treaty recognizing Norwegian sovereignty over Svalbard. There is a NATO base on Jan Mayen. Bouvet Island is a nature reserve.

Bouvet Island
South Atlantic
Status: Dependency
Claimed: 1929
Capital: Not applicable
Population: None
Area: 19 sq miles (49 sq km)

Jan Mayen
North Atlantic
Status: Dependency
Claimed: 1929
Capital: Not applicable
Population: 35 (military personnel / scientists)
Area: 146 sq miles (377 sq km)

Peter I. Island
Southern Ocean
Status: Dependency
Claimed: 1931
Capital: Not applicable
Population: None
Area: 60 sq miles (156 sq km)

Svalbard
Arctic Ocean
Status: Dependency
Claimed: 1920
Capital: Longyearbyen
Population: 2926
Area: 23,956 sq miles (62,045 sq km)

UNITED KINGDOM

The UK still has the largest number of overseas territories. These are locally-governed by a mixture of elected representatives and appointed officials, and they all enjoy a large measure of internal self-government, but certain powers, such as foreign affairs and defense, are reserved for Governors of the British Crown.

Anguilla
West Indies
Status: Overseas Territory
Claimed: 1650
Capital: The Valley
Population: 18,741
Area: 35 sq miles (91 sq km)

Ascension Island
South Atlantic
Status: Overseas Territory
Claimed: 1815
Capital: Georgetown
Population: 880
Area: 34 sq miles (88 sq km)

Bermuda
North Atlantic
Status: Overseas Territory
Claimed: 1612
Capital: Hamilton
Population: 72,337
Area: 21 sq miles (54 sq km)

British Indian Ocean Territory
Status: Overseas Territory
Claimed: 1814
Capital: Diego Garcia
Population: 4000 (UK/US air base)
Area: 23 sq miles (60 sq km)

British Virgin Islands
West Indies
Status: Overseas Territory
Claimed: 1672
Capital: Road Town
Population: 38,632
Area: 58 sq miles (151 sq km)

Cayman Islands
West Indies
Status: Overseas Territory
Claimed: 1670
Capital: George Town
Population: 64,309
Area: 102 sq miles (264 sq km)

Falkland Islands
South Atlantic
Status: Overseas Territory
Claimed: 1833
Capital: Stanley
Population: 3780
Area: 4699 sq miles (12,173 sq km)

Gibraltar
Southwest Europe
Status: Overseas Territory
Claimed: 1713
Capital: Gibraltar
Population: 29,573
Area: 3 sq miles (7 sq km)

Guernsey
Channel Islands
Status: Crown Dependency
Claimed: 1066
Capital: St. Peter Port
Population: 67,491
Area: 30 sq miles (78 sq km)

Isle of Man
British Isles
Status: Crown Dependency
Claimed: 1765
Capital: Douglas
Population: 91,382
Area: 221 sq miles (572 sq km)

Jersey
Channel Islands
Status: Crown Dependency
Claimed: 1066
Capital: St. Helier
Population: 102,146
Area: 45 sq miles (116 sq km)

Montserrat
West Indies
Status: Overseas Territory
Claimed: 1632
Capital: Plymouth (de jure), Brades (de facto)
Population: 5414
Area: 40 sq miles (102 sq km)

Pitcairn Group of Islands
South Pacific
Status: Overseas Territory
Claimed: 1838
Capital: Adamstown
Population: 50
Area: 18 sq miles (47 sq km)

St. Helena
South Atlantic
Status: Overseas Territory
Claimed: 1657
Capital: Jamestown
Population: 4217
Area: 47 sq miles (122 sq km)

South Georgia & The South Sandwich Islands
South Atlantic
Status: Overseas Territory
Claimed: 1775
Capital: Not applicable
Population: None
Area: 1507 sq miles
(3903 sq km)

Tristan da Cunha
South Atlantic
Status: Overseas territory
Claimed: 1816
Capital: Edinburgh
Population: 242
Area: 38 sq miles (98 sq km)

Turks & Caicos Islands
West Indies
Status: Overseas territory
Claimed: 1799
Capital: Cockburn Town
Population: 58,286
Area: 366 sq miles (948 sq km)

UNITED STATES OF AMERICA

America's overseas territories have been seen as strategically useful, if expensive, links with its "backyards." The US has, in most cases, given the local population a say in deciding their own status. A US Commonwealth territory, such as Puerto Rico, has a greater level of independence than that of a US unincorporated territory.

American Samoa
South Pacific
Status: Unincorporated territory
Claimed: 1900
Capital: Pago Pago
Population: 45,443
Area: 86 sq miles (224 sq km)

Baker Island
Central Pacific
Status: Unincorporated territory
Claimed: 1856
Capital: Not applicable
Population: None
Area: 0.8 sq miles (1.2 sq km)

Guam
West Pacific
Status: Unincorporated territory
Claimed: 1898
Capital: Hagåtña
Population: 169,086
Area: 210 sq miles
(544 sq km)

Howland Island
Central Pacific
Status: Unincorporated territory
Claimed: 1856
Capital: Not applicable
Population: None
Area: 1 sq mile (2.6 sq km)

Jarvis Island
Central Pacific
Status: Unincorporated territory
Claimed: 1935
Capital: Not applicable
Population: None
Area: 2 sq miles (5 sq km)

Johnston Atoll
Central Pacific
Status: Unincorporated territory
Claimed: 1858
Capital: Not applicable
Population: None
Area: 1 sq mile (2.8 sq km)

Kingman Reef
Central Pacific
Status: Unincorporated territory
Claimed: 1856
Capital: Not applicable
Population: None
Area: 0.4 sq mile (1 sq km)

Midway Atoll
Central Pacific
Status: Unincorporated territory
Claimed: 1867
Capital: Not applicable
Population: 40 (US air base)
Area: 2 sq miles (5.2 sq km)

Navassa Island
West Indies
Status: Unincorporated territory
Claimed: 1856
Capital: Not applicable
Population: None
Area: 2 sq miles
(5.2 sq km)

Northern Mariana Islands
West Pacific
Status: Commonwealth territory
Claimed: 1944
Capital: Saipan
Population: 55,020
Area: 179 sq miles
(464 sq km)

Palmyra Atoll
Central Pacific
Status: Incorporated territory
Claimed: 1898
Capital: Not applicable
Population: 20
Area: 2 sq miles
(4 sq km)

Puerto Rico
West Indies
Status: Commonwealth territory
Claimed: 1898
Capital: San Juan
Population: 3.1 million
Area: 3515 sq miles
(9104 sq km)

Virgin Islands
West Indies
Status: Unincorporated territory
Claimed: 1917
Capital: Charlotte Amalie
Population: 105,413
Area: 134 sq miles
(346 sq km)

Wake Island
Central Pacific
Status: Unincorporated territory
Claimed: 1899
Capital: Not applicable
Population: 150 (US air base)
Area: 3 sq miles
(7 sq km)

Glossary of geographical terms

The following glossary lists all geographical terms occuring on the maps and in the main-entry names in the Index–Gazetteer. These terms may precede, follow or be run together with the proper elements of the name; where they precede it the term is reversed for indexing purposes – thus Poluostov Yamal is indexed as Yamal, Poluostrov.

A

Å *Danish, Norwegian,* River
Alpen *German,* Alps
Altiplanicie *Spanish,* Plateau
Älv(en) *Swedish,* River
Anse *French,* Bay
Archipiélago *Spanish,* Archipelago
Arcipelago *Italian,* Archipelago
Arquipélago *Portuguese,* Archipelago
Aukštuma *Lithuanian,* Upland
Aýlag Aýlagy *Turkmen,* Gulf

B

Bahía *Spanish,* Bay
Baía *Portuguese,* Bay
Baḥr *Arabic,* River
Baie *French,* Bay
Bandao *Chinese,* Peninsula
Banjaran *Malay,* Mountain range
Batang *Malay,* Stream
-berg *Afrikaans, Norwegian,* Mountain
Birket *Arabic,* Lake
Boğazı *Turkish,* Strait
Bucht *German,* Bay
Bugten *Danish,* Bay
Buḥayrat *Arabic,* Lake, reservoir
Buḥeiret *Arabic,* Lake
Bukit *Malay,* Mountain
-bukta *Norwegian,* Bay
bukten *Swedish,* Bay
Burnu *Turkish,* Cape, point
Buuraha *Somali,* Mountains

C

Cabo *Portuguese,* Cape
Cap *French,* Cape
Cascada *Portuguese,* Waterfall
Cerro *Spanish,* Hill
Chaîne *French,* Mountain range
Chau *Cantonese,* Island
Chāy *Turkish,* Stream
Chhâk *Cambodian,* Bay
Chhu *Tibetan,* River
-chôsuji *Korean,* Reservoir

Chott *Arabic,* Salt lake, depression
Ch'ün-tao *Chinese,* Island group
Chuŏr Phnum *Cambodian,* Mountains
Cordillera *Spanish,* Mountain range
Costa *Spanish,* Coast
Côte *French,* Coast
Cuchilla *Spanish,* Mountains

D

Dağı *Azerbaijani, Turkish,* Mountain
Dağları *Azerbaijani, Turkish,* Mountains
-dake *Japanese,* Peak
Danau *Indonesian,* Lake
Dao *Chinese,* Island
Đao *Vietnamese,* Island
Daryā *Persian,* River
Daryācheh *Persian,* Lake
Dasht *Persian,* Plain, desert
Dawḥat *Arabic,* Bay
Dere *Turkish,* Stream
Dili *Azerbaijani,* Spit
-do *Korean,* Island
Dooxo *Somali,* Valley
Düzü *Azerbaijani,* Steppe
-dwīp *Bengali,* Island

E

Embalse *Spanish,* Reservoir
Erg *Arabic,* Dunes
Estany *Catalan,* Lake
Estrecho *Spanish,* Strait
-ey *Icelandic,* Island
Ezero *Bulgarian, Macedonian,* Lake

F

Fjord *Danish,* Fjord
-fjorden *Norwegian,* Fjord
-fjørdhur *Faroese,* Fjord
Fleuve *French,* River
Fliegu *Maltese,* Channel
-fljót *Icelandic,* River

G

-gang *Korean,* River
Ganga *Nepali, Sinhala,* River
Gaoyuan *Chinese,* Plateau
-gawa *Japanese,* River

Gebel *Arabic,* Mountain
-gebirge *German,* Mountains
Ghubbat *Arabic,* Bay
Gjiri *Albanian,* Bay
Gol *Mongolian,* River
Göl, Gölü *Turkish,* Lake
Golfe *French,* Gulf
Golfo *Italian, Spanish,* Gulf
Gora *Russian, Serbian,* Mountain
Gory *Russian,* Mountains
Guba *Russian,* Bay
Gunung *Malay,* Mountain

H

Ḥadd *Arabic,* Spit
-haehyôp *Korean,* Strait
Haff *German,* Lagoon
Hai *Chinese,* Sea, bay
Ḥammādat *Arabic,* Plateau
Hāmūn *Persian,* Lake
Hawr *Arabic,* Lake
Hāyk' *Amharic,* Lake
He *Chinese,* River
Helodrano *Malagasy,* Bay
-hegység *Hungarian,* Mountain range
Hka *Burmese,* River
-ho *Korean,* Lake
Hô *Korean,* Reservoir
Ḥolot *Hebrew,* Dunes
Hora *Belarusian,* Mountain
Hrada *Belarusian,* Mountains, ridge
Hsi *Chinese,* River
Hu *Chinese,* Lake

I

Île(s) *French,* Island(s)
Ilha(s) *Portuguese,* Island(s)
Ilhéu(s) *Portuguese,* Islet(s)
Irmak *Turkish,* River
Isla(s) *Spanish,* Island(s)
Isola (Isole) *Italian,* Island(s)

J

Jabal *Arabic,* Mountain
Jāl *Arabic,* Ridge
-järvi *Finnish,* Lake
Jazīrat *Arabic,* Island

Jazīreh *Persian,* Island
Jebel *Arabic,* Mountain
Jezero *Serbian/Croatian,* Lake
Jiang *Chinese,* River
-joki *Finnish,* River
-jökull *Icelandic,* Glacier
Juzur *Arabic,* Islands

K

Kaikyō *Japanese,* Strait
-kaise *Sámi,* Mountain
Kali *Nepali,* River
Kalnas *Lithuanian,* Mountain
Kalns *Latvian,* Mountain
Kang *Chinese,* Harbor
Kangri *Tibetan,* Mountain(s)
Kaôh *Cambodian,* Island
Kapp *Norwegian,* Cape
Kavir *Persian,* Desert
Kedi *Georgian,* Mountain range
Kediet *Arabic,* Mountain
Kepulauan *Indonesian, Malay,* Island group
Khalīg, Khalīj *Arabic,* Gulf
Khawr *Arabic,* Inlet
Khola *Nepali,* River
Khrebet *Russian,* Mountain range
Ko *Thai,* Island
Köl, Köli *Kazakh, Kyrgyz,* Lake
Kólpos *Greek,* Bay
-kopf *German,* Peak
Körfäzi *Azerbaijani,* Bay
Körfezi *Turkish,* Bay
Kõrgustik *Estonian,* Upland
Koshi *Nepali,* River
Kowtal *Persian (Dari),* Pass
Kūh(hā) *Persian (Dari),* Mountain(s)
-kundo *Korean,* Island group
-kysten *Norwegian,* Coast
Kyun *Burmese,* Island

L

Laaq *Somali,* Watercourse
Lac *French,* Lake
Lacul *Romanian,* Lake
Lago *Italian, Portuguese, Spanish,* Lake

Laguna *Spanish,*
Lagoon, Lake
Laht *Estonian,* Bay
Laut *Indonesian,* Sea
Lembalemba *Malagasy,*
Plateau
Lerr *Armenian,*
Mountain
Lerrnashght'a *Armenian,*
Mountain range
Les *Czech,* Forest
Lich *Armenian,* Lake
Liqeni *Albanian,* Lake
Lumi *Albanian,* River
Lyman *Ukrainian,*
Estuary

M

Mae Nam *Thai,* River
-mägi *Estonian,* Hill
Maja *Albanian,* Mountain
-man *Korean,* Bay
Marios *Lithuanian,* Lake
-meer *Dutch,* Lake
Melkosopochnik *Russian,*
Plain
-meri *Estonian,* Sea
Mifrats *Hebrew,* Bay
Monkhafad *Arabic,*
Depression
Mont(s) *French,*
Mountain(s)
Monte *Italian,*
Portuguese, Mountain
More *Russian,* Sea
Mörön *Mongolian,* River

N

Nagor'ye *Russian,*
Upland
Nahal *Hebrew,* River
Nahr *Arabic,* River
Nam *Laotian,* River
Nehri *Turkish,* River
Nevado *Spanish,*
Mountain (snow-
capped)
Nisiá, Nísoi *Greek,*
Islands
Nizmennost' *Russian,*
Lowland, plain
Nosy *Malagasy,* Island
Nur *Mongolian,* Lake
Nuruu *Mongolian,*
Mountains
Nuur *Mongolian,* Lake
Nyzovyna *Ukrainian,*
Lowland, plain

O

Ostrov(a) *Russian,*
Island(s)
Oued *Arabic,*
Watercourse
-oy *Faroese,* Island
-øy(a) *Norwegian,* Island
Oya *Sinhala,* River

Ozero *Russian,*
Ukrainian, Lake

P

Passo *Italian,* Pass
Pegunungan *Indonesian,*
Malay, Mountain range
Pélagos *Greek,* Sea
Penisola *Italian,*
Peninsula
Peski *Russian,* Sands
Phanom *Thai,* Mountain
Phou *Laotian,* Mountain
Pic *Catalan,* Peak
Pico *Portuguese,*
Spanish, Peak
Pik *Russian,* Peak
Planalto *Portuguese,*
Plateau
Planina, Planini
Bulgarian, Macedonian,
Serbian, Croatian,
Mountain range
Ploskogor'ye *Russian,*
Upland
Poluostrov *Russian,*
Peninsula
Potamós *Greek,* River
Proliv *Russian,* Strait
Pulau *Indonesian,*
Malay, Island
Pulu *Malay,* Island
Punta *Portuguese,*
Spanish, Point

Q

Qā' *Arabic,* Depression
Qolleh *Persian,*
Mountain
Qullai *Tajik,* Mountain
Qundao *Chinese,* Islands

R

Raas *Somali,* Cape
-rags *Latvian,* Cape
Ramlat *Arabic,* Sands
Ra's *Arabic,* Cape,
point, headland
Ravnina *Bulgarian,*
Russian, Plain
Récif *French,* Reef
Represa (Rep.) *Spanish,*
Portuguese, Reservoir
-rettō *Japanese,* Island
chain
Riacho *Spanish,* Stream
Riban' *Malagasy,*
Mountains
Rio *Portuguese,* River
Río *Spanish,* River
Riu *Catalan,* River
Rivier *Dutch,* River
Rivière *French,* River
Rōd *Pashtu (Dari),*
River
Rúd *Persian,* River

Rudohorie *Slovak,*
Mountains
Ruisseau *French,* Stream

S

Sabkhat *Arabic,* Salt
marsh
Şaḥrā' *Arabic,* Desert
Samudra *Sinhala,*
Reservoir
-san *Japanese, Korean,*
Mountain
-sanchi *Japanese,*
Mountains
-sanmaek *Korean,*
Sarīr *Arabic,* Desert
Sebkha, Sebkhet *Arabic,*
Salt marsh, depression
See *German,* Lake
Selat *Indonesian,* Strait
-selkä *Finnish,* Ridge
Selseleh *Persian,*
Mountain range
Serra *Portuguese,*
Mountain
Serranía *Spanish,*
Mountain
Sha'īb *Arabic,*
Watercourse
Shamo *Chinese,*
Desert
Shan *Chinese,*
Mountain(s)
Shan-mo *Chinese,*
Mountain range
Shaṭṭ *Arabic,*
Distributary
-shima *Japanese,* Island
Shui-tao *Chinese,*
Channel
Sierra *Spanish,*
Mountains
Silsilah *Persian (Dari),*
Mountain range
Sòn *Vietnamese,*
Mountain
Sông *Vietnamese,* River
-spitze *German,* Peak
Štít *Slovak,* Peak
Stoeng *Cambodian,*
River
Stretto *Italian,* Strait
Su Anbarı *Azerbaijani,*
Reservoir
Sungai *Indonesian,*
Malay, River
Suu *Turkish,* River

T

Tal *Mongolian,* Plain
Tandavan' *Malagasy,*
Mountain range
Tangorombohitr'
Malagasy, Mountain
massif
Tao *Chinese,* Island

Tassili *Tamazight,*
Plateau, mountain
Tau *Russian,*
Mountain(s)
Taungdan *Burmese,*
Mountain range
Teluk *Indonesian,*
Malay, Bay
Terara *Amharic,*
Mountain
Tog *Somali,* Valley
Tônlé *Khmer,* Lake
Too *Kyrgyz,* Mountain,
Mountain range
Top *Dutch,* Peak
-tunturi *Finnish,*
Mountain
Tur'at *Arabic,* Channel

V

Väin *Estonian,* Strait
-vatn *Icelandic,* Lake
-vesi *Finnish,* Lake
Vinh *Vietnamese,* Bay
**Vodokhranilishche
(Vdkhr.)** *Russian,*
Reservoir
**Vodoskhovyshche
(Vdskh.)** *Ukrainian,*
Reservoir
Volcán *Spanish,* Volcano
Vozvyshennost' *Russian,*
Upland, plateau
Vrh *Macedonian,* Peak
Vysochyna *Ukrainian,*
Upland
Vysočina *Czech,* Upland

W

Waadi *Somali,*
Watercourse
Wādī *Arabic,*
Watercourse
Wāḥāt *Arabic,* Oasis
Wald *German,* Forest
Wan *Chinese,* Bay
Wyżyna *Polish,* Upland

X

Xé *Laotian,* River
Xi *Chinese,* River

Y

Yarımadası *Azerbaijani,*
Peninsula
Yazovir *Bulgarian,*
Reservoir
Yoma *Burmese,*
Mountains
Yu *Chinese,* Islet

Z

Zaliv *Bulgarian, Russian,*
Bay
Zatoka *Ukrainian,* Bay
Zemlya *Russian,* Land

Continental factfile

North & Central America

Total area: 9,400,000 sq miles (24,346,000 sq km)

Total number of countries: 23

Total population: 590 million

Largest city with population: Mexico City, Mexico 24.7 million

Country with highest population density: Barbados 1730 people per sq mile (669 people per sq km)

Largest country: Canada 3,855,171 sq miles (9,984,670 sq km)

Smallest country: St Kittws & Nevis 101 sq miles (261 sq km)

Largest lake: Lake Superior, Canada/ USA 31,700 sq miles (82,100 sq km)

Longest river: Mississippi-Missouri, USA 3902 miles (6280 km)

Highest point: Denali (Mt. McKinley), Alaska, USA 20,310 ft (6190 m)

Lowest point: Death Valley, California, USA -282 ft (-86 m) below sea level

South America

Total area: 6,880,000 sq miles (17,819,000 sq km)

Total number of countries: 12

Total population: 434 million

Largest city with population: São Paulo, Brazil 22.7 million

Country with highest population density: Ecuador 157 people per sq mile (65 people per sq km)

Largest country: Brazil 3,287,957 sq miles (8,515,770 sq km)

Smallest country: Suriname 63,251 sq miles (163,820 sq km)

Largest lake: Lake Titicaca, Bolivia/Peru 3220 sq miles (8340 sq km)

Longest river: Amazon, Brazil 3976 miles (6400 km)

Highest point: Cerro Aconcagua, Argentina 22,838 ft (6961 m)

Lowest point: Laguna del Carbón, Argentina -344 ft (-105 m) below sea level

Africa

Total area: 11,677,250 sq miles (30,244,050 sq km)

Total number of countries: 54

Total population: 1372 million

Largest city with population: Cairo, Egypt 21.9 million

Country with highest population density: Mauritius 1616 people per sq mile (624 people per sq km)

Largest country: Algeria 919,590 sq miles (2,381,740 sq km)

Smallest country: Seychelles 176 sq miles (455 sq km)

Largest lake: Lake Victoria, Uganda, Kenya, Tanzania 26,590 sq miles (68,870 sq km)

Longest river: Nile, Uganda/Sudan/Egypt 4130 miles (6650 km)

Highest point: Kilimanjaro, Tanzania 19,340 ft (5895 m)

Lowest point: Lac', Assal, Djibouti -512 ft (-156 m) below sea level

Europe

Total area: 3,743,246 sq miles (9,694,996 sq km)

Total number of countries: 47

Total population: 723 million

Largest city with population: Moscow, European Russia 17.4 million

Country with highest population density: Monaco 50,145 people per sq mile (19,497 people per sq km)

Largest country: European Russia 1,527,341 sq miles (3,955,818 sq km)

Smallest country: Vatican City, Italy 0.17 sq miles (0.44 sq km)

Largest lake: Ladoga, European Russia 6800 sq miles (17,700 sq km)

Longest river: Volga, European Russia 2194 miles (3531 km)

Highest point: El'brus, Caucasus Mts, European Russia 18,510 ft (5642 m)

Lowest point: Volga Delta, Caspian Sea, European Russia -92 ft (-28 m) below sea level

North & West Asia

 Total area:
9,585,550 sq miles
(24,826,600 sq km)

 Total number of
countries: 25

 Total population:
506 million

 Largest city with
population: Istanbul,
Turkey 16.5 million

 Country with highest
population density: Bahrain
5651 people per sq mile
(2241 people per sq km)

 Largest country: Asiatic Russia
5,065,471 sq miles
(13,119,582 sq km)

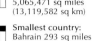 Smallest country:
Bahrain 293 sq miles
(760 sq km)

 Largest lake:
Caspian Sea 142,240 sq miles
(371,000 sq km)

 Longest river: Yenisey, Asiatic
Russia 3445 miles
(5544 km)

 Highest point: Tömür Feng,
Kyrgyzstan/China 24,406 ft
(7439 m)

 Lowest point: Dead Sea,
Israel/Jordan -1411 ft
(-430 m) below sea level

South & East Asia

 Total area:
7,936,200 sq miles
(20,554,700 sq km)

 Total number of
countries: 24

 Total population:
4156 million

 Largest city with
population: Guangzhou,
China 65.1 million

 Country with highest
population density: Singapore
20,769 people per sq mile
(7692 people per sq km)

 Largest country:
China 3,705,386 sq miles
(9,596,960 sq km)

 Smallest country:
Maldives 120 sq miles
(300 sq km)

 Largest lake: Tonle Sap,
Cambodia 1042 sq miles
(2850 sq km)

 Longest river: Chang Jiang
(Yangtze) 3917 miles
(6300 km)

 Highest point:
Mount Everest, Nepal
29,032 ft (8849 m)

 Lowest point: Turpan Hami,
(Turfan basin), China -505 ft
(-154 m) below sea level

Australasia & Oceania

 Total area:
3,376,700 sq miles
(8,745,750 sq km)

 Total number of
countries: 14

 Total population:
42 million

 Largest city with
population: Sydney,
Australia 5.5 million

 Country with highest
population density: Nauru
1211 people per sq mile
(603 people per sq km)

 Largest country:
Australia 2,969,907 sq miles
(7,692,024 sq km)

 Smallest country:
Nauru 8 sq miles
(21 sq km)

 Largest lake: Lake Eyre,
Australia 3700 sq miles
(9583 sq km)

 Longest river: Murray-
Darling, Australia
2330 miles (3750 km)

 Highest point: Mt. Wilhelm,
Papua New Guinea 14,794 ft
(4509 m)

 Lowest point: Lake Eyre,
Australia -49 ft
(-15 m) below sea level

Antarctica

 Total area: 5,500,000 sq miles (14,200,000 sq km)
of which approx. 324,300 sq miles
(840,000 sq km) is ice-free.

 Total number of countries: The Antarctic Treaty has
29 participating nations and 24 with observer status.
Claims by Australia, France, New Zealand, Norway,
Argentina, Chile, and the UK are not recognized by
other member states.

 Total Population: No indigenous population.
70 permanent research stations which represent 29
nations. Population varies between about 1000 (winter)
and 4000 (summer).

 Total volume of ice:
7,200,000 cu miles (30,000,000 cu km):
contains 90% of Earth's fresh water

 Sea ice: 1,158,300 sq miles (3,000,000
sq km) in February. 7,722,000 sq miles
(20,000,000 sq km) in October

 Lowest temperature: Vostok station
-89.5°C (-129°F)

 Highest point: Vinson Massif
16,072 ft (4897 m)

 Lowest Point: Coastline 0ft/m

Geographical comparisons

Largest countries

Russia ....6,601,668 sq miles ... (17,098,242 sq km)
Canada ....3,855,171 sq miles .....(9,984,670 sq km)
China ....3,705,386 sq miles .... (9,596,960 sq km)
USA .... 3,677,649 sq miles .... (9,525,067 sq km)
Brazil .... 3,287,957 sq miles .... (8,515,770 sq km)
Australia .... 2,969,907 sq miles .... (7,692,024 sq km)
India.................. 1,269,219 sq miles .... (3,287,263 sq km)
Argentina .... 1,073,518 sq miles .... (2,780,400 sq km)
Kazakhstan ...... 1,052,089 sq miles .....(2,717,300 sq km)
Algeria .... 919,590 sq miles .... (2,724,900 sq km)

Smallest countries

Vatican City .... 0.17 sq miles ....(0.44 sq km)
Monaco.................... 0.77 sq miles ....(2 sq km)
Nauru .... 8 sq miles ....(21 sq km)
Tuvalu .... 10 sq miles ....(26 sq km)
San Marino .... 24 sq miles ....(61 sq km)
Liechtenstein.............. 62 sq miles ....(160 sq km)
Marshall Islands......... 70 sq miles ....(181 sq km)
St. Kitts & Nevis .... 101 sq miles ....(261 sq km)
Maldives................... 120 sq miles ....(300 sq km)
Malta........................ 122 sq miles ....(316 sq km)

Largest islands

Greenland...............822,700 sq miles (2,130,783 sq km)
New Guinea ....303,381 sq miles (785,753 sq km)
Borneo ....288,869 sq miles (748,167 sq km)
Madagascar ....226,658 sq miles (587,042 sq km)
Baffin Island ....195,928 sq miles (507,451 sq km)
Sumatra ....171,068 sq miles (443,064 sq km)
Honshu ....87,200 sq miles (225,847 sq km)
Victoria Island............83,897 sq miles (217,292 sq km)
Britain........................80,823 sq miles (209,331 sq km)
Ellesmere Island ....75,767 sq miles (196,236 sq km)

Richest countries (GNI per capita, in US$)

Liechtenstein 116,540
Switzerland 90,360
Norway... 84,490
Luxembourg 81,110
Ireland.. 74,520
USA 70,430
Denmark 68,110
Iceland 64,410
Singapore.. 64,010
Sweden 58,890

Poorest countries (GNI per capita, in US$)

Burundi 240
Somalia 450
Mozambique.. 480
Madagascar... 500
Afghanistan... 500
Sierra Leone 510
Central African Republic 530
Congo, Dem. Republic.. 580
Niger 590
Eritrea.. 600

Most populous countries

China.. 1.41 billion
India .... 1.39 billion
USA .... 332 million
Indonesia .... 276 million
Pakistan .... 225 million
Brazil .... 214 million
Nigeria.. 211 million
Bangladesh... 166 million
Russia .... 143 million
Mexico .... 130 million

Least populous countries

Vatican City 825
Nauru 10,870
Tuvalu 11,930
Palau 18,170
San Marino 34,010
Liechtenstein 38,250
Monaco 39,550
St. Kitts & Nevis.. 53,550
Marshall Islands 59,620
Dominica... 72,170

Most densely populated countries

Monaco ..50,145 people per sq mile (19,497 per sq km)
Singapore ..20,769 people per sq mile (7692 per sq km)
Bahrain ....5651 people per sq mile (2241 per sq km)
Vatican City ..4852 people per sq mile (1875 per sq km)
Maldives.......4667 people per sq mile (1812 per sq km)
Malta ....4252 people per sq mile (1615 per sq km)
Bangladesh...3276 people per sq mile (1278 per sq km)
Barbados ....1730 people per sq mile (669 per sq km)
Lebanon ....1728 people per sq mile (662 per sq km)
Mauritius........1616 people per sq mile (624 per sq km)

Most sparsely populated countries

Mongolia......... 5 people per sq mile......... (2 per sq km)
Namibia.......... 8 people per sq mile......... (3 per sq km)
Australia.......... 8 people per sq mile......... (3 per sq km)
Iceland.......... 10 people per sq mile......... (4 per sq km)
Guyana......... 10 people per sq mile......... (4 per sq km)
Suriname....... 10 people per sq mile......... (4 per sq km)
Canada....... 10 people per sq mile......... (4 per sq km)
Libya.............. 11 people per sq mile......... (4 per sq km)
Botswana....... 11 people per sq mile......... (4 per sq km)
Mauritania..... 11 people per sq mile......... (4 per sq km)

Most spoken languages (Native speakers)

1. Chinese (Mandarin)	6. Bengali
2. Spanish	7. Portuguese
3. English	8. Russian
4. Arabic	9. Japanese
5. Hindi	10. Western Punjabi

Largest conurbations

Guangzhou (China)..................................... 65,100,000
Tokyo (Japan) 40,700,000
Shanghai (China)....................................... 39,300,000
Delhi (India) 32,400,000
Jakarta (Indonesia).................................... 28,600,000
Manila (Philippines) 26,400,000
Mumbai (India) 26,100,000
Seoul (South Korea)................................... 24,800,000
Mexico City (Mexico)................................. 24,700,000
New York (USA) 23,000,000
São Paulo (Brazil) 22,700,000
Cairo (Egypt) 21,900,000
Dhaka (Bangladesh) 20,900,000
Lagos (Nigeria).. 20,700,000
Beijing (China) 20,500,000
Bangkok (Thailand) 19,900,000
Karachi (Pakistan)..................................... 18,600,000
Osaka (Japan).. 17,700,000
Los Angeles (USA)..................................... 17,500,000
Moscow (Russia) 17,400,000
Kolkata (India)... 17,200,000
Buenos Aires (Argentina) 16,800,000
Istanbul (Turkey) 16,500,000
Tehran (Iran) 15,800,000
Chengdu (China).. 15,200,000

Longest rivers

Nile (Northeast Africa) .... 4130 miles ...... (6650 km)
Amazon (South America)........... 3976 miles ...... (6400 km)
Yangtze (China).......................... 3917 miles ...... (6300 km)
Mississippi/Missouri (USA).......... 3902 miles........ (6280 km)
Yenisey (Russia) .... 3445 miles ...... (5544 km)
Yellow River (China) .... 3395 miles ...... (5464 km)
Ob (Russia) .... 3364 miles....... (5414 km)
Parana (South America) .... 3030 miles...... (4876 km)
Congo (Central Africa) .... 2922 miles ...... (4703 km)
Amur (East Asia)....................... 2763 miles...... (4447 km)

Highest mountains (Height above sea level)

Everest...................................... 29,032 ft....... (8849 m)
K2 .... 28,253 ft....... (8611 m)
Kanchenjunga I...................... 28,169 ft....... (8586 m)
Lhotse...................................... 27,940 ft....... (8516 m)
Makalu .... 27,838 ft....... (8485 m)
Cho Oyu .... 26,864 ft....... (8188 m)
Dhaulagiri I............................. 26,795 ft....... (8167 m)
Manaslu .... 26,781 ft....... (8163 m)
Nanga Parbat.......................... 26,660 ft....... (8126 m)
Annapurna............................... 26,545 ft....... (8091 m)

Largest bodies of inland water (Area & depth)

Caspian Sea
    143,240 sq miles (371,000 sq km).....3363 ft (1025 m)
Lake Superior
    31,700 sq miles (82,100 sq km).......1332 ft (406 m)
Lake Victoria
    26,590 sq miles (68,870 sq km)...........276 ft (84 m)
Lake Huron
    23,000 sq miles (59,600 sq km).........751 ft (229 m)
Lake Michigan
    22,400 sq miles (58,000 sq km).........922 ft (281 m)
Lake Tanganyika
    12,600 sq miles (32,600 sq km).... 4820 ft (1470 m)
Lake Baikal
    12,200 sq miles (31,500 sq km).....5371 ft (1637 m)
Great Bear Lake
    12,000 sq miles (31,150 sq km)...... 1463 ft (446 m)
Lake Malawi
    11,400 sq miles (29,500 sq km)...... 2316 ft (706 m)
Great Slave Lake
    10,000 sq miles (27,000 sq km)...... 2014 ft (614 m)

......continued on page 166

Geographical comparisons continued

Deepest ocean features

Challenger Deep, Mariana Trench (Pacific)
36,197 ft .... (11,034 m)
Horizon Deep, Tonga Trench (Pacific)
35,702 ft .... (10,882 m)
Galathea Depth, Philippine Trench (Pacific)
34,580 ft .... (10,545 m)
Kuril-Kamchatka Trench (Pacific)
34,449 ft .... (10,542 m)
Kermadec Trench (Pacific)
32,963 ft .... (10,047 m)
Izu–Ogasawara Trench (Pacific)
32,087 ft .... (9810 m)
Japan Trench (Pacific)
29,527 ft .... (9000 m)
Milwaukee Deep, Puerto Rico Trench (Atlantic)
28,232 ft .... (8605 m)
Yap Trench (Pacific)
27,976 ft .... (8527 m)
Meteor Deep, South Sandwich Trench (Atlantic)
27,651 ft .... (8428 m)

Greatest waterfalls   (Mean flow of water)

Boyoma (Congo, D.R.).... 600,000 cu. ft/sec (16,990 cu.m/sec)
Khône (Laos/Cambodia) ... 410,000 cu. ft/sec (11,600 cu.m/sec)
Pará (Venezuela) .... 125,000 cu. ft/sec (3540 cu.m/sec)
Paulo Afonso (Brazil) .... 100,000 cu. ft/sec (2800 cu.m/sec)
Niagara (USA/Canada).......... 85,000 cu. ft/sec (2407 cu.m/sec)
Vermilion (Canada)................ 64,000 cu. ft/sec (1800 cu.m/sec)
Iguaçu (Argentina/Brazil) .... 6,800 cu. ft/sec (1750 cu.m/sec)
Limestone (Canada).............51,600 cu. ft/sec (1460 cu.m/sec)
Pyrite (Canada)....................51,600 cu. ft/sec (1460 cu.m/sec)
Victoria (Zimbabwe)................ 39,000 cu. ft/sec (1100 cu.m/sec)
Virginia (Canada)................... 35,300 cu. ft/sec (1000 cu.m/sec)
Shivanasamudra (India) .... 33,000 cu. ft/sec (930 cu.m/sec)

Highest waterfalls

Angel (Venezuela) .... 3212 ft .... (979 m)
Tugela (South Africa) .... 3110 ft .... (948 m)
Tres Hermanas (Peru)................. 2999 ft .... (914 m)
Olo'upena (USA) .... 2953 ft .... (900 m)
Yumbilla (Peru) .... 2940 ft .... (896 m)
Skorga (Norway)........................... 2871 ft .... (875 m)
Vinnufossen (Norway).................. 2822 ft .... (860 m)
Baláifossen (Norway)................... 2789 ft .... (850 m)
Mattenbachfall (Switzerland)...... 2756 ft .... (840 m)
Pu'uka'oku (USA) .... 2756 ft .... (840 m)
James Bruce (Canada)................. 2756 ft .... (840 m)
Browne (New Zealand) .... 2743 ft .... (836 m)

Largest deserts

Sahara................ 3,552,140 sq miles (9,200,000 sq km)
Gobi...................... 500,000 sq miles (1,295,000 sq km)
Kalahari .... 347,492 sq miles (900,000 sq km)
Patagonian .... 259,847 sq miles (673,000 sq km)
Ar Rub al Khali .... 250,000 sq miles (650,000 sq km)
Great Basin............. 190,000 sq miles (492,100 sq km)
Chihuahuan............. 175,000 sq miles (453,250 sq km)
Karakum .... 135,136 sq miles (350,000 sq km)
Great Victorian ...... 134,653 sq miles (348,750 sq km)
Sonoran.................. 130,116 sq miles (337,000 sq km)

*NB – Most of Antarctica is a polar desert, with only
2 inches (50 mm) of precipitation annually*

Hottest inhabited places

Abéché (Chad) .... 90.0°F .... (32.2°C)
Mecca (Saudi Arabia) .... 89.8°F .... (32.1°C)
Kaédi (Mauritania) .... 89.2°F .... (31.7°C)
Yélimané (Mali)............................... 88.7°F .... (31.5°C)
Jizan (Saudi Arabia) .... 88.0°F .... (31.1°C)
Kiffa (Mauritania) .... 87.9°F .... (31.0°C)
Atbara (Sudan) .... 87.7°F .... (30.9°C)
Matam (Senegal) .... 87.6°F .... (30.8°C)
Ayoun al Atrous (Mauritania) ...... 87.2°F .... (30.6°C)

Driest inhabited places

Al Jawf (Libya) .... <0.10 in ..... (<2.5 mm)
Chimbote (Peru)......................... <0.10 in ..... (<2.5 mm)
Kaktovik (AK, USA).................... <0.10 in ..... (<2.5 mm)
Pisco (Peru) .... <0.10 in ..... (<2.5 mm)
Wadi Halfa (Sudan) .... <0.10 in ..... (<2.5 mm)
Siwa Oasis (Egypt) .... <0.10 in ..... (<2.5 mm)
Kharga Oasis (Egypt)................. 0.10 in .... (2.5 mm)
Aswan (Egypt) .... 0.10 in .... (2.5 mm)
Nok Kundi (Pakistan) .... 0.10 in .... (2.5 mm)
Altos del Mar (Chile)................... 0.10 in .... (2.5 mm)

Wettest inhabited places

Mawsynram (India) .... 467 in .. (11,871 mm)
Cherrapunji (India) .... 464 in .. (11,777 mm)
Tutunendo (Colombia).................. 454 in .. (11,770 mm)
San Ant. de Ureca (Eq. Guin.) ...... 418 in .. (10,450 mm)
Debundscha (Cameroon) .... 405 in . (10,299 mm)
Quibdó City (Colombia) .... 289 in .. (7328 mm)
Buenaventura (Colombia) .... 247 in .... (6276 mm)
Mawlamyine (Myanmar)............... 188 in .... (4772 mm)
Monrovia (Liberia) .... 179 in .... (4540 mm)
Hilo (Hawaii) .... 127 in .... (3219 mm)

Country profiles

These country profiles are intended as a guide to a world that is continually changing. All the data has been researched from the most up-to-date and authoritative sources to give an overview of the geographical, political, and social characteristics that make each country so unique.

There are currently 196 independent countries in the world - more than at any previous time - and over 50 dependencies. Antarctica is the only land area on Earth that is not officially part of, and does not belong to, any single country.

Country profile key

Formation Date of formation denotes the date of political origin or independence of a state, i.e. its emergence as a recognizable entity in the modern political world / date current borders were established

Population Total population / population density – based on total land area

Languages An asterisk (*) denotes the official language(s)

Life expectancy Average life expectancy at birth for males and females combined

---

AFGHANISTAN
Central Asia

Official name Islamic Republic of Afghanistan
Formation 1919 / 1919
Capital Kabul
Population 38.3 million / 152 people per sq mile (59 people per sq km)
Total area 251,827 sq miles (652,230 sq km)
Languages Pashto*, Tajik, Dari*, Farsi, Uzbek, Turkmen, Urdu
Religions Muslim 99% (Sunni 87%, Shi'a 12%), Other 1%
Demographics Pashtun 38%, Tajik 25%, Hazara 19%, Uzbek and Turkmen 15%, Other 3%
Government Islamic theocracy
Currency Afghani = 100 puls
Literacy rate 37%
Life expectancy 62 years

ANDORRA
Southwest Europe

Official name Principality of Andorra
Formation 1278 / 1278
Capital Andorra la Vella
Population 85,560 / 473 people per sq mile (183 people per sq km)
Total area 181 sq miles (468 sq km)
Languages Spanish, Catalan*, French, Portuguese, Castilian
Religions Christian (mainly Roman Catholic) 90%, Other 10%
Demographics Andorran 48%, Spanish 25%, Portuguese 11%, French 5%, Other 11%
Government Parliamentary system
Currency Euro = 100 cents
Life expectancy 84 years

ARGENTINA
South America

Official name Argentine Republic
Formation 1816 / 1898
Capital Buenos Aires
Population 46.2 million / 43 people per sq mile (17 people per sq km)
Total area 1,073,518 sq miles (2,780,400 sq km)
Languages Spanish*, Italian, English, German, French, Indigenous (Mapudungun, Quechua)
Religions Roman Catholic 63%, Evangelical 15%, Nonreligious 19%, Other 3%
Demographics European and mixed race 97%, Indigenous 2%, Other 1%
Government Presidential system
Currency Argentine peso = 100 centavos
Literacy rate 99%
Life expectancy 75 years

AUSTRIA
Central Europe

Official name Republic of Austria
Formation 1918 / 1920
Capital Vienna
Population 8.9 million / 275 people per sq mile (106 people per sq km)
Total area 32,383 sq miles (83,871 sq km)
Languages German*, Turkish, Serbian, Croatian, Slovenian, Hungarian (Magyar)
Religions Roman Catholic 55%, None 26%, Other Christian 9%, Muslim 8%, Other 2%
Demographics Austrian 81%, German 3%, Turkish 2%, Other 14%
Government Parliamentary system
Currency Euro = 100 cents
Literacy rate 98%
Life expectancy 81 years

ALBANIA
Southeast Europe

Official name Republic of Albania
Formation 1912 / 1912
Capital Tirana
Population 3 million / 270 people per sq mile (104 people per sq km)
Total area 11,100 sq miles (28,748 sq km)
Languages Albanian*, Greek
Religions Muslim (mainly Sunni) 68%, Roman Catholic 12%, Albanian Orthodox 8%, Nonreligious 6%, Other 6%
Demographics Albanian 83%, Other 17%
Government Parliamentary system
Currency Lek = 100 qindarka (qintars)
Literacy rate 98%
Life expectancy 76 years

ANGOLA
Southern Africa

Official name Republic of Angola
Formation 1975 / 1975
Capital Luanda
Population 34.7 million / 72 people per sq mile (28 people per sq km)
Total area 481,354 sq miles (1,246,700 sq km)
Languages Portuguese*, Kimbundu, Umbundu, Chokwe, Kikongo
Religions Roman Catholic 40%, Protestant 38%, Nonreligious 12%, Other (including animist) 10%
Demographics Ovimbundu 37%, Ambundu 25%, Bakongo 13%, Other African 21%, Other 4%
Government Presidential system
Currency Kwanza = 100 centimos
Literacy rate 71%
Life expectancy 62 years

ARMENIA
Southwest Asia

Official name Republic of Armenia
Formation 1991 / 1991
Capital Yerevan
Population 3 million / 261 people per sq mile (101 people per sq km)
Total area 11,484 sq miles (29,743 sq km)
Languages Armenian*, Azeri, Russian, Kurdish
Religions Orthodox Christian 89%, Nonreligious 2%, Armenian Catholic Church 1%, Other 8%
Demographics Armenian 98%, Yezidi 1%, Other 1%
Government Parliamentary system
Currency Dram = 100 luma
Literacy rate 100%
Life expectancy 72 years

AZERBAIJAN
Southwest Asia

Official name Republic of Azerbaijan
Formation 1991 / 1991
Capital Baku
Population 10.3 million / 308 people per sq mile (119 people per sq km)
Total area 33,436 sq miles (86,600 sq km)
Languages Azeri*, Russian, Armenian
Religions Muslim (mainly Shi'a) 97%, Christian 3%
Demographics Azeri 92%, Lezgin 2%, Other 6%
Government Presidential system
Currency Manat = 100 gopik
Literacy rate 100%
Life expectancy 69 years

ALGERIA
North Africa

Official name People's Democratic Republic of Algeria
Formation 1962 / 1962
Capital Algiers
Population 44.1 million / 48 people per sq mile (19 people per sq km)
Total area 919,590 sq miles (2,381,740 sq km)
Languages Arabic*, Tamazight* (Kabyle, Shawia, Tamashek), French
Religions Muslim (mainly Sunni) 99%, Other 1%
Demographics Arab 75%, Amazigh 24%, European & Jewish 1%
Government Presidential system
Currency Algerian dinar = 100 centimes
Literacy rate 81%
Life expectancy 76 years

ANTIGUA & BARBUDA
West Indies

Official name Antigua and Barbuda
Formation 1981 / 1981
Capital St. John's
Population 100,335 / 587 people per sq mile (226 people per sq km)
Total area 171 sq miles (443 sq km)
Languages English*, English patois
Religions Other Christian 49%, Anglican 19%, Seventh-day Adventist 13%, Other 19%
Demographics Black 87%, Mixed race 5%, Hispanic 3%, White 2%, Other 3%
Government Parliamentary system
Currency East Caribbean dollar = 100 cents
Literacy rate 98%
Life expectancy 78 years

AUSTRALIA
Australasia & Oceania

Official name Commonwealth of Australia
Formation 1901 / 1901
Capital Canberra
Population 26.1 million / 8 people per sq mile (3 people per sq km)
Total area 2,969,907 sq miles (7,692,024 sq km)
Languages English*, Italian, Cantonese, Greek, Arabic, Vietnamese, Mandarin
Religions Roman Catholic 28%, Nonreligious 24%, Other Christian 20%, Anglican 19%, Other 9%,
Demographics British 33%, Australian 30%, Irish 9%, Scottish 9%, Chinese 5%, Indigenous 3%, Other 11%
Government Parliamentary system
Currency Australian dollar = 100 cents
Literacy rate 99%
Life expectancy 83 years

BAHAMAS, THE
West Indies

Official name Commonwealth of The Bahamas
Formation 1973 / 1973
Capital Nassau
Population 355,608 / 66 people per sq mile (26 people per sq km)
Total area 5359 sq miles (13,880 sq km)
Languages English*, English Creole, French Creole
Religions Baptist 36%, Anglican 14%, Roman Catholic 12%, Pentecostal 9%, Seventh-day Adventist 5%, Methodist 4%, Other 20%
Demographics Black 90%, White 5%, Mixed race 2%, Other 3%
Government Parliamentary system
Currency Bahamian dollar = 100 cents
Literacy rate 96%
Life expectancy 72 years

BAHRAIN
Southwest Asia

Official name Kingdom of Bahrain
Formation 1971 / 2001
Capital Manama
Population 1.5 million / 5651 people per sq mile (2241 people per sq km)
Total area 293 sq miles (760 sq km)
Languages Arabic*, English, Farsi, Urdu
Religions Muslim (mainly Shi'a) 74%, Christian 9%, Other 17%
Demographics Bahraini 46%, Asian 46%, Other Arab 5%, Other 3%
Government Monarchical / parliamentary system
Currency Bahraini dinar = 1000 fils
Literacy rate 100%
Life expectancy 79 years

BELGIUM
Northwest Europe

Official name Kingdom of Belgium
Formation 1830 / 1919
Capital Brussels
Population 11.8 million / 1001 people per sq mile (387 people per sq km)
Total area 11,787 sq miles (30,528 sq km)
Languages Dutch*, French*, German*
Religions Roman Catholic 57%, Nonreligious 30%, Muslim 7%, Other Christian 4%, Other 2%
Demographics Belgian 75%, Italian 4%, Moroccan 4%, French 2%, Turkish 2%, Dutch 2%, Other 11%
Government Parliamentary system
Currency Euro = 100 cents
Literacy rate 99%
Life expectancy 82 years

BOLIVIA
South America

Official name Plurinational State of Bolivia
Formation 1825 / 1938
Capital La Paz (admin.); Sucre (judicial)
Population 12 million / 28 people per sq mile (11 people per sq km)
Total area 424,164 sq miles (1,098,581 sq km)
Languages Aymara*, Quechua*, Spanish*, Guarani
Religions Roman Catholic 70%, Evangelical 15%, Adventist 2%, Church of Jesus Christ 1%, Other 12%
Demographics Mixed race 70%, Indigenous 20%, White 5%, African 1%, Other 4%
Government Presidential system
Currency Boliviano = 100 centavos
Literacy rate 93%
Life expectancy 64 years

BRUNEI
Southeast Asia

Official name Brunei Darussalam
Formation 1984 / 1984
Capital Bandar Seri Begawan
Population 474,054 / 213 people per sq mile (82 people per sq km)
Total area 2226 sq miles (5765 sq km)
Languages Malay*, English, Chinese dialects
Religions Muslim (mainly Sunni) 79%, Christian 9%, Buddhist 8%, Other 4%
Demographics Malay 66%, Chinese 10%, Indigenous 4%, Other 20%
Government Monarchy
Currency Bruneian dollar = 100 cents
Literacy rate 97%
Life expectancy 75 years

BANGLADESH
South Asia

Official name People's Republic of Bangladesh
Formation 1971 / 2015
Capital Dhaka
Population 166 million / 3276 people per sq mile (1278 people per sq km)
Total area 57,321 sq miles (148,460 sq km)
Languages Bengali*, Urdu, Chakma, Marma (Magh), Garo, Khasi, Santhali, Tripuri, Mro
Religions Muslim (mainly Sunni) 91%, Hindu 8%, Other 1%
Demographics Bengali 98%, Other indigenous ethnic groups 2%
Government Parliamentary system
Currency Taka = 100 poisha
Literacy rate 78%
Life expectancy 72 years

BELIZE
Central America

Official name Belize
Formation 1981 / 1981
Capital Belmopan
Population 412,387 / 47 people per sq mile (18 people per sq km)
Total area 8867 sq miles (22,966 sq km)
Languages English Creole, Spanish, English*, Mayan, Garifuna (Carib), German
Religions Roman Catholic 40%, Other Christian 34%, Nonreligious 16%, Other 10%
Demographics Mixed race 40%, Creole 24%, Maya 10%, Garifuna 6%, Asian Indian 4%, Other 7%
Government Parliamentary system
Currency Belizean dollar = 100 cents
Literacy rate 75%
Life expectancy 70 years

BOSNIA & HERZEGOVINA
Southeast Europe

Official name Bosnia and Herzegovina
Formation 1992 / 1992
Capital Sarajevo
Population 3.8 million / 192 people per sq mile (74 people per sq km)
Total area 19,767 sq miles (51,197 sq km)
Languages Bosnian*, Serbian*, Croatian*
Religions Muslim (mainly Sunni) 53%, Orthodox Christian 35%, Roman Catholic 8%, Nonreligious 3%, Other 1%
Demographics Bosniak 50%, Serb 31%, Croat 15%, Other 4%
Government Parliamentary system
Currency Marka = 100 pfeninga
Literacy rate 99%
Life expectancy 75 years

BULGARIA
Southeast Europe

Official name Republic of Bulgaria
Formation 1908 / 1947
Capital Sofia
Population 6.8 million / 159 people per sq mile (61 people per sq km)
Total area 42,811 sq miles (110,879 sq km)
Languages Bulgarian*, Turkish, Romani
Religions Orthodox Christian 75%, Muslim 15%, Nonreligious 5%, Protestant 1%, Roman Catholic 1%, Other 3%
Demographics Bulgarian 85%, Turkish 9%, Roma 5%, Other 1%
Government Parliamentary system
Currency Lev = 100 stotinki
Literacy rate 98%
Life expectancy 72 years

BARBADOS
West Indies

Official name Barbados
Formation 1966 / 1966
Capital Bridgetown
Population 302,674 / 1730 people per sq mile (669 people per sq km)
Total area 166 sq miles (430 sq km)
Languages Bajan (Barbadian English), English*
Religions Anglican 24%, Nonreligious 21%, Pentecostal 20%, Seventh-day Adventist 6%, Methodist 4%, Roman Catholic 4%, Other 21%
Demographics Black 93%, Mixed race 3%, White 3%, Other 1%
Government Parliamentary system
Currency Barbados dollar = 100 cents
Literacy rate 100%
Life expectancy 78 years

BENIN
West Africa

Official name Republic of Benin
Formation 1960 / 1960
Capital Porto-Novo; Cotonou
Population 13.7 million / 315 people per sq mile (122 people per sq km)
Total area 43,484 sq miles (112,622 sq km)
Languages Fon, Bariba, Yoruba, Adja, Houeda, Somba, French*
Religions Muslim 28%, Roman Catholic 26%, Other Christian 24%, Vodoun 12%, Other 10%
Demographics Fon 38%, Adja 15%, Yoruba 12%, Bariba 10%, Fulani 9%, Other 16%
Government Presidential system
Currency CFA franc = 100 centimes
Literacy rate 42%
Life expectancy 60 years

BOTSWANA
Southern Africa

Official name Republic of Botswana
Formation 1966 / 1966
Capital Gaborone
Population 2.3 million / 13 people per sq mile (5 people per sq km)
Total area 224,607 sq miles (581,730 sq km)
Languages Setswana, English*, Sekalanga, Shona, San, Khoikhoi, isiNdebele
Religions Christian (mainly Protestant) 80%, Nonreligious 15%, Traditional beliefs 4%, Other (including Muslim) 1%
Demographics Tswana (or Setswana) 79%, Kalanga 11%, Basarwa 3%, Other 7%
Government Parliamentary system
Currency Pula = 100 thebe
Literacy rate 89%
Life expectancy 61 years

BURKINA FASO
West Africa

Official name Burkina Faso
Formation 1960 / 1960
Capital Ouagadougou
Population 21.9 million / 207 people per sq mile (80 people per sq km)
Total area 105,869 sq miles (274,200 sq km)
Languages Mossi, Fulani, French*, Tuareg,Dyula, Songhai
Religions Muslim 60% Christian 23%, Indigenous beliefs 15%, Other 2%
Demographics Mossi 52%, Fulani 8%, Gurma 7%, Bobo 5%, Gurunsi 5%, Senufo 5%, Bissa 4%, Lobi 2%, Dagara 2%, Other 10%
Government Presidential system
Currency CFA franc = 100 centimes
Literacy rate 39%
Life expectancy 59 years

BELARUS
Eastern Europe

Official name Republic of Belarus
Formation 1991 / 1991
Capital Minsk
Population 9.4 million / 117 people per sq mile (45 people per sq km)
Total area 80,155 sq miles (207,600 sq km)
Languages Belarusian*, Russian*
Religions Orthodox Christian 73%, Roman Catholic 12%, Nonreligious 3%, Other 12%
Demographics Belarussian 86%, Russian 8%, Polish 3%, Ukrainian 1%, Other 2%
Government Presidential system
Currency Belarussian ruble = 100 kopeks
Literacy rate 100%
Life expectancy 72 years

BHUTAN
South Asia

Official name Kingdom of Bhutan
Formation 1907 / 2006
Capital Thimphu
Population 867,775 / 59 people per sq mile (23 people per sq km)
Total area 14,824 sq miles (38,394 sq km)
Languages Dzongkha*, Sharchopkha, Lhotshamkha
Religions Mahayana Buddhist 75%, Hindu 22%, Other 3%
Demographics Ngalop 50%, Nepali 35%, Tribal groups 15%
Government Monarchical / parliamentary system
Currency Ngultrum = 100 chetrum
Literacy rate 67%
Life expectancy 72 years

BRAZIL
South America

Official name Federative Rep. of Brazil
Formation 1822 / 1909
Capital Brasília
Population 214 million / 65 people per sq mile (25 people per sq km)
Total area 3,287,957 sq miles (8,515,770 sq km)
Languages Portuguese*, German, Italian, Spanish, Polish, Japanese, Amerindian languages
Religions Roman Catholic 61%, Protestant 26%, Nonreligious 8%, Other 5%
Demographics White 48%, Mixed race 43%, Black 8%, Other 1%
Government Presidential system
Currency Real = 100 centavos
Literacy rate 93%
Life expectancy 77 years

BURUNDI
Central Africa

Official name Republic of Burundi
Formation 1962 / 1962
Capital Bujumbura; Gitega
Population 12.6 million / 1173 people per sq mile (453 people per sq km)
Total area 10,745 sq miles (27,830 sq km)
Languages Kirundi*, French*, Kiswahili, English
Religions Roman Catholic 65%, Protestant 23%, Muslim 3%, Seventh-day Adventist 2%, Other 7%
Demographics Hutu 85%, Tutsi 14%, Twa 1%
Government Presidential system
Currency Burundian franc = 100 centimes
Literacy rate 68%
Life expectancy 62 years

CAMBODIA
Southeast Asia

Official name Kingdom of Cambodia
Formation 1953 / 1953
Capital Phnom Penh
Population 16.7 million/ 239 people per sq mile (92 people per sq km)
Total area 69,898 sq miles (181,035 sq km)
Languages Khmer*, French, Chinese, Vietnamese, Cham
Religions Buddhist 97%, Muslim 2%, Other (mostly Christian) 1%
Demographics Khmer 95%, Cham 2%, Chinese 2%, Other 1%
Government Parliamentary system
Currency Riel = 100 sen
Literacy rate 81%
Life expectancy 70 years

CAMEROON
Central Africa

Official name Republic of Cameroon
Formation 1960 / 2006
Capital Yaoundé
Population 29.3 million / 160 people per sq mile (62 people per sq km)
Total area 183,568 sq miles (475,440 sq km)
Languages Bamileke, Fang, Fulani, French*, English*
Religions Roman Catholic 38%, Protestant 26%, Other Christian 7%, Muslim 24%, Other 5%
Demographics Bamileke-Bamu 24%, Beti/Bassa, Mbam 21%, Biu-Mandara 14%, Arab-Choa/Hausa/Kanuri 11%, Adamawa-Ubangi 9%, Other 21%
Government Presidential system
Currency CFA franc = 100 centimes
Literacy rate 77%
Life expectancy 60 years

CANADA
North America

Official name Canada
Formation 1867 / 1949
Capital Ottawa
Population 38.2 million / 10 people per sq mile (4 people per sq km)
Total area 3,855,171 sq miles (9,984,670 sq km)
Languages English*, French*, Pubjabi, Cantonese, Spanish, Arabic,Tagalog, Italian, German
Religions Roman Catholic 39%, Other Christian 28%, Nonreligious 24%, Muslim 3%, Other 6%
Demographics European descent 80%, Asian 15%, First Nations, Métis, and Inuit 5%
Government Parliamentary system
Currency Canadian dollar = 100 cents
Literacy rate 99%
Life expectancy 82 years

CAPE VERDE (CABO VERDE)
Atlantic Ocean

Official name Republic of Cabo Verde
Formation 1975 / 1975
Capital Praia
Population 596,707 / 383 people per sq mile (148 people per sq km)
Total area 1557 sq miles (4033 sq km)
Languages Portuguese Creole, Portuguese*
Religions Roman Catholic 77%, Protestant 5%, other Christian 3% , Muslim 2%, Other 13%
Demographics Mixed race 71%, African 28%, European 1%
Government Presidential / parliamentary system
Currency Escudo = 100 centavos
Literacy rate 87%
Life expectancy 74 years

CENTRAL AFRICAN REPUBLIC
Central Africa

Official name Central African Republic
Formation 1960 / 1960
Capital Bangui
Population 5.4 million / 22 people per sq mile (9 people per sq km)
Total area 240,535 sq miles (622,984 sq km)
Languages Sango, Banda, Gbaya, French*
Religions Christian 89%, Muslim 9%, Folk Religion 1%, Unaffiliated 1%
Demographics Baya 29%, Banda 23%, Mandjia 10%, Sara 8%, M'Baka-Bantu 8%, Mbum 6%, Arab-Fulani 6%, Ngbanki 6%, Other 4%
Government Presidential system
Currency CFA franc = 100 centimes
Literacy rate 37%
Life expectancy 54 years

CHAD
Central Africa

Official name Republic of Chad
Formation 1960 / 1960
Capital N'Djaména
Population 17.9 million / 36 people per sq mile (14 people per sq km)
Total area 495,755 sq miles (1,284,000 sq km)
Languages French*, Sara, Arabic*, Maba
Religions Muslim 52%, Protestant 24%, Roman Catholic 20%, None 3%, Other 1%
Demographics Sara 31%, Arab 10%, Kanembu 9%, Masalit 7%, Gorane 6%, Other Indigenous groups 34%, Other 3%
Government Presidential system
Currency CFA franc = 100 centimes
Literacy rate 22%
Life expectancy 53 years

CHILE
South America

Official name Republic of Chile
Formation 1810 / 1898
Capital Santiago
Population 18.4 million / 63 people per sq mile (24 people per sq km)
Total area 291,933 sq miles (756,120 sq km)
Languages Spanish*, Amerindian languages
Religions Roman Catholic 60%, Evangelical 18%, Atheist or Agnostic 4%, None 17%
Demographics White and mixed race 89%, Mapuche 9%, Aymara 1%, Other Indigenous 1%
Government Presidential system
Currency Chilean peso = 100 centavos
Literacy rate 96%
Life expectancy 79 years

CHINA
East Asia

Official name People's Republic of China
Formation 1949 / 2011
Capital Beijing
Population 1.41 billion / 381 people per sq mile (147 people per sq km)
Total area 3,705,386 sq miles (9,596,960 sq km)
Languages Mandarin*, Wu, Cantonese, Hsiang, Min, Hakka, Kan
Religions Folk religion 22%, Buddhist 18%, Christian 5%, Muslim 2%, Unaffiliated 52%, Hindu, Jewish, & other 1%
Demographics Han Chinese 91%, Ethnic minorities 9%
Government One-party state
Currency Renminbi (or yuan) = 10 jiao = 100 fen
Literacy rate 97%
Life expectancy 78 years

COLOMBIA
South America

Official name Republic of Colombia
Formation 1810 / 1903
Capital Bogotá
Population 49.1 million / 112 people per sq mile (43 people per sq km)
Total area 439,736 sq miles (1,138,910 sq km)
Languages Spanish*, Amerindian languages
Religions Christian 92% (mainly Roman Catholic), Unspecified 7%, Other 1%
Demographics Mixed race and White 88%, Indigenous 4%, Afro-Colombian 7%, Unspecified 1%
Government Presidential system
Currency Colombian peso = 100 centavos
Literacy rate 96%
Life expectancy 73 years

COMOROS
Indian Ocean

Official name Union of the Comoros
Formation 1975 / 1975
Capital Moroni
Population 876,437 / 1016 people per sq mile (392 people per sq km)
Total area 863 sq miles (2235 sq km)
Languages Arabic*, Comoran*, French*
Religions Sunni Muslim 98%, Other 2%
Demographics Comoran 97%, Other 3%
Government Presidential system
Currency Comoros franc = 100 centimes
Literacy rate 59%
Life expectancy 63 years

CONGO
Central Africa

Official name Republic of the Congo
Formation 1960 / 1960
Capital Brazzaville
Population 5.5 million/ 42 people per sq mile (16 people per sq km)
Total area 132,047 sq miles (342,000 sq km)
Languages Kongo, Teke, Lingala, French*
Religions Roman Catholic 34%, Awakening Churches/Christian Revival 22%, Protestant 20%, None 12%, Other 12%
Demographics Kongo 41%, Teke 17%, Mbochi 13%, Mbere/Mbeti/Kele 4%, Punu 4%, Other 15%
Government Presidential system
Currency CFA franc = 100 centimes
Literacy rate 80%
Life expectancy 64 years

CONGO, DEM. REP.
Central Africa

Official name Democratic Republic of the Congo
Formation 1960 / 1960
Capital Kinshasa
Population 108.4 million/ 120 people per sq mile (46 people per sq km)
Total area 905,355 sq miles (2,344,858 sq km)
Languages Kiswahili, Tshiluba, Kikongo, Lingala, French*
Religions Roman Catholic 31%, Protestant 28%, Other Christian 37%, Muslim 1%, Other 3%
Demographics Mongo, Luba, Kongo, & Mangbetu-Azande 45%, Other African ethnic groups 55%
Government Semi-Presidential system
Currency Congolese franc = 100 centimes
Literacy rate 77%
Life expectancy 59 years

COSTA RICA
Central America

Official name Republic of Costa Rica
Formation 1821 / 1941
Capital San José
Population 5.2 million / 264 people per sq mile (102 people per sq km)
Total area 19,730 sq miles (51,100 sq km)
Languages Spanish*, English Creole, Bribri, Cabecar
Religions Roman Catholic 48%, Evangelical and Pentecostal 20%, Jehovah's Witness 1%, Other Protestant 1%, None 27%, Other 3%
Demographics Mixed race and European 96%, Indigenous 3%, Black 1%
Government Presidential system
Currency Costa Rican colón = 100 céntimos
Literacy rate 98%
Life expectancy 77 years

CROATIA
Southeast Europe

Official name Republic of Croatia
Formation 1991 / 1991
Capital Zagreb
Population 4.1 million / 188 people per sq mile (72 people per sq km)
Total area 21,851 sq miles (56,594 sq km)
Languages Croatian*
Religions Roman Catholic 84%, Nonreligious 7%, Orthodox Christian 4%, Muslim 2%, Other 3%
Demographics Croat 92%, Serb 4%, Bosniak 1%, Other 3%
Government Parliamentary system
Currency Kuna = 100 lipa
Literacy rate 99%
Life expectancy 76 years

CUBA
West Indies

Official name Republic of Cuba
Formation 1902 / 1902
Capital Havana
Population 11 million / 257 people per sq mile (99 people per sq km)
Total area 42,803 sq miles (110,860 sq km)
Languages Spanish
Religions Christian 58%, Folk religion 16%, Buddhist 1%, Hindu 1%, Jewish 1%, Muslim 1%, None 22%
Demographics White 65%, Mixed race 25%, Black 10%
Government One-party state
Currency Cuban peso = 100 centavos
Literacy rate 100%
Life expectancy 74 years

CYPRUS
Southeast Europe

Official name Republic of Cyprus
Formation 1960 / 1960
Capital Nicosia
Population 1.2 million / 336 people per sq mile (130 people per sq km)
Total area 3572 sq miles (9251 sq km)
Languages Greek*, Turkish*, English, Romanian, Russian, Bulgarian, Arabic, Filipino
Religions Orthodox Christian 89%, Roman Catholic 3%, Protestant/Anglican 2%, Muslim 2%, Other 4%
Demographics Greek 99%, Other 1%
Government Presidential system
Currency Euro = 100 cents (In TRNC, Turkish lira = 100 kurus)
Literacy rate 99%
Life expectancy 81 years

CZECHIA
Central Europe

Official name Czech Republic
Formation 1993 / 1993
Capital Prague
Population 10.7 million / 351 people per sq mile (136 people per sq km)
Total area 30,450 sq miles (78,867 sq km)
Languages Czech*, Slovak, Hungarian
Religions Roman Catholic 9%, Other Christian 2%, Nonreligious 57%, Unspecified 32%
Demographics Czech 57%, Moravian 3%, Unspecified 32%, Other 8%
Government Parliamentary system
Currency Czech koruna = 100 haleru
Literacy rate 99%
Life expectancy 77 years

DENMARK
Northern Europe

Official name Kingdom of Denmark
Formation 965 / 1944
Capital Copenhagen
Population 5.9 million / 355 people per sq mile (137 people per sq km)
Total area 16,639 sq miles (43,094 sq km)
Languages Danish*, English, Faroese
Religions Evangelical Lutheran 75%, Muslim 5%, Other 20%,
Demographics Danish 86%, Turkish 1%, Other 13%
Government Parliamentary system
Currency Danish krone = 100 øre
Literacy rate 99%
Life expectancy 81 years

DJIBOUTI
East Africa

Official name Republic of Djibouti
Formation 1977 / 1977
Capital Djibouti
Population 957,273 / 107 people per sq mile (41 people per sq km)
Total area 8958 sq miles (23,200 sq km)
Languages Somali, Afar, French*, Arabic*
Religions Muslim (mainly Sunni) 94%, Other 6%
Demographics Somali 60%, Afar 35%, Other 5%
Government Presidential system
Currency Djibouti franc = 100 centimes
Literacy rate 68%
Life expectancy 62 years

DOMINICA
West Indies

Official name Commonwealth of Dominica
Formation 1978 / 1978
Capital Roseau
Population 72,170 / 249 people per mile (96 people per sq km)
Total area 290 sq miles (751 sq km)
Languages French Creole, English*
Religions Roman Catholic 62%, Protestant 30%, Nonreligious 6%, Other 2%
Demographics African 84%, Mixed 9%, Indigenous 4%, Other 3%
Government Parliamentary system
Currency East Caribbean dollar = 100 cents
Literacy rate 94%
Life expectancy 73 years

DOMINICAN REPUBLIC
West Indies

Official name Dominican Republic
Formation 1844 / 1936
Capital Santo Domingo
Population 10.6 million / 564 people per sq mile (218 people per sq km)
Total area 18,792 sq miles (48,670 sq km)
Languages Spanish*, French Creole
Religions Roman Catholic 44%, Evangelical 14%, Protestant 8%, None 29%, Unspecified 2%, Other 3%
Demographics Mixed race 70%, Black 16%, White 13%, Other 1%
Government Presidential system
Currency Dominican Republic peso = 100 centavos
Literacy rate 94%
Life expectancy 73 years

EAST TIMOR (TIMOR-LESTE)
Southeast Asia

Official name Democratic Republic of Timor-Leste
Formation 2002 / 2002
Capital Dili
Population 1.4 million / 244 people per sq mile (94 people per sq km)
Total area 5743 sq miles (14,874 sq km)
Languages Tetum* (Portuguese/Austronesian), Bahasa Indonesia, Portuguese*
Religions Roman Catholic 96%, Protestant/Evangelical 2%, Other 2%
Demographics Papuan groups approx. 85%, Indonesian groups approx. 13%, Chinese 2%
Government Semi-presidential system
Currency US dollar = 100 cents
Literacy rate 58%
Life expectancy 68 years

ECUADOR
South America

Official name Republic of Ecuador
Formation 1822 / 1942
Capital Quito
Population 17.2 million / 157 people per sq mile (61 people per sq km)
Total area 109,484 sq miles (283,561 sq km)
Languages Spanish*, Quechua, Other Amerindian languages
Religions Roman Catholic 69%, Evangelical 16%, Adventist 1%, Agnostic or Atheist 2%, None 10%, Other 2%
Demographics Mixed race 79%, Black 7%, Indigenous 7%, White 6%, Other 1%
Government Presidential system
Currency US dollar = 100 cents
Literacy rate 94%
Life expectancy 74 years

EGYPT
North Africa

Official name Arab Republic of Egypt
Formation 1953 / 2017
Capital Cairo
Population 107.7 million / 279 people per sq mile (108 people per sq km)
Total area 386,662 sq miles (1,001,450 sq km)
Languages Arabic*, French, English, Berber
Religions Muslim (mainly Sunni) 90%, Coptic Christian 9%, Other Christian 1%
Demographics Egyptian 99%, Other 1%
Government Presidential system
Currency Egyptian pound = 100 piastres
Literacy rate 71%
Life expectancy 70 years

EL SALVADOR
Central America

Official name Republic of El Salvador
Formation 1821 / 1998
Capital San Salvador
Population 6.5 million / 800 people per sq mile (309 people per sq km)
Total area 8124 sq miles (21,041 sq km)
Languages Spanish*, Nawat
Religions Roman Catholic 50%, Protestant 36%, Nonreligious 12%, Other 2%
Demographics Mixed race 86%, White 13%, Indigenous and other 1%
Government Presidential system
Currency Salvadorean colón = 100 centavos; US dollar = 100 cents
Literacy rate 89%
Life expectancy 71 years

EQUATORIAL GUINEA
Central Africa

Official name Republic of Equatorial Guinea
Formation 1968 / 1968
Capital Malabo
Population 1.6 million / 148 people per sq mile (57 people per sq km)
Total area 10,830 sq miles (28,051 sq km)
Languages Spanish*, Fang, Bubi, French*
Religions Roman Catholic 88%, Protestant 5%, Muslim 2%, Other 5%
Demographics Fang 86%, Bubi 6%, Other 8%
Government Presidential system
Currency CFA franc = 100 centimes
Literacy rate 95%
Life expectancy 61 years

ERITREA
East Africa

Official name State of Eritrea
Formation 1993 / 2002
Capital Asmera
Population 6.2 million / 137 people per sq mile (53 people per sq km)
Total area 45,406 sq miles (117,600 sq km)
Languages Tigrinya*, English*, Tigre, Afar, Arabic*, Saho, Bilen, Kunama, Nara, Hadareb
Religions Christian 50%, Muslim 48%, Other 2%
Demographics Tigrinya 50%, Tigre 30%, Saho 4%, Afar 4%, Kunama 4%, Bilen 3%, Other 5%
Government Presidential system
Currency Nakfa = 100 cents
Literacy rate 77%
Life expectancy 67 years

ESTONIA
Northeast Europe

Official name Republic of Estonia
Formation 1991 / 1991
Capital Tallinn
Population 1.2 million/ 69 people per sq mile (27 people per sq km)
Total area 17,463 sq miles (45,228 sq km)
Languages Estonian*, Russian, Ukrainian
Religions Nonreligious 45%, Orthodox Christian 25%, Lutheran 20%, Other 10%
Demographics Estonian 70%, Russian 25%, Ukrainian 2%, Belarussian 1%, Other 2%
Government Parliamentary system
Currency Euro = 100 cents
Literacy rate 100%
Life expectancy 77 years

ESWATINI
Southern Africa

Official name Kingdom of Eswatini
Formation 1968 / 1968
Capital Mbabane; Lobamba
Population 1.1 million / 164 people per sq mile (63 people per sq km)
Total area 6704 sq miles (17,364 sq km)
Languages English*, siSwati*, isiZulu, Xitsonga
Religions Christian 90%, Muslim 2%, Other 8%
Demographics Swazi 97%, Other 3%
Government Absolute monarchy
Currency Swazi lilangeni = 100 cents; South African rand = 100 cents
Literacy rate 83%
Life expectancy 57 years

ETHIOPIA
East Africa

Official name Federal Democratic Republic of Ethiopia
Formation 1896 / 2002
Capital Addis Ababa
Population 113 million / 265 people per sq mile (102 people per sq km)
Total area 426,373 sq miles (1,104,300 sq km)
Languages Amharic*, Oromo, Tigrinya, Galla, Sidamo, Somali, English, Arabic
Religions Christian 62%, Muslim 34%, Other 4%
Demographics Oromo 36%, Amhara 24%, Somali 7%, Tigray 6%, Sidama 4%, Guragie 3%, Welaita 2%, Afar 2%, Other 16%
Government Parliamentary system
Currency Birr = 100 santim
Literacy rate 52%
Life expectancy 65 years

FIJI
Australasia & Oceania

Official name Republic of Fiji
Formation 1970 / 1970
Capital Suva
Population 943,737 / 134 people per sq mile (52 people per sq km)
Total area 7056 sq miles (18,274 sq km)
Languages Fijian, English*, Hindi, Urdu, Tamil, Telugu
Religions Methodist 35%, Hindu 28%, Other Christian 21%, Roman Catholic 9%, Muslim 6%, Other and nonreligious 1%
Demographics Melanesian 57%, Indian 38%, Other 5%
Government Parliamentary system
Currency Fiji dollar = 100 cents
Literacy rate 99%
Life expectancy 67 years

FINLAND
Northern Europe

Official name Republic of Finland
Formation 1917 / 1947
Capital Helsinki / Helsingfors
Population 5.6 million / 43 people per sq mile (17 people per sq km)
Total area 130,127 sq miles (338,145 sq km)
Languages Finnish*, Swedish*, Russian
Religions Lutheran 67%, Greek Orthodox 1%, None 30%, Other 2%
Demographics Finnish 93%, Other (including Sámi) 7%
Government Parliamentary system
Currency Euro = 100 cents
Literacy rate 100%
Life expectancy 82 years

FRANCE
Western Europe

Official name French Republic
Formation 987 / 1947
Capital Paris
Population 68.3 million / 321 people per sq mile (124 people per sq km)
Total area 212,935 sq miles (551,500 sq km)
Languages French*, Provençal, German, Breton, Catalan, Basque
Religions Roman Catholic 47%, Muslim 4%, None 33%, Protestant 2%, Unspecified 9%, Other 5%
Demographics French 86%, Black 10%, German (Alsace) 2%, Breton 1%, Other 1%
Government Presidential / Parliamentary system
Currency Euro = 100 cents
Literacy rate 99%
Life expectancy 82 years

GERMANY
Northern Europe

Official name Federal Republic of Germany
Formation 1871 / 1990
Capital Berlin
Population 84.3 million / 612 people per sq mile (236 people per sq km)
Total area 137,847 sq miles (357,022 sq km)
Languages German*, Turkish
Religions Roman Catholic 27%, Protestant 24%, Muslim 3%, None 41%, Other 5%
Demographics German 86%, Turkish 2%, Polish 1%, Syrian 1%, Romanian 1%, Other 9%
Government Parliamentary system
Currency Euro = 100 cents
Literacy rate 99%
Life expectancy 81 years

GUATEMALA
Central America

Official name Republic of Guatemala
Formation 1821 / 1838
Capital Guatemala City
Population 17.7 million / 421 people per sq mile (163 people per sq km)
Total area 42,042 sq miles (108,889 sq km)
Languages Quiché, Mam, Cakchiquel, Kekchí, Spanish*, Maya Languages
Religions Roman Catholic 42%, Evangelical 39%, None 14%, Unspecified 2%, Other 3%
Demographics Mixed race 56%, Maya 42%, Xinca 1%, Other 1%
Government Presidential system
Currency Quetzal = 100 centavos
Literacy rate 81%
Life expectancy 69 years

HAITI
West Indies

Official name Republic of Haiti
Formation 1804 / 1936
Capital Port-au-Prince
Population 11.3 million / 1055 people per sq mile (407 people per sq km)
Total area 10,714 sq miles (27,750 sq km)
Languages French Creole*, French*
Religions Catholic 55%, Protestant 28%, Vodou 2%, None 10%, Other 5%
Demographics Black 95%, Mixed race 5%
Government Semi-presidential system
Currency Gourde = 100 centimes
Literacy rate 62%
Life expectancy 63 years

GABON
Central Africa

Official name Gabonese Republic
Formation 1960 / 1960
Capital Libreville
Population 2.3 million / 22 people per sq mile (9 people per sq km)
Total area 103,347 sq miles (267,667 sq km)
Languages Fang, French*, Punu, Sira, Nzebi, Mpongwe, Bandjabi
Religions Roman Catholic 42%, Protestant 12%, other Christian 27%, Muslim 10%, Animist 1%, None 7%, Other 1%
Demographics Gabonese 80%, Cameroonian 5%, Malian 2%, Beninese 2%, Togolese 2%, Other 9%
Government Presidential system
Currency CFA franc = 100 centimes
Literacy rate 85%
Life expectancy 66 years

GHANA
West Africa

Official name Republic of Ghana
Formation 1957 / 1957
Capital Accra
Population 33.1 million/ 359 people per sq mile (139 people per sq km)
Total area 92,098 sq miles (238,533 sq km)
Languages Twi, Fanti, Ewe, Ga, Adangbe, Gurma, Dagomba (Dagbani), English*, Asante
Religions Christian 71%, Muslim 20%, Traditionalist 3%, None 1%, Other 5%
Demographics Akan 46%, Mole-Dagbani 18%, Ewe 13%, Ga-Dangme 7%, Other 16%
Government Presidential system
Currency Cedi = 100 pesewas
Literacy rate 79%
Life expectancy 64 years

GUINEA
West Africa

Official name Republic of Guinea
Formation 1958 / 1958
Capital Conakry
Population 13.2 million / 139 people per sq mile (54 people per sq km)
Total area 94,926 sq miles (245,857 sq km)
Languages Pulaar, Malinké, Soussou, French*
Religions Muslim 89%, Christian 7%, Nonreligious 2%, Traditional beliefs and other 2%
Demographics Fulani (Peuhl) 33%, Malinke 30%, Susu 21%, Guerze 8%, Kissi 6%, Other 2%
Government Presidential system
Currency Guinea franc = 100 centimes
Literacy rate 40%
Life expectancy 59 years

HONDURAS
Central America

Official name Republic of Honduras
Formation 1821 / 1998
Capital Tegucigalpa
Population 9.4 million / 217 people per sq mile (84 people per sq km)
Total area 43,278 sq miles (112,090 sq km)
Languages Spanish*, Garífuna (Carib), English Creole
Religions Evangelical/Protestant 48%, Roman Catholic 34%, None 17%, Other 1%
Demographics Mixed race 90%, Black 2%, Indigenous 7%, White 1%
Government Presidential system
Currency Lempira = 100 centavos
Literacy rate 89%
Life expectancy 70 years

GAMBIA, THE
West Africa

Official name Republic of The Gambia
Formation 1965 / 1965
Capital Banjul
Population 2.4 million / 550 people per sq mile (212 people per sq km)
Total area 4363 sq miles (11,300 sq km)
Languages Mandinka, Fulani, Wolof, Jola, Soninke, English*
Religions Muslim 96%, Christian 3%, Other 1%
Demographics Mandinka/Jahanka 33%, Fulani/Tukulur/Lorobo 18%, Wolof 13%, Jola/Karoninka 11%, Non-Gambian 10%, Other 15%
Government Presidential system
Currency Dalasi = 100 butut
Literacy rate 51%
Life expectancy 62 years

GREECE
Southeast Europe

Official name Hellenic Republic
Formation 1830 / 1947
Capital Athens
Population 10.5 million / 206 people per sq mile (80 people per sq km)
Total area 50,949 sq miles (131,957 sq km)
Languages Greek*, Turkish, Macedonian, Albanian
Religions Orthodox Christian 90%, Nonreligious 4%, Muslim 2%, Other 4%
Demographics Greek 92%, Albanian 4%, Other 4%
Government Parliamentary system
Currency Euro = 100 cents
Literacy rate 98%
Life expectancy 80 years

GUINEA-BISSAU
West Africa

Official name Republic of Guinea-Bissau
Formation 1974 / 1974
Capital Bissau
Population 2 million/ 143 people per sq mile (55 people per sq km)
Total area 13,948 sq miles (36,125 sq km)
Languages Portuguese Creole, Balante, Fulani, Malinké, Portuguese*
Religions Muslim 46%, Folk religions 31%, Christian 19%, Unaffiliated or Other 4%
Demographics Balante 30%, Fulani 30%, Papel 7%, Mandyako 14%, Mandinka 13%, Other 6%
Government Semi-presidential system
Currency CFA franc = 100 centimes
Literacy rate 60%
Life expectancy 60 years

HUNGARY
Central Europe

Official name Hungary
Formation 1918 / 1947
Capital Budapest
Population 9.7 million / 270 people per sq mile (104 people per sq km)
Total area 35,922 sq miles (93,038 sq km)
Languages Hungarian* (Magyar), English, German, Russian
Religions Roman Catholic 56%, Nonreligious 21%, Presbyterian 13%, Other 10%
Demographics Magyar 92%, Roma 3%, German 2%, Other 3%
Government Parliamentary system
Currency Forint = 100 fillér
Literacy rate 99%
Life expectancy 74 years

GEORGIA
Southwest Asia

Official name Georgia
Formation 1991 / 1991
Capital Tbilisi
Population 4.9 million / 182 people per sq mile (70 people per sq km)
Total area 26,911 sq miles (69,700 sq km)
Languages Georgian*, Russian, Azeri, Armenian, Mingrelian, Ossetian, Abkhazian (* in Abkhazia)
Religions Orthodox Christian 89%, Muslim 9%, Roman Catholic 1%, Other 1%
Demographics Georgian 87%, Azeri 6%, Armenian 4%, Russian 1%, Other 2%
Government Presidential / Parliamentary system
Currency Lari = 100 tetri
Literacy rate 100%
Life expectancy 72 years

GRENADA
West Indies

Official name Grenada
Formation 1974 / 1974
Capital St. George's
Population 113,949 / 857 people per sq mile (331 people per sq km)
Total area 133 sq miles (344 sq km)
Languages English*, English Creole
Religions Protestant 49%, Roman Catholic 36%, Jehovah's Witness 1%, Rastafarian 1%, None 6%, Unspecified 1%, Other 6%
Demographics African 82%, Mixed 14%, East Indian 2%, Unspecified 1%, Other 1%
Government Parliamentary system
Currency East Caribbean dollar = 100 cents
Literacy rate 99%
Life expectancy 75 years

GUYANA
South America

Official name Cooperative Republic of Guyana
Formation 1966 / 1966
Capital Georgetown
Population 789,683 / 10 people per sq mile (4 people per sq km)
Total area 83,000 sq miles (214,969 sq km)
Languages English Creole, Hindi, Tamil, Amerindian languages, English*
Religions Protestant 35%, Hindu 25%, Roman Catholic 7%, Muslim 7%, Other Christian 20%, None 3%, Other 3%
Demographics East Indian 40%, African 29%, Mixed race 20%, Indigenous 11%
Government Parliamentary system
Currency Guyanese dollar = 100 cents
Literacy rate 89%
Life expectancy 66 years

ICELAND
Northwest Europe

Official name Iceland
Formation 1944 / 1944
Capital Reykjavík
Population 357,603 / 10 people per sq mile (4 people per sq km)
Total area 39,769 sq miles (103,000 sq km)
Languages Icelandic*, English, Nordic languages
Religions The Evangelical Lutheran Church of Iceland 63%, Other Christian 8%, Nonreligious 7%, Roman Catholic 4%, Other 18%
Demographics Icelandic 81%, Polish 6%, Danish 1%, Other 12%
Government Parliamentary system
Currency Icelandic króna = 100 aurar
Literacy rate 99%
Life expectancy 83 years

INDIA
South Asia

Official name Republic of India
Formation 1947 / 2015
Capital New Delhi
Population 1.39 billion / 1095 people per sq mile (423 people per sq km)
Total area 1,269,219 sq miles (3,287,263 sq km)
Languages Hindi*, English*, Urdu, Bengali, Marathi, Telugu, Tamil, Bihari, Gujarati
Religions Hindu 80%, Muslim 14%, Christian 2%, Sikh 2%, Unspecified or Other 2%
Demographics Indo-Aryan 72%, Dravidian 25%, Other 3%
Government Parliamentary system
Currency Indian rupee = 100 paise
LLiteracy rate 74%
Life expectancy 67 years

IRELAND
Northwest Europe

Official name Ireland
Formation 1922 / 1922
Capital Dublin
Population 5.2 million / 192 people per sq mile (74 people per sq km)
Total area 27,133 sq miles (70,273 sq km)
Languages English*, Irish*
Religions Roman Catholic 86%, Other Christian 6%, Nonreligious 6%, Muslim 1%, Other 1%
Demographics Irish 86%, Other White 9%, Asian 2%, Black 1%, Other 2%
Government Parliamentary system
Currency Euro = 100 cents
Literacy rate 99%
Life expectancy 82 years

JAMAICA
West Indies

Official name Jamaica
Formation 1962 / 1962
Capital Kingston
Population 2.8 million / 660 people per sq mile (255 people per sq km)
Total area 4244 sq miles (10,991 sq km)
Languages English Creole, English*
Religions Church of God 26%, Nonreligious 22%, Other Christian 21%, Seventh-day Adventist 12%, Pentecostal 11%, Other 8%
Demographics Black 92%, Mixed race 6%, East Indian 1%, Other 1%
Government Parliamentary system
Currency Jamaican dollar = 100 cents
Literacy rate 89%
Life expectancy 71 years

KENYA
East Africa

Official name Republic of Kenya
Formation 1963 / 1963
Capital Nairobi
Population 55.8 million / 249 people per sq mile (96 people per sq km)
Total area 224,081 sq miles (580,367 sq km)
Languages Kiswahili*, English*, Kikuyu, Luo, Kalenjin, Kamba
Religions Christian 86%, Muslim 11%, Other 3%
Demographics Kikuyu 17%, Luhya 14%, Kalenjin 13%, Luo 11%, Kamba 10%, Somali 6%, Kisii 6%, Mijikenda 5%, Meru 4%, Maasai 3%, Turkana 2%, Non-Kenyan 1%, Other 8%
Government Presidential system
Currency Kenya shilling = 100 cents
Literacy rate 82%
Life expectancy 61 years

INDONESIA
Southeast Asia

Official name Republic of Indonesia
Formation 1945 / 1999
Capital Jakarta
Population 276 million / 375 people per sq mile (145 people per sq km)
Total area 735,358 sq miles (1,904,569 sq km)
Languages Javanese, Sundanese, Madurese, Bahasa Indonesia*, Dutch, English
Religions Sunni Muslim 87%, Protestant 7%, Roman Catholic 3%, Hindu 2%, Buddhist 1%
Demographics Javanese 40%, Sundanese 16%, Malay 4%, Batak 4%, Madurese 3%, Betawi 3%, Minangkabau 3%, Buginese 3%, Other 24%
Government Presidential system
Currency Rupiah = 100 sen
Literacy rate 96%
Life expectancy 68 years

ISRAEL
Southwest Asia

Official name State of Israel
Formation 1948 / 1994
Capital Jerusalem (not recognized by all nations)
Population 8.9 million / 1051 people per sq mile (406 people per sq km)
Total area 8470 sq miles (21,937 sq km)
Languages Hebrew*, Arabic*, Yiddish, German, Russian, Polish, Romanian, Persian, English
Religions Jewish 74%, Muslim 18%, Christian 2%, Druze 2%, Other 4%
Demographics Jewish 74%, Arab 21%, Other 5%
Government Parliamentary system
Currency Shekel = 100 agorot
Literacy rate 98%
Life expectancy 83 years

JAPAN
East Asia

Official name Japan
Formation 1890 / 1972
Capital Tokyo
Population 124 million / 850 people per sq mile (328 people per sq km)
Total area 145,914 sq miles (377,915 sq km)
Languages Japanese*, Korean, Chinese
Religions Shintoism 71%, Buddhism 67%, Christianity 2%, Other 6%
Demographics Japanese 98%, Chinese and Korean 1%, Other 1%
Government Parliamentary system
Currency Yen = 100 sen
Literacy rate 99%
Life expectancy 84 years

KIRIBATI
Australasia & Oceania

Official name Republic of Kiribati
Formation 1979 / 2012
Capital Tarawa Atoll
Population 114,189 / 365 people per sq mile (141 people per sq km)
Total area 313 sq miles (811 sq km)
Languages English*, Kiribati
Religions Roman Catholic 59%, Kiribati Uniting Church 21%, Kiribati Protestant Church 8%, Church of Jesus Christ 6%, Other 6%
Demographics I-Kiribati 95%, Mixed race 3%, Other 2%
Government Presidential system
Currency Australian dollar = 100 cents
Literacy rate 99%
Life expectancy 68 years

IRAN
Southwest Asia

Official name Islamic Republic of Iran
Formation 1979 / 1990
Capital Tehran
Population 86.7 million / 136 people per sq mile (53 people per sq km)
Total area 636,372 sq miles (1,648,195 sq km)
Languages Farsi*, Azeri, Luri, Gilaki, Mazanderani, Kurdish, Turkmen, Arabic, Baluchi
Religions Shi'a Muslim 90%, Sunni Muslim 9%, Other 1%
Demographics Persian 51%, Azari 24%, Lur and Bakhtiari 8%, Kurdish 7%, Other 10%
Government Islamic theocracy
Currency Iranian rial = 100 dinars
Literacy rate 86%
Life expectancy 74 years

ITALY
Southern Europe

Official name Italian Republic
Formation 1861 / 1954
Capital Rome
Population 61 million / 524 people per sq mile (202 people per sq km)
Total area 116,348 sq miles (301,340 sq km)
Languages Italian*, German, French, Rhaeto-Romanic, Sardinian, Slovene
Religions Christian 81%, Muslim 5%, Unaffiliated 13%, Other 1%
Demographics Italian 92%, Other European 5%, African 1%, Other 2%
Government Parliamentary system
Currency Euro = 100 cents
Literacy rate 99%
Life expectancy 83 years

JORDAN
Southwest Asia

Official name Hashemite Kingdom of Jordan
Formation 1946 / 1967
Capital Amman
Population 10.9 million / 316 people per sq mile (122 people per sq km)
Total area 34,495 sq miles (89,342 sq km)
Languages Arabic*, English
Religions Muslim 97% (official; predominantly Sunni), Christian 2%, Other 1%
Demographics Jordanian 69%, Syrian 13%, Palestinian 7%, Egyptian 7%, Iraqi 1%, Other 3%
Government Monarchy
Currency Jordanian dinar = 1000 fils
Literacy rate 98%
Life expectancy 74 years

KOSOVO
Southeast Europe

Official name Republic of Kosovo
Formation 2008 / 2008
Capital Pristina
Population 1.9 million / 452 people per sq mile (175 people per sq km)
Total area 4203 sq miles (10,887 sq km)
Languages Albanian*, Serbian*, Bosniak, Gorani, Roma, Turkish
Religions Muslim 96%, Roman Catholic 2%, Orthodox 1%, Other 1%
Demographics Albanians 93%, Bosniaks 1%, Serbs 1%, Turk 1%, Other 4%
Government Parliamentary system
Currency Euro = 100 cents
Literacy rate 92%
Life expectancy 80 years

IRAQ
Southwest Asia

Official name Republic of Iraq
Formation 1932 / 1991
Capital Baghdad
Population 40.4 million / 239 people per sq mile (92 people per sq km)
Total area 169,235 sq miles (438,317 sq km)
Languages Arabic*, Kurdish*, Turkic languages, Armenian, Assyrian
Religions Muslim 97%, Christian 1%, Other 2%
Demographics Arab 80%, Kurdish 15%, Turkmen 3%, Other 2%
Government Parliamentary system
Currency New Iraqi dinar = 1000 fils
Literacy rate 86%
Life expectancy 70 years

IVORY COAST (CÔTE D'IVOIRE)
West Africa

Official name Republic of Côte d'Ivoire
Formation 1960 / 1960
Capital Yamoussoukro
Population 28.7 million / 231 people per sq mile (89 people per sq km)
Total area 124,504 sq miles (322,463 sq km)
Languages Akan, French*, Krou, Voltaïque, Dioula
Religions Muslim 43%, Nonreligious or traditional beliefs 23%, Roman Catholic 17%, Evangelical 12%, Other 5%
Demographics Akan 29%, Voltaique or Gur 16%, Northern Mande 15%, Kru 8%, Southern Mande 7%, Non-Ivoirian 24%, Other 1%
Government Presidential system
Currency CFA franc = 100 centimes
Literacy rate 90%
Life expectancy 59 years

KAZAKHSTAN
Central Asia

Official name Republic of Kazakhstan
Formation 1991 / 1991
Capital Astana.
Population 19.3 million / 18 people per sq mile (7 people per sq km)
Total area 1,052,089 sq miles (2,724,900 sq km)
Languages Kazakh*, Russian*, Ukrainian, German, Uzbek, Tatar, Uighur
Religions Muslim (mainly Sunni) 71%, Christian (mainly Orthodox) 26%, Nonreligious
Demographics Kazakh 68%, Russian 19%, Uzbek 3%, Ukrainian 2%, Uighur 2%, Other 6%
Government Presidential system
Currency Tenge = 100 tiyn
Literacy rate 100%
Life expectancy 70 years

KUWAIT
Southwest Asia

Official name State of Kuwait
Formation 1961 / 1969
Capital Kuwait City
Population 4.1 million / 596 people per sq mile (230 people per sq km)
Total area 6880 sq miles (17,818 sq km)
Languages Arabic*, English
Religions Muslim (official) 75%, Christian 18%, Unspecified or Other 7%
Demographics Kuwaiti 30%, Other Arab 28%, Asian 40%, African 1%, Other 1%
Government Monarchy
Currency Kuwaiti dinar = 1000 fils
Literacy rate 97%
Life expectancy 79 years

KYRGYZSTAN
Central Asia

Official name Kyrgyz Republic
Formation 1991 / 1991
Capital Bishkek
Population 6 million / 78 people per sq mile (30 people per sq km)
Total area 77,202 sq miles (199,951 sq km)
Languages Kyrgyz*, Russian*, Uzbek, Tatar, Ukrainian
Religions Muslim 90% (majority Sunni), Christian 7%, Other 3%
Demographics Kyrgyz 74%, Uzbek 15%, Russian 5%, Dungan 1%, Other 5%
Government Presidential / Parliamentary system
Currency Som = 100 tyiyn
Literacy rate 100%
Life expectancy 72 years

LAOS
Southeast Asia

Official name Lao People's Democratic Republic
Formation 1953 / 1953
Capital Viangchan (Vientiane)
Population 7.7 million / 84 people per sq mile (28 people per sq km)
Total area 91,429 sq miles (236,800 sq km)
Languages Lao*, Mon-Khmer, Yao, Vietnamese, Chinese, French, English
Religions Buddhist 65%, Christian 2%, None 31%, Other 2%
Demographics Lao 53%, Khmou 11%, Hmong 9%, Phouthay 3%, Tai 3%, Makong 3%, Katong 2%, Lue 2%, Akha 2%, Other 12%
Government One-party state
Currency Kip = 100 att
Literacy rate 85%
Life expectancy 68 years

LATVIA
Northeast Europe

Official name Republic of Latvia
Formation 1991 / 1991
Capital Riga
Population 1.8 million/ 72 people per sq mile (28 people per sq km)
Total area 24,938 sq miles (64,589 sq km)
Languages Latvian*, Russian
Religions Lutheran 36%, Roman Catholic 19%, Orthodox 19%, Other Christian 2%, Unspecified 24%
Demographics Latvian 63%, Russian 25%, Belarusian 3%, Ukrainian 2%, Polish 2%, Unspecified 2%, Other 3%
Government Parliamentary system
Currency Euro = 100 cents
Literacy rate 100%
Life expectancy 73 years

LEBANON
Southwest Asia

Official name Lebanese Republic
Formation 1943 / 1941
Capital Beirut
Population 5.2 million/ 1728 people per sq mile (662 people per sq km)
Total area 4015 sq miles (10,400 sq km)
Languages Arabic*, French, Armenian, Assyrian, English
Religions Muslim 64%, Christian 31%, Druze 4%, Other 1%
Demographics Arab 95%, Armenian 4%, Other 1%
Government Parliamentary system
Currency Lebanese pound = 100 piastres
Literacy rate 95%
Life expectancy 75 years

LESOTHO
Southern Africa

Official name Kingdom of Lesotho
Formation 1966 / 1966
Capital Maseru
Population 2 million / 188 people per sq mile (72 people per sq km)
Total area 11,720 sq miles (30,355 sq km)
Languages English*, Sesotho*, isiZulu, Xhosa
Religions Protestant 48%, Roman Catholic 40%, Other Christian 9%, Non-Christian 1%, None 2%
Demographics Sotho 99%, European and Asian 1%
Government Parliamentary system
Currency Loti = 100 lisente; South African rand = 100 cents
Literacy rate 79%
Life expectancy 53 years

LIBERIA
West Africa

Official name Republic of Liberia
Formation 1847 / 1911
Capital Monrovia
Population 5.3 million / 123 people per sq mile (48 people per sq km)
Total area 43,000 sq miles (111,369 sq km)
Languages Kpelle, Vai, Bassa, Kru, Grebo, Kissi, Gola, Loma, English*
Religions Christian 86%, Muslim 12%, Nonreligious 1%, Traditional beliefs and other 1%
Demographics Indigenous tribes (12 groups) 40%, Kpellé 20%, Bassa 14%, Grebo 10%, Gio 8%, Krou 6%, Other 2%
Government Presidential system
Currency Liberian dollar = 100 cents
Literacy rate 48%
Life expectancy 61 years

LIBYA
North Africa

Official name State of Libya
Formation 1951 / 1951
Capital Tripoli
Population 7.1 million / 11 people per sq mile (4 people per sq km)
Total area 679,362 sq miles (1,759,540 sq km)
Languages Arabic*, Tuareg, English, Italian
Religions Muslim 96%, Christian 2%, Other 2%
Demographics Arab and Berber 97%, Other 3%
Government Transitional regime
Currency Libyan dinar = 1000 dirhams
Literacy rate 91%
Life expectancy 72 years

LIECHTENSTEIN
Central Europe

Official name Principality of Liechtenstein
Formation 1719 / 1719
Capital Vaduz
Population 38,250 / 617 people per sq mile (239 people per sq km)
Total area 62 sq miles (160 sq km)
Languages German*, Alemannish dialect, Italian, Turkish, Portuguese
Religions Roman Catholic 78%, Protestant 9%, Muslim 6%, Nonreligious 5%, Orthodox Christian 1%, Other 1%
Demographics Liechtensteiner 65%, German 5% Austrian 6%, Swiss 10%, Italian 3%, Other 11%
Government Monarchy
Currency Swiss franc = 100 rappen/centimes
Literacy rate 100%
Life expectancy 84 years

LITHUANIA
Northeast Europe

Official name Republic of Lithuania
Formation 1991 / 2003
Capital Vilnius
Population 2.6 million / 103 people per sq mile (40 people per sq km)
Total area 25,212 sq miles (65,300 sq km)
Languages Lithuanian*, Russian, Polish
Religions Roman Catholic 74%, Russian Orthodox 4%, None 6%, Unspecified 14%, Other 2%
Demographics Lithuanian 85%, Polish 6%, Russian 5%, Belarusian 1%, Other 3%
Government Semi-presidential system
Currency Euro = 100 cents
Literacy rate 100%
Life expectancy 74 years

LUXEMBOURG
Northwest Europe

Official name Grand Duchy of Luxembourg
Formation 1839 / 1867
Capital Luxembourg
Population 650,364 / 652 people per sq mile (251 people per sq km)
Total area 998 sq miles (2586 sq km)
Languages Luxembourgish*, Portuguese, Italian, German*, French*, English
Religions Christian 71%, Muslim 2%, Unaffiliated 27%
Demographics Luxembourger 53%, Portuguese 15%, French 8%, Italian 4%, Belgian 3%, German 2%, Spanish 1%, Romania 1%, Other 13%
Government Monarchy
Currency Euro = 100 cents
Literacy rate 100%
Life expectancy 83 years

MADAGASCAR
Indian Ocean

Official name Republic of Madagascar
Formation 1960 / 1960
Capital Antananarivo
Population 28 million / 124 people per sq mile (48 people per sq km)
Total area 226,658 sq miles (587,042 sq km)
Languages Malagasy*, French*, English*
Religions Traditional beliefs 52%, Christian 41%, Muslim 7%
Demographics Malay 46%, Merina 26%, Betsimisaraka 15%, Betsileo 12%, Other 1%
Government Semi-presidential system
Currency Ariary = 5 iraimbilanja
Literacy rate 77%
Life expectancy 64 years

MALAWI
Southern Africa

Official name Republic of Malawi
Formation 1964 / 1964
Capital Lilongwe
Population 21 million / 459 people per sq mile (177 people per sq km)
Total area 45,745 sq miles (118,484 sq km)
Languages Chewa, Lomwe, Yao, Ngoni, English*
Religions Christian (mainly Protestant) 50%, Other Christian 27%, Muslim 14%, Traditionalist 1%, none 2%, Other 6%
Demographics Chewa 34%, Lomwe 19%, Yao 13%, Ngoni 10%, Tumbuka 9%, Sena 4%, Mang'anja 3%, Tonga 2%, Nyanja 2%, Nkhonde 1%, Other 3%
Government Presidential system
Currency Malawi kwacha = 100 tambala
Literacy rate 62%
Life expectancy 63 years

MALAYSIA
Southeast Asia

Official name Malaysia
Formation 1957 / 1965
Capital Kuala Lumpur
Population 33.1 million / 259 people per sq mile (100 people per sq km)
Total area 127,723 sq miles (330,803 sq km)
Languages Bahasa Malaysia*, Malay, Chinese, Tamil, English
Religions Muslim (mainly Sunni) 61%, Buddhist 20%, Christian 9%, Hindu 6%, Other 3%
Demographics Malay 50%, Chinese 22%, Indigenous tribes 11%, Indian 7%, Other 9%
Government Parliamentary system
Currency Ringgit = 100 sen
Literacy rate 95%
Life expectancy 75 years

MALDIVES
Indian Ocean

Official name Republic of Maldives
Formation 1965 / 1965
Capital Maale (Male')
Population 390,164 / 4667 people per sq mile (1812 people per sq km)
Total area 120 sq miles (300 sq km)
Languages Dhivehi* (Maldivian), Sinhala, Tamil, Arabic, English
Religions Sunni Muslim 94%, Hindu 3%, Christian 2%, Buddhist 1%
Demographics Arab-Sinhalese–Malay 100%
Government Presidential system
Currency Rufiyaa = 100 laari
Literacy rate 98%
Life expectancy 80 years

MALI
West Africa

Official name Republic of Mali
Formation 1960 / 1986
Capital Bamako
Population 20.7 million / 43 people per sq mile (17 people per sq km)
Total area 478,840 sq miles (1,240,192 sq km)
Languages Bambara, Fulani, Senufo, Soninke, French*
Religions Muslim (mainly Sunni) 93%, Traditional beliefs 6%, Christian 4%
Demographics Bambara 33%, Fulani 13%, Sarakole/Soninke/Marka 10%, Senufo/Manianka 10%, Malinke 9%, Dogon 9%, Sonrai 6%, Other 10%
Government Presidential system
Currency CFA franc = 100 centimes
Literacy rate 36%
Life expectancy 59 years

MALTA
Southern Europe

Official name Republic of Malta
Formation 1964 / 1964
Capital Valletta
Population 464,186/ 4252 people per sq mile (1615 people per sq km)
Total area 122 sq miles (316 sq km)
Languages Maltese*, English*
Religions Roman Catholic 98%, Nonreligious and other 2%
Demographics Maltese 96%, Other 4%
Government Parliamentary system
Currency Euro = 100 cents
Literacy rate 95%
Life expectancy 83 years

MARSHALL ISLANDS
Australasia & Oceania

Official name Republic of the
Marshall Islands
Formation 1986 / 1986
Capital Majuro Atoll
Population 59,620 / 852 people per sq
mile (329 people per sq km)
Total area 70 sq miles (181 sq km)
Languages Marshallese*, English*,
Japanese, German
Religions Protestant 81%, Roman
Catholic 8%, Other 11%
Demographics Micronesian 90%,
Other 10%
Government Presidential system
Currency US dollar = 100 cents
Literacy rate 98%
Life expectancy 65 years

MAURITANIA
West Africa

Official name Islamic Republic of
Mauritania
Formation 1960 / 1960
Capital Nouakchott
Population 4.1 million / 13 people per
sq mile (5 people per sq km)
Total area 397,953 sq miles
(1,030,700 sq km)
Languages Arabic*, Hassaniyah Arabic,
Wolof, Pular, Soninke, French
Religions Sunni Muslim 100%
Demographics Maure 81%, Wolof 7%,
Tukolor 5%, Soninka 3%, Other 4%
Government Presidential system
Currency Ouguiya = 5 khoums
Literacy rate 52%
Life expectancy 64 years

MAURITIUS
Indian Ocean

Official name Republic of Mauritius
Formation 1968 / 1968
Capital Port Louis
Population 1.3 million / 1616 people per
sq mile (624 people per sq km)
Total area 787 sq miles (2040 sq km)
Languages French Creole, Hindi, Urdu,
Tamil, Chinese, English*, French
Religions Hindu 48%, Roman Catholic
26%, Muslim 17%, Other Christian 7%,
Other 2%
Demographics Indo-Mauritian 68%,
Creole 27%, Sino-Mauritian 3%,
Franco-Mauritian 2%
Government Parliamentary system
Currency Mauritian rupee = 100 cents
Literacy rate 91%
Life expectancy 74 years

MEXICO
North America

Official name United Mexican States
Formation 1821 / 1848
Capital Mexico City
Population 130 million / 171 people per
sq mile (66 people per sq km)
Total area 758,449 sq miles
(1,964,375 sq km)
Languages Spanish*, Nahuatl, Mayan,
Zapotec, Mixtec, Otomi, Totonac,
Tzotzil, Tzeltal
Religions Roman Catholic 78%,
Protestant 11%, Nonreligious 11%,
Other 1%
Demographics Mixed race 62%,
Indigenous 21%, Other 10%
Government Presidential system
Currency Mexican peso = 100 centavos
Literacy rate 95%
Life expectancy 70 years

MICRONESIA, FED. STATES OF
Australasia & Oceania

Official name Federated States of
Micronesia
Formation 1986 / 1986
Capital Palikir (Pohnpei Island)
Population 101,009 / 373 people per sq
mile (144 people per sq km)
Total area 271 sq miles (702 sq km)
Languages English, Chuukese, Kosraean,
Pohnpeian, Yapese, Ulithian, Woleaian,
Nukuoro, Kapingamarangi
Religions Roman Catholic 55%,
Protestant 41%, Other 4%
Demographics Chuukese 49%,
Pohnpeian 30%, Kosraean 6%,
Yapese 5%, Asian 2%, Other 14%
Government Nonparty system
Currency US dollar = 100 cents
Literacy rate 72%
Life expectancy 71 years

MOLDOVA
Southeast Europe

Official name Republic of Moldova
Formation 1991 / 1991
Capital Chisinau
Population 3.2 million / 245 people per
sq mile (95 people per sq km)
Total area 13,069 sq miles (33,851 sq km)
Languages Moldovan*, Ukrainian,
Russian
Religions Orthodox Christian 90%,
Nonreligious 2%, Other 8%
Demographics Moldovan 75%,
Ukrainian 7%, Russian 4%, Gagauz 5%,
Romanian 7%, Bulgarian 2%, Other 1%
Government Parliamentary system
Currency Moldovan leu = 100 bani
Literacy rate 99%
Life expectancy 69 years

MONACO
Southern Europe

Official name Principality of Monaco
Formation 1419 / 1861
Capital Monaco
Population 39,520 / 50,145 people per
sq mile (19,497 people per sq km)
Total area 0.77 sq miles (2 sq km)
Languages French*, Italian,
Monégasque, English
Religions Roman Catholic 89%,
Protestant 6%, Other 5%
Demographics French 20%, Italian 15%,
Monégasque 32%, British 5%, Belgian
2%, Swiss 2%, German 2%, Russian 2%,
American 1%, Dutch 1%, Other 18%
Government Monarchical /
parliamentary system
Currency Euro = 100 cents
Literacy rate 99%
Life expectancy 90 years

MONGOLIA
East Asia

Official name Mongolia
Formation 1921 / 1924
Capital Ulaanbaatar
Population 3.3 million / 5 people per
sq mile (2 people per sq km)
Total area 603,908 sq miles
(1,564,116 sq km)
Languages Khalkha Mongolian*, Kazakh,
Chinese, Russian
Religions Tibetan Buddhist 52%,
Nonreligious 41%, Muslim 3%, Shamanist
2%, Christian 1%, Other 1%
Demographics Khalkh 84%, Kazakh 4%,
Dorvod 3%, Bayad 2%, Other 4%
Government Presidential /
Parliamentary system
Currency Tugrik (tögrög) = 100 möngö
Literacy rate 99%
Life expectancy 71 years

MONTENEGRO
Southeast Europe

Official name Montenegro
Formation 2006 / 2006
Capital Podgorica
Population 604,966 / 113 people per sq
mile (44 people per sq km)
Total area 5332 sq miles (13,812 sq km)
Languages Montenegrin*, Serbian,
Albanian, Bosnian, Croatian
Religions Orthodox Christian 74%,
Muslim 20%, Roman Catholic 4%,
Nonreligious 1%, Other 1%
Demographics Montenegrin 45%,
Serb 29%, Bosniak 9%, Albanian 5%,
Other 12%
Government Parliamentary system
Currency Euro = 100 cents
Literacy rate 99%
Life expectancy 74 years

MOROCCO
North Africa

Official name Kingdom of Morocco
Formation 1956 / 1969
Capital Rabat
Population 36.7 million / 133 people per
sq mile (51 people per sq km)
Total area 276,661 sq miles
(716,550 sq km)
Languages Arabic*, Tamazight*
(Amazigh), French, Spanish
Religions Muslim (mainly Sunni) 99%,
Other 1%
Demographics Arab 70%, Amazigh 29%,
European 1%
Government Monarchical /
parliamentary system
Currency Moroccan dirham =
100 centimes
Literacy rate 74%
Life expectancy 74 years

MOZAMBIQUE
Southern Africa

Official name Republic of Mozambique
Formation 1975 / 1975
Capital Maputo
Population 31.6 million / 102 people per
sq mile (40 people per sq km)
Total area 308,642 sq miles (799,380 sq km)
Languages Makua, Xitsonga, Sena,
Nyanja, Chuwabo, Ndau, Tswa, Lomwe,
Portuguese*
Religions Roman Catholic 27%, Muslim
19%, Christian 16%, Evangelical/
Pentecostal 15%, None 14%, Other 9%
Demographics Makua Lomwe 47%,
Tsonga 23%, Malawi 12%, Shona 11%,
Yao 4%, Other 3%
Government Presidential system
Currency New metical = 100 centavos
Literacy rate 61%
Life expectancy 59 years

MYANMAR (BURMA)
Southeast Asia

Official name Republic of the Union
of Myanmar
Formation 1948 / 1948
Capital Nay Pyi Taw
Population 57.5 million / 220 people per
sq mile (85 people per sq km)
Total area 261,228 sq miles (676,578 sq km)
Languages Burmese* (Myanmar), Shan,
Karen, Rakhine, Chin, Yangbye,
Kachin, Mon
Religions Buddhist 88%, Christian 6%,
Muslim 4%, Animist 1%, Other 1%
Demographics Burman (Bamah) 68%,
Chinese 3%, Shan 9%, Karen 7%, Rakhine
4%, Indian 2%, Mon 2%, Other 5%
Government Parliamentary system
Currency Kyat = 100 pyas
Literacy rate 89%
Life expectancy 66 years

NAMIBIA
Southern Africa

Official name Republic of Namibia
Formation 1990 / 1994
Capital Windhoek
Population 2.7 million / 8 people per sq
mile (3 people per sq km)
Total area 318,260 sq miles
(824,292 sq km)
Languages Ovambo, Kavango, English*,
Bergdama, German, Afrikaans
Religions Christian 98%, Traditional
beliefs 2%
Demographics Ovambo 50%, Other
tribes 22%, Kavango 9%, Herero 7%,
Damara 7%, Other 5%
Government Presidential system
Currency Namibian dollar = 100 cents;
South African rand = 100 cents
Literacy rate 92%
Life expectancy 59 years

NAURU
Australasia & Oceania

Official name Republic of Nauru
Formation 1968 / 1968
Capital None (Yaren *de facto* capital)
Population 10,870 / 1359 people per sq
mile (518 people per sq km)
Total area 8 sq miles (21 sq km)
Languages Nauruan*, Kiribati, Chinese,
Tuvaluan, English
Religions Nauruan Congregational
Church 60%, Roman Catholic 35%,
Other 5%
Demographics Nauruan 93%,
Chinese 5%, Other Pacific islanders 1%,
European 1%
Government Parliamentary system
Currency Australian dollar = 100 cents
Literacy rate 95%
Life expectancy 64 years

NEPAL
South Asia

Official name Nepal
Formation 1768 / 1768
Capital Kathmandu
Population 30 million / 528 people per
sq mile (204 people per sq km)
Total area 56,826 sq miles
(147,181 sq km)
Languages Nepali*, Maithili, Bhojpuri
Religions Hindu 81%, Buddhist 9%,
Muslim 4%, Other 5%
Demographics Chhetri 17%, Hill
Brahman 12%, Magar 7%, Tharu 7%,
Tamang 6%, Newar 5%, Kami 5%,
Muslim 4%, Yadav 4%, Other 33%
Government Parliamentary system
Currency Nepalese rupee = 100 paisa
Literacy rate 68%
Life expectancy 68 years

NETHERLANDS
Northwest Europe

Official name Kingdom of the
Netherlands
Formation 1648 / 1839
Capital Amsterdam; The Hague
(administrative)
Population 17.4 million / 1085 people
per sq mile (419 people per sq km)
Total area 16,039 sq miles (41,543 sq km)
Languages Dutch*, Frisian
Religions Roman Catholic 20%,
Protestant 15%, Muslim 5%, None 54%,
Other 6%
Demographics Dutch 82%, Surinamese
2%, Turkish 2%, Moroccan 2%,
Other 12%
Government Parliamentary system
Currency Euro = 100 cents
Literacy rate 99%
Life expectancy 81 years

NEW ZEALAND
Australasia & Oceania

Official name New Zealand
Formation 1907 / 1947
Capital Wellington
Population 5 million / 48 people per sq mile (19 people per sq km)
Total area 103,798 sq miles (268,838 sq km)
Languages English*, Maori*
Religions Nonreligious 48%, Christian 15%, Anglican 7%, Roman Catholic 10%, Presbyterian 5%, Other 8%
Demographics European 64%, Maori 17%, Chinese 5%, Samoan 4%, Other 14%
Government Parliamentary system
Currency New Zealand dollar = 100 cents
Literacy rate 99%
Life expectancy 82 years

NICARAGUA
Central America

Official name Republic of Nicaragua
Formation 1821 / 1838
Capital Managua
Population 6.3 million / 125 people per sq mile (48 people per sq km)
Total area 50,336 sq miles (130,370 sq km)
Languages Spanish*, English Creole, Miskito
Religions Roman Catholic 50%, Protestant 33%, Nonreligious 1%, Other 16%
Demographics Mixed race 69%, White 17%, Black 9%, Indigenous 5%
Government Presidential system
Currency Córdoba oro = 100 centavos
Literacy rate 83%
Life expectancy 74 years

NIGER
West Africa

Official name Republic of Niger
Formation 1960 / 2016
Capital Niamey
Population 24.4 million / 50 people per sq mile (19 people per sq km)
Total area 489,188 sq miles (1,267,000 sq km)
Languages Hausa, Djerma, Fulani, Tuareg, Teda, French*
Religions Muslim 99%, Other 1%
Demographics Hausa 53%, Djerma and Songhai 21%, Tuareg 11%, Peul 7%, Kanuri 6%, Other 1%
Government Semi-presidential system
Currency CFA franc = 100 centimes
Literacy rate 35%
Life expectancy 62 years

NIGERIA
West Africa

Official name Federal Republic of Nigeria
Formation 1960 / 2006
Capital Abuja
Population 211 million / 592 people per sq mile (228 people per sq km)
Total area 356,667 sq miles (923,768 sq km)
Languages Hausa, English*, Yoruba, Ibo
Religions Muslim 54%, Christian 45%, Traditional beliefs 1%
Demographics Hausa 30%, Yoruba 16%, Ibo 15%, Fulani 6%, Tiv 2%, Kanuri/Beriberi 2%, Ibibio 2%, Ijaw/Izon 2%, Other 25%
Government Presidential system
Currency Naira = 100 kobo
Literacy rate 62%
Life expectancy 53 years

NORTH KOREA
East Asia

Official name Democratic People's Republic of Korea
Formation 1945 / 1953
Capital Pyongyang
Population 25.9 million / 557 people per sq mile (215 people per sq km)
Total area 46,539 sq miles (120,538 sq km)
Languages Korean*
Religions Atheist 100%
Demographics Korean 100%
Government One-party state
Currency North Korean won = 100 chon
Literacy rate 100%
Life expectancy 73 years

NORTH MACEDONIA
Southeast Europe

Official name Republic of North Macedonia
Formation 1991 / 1991
Capital Skopje
Population 2.1 million / 212 people per sq mile (82 people per sq km)
Total area 9928 sq miles (25,713 sq km)
Languages Macedonian*, Albanian*, Turkish, Romani, Serbian
Religions Orthodox Christian 65%, Muslim 33%, Other 2%
Demographics Macedonian 64%, Albanian 25%, Turkish 4%, Roma 3%, Serb 2%, Other 2%
Government Parliamentary system
Currency Macedonian denar = 100 deni
Literacy rate 98%
Life expectancy 75%

NORWAY
Northern Europe

Official name Kingdom of Norway
Formation 1905 / 1905
Capital Oslo
Population 5.5 million / 44 people per sq mile (17 people per sq km)
Total area 125,020 sq miles (323,802 sq km)
Languages Norwegian* (Bokmål "book language" and Nynorsk "new Norsk"), Sámi
Religions Evangelical Lutheran 68%, Roman Catholic 3%, Other Christian 4%, Muslim 3%, Unspecified 20%, Other 2%
Demographics Norwegian 81%, Other European 19%
Government Parliamentary system
Currency Norwegian krone = 100 øre
Literacy rate 99%
Life expectancy 83 years

OMAN
Southwest Asia

Official name Sultanate of Oman
Formation 1650 / 1955
Capital Muscat
Population 3.7 million / 31 people per sq mile (12 people per sq km)
Total area 119,498 sq miles (309,500 sq km)
Languages Arabic*, Baluchi, Farsi, Hindi, Punjabi
Religions Ibadi Muslim 86%, Other 14%
Demographics Arab 88%, Baluchi 4%, Indian and Pakistani 3%, Persian 3%, African 2%
Government Monarchy
Currency Omani rial = 1000 baisa
Literacy rate 96%
Life expectancy 73 years

PAKISTAN
South Asia

Official name Islamic Republic of Pakistan
Formation 1947 / 1971
Capital Islamabad
Population 225 million / 732 people per sq mile (283 people per sq km)
Total area 307,373 sq miles (796,095 sq km)
Languages Punjabi, Sindhi, Pashto, Urdu*, Baluchi, Brahui
Religions Sunni Muslim 77%, Shi'a Muslim 20%, Hindu 2%, Christian 1%
Demographics Punjabi 45%, Pathan (Pashtun) 15%, Sindhi 14%, Mohajir 8%, Baluchi 4%, Other 18%
Government Parliamentary system
Currency Pakistani rupee = 100 paisa
Literacy rate 58%
Life expectancy 66 years

PALAU
Australasia & Oceania

Official name Republic of Palau
Formation 1994 / 1994
Capital Ngerulmud
Population 18,170 / 103 people per sq mile (40 people per sq km)
Total area 177 sq miles (459 sq km)
Languages Palauan*, English*, Japanese, Angaur, Tobi, Sonsorolese
Religions Roman Catholic 45%, Protestant 35%, Modekngei 6%, Muslim 3%, Other 11%
Demographics Palauan 73%, Filipino 16%, Other Asian 7%, Other Micronesian 3%, Other 2%
Government Nonparty system
Currency US dollar = 100 cents
Literacy rate 97%
Life expectancy 69 years

PANAMA
Central America

Official name Republic of Panama
Formation 1903 / 1941
Capital Panama City
Population 4.3 million / 148 people per sq mile (57 people per sq km)
Total area 29,119 sq miles (75,420 sq km)
Languages English Creole, Spanish*, Amerindian languages, Chibchan languages
Religions Roman Catholic 49%, Protestant 30%, Nonreligious 12%, Other 9%
Demographics Mixed race 64%, Black 9%, White 7%, Indigenous 12%, Other 1%
Government Presidential system
Currency Balboa = 100 centésimos; US dollar
Literacy rate 96%
Life expectancy 76 years

PAPUA NEW GUINEA
Australasia & Oceania

Official name Independent State of Papua New Guinea
Formation 1975 / 1975
Capital Port Moresby
Population 9.5 million / 53 people per sq mile (21 people per sq km)
Total area 178,703 sq miles (462,840 sq km)
Languages Pidgin English, Papuan, English*, Motu, 800 (est.) native languages
Religions Protestant 64%, Roman Catholic 26%, Other 9%
Demographics Melanesian and mixed race 100%
Government Parliamentary system
Currency Kina = 100 toea
Literacy rate 64%
Life expectancy 65 years

PARAGUAY
South America

Official name Republic of Paraguay
Formation 1811 / 1938
Capital Asunción
Population 7.3 million / 46 people per sq mile (18 people per sq km)
Total area 157,047 sq miles (406,752 sq km)
Languages Guaraní*, Spanish*, German
Religions Roman Catholic 90%, Protestant 6%, Other Christian 1%, Other 3%
Demographics Mixed race 95%, Other 5%
Government Presidential system
Currency Guaraní = 100 céntimos
Literacy rate 95%
Life expectancy 70 years

PERU
South America

Official name Republic of Peru
Formation 1821 / 1941
Capital Lima
Population 32.2 million / 65 people per sq mile (25 people per sq km)
Total area 496,224 sq miles (1,285,216 sq km)
Languages Spanish*, Quechua*, Aymara
Religions Roman Catholic 60%, Protestant 15%, Nonreligious 4%, Other 21%
Demographics Indigenous 26%, Mixed race 60%, White 6%, Black 4%, Other 4%
Government Presidential system
Currency New sol = 100 céntimos
Literacy rate 95%
Life expectancy 72 years

PHILIPPINES
Southeast Asia

Official name Republic of the Philippines
Formation 1946 / 1946
Capital Manila
Population 114 million / 984 people per sq mile (380 people per sq km)
Total area 115,830 sq miles (300,000 sq km)
Languages Filipino*, English*, Tagalog, Cebuano, Ilocano, Hiligaynon, many other local languages
Religions Roman Catholic 80%, Other Christian 6%, Muslim 6%, Other 8%
Demographics Tagalog 24%, Cebuano 10%, Ilocano 9%, Hiligaynon 8%, Bisaya 11%, Bikol 7%, Waray 4%, Other 27%
Government Presidential system
Currency Philippine peso = 100 centavos
Literacy rate 96%
Life expectancy 69 years

POLAND
Northern Europe

Official name Republic of Poland
Formation 1918 / 1945
Capital Warsaw
Population 38 million / 315 people per sq mile (122 people per sq km)
Total area 120,728 sq miles (312,685 sq km)
Languages Polish*, Silesian
Religions Roman Catholic 85%, Orthodox Christian 1%, Protestant 1%, Other 13%
Demographics Polish 97%, Silesian 1%, Other 2%
Government Parliamentary system
Currency Zloty = 100 groszy
Literacy rate 100%
Life expectancy 76 years

PORTUGAL
Southwest Europe

Official name Portuguese Republic
Formation 1143 / 1640
Capital Lisbon
Population 10.2 million / 287 people per sq mile (111 people per sq km)
Total area 35,556 sq miles (92,090 sq km)
Languages Portuguese*, Mirandese
Religions Roman Catholic 81%, Nonreligious 7%, Other Christian 3%, Unspecified 8%, Other 1%
Demographics Portuguese 95%, Other 5%
Government Parliamentary system
Currency Euro = 100 cents
Literacy rate 96%
Life expectancy 81 years

QATAR
Southwest Asia

Official name State of Qatar
Formation 1971 / 2001
Capital Doha
Population 2.5 million / 559 people per sq mile (216 people per sq km)
Total area 4473 sq miles (11,586 sq km)
Languages Arabic*
Religions Muslim (mainly Sunni) 65%, Christian 14%, Other 14%
Demographics Qatari 12%, Other Arab 20%, Indian 20%, Nepalese 13%, Filipino 10%, Pakistani 7%, Other 10%
Government Monarchy
Currency Qatar riyal = 100 dirhams
Literacy rate 94%
Life expectancy 79 years

ROMANIA
Southeast Europe

Official name Romania
Formation 1878 / 2009
Capital Bucharest
Population 18.5 million / 201 people per sq mile (78 people per sq km)
Total area 92,043 sq miles (238,391 sq km)
Languages Romanian*, Hungarian (Magyar), Romani, German
Religions Orthodox Christian 82%, Roman Catholic 4%, Nonreligious 1% , Other 13%
Demographics Romanian 83%, Magyar 6%, Roma 3%, Other 7%
Government Presidential / Parliamentary system
Currency New Romanian leu = 100 bani
Literacy rate 99%
Life expectancy 73 years

RUSSIA
Europe / Asia

Official name Russian Federation
Formation 1547 / 2008
Capital Moscow
Population 143 million / 22 people per sq mile (8 people per sq km)
Total area 6,601,668 sq miles (17,098,242 sq km)
Languages Russian*, Tatar, Ukrainian, Chavash, other national languages
Religions Russian Orthodox 20%, Muslim 15%, Other Christian 2%
Demographics Russian 78%, Tatar 4%, Ukrainian 1%, Bashkir 1%, Chavash 1%, Chechen 1%, Other 10%
Government Presidential / Parliamentary system
Currency Russian rouble = 100 kopeks
Literacy rate 100%
Life expectancy 69 years

RWANDA
Central Africa

Official name Republic of Rwanda
Formation 1962 / 1962
Capital Kigali
Population 13.1 million / 1288 people per sq mile (497 people per sq km)
Total area 10,169 sq miles (26,338 sq km)
Languages Kinyarwanda*, French*, Kiswahili, English*
Religions Roman Catholic 38%, Protestant 38%, Seventh-day Adventist 13%, Muslim 2%, Other 2%
Demographics Hutu 85%, Tutsi 14%, Other 1%
Government Presidential system
Currency Rwanda franc = 100 centimes
Literacy rate 73%
Life expectancy 66 years

ST KITTS AND NEVIS
West Indies

Official name Federation of Saint Christopher and Nevis
Formation 1983 / 1983
Capital Basseterre
Population 53,550 / 530 people per sq mile (205 people per sq km)
Total area 101 sq miles (261 sq km)
Languages English*, English Creole
Religions Protestant 76%, Roman Catholic 6%, Hindu 2%, Jehovah's Witness 1%, Rastafarian 1%, None 9%, Other 5%
Demographics Black 93%, Mixed race 3%, White 2%, Indigenous and other 2%
Government Parliamentary system
Currency East Caribbean dollar = 100 cents
Literacy rate 98%
Life expectancy 72 years

ST LUCIA
West Indies

Official name Saint Lucia
Formation 1979 / 1979
Capital Castries
Population 167,122 / 705 people per sq mile (271 people per sq km)
Total area 237 sq miles (616 sq km)
Languages English*, French Creole
Religions Roman Catholic 62%, Seventh-day Adventist 10%, Other Christian 3%, Pentecostal 9%, Nonreligious 6%, Rastafarian 2%, Other 1%
Demographics Black 85%, Mixed race 11%, Asian 2%, Other 2%
Government Parliamentary system
Currency East Caribbean dollar = 100 cents
Literacy rate 95%
Life expectancy 71 years

ST VINCENT & THE GRENADINES
West Indies

Official name Saint Vincent and the Grenadines
Formation 1979 / 1979
Capital Kingstown
Population 100,969 / 673 people per sq mile (260 people per sq km)
Total area 150 sq miles (389 sq km)
Languages English*, English Creole
Religions Other Christian 48%, Anglican 18%, Pentecostal 18%, Nonreligious 9%, Other 7%
Demographics Black 71%, Mixed race 23%, Carib 4%, Asian 2%, Other 1%
Government Parliamentary system
Currency East Caribbean dollar = 100 cents
Literacy rate 96%
Life expectancy 77 years

SAMOA
Australasia & Oceania

Official name Independent State of Samoa
Formation 1962 / 1962
Capital Apia
Population 206,179 / 189 people per sq mile (73 people per sq km)
Total area 1093 sq miles (2,831 sq km)
Languages Samoan*, English*
Religions Other Christian 78%, Roman Catholic 19%, Other 3%
Demographics Samoan 96%, Samoan/New Zealander 2%, Other 2%
Government Parliamentary system
Currency Tala = 100 sene
Literacy rate 99%
Life expectancy 73 years

SAN MARINO
Southern Europe

Official name Republic of San Marino
Formation 301 / 1631
Capital San Marino
Population 34,010 / 1417 people per sq mile (558 people per sq km)
Total area 24 sq miles (61 sq km)
Languages Italian*
Religions Roman Catholic 93%, Nonreligious and other 7%
Demographics Sammarinese 88%, Italian 10%, Other 2%
Government Parliamentary system
Currency Euro = 100 cents
Literacy rate 100%
Life expectancy 84 years

SÃO TOMÉ & PRÍNCIPE
West Africa

Official name Democratic Republic of São Tomé and Príncipe
Formation 1975 / 1975
Capital São Tomé
Population 217,164 / 584 people per sq mile (225 people per sq km)
Total area 372 sq miles (964 sq km)
Languages Portuguese Creole, Portuguese*
Religions Roman Catholic 56%, Nonreligious 21%, Other Christian 16%, Other 7%
Demographics Black 90%, Portuguese and Creole 10%
Government Parliamentary system
Currency Dobra = 100 céntimos
Literacy rate 93%
Life expectancy 68 years

SAUDI ARABIA
Southwest Asia

Official name Kingdom of Saudi Arabia
Formation 1932 / 2017
Capital Riyadh
Population 35.3 million / 43 people per sq mile (16 people per sq km)
Total area 829,999 sq miles (2,149,690 sq km)
Languages Arabic*
Religions Sunni Muslim 85%, Shi'a Muslim 15%
Demographics Arab 90%, Afro-Asian 10%
Government Monarchy
Currency Saudi riyal = 100 halalat
Literacy rate 98%
Life expectancy 77 years

SENEGAL
West Africa

Official name Republic of Senegal
Formation 1960 / 1960
Capital Dakar
Population 17.9 million / 236 people per sq mile (91 people per sq km)
Total area 75,954 sq miles (196,722 sq km)
Languages Wolof, Pulaar, Serer, Diola, Mandinka, Malinké, Soninké, French*
Religions Sunni Muslim 97%, Christian 3%
Demographics Wolof 40%, Serer 16%, Peul 14%, Toucouleur 9%, Diola 5%, Other 14%
Government Presidential system
Currency CFA franc = 100 centimes
Literacy rate 52%
Life expectancy 67 years

SERBIA
Southeast Europe

Official name Republic of Serbia
Formation 2006 / 2008
Capital Belgrade
Population 6.7 million / 224 people per sq mile (86 people per sq km)
Total area 29,912 sq miles (77,474 sq km)
Languages Serbian*, Hungarian (Magyar)
Religions Orthodox Christian 85%, Roman Catholic 5%, Nonreligious 5%, Muslim 3%, Other 1%
Demographics Serb 83%, Magyar 4%, Roma 2%, Bosniak 2%, Croat 1%, Slovak 1%, Other 3%
Government Parliamentary system
Currency Serbian dinar = 100 para
Literacy rate 99%
Life expectancy 73 years

SEYCHELLES
Indian Ocean

Official name Republic of Seychelles
Formation 1976 / 1976
Capital Victoria
Population 97,017 / 551 people per sq mile (213 people per sq km)
Total area 176 sq miles (455 sq km)
Languages French Creole*, English*, French*
Religions Roman Catholic 76%, Anglican 6%, Other Christian 7%, Hindu 2%, Muslim 2%, Nonreligious and other 7%
Demographics Creole 89%, Indian 5%, Chinese 2%, Other 4%
Government Presidential system
Currency Seychelles rupee = 100 cents
Literacy rate 96%
Life expectancy 73 years

SIERRA LEONE
West Africa

Official name Republic of Sierra Leone
Formation 1961 / 1961
Capital Freetown
Population 8.6 million / 310 people per sq mile (120 people per sq km)
Total area 27,698 sq miles (71,740 sq km)
Languages Mende, Temne, Krio, English*
Religions Muslim 77%, Christian 23%,
Demographics Mende 31%, Temne 35%, Limba 9%, Kono 4%, Kurankoh 4%, Fullah 4%, Other 13%
Government Presidential system
Currency Leone = 100 cents
Literacy rate 32%
Life expectancy 60 years

SINGAPORE
Southeast Asia

Official name Republic of Singapore
Formation 1965 / 1965
Capital Singapore
Population 5.9 million / 20,769 people per sq mile (7692 people per sq km)
Total area 277 sq miles (719 sq km)
Languages Mandarin*, Malay*, Tamil*, English*
Religions Christian 19%, Buddhist 31%, Nonreligious 20%, Muslim 15%, Taoist 9%, Hindu 5%, Other 1%
Demographics Chinese 74%, Malay 14%, Indian 9%, Other 3%
Government Parliamentary system
Currency Singapore dollar = 100 cents
Literacy rate 98%
Life expectancy 83 years

SLOVAKIA
Central Europe

Official name Slovak Republic
Formation 1993 / 1993
Capital Bratislava
Population 5.4 million / 285 people per sq mile (110 people per sq km)
Total area 18,932 sq miles (49,035 sq km)
Languages Slovak*, Hungarian (Magyar), Czech
Religions Roman Catholic 56%, Nonreligious 24%, Other Christian 7%, Greek Catholic (Uniate) 4%, Other 8%
Demographics Slovak 84%, Magyar 9%, Roma 2%, Other 5%
Government Parliamentary system
Currency Euro = 100 cents
Literacy rate 99%
Life expectancy 75 years

SLOVENIA
Central Europe

Official name Republic of Slovenia
Formation 1991 / 1991
Capital Ljubljana
Population 2.1 million / 268 people per sq mile (104 people per sq km)
Total area 7827 sq miles (20,273 sq km)
Languages Slovenian*
Religions Roman Catholic 58%, Nonreligious 36%, Muslim 2%, Orthodox Christian 2%, Other 1%
Demographics Slovene 83%, Serb 2%, Croat 2%, Bosniak 1%, Other 12%
Government Parliamentary system
Currency Euro = 100 cents
Literacy rate 100%
Life expectancy 81 years

SOLOMON ISLANDS
Australasia & Oceania

Official name Solomon Islands
Formation 1978 / 1978
Capital Honiara
Population 702,694 / 63 people per sq mile (24 people per sq km)
Total area 11,156 sq miles (28,896 sq km)
Languages English*, Pidgin English, Melanesian Pidgin, around 120 native languages
Religions Protestant 73%, Roman Catholic 20%, Other Christian 3%, Other 4%
Demographics Melanesian 95%, Polynesian 3%, Micronesian 1%, Other 1%
Government Parliamentary system
Currency Solomon Islands dollar = 100 cents
Literacy rate 77%
Life expectancy 70 years

SOMALIA
East Africa

Official name Federal Republic of Somalia
Formation 1960 / 1960
Capital Mogadishu
Population 12.3 million / 50 people per sq mile (19 people per sq km)
Total area 246,199 sq miles (637,657 sq km)
Languages Somali*, Arabic*, English, Italian
Religions Sunni Muslim 99%, Christian 1%
Demographics Somali 85%, Other 15%
Government Parliamentary system
Currency Somali shilin = 100 senti
Literacy rate 38%
Life expectancy 55 years

SOUTH AFRICA
Southern Africa

Official name Republic of South Africa
Formation 1910 / 1994
Capital Pretoria; Cape Town; Bloemfontein
Population 57.5 million / 122 people per sq mile (47 people per sq km)
Total area 470,693 sq miles (1,219,090 sq km)
Languages English*, isiZulu*, isiXhosa*, Afrikaans*, Sepedi*, Setswana*, Sesotho*, Xitsonga*, siSwati*, Tshivenda*, isiNdebele*
Religions Christian 86%, Nonreligious 11%, Muslim 2%, Other 1%
Demographics Black 81%, White 8%, Colored 9%, Asian 3%
Government Presidential system
Currency Rand = 100 cents
Literacy rate 95%
Life expectancy 62 years

SOUTH KOREA
East Asia

Official name Republic of Korea
Formation 1948 / 1953
Capital Seoul
Population 51 million / 1325 people per sq mile (511 people per sq km)
Total area 38,502 sq miles (99,720 sq km)
Languages Korean*, English
Religions Nonreligious 57%, Mahayana Buddhist 16%, Other Christian 20%, Roman Catholic 8%, Other 1%
Demographics Korean 100%
Government Presidential system
Currency South Korean won = 100 chon
Literacy rate 99%
Life expectancy 84 years

SOUTH SUDAN
East Africa

Official name Republic of South Sudan
Formation 2011 / 2011
Capital Juba
Population 11.5 million / 46 people per sq mile (18 people per sq km)
Total area 248,777 sq miles (644,329 sq km)
Languages Arabic, Dinka, Nuer, Zande, Bari, Shilluk, Lotuko, English*
Religions Christian 60% traditional beliefs 33%, Muslim 6%, Other 1%
Demographics Dinka 40%, Nuer 15%, Shilluk 10%, Azande 10%, Arab 10%, Bari 10%, Other 5%
Government Presidential system
Currency South Sudan Pound = 100 piastres
Literacy rate 35%
Life expectancy 55 years

SPAIN
Southwest Europe

Official name Kingdom of Spain
Formation 1492 / 1713
Capital Madrid
Population 47.1 million / 241 people per sq mile (93 people per sq km)
Total area 195,124 sq miles (505,370 sq km)
Languages Spanish*, Catalan*, Galician*, Basque*
Religions Roman Catholic 58%, Nonreligious 16%, Other 3%
Demographics Castilian Spanish 72%, Catalan 17%, Galician 6%, Basque 2%, Roma 1%, Other 2%
Government Parliamentary system
Currency Euro = 100 cents
Literacy rate 99%
Life expectancy 83 years

SRI LANKA
South Asia

Official name Democratic Socialist Republic of Sri Lanka
Formation 1948 / 1948
Capital Colombo; Sri Jayawardenapura Kotte
Population 23.1 million / 912 people per sq mile (352 people per sq km)
Total area 25,332 sq miles (65,610 sq km)
Languages Sinhala*, Tamil*, Sinhala-Tamil, English
Religions Buddhist 70%, Hindu 13%, Muslim 10%, Christian (mainly Roman Catholic) 7%
Demographics Sinhalese 75%, Tamil 15%, Moor 9%, Other 1%
Government Presidential system
Currency Sri Lankan rupee = 100 cents
Literacy rate 92%
Life expectancy 76 years

SUDAN
East Africa

Official name Republic of the Sudan
Formation 1956 / 2011
Capital Khartoum
Population 47.9 million / 67 people per sq mile (26 people per sq km)
Total area 718,723 sq miles (1,861,484 sq km)
Languages Arabic*, English, Nubian, Beja, Fur
Religions Muslim (mainly Sunni) 99%, Other 1%
Demographics Arab 70%, Nubian 3%, Beja 6%, Fur 2%, Egyptian 1%, Fulani 1%, Coptic 1%, Other 16%
Government Presidential system
Currency Sudanese pound = 100 piastres
Literacy rate 60%
Life expectancy 65 years

SURINAME
South America

Official name Republic of Suriname
Formation 1975 / 1975
Capital Paramaribo
Population 632,000 / 10 people per sq mile (4 people per sq km)
Total area 63,251 sq miles (163,820 sq km)
Languages Sranang Tongo (creole), Dutch*, Caribbean Hindustani , Javanese
Religions Christian 50%, Hindu 23%, Muslim 14%, Other 13%
Demographics East Indian 27%, Creole 16%, Black 37%, Javanese 14%, Mixed race 13%, Other 8%
Government Presidential system
Currency Surinamese dollar = 100 cents
Literacy rate 94%
Life expectancy 70 years

SWEDEN
Northern Europe

Official name Kingdom of Sweden
Formation 1523 / 1921
Capital Stockholm
Population 10.4 million / 60 people per sq mile (23 people per sq km)
Total area 173,860 sq miles (450,295 sq km)
Languages Swedish*, Finnish, Sámi
Religions Evangelical Lutheran 58%, Muslim 2%, Other 40%
Demographics Swedish 80%, Syrian 2%, Iraqi 2%, Finnish 1%, Other 15%
Government Parliamentary constitutional monarchy
Currency Swedish krona = 100 öre
Literacy rate 99%
Life expectancy 83 years

SWITZERLAND
Central Europe

Official name Swiss Confederation
Formation 1291 / 1857
Capital Bern
Population 8.5 million / 533 people per sq mile (206 people per sq km)
Total area 15,937 sq miles (41,277 sq km)
Languages German*, Swiss-German, French*, Italian*, Portuguese, Romansch*
Religions Roman Catholic 34%, Protestant 23%, Muslim 5%, Other Christian 6%, Other 32%
Demographics Swiss 69%, German 4%, French 2%, Italian 3%, Portuguese 3%, Kosovo 1%, Turkish 1%, Other 17%
Government Federal republic
Currency Swiss franc = 100 rappen/centimes
Literacy rate 99%
Life expectancy 84 years

SYRIA
Southwest Asia

Official name Syrian Arab Republic
Formation 1946 / 1967
Capital Damascus
Population 21.5 million / 297 people per sq mile (117 people per sq km)
Total area 72,370 sq miles (184,437 sq km)
Languages Arabic*, French, Kurdish, Armenian, Circassian, Turkic languages, Assyrian, Aramaic
Religions Muslim 87%, Christian 10%, Druze 3%
Demographics Arab 50%, Kurd 10%, Alawite 15%, Levantine 10%, Other 15%
Government Presidential system
Currency Syrian pound = 100 piastres
Literacy rate 86%
Life expectancy 72 years

TAIWAN
East Asia

Official name Republic of China (ROC)
Formation 1945 / 1945
Capital Taipei
Population 23.5 million / 1692 people per sq mile (653 people per sq km)
Total area 13,892 sq miles (35,980 sq km)
Languages Amoy Chinese, Mandarin Chinese*, Hakka Chinese
Religions Buddhist 35%, Taoist 33%, Christian 4%, Indigenous 10%, Other 18%
Demographics Han Chinese 95%, Indigenous 2%, Other 3%
Government Semi-presidential system
Currency New Taiwan dollar = 100 cents
Literacy rate 99%
Life expectancy 81 years

TAJIKISTAN
Central Asia

Official name Republic of Tajikistan
Formation 1991 / 2011
Capital Dushanbe
Population 9.1 million / 164 people per sq mile (63 people per sq km)
Total area 55,637 sq miles (144,100 sq km)
Languages Tajik*, Uzbek, Russian
Religions Sunni Muslim 95%, Shi'a Muslim 3%, Other 2%
Demographics Tajik 84%, Uzbek 14%, Other 2%
Government Presidential system
Currency Somoni = 100 diram
Literacy rate 100%
Life expectancy 72 years

TONGA
Australasia & Oceania

Official name Kingdom of Tonga
Formation 1970 / 1970
Capital Nuku'alofa
Population 100,000/ 347 people per sq mile (134 people per sq km)
Total area 288 sq miles (747 sq km)
Languages English*, Tongan*, Other
Religions Free Wesleyan 38%, Roman Catholic 16%, Church of Jesus Christ of Latter-day Saints 17%, Other Christian 16%, Free Church of Tonga 12%, Other 1%
Demographics Tongan 97%, Other 3%
Government Constitutional monarchy
Currency Pa'anga = 100 seniti
Literacy rate 99%
Life expectancy 71 years

TURKMENISTAN
Central Asia

Official name Turkmenistan
Formation 1991 / 1991
Capital Ashgabat
Population 5.6 million / 30 people per sq mile (11 people per sq km)
Total area 188,456 sq miles (488,100 sq km)
Languages Turkmen*, Uzbek, Russian, Kazakh, Tatar
Religions Muslim 93%, Christian 6%, Other 2%
Demographics Turkmen 85%, Uzbek 5%, Russian 4%, Other 6%
Government Presidential system / authoritarian
Currency Manat = 100 tenge
Literacy rate 100%
Life expectancy 69 years

UNITED ARAB EMIRATES
Southwest Asia

Official name United Arab Emirates
Formation 1971 / 1974
Capital Abu Dhabi
Population 9.9 million / 307 people per sq mile (118 people per sq km)
Total area 32,278 sq miles (83,600 sq km)
Languages Arabic*, Farsi, Indian and Pakistani languages, English
Religions Muslim (mainly Sunni) 76%, Christian 9%, Other 15%
Demographics South Asian 59%, Emirati 12%, Filipino 6%, Egyptian 10%, Other 13%
Government Federation of monarchies
Currency UAE dirham = 100 fils
Literacy rate 98%
Life expectancy 79 years

TANZANIA
East Africa

Official name United Republic of Tanzania
Formation 1964 / 1964
Capital Dodoma
Population 63.8 million / 174 people per sq mile (67 people per sq km)
Total area 365,754 sq miles (947,300 sq km)
Languages Kiswahili*, Sukuma, Chagga, Nyamwezi, Hehe, Makonde, Yao, Sandawe, English*
Religions Christian 63%, Muslim 34%, Other 3%
Demographics Indigenous tribes 99%, Other 1%
Government Presidential system
Currency Tanzanian shilling = 100 cents
Literacy rate 78%
Life expectancy 66 years

TRINIDAD AND TOBAGO
West Indies

Official name Rep. of Trinidad & Tobago
Formation 1962 / 1962
Capital Port of Spain
Population 1.4 million / 707 people per sq mile (273 people per sq km)
Total area 1980 sq miles (5128 sq km)
Languages English Creole, Trinidadian Creole, English*, Caribbean Hindustani, French, Spanish
Religions Protestant 32%, Roman Catholic 22%, Hindu 18%, Muslim 5%, Jehovah's Witness 2%, Other 21%
Demographics East Indian 38%, Black 36%, Mixed race 24%, Other 2%
Government Parliamentary system
Currency Trinidad and Tobago dollar = 100 cents
Literacy rate 99%
Life expectancy 73 years

TUVALU
Australasia & Oceania

Official name Tuvalu
Formation 1978 / 2012
Capital Funafuti Atoll
Population 11,930 / 1193 people per sq mile (459 people per sq km)
Total area 10 sq miles (26 sq km)
Languages Tuvaluan, Kiribati, Samoan, English*
Religions Church of Tuvalu 91%, Seventh-day Adventist 2%, Baha'i 2%, Other 5%
Demographics Tuvaluan 97%, Other 3%
Government Parliamentary democracy
Currency Australian dollar = 100 cents; Tuvaluan dollar = 100 cents
Literacy rate 95%
Life expectancy 65 years

UNITED KINGDOM
Northwest Europe

Official name United Kingdom of Great Britain and Northern Ireland
Formation 1707 / 1922
Capital London
Population 67.7 million / 720 people per sq mile (278 people per sq km)
Total area 94,058 sq miles (243,610 sq km)
Languages English*, Welsh (* in Wales), Gaelic, Irish
Religions Christian 64%, Nonreligious 28%, Muslim 5%, Hindu 1%, Other 3%
Demographics White 87%, Indian and Pakistani 4%, Black 3%, Other Asian 2%, Bengali 1%, Other 3%
Government Parliamentary constitutional monarchy
Currency Pound sterling = 100 pence
Literacy rate 99%
Life expectancy 81 years

THAILAND
Southeast Asia

Official name Kingdom of Thailand
Formation 1238 / 1907
Capital Bangkok
Population 69.6 million / 351 people per sq mile (136 people per sq km)
Total area 198,117 sq miles (513,120 sq km)
Languages Thai*, Chinese, Malay, Khmer, Mon, Karen, Miao
Religions Buddhist 94%, Muslim 4%, Other 2%
Demographics Thai 98%, Burmese 1%, Other 1%
Government Constitutional monarchy
Currency Baht = 100 satang
Literacy rate 94%
Life expectancy 79 years

TUNISIA
North Africa

Official name Republic of Tunisia
Formation 1956 / 1956
Capital Tunis
Population 11.8 million / 187 people per sq mile (72 people per sq km)
Total area 63,170 sq miles (163,610 sq km)
Languages Arabic*, French, Berber
Religions Muslim (mainly Sunni) 99%, Other 1%
Demographics Arab 98%, European 1%, Jewish and other 1%
Government Parliamentary system
Currency Tunisian dinar = 1000 millimes
Literacy rate 90%
Life expectancy 74 years

UGANDA
East Africa

Official name Republic of Uganda
Formation 1962 / 1962
Capital Kampala
Population 46.2 million / 496 people per sq mile (192 people per sq km)
Total area 93,065 sq miles (241,038 sq km)
Languages Luganda, Nkole, Chiga, Lango, Acholi, Teso, Lugbara, English*
Religions Roman Catholic 42%, Protestant 42%, Muslim (mainly Sunni) 12%, Other 4%
Demographics Baganda 17%, Banyakole 10%, Basoga 9%, Bakiga 7%, Iteso 7%, Langi 6%, Bagisu 5%, Acholi 4%, Lugbara 3%, Other 32%
Government Presidential system
Currency Ugandan shilling
Literacy rate 77%
Life expectancy 63 years

UNITED STATES
North America

Official name United States of America
Formation 1776 / 1959
Capital Washington D.C.
Population 332 million / 90 people per sq mile (35 people per sq km)
Total area 3,677,649 sq miles (9,525,067 sq km)
Languages English*, Spanish, Chinese, French, Polish, German, Tagalog, Vietnamese, Italian, Korean, Russian
Religions Protestant 47%, Nonreligious 23%, Roman Catholic 21%, Jewish 2%, Muslim 1%, Other 6%
Demographics White 62%, Black 12%, Asian 6%, Indigenous 2%, Mixed race 10%, Other 8%
Government Constitutional federal republic
Currency US dollar = 100 cents
Literacy rate 99%
Life expectancy 76 years

TOGO
West Africa

Official name Togolese Republic
Formation 1960 / 1960
Capital Lomé
Population 8.4 million / 383 people per sq mile (148 people per sq km)
Total area 21,925 sq miles (56,785 sq km)
Languages Ewe and Mina, Kabye, Gurma, French*
Religions Christian 42%, Indigenous 37%, Muslim 14%, Other 7%
Demographics Ewe 42%, Other African 41%, Kabye 12%, Foreigners 5%
Government Presidential system
Currency West African CFA franc = 100 centimes
Literacy rate 67%
Life expectancy 62 years

TURKEY (TÜRKIYE)
Asia / Europe

Official name Republic of Türkiye
Formation 1923 / 1939
Capital Ankara
Population 83 million / 274 people per sq mile (106 people per sq km)
Total area 302,535 sq miles (783,562 sq km)
Languages Turkish*, Kurdish, Arabic, Circassian, Armenian, Greek, Georgian, Ladino
Religions Muslim (mainly Sunni) 99%, Other 1%
Demographics Turkish 73%, Kurdish 17%, Other 10%
Government Presidential system
Currency Turkish lira = 100 kuruş
Literacy rate 97%
Life expectancy 76 years

UKRAINE
Eastern Europe

Official name Ukraine
Formation 1991 / 1991
Capital Kyiv
Population 43.5 million / 187 people per sq mile (72 people per sq km)
Total area 233,032 sq miles (603,550 sq km)
Languages Ukrainian*, Russian, Tatar, Other
Religions Orthodox Christian 78%, Roman Catholic 10%, Nonreligious 7%, Other 5%
Demographics Ukrainian 78%, Russian 17%, Belarussian 1%, Other 4%
Government Semi-presidential system
Currency Hryvnia = 100 kopiykas
Literacy rate 100%
Life expectancy 70 years

URUGUAY
South America

Official name Oriental Republic of Uruguay
Formation 1828 / 1828
Capital Montevideo
Population 3.4 million / 50 people per sq mile (19 people per sq km)
Total area 68,037 sq miles (176,215 sq km)
Languages Spanish*
Religions Roman Catholic 42%, Nonreligious 37%, Protestant 15%, Other 6%
Demographics White 87%, Black 7%, Mixed race 5%, Other 1%
Government Presidential system
Currency Uruguayan peso = 100 centésimos
Literacy rate 99%
Life expectancy 75 years

UZBEKISTAN
Central Asia

Official name Republic of Uzbekistan
Formation 1991 / 1991
Capital Tashkent
Population 31.1million / 180 people per sq mile (70 people per sq km)
Total area 172,742 sq miles (447,400 sq km)
Languages Uzbek*, Russian, Tajik, Kazakh
Religions Sunni Muslim 88%, Orthodox Christian 9%, Other 3%
Demographics Uzbek 84%, Russian 2%, Tajik 5%, Kazakh 3%, Other 6%
Government Presidential system / authoritarian
Currency So'm = 100 tiyin
Literacy rate 100%
Life expectancy 71 years

VANUATU
Australasia & Oceania

Official name Republic of Vanuatu
Formation 1980 / 1980
Capital Port Vila
Population 300,000 / 63 people per sq mile (25 people per sq km)
Total area 4706 sq miles (12,189 sq km)
Languages Bislama*, English*, French*, Other indigenous languages
Religions Protestant 45%, Presbyterian 35%, Roman Catholic 14%. Indigenous 4%, Other 2%
Demographics ni-Vanuatu 99%, Other 1%
Government Parliamentary system
Currency Vanuatu Vatu
Literacy rate 88%
Life expectancy 70 years

VATICAN CITY
Southern Europe

Official name Vatican City State
Formation 1929 / 1929
Capital Vatican City
Population 825 / 4852 people per sq mile (1875 people per sq km)
Total area 0.17 sq miles (0.44 sq km)
Languages Italian*, Latin*, French
Religions Roman Catholic 100%
Demographics Clergy 9%, Diplomats 39%, Swiss Guard 13%, Other 39%
Government Papal state
Currency Euro = 100 cents
Literacy rate 99%
Life expectancy 78 years

VENEZUELA
South America

Official name Bolivarian Republic of Venezuela
Formation 1811 / 1811
Capital Caracas
Population 29.7 million / 84 people per sq mile (33 people per sq km)
Languages Spanish*, Indigenous languages
Religions Roman Catholic 73%, Protestant 17%, Nonreligious 7%, Other 3%
Demographics Mixed race 69%, White 20%, Black 9%, Indigenous 2%
Government Federal presidential republic
Currency Venezuelan bolívar = 100 céntimos
Literacy rate 97%
Life expectancy 71 years

VIETNAM
Southeast Asia

Official name Socialist Republic of Viet Nam
Formation 1945 / 1976
Capital Hanoi
Population 103.8 million / 812 people per sq mile (313 people per sq km)
Total area 127,881 sq miles (331,210 sq km)
Languages Vietnamese*, English, French, Chinese, Thai, Khmer
Religions Catholic 6%, Buddhist 6%, Protestant 1%, None 86%, Other 1%
Demographics Vietnamese 85%, Tay 2%, Thai 2%, Muong 2%, Other 8%
Government Communist state
Currency Đông
Literacy rate 96%
Life expectancy 74 years

YEMEN
Southwest Asia

Official name Republic of Yemen
Formation 1990 / 2000
Capital Sanaa
Population 30.9 million / 152 people per sq mile (59 people per sq km)
Total area 203,850 sq miles (527,968 sq km)
Languages Arabic*
Religions Muslim 99% (Sunni 65%, Shi'a 35%), Other 1%
Demographics Arab 99%, Other 1%
Government Transitional regime
Currency Yemeni rial = 100 fils
Literacy rate 70%
Life expectancy 64 years

ZAMBIA
Southern Africa

Official name Republic of Zambia
Formation 1964 / 1964
Capital Lusaka
Population 19.6 million / 67 people per sq mile (26 people per sq km)
Total area 290,587 sq miles (752,618 sq km)
Languages Bemba, Tonga, Nyanja, Lozi, Lala-Bisa, Nsenga, English*
Religions Protestant 75%, Roman Catholic 20%, Nonreligious 2%, Other 3%
Demographics Bemba 21%, Tonga 14%, Chewa 7%, Lozi 6%, Nsenga 5%, Tumbuka 4%, Ngoni 4%, Lala 3%, Kaonde 3%, Lunda 3%, Other 30%
Government Presidential system
Currency New Zambian kwacha = 100 ngwee
Literacy rate 87%
Life expectancy 61 years

ZIMBABWE
Southern Africa

Official name Republic of Zimbabwe
Formation 1980 / 1980
Capital Harare
Population 15.1 million / 100 people per sq mile (39 people per sq km)
Total area 150,872 sq miles (390,757 sq km)
Languages Shona, isiNdebele, English*
Religions Protestant 75%, Roman Catholic 7%, Other Christian 5%, Traditional 2%, None 10%, Other 1%
Demographics Shona 99%, Other 1%
Government Presidential system
Currency RTGS dollar = 100 cents; multiple foreign currencies
Literacy rate 87%
Life expectancy 59 years

GLOSSARY OF ABBREVIATIONS

This Glossary provides a comprehensive guide to the abbreviations used in this Atlas, and in the Index.

A
abbrev. abbreviated
admin. administrative
Afr. Afrikaans
Alb. Albanian
Amh. Amharic
anc. ancient
Ar. Arabic
Arm. Armenian
Az. Azerbaijani

B
Basq. Basque
Bel. Belarusian
Ben. Bengali
Bibl. Biblical
Bret. Breton
Bul. Bulgarian
Bur. Burmese

C
Cant. Cantonese
Cast. Castilian
Cat. Catalan
Chin. Chinese
Cro. Croatian
Cz. Czech

D
Dan. Danish
Dut. Dutch

E
E. East
Eng. English
Est. Estonian
est. estimated

F
Far. Faroese
Fij. Fijian
Fin. Finnish
Flem. Flemish
Fr. French
Fris. Frisian

G
Geor. Georgian
Ger. German
Gk. Greek
Guj. Gujarati

H
Haw. Hawaiian
Heb. Hebrew
Hind. Hindi
hist. historical
Hung. Hungarian

I
Icel. Icelandic
Ind. Indonesian
In. Inuit
Ir. Irish
It. Italian

J
Jap. Japanese

K
Kaz. Kazakh
Khm. Khmer
Kir. Kyrgyz
Kor. Korean
Kurd. Kurdish

L
Lao. Laotian
Lat. Latin
Latv. Latvian
Lith. Lithanian
Lus. Lusatian

M
Mac. Macedonian
Mal. Malay
Malg. Malagasy
Malt. Maltese
Mon. Montenegro
Mong. Mongolian

N
Nepali. Nepali
Nor. Norwegian

O
off. officially

P
Pash. Pashto
Per. Persian
Pol. Polish
Port. Portuguese
prev. previously

R
Rmsch. Romansch
Roman. Romanian
Rus. Russian

S
Smi. Sámi
SCr. Serbo - Croatian
Serb. Serbian
Slvk. Slovak
Slvn. Slovene
Som. Somali
Sp. Spanish
Swa. Swahili
Swe. Swedish

T
Taj. Tajik
Th. Thai
Tib. Tibetan
Turk. Turkish
Turkm. Turkmenistan

U
Ukr. Ukrainian
Uygh. Uyghur
Uzb. Uzbek

V
var. variant
Vtn. Vietnamese

W
Wel. Welsh

X
Xh. Xhosa

A

Aa *see* Gauja
Aachen *94 A4 Dut.* Aken, *Fr.*
Aix-la-Chapelle; *anc.* Aquae Grani,
Aquisgranum. Nordrhein-Westfalen,
W Germany
Aaiún *see* Laâyoune
Aalborg *85 B7 var.* Ålborg,
Ålborg-Nørresundby; *anc.* Alburgum.
Nordjylland, N Denmark
Aalen *95 B6* Baden-Württemberg,
S Germany
Aalsmeer *86 C3* Noord-Holland,
C Netherlands
Aalst *87 B6* Oost-Vlaanderen, C Belgium
Aalten *86 E4* Gelderland, E Netherlands
Aalter *87 B5* Oost-Vlaanderen,
NW Belgium
Aanaarjävri *see* Inarijärvi
Äänekoski *85 D5* Keski-Suomi,
W Finland
Aar *see* Aare
Aare *95 A7 var.* Aar. *river* W Switzerland
Aarhus *85 B7 var.* Århus, C Denmark
Aarlen *see* Arlon
Aassi, Nahr El *see* Orontes
Aat *see* Ath
Aba *77 E5* Orientale, NE Dem. Rep.
Congo
Aba *75 G5* Abia, S Nigeria
Abā as Su'ūd *see* Najrān
Abaco Island *see* Great Abaco, N The
Bahamas
Ābādān *120 C4* Khūzestān, SW Iran
Abadan *122 D3 prev.* Bezmein,
Büzmeýin, *Rus.* Byuzmeyin. Ahal
Welaýaty, C Turkmenistan
Abai *see* Blue Nile
Abakan *114 D4* Respublika Khakasiya,
S Russia
Abancay *60 D4* Apurímac, SE Peru
Abariringa *see* Kanton
Abashiri *130 D2 var.* Abasiri. Hokkaidō,
NE Japan
Abasiri *see* Abashiri
Ābay Wenz *see* Blue Nile
Abbaia *see* Åbaya Häyk'
Abbatis Villa *see* Abbeville
Abbazia *see* Opatija
Abbeville *90 C2 anc.* Abbatis Villa.
Somme, N France
'Abd al 'Azīz, Jabal *118 D2 mountain
range* NE Syria
Abéché *76 C3 var.* Abécher, Abeshr.
Ouaddaï, SE Chad
Abécher *see* Abéché
Abela *see* Ávila
Abellinum *see* Avellino
Abemama *144 D2 var.* Apamama; *prev.*
Roger Simpson Island. *atoll* Tungaru,
W Kiribati
Abengourou *75 E5* E Ivory Coast
Aberbrothock *see* Arbroath
Abercorn *see* Mbala
Aberdeen *88 D3 anc.* Devana.
NE Scotland, United Kingdom
Aberdeen *45 E2* South Dakota, N USA
Aberdeen *46 B2* Washington,
NW USA
Abergwaun *see* Fishguard
Abertawe *see* Swansea
Aberystwyth *89 C6* W Wales,
United Kingdom
Abeshr *see* Abéché
Abhā *121 B6* 'Asīr, SW Saudi Arabia
Abidavichy *107 D7 Rus.* Obidovichi.
Mahilyowskaya Voblasts', E Belarus
Abidjan *75 E5* S Ivory Coast
Abilene *49 F3* Texas, SW USA
Abingdon *see* Pinta, Isla
Abkhazia *117 E1 var.* Apkhazeti.
autonomous republic NW Georgia
Åbo *see* Turku
Aboisso *75 E5* SE Ivory Coast
Abo, Massif d' *76 B1 mountain range*
NW Chad
Abomey *75 F5* S Benin
Abou-Déïa *76 C3* Salamat, SE Chad
Aboudouhour *see* Abū aḑ Ḑuhūr
Abou Kémal *see* Abū Kamāl
Abrantes *92 B3 var.* Abrántes. Santarém,
C Portugal
Abrashlare *see* Brezovo
Abrolhos Bank *56 E4 undersea bank*
W Atlantic Ocean
Abrova *107 B6 Rus.* Obrovo. Brestskaya
Voblasts', SW Belarus

Abrud *108 B4 Ger.* Gross-Schlatten,
Hung. Abrudbánya. Alba, SW Romania
Abrudbánya *see* Abrud
Abruzzese, Appennino *96 C4 mountain
range* C Italy
Absaroka Range *44 B2 mountain range*
Montana/Wyoming, NW USA
Abū aḑ Ḑuhūr *118 B3 Fr.* Aboudouhour.
Idlib, NW Syria
Abu Dhabi *see* Abū Ẓaby
Abu Hamed *72 C3* River Nile, N Sudan
Abū Ḩardān *see* Hajîn
Abuja *75 G4 country capital* (Nigeria)
Federal Capital District, C Nigeria
Abū Kamāl *118 E3 Fr.* Abou Kémal.
Dayr az Zawr, E Syria
Abula *see* Ávila
Abunã, Rio *62 C2 var.* Río Abuná. *river*
Bolivia/Brazil
Abut Head *151 B6 headland* South
Island, New Zealand
Abuye Meda *72 D4 mountain* C Ethiopia
Abū Ẓaby *121 C5 var.* Abū Ẓabī, *Eng.*
Abu Dhabi. *country capital* (United
Arab Emirates)
Abū Ẓaby, C United Arab Emirates
Abū Ẓabī *see* Abū Ẓaby
Abyad, Al Baḩr al *see* White Nile
Abyei Area *73 B5 disputed region* Sudan/
South Sudan
Abyla *see* Ávila
Abyssinia *see* Ethiopia
Acalayong *77 A5* SW Equatorial Guinea
Acaponeta *50 D4* Nayarit, C Mexico
Acapulco *51 E5 var.* Acapulco de Juárez.
Guerrero, S Mexico
Acapulco de Juárez *see* Acapulco
Acaraí Mountains *59 F4 Sp.* Serra
Acaraí. *mountain range* Brazil/Guyana
Acaraí, Serra *see* Acaraí Mountains
Acarigua *58 D2* Portuguesa, N Venezuela
Accra *75 E5 country capital* (Ghana)
SE Ghana
Achacachi *61 E4* La Paz, W Bolivia
Ach'ara *see* Ajaria
Acklins Island *54 C2 island* SE
The Bahamas
Aconcagua, Cerro *64 B4 mountain*
W Argentina
**Açores/Açores, Arquipélago dos/
Açores, Ilhas dos** *see* Azores
A Coruña *92 B1 Cast.* La Coruña, *Eng.*
Corunna; *anc.* Caronium. Galicia,
NW Spain
Acre *62 C2 off.* Estado do Acre. *state/
region* W Brazil
Açu *see* Assu
Acunum Acusio *see* Montélimar
Ada *100 D3* Vojvodina, N Serbia
Ada *49 G2* Oklahoma, C USA
Ada Bazar *see* Adapazarı
Adalia *see* Antalya
Adalia, Gulf of *see* Antalya Körfezi
Adama *see* Nazrēt
'Adan *121 B7 Eng.* Aden. SW Yemen
Adana *116 D4 var.* Seyhan. Adana,
S Turkey (Türkiye)
Adâncata *see* Horlivka
Adapazarı *116 B2 prev.* Ada Bazar.
Sakarya, NW Turkey (Türkiye)
Adare, Cape *154 B4 cape* Antarctica
Ad Dahna *120 C4 desert* E Saudi Arabia
Ad Dakhla *70 A4 var.* Dakhla.
SW Western Sahara
Ad Dalanj *see* Dilling
Ad Damar *see* Ed Damer
Ad Damazin *see* Ed Damazin
Ad Dāmir *see* Ed Damer
Ad Dammām *120 C4* Ash Sharqīyah,
NE Saudi Arabia
Ad Dāmūr *see* Damoûr
Ad Dawḩah *120 C4 Eng.* Doha. *country
capital* (Qatar) C Qatar
Ad̲ Ḑiffah *see* Libyan Plateau
Addis Ababa *see* Ādīs Ābeba
Addoo Atoll *see* Addu Atoll
Addu Atoll *132 A5 var.* Addoo Atoll,
Seenu Atoll. *atoll* S Maldives
Adelaide *149 B6 state capital* South
Australia
Adelsberg *see* Postojna
Aden *see* 'Adan
Aden, Gulf of *121 C7 gulf* SW
Arabian Sea
Adige *96 C2 Ger.* Etsch. *river* N Italy
Adirondack Mountains *41 F2 mountain
range* New York, NE USA
Ādīs Ābeba *73 C5 Eng.* Addis Ababa.
country capital (Ethiopia) Ādīs Ābeba,
C Ethiopia

Adıyaman *117 E4* Adıyaman,
SE Turkey (Türkiye)
Adjud *108 C4* Vrancea, E Romania
Admiralty Islands *144 B3 island group*
N Papua New Guinea
Adra *93 E5* Andalucía, S Spain
Adrar *70 D3* C Algeria
Adrian *40 C3* Michigan, N USA
Adrianople/Adrianopolis *see* Edirne
Adriatico, Mare *see* Adriatic Sea
Adriatic Sea *103 E2 Alb.* Deti Adriatik,
It. Mare Adriatico, *Croatian* Jadransko
More, *Slvn.* Jadransko Morje. *sea*
N Mediterranean Sea
Adriatik, Deti *see* Adriatic Sea
Adycha *115 F2 river* NE Russia
Aegean Sea *105 C5 Gk.* Aigaíon Pelagos,
Aigaío Pélagos, *Turk.* Ege Denizi. *sea*
NE Mediterranean Sea
Aegviidu *106 D2 Ger.* Charlottenhof.
Harjumaa, NW Estonia
Aegyptus *see* Egypt
Aelana *see* Al 'Aqabah
Aelok *see* Ailuk Atoll
Aelönlaplap *see* Ailinglaplap Atoll
Aemona *see* Ljubljana
Aeolian Islands *97 C6 var.* Isole Lipari,
Eng. Aeolian Islands, Lipari Islands.
island group S Italy
Aeolian Islands *see* Eolie, Isole
Æsernia *see* Isernia
Afar Depression *see* Danakil Desert
**Afars et des Issas, Territoire Français
des** *see* Djibouti
Afghanistan *122 C4 off.* Islamic Republic
of Afghanistan; *var. Dari,* Jamhūrī-ye
Islāmī-ye Afghānistān, *Pash.* Dĕ
Afghānistān Islāmī Jumhūriyat; *prev.*
Republic of Afghanistan. *country*
C Asia
Afghānistān, Jamhūrī-ye Islāmī-ye *see*
Afghanistan
Afmadow *73 D6* Jubbada Hoose,
S Somalia
Africa *68 continent*
Africa, Horn of *68 E4 physical region*
Ethiopia/Somalia
Africana Seamount *141 A6 seamount*
SW Indian Ocean
'Afrīn *118 B2 Ḩalab,* N Syria
Afyon *116 B3 prev.* Afyonkarahisar.
Afyon, W Turkey (Türkiye)
Agadès *see* Agadez
Agadez *75 G3 prev.* Agadès. Agadez,
C Niger
Agadir *70 B3* SW Morocco
Agana/Agaña *see* Hagåtña
Agaro *73 C5* Oromiya, C Ethiopia
Agassiz Fracture Zone *143 G5 fracture
zone* S Pacific Ocean
Agatha *see* Agde
Agathónisi *105 D6 island* Dodékánisa,
Greece, Aegean Sea
Agde *91 C6 anc.* Agatha. Hérault,
S France
Agedabia *see* Ajdābiyā
Agen *91 B5 anc.* Aginnum.
Lot-et-Garonne, SW France
Agendicum *see* Sens
Agiá *104 B4 var.* Ayiá. Thessalía,
C Greece
Agía Marína *105 E6* Léros, Dodékánisa,
Greece, Aegean Sea
Aginnum *see* Agen
Ágios Efstrátios *104 D4 var.* Áyios
Evstrátios, Hagios Evstrátios. *island*
E Greece
Ágios Nikólaos *105 D8 var.* Áyios
Nikólaos. Kríti, Greece,
E Mediterranean Sea
Āgra *134 D3* Uttar Pradesh, N India
Agra and Oudh, United Provinces of *see*
Uttar Pradesh
Agram *see* Zagreb
Ağrı *117 F3 var.* Karaköse; *prev.*
Karakılısse. Ağrı, NE Turkey (Türkiye)
Ağrı Dağı *see* Büyük Ağrı Dağı
Agrigento *97 C7 Gk.* Akragas; *prev.*
Girgenti. Sicilia, Italy,
C Mediterranean Sea
Agrínio *105 C5* Évvoia, C Greece
Agropoli *97 D5* Campania, S Italy
Aguachica *58 B2* Cesar, N Colombia
Aguadulce *53 F5* Coclé, S Panama
Agua Prieta *50 B1* Sonora, NW Mexico
Aguascalientes *50 D4* Aguascalientes,
C Mexico
Aguaytía *60 C3* Ucayali, C Peru
Aguilas *93 E4* Murcia, SE Spain
Aguililla *50 D4* Michoacán, SW Mexico

Agulhas Basin *69 D8 undersea basin*
SW Indian Ocean
Agulhas, Cape *78 C5 headland*
SW South Africa
Agulhas Plateau *67 D6 undersea plateau*
SW Indian Ocean
Ahaggar *75 F2 high plateau region*
SE Algeria
Ahlen *94 B4* Nordrhein-Westfalen,
W Germany
Ahmadābād *see* Ahmedābād
Ahmadnagar *134 C5 var.* Ahmednagar.
Mahārāshtra, W India
Ahmedābād *134 C4 var.* Ahmadābād.
Gujarāt, W India
Ahmednagar *see* Ahmadnagar
Ahuachapán *52 B3* Ahuachapán,
W El Salvador
Ahvāz *120 C3 var.* Ahwāz; *prev.* Nāsiri.
Khūzestān, SW Iran
Ahvenanmaa *see* Åland
Ahwāz *see* Ahvāz
Aigaíon Pelagos/Aigaío Pélagos *see*
Aegean Sea
Aigialoúsa *see* Yenierenköy
Aígina *105 C6 var.* Aíyina, Egina.
Aígina, C Greece
Aígio *105 B5 var.* Egio; *prev.* Aíyion.
Dytikí Ellás, S Greece
Aiken *43 E2* South Carolina, SE USA
Ailinglaplap Atoll *144 D2 var.*
Aelönlaplap. *atoll* Ralik Chain,
S Marshall Islands
Ailuk Atoll *144 D1 var.* Aelok. *atoll*
Ratak Chain, N Marshall Islands
Aináži *106 D3 Est.* Heinaste, *Ger.*
Hainasch. Limbaži, N Latvia
'Aïn Ben Tili *74 D1* Tiris Zemmour,
N Mauritania
Aintab *see* Gaziantep
Aïoun el Atrous/Aïoun el Atroûss *see*
'Ayoûn el 'Atroûs
Aiquile *61 F4* Cochabamba, C Bolivia
Aïr *see* Aïr, Massif de l'
Air du Azbine *see* Aïr, Massif de l'
Aïr, Massif de l' *75 G2 var.* Aïr, Air du
Azbine, Asben. *mountain range*
NC Niger
Aiud *108 B4 Ger.* Strassburg, *Hung.*
Nagyenyed; *prev.* Engeten. Alba,
SW Romania
Aix *see* Aix-en-Provence
Aix-en-Provence *91 D6 var.* Aix; *anc.*
Aquae Sextiae. Bouches-du-Rhône,
SE France
Aix-la-Chapelle *see* Aachen
Aiyina *see* Aígina
Aíyion *see* Aígio
Aizkraukle *106 C4* Aizkraukle, S Latvia
Ajaccio *91 E7* Corse, France,
C Mediterranean Sea
Ajaria *117 F2 var.* Ach'ara. *autonomous
republic* SW Georgia
Aj Bogd Uul *126 D2 mountain*
SW Mongolia
Ajdābiyā *71 G2 var.* Agedabia,
Ajdābiyah. NE Libya
Ajdābiyah *see* Ajdābiyā
Ajjinena *see* El Geneina
Ajmer *134 D3 var.* Ajmere. Rājasthān,
N India
Ajo *48 A3* Arizona, SW USA
Akaba *see* Al 'Aqabah
Akamagaseki *see* Shimonoseki
Akasha *72 B3* Northern, N Sudan
Akchâr *74 C2 desert* W Mauritania
Aken *see* Aachen
Akermanceaster *see* Bath
Akhalts'ikhe *117 F2 prev.* Akhalts'ikhe.
SW Georgia
Akhalts'ikhe *see* Akhaltsikhe
Akhisar *116 A3* Manisa, W Turkey
(Türkiye)
Akhmîm *72 B2 var.* Akhmîm; *anc.*
Panopolis. C Egypt
Akhtubinsk *111 C7* Astrakhanskaya
Oblast', SW Russia
Akhtyrka *see* Okhtyrka
Akimiski Island *38 C3 island* Nunavut,
C Canada
Akinovka *109 F4* Zaporiz'ka Oblast',
S Ukraine
Akita *130 D4* Akita, Honshū, C Japan
Akjoujt *74 C2 prev.* Fort-Repoux.
Inchiri, W Mauritania
Akkerman *see* Bilhorod-Dnistrovskyi
Akkeshi *130 E2* Hokkaidō, NE Japan
Aklavik *36 D3* Northwest Territories,
NW Canada
Akmola *see* Astana

Akmolinsk *see* Astana
Aknavásár *see* Târgu Ocna
Akpatok Island *39 E1 island* Nunavut, E Canada
Akragas *see* Agrigento
Akron *40 D4* Ohio, N USA
Akrotiri Sovereign Base Area *102 C5 var.* Akrotírion. *UK Sovereign Base Area* S Cyprus
Akrotírion *see* Akrotiri Sovereign Base Area
Aksai Chin *124 B2 Chin.* Aksayqin. *disputed region* China/India
Aksaray *116 C4* Aksaray, C Turkey (Türkiye)
Akşehir *116 B4* Konya, W Turkey (Türkiye)
Aksayqin *see* Aksai Chin
Aktash *see* Oqtosh
Aktau *see* Aqtaū
Aktjubinsk/Aktyubinsk *see* Aqtöbe
Aktobe *see* Aqtöbe
Aktsyabrski *107 C7 Rus.* Oktyabr'skiy; *prev.* Karpilovka. Homyel'skaya Voblasts', SE Belarus
Aktyubinsk *see* Aqtöbe
Akula *77 C5* Equateur, NW Dem. Rep. Congo
Akureyri *83 E4* Nordhurland Eystra, N Iceland
Akyab *see* Sittwe
Alabama *42 C2 off.* State of Alabama, *also known as* Camellia State, Heart of Dixie, The Cotton State, Yellowhammer State. *state* S USA
Alabama River *42 C3 river* Alabama, S USA
Alaca *116 C3* Çorum, N Turkey (Türkiye)
Alacant *93 F4 var.* Alicante, *Lat.* Lucentum. Comunitat Valenciana, SE Spain
Alagoas *63 G2 off.* Estado de Alagoas. *state/region* E Brazil
Alais *see* Alès
Alajuela *53 E4* Alajuela, C Costa Rica
Alakanuk *36 C2* Alaska, USA
Al 'Alamayn *72 B1 var.* El 'Alamein. N Egypt
Al 'Amārah *120 C3 var.* Amara. Maysān, E Iraq
Alamo *47 D6* Nevada, W USA
Alamogordo *48 D3* New Mexico, SW USA
Alamosa *44 C5* Colorado, C USA
Åland *85 C6 var.* Aland Islands, *Fin.* Ahvenanmaa. *island group* SW Finland
Aland Islands *see* Åland
Aland Sea *see* Ålands Hav
Ålands Hav *85 C6 var.* Aland Sea. *strait* Baltic Sea/Gulf of Bothnia
Alanya *116 C4* Antalya, S Turkey (Türkiye)
Alappuzha *132 C3 var.* Alleppey. Kerala, SW India
Al 'Aqabah *119 B8 var.* Akaba, Aqaba, 'Aqaba; *anc.* Aelana, Elath. Al 'Aqabah, SW Jordan
Al 'Arabiyah as Su'ūdīyah *see* Saudi Arabia
Alasca, Golfo de *see* Alaska, Gulf of
Alaşehir *116 A4* Manisa, W Turkey (Türkiye)
Al 'Ashārah *118 E3 var.* Ashara. Dayr az Zawr, E Syria
Alaska *36 C3 off.* State of Alaska, *also known as* Land of the Midnight Sun, The Last Frontier, Seward's Folly; *prev.* Russian America. *state* NW USA
Alaska, Gulf of *36 C4 var.* Golfo de Alasca. *gulf* Canada/USA
Alaska Peninsula *36 C3 peninsula* Alaska, USA
Alaska Range *34 B2 mountain range* Alaska, USA
Al-Asnam *see* Chlef
Alattio *see* Alta
Al Awaynāt *see* Al 'Uwaynāt
Al 'Awjā *119 E7* E West Bank, SW Asia
Al 'Aynā *119 B7* Al Karak, W Jordan
Alazeya *115 G2 river* NE Russia
Al Bāb *118 B2* Ḥalab, N Syria
Albacete *93 E3* Castilla-La Mancha, C Spain
Al Baghdādī *120 B3 var.* Ḥadīthah. Al Anbār, SW Iraq
Al Bāha *see* Al Bāḥah
Al Bāḥah *121 B5 var.* Al Bāha. Al BāḤah, SW Saudi Arabia

Al Baḥrayn *see* Bahrain
Alba Iulia *108 B4 Ger.* Weissenburg, *Hung.* Gyulafehérvár; *prev.* Bálgrad, Karlsburg, Károly-Fehérvár. Alba, W Romania
Albania *101 C7 off.* Republic of Albania, *Alb.* Republika e Shqipërisë, Shqipëria; *prev.* People's Socialist Republic of Albania. *country* SE Europe
Albania *see* Aubagne
Albany *147 B7* Western Australia
Albany *42 D3* Georgia, SE USA
Albany *41 F3 state capital* New York, NE USA
Albany *46 B3* Oregon, NW USA
Albany *38 C3 river* Ontario, S Canada
Alba Regia *see* Székesfehérvár
Al Bāridah *118 C4 var.* Bāridah. Ḥimṣ, C Syria
Al Baṣrah *120 C3 Eng.* Basra, *hist.* Busra, Bussora. Al Baṣrah, SE Iraq
Al Batrūn *see* Batroûn
Al Bawītī *72 B2 var.* Bawīti C Egypt
Al Baydā' *71 G2 var.* Beida. NE Libya
Albemarle Island *see* Isabela, Isla
Albemarle Sound *43 G1 inlet* W Atlantic Ocean
Albergaria-a-Velha *92 B2* Aveiro, N Portugal
Albert *90 C3* Somme, N France
Alberta *37 E4 province* SW Canada
Albert Edward Nyanza *see* Edward, Lake
Albert, Lake *73 B6 var.* Albert Nyanza, Lac Mobutu Sese Seko. *lake* Uganda/ Dem. Rep. Congo
Albert Lea *45 F3* Minnesota, N USA
Albert Nyanza *see* Albert, Lake
Albertville *see* Kalemie
Albi *91 C6 anc.* Albiga. Tarn, S France
Albiga *see* Albi
Ålborg *see* Aalborg
Ålborg-Nørresundby *see* Aalborg
Alborz, Reshteh-ye Kūhhā-ye *120 C2 Eng.* Elburz Mountains. *mountain range* N Iran
Albuquerque *48 D2* New Mexico, SW USA
Al Burayqah *see* Marsá al Burayqah
Alburgum *see* Aalborg
Albury *149 C7* New South Wales, SE Australia
Al Buşayrah *118 D3* Dayr az Zawr, E Syria
Alcácer do Sal *92 B4* Setúbal, W Portugal
Alcalá de Henares *93 E3 Ar.* Alkal'a; *anc.* Complutum. Madrid, C Spain
Alcamo *97 C7* Sicilia, Italy, C Mediterranean Sea
Alcañiz *93 F2* Aragón, NE Spain
Alcántara, Embalse de *72 C3 reservoir* W Spain
Alcaudete *92 D4* Andalucía, S Spain
Alcázar *see* Ksar-el-Kebir
Alcazarquivir *see* Ksar-el-Kebir
Alcoi *93 F4 var.* Alcoy. Comunitat Valenciana, E Spain
Alcoy *see* Alcoi
Aldabra Group *79 G2 island group* SW Seychelles
Aldan *115 F3 river* NE Russia
al Dar al Baida *see* Rabat
Alderney *90 A2 island* Channel Islands
Aleg *74 C3* Brakna, SW Mauritania
Aleksandriya *see* Oleksandriia
Aleksandropol' *see* Gyumri
Aleksandrovka *see* Oleksandrivka
Aleksandrovsk *see* Zaporizhzhia
Aleksin *111 B5* Tul'skaya Oblast', W Russia
Aleksinac *100 E4* Serbia, SE Serbia
Alençon *90 B3* Orne, N France
Alenquer *63 E2* Pará, NE Brazil
Alep/Aleppo *see* Ḥalab
Alert *37 F1* Ellesmere Island, Nunavut, N Canada
Alès *91 C6 prev.* Alais. Gard, S France
Aleşd *108 B3 Hung.* Élesd. Bihor, SW Romania
Alessandria *96 B2 Fr.* Alexandrie. Piemonte, N Italy
Ålesund *85 A5* Møre og Romsdal, S Norway
Aleutian Basin *113 G3 undersea basin* Bering Sea
Aleutian Islands *36 A3 island group* Alaska, USA
Aleutian Range *34 A2 mountain range* Alaska, USA
Aleutian Trench *113 H3 trench* S Bering Sea

Alexander Archipelago *36 D4 island group* Alaska, USA
Alexander City *42 D2* Alabama, S USA
Alexander Island *154 A3 island* Antarctica
Alexander Range *see* Kirghiz Range
Alexandra *151 B7* Otago, South Island, New Zealand
Alexándreia *104 B4 var.* Alexándria. Kentrikí Makedonía, N Greece
Alexandretta *see* İskenderun
Alexandretta, Gulf of *see* İskenderun Körfezi
Alexandria *see* Al Iskandariyah
Alexandria *108 C5* Teleorman, S Romania
Alexandria *42 B3* Louisiana, S USA
Alexandria *45 F2* Minnesota, N USA
Alexándria *see* Alexándreia
Alexandrie *see* Alessandria
Alexandroúpoli *104 D3 var.* Alexandroúpolis, *Turk.* Dedeagaç, Dedeagach. Anatolikí Makedonía kai Thráki, NE Greece
Alexandroúpolis *see* Alexandroúpoli
Al Fāshir *see* El Fasher
Alfatar *104 E1* Silistra, NE Bulgaria
Alfeiós *105 B6 var.* Alfiós; *anc.* Alpheius, Alpheus. *river* S Greece
Alfiós *see* Alfeiós
Alföld *see* Great Hungarian Plain
Al-Furāt *see* Euphrates
Alga *114 B4 Kaz.* Algha. Aktyubinsk, NW Kazakhstan
Algarve *92 B4 cultural region* S Portugal
Algeciras *92 C5* Andalucía, SW Spain
Algemesí *93 F3* Comunitat Valenciana, E Spain
Alger *71 E1 var.* Algiers, El Djazaïr, Al Jazair. *country capital* (Algeria) N Algeria
Algeria *70 C3 off.* People's Democratic Republic of Algeria, *prev.* Democratic and Popular Republic of Algeria. *country* N Africa
Algeria, Democratic and Popular Republic of *see* Algeria
Algeria, People's Democratic Republic of *see* Algeria
Algerian Basin *80 C5 var.* Balearic Plain. *undersea basin* W Mediterranean Sea
Algha *see* Alga
Al Ghābah *121 E5 var.* Ghaba. C Oman
Alghero *97 A5* Sardegna, Italy, C Mediterranean Sea
Al Ghardaqah *72 C2 var.* Hurghada, Ghurdaqah. E Egypt
Algiers *see* Alger
Al Golea *see* El Goléa
Algona *45 F3* Iowa, C USA
Al Ḥajar al Gharbī *121 D5 mountain range* N Oman
Al Hamad *see* Syrian Desert
Al Ḥasakah *118 D2 var.* Al Hasijah, El Haseke, Fr. Hassetché. Al Ḥasakah, NE Syria
Al Hasijah *see* Al Ḥasakah
Al Ḥillah *120 B3 var.* Hilla. Bābil, C Iraq
Al Ḥisā *119 B7* Aṭ Ṭafīlah, W Jordan
Al Ḥudayda *121 B6 var.* Hodeidah. W Yemen
Al Ḥufūf *120 C4 var.* Hofuf. Ash Sharqīyah, NE Saudi Arabia
Al Imārāt al 'Arabīyah Muttaḥidah *see* United Arab Emirates
Alindao *76 C4* Basse-Kotto, S Central African Republic
Aliquippa *40 D4* Pennsylvania, NE USA
Al Iskandariyah *72 B1 var.* Alexandria. N Egypt
Al Ismā'īliya *72 B1 var.* Ismailia, Ismâ'iliya. N Egypt
Alistráti *104 C3* Kentrikí Makedonía, NE Greece
Alivéri *105 C5 var.* Alivérion. Évvoia, C Greece

Alivérion *see* Alivéri
Al Jabal al Akhḍar *71 G2 mountain range* NE Libya
Al Jafr *119 B7* Ma'ān, S Jordan
Al Jaghbūb *71 H3* NE Libya
Al Jahrā' *120 C4 var.* Al Jahrah, Jahra. C Kuwait
Al Jahrah *see* Al Jahrā'
Al Jawf *120 B4 off.* Jauf. Al Jawf, NW Saudi Arabia
Al Jawlān *see* Golan Heights
Al Jazair *see* Alger
Al Jizah *118 E2 physical region* Iraq/Syria
Al Jīzah *72 B1 var.* El Gîza, Gizeh; Eng. Giza. N Egypt
Al Junaynah *see* El Geneina
Alkal'a *see* Alcalá de Henares
Al Karak *119 B7 var.* El Kerak, Karak, Kerak; *anc.* Kir Moab, Kir of Moab. Al Karak, W Jordan
Al-Kasr al-Kebir *see* Ksar-el-Kebir
Al Khalīl *see* Hebron
Al Khārjah *72 B2 var.* El Khârga; Eng. Kharga. C Egypt
Al Khums *71 F2 var.* Homs, Khoms, Khums. NW Libya
Alkmaar *86 C2* Noord-Holland, NW Netherlands
Al Kufrah *71 H4* SE Libya
Al Kūt *120 C3 var.* Kūt al 'Amārah, Kut al Imara. Wāsiṭ, E Iraq
Al-Kuwait *see* Al Kuwayt
Al Kuwayt *120 C4 var.* Al-Kuwait, Madīnat al Kuwayt, Eng. Kuwait, Kuwait City; *prev.* Qurein. *country capital* (Kuwait) E Kuwait
Al Lādhiqīyah *118 A3 Eng.* Latakia, Fr. Lattaquié; *anc.* Laodicea, Laodicea ad Mare. Al Lādhiqīyah, W Syria
Allahābād *see* Prayagraj
Allanmyo *see* Aunglan
Allegheny Plateau *41 E3 mountain range* New York/Pennsylvania, NE USA
Allenstein *see* Olsztyn
Allentown *41 F4* Pennsylvania, NE USA
Alleppey *see* Alappuzha
Alliance *44 D3* Nebraska, C USA
Al Līth *121 B5* Makkah, SW Saudi Arabia
Alma-Ata *see* Almaty
Almada *92 B4* Setúbal, W Portugal
Al Madīnah *121 A5 Eng.* Medina. Al Madīnah, W Saudi Arabia
Al Mafraq *119 B6 var.* Mafraq. Al Mafraq, N Jordan
Al Maghrib *see* Morocco
Al Mahdīyah *see* Mahdia
Al Mahrah *121 C6 mountain range* E Yemen
Al Majma'ah *120 B4* Ar Riyāḍ, C Saudi Arabia
Al Mālikīyah *118 E1 var.* Malkiye. Al Ḥasakah, N Syria
Almalyk *see* Olmaliq
Al Mamlaka al Urdunīya al Hashemīyah *see* Jordan
Al Manāmah *120 C4 Eng.* Manama. *country capital* (Bahrain) N Bahrain
Al Manāṣif *118 E3 mountain range* E Syria
Almansa *93 F4* Castilla-La Mancha, C Spain
Al-Mariyya *see* Almería
Al Marj *71 G2 var.* Barka, It. Barce. NE Libya
Almaty *114 C5 var.* Alma-Ata. Almaty, SE Kazakhstan
Al Mawṣil *120 B2 Eng.* Mosul. Nīnawá, N Iraq
Al Mayādīn *118 D3 var.* Mayadin, Fr. Meyadine. Dayr az Zawr, E Syria
Al Mazra'a *see* Al Mazra'ah
Al Mazra'ah *119 B6 var.* Al Mazra', Mazra'a. Al Karak, W Jordan
Almelo *86 E3* Overijssel, E Netherlands
Almendra, Embalse de *92 C2 reservoir* Castilla y León, NW Spain
Almendralejo *92 C4* Extremadura, W Spain
Almere *86 C3 var.* Almere-stad. Flevoland, C Netherlands
Almere-stad *see* Almere
Almería *93 E4 Ar.* Al-Mariyya; *anc.* Unci, Lat. Portus Magnus. Andalucía, S Spain
Al Mīnā' *see* El Mina
Al Minyā *72 B2 var.* El Minya, Minya. C Egypt

Almirante 53 E4 Bocas del Toro, NW Panama
Al Mudawwarah 119 B8 Ma'ān, SW Jordan
Al Mukallā 121 C6 var. Mukalla. SE Yemen
Al Obayyid see El Obeid
Alofi 145 E4 dependent territory capital (Niue) W Niue
Aloha State see Hawaii
Aloja 106 D3 Limbaži, N Latvia
Alónnisos 105 C5 island Vóreies Sporádes, Greece, Aegean Sea
Álora 92 D5 Andalucía, S Spain
Alor, Kepulauan 139 E5 island group E Indonesia
Al Oued see El Oued
Alpen see Alps
Alpena 40 D2 Michigan, N USA
Alpes see Alps
Alpha Cordillera 155 B3 var. Alpha Ridge. seamount range Arctic Ocean
Alpha Ridge see Alpha Cordillera
Alpheius see Alfeiós
Alphen see Alphen aan den Rijn
Alphen aan den Rijn 86 C3 var. Alphen. Zuid-Holland, C Netherlands
Alpheus see Alfeiós
Alpi see Alps
Alpine 49 E4 Texas, SW USA
Alps 102 C1 Fr. Alpes, Ger. Alpen, It. Alpi. mountain range C Europe
Al Qadārif see Gedaref
Al Qāhirah 72 B2 var. El Qâhira, Eng. Cairo. country capital (Egypt) N Egypt
Al Qāmishli 118 E1 var. Kamishli, Qamishly.
Al Ḥasakah, NE Syria
Al Qaşrayn see Kasserine
Al Qayrawān see Kairouan
Al-Qṣar al-Kbir see Ksar-el-Kebir
Al Qubayyāt see Qoubaïyât
Al Quds/Al Quds ash Sharif see Jerusalem
Alqueva, Barragem do 92 C4 reservoir Portugal/Spain
Al Qunayṭirah 119 B5 var. El Kuneitra, El Quneitra, Kuneitra, Qunaytra. Al Qunayṭirah, SW Syria
Al Quşayr 118 B4 var. El Quseir, Quşayr, Fr. Kousseir. Ḥimṣ, W Syria
Al Quwayrah 119 B8 var. El Quweira. Al 'Aqabah, SW Jordan
Alsace 91 E3 Ger. Elsass; anc. Alsatia. cultural region NE France
Alsatia see Alsace
Alsdorf 94 A4 Nordrhein-Westfalen, W Germany
Alt see Olt
Alta 84 D2 Fin. Alattio. Finnmark, N Norway
Altai see Altai Mountains
Altai Mountains 126 C2 var. Altai, Chin. Altay Shan, Rus. Altay. mountain range Asia/Europe
Altamaha River 43 E3 river Georgia, SE USA
Altamira 63 E2 Pará, NE Brazil
Altamura 97 E5 anc. Lupatia. Puglia, SE Italy
Altar, Desierto de 50 A1 var. Sonoran Desert. desert Mexico/USA
Altar, Desierto de see Sonoran Desert
Altay 126 C2 Xinjiang Uygur Zizhiqu, NW China
Altay 126 D2 prev. Yösönbulag. Govĭ-Altay, W Mongolia
Altay see Altai Mountains
Altay Shan see Altai Mountains
Altbetsche see Bečej
Altenburg see Bucureşti, Romania
Altin Köprü see Altūn Kawbrī
Altiplano 61 F4 physical region W South America
Altkanischa see Kanjiža
Alton 40 B4 Illinois, C USA
Altoona 41 E4 Pennsylvania, NE USA
Alto Paraná see Paraná
Altpasua see Stara Pazova
Alt-Schwanenburg see Gulbene
Altsohl see Zvolen
Altūn Kawbrī 120 B3 var. Altin Köprü, Altun Kupri. Kirkūk, N Iraq
Altun Kupri see Altūn Kawbrī
Altun Shan 126 C3 var. Altyn Tagh. mountain range NW China
Altus 49 F2 Oklahoma, C USA
Altyn Tagh see Altun Shan
Al Ubayyiḍ see El Obeid

Alūksne 106 D3 Ger. Marienburg. Alūksne, NE Latvia
Al 'Ulā 120 A4 Al Madīnah, NW Saudi Arabia
Al 'Umari 119 C6 'Ammān, E Jordan
Alupka 109 F5 Respublika Krym, S Ukraine
Al Uqsur 72 B2 Eng. Luxor. E Egypt
Al Urdunn see Jordan
Alushta 109 F5 Respublika Krym, S Ukraine
Al 'Uwaynāt 71 F4 var. Al Awaynāt. SW Libya
Alva 49 F1 Oklahoma, C USA
Alvarado 51 F4 Veracruz-Llave, E Mexico
Alvin 49 H4 Texas, SW USA
Al Wajh 120 A4 Tabūk, NW Saudi Arabia
Alwar 134 D3 Rājasthān, N India
Al Wari'ah 120 C4 Ash Sharqīyah, N Saudi Arabia
Al Yaman see Yemen
Alykí 104 C4 var. Aliki. Thásos, N Greece
Alytus 107 B5 Pol. Olita. Alytus, S Lithuania
Alzette 87 D8 river S Luxembourg
Amadeus, Lake 147 D5 seasonal lake Northern Territory, C Australia
Amadi 73 B5 W Equatoria, S South Sudan
Amadjuak Lake 37 G3 lake Baffin Island, Nunavut, N Canada
Amakusa-nada 131 A7 gulf SW Japan
Åmål 85 B6 Västra Götaland, S Sweden
Amami-gunto 130 A3 island group SW Japan
Amami-o-shima 130 A3 island S Japan
Amantea 97 D6 Calabria, SW Italy
Amapá 63 E1 off. Estado de Amapá; prev. Território de Amapá. state/region NE Brazil
Amapá, Estado de see Amapá
Amapá, Território de see Amapá
Amara see Al 'Amārah
Amarapura 136 B3 Mandalay, C Myanmar (Burma)
Amarillo 49 E2 Texas, SW USA
Amay 87 C6 Liège, E Belgium
Amazon 63 E1 Sp. Amazonas. river Brazil/Peru
Amazonas see Amazon
Amazon Basin 62 D2 basin N South America
Amazonía 58 B4 region S Colombia
Amazon, Mouths of the 63 F1 delta NE Brazil
Ambam 77 B5 Sud, S Cameroon
Ambanja 79 G2 Antsiranana, N Madagascar
Ambarchik 115 G2 Respublika Sakha (Yakutiya), NE Russia
Ambato 60 B1 Tungurahua, C Ecuador
Ambérieu-en-Bugey 91 D5 Ain, E France
Ambianum see Amiens
Amboasary 79 F4 Toliara, S Madagascar
Amboina see Ambon
Ambon 139 F4 prev. Amboina, Amboyna. Pulau Ambon, E Indonesia
Ambositra 79 G3 Fianarantsoa, SE Madagascar
Amboyna see Ambon
Ambracia see Árta
Ambre, Cap d' see Bobaomby, Tanjona
Ambrim see Ambrym
Ambriz 78 A1 Bengo, NW Angola
Ambrym 144 D4 var. Ambrim. island C Vanuatu
Amchitka Island 36 A2 island Aleutian Islands, Alaska, USA
Amdo 126 C5 Xizang Zizhiqu, W China
Ameland 86 D1 Fris. It Amelân. island Waddeneilanden, N Netherlands
Amelân, It see Ameland
America see United States of America
America-Antarctica Ridge 67 C7 undersea ridge S Atlantic Ocean
America in Miniature see Maryland
American Falls Reservoir 46 E4 reservoir Idaho, NW USA
American Samoa 145 E4 US unincorporated territory W Polynesia
Amersfoort 86 D3 Utrecht, C Netherlands
Ames 45 F3 Iowa, C USA
Amfilochía 105 A5 var. Amfilokhía. Dytikí Ellás, C Greece
Amfilokhía see Amfilochía

Amga 115 F3 river NE Russia
Amherst 39 F4 Nova Scotia, SE Canada
Amherst see Kyaikkami
Amida see Diyarbakır
Amiens 90 C3 anc. Ambianum, Samarobriva. Somme, N France
Amíndaion/Amindeo see Amýntaio
Amindivi Islands 132 A2 island group Lakshadweep, India, N Indian Ocean
Amirante Islands 79 G1 var. Amirantes Group. island group C Seychelles
Amirantes Group see Amirante Islands
Amistad, Presa de la see Amistad Reservoir
Amistad Reservoir 49 F4 var. Presa de la Amistad. reservoir Mexico/USA
Amisus see Samsun
Ammaia see Portalegre
'Ammān 119 B6 anc. Philadelphia, Bibl. Rabbah Ammon, Rabbath Ammon. country capital (Jordan) 'Ammān, NW Jordan
Ammassalik see Tasiilaq
Ammóchostos see Gazimağusa
Ammóchostos, Kólpos see Gazimağusa Körfezi
Amnok-kang see Yalu
Amoea see Portalegre
Amoentai see Amuntai
Åmol 120 D2 var. Amul. Māzandarān, N Iran
Amorgós 105 D6 Amorgós, Kykládes, Greece, Aegean Sea
Amorgós 105 D6 island Kykládes, Greece, Aegean Sea
Amos 38 D4 Québec, SE Canada
Amourj 74 D3 Hodh ech Chargui, SE Mauritania
Amoy see Xiamen
Ampato, Nevado 61 E4 mountain S Peru
Amposta 93 F2 Cataluña, NE Spain
Amraoti see Amrāvati
Amrāvati 134 D4 prev. Amraoti. Mahārāshtra, C India
Amritsar 134 D2 Punjab, N India
Amstelveen 86 C3 Noord-Holland, C Netherlands
Amsterdam 86 C3 country capital (Netherlands) Noord-Holland, C Netherlands
Amsterdam Island 141 C6 island NE French Southern and Antarctic Lands
Am Timan 76 C3 Salamat, SE Chad
Amu Darya 122 D2 Rus. Amudar'ya, Taj. Dar''yoi Amu, Turkm. Amyderya, Uzb. Amudaryo; anc. Oxus. river C Asia
Amu-Dar'ya see Amyderýa
Amudar'ya/Amudaryo/Amu, Dar''yoi see Amu Darya
Amul see Āmol
Amund Ringnes Island 37 F2 island Nunavut, N Canada
Amundsen Basin see Fram Basin
Amundsen Plain 154 A4 abyssal plain S Pacific Ocean
Amundsen-Scott 154 B3 US research station Antarctica
Amundsen Sea 154 A4 sea S Pacific Ocean
Amuntai 138 D4 prev. Amoentai. Borneo, C Indonesia
Amur 115 G4 Chin. Heilong Jiang. river China/Russia
Amvrosiivka 109 H3 prev. Amvrosiyivka, Rus. Amvrosiyevka. Donets'ka Oblast', SE Ukraine
Amvrosiyevka see Amvrosiivka
Amvrosiyivka see Amvrosiivka
Amyderýa 123 E3 Rus. Amu-Dar'ya. Lebap Welaýaty, NE Turkmenistan
Amyderya see Amu Darya
Amýntaio 104 B4 var. Amindeo; prev. Amíndaion. Dytikí Makedonía, N Greece
Anabar 115 E2 river NE Russia
An Abhainn Mhór see Blackwater
Anaco 59 E2 Anzoátegui, NE Venezuela
Anaconda 44 B2 Montana, NW USA
Anacortes 46 B1 Washington, NW USA
Anadolu Dağları see Doğu Karadeniz Dağları
Anadyr' 115 H1 Chukotskiy Avtonomnyy Okrug, NE Russia
Anadyr' 115 H1 Eng. Gulf of Anadyr. gulf NE Russia
Anadyr, Gulf of see Anadyrskiy Zaliv
Anadyrskiy Zaliv 115 H1 Eng. Gulf of Anadyr. gulf NE Russia

Anáfi 105 D7 anc. Anaphe. island Kykládes, Greece, Aegean Sea
'Ānah see 'Annah
Anaheim 46 E2 California, W USA
Anaiza see 'Unayzah
Analalava 79 G2 Mahajanga, NW Madagascar
Anamur 116 C5 İçel, S Turkey (Türkiye)
Anantapur 132 C2 Andhra Pradesh, S India
Anaphe see Anáfi
Anápolis 63 F3 Goiás, C Brazil
Anār 120 D3 Kermān, C Iran
Anatolia 116 C4 plateau C Turkey (Türkiye)
Anatom see Aneityum
Añatuya 64 C3 Santiago del Estero, N Argentina
An Bhearú see Barrow
Anchorage 36 C3 Alaska, USA
Ancona 96 C3 Marche, C Italy
Ancud 65 B6 prev. San Carlos de Ancud. Los Lagos, S Chile
Ancyra see Ankara
Åndalsnes 85 A5 Møre og Romsdal, S Norway
Andalucía 92 D4 cultural region S Spain
Andalusia 42 C3 Alabama, S USA
Andaman Islands 124 B4 island group India, NE Indian Ocean
Andaman Sea 124 C4 sea NE Indian Ocean
Andenne 87 C6 Namur, SE Belgium
Anderlues 87 B7 Hainaut, S Belgium
Anderson 40 C4 Indiana, N USA
Andes 64 B3 mountain range W South America
Andhra Pradesh 135 E5 cultural region E India
Andijon 123 F2 Rus. Andizhan. Andijon Viloyati, E Uzbekistan
Andikíthira see Antikýthira
Andípaxi see Antípaxoi
Andípsara see Antípsara
Ándissa see Ántissa
Andizhan see Andijon
Andkhvoy 122 D3 Fāryāb, N Afghanistan
Andorra 91 A7 off. Principality of Andorra, Cat. Valls d'Andorra, Fr. Vallée d'Andorre. country SW Europe
Andorra la Vella 91 A8 var. Andorra, Fr. Andorre la Vieille, Sp. Andorra la Vieja. country capital (Andorra) C Andorra
Andorra la Vieja see Andorra la Vella
Andorra, Principality of see Andorra
Andorra, Valls d'/Andorra, Vallée d' see Andorra
Andorre la Vielle see Andorra la Vella
Andover 89 D7 S England, United Kingdom
Andøya 84 C2 island C Norway
Andreanof Islands 36 A3 island group Aleutian Islands, Alaska, USA
Andrews 49 E3 Texas, SW USA
Andrew Tablemount 140 A4 var. Gora Andryu. seamount N Indian Ocean
Andria 97 D5 Puglia, SE Italy
An Droichead Nua see Newbridge
Andropov see Rybinsk
Ándros 105 D6 Ándros, Kykládes, Greece, Aegean Sea
Ándros 105 C6 island Kykládes, Greece, Aegean Sea
Andros Island 54 B2 island NW The Bahamas
Andros Town 54 C1 Andros Island, NW The Bahamas
Andryu, Gora see Andrew Tablemount
Aneityum 144 D5 var. Anatom; prev. Kéamu. island S Vanuatu
Ånewetak see Enewetak Atoll
Angara 115 E4 river C Russia
Angarsk 115 E4 Irkutskaya Oblast', S Russia
Ånge 85 C5 Västernorrland, C Sweden
Ángel de la Guarda, Isla 50 B2 island NW Mexico
Angeles 139 E1 off. Angeles City. Luzon, N Philippines
Angeles City see Angeles
Angel Falls 59 E3 Eng. Angel Falls. waterfall E Venezuela
Angel Falls see Ángel, Salto
Angerburg see Węgorzewo
Ångermanälven 84 C4 river N Sweden

Angermünde *94 D3* Brandenburg,
NE Germany
Angers *90 B4* *anc.* Juliomagus.
Maine-et-Loire, NW France
Anglesey *89 C5* *island* NW Wales,
United Kingdom
Anglet *91 A6* Pyrénées-Atlantiques,
SW France
Angleton *49 H4* Texas, SW USA
Anglia *see* England
Anglo-Egyptian Sudan *see* Sudan
Angmagssalik *see* Tasiilaq
Ang Nam Ngum *136 C4* *lake* C Laos
Angola *78 B2* *off.* Republic of Angola;
prev. People's Republic of Angola,
Portuguese West Africa. *country*
SW Africa
Angola Basin *69 B5* *undersea basin*
E Atlantic Ocean
Angola, People's Republic of *see*
Angola
Angola, Republic of *see* Angola
Angostura *see* Ciudad Bolívar
Angostura, Presa de la *51 G5* *reservoir*
SE Mexico
Angoulême *91 B5* *anc.* Iculisma.
Charente, W France
Angoumois *91 B5* *cultural region*
W France
Angra Pequena *see* Lüderitz
Angren *123 F2* Toshkent Viloyati,
E Uzbekistan
Anguilla *55 G3* *UK Overseas Territory*
E West Indies
Anguilla Cays *54 B2* *islets* SW The
Bahamas
Anhui *129 C5* *var.* Anhui Sheng,
Anhwei, Wan. *province* E China
AnhuiSheng/Anhwei Wan *see* Anhui
Anicium *see* le Puy
Anina *108 A4* *Ger.* Steierdorf, *Hung.*
Stájerlakanina; *prev.* Ştaierdorf-Anina,
Steierdorf-Anina, Steyerlak-Anina.
Caraş-Severin, SW Romania
Anjou *90 B4* *cultural region* NW France
Anjouan *79 F2* *var.* Ndzouani, Nzwani.
island SE Comoros
Ankara *116 C3* *prev.* Angora; *anc.*
Ancyra. *country capital* (Turkey)
Ankara, C Turkey (Türkiye)
Ankeny *45 F3* Iowa, C USA
Anklam *94 D2* Mecklenburg-
Vorpommern, NE Germany
An Mhuir Cheilteach *see* Celtic Sea
Annaba *71 E1* *prev.* Bône. NE Algeria
An Nafud *120 B4* *desert*
NW Saudi Arabia
'Annah *120 B3* *var.* 'Ānah. Al Anbār,
NW Iraq
An Najaf *120 B3* *var.* Najaf. An Najaf,
S Iraq
Annamite Mountains *136 D4* *Fr.*
Annamitique, Chaîne. *mountain range*
C Laos
Annamitique, Chaîne *see* Annamite
Mountains
Annapolis *41 F4* *state capital* Maryland,
NE USA
Annapurna *135 E3* *mountain* C Nepal
An Nâqûrah *see* En Nâqoûra
Ann Arbor *40 C3* Michigan, N USA
An Nāşirīyah *120 C3* *var.* Nasiriya. Dhī
Qār, SE Iraq
Anneciacum *see* Annecy
Annecy *91 D5* *anc.* Anneciacum.
Haute-Savoie, E France
An Nil al Abyaḍ *see* White Nile
An Nil al Azraq *see* Blue Nile
Anniston *42 D2* Alabama, S USA
Annotto Bay *54 B4* C Jamaica
An Nuway'imah *119 E7* E West Bank,
SW Asia
An Omaigh *see* Omagh
Anqing *129 D5* Anhui, E China
Anse La Raye *55 F1* NW Saint Lucia
Anshun *129 B6* Guizhou, S China
Ansongo *75 E3* Gao, E Mali
An Srath Bán *see* Strabane
Antakya *116 D4* *anc.* Antioch,
Antiochia. Hatay, S Turkey (Türkiye)
Antalaha *79 G2* Antsiranana,
NE Madagascar
Antalya *116 B4* *prev.* Adalia; *anc.*
Attaleia, *Bibl.* Attalia. Antalya,
SW Turkey (Türkiye)
Antalya, Gulf of *116 B4* *var.* Gulf of
Adalia, *Eng.* Gulf of Antalya. *gulf*
SW Turkey (Türkiye)
Antalya, Gulf of *see* Antalya Körfezi

Antananarivo *79 G3* *prev.* Tananarive.
country capital (Madagascar)
Antananarivo, C Madagascar
Antarctica *154 B3* *continent*
Antarctic Peninsula *154 A2* *peninsula*
Antarctica
Antep *see* Gaziantep
Antequera *92 D5* *anc.* Anticaria,
Antiquaria. Andalucía, S Spain
Antequera *see* Oaxaca
Antibes *91 D6* *anc.* Antipolis.
Alpes-Maritimes, SE France
Anticaria *see* Antequera
Anticosti, Île d' *39 F3* *Eng.* Anticosti
Island. *island* Québec, E Canada
Anticosti Island *see* Anticosti, Île d'
Antigua *55 G3* *island* S Antigua and
Barbuda, Leeward Islands
Antigua and Barbuda *55 G3* *country*
E West Indies
Antikythira *105 B7* *var.* Andikíthira.
island S Greece
Anti-Lebanon *118 B4* *var.* Jebel esh
Sharqi, *Ar.* Al Jabal ash Sharqī, *Fr.*
Anti-Liban. *mountain range* Lebanon/
Syria
Anti-Liban *see* Anti-Lebanon
Antioch *see* Antakya
Antiochia *see* Antakya
Antípaxoi *105 A5* *var.* Andipaxi. *island*
Iónia Nísiá, Greece,
C Mediterranean Sea
Antipodes Islands *142 D5* *island group*
S New Zealand
Antipolis *see* Antibes
Antípsara *105 D5* *var.* Andípsara. *island*
E Greece
Antiquaria *see* Antequera
Ántissa *105 D5* *var.* Ándissa. Lésvos,
E Greece
An tIúr *see* Newry
Antivari *see* Bar
Antofagasta *64 B2* Antofagasta, N Chile
Antony *90 B2* Hauts-de-Seine, N France
An tSionainn *see* Shannon
Antsirañana *79 G2* *province*
N Madagascar
Antsohihy *79 G2* Mahajanga,
NW Madagascar
An-tung *see* Dandong
Antwerpen *87 C5* *Eng.* Antwerp, *Fr.*
Anvers. Antwerpen, N Belgium
Anuradhapura *132 D3* North Central
Province, C Sri Lanka
Anvers *see* Antwerpen
Anyang *128 C4* Henan, C China
A'nyêmaqên Shan *126 D4* *mountain
range* C China
Anykščiai *106 C4* Utena, E Lithuania
Anzio *97 C5* Lazio, C Italy
Aomori *130 D3* Aomori, Honshū,
C Japan
Aóos *see* Vjosës, Lumi i
Aoraki *151 B6* *var.* Mount Cook. South
Island, New Zealand
Aoraki *151 B6* *prev.* Aorangi, *var.* Mount
Cook. *mountain* South Island, New
Zealand
Aorangi *see* Aoraki
Aosta *96 A1* *anc.* Augusta Praetoria.
Valle d'Aosta, NW Italy
Aotearoa *see* New Zealand
Aoukâr *74 D3* *var.* Aouker. *plateau*
C Mauritania
Aouk, Bahr *76 C4* *river* Central African
Republic/Chad
Aouker *see* Aoukâr
Aozou *76 C1* Borkou-Ennedi-Tibesti,
N Chad
Apalachee Bay *42 D3* *bay* Florida,
SE USA
Apalachicola River *42 D3* *river* Florida,
SE USA
Apamama *see* Abemama
Apaporis, Río *58 C4* *river* Brazil/
Colombia
Aparima *see* Riverton
Apatity *112 C2* Murmanskaya Oblast',
NW Russia
Ape *106 D3* Alûksne, NE Latvia
Apeldoorn *86 D3* Gelderland,
E Netherlands
Apennines *96 E2* *Eng.* Apennines.
mountain range Italy/San Marino
Apennines *see* Appennino
Äpia *145 F4* *country capital* (Samoa)
Upolu, SE Samoa
Apkhazeti *see* Abkhazia
Apoera *59 G3* Sipaliwini,
NW Surinam

Apostle Islands *40 B1* *island group*
Wisconsin, N USA
Appalachian Mountains *35 D5*
mountain range E USA
Appingedam *86 E1* Groningen,
NE Netherlands
Appleton *40 B2* Wisconsin, N USA
Apulia *see* Puglia
Apure, Río *58 C2* *river* W Venezuela
Apurímac, Río *60 D3* *river* S Peru
Apuseni, Munţii *108 A4* *mountain
range* W Romania
Aqaba/'Aqaba *see* Al 'Aqabah
Aqaba, Gulf of *120 A4* *var.* Gulf of Elat,
Ar. Khalīj al 'Aqabah; *anc.* Sinus
Aelaniticus. *gulf* NE Red Sea
'Aqabah, Khalīj al *see* Aqaba, Gulf of
Āqchah *123 E3* *var.* Āqcheh. Jowzjān,
N Afghanistan
Āqcheh *see* Āqchah
Aqmola *see* Astana
Aqtaū *114 A4* *var.* Aktau; *prev.*
Shevchenko. Mangistau, W Kazakhstan
Aqtöbe *114 B4* *var.* Aktobe; *prev.*
Aktjubinsk, Aktyubinsk. Aktyubinsk,
NW Kazakhstan
Aquae Augustae *see* Dax
Aquae Calidae *see* Bath
Aquae Flaviae *see* Chaves
Aquae Grani *see* Aachen
Aquae Sextiae *see* Aix-en-Provence
Aquae Solis *see* Bath
Aquae Tarbelicae *see* Dax
Aquidauana *63 E4* Mato Grosso do Sul,
S Brazil
Aquila/Aquila degli Abruzzi *see*
L'Aquila
Aquisgranum *see* Aachen
Aquitaine *91 B6* *cultural region*
SW France
'Arabah, Wadi al *119 B7* *Heb.* Ha'Arava.
dry watercourse Israel/Jordan
Arabian Basin *124 A4* *undersea basin*
N Arabian Sea
Arabian Desert *see* Sahara el Sharqīya
Arabian Peninsula *121 B5* *peninsula*
SW Asia
Arabian Sea *124 A3* *sea*
NW Indian Ocean
Arabicus, Sinus *see* Red Sea
'Arabi, Khalīj al *see* Persian Gulf
**'Arabīyah as Su'ūdīyah, Al Mamlakah
al** *see* Saudi Arabia
'Arabīyah Jumhūrīyat, Mişr al *see*
Egypt
Arab Republic of Egypt *see* Egypt
Aracaju *63 G3* *state capital* Sergipe,
E Brazil
Araçuai *63 F3* Minas Gerais, SE Brazil
Arad *119 B7* Southern, S Israel
Arad *108 A4* Arad, W Romania
Arafura Sea *142 A3* *Ind.* Laut Arafura.
sea W Pacific Ocean
Arafuru, Laut *see* Arafura Sea
Aragón *93 E2* *autonomous community*
E Spain
Araguaia, Río *63 E3* *var.* Araguaya. *river*
C Brazil
Araguari *63 F3* Minas Gerais, SE Brazil
Araguaya *see* Araguaia, Río
Ara Jovis *see* Aranjuez
Arāk *120 C3* *prev.* Sultānābād. Markazī,
W Iran
Arakan Yoma *136 A3* *mountain range*
W Myanmar (Burma)
Araks/Arak's *see* Aras
Aral *114 B4* *var.* Aral'sk. Kzylorda,
SW Kazakhstan
Aral Sea *114 C1* *Kaz.* Aral Tengizi, *Rus.*
Aral'skoye More, *Uzb.* Orol Dengizi.
inland sea Kazakhstan/Uzbekistan
Aral'sk *see* Aral
Aral'skoye More/Aral Tengizi *see*
Aral Sea
Aranda de Duero *92 D2* Castilla y León,
N Spain
Arandelovac *100 D4* *prev.* Arandjelovac.
Serbia, C Serbia
Arandjelovac *see* Arandelovac
Aranjuez *92 D3* *anc.* Ara Jovis. Madrid,
C Spain
Araouane *75 E2* Tombouctou, N Mali
'Ar'ar *120 B3* Al Ḥudūd ash Shamālīyah,
NW Saudi Arabia
Ararat, Mount *see* Büyük Ağrı Dağı
Aras *117 E3* *var.* Araxes, *Arm.* Araz
Nehri, *Per.* Rūd-e Aras, *Rus.* Araks;
prev. Araxes. *river* SW Asia

Aras, Rūd-e *see* Aras
Arauca *58 C2* Arauca, NE Colombia
Arauca, Río *58 C2* *river* Colombia/
Venezuela
Arausio *see* Orange
Araxes *see* Aras
Araz Nehri *see* Aras
Arbela *see* Arbīl
Arbīl *120 B2* *var.* Erbil, Irbil, *Kurd.*
Hewlêr; *anc.* Arbela. Arbīl/Hewlêr,
N Iraq
Arbroath *88 D3* *anc.* Aberbrothock.
E Scotland, United Kingdom
Arbuzinka *see* Arbuzynka
Arbuzynka *109 E3* *Rus.* Arbuzinka.
Mykolayivs'ka Oblast', S Ukraine
Arcachon *91 B5* Gironde, SW France
Arcae Remorum *see*
Châlons-en-Champagne
Arcata *46 A4* California, W USA
Archangel *see* Arkhangel'sk
Archangel Bay *see* Chëshskaya Guba
Archidona *92 D5* Andalucía, S Spain
Arco *96 C2* Trentino-Alto Adige, N Italy
Arctic Mid Oceanic Ridge *see* Nansen
Cordillera
Arctic Ocean *155 B3* *ocean*
Arda *104 C3* *var.* Ardas, *Gk.* Ardas.
river Bulgaria/Greece
Ardabīl *120 C2* *var.* Ardebil. Ardabīl,
NW Iran
Ardakān *120 D3* Yazd, C Iran
Ardas *104 D3* *var.* Ardhas, *Bul.* Arda.
river Bulgaria/Greece
Arḍ aş Şawwān *119 C7* *var.* Ardh es
Suwwān. *plain* S Jordan
Ardeal *see* Transylvania
Ardebil *see* Ardabīl
Ardèche *91 C5* *cultural region*
E France
Ardennes *87 C8* *physical region* Belgium/
France
Ardhas *see* Arda/Ardas
Ardh es Suwwān *see* Arḍ aş Şawwān
Ardino *104 D3* Kardzhali, S Bulgaria
Ard Mhacha *see* Armagh
Ardmore *49 G2* Oklahoma, C USA
Arel *see* Arlon
Arelas/Arelate *see* Arles
Arendal *85 A6* Aust-Agder, S Norway
Arensburg *see* Kuressaare
Arenys de Mar *93 G2* Cataluña,
NE Spain
Areópoli *105 B7* *prev.* Areópolis.
Pelopónnisos, S Greece
Areópolis *see* Areópoli
Arequipa *61 E4* Arequipa, SE Peru
Arezzo *96 C3* *anc.* Arretium. Toscana,
C Italy
Argalastí *105 C5* Thessalía, C Greece
Argenteuil *90 D1* Val-d'Oise, N France
Argentina *65 B5* *off.* Argentine Republic.
country S South America
Argentina Basin *see* Argentine Basin
Argentine Basin *57 C7* *var.* Argentina
Basin. *undersea basin* SW Atlantic
Ocean
Argentine Republic *see* Argentina
Argentine Rise *see* Falkland Plateau
Argentoratum *see* Strasbourg
Arghandāb, Darya-ye *see*
Arghandāb Röd
Arghandāb Röd *123 E5* *var.* Darya-ye
Arghandāb. *river* SE Afghanistan
Argirocastro *see* Gjirokastër
Argo *72 B3* Northern, N Sudan
Argo Fracture Zone *141 C5* *tectonic
Feature* C Indian Ocean
Árgos *105 B6* Pelopónnisos, S Greece
Argostóli *105 A5* *var.* Argostólion.
Kefalloniá, Iónia Nísiá, Greece,
C Mediterranean Sea
Argostólion *see* Argostóli
Argun *125 E1* *Chin.* Ergun He, *Rus.*
Argun'. *river* China/Russia
Argyrokastron *see* Gjirokastër
Århus *see* Aarhus
Aria *see* Herāt
Ari Atoll *132 A4* *var.* Alifu Atoll. *atoll*
C Maldives
Arica *64 B1* *hist.* San Marcos de Arica.
Tarapacá, N Chile
Aridaia *104 B3* *var.* Aridea, Aridhaía.
Dytikí Makedonía, N Greece
Aridea *see* Aridaía
Aridhaía *see* Aridaía
Arīḥā *118 B3* Al Karak, W Jordan
Arīḥā *see* Jericho
Ariminum *see* Rimini
Arinsal *91 A7* NW Andorra

Arizona *48 A2 off.* State of Arizona, *also known as* Copper State, Grand Canyon State. *state* SW USA
Arkadag *122 C3* Ahal Welaýaty, C Turkmenistan
Arkansas *42 A1 off.* State of Arkansas, *also known as* The Land of Opportunity. *state* S USA
Arkansas City *45 F5* Kansas, C USA
Arkansas River *49 G1 river* C USA
Arkhangel'sk *114 B2 Eng.* Archangel. Arkhangel'skaya Oblast', NW Russia
Arkoí *105 E6 island* Dodekánisa, Greece, Aegean Sea
Arles *91 D6 var.* Arles-sur-Rhône; *anc.* Arelas, Arelate. Bouches-du-Rhône, SE France
Arles-sur-Rhône *see* Arles
Arlington *49 G2* Texas, SW USA
Arlington *41 E4* Virginia, NE USA
Arlon *87 D8 Dut.* Aarlen, *Ger.* Arel, *Lat.* Orolaunum. Luxembourg, SE Belgium
Armagh *89 B5 Ir.* Ard Mhacha. S Northern Ireland, United Kingdom
Armagnac *91 B6 cultural region* S France
Armenia *58 B3* Quindío, W Colombia
Armenia *117 F3 off.* Republic of Armenia, *var.* Hayastan, *Arm.* Hayastani Hanrapetut'yun; *prev.* Armenian Soviet Socialist Republic. *country* SW Asia
Armenian Soviet Socialist Republic *see* Armenia
Armenia, Republic of *see* Armenia
Armiansk *109 F4 prev.* Armyans'k, *Rus.* Armyansk. Respublika Krym, S Ukraine
Armidale *149 D6* New South Wales, SE Australia
Armstrong *38 B3* Ontario, S Canada
Armyansk/ Armyans'k *see* Armiansk
Arnaía *104 C4 Cont.* Arnea. Kentrikí Makedonía, N Greece
Arnaud *82 A3 river* Québec, E Canada
Arnea *see* Arnaía
Arnedo *93 E2* La Rioja, N Spain
Arnhem *86 D4* Gelderland, SE Netherlands
Arnhem Land *148 A2 physical region* Northern Territory, N Australia
Arno *96 B3 river* C Italy
Arnold *45 G4* Missouri, C USA
Arnswalde *see* Choszczno
Aroe Islands *see* Aru, Kepulauan
Arorae *145 E3 atoll* Tungaru, W Kiribati
Arrabona *see* Győr
Ar Rahad *see* Er Rahad
Ar Ramādī *120 B3 var.* Ramadi, Rumadiya. Al Anbār, SW Iraq
Ar Rāmī *118 C4* Ḥimṣ, C Syria
Ar Ramtha *119 B5 var.* Ramtha. Irbid, N Jordan
Arran, Isle of *88 C4 island* SW Scotland, United Kingdom
Ar Raqqah *118 C2 var.* Rakka; *anc.* Nicephorium. Ar Raqqah, N Syria
Arras *90 C2 anc.* Nemetocenna. Pas-de-Calais, N France
Ar Rawḍatayn *120 C4 var.* Raudhatain. N Kuwait
Arretium *see* Arezzo
Arriaca *see* Guadalajara
Arriaga *51 G5* Chiapas, SE Mexico
Ar Riyāḍ *121 C5 Eng.* Riyadh. *country capital* (Saudi Arabia) Ar Riyāḍ, C Saudi Arabia
Ar Rub 'al Khali *121 C6 Eng.* Empty Quarter, Great Sandy Desert. *desert* SW Asia
Ar Rustāq *121 E5 var.* Rostak, Rustaq. N Oman
Ar Ruţbah *120 B3 var.* Rutba. Al Anbār, SW Iraq
Árta *105 A5 anc.* Ambracia. Ípeiros, W Greece
Artashat *117 F3* S Armenia
Artemisa *54 B2* La Habana, W Cuba
Artesia *48 D3* New Mexico, SW USA
Arthur's Pass *151 C6 pass* South Island, New Zealand
Artigas *64 D3 prev.* San Eugenio, San Eugenio del Cuareim. Artigas, N Uruguay
Art'ik *117 F2* W Armenia
Artois *90 C2 cultural region* N France
Artsiz *see* Artsyz
Artsyz *108 D4 Rus.* Artsiz. Odes'ka Oblast', SW Ukraine

Arturo Prat *154 A2 Chilean research station* South Shetland Islands, Antarctica
Artvin *117 F2* Artvin, NE Turkey (Türkiye)
Arua *73 B6* NW Uganda
Aruba *58 C1 var.* Oruba. *Dutch self-governing territory* S West Indies
Aru Islands *see* Aru, Kepulauan
Aru, Kepulauan *139 G4 Eng.* Aru Islands; *prev.* Aroe Islands. *island group* E Indonesia
Arunāchal Pradesh *135 G3 prev.* North East Frontier Agency, North East Frontier Agency of Assam. *cultural region* NE India
Arusha *73 C7* Arusha, N Tanzania
Arviat *37 G4 prev.* Eskimo Point. Nunavut, C Canada
Arvidsjaur *84 C4* Norrbotten, N Sweden
Arys' *114 B5 Kaz.* Arys. Türkistan/ Turkestan, S Kazakhstan
Arys *see* Arys'
Asadābād *123 F4 var.* Asaḍābād; *prev.* Chaghasarāy. Konar, E Afghanistan
Asaḍābād *see* Asadābād
Asad, Buḩayrat al *118 C2 Eng.* Lake Assad. *lake* N Syria
Asahi-dake *130 D2 mountain* Hokkaidō, N Japan
Asahikawa *130 D2* Hokkaidō, N Japan
Asamankese *75 E5* SE Ghana
Asānsol *135 F4* West Bengal, NE India
Asben *see* Aïr, Massif de l'
Ascension Fracture Zone *69 A5 tectonic Feature* C Atlantic Ocean
Ascension Island *see* St Helena, Ascension and Tristan da Cunha
Ascoli Piceno *96 C4 anc.* Asculum Picenum. Marche, C Italy
Asculum Picenum *see* Ascoli Piceno
'Aseb *72 D4 var.* Assab, *Amh.* Āseb. SE Eritrea
Assen *86 E2* Drenthe, NE Netherlands
Aşgabat *122 C3 prev.* Ashgabat, Ashkhabad, Poltoratsk. *country capital* (Turkmenistan) Ahal Welaýaty, C Turkmenistan
Ashara *see* Al 'Ashārah
Ashburton *151 C6* Canterbury, South Island, New Zealand
Ashburton River *146 A4 river* Western Australia
Ashdod *119 A6 anc.* Azotos, *Lat.* Azotus. Central, W Israel
Asheville *43 E1* North Carolina, SE USA
Ashgabat *see* Aşgabat
Ashkelon *119 A6 prev.* Ashqelon. Southern, C Israel
Ashkhabad *see* Aşgabat
Ashland *46 B4* Oregon, NW USA
Ashland *40 B1* Wisconsin, N USA
Ashmore and Cartier Islands *142 A3 Australian external territory* E Indian Ocean
Ashmyany *107 C5 Rus.* Oshmyany. Hrodzyenskaya Voblasts', W Belarus
Ashqelon *see* Ashkelon
Ash Shadādāh *118 D2 var.* Ash Shaddādah, Jisr ash Shadadi, Shaddādī, Shedadi, Tell Shedadi. Al Ḩasakah, NE Syria
Ash Shaddādah *see* Ash Shadādāh
Ash Sharah *119 B7 var.* Esh Sharā. *mountain range* W Jordan
Ash Shāriqah *120 D4 Eng.* Sharjah. Ash Shāriqah, NE United Arab Emirates
Ash Shawbak *119 B7* Ma'ān, W Jordan
Ash Shiḩr *121 C6* SE Yemen
'Aşi, Nahr Al *see* Orontes
Asinara *96 A4 island* W Italy
Asi Nehri *see* Orontes
Asipovichy *107 D6 Rus.* Osipovichi. Mahilyowskaya Voblasts', C Belarus
Aşkale *117 E3* Erzurum, NE Turkey (Türkiye)
Askersund *85 C6* Örebro, C Sweden
Asmara *see* Asmera
Asmera *72 C4 var.* Asmara. *country capital* (Eritrea) C Eritrea
Aspadana *see* Eşfahān
Asphaltites, Lacus *see* Dead Sea
Aspinwall *see* Colón
Assab *see* 'Aseb
As Sabkhah *118 D2 var.* Sabkha. Ar Raqqah, NE Syria
Assad, Lake *see* Asad, Buḩayrat al
Aş Şafāwī *119 C6* Al Mafraq, N Jordan

Aş Şaḩrā' ash Sharqīyah *see* Sahara el Sharqīya
As Salamīyah *see* Salamīyah
'Assal, Lac *68 E4 lake* C Djibouti
As Salţ *119 B6 var.* Salt. Al Balqā', C Jordan
Assamakka *see* Assamakka
Assamakka *75 F2 var.* Assamaka. Agadez, NW Niger
As Samāwah *120 B3 var.* Samawa. Al Muthanná, S Iraq
Assenede *87 B5* Oost-Vlaanderen, NW Belgium
Assiout *see* Asyūţ
Assiut *see* Asyūţ
Assling *see* Jesenice
Assouan *see* Aswān
Assu *63 G2 var.* Açu. Rio Grande do Norte, E Brazil
Assuan *see* Aswān
As Sukhnah *118 C3 var.* Sukhne, *Fr.* Soukhné. Ḩimṣ, C Syria
As Sulaymānīyah *120 C3 var.* Sulaimaniya, *Kurd.* Slêmanî. As Sulaymānīyah/Slêmanî, NE Iraq
As Sulayyil *121 B5* Ar Riyāḍ, S Saudi Arabia
As Suwaydā' *119 B5 var.* El Suweida, Es Suweida, Suweida, *Fr.* Soueida. As Suwayda', SW Syria
As Suways *72 B1 var.* Suez; *Ar.* El Suweis. NE Egypt
Asta Colonia *see* Asti
Astacus *see* İzmit
Astana *114 C4 prev.* Nur-Sultan, Akmola, Akmolinsk, Tselinograd, Aqmola. *country capital* (Kazakhstan) Akmola, N Kazakhstan
Asta Pompeia *see* Asti
Astarabad/Asterābād *see* Gorgān
Asti *96 A2 anc.* Asta Colonia, Asta Pompeia, Hasta Colonia, Hasta Pompeia. Piemonte, NW Italy
Astigi *see* Écija
Astipálaia *see* Astypálaia
Astorga *92 C1 anc.* Asturica Augusta. Castilla y León, N Spain
Astrabad *see* Gorgān
Astrakhan' *111 C7* Astrakhanskaya Oblast', SW Russia
Asturias *92 C1 autonomous community* NW Spain
Asturias *see* Oviedo
Asturica Augusta *see* Astorga
Astypálaia *105 D7 var.* Astipálaia, *It.* Stampalia. *island* Kykládes, Greece, Aegean Sea
Asunción *64 D2 country capital* (Paraguay) Central, S Paraguay
Asuka *154 C2 Japanese research station* Antarctica
Aswān *72 B2 var.* Assouan, Assuan, Aswân; *anc.* Syene. SE Egypt
Aswân *see* Aswān
Asyūţ *72 B2 var.* Assiout, Assiut, Asyût, Siut; *anc.* Lycopolis. C Egypt
Asyût *see* Asyūţ
Atacama Desert *64 B2 Eng.* Atacama Desert. *desert* N Chile
Atacama Desert *see* Atacama, Desierto de
Atafu Atoll *145 E3 island* NW Tokelau
Atamyrat *see* Kerki
Atār *74 C2* Adrar, W Mauritania
Atas Bogd *126 D3 mountain* SW Mongolia
Atascadero *47 B7* California, W USA
Atatürk Baraji *117 E4 reservoir* S Turkey (Türkiye)
Atbara *72 C3 var.* 'Aţbārah. River Nile, NE Sudan
'Aţbarah/'Aţbarah, Nahr *see* Atbara
Atbasar *114 C4* Akmola, N Kazakhstan
Atchison *45 F4* Kansas, C USA
Aternum *see* Pescara
Ath *87 B6 var.* Aat. Hainaut, SW Belgium
Athabasca *37 E5* Alberta, SW Canada
Athabasca *37 E5 var.* Athabaska. *river* Alberta, SW Canada
Athabaska *see* Athabasca
Athenae *see* Athína
Athens *43 E2* Georgia, SE USA
Athens *40 D4* Ohio, N USA
Athens *49 G3* Texas, SW USA
Athens *see* Athína
Atherton *148 D3* Queensland, NE Australia

Athína *105 C6 Eng.* Athens, *prev.* Athínai; *anc.* Athenae. *country capital* (Greece) Attikí, C Greece
Athínai *see* Athína
Athlone *89 B5 Ir.* Baile Átha Luain. C Ireland
Ath Thawrah *118 C2 var.* Madīnat ath Thawrah. Ar Raqqah, N Syria
Ati *76 C3* Batha, C Chad
Atikokan *38 B4* Ontario, S Canada
Atka *115 G3* Magadanskaya Oblast', E Russia
Atka *36 A3* Atka Island, Alaska, USA
Atlanta *42 D2 state capital* Georgia, SE USA
Atlanta *49 H2* Texas, SW USA
Atlantic City *41 F4* New Jersey, NE USA
Atlantic-Indian Basin *67 D7 undersea basin* SW Indian Ocean
Atlantic-Indian Ridge *69 B8 undersea ridge* SW Indian Ocean
Atlantic Ocean *66 B4 ocean*
Atlas Mountains *70 C2 mountain range* NW Africa
Atlasovo *115 H3* Kamchatskaya Oblast', E Russia
Atlas Saharien *70 D2 var.* Saharan Atlas. *mountain range* Algeria/Morocco
Atlas, Tell *see* Atlas Tellien
Atlas Tellien *102 C3 Eng.* Tell Atlas. *mountain range* N Algeria
Atlin *36 D4* British Columbia, W Canada
At Tafīlah *119 B7 var.* Et Tafila, Tafila. Aţ Ţafīlah, W Jordan
Aţ Ţā'if *121 B5* Makkah, W Saudi Arabia
Attalea/Attalia *see* Antalya
At Tall al Abyaḍ *see* Tall Abyaḍ
Aţ Ţanf *118 D4* Ḩimṣ, S Syria
Attapu *137 E5 var.* Samakhixai, Attopeu. Attapu, S Laos
Attawapiskat *38 C3* Ontario, C Canada
Attawapiskat *38 C3 river* Ontario, S Canada
At Tibnī *118 D2 var.* Tibni. Dayr az Zawr, NE Syria
Attopeu *see* Attapu
Attu Island *36 A2 island* Aleutian Islands, Alaska, USA
Atyraū *114 B4 var.* Atyrau, *prev.* Gur'yev. Atyrau, W Kazakhstan
Atyrau *see* Atyraū
Aubagne *91 D6 anc.* Albania. Bouches-du-Rhône, SE France
Aubange *87 D8* Luxembourg, SE Belgium
Aubervilliers *90 E1* Seine-St-Denis, Île-de-France, N France Europe
Auburn *46 B2* Washington, NW USA
Auch *91 B6 Lat.* Augusta Auscorum, Elimberrum. Gers, S France
Auckland *150 D2* Auckland, North Island, New Zealand
Auckland Islands *142 C5 island group* S New Zealand
Audern *see* Audru
Audincourt *90 E4* Doubs, E France
Audru *106 D2 Ger.* Audern. Pärnumaa, SW Estonia
Augathella *149 D5* Queensland, E Australia
Augsbourg *see* Augsburg
Augsburg *95 C6 Fr.* Augsbourg; *anc.* Augusta Vindelicorum. Bayern, S Germany
Augusta *147 A7* Western Australia
Augusta *43 E2* Georgia, SE USA
Augusta *41 G2 state capital* Maine, NE USA
Augusta *see* London
Augusta Auscorum *see* Auch
Augusta Emerita *see* Mérida
Augusta Praetoria *see* Aosta
Augusta Trajana *see* Stara Zagora
Augusta Treverorum *see* Trier
Augusta Vangionum *see* Worms
Augusta Vindelicorum *see* Augsburg
Augustobona Tricassium *see* Troyes
Augustodurum *see* Bayeux
Augustoritum Lemovicensium *see* Limoges
Augustów *98 E2 Rus.* Avgustov. Podlaskie, NE Poland
Aulie Ata/Auliye-Ata *see* Taraz
Aunglan *136 B4 var.* Allanmyo, Myaydo. Magway, C Myanmar (Burma)
Auob *78 C4 var.* Oup. *river* Namibia/ South Africa
Aurangābād *134 D5* Mahārāshtra, C India
Auray *90 A3* Morbihan, NW France

Aurelia Aquensis see Baden-Baden
Aurelianum see Orléans
Aurès, Massif de l' *102 C4 mountain range* NE Algeria
Aurillac *91 C5* Cantal, C France
Aurium see Ourense
Aurora *59 F2* NW Guyana
Aurora *44 D4* Colorado, C USA
Aurora *40 B3* Illinois, N USA
Aurora *45 G5* Missouri, C USA
Aurora see Maéwo, Vanuatu
Aus *78 B4* Karas, SW Namibia
Ausa see Vic
Aussig see Ústí nad Labem
Austin *45 G3* Minnesota, N USA
Austin *49 G3 state capital* Texas, SW USA
Australes, Archipel des see Australes, Îles
Australes et Antarctiques Françaises, Terres see French Southern and Antarctic Lands
Australes, Îles *143 F4 var.* Archipel des Australes, Îles Tubuai, Tubuai Islands, *Eng.* Austral Islands. *island group* SW French Polynesia
Austral Fracture Zone *143 H4 tectonic feature* S Pacific Ocean
Australia *142 A4 off.* Commonwealth of Australia. *country*
Australia, Commonwealth of see Australia
Australian Alps *149 C7 mountain range* SE Australia
Australian Capital Territory *149 D7 prev.* Federal Capital Territory. *territory* SE Australia
Australie, Bassin Nord de l' see North Australian Basin
Austral Islands see Australes, Îles
Austrava see Ostrov
Austria *95 D7 off.* Republic of Austria, *Ger.* Österreich. *country* C Europe
Austria, Republic of see Austria
Autesiodorum see Auxerre
Autissiodorum see Auxerre
Autricum see Chartres
Auvergne *91 C5 cultural region* C France
Auxerre *90 C4 anc.* Autesiodorum, Autissiodorum. Yonne, C France
Avaricum see Bourges
Avarua *145 G5 dependent territory capital* (Cook Islands) Rarotonga, S Cook Islands
Avasfelsőfalu see Negreşti-Oaş
Ávdira *104 C3* Anatolikí Makedonía kai Thráki, NE Greece
Aveiro *92 B2 anc.* Talabriga. Aveiro, W Portugal
Avela see Ávila
Avellino *97 D5 anc.* Abellinum. Campania, S Italy
Avenio see Avignon
Avesta *85 C6* Dalarna, C Sweden
Aveyron *91 C6 river* S France
Avezzano *96 C4* Abruzzo, C Italy
Avgustov see Augustów
Aviemore *88 C3* N Scotland, United Kingdom
Avignon *91 D6 anc.* Avenio. Vaucluse, SE France
Ávila *92 D3 var.* Avila; *anc.* Abela, Abula, Abyla, Avela. Castilla y León, C Spain
Avilés *92 C1* Asturias, NW Spain
Avranches *90 B3* Manche, N France
Avveel see Ivalo, Finland
Avvil see Ivalo
Awaji-shima *131 C6 island* SW Japan
Āwash *73 D5* Āfar, NE Ethiopia
Āwbārī *71 F3* SW Libya
Ax see Dax
Axel *87 B5* Zeeland, SW Netherlands
Axel Heiberg Island *37 E1 var.* Axel Heiburg. *island* Nunavut, N Canada
Axel Heiburg see Axel Heiberg Island
Áxios see Vardar
Ayacucho *60 D4* Ayacucho, S Peru
Ayagoz *114 C5 var.* Ayaguz, *Kaz.* Ayakoz. *river* E Kazakhstan
Ayamonte *92 C4* Andalucía, S Spain
Ayaviri *61 E4* Puno, S Peru
Aydarken *123 F2 prev.* Khaydarken. Batken Oblusu/Batkenskaya Oblast', SW Kyrgyzstan
Aydarko'l Ko'li *123 E2 Rus.* Ozero Aydarkul'. *lake* C Uzbekistan
Aydarkul', Ozero see Aydarko'l Ko'li
Aydın *116 A4 var.* Aidin; *anc.* Tralles Aydın. Aydın, SW Turkey (Türkiye)
Ayers Rock see Uluru

Ayeyarwady *136 B2 var.* Irrawaddy. *river* W Myanmar (Burma)
Ayiá see Agiá
Áyios Evstrátios see Ágios Efstrátios
Áyios Nikólaos see Ágios Nikólaos
Ayorou *75 E3* Tillabéri, W Niger
'Ayoûn el 'Atroûs *74 D3 var.* Aïoun el Atrous, Aioun el Atroûss. Hodh el Gharbi, SE Mauritania
Ayr *88 C4* W Scotland, United Kingdom
Ayteke Bi *114 B4 Kaz.* Zhangaqazaly; *prev.* Novokazalinsk. Kzylorda, SW Kazakhstan
Aytos *104 E2* Burgas, E Bulgaria
Ayutthaya *137 C5 var.* Phra Nakhon Si Ayutthaya. Phra Nakhon Si Ayutthaya, C Thailand
Ayvalık *116 A3* Balıkesir, W Turkey (Türkiye)
Azahar, Costa del *93 F3 coastal region* E Spain
Azaouâd *75 E3 desert* C Mali
Azärbaycan/Azärbaycan Respublikasi see Azerbaijan
A'zäz see I'zaz
Azerbaijan *117 G2 off.* Republic of Azerbaijan, *Az.* Azärbaycan, Azärbaycan Respublikasi; *prev.* Azerbaijan SSR. *country* SE Asia
Azerbaijan, Republic of see Azerbaijan
Azerbaijan SSR see Azerbaijan
Azimabad see Patna
Azizie see Telish
Azogues *60 B2* Cañar, S Ecuador
Azores *92 A4 var.* Açores, Ilhas dos Açores, *Port.* Arquipélago dos Açores. *island group* Portugal, NE Atlantic Ocean
Azores-Biscay Rise *80 A3 undersea rise* E Atlantic Ocean
Azotos/Azotus see Ashdod
Azoum, Bahr *76 C3 seasonal river* SE Chad
Azov, Sea of *103 H1 Rus.* Azovskoye More, *Ukr.* Azovske More. *sea* NE Black Sea
Azovskoye More/Azovskoye More see Azov, Sea of
Azraq, Wāḥat al *119 C6 oasis* N Jordan
Aztec *48 C1* New Mexico, SW USA
Azuaga *92 C4* Extremadura, W Spain
Azuero, Península de *53 F5 peninsula* S Panama
Azul *65 D5* Buenos Aires, E Argentina
Azur, Côte d' *91 E6 coastal region* SE France
'Azza see Gaza
Az Zaqāzīq *72 B1 var.* Zagazig. N Egypt
Az Zarqā' *119 B6* NW Jordan
Az Zāwiyah *71 F2 var.* Zawia. NW Libya
Az Zilfī *120 B4* Ar Riyāḍ, N Saudi Arabia

B

Baalbek *118 B4 var.* Ba'labakk; *anc.* Heliopolis. E Lebanon
Baardheere *73 D6 var.* Bardere, *It.* Bardera. Gedo, SW Somalia
Baarle-Hertog *87 C5* Antwerpen, N Belgium
Baarn *86 C3* Utrecht, C Netherlands
Babadag *108 D5* Tulcea, SE Romania
Babahoyo *60 B2 prev.* Bodegas. Los Ríos, C Ecuador
Bābā, Köh-e *123 E4 mountain range* C Afghanistan
Babayevo *110 B4* Vologodskaya Oblast', NW Russia
Babeldaob *144 A1 var.* Babeldaop, Babelthuap. *island* N Palau
Babeldaop see Babeldaob
Bab el Mandeb *121 B7 strait* Gulf of Aden/Red Sea
Babelthuap see Babeldaob
Babian Jiang see Black River
Babruysk *107 D7 Rus.* Bobruysk. Mahilyowskaya Voblasts', E Belarus
Babuyan Channel *139 E1 channel* N Philippines
Babuyan Islands *139 E1 island group* N Philippines
Bacabal *63 F2* Maranhão, E Brazil
Bacău *108 C4 Hung.* Bakó. Bacău, NE Romania
Bắc Bộ, Vịnh see Tonkin, Gulf of
Bắc Giang *136 D3* Ha Bắc, N Vietnam
Bacheykava *107 D5 Rus.* Bocheykovo. Vitsyebskaya Voblasts', N Belarus
Back *37 F3 river* Nunavut, N Canada

Bačka Palanka *100 D3 prev.* Palanka. Serbia, NW Serbia
Bačka Topola *100 D3 Hung.* Topolya; *prev. Hung.* Bácstopolya. Vojvodina, N Serbia
Bạc Liêu *137 D6 var.* Vinh Loi. Minh Hai, S Vietnam
Bacolod *125 E4 off.* Bacolod City. Negros, C Philippines
Bacolod City see Bacolod
Bácsszenttamás see Srbobran
Bácstopolya see Bačka Topola
Bactra see Balkh
Badajoz *92 C4 anc.* Pax Augusta. Extremadura, W Spain
Baden-Baden *95 B6 anc.* Aurelia Aquensis. Baden-Württemberg, SW Germany
Bad Freienwalde *94 D3* Brandenburg, NE Germany
Badger State see Wisconsin
Bad Hersfeld *94 B4* Hessen, C Germany
Bad Homburg see Bad Homburg vor der Höhe
Bad Homburg vor der Höhe *95 B5 var.* Bad Homburg. Hessen, W Germany
Bad Ischl *95 D7* Oberösterreich, N Austria
Bad Krozingen *95 A6* Baden-Württemberg, SW Germany
Badlands *44 D2 physical region* North Dakota/South Dakota, N USA
Badu Island *148 C1 island* Queensland, NE Australia
Bad Vöslau *95 E6* Niederösterreich, NE Austria
Baeterrae/Baeterrae Septimanorum see Béziers
Baetic Cordillera/Baetic Mountains see Béticos, Sistemas
Bafatá *74 C4* C Guinea-Bissau
Baffin Bay *37 G2 bay* Canada/Greenland
Baffin Island *37 G2 island* Nunavut, NE Canada
Bafing *74 C3 river* W Africa
Bafoussam *76 A4* West, W Cameroon
Bafra *116 D2* Samsun, N Turkey (Türkiye)
Bagdad see Baghdād
Bagé *63 E5* Rio Grande do Sul, S Brazil
Baghdād *120 B3 var.* Bagdad, *country capital* (Iraq) Baghdād, C Iraq
Baghlān *123 E3* Baghlān, NE Afghanistan
Bago *136 B4 var.* Pegu. Bago, SW Myanmar (Burma)
Bagoé *74 D4 river* Ivory Coast/Mali
Bagrationovsk *106 A4 Ger.* Preussisch Eylau. Kaliningradskaya Oblast', W Russia
Bagrax Hu see Bosten Hu
Baguio *139 E1 off.* Baguio City. Luzon, N Philippines
Baguio City see Baguio
Bagzane, Monts *75 F3 mountain* N Niger
Bahama Islands see Bahamas, The
Bahamas, The *54 C2 off.* Commonwealth of The Bahamas. *country* W Indies
Bahamas *35 D6 var.* Bahama Islands. *island group* N West Indies
Bahamas, Commonwealth of The see Bahamas, The
Baharly see Bäherden
Bahawalpur *134 C2* Punjab, E Pakistan
Bäherden *122 C3 prev.* Baharly, *Rus.* Bakharden; *prev.* Bakherden. Ahal Welayaty, C Turkmenistan
Bahia *63 F3 off.* Estado da Bahia. *state/region* E Brazil
Bahía Blanca *65 C5* Buenos Aires, E Argentina
Bahia, Estado da see Bahia
Bahir Dar *72 C4 var.* Bahr Dar, Bahrdar Giyorgis. Āmara, N Ethiopia
Bahraich *135 E3* Uttar Pradesh, N India
Bahrain *120 C4 off.* Kingdom of Bahrain, Mamlakat al Baḥrayn, *Ar.* Al Baḥrayn, *prev.* Bahrein; *anc.* Tylos, Tyros. *country* SW Asia
Bahrain, Kingdom of see Bahrain
Bahrayn, Mamlakat al see Bahrain
Bahr Dar/Bahrdar Giyorgis see Bahir Dar
Bahrein see Bahrain
Bahr el, Azraq see Blue Nile
Bahr Tabariya, Sea of see Kinneret, Yam

Bahushewsk *107 E6 Rus.* Bogushëvsk. Vitsyebskaya Voblasts', NE Belarus
Baia Mare *108 B3 Ger.* Frauenbach, *Hung.* Nagybánya; *prev.* Neustadt. Maramureş, NW Romania
Baia Sprie *108 B3 Ger.* Mittelstadt, *Hung.* Felsőbánya. Maramureş, NW Romania
Baïbokoum *76 B4* Logone-Oriental, SW Chad
Baidoa see Baydhabo
Baie-Comeau *39 E3* Québec, SE Canada
Baikal, Lake *115 E4 Eng.* Lake Baikal. *lake* S Russia
Baikal, Lake see Baykal, Ozero
Baile Átha Cliath see Dublin
Baile Átha Luain see Athlone
Bailén *92 D4* Andalucía, S Spain
Baile na Mainistreach see Newtownabbey
Băileşti *108 B5* Dolj, SW Romania
Ba Illi *76 B3* Chari-Baguirmi, SW Chad
Bainbridge *42 D3* Georgia, SE USA
Bā'ir see Bāyir
Baireuth see Bayreuth
Bairnsdale *149 C7* Victoria, SE Australia
Baishan *129 E3 prev.* Hunjiang. Jilin, NE China
Baiyin *128 B4* Gansu, C China
Baja *99 C7* Bács-Kiskun, S Hungary
Baja California *48 A4 Eng.* Lower California. *peninsula* NW Mexico
Baja California Norte *50 B2 state* NW Mexico
Bajo Boquete see Boquete
Bajram Curri *101 D5* Kukës, N Albania
Bakala *76 C4* Ouaka, C Central African Republic
Bakan see Shimonoseki
Bakay-Ata *123 F2 prev.* Leninpol' Talas Oblusu/Talasskaya Oblast', NW Kyrgyzstan
Baker *46 C3* Oregon, NW USA
Baker Island *145 E2 US unincorporated territory* W Polynesia
Baker Lake *37 F3* Nunavut, N Canada
Bakersfield *47 C7* California, W USA
Bakharden see Bäherden
Bakhchisarai see Bakhchysaray
Bakhchysarai *109 F5 prev.* Bakhchisaray, *Rus.* Bakhchisaray. Respublika Krym, S Ukraine
Bakhchysaray see Bakhchysaray
Bakherden see Bäherden
Bakhmach *109 F1* Chernihivs'ka Oblast', N Ukraine
Bäkhtärän see Kermānshāh
Bakı *117 H2 Eng.* Baku. *country capital* (Azerbaijan) E Azerbaijan
Bákó see Bacău
Bakony *99 C7 Eng.* Bakony Mountains, *Ger.* Bakonywald. *mountain range* W Hungary
Bakony Mountains/Bakonywald see Bakony
Baku see Bakı
Bakwanga see Mbuji-Mayi
Balabac Island *129 C8 island* W Philippines
Balabac, Selat see Balabac Strait
Balabac Strait *138 D2 var.* Selat Balabac. *strait* Malaysia/Philippines
Ba'labakk see Baalbek
Balaguer *93 F2* Cataluña, NE Spain
Balakovo *111 C6* Saratovskaya Oblast', W Russia
Bālā Morghāb see Bālā Murghāb
Bālā Murghāb *122 D3 prev.* Bālā Morghāb. Laghmān, NW Afghanistan
Balashov *111 B6* Saratovskaya Oblast', W Russia
Balasore see Bāleshwar
Balaton, Lake *149 C7 var.* Lake Balaton, *Ger.* Plattensee. *lake* W Hungary
Balaton, Lake see Balaton
Balbina, Represa *62 D1 reservoir* NW Brazil
Balboa *53 G4* Panamá, C Panama
Balcarce *65 D5* Buenos Aires, E Argentina
Balclutha *151 B7* Otago, South Island, New Zealand
Baldy Mountain *44 C1 mountain* Montana, NW USA
Bâle see Basel
Balearic Plain see Algerian Basin
Baleares Major see Mallorca
Baleares, Islas *93 G3 Eng.* Balearic Islands. *island group* Spain, W Mediterranean Sea
Balearic Islands see Baleares, Islas

Balearis Minor see Menorca
Baleine, Rivière à la 39 E2 river Québec,
  E Canada
Balen 87 C5 Antwerpen, N Belgium
Bāleshwar 135 F4 prev. Balasore.
  Odisha, E India
Bālgrad see Alba Iulia
Bali 138 D5 island C Indonesia
Balıkesir 116 A3 Balıkesir,
  W Turkey (Türkiye)
Balikh, Nahr 118 C2 river N Syria
Balikpapan 138 D4 Borneo,
  C Indonesia
Balkanabat 122 B2 Rus. Nebitdag.
  Balkan Welaýaty, W Turkmenistan
Balkan Mountains 104 C2 Bul./
  Croatian Stara Planina. mountain
  range Bulgaria/Serbia
Balkash see Balqash
Balkash, Ozero see Balqash Köli
Balkh 123 E3 anc. Bactra. Balkh,
  N Afghanistan
Balkhash, Lake see Balqash Köli
Balladonia 147 C6 Western Australia
Ballarat 149 C7 Victoria, SE Australia
Balleny Islands 154 B5 island group
  Antarctica
Ballinger 49 F3 Texas, SW USA
Balochistan see Baluchistan
Balqash 114 C5 Rus. Balkash.
  Karagandy, SE Kazakhstan see Balkhash
Balqash Köli 114 C5 Eng. Lake Balkhash,
  Rus. Ozero Balkash. lake
  SE Kazakhstan
Balş 108 B5 Olt, S Romania
Balsas 63 F2 Maranhão, E Brazil
Balsas, Río 51 E5 var. Río Mexcala. river
  S Mexico
Bal'shavik 107 D7 Rus. Bol'shevik.
  Homyel'skaya Voblasts', SE Belarus
Balta 108 D3 Odes'ka Oblast',
  SW Ukraine
Bălţi 108 D3 Rus. Bel'tsy. N Moldova
Baltic Port see Paldiski
Baltic Sea 85 C7 Ger. Ostee, Rus.
  Baltiskoye More. sea N Europe
Baltimore 41 F4 Maryland, NE USA
Baltischport/Baltiski see Paldiski
Baltiskoye More see Baltic Sea
Baltkrievija see Belarus
Baluchistan 134 B3 var. Balochistan,
  Beluchistan. province SW Pakistan
Balvi 106 D4 Balvi, NE Latvia
Balykchy 123 G2 Kir. Ysyk-Köl; prev.
  Issyk-Kul', Rybach'ye. Ysyk-Köl
  Oblusu/Issyk-Kul'skaya Oblast',
  NE Kyrgyzstan
Balzers 94 E2 S Liechtenstein
Bam 120 E4 Kermān, SE Iran
Bamako 74 D4 country capital (Mali)
  Capital District, SW Mali
Bambari 76 C4 Ouaka, C Central
  African Republic
Bamberg 95 C5 Bayern, SE Germany
Bamenda 76 A4 Nord-Ouest,
  W Cameroon
Banaba 144 D2 var. Ocean Island. island
  Tungaru, W Kiribati
Banaras see Vārānasi
Bandaaceh 138 A3 var. Banda Atjeh;
  prev. Koetaradja, Kutaradja, Kutaraja.
  Sumatera, W Indonesia
Banda Atjeh see Bandaaceh
Banda, Laut see Banda Sea
Bandama 74 D5 var. Bandama Fleuve.
  river S Ivory Coast
Bandama Fleuve see Bandama
Bandar 'Abbās see Bandar-e 'Abbās
Bandarbeyla 73 E5 var. Bender Beila,
  Bender Beyla. Bari, NE Somalia
Bandar-e 'Abbās 120 D4 var. Bandar
  'Abbās; prev. Gombroon. Hormozgān,
  S Iran
Bandar-e Büshehr 120 C4 var. Büshehr,
  Eng. Bushire. Büshehr, S Iran
Bandar-e Kangān 120 D4 var. Kangān.
  Büshehr, S Iran
Bandar-e Khamir 120 D4 Hormozgān,
  S Iran
Bandar-e Langeh see Bandar-e Lengeh
Bandar-e Lengeh 120 D4 var. Bandar-e
  Langeh, Lingeh. Hormozgān,
  S Iran
Bandar Kassim see Boosaaso
Bandar Lampung 138 C4 var.
  Bandarlampung, Tanjungkarang-
  Telukbetung; prev. Tandjoengkarang,
  Tanjungkarang, Teloekbetoeng,
  Telukbetung. Sumatera, W Indonesia
Bandarlampung see Bandar Lampung

Bandar Maharani see Muar
Bandar Masulipatnam see
  Machilipatnam
Bandar Penggaram see Batu Pahat
Bandar Seri Begawan 138 D3 prev.
  Brunei Town. country capital (Brunei)
  N Brunei
Banda Sea 139 F5 var. Laut Banda. sea
  E Indonesia
Bandiagara 75 E3 Mopti, C Mali
Bandırma 116 A3 var. Penderma.
  Balıkesir, NW Turkey (Türkiye)
Bandjarmasin see Banjarmasin
Bandoeng see Bandung
Bandundu 77 C6 prev. Banningville.
  Bandundu, W Dem. Rep. Congo
Bandung 138 C5 prev. Bandoeng. Jawa,
  C Indonesia
Bangalore see Bengalūru
Bangassou 76 D4 Mbomou, SE Central
  African Republic
Banggai, Kepulauan 139 E4 island
  group C Indonesia
Banghāzī 71 G2 Eng. Bengazi, Benghazi,
  It. Bengasi. NE Libya
Bangka, Pulau 138 C4 island
  W Indonesia
Bangkok see Krung Thep
Bangkok, Bight of see Krung Thep, Ao
Bangladesh 135 G3 off. People's
  Republic of Bangladesh; prev. East
  Pakistan. country S Asia
Bangladesh, People's Republic of see
  Bangladesh
Bangor 89 C6 NW Wales,
  United Kingdom
Bangor 89 B5 Ir. Beannchar. E Northern
  Ireland, United Kingdom
Bangor 41 G2 Maine, NE USA
Bang Pla Soi see Chon Buri
Bangui 76 C4 country capital (Central
  African Republic) Ombella-Mpoko,
  SW Central African Republic
Bangweulu, Lake 73 B8 var. Lake
  Bengweulu. lake N Zambia
Ban Hat Yai see Hat Yai
Ban Hin Heup 136 C4 Viangchan,
  C Laos
Ban Houayxay/Ban Houei Sai see
  Houayxay
Ban Hua Hin 137 C6 var. Hua Hin.
  Prachuap Khiri Khan, SW Thailand
Bani 74 D3 river S Mali
Banias see Bāniyās
Banijska Palanka see Glina
Banī Suwayf 72 B2 var. Beni Suef.
  N Egypt
Bāniyās 118 B3 var. Banias, Baniyas,
  Paneas. Ţarţūs, W Syria
Banjak, Kepulauan see Banyak,
  Kepulauan
Banja Luka 100 B3 Republika Srpska,
  NW Bosnia and Herzegovina
Banjarmasin 138 D4 prev.
  Bandjarmasin. Borneo, C Indonesia
Banjul 74 B3 prev. Bathurst. country
  capital (The Gambia) W The Gambia
Banks, Îles see Banks Islands
Banks Island 37 E2 island Northwest
  Territories, NW Canada
Banks Islands 144 D4 Fr. Îles Banks.
  island group N Vanuatu
Banks Lake 46 B1 reservoir Washington,
  NW USA
Banks Peninsula 151 C6 peninsula
  South Island, New Zealand
Banks Strait 149 C8 strait SW
  Tasman Sea
Bānkura 135 F4 West Bengal, NE India
Ban Mak Khaeng see Udon Thani
Banmo see Bhamo
Bannau Brycheiniog 89 C6 Eng. Brecon
  Beacons. mountain range S Wales,
  United Kingdom
Banningville see Bandundu
Bañolas see Banyoles
Ban Pak Phanang see Pak Phanang
Ban Sichon see Sichon
Banská Bystrica 99 C6 Ger. Neusohl,
  Hung. Besztercebánya.
  Banskobystricky Kraj, C Slovakia
Bantry Bay 89 A7 Ir. Bá Bheanntraí. bay
  SW Ireland
Banya 104 E2 Burgas, E Bulgaria
Banyak, Kepulauan 138 A3 prev.
  Kepulauan Banjak. island group
  NW Indonesia
Banyo 76 B4 Adamaoua, NW Cameroon
Banyoles 93 G2 var. Bañolas. Cataluña,
  NE Spain

Banzare Seamounts 141 C7 seamount
  range S Indian Ocean
Banzart see Bizerte
Baoji 128 B4 var. Pao-chi, Paoki.
  Shaanxi, C China
Baoro 76 B4 Nana-Mambéré, W Central
  African Republic
Baoshan 129 A6 var. Pao-shan. Yunnan,
  SW China
Baotou 127 F3 var. Pao-t'ou, Paotow.
  Nei Mongol Zizhiqu, N China
Ba'qūbah 120 B3 var. Qubba. Diyālá,
  C Iraq
Baquerizo Moreno see Puerto Baquerizo
  Moreno
Bar 101 C5 It. Antivari. S Montenegro
Baraawe 73 D6 It. Brava. Shabeellaha
  Hoose, S Somalia
Bārāmati 134 C5 Mahārāshtra, W India
Baranavichy 107 B6 Pol. Baranowicze,
  Rus. Baranovichi. Brestskaya Voblasts',
  SW Belarus
Baranovichi/Baranowicze see
  Baranavichy
Barbados 55 G1 country SE West Indies
Barbastro 93 F2 Aragón,
  NE Spain
Barbate de Franco 92 C5 Andalucía,
  S Spain
Barbuda 55 G3 island N Antigua
  and Barbuda
Barcaldine 148 C4 Queensland,
  E Australia
Barcarozsnyó see Râşnov
Barcău see Berettyó
Barce see Al Marj
Barcelona 93 G2 anc. Barcino,
  Barcinona. Cataluña, E Spain
Barcelona 59 E2 Anzoátegui,
  NE Venezuela
Barcino/Barcinona see Barcelona
Barcoo see Cooper Creek
Barcs 99 C7 Somogy, SW Hungary
Bardaï 76 C1 Borkou-Ennedi-Tibesti,
  N Chad
Bardejov 99 D5 Ger. Bartfeld, Hung.
  Bártfa. Presovský Kraj, E Slovakia
Bardera/Bardere see Baardheere
Barduli see Barletta
Bareilly 135 E3 var. Bareli. Uttar
  Pradesh, N India
Bareli see Bareilly
Barendrecht 86 C4 Zuid-Holland,
  SW Netherlands
Barentin 90 C3 Seine-Maritime,
  N France
Barentsburg 83 G2 Spitsbergen,
  W Svalbard
Barentsevo More/Barents Havet see
  Barents Sea
Barentsøya 83 G2 island E Svalbard
Barents Sea 110 C2 Nor. Barents Havet,
  Rus. Barentsevo More. sea
  Arctic Ocean
Bar Harbor 41 H2 Mount Desert Island,
  Maine, NE USA
Bari 97 E5 var. Bari delle Puglie; anc.
  Barium. Puglia, SE Italy
Bāridah see Al Bāridah
Bari delle Puglie see Bari
Barikot see Barī Kōţ
Barī Kōţ 123 F4 var. Barikot, Barīkowţ.
  Konar, NE Afghanistan
Barīkowţ see Barī Kōţ
Barillas 52 A2 var. Santa Cruz Barillas.
  Huehuetenango, NW Guatemala
Barinas 58 C2 Barinas, W Venezuela
Barisal see Barishal
Barisan, Pegunungan 138 B4 mountain
  range Sumatera, W Indonesia
Barishal 135 G4 prev. Barisal. Barishal,
  S Bangladesh
Barito, Sungai 138 D4 river Borneo,
  C Indonesia
Barium see Bari
Barka see Al Marj
Barkly Tableland 148 B3 plateau
  Northern Territory/Queensland,
  N Australia
Bârlad 108 D4 prev. Bîrlad. Vaslui,
  E Romania
Barlavento, Ilhas de 74 A2 var.
  Windward Islands. island group
  N Cape Verde (Cabo Verde)
Bar-le-Duc 90 D3 var. Bar-sur-Ornain.
  Meuse, NE France
Barlee, Lake 147 B6 lake
  Western Australia
Barlee Range 146 A4 mountain range
  Western Australia

Barletta 97 D5 anc. Barduli. Puglia,
  SE Italy
Barlinek 98 B3 Ger. Berlinchen.
  Zachodnio-pomorskie, NW Poland
Barmen-Elberfeld see Wuppertal
Barmouth 89 C6 NW Wales,
  United Kingdom
Barnaul 114 D4 Altayskiy Kray, C Russia
Barnet 89 A7 United Kingdom
Barnstaple 89 C7 SW England,
  United Kingdom
Baroda see Vadodara
Baroghil Pass 123 F3 var. Kowtal-e
  Barowghil. pass Afghanistan/Pakistan
Baron'ki 107 E7 Rus. Boron'ki.
  Mahilyowskaya Voblasts', E Belarus
Barowghil, Kowtal-e see Baroghil Pass
Barquisimeto 58 C2 Lara,
  NW Venezuela
Barra 88 B3 island NW Scotland,
  United Kingdom
Barra de Río Grande 53 E3 Región
  Autónoma Atlántico Sur, E Nicaragua
Barranca 61 C3 Lima, W Peru
Barrancabermeja 58 B2 Santander,
  N Colombia
Barranquilla 58 B1 Atlántico,
  N Colombia
Barreiro 92 B4 Setúbal, W Portugal
Barrier Range 149 C6 hill range New
  South Wales, SE Australia
Barrow 36 D2 Alaska, USA
Barrow 89 B6 Ir. An Bhearú. river
  SE Ireland
Barrow-in-Furness 89 C5 NW England,
  United Kingdom
Barrow Island 146 A4 island Western
  Australia
Barstow 47 C7 California, W USA
Bar-sur-Ornain see Bar-le-Duc
Bartang 123 F3 river SE Tajikistan
Bartenstein see Bartoszyce
Bártfa/Bartfeld see Bardejov
Bartica 59 F3 N Guyana
Bartın 116 C2 Bartın, NW
  Turkey (Türkiye)
Bartlesville 49 G1 Oklahoma,
  C USA
Bartoszyce 98 D2 Ger. Bartenstein.
  Warmińsko-mazurskie, NE Poland
Baruun-Urt 127 F2 Sühbaatar,
  E Mongolia
Barú, Volcán 53 E5 var. Volcán de
  Chiriquí. volcano W Panama
Barwon River 149 D5 river New South
  Wales, SE Australia
Barysaw 107 D6 Rus. Borisov. Minskaya
  Voblasts', NE Belarus
Basarabeasca 108 D4 Rus. Bessarabka.
  SE Moldova
Basel 95 A7 Eng. Basle, Fr. Bâle.
  Basel-Stadt, NW Switzerland
Basilan 139 E3 island Sulu Archipelago,
  SW Philippines
Basle see Basel
Basra see Al Başrah
Bassano del Grappa 96 C2 Veneto,
  NE Italy
Bassein see Pathein
Basseterre 55 G4 country capital (Saint
  Kitts and Nevis) Saint Kitts
  and Nevis
Basse-Terre 55 G3 dependent territory
  capital (Guadeloupe) Basse Terre,
  SW Guadeloupe
Basse Terre 55 G4 island W Guadeloupe
Bassikounou 74 D3 Hodh ech Chargui,
  SE Mauritania
Bass, Îlots de see Marotiri
Bass Strait 149 C7 strait SE Australia
Bassum 94 B3 Niedersachsen,
  NW Germany
Bastia 91 E7 Corse, France,
  C Mediterranean Sea
Bastogne 87 D7 Luxembourg,
  SE Belgium
Bastrop 42 B2 Louisiana, S USA
Bastyn' 107 B7 Rus. Bostyn'. Brestskaya
  Voblasts', SW Belarus
Basuo see Dongfang
Basutoland see Lesotho
Bata 77 A5 NW Equatorial Guinea
Batae Coritanorum see Leicester
Batajnica 100 D3 Vojvodina, N Serbia
Batangas 139 E2 off. Batangas City.
  Luzon, N Philippines
Batangas City see Batangas
Batavia see Jakarta
Bătdâmbâng see Battambang
Batéké, Plateaux 77 B6 plateau S Congo

Bath *89 D7 hist.* Akermanceaster; *anc.*
Aquae Calidae, Aquae Solis.
SW England, United Kingdom
Bathinda *134 D2* Punjab, NW India
Bathsheba *55 G1* E Barbados
Bathurst *149 D6* New South Wales,
SE Australia
Bathurst *39 F4* New Brunswick,
SE Canada
Bathurst *see* Banjul
Bathurst Island *146 D2 island* Northern
Territory, N Australia
Bathurst Island *37 F2 island* Parry
Islands, Nunavut, N Canada
Batin, Wadi al *120 C4 dry watercourse*
SW Asia
Batman *117 E4 var.* İluh. Batman,
SE Turkey (Türkiye)
Batna *71 E2* NE Algeria
Baton Rouge *42 B3 state capital*
Louisiana, S USA
Batroûn *118 A4 var.* Al Batrūn.
N Lebanon
Battambang *137 C5* Khmer.
Bătdâmbâng, NW Cambodia
Batticaloa *132 D3* Eastern Province,
E Sri Lanka
Battipaglia *97 D5* Campania, S Italy
Battle Born State *see* Nevada
Batumi *117 F2 prev.* Bat'umi. W Georgia
Bat'umi *see* Batumi
Batu Pahat *138 B3 prev.* Bandar
Penggaram. Johor, Peninsular Malaysia
Bauchi *75 G4* Bauchi, NE Nigeria
Bauer Basin *153 F3 undersea basin*
E Pacific Ocean
Bauska *106 C3 Ger.* Bauske. Bauska,
S Latvia
Bauske *see* Bauska
Bautzen *94 D4 Lus.* Budyšin. Sachsen,
E Germany
Bauzanum *see* Bolzano
Bavaria *see* Bayern
Bavarian Alps *95 C7 Ger.* Bayrische
Alpen. *mountain range* Austria/
Germany
Bavière *see* Bayern
Bavispe, Río *50 C2 river* NW Mexico
Bawiti *see* Al Bawîtî
Bawku *75 E4* N Ghana
Bayamo *54 C3* Granma, E Cuba
Bayan Har Shan *126 D4 var.* Bayan
Khar. *mountain range* C China
Bayanhongor *126 D2* Bayanhongor,
C Mongolia
Bayan Khar *see* Bayan Har Shan
Bayano, Lago *53 G4 lake* E Panama
Bay City *40 C3* Michigan, N USA
Bay City *49 G4* Texas, SW USA
Baydhabo *73 D6 var.* Baydhowa, Isha
Baydhabo, *It.* Baidoa. Bay, SW Somalia
Baydhowa *see* Baydhabo
Bayern *95 C6 Eng.* Bavaria, *Fr.* Bavière.
state SE Germany
Bayeux *90 B3 anc.* Augustodurum.
Calvados, N France
Bäyir *119 C7 var.* Ba'ir. Ma'ān,
S Jordan
Bay Islands *52 C1 Eng.* Bay Islands.
island group N Honduras
Bay Islands *see* Bahía, Islas de la
Baymak *111 D6* Respublika
Bashkortostan, W Russia
Bayonne *91 A6 anc.* Lapurdum.
Pyrénées-Atlantiques, SW France
Bayou State *see* Mississippi
Bayram-Ali *see* Baýramaly
Baýramaly *122 D3 var.* Bayramaly; *prev.*
Bayram-Ali. Mary Welaýaty,
S Turkmenistan
Bayreuth *95 C5 var.* Baireuth. Bayern,
SE Germany
Bayrische Alpen *see* Bavarian Alps
Bayrūt *see* Beyrouth
Bay State *see* Massachusetts
Baysun *see* Boysun
Bayt Lahm *see* Bethlehem
Baytown *49 H4* Texas, SW USA
Baza *93 E4* Andalucía, S Spain
Bazargic *see* Dobrich
Bazin *see* Pezinok
Beagle Channel *65 C8 channel*
Argentina/Chile
Béal Feirste *see* Belfast
Beannchar *see* Bangor, Northern Ireland,
United Kingdom
Bear Island *see* Bjørnøya
Bear Lake *46 E4 lake* Idaho/Utah,
NW USA
Beas de Segura *93 E4* Andalucía, S Spain

Beata, Isla *55 E3 island* SW
Dominican Republic
Beatrice *45 F4* Nebraska, C USA
Beaufort Sea *36 D2 sea* Arctic Ocean
Beaufort-Wes *see* Beaufort West
Beaufort West *78 C5 Afr.* Beaufort-Wes.
Western Cape, SW South Africa
Beaumont *49 H3* Texas, SW USA
Beaune *90 D4* Côte d'Or, C France
Beauvais *90 C3 anc.* Bellovacum,
Caesaromagus. Oise, N France
Beaver Island *40 C2 island* Michigan,
N USA
Beaver Lake *49 H1 reservoir* Arkansas,
C USA
Beaver River *49 F1 river* Oklahoma,
C USA
Beaver State *see* Oregon
Beāwar *134 C3* Rājasthān, N India
Bečej *100 D3 Ger.* Altbetsche, *Hung.*
Óbecse, Rácz-Becse; *prev.*
Magyar-Becse, Stari Bečej. Vojvodina,
N Serbia
Béchar *70 D2 prev.* Colomb-Béchar.
W Algeria
Beckley *40 D5* West Virginia, NE USA
Bécs *see* Wien
Bedford *89 D6* E England,
United Kingdom
Bedum *86 E1* Groningen,
NE Netherlands
Beehive State *see* Utah
Be'er Menuha *119 B7 prev.* Be'ér
Menuḥa. Southern, S Israel
Be'ér Menuḥa *see* Be'er Menuha
Beernem *87 A5* West-Vlaanderen,
NW Belgium
Beersheba *see* Be'er Sheva
Be'er Sheva *119 A7 var.* Beersheba, *Ar.*
Bir es Saba; *prev.* Be'ér Sheva'.
Southern, S Israel
Be'ér Sheva' *see* Be'er Sheva
Beesel *87 D5* Limburg, SE Netherlands
Beeville *49 G4* Texas, SW USA
Bega *149 D7* New South Wales,
SE Australia
Begoml' *see* Byahoml'
Begovat *see* Bekobod
Behagle *see* Laï
Behar *see* Bihār
Beibu Wan *see* Tonkin, Gulf of
Beida *see* Al Bayḍā'
Beihai *129 B6* Guangxi Zhuangzu
Zizhiqu, S China
Beijing *128 C3 var.* Pei-ching, Peking;
prev. Pei-p'ing. *country capital* (China)
Beijing Shi, E China
Beilen *86 E2* Drenthe, NE Netherlands
Beira *79 E3* Sofala, C Mozambique
Beirut *see* Beyrouth
Beit Lekhem *see* Bethlehem
Beiuş *108 B3 Hung.* Belényes. Bihor,
NW Romania
Beja *92 B4 anc.* Pax Julia. Beja,
SE Portugal
Béjar *92 C3* Castilla y León, N Spain
Bejraburi *see* Phetchaburi
Bekabad *see* Bekobod
Békás *see* Bicaz
Bek-Budi *see* Qarshi
Békéscsaba *99 D7 Rom.* Bichiş-Ciaba.
Békés, SE Hungary
Bekobod *123 E2 Rus.* Bekabad; *prev.*
Begovat. Toshkent Viloyati,
E Uzbekistan
Bela Crkva *100 E3 Ger.* Weisskirchen,
Hung. Fehértemplom. Vojvodina,
W Serbia
Belagāvi *132 B1 prev.* Belgaum.
Karnātaka, W India
Belarus *107 B6 off.* Republic of Belarus,
var. Belorussia, *Latv.* Baltkrievija; *prev.*
Belorussian SSR, *Rus.* Belorusskaya
SSR. *country* E Europe
Belaya Tserkov' *see* Bila Tserkva
Belchatów *98 C4 var.* Bełchatow.
Łódzski, C Poland
Belchatow *see* Bełchatów
Belcher, Îles *see* Belcher Islands
Belcher Islands *38 C2 Fr.* Îles Belcher.
island group Nunavut, SE Canada
Beledweyne *73 D5 var.* Belet Huen, *It.*
Belet Uen. Hiiraan, C Somalia
Belém *63 F1 var.* Pará. *state capital* Pará,
N Brazil
Belén *52 D4* Rivas, SW Nicaragua
Belen *48 D2* New Mexico, SW USA
Belényes *see* Beiuş
Belet Huen/Belet Uen *see* Beledweyne

Belfast *89 B5 Ir.* Béal Feirste. *national
capital* E Northern Ireland,
United Kingdom
Belfast *41 D2* North Dakota, N USA
Belfort *90 E4* Territoire-de-Belfort,
E France
Belgard *see* Białogard
Belgaum *see* Belagāvi
Belgian Congo *see* Congo (Democratic
Republic of)
België/Belgique *see* Belgium
Belgium *87 B6 off.* Kingdom of Belgium,
Dut. België, *Fr.* Belgique. *country*
NW Europe
Belgium, Kingdom of *see* Belgium
Belgorod *111 A6* Belgorodskaya Oblast',
W Russia
Belgorod-Dnestrovskiy *see*
Bilhorod-Dnistrovskyi
Belgrano II *154 A2* Argentinian research
station Antarctica
Belice *see* Belize/Belize City
Beligrad *see* Berat
Beli Manastir *100 C3 Hung.*
Pélmonostor; *prev.* Monostor.
Osijek-Baranja, NE Croatia
Belinga *77 B5* Ogooué-Ivindo, NE Gabon
Belitung, Pulau *138 C4 island*
W Indonesia
Belize *52 B1 Sp.* Belice; *prev.* British
Honduras, Colony of Belize. *country*
Central America
Belize *see* Belize City
Belize City *52 C1 var.* Belize, *Sp.* Belice.
Belize, NE Belize
Belize, Colony of *see* Belize
Beljak *see* Villach
Belkofski *36 B3* Alaska, USA
Belle Île *90 A4 island* NW France
Belle Isle, Strait of *39 G3 strait*
Newfoundland and Labrador, E Canada
Bellenz *see* Bellinzona
Belleville *40 B4* Illinois, N USA
Bellevue *45 F4* Iowa, C USA
Bellevue *46 B2* Washington, NW USA
Bellingham *46 B1* Washington, NW USA
Belling Hausen Mulde *see* Southeast
Pacific Basin
Bellingshausen Abyssal Plain *see*
Bellingshausen Plain
Bellingshausen Plain *153 F5 var.*
Bellingshausen Abyssal Plain. *abyssal
plain* SE Pacific Ocean
Bellingshausen Sea *154 A3 sea*
Antarctica
Bellinzona *95 B8 Ger.* Bellenz. Ticino,
S Switzerland
Bello *58 B2* Antioquia, W Colombia
Bello Horizonte *see* Belo Horizonte
Bellovacum *see* Beauvais
Bellville *78 B5* Western Cape, SW South
Africa
Belmopan *52 C1 country capital* (Belize)
Cayo, C Belize
Belogradchik *104 B1* Vidin,
NW Bulgaria
Belo Horizonte *63 F4 prev.* Bello
Horizonte. *state capital* Minas Gerais,
SE Brazil
Belomorsk *110 B3* Respublika Kareliya,
NW Russia
Beloretsk *111 D6* Respublika
Bashkortostan, W Russia
Belorussia/Belorussian SSR *see* Belarus
Belorusskaya Gryada *see* Byelaruskaya
Hrada
Belorussian SSR *see* Belarus
Beloshchel'ye *see* Nar'yan-Mar
Belostok *see* Białystok
Belovár *see* Bjelovar
Beloye More *110 C3 Eng.* White Sea. *sea*
NW Russia
Belozërsk *110 B4* Vologodskaya Oblast',
NW Russia
Belton *49 G3* Texas, SW USA
Bel'tsy *see* Bălţi
Beluchistan *see* Baluchistan
Belukha, Gora *114 D5 mountain*
Kazakhstan/Russia
Belynichi *see* Byalynichy
Belyy, Ostrov *114 D2 island* N Russia
Bemaraha, Plateau du *see* Bemaraha
Bemidji *45 F1* Minnesota, N USA
Bemmel *86 D4* Gelderland,
SE Netherlands
Benaco *see* Garda, Lago di
Benares *see* Vārānasi

Benavente *92 D2* Castilla y León, N Spain
Bend *46 B3* Oregon, NW USA
Bender *108 D4 Rus.* Bendery; *var.*
Tighina. E Moldova
Bender Beila/Bender Beyla *see*
Bandarbeyla
Bender Cassim/Bender Qaasim *see*
Boosaaso
Bendern *94 E1* NW Liechtenstein Europe
Bendery *see* Bender
Bendigo *149 C7* Victoria, SE Australia
Beneschau *see* Benešov
Benešov *99 B5 Ger.* Beneschau.
Středočeský Kraj, W Czechia
(Czech Republic)
Benevento *97 D5 anc.* Beneventum,
Malventum. Campania, S Italy
Beneventum *see* Benevento
Bengal, Bay of *124 C4 bay*
N Indian Ocean
Bengalūru *132 C2 prev.* Bangalore. *state
capital* Karnātaka, S India
Bengasi *see* Banghāzi
Bengbu *129 D5 var.* Peng-pu. Anhui,
E China
Benghazi *see* Banghāzi
Bengkulu *138 B4 prev.* Bengkoeloe,
Benkoelen, Benkulen. Sumatera,
W Indonesia
Benguela *78 A2 var.* Benguella.
Benguela, W Angola
Benguella *see* Benguela
Bengweulu, Lake *see* Bangweulu, Lake
Ben Hope *88 B2 mountain* N Scotland,
United Kingdom
Beni *77 E5* Nord-Kivu, NE Dem.
Rep. Congo
Benidorm *93 F4* Comunitat Valenciana,
E Spain
Beni-Mellal *70 C2* C Morocco
Benin *75 F4 off.* Republic of Benin; *prev.*
Dahomey. *country* W Africa
Benin, Bight of *75 F5 gulf* W Africa
Benin City *75 F5* Edo, SW Nigeria
Benin, Republic of *see* Benin
Beni, Río *61 E3 river* N Bolivia
Beni Suef *see* Banī Suwayf
Ben Nevis *88 C3 mountain* N Scotland,
United Kingdom
Bénoué *see* Benue
Benson *48 B3* Arizona, SW USA
Bent Jbaïl *119 A5 var.* Bint Jubayl.
S Lebanon
Benton *42 B1* Arkansas, C USA
Benue *76 B4 Fr.* Bénoué. *river*
Cameroon/Nigeria
Beograd *100 D3 Eng.* Belgrade, *Serb.*
N Serbia
Berane *101 D5 prev.* Ivangrad.
E Montenegro
Berat *101 C6 var.* Berati, Croatian
Beligrad. Berat, C Albania
Berātău *see* Berettyó
Berati *see* Berat
Berau, Teluk *139 G4 var.* MacCluer Gulf.
bay Papua, E Indonesia
Berbera *72 D4* Sahil, NW Somalia
Berbérati *77 B5* Mambéré-Kadéï,
SW Central African Republic
Berck-Plage *90 C2* Pas-de-Calais,
N France
Berdiansk *109 G4 var.* Berdyans'k, *Rus.*
Berdyansk; *prev.* Osipenko. Zaporiz'ka
Oblast', SE Ukraine
Berdichev *see* Berdychiv
Berdyans'k/Berdyansk *see* Berdiansk
Berdychiv *108 D2 Rus.* Berdichev.
Zhytomyrs'ka Oblast', N Ukraine
Beregovo/Beregszász *see* Berehove
Bereket *122 B2 prev. Rus.* Gazandzhyk,
Kazandzhik, *Turkm.* Gazanjyk. Balkan
Welaýaty, W Turkmenistan
Beretău *see* Berettyó
Berettyóújfalu *99 D6* Hajdú-Bihar,
E Hungary
Berettyó *108 C2 Pol.* Brzeżany.
Ternopil's'ka Oblast', W Ukraine
Berezina *see* Byerezino
Berezniki *111 D5* Permskaya Oblast',
NW Russia
Berga *93 G2* Cataluña, NE Spain
Bergara *93 E1* País Vasco, N Spain

Bergen 94 D2 Mecklenburg-
Vorpommern, NE Germany
Bergen 86 C2 Noord-Holland,
NW Netherlands
Bergen 85 A5 Hordaland, S Norway
Bergen *see* Mons
Bergerac 91 B5 Dordogne,
SW France
Bergeyk 87 C5 Noord-Brabant,
S Netherlands
Bergomum *see* Bergamo
Bergse Maas 86 D4 *river*
S Netherlands
Beringen 87 C5 Limburg, NE Belgium
Beringov Proliv *see* Bering Strait
Bering Sea 36 A2 *sea* N Pacific Ocean
Bering Strait 36 C2 *Rus.* Beringov Proliv.
strait Bering Sea/Chukchi Sea
Berja 93 E5 Andalucía, S Spain
Berkeley 47 B6 California, W USA
Berkner Island 154 A2 *island* Antarctica
Berkovitsa 104 C2 Montana,
NW Bulgaria
Berlin 94 D3 *country capital* (Germany)
Berlin, NE Germany
Berlin 41 G2 New Hampshire, NE USA
Berlinchen *see* Barlinek
Bermejo, Río 64 C2 *river* N Argentina
Bermeo 93 E1 País Vasco, N Spain
Bermuda 35 D6 *var.* Bermuda Islands,
Bermudas; *prev.* Somers Islands. *UK*
Overseas Territory NW Atlantic Ocean
Bermuda Islands *see* Bermuda
Bermuda Rise 35 E6 *undersea rise*
C Sargasso Sea
Bermudas *see* Bermuda
Bern 95 A7 *Fr.* Berne. *country capital*
(Switzerland) Bern, W Switzerland
Bernau 94 D3 Brandenburg,
NE Germany
Bernburg 94 C4 Sachsen-Anhalt,
C Germany
Berne *see* Bern
Berner Alpen 95 A7 *var.* Berner
Oberland, *Eng.* Bernese Oberland.
mountain range SW Switzerland
Berner Oberland/Bernese Oberland *see*
Berner Alpen
Bernier Island 147 A5 *island*
Western Australia
Beroea *see* Ḥalab
Berry 90 C4 *cultural region* C France
Berry Islands 54 C1 *island group*
N The Bahamas
Bertoua 77 B5 Est, E Cameroon
Beru 145 E2 *var.* Peru. *atoll* Tungaru,
W Kiribati
Berwick-upon-Tweed 88 D4 N England,
United Kingdom
Berytus *see* Beyrouth
Besançon 90 D4 *anc.* Besontium,
Vesontio. Doubs, E France
Beskra *see* Biskra
Besontium *see* Besançon
Bessarabka *see* Basarabeasca
Besztercze *see* Bistrița
Besztercebánya *see* Banská Bystrica
Betafo 79 G3 Antananarivo,
C Madagascar
Betanzos 92 B1 Galicia, NW Spain
Bethlehem 78 D4 Free State,
C South Africa
Bethlehem 119 B6 *var.* Beit Leḥem, *Ar.*
Bayt Laḥm, *Heb.* Bet Leḥem. C West
Bank, SW Asia
Béticos, Sistemas 92 D4 *var.* Sistema
Penibético, *Eng.* Baetic Cordillera,
Baetic Mountains. *mountain range*
S Spain
Bet Leḥem *see* Bethlehem
Bétou 77 C5 Likouala, N Congo
Bette, Picco 71 G4 *var.* Bikkū Bīttī, *It.*
Picco Bette. *mountain* S Libya
Bette, Picco *see* Bette, Picco
Beulah 40 C2 Michigan, N USA
Beuthen *see* Bytom
Beveren 87 B5 Oost-Vlaanderen,
N Belgium
Beverley 89 D5 E England,
United Kingdom
Bexley 89 B8 Bexley, SE England,
United Kingdom
Beyla 74 D4 SE Guinea
Beyrouth 118 A4 *var.* Bayrūt, *Eng.*
Beirut; *anc.* Berytus. *country capital*
(Lebanon) W Lebanon
Beyşehir 116 B4 Konya,
SW Turkey (Türkiye)
Beyşehir Gölü 116 B4 *lake*
C Turkey (Türkiye)

Béziers 91 C6 *anc.* Baeterrae, Baeterrae
Septimanorum, Julia Beterrae. Hérault,
S France
Bezmein *see* Abadan
Bezwada *see* Vijayawāda
Bhadrāvati 132 C2 Karnātaka, SW India
Bhāgalpur 135 F3 Bihār, NE India
Bhaktapur 135 F3 Central, C Nepal
Bhamo 136 B2 *var.* Banmo. Kachin
State, N Myanmar (Burma)
Bhārat *see* India
Bharati 154 D3 *Indian research station.*
Antarctica
Bharūch 134 C4 Gujarāt, W India
Bhaunagar *see* Bhāvnagar
Bhāvnagar 134 C4 *prev.* Bhaunagar.
Gujarāt, W India
Bheanntraí, Bá *see* Bantry Bay
Bhopāl 134 D4 *state capital* Madhya
Pradesh, C India
Bhubaneshwar 135 F5 *prev.*
Bhubaneswar, Bhuvaneshwar. *state
capital* Odisha, E India
Bhubaneswar *see* Bhubaneshwar
Bhuket *see* Phuket
Bhusaval *see* Bhusāwal
Bhusāwal 134 D4 *prev.* Bhusaval.
Mahārāshtra, C India
Bhutan 135 G3 *off.* Kingdom of Bhutan,
var. Druk-yul. *country* S Asia
Bhutan, Kingdom of *see* Bhutan
Bhuvaneshwar *see* Bhubaneshwar
Biak, Pulau 139 G4 *island* E Indonesia
Biała Podlaska 98 E3 Lubelskie,
E Poland
Białogard 98 B2 *Ger.* Belgard.
Zachodnio-pomorskie, NW Poland
Białystok 98 E3 *Rus.* Belostok, Bielostok.
Podlaskie, NE Poland
Bianco, Monte *see* Blanc, Mont
Biarritz 91 A6 Pyrénées-Atlantiques,
SW France
Bicaz 108 C3 *Hung.* Békás. Neamț,
NE Romania
Bichiş-Ciaba *see* Békéscsaba
Biddeford 41 G2 Maine, NE USA
Bideford 89 C7 SW England,
United Kingdom
Biel 95 A7 *Fr.* Bienne. Bern,
W Switzerland
Bielefeld 94 B4 Nordrhein-Westfalen,
NW Germany
Bielitz/Bielitz-Biala *see* Bielsko-Biała
Bielostok *see* Białystok
Bielsko-Biała 99 C5 *Ger.* Bielitz,
Bielitz-Biala. Śląskie, S Poland
Bielsk Podlaski 98 E3 Białystok,
E Poland
Bien Bien *see* Diên Biên
Biên Hoa 137 E6 Đông Nai, S Vietnam
Bienne *see* Biel
Bienville, Lac 38 D2 *lake* Québec,
C Canada
Bié, Planalto do 78 B2 *var.* Bié Plateau.
plateau C Angola
Bié Plateau *see* Bié, Planalto do
Big Cypress Swamp 43 E5 *wetland*
Florida, SE USA
Bigge Island 146 C2 *island*
Western Australia
Bighorn Mountains 44 C2 *mountain
range* Wyoming, C USA
Bighorn River 44 C2 *river* Montana/
Wyoming, NW USA
Bignona 74 B3 SW Senegal
Bigorra *see* Tarbes
Bigosovo *see* Bihosava
Big Sioux River 45 E2 *river* Iowa/South
Dakota, N USA
Big Spring 49 E3 Texas, SW USA
Bihać 100 B3 Federacija Bosna I
Hercegovina, NW Bosnia and
Herzegovina
Bihār 135 F3 *prev.* Behar. *cultural region*
N India
Bihār *see* Bihār Sharīf
Biharamulo 73 B7 Kagera, NW Tanzania
Bihār Sharīf 135 F3 *var.* Bihār. Bihār,
N India
Bihosava 107 D5 *Rus.* Bigosovo.
Vitsyebskaya Voblasts', NW Belarus
Bijeljina 100 C3 Republika Srpska,
NE Bosnia and Herzegovina
Bijelo Polje 101 D5 E Montenegro
Bikāner 134 C3 Rājasthān, NW India
Bikin 115 G4 Khabarovskiy Kray,
SE Russia
Bikini Atoll 144 C1 *var.* Pikinni. *atoll*
Ralik Chain, NW Marshall Islands
Bīkkū Bīttī *see* Bette, Picco

Biläspur 135 E4 Chhattisgarh, C India
Biläsuvar 117 H3 *Rus.* Bilyasuvar; *prev.*
Pushkino. SE Azerbaijan
Bila Tserkva 109 E2 *Rus.* Belaya
Tserkov'. Kyivska Oblast, N Ukraine
Bilauktaung Range 137 C6 *var.*
Thanintari Taungdan. *mountain range*
Myanmar (Burma)/Thailand
Bilbao 93 E1 *Basq.* Bilbo. País Vasco,
N Spain
Bilbo *see* Bilbao
Bilecik 116 B3 Bilecik,
NW Turkey (Türkiye)
Bilhorod-Dnistrovskyi 109 E4 *prev.*
Akkerman, *Rus.* Belgorod-
Dnestrovsky, *Rom.* Cetatea Albă, *anc.*
Tyras. SW Ukraine
Billings 44 C2 Montana, NW USA
Bilma, Grand Erg de 75 H3 *desert*
NE Niger
Biloela 148 D4 Queensland, E Australia
Biloxi 42 C3 Mississippi, S USA
Biltine 76 C3 Biltine, E Chad
Bilwi *see* Puerto Cabezas
Bilyasuvar *see* Biläsuvar
Bilzen 87 D6 Limburg, NE Belgium
Bimini Islands 54 C1 *island group*
W The Bahamas
Binche 87 B7 Hainaut, S Belgium
Bindloe Island *see* Marchena, Isla
Bin Ghalfān, Jazā'ir *see* Ḥalāniyāt,
Juzur al
Binghamton 41 F3 New York, NE USA
Bingöl 117 E3 Bingöl,
E Turkey (Türkiye)
Bint Jubayl *see* Bent Jbaïl
Bintulu 138 D3 Sarawak, East Malaysia
Binzhou 128 D4 Shandong, E China
Bío Bío, Río 65 B5 *river* C Chile
Bioco, Isla de *see* Bioko
Bioko 77 A5 *var.* Isla de Bioco, *Eng.*
Fernando Po, *Sp.* Fernando Póo; *prev.*
Macías Nguema Biyogo. *island*
NW Equatorial Guinea
Birāk 71 F3 *var.* Brak. C Libya
Birao 76 D3 Vakaga, NE Central
African Republic
Birātnagar 135 F3 Eastern, SE Nepal
Bir es Saba *see* Be'er Sheva
Bîrjand 120 E3 Khorāsān-e Janūbī,
E Iran
Birkenfeld 95 A5 Rheinland-Pfalz,
SW Germany
Birkenhead 89 C5 NW England,
United Kingdom
Bîrlad *see* Bârlad
Birmingham 89 C6 C England,
United Kingdom
Birmingham 42 C2 Alabama, S USA
Bir Moghrein *see* Bîr Mogreïn
Bîr Mogreïn 74 C1 *var.* Bir Moghrein;
prev. Fort-Trinquet. Tiris Zemmour,
N Mauritania
Birnie Island 145 E3 *atoll* Phoenix
Islands, C Kiribati
Birnin Konni 75 F3 *var.* Birni-Nkonni.
Tahoua, SW Niger
Birni-Nkonni *see* Birnin Konni
Birobidzhan 115 G4 Yevreyskaya
Avtonomnaya Oblast', SE Russia
Birsen *see* Biržai
Birsk 111 D5 Respublika Bashkortostan,
W Russia
Biržai 106 C4 *Ger.* Birsen. Panevėžys,
NE Lithuania
Birżebbuġa 102 B5 SE Malta
Bisanthe *see* Tekirdağ
Bisbee 48 B3 Arizona, SW USA
Biscaia, Baía de *see* Biscay, Bay of
Biscay, Bay of 80 B4 *Sp.* Golfo de
Vizcaya, *Port.* Baía de Biscaia. *bay*
France/Spain
Biscay Plain 80 B3 *abyssal plain* SE Bay
of Biscay
Bischofsburg *see* Biskupiec
Bishah, Wadi 121 B5 *dry watercourse*
C Saudi Arabia
Bishkek 123 G2 *var.* Pishpek; *prev.*
Frunze. *country capital* (Kyrgyzstan)
Bishkek Shaary/Gorod Bishkek,
N Kyrgyzstan
Bishop's Lynn *see* King's Lynn
Bishri, Jabal 118 D3 *mountain range*
E Syria
Biskra 71 E2 *var.* Beskra, Biskara.
NE Algeria
Biskupiec 98 D2 *Ger.* Bischofsburg.
Warmińsko-Mazurskie, NE Poland
Bislig 139 F2 Mindanao, S Philippines

Bismarck 45 E2 *state capital* North
Dakota, N USA
Bismarck Archipelago 144 B3 *island
group* NE Papua New Guinea
Bismarck Sea 144 B3 *sea*
W Pacific Ocean
Bisnulok *see* Phitsanulok
Bissau 74 B4 *country capital*
(Guinea-Bissau) W Guinea-Bissau
Bistrița 108 B3 *Ger.* Bistritz, *Hung.*
Beszterce; *prev.* Nösen.
Bistrița-Năsăud, N Romania
Bistritz *see* Bistrița
Bitam 77 B5 Woleu-Ntem, N Gabon
Bitburg 95 A5 Rheinland-Pfalz,
SW Germany
Bitlis 117 F3 Bitlis, SE Turkey (Türkiye)
Bitoeng *see* Bitung
Bitola 101 D6 *Turk.* Monastir; *prev.*
Bitolj. S North Macedonia
Bitolj *see* Bitola
Bitonto 97 D5 *anc.* Butuntum. Puglia,
SE Italy
Bitterroot Range 46 D2 *mountain range*
Idaho/Montana, NW USA
Bitung 139 F3 *prev.* Bitoeng. Sulawesi,
C Indonesia
Biu 75 H4 Borno, E Nigeria
Biwa-ko 131 C6 *lake* Honshū, SW Japan
Bizerta *see* Bizerte
Bizerte 71 E1 *Ar.* Banzart, *Eng.* Bizerta.
N Tunisia
Bjelovar 100 B2 *Hung.* Belovár.
Bjelovar-Bilogora, N Croatia
Bjeshkët e Namuna *see* North
Albanian Alps
Björneborg *see* Pori
Bjørnøya 83 F3 *Eng.* Bear Island. *island*
N Norway
Blackall 148 C4 Queensland, E Australia
Black Drin 101 D6 *Alb.* Lumi i Drinit të
Zi, *Croatian* Crni Drim. *river* Albania/
North Macedonia
Blackfoot 46 E4 Idaho, NW USA
Black Forest 95 B6 *Eng.* Black Forest.
mountain range SW Germany
Black Forest *see* Schwarzwald
Black Hills 44 D3 *mountain range* South
Dakota/Wyoming, N USA
Blackpool 89 C5 NW England,
United Kingdom
Black Range 48 C2 *mountain range*
New Mexico, SW USA
Black River 54 A5 W Jamaica
Black River 136 C3 *Chin.* Babian Jiang,
Lixian Jiang, *Fr.* Rivière Noire, *Vtn.*
Sông Đa. *river* China/Vietnam
Black Rock Desert 47 C5 *desert* Nevada,
W USA
Black Sand Desert *see* Garagum
Black Sea 116 B1 *var.* Euxine Sea, *Bul.*
Cherno More, *Rom.* Marea Neagră, *Rus.*
Chernoye More, *Turk.* Karadeniz, *Ukr.*
Chorne More. *sea* Asia/Europe
Black Sea Lowland 109 E4 *Ukr.*
Prychornomor'ska Nyzovyna.
depression SE Europe
Black Volta 75 E4 *var.* Borongo,
Mouhoun, Moun Hou, *Fr.* Volta Noire.
river W Africa
Blackwater 89 A6 *Ir.* An Abhainn Mhór.
river S Ireland
Blackwater State *see* Nebraska
Blagoevgrad 104 C3 *prev.* Gorna
Dzhumaya. Blagoevgrad, W Bulgaria
Blagoveshchensk 115 G4 Amurskaya
Oblast', SE Russia
Blahovishchenske 109 E3 *prev.*
Ulianovka, *Rus.* Ulyanovka.
Kirovohrads'ka Oblast', C Ukraine
Blake Plateau 35 D6 *var.* Blake Terrace.
undersea plateau W Atlantic Ocean
Blake Terrace *see* Blake Plateau
Blanca, Bahía 65 C5 *bay* E Argentina
Blanca, Costa 93 F4 *physical region*
SE Spain
Blanche, Lake 149 B5 *lake*
South Australia
Blanc, Mont 91 D5 *It.* Monte Bianco.
mountain France/Italy
Blanco, Cape 46 A4 *headland* Oregon,
NW USA
Blanes 93 G2 Cataluña, NE Spain
Blankenberge 87 A5 West-Vlaanderen,
NW Belgium
Blankenheim 95 A5 Nordrhein-
Westfalen, W Germany
Blanquilla, La 59 E1 *var.* La Blanquilla.
island N Venezuela
Blanquilla, La *see* Blanquilla, Isla

Blantyre 79 E2 var. Blantyre-Limbe.
  Southern, S Malawi
Blantyre-Limbe see Blantyre
Blaricum 86 C3 Noord-Holland,
  C Netherlands
Blatnitsa see Durankulak
Blenheim 151 C5 Marlborough, South
  Island, New Zealand
Blesae see Blois
Blida 70 D2 var. El Boulaida, El Boulaïda.
  N Algeria
Bloemfontein 78 C4 var. Mangaung.
  country capital (South Africa-judicial
  capital) Free State, C South Africa
Blois 90 C4 anc. Blesae. Loir-et-Cher,
  C France
Bloomfield 48 C1 New Mexico, SW USA
Bloomington 40 B4 Illinois, N USA
Bloomington 40 C4 Indiana, N USA
Bloomington 45 F2 Minnesota, N USA
Bloomsbury 148 D3 Queensland,
  NE Australia
Bluefield 40 D5 West Virginia, NE USA
Bluefields 53 E3 Región Autónoma
  Atlántico Sur, SE Nicaragua
Bluegrass State see Kentucky
Blue Hen State see Delaware
Blue Law State see Connecticut
Blue Mountain Peak 54 B5 mountain
  E Jamaica
Blue Mountains 46 C3 mountain range
  Oregon/Washington, NW USA
Blue Nile 72 C4 var. Abai, Bahr el,
  Azraq, Amh. Ābay Wenz, Ar. An Nīl al
  Azraq. river Ethiopia/Sudan
Blumenau 63 E5 Santa Catarina, S Brazil
Blythe 47 D8 California, W USA
Blytheville 42 C1 Arkansas, C USA
Bo 74 C4 S Sierra Leone
Boaco 52 D3 Boaco, S Nicaragua
Boa Vista 62 D1 state capital Roraima,
  NW Brazil
Boa Vista 74 A3 var. Ilhas de
  Barlavento, E Cape Verde (Cabo Verde)
Bobaomby, Tanjona 79 G2 Fr. Cap
  d'Ambre. headland N Madagascar
Bobigny 90 E1 Seine-St-Denis, N France
Bobo-Dioulasso 74 D4 SW Burkina
Bobrinets see Bobrynets
Bobruysk see Babruysk
Bobrynets 109 E3 Rus. Bobrinets.
  Kirovohrads'ka Oblast', C Ukraine
Boca Raton 43 F5 Florida, SE USA
Bocay 52 D2 Jinotega, N Nicaragua
Bocheykovo see Bacheykava
Bocholt 94 A4 Nordrhein-Westfalen,
  W Germany
Bochum 94 A4 Nordrhein-Westfalen,
  W Germany
Bocşa 108 A4 Ger. Bokschen, Hung.
  Boksánbánya. Caraş-Severin,
  SW Romania
Bodaybo 115 F4 Irkutskaya Oblast',
  E Russia
Bodegas see Babahoyo
Boden 84 D4 Norrbotten, N Sweden
Bodensee see Constance, Lake, C Europe
Bodmin 89 C7 SW England,
  United Kingdom
Bodø 84 C3 Nordland, C Norway
Bodrum 116 A4 Muğla,
  SW Turkey (Türkiye)
Boeloekoemba see Bulukumba
Boende 77 C5 Equateur, C Dem.
  Rep. Congo
Boeroe see Buru, Pulau
Boetoeng see Buton, Pulau
Bogale 136 B4 Ayeyarwady, SW Myanmar
  (Burma)
Bogalusa 42 B3 Louisiana, S USA
Bogatynia 98 B4 Ger. Reichenau.
  Dolnośląskie, SW Poland
Bogendorf see Łuków
Bogor 138 C5 Dut. Buitenzorg. Jawa,
  C Indonesia
Bogotá 58 B3 prev. Santa Fe, Santa Fe de
  Bogotá. country capital (Colombia)
  Cundinamarca, C Colombia
Bogushëvsk see Bahushewsk
Boguslav see Bohuslav
Bo Hai 128 D4 var. Gulf of Chihli. gulf
  NE China
Bohemia 99 A5 Cz. Čechy, Ger. Böhmen.
  W Czechia (Czech Republic)
Bohemian Forest 95 C5 Cz. Český Les,
  Šumava, Ger. Böhmerwald. mountain
  range C Europe
Böhmen see Bohemia

Böhmerwald see Bohemian Forest
Böhmisch-Krumau see Český Krumlov
Bohol Sea 139 E2 var. Mindanao Sea. sea
  S Philippines
Bohuslav 109 E2 Rus. Boguslav. Kyivska
  Oblast', N Ukraine
Boise 46 D3 var. Boise City. state capital
  Idaho, NW USA
Boise City 49 E1 Oklahoma, C USA
Boise City see Boise
Bois, Lac des see Woods, Lake of the
Bois-le-Duc see 's-Hertogenbosch
Boiarka 109 E2 var. Boyarka. Kyivska
  Oblast', N Ukraine
Boizenburg 94 C3 Mecklenburg-
  Vorpommern, N Germany
Bojador see Boujdour
Bojnūrd 120 D2 var. Bujnurd.
  Khorāsān-e Shemālī, N Iran
Bokāro 135 F4 Jhārkhand, N India
Boké 74 C4 W Guinea
Bokhara see Buxoro
Bokhtar 123 E3 prev. Qürghonteppa,
  Rus. Kurgan-Tyube. SW Tajikistan
Boksánbánya/Bokschen see Bocşa
Bol 76 B3 Lac, W Chad
Bolgatanga 75 E4 N Ghana
Bolgrad see Bolhrad
Bolhrad 108 D4 Rus. Bolgrad. Odes'ka
  Oblast', SW Ukraine
Bolívar, Pico 58 C2 mountain
  W Venezuela
Bolivia 61 F3 off. Plurinational State of
  Bolivia. country W South America
Bolivia, Plurinational Republic of see
  Bolivia
Bollène 91 D6 Vaucluse, SE France
Bollnäs 85 C5 Gävleborg, C Sweden
Bollon 149 D5 Queensland, C Australia
Bologna 96 C3 Emilia-Romagna,
  N Italy
Bol'shevik see Bal'shavik
Bol'shevik, Ostrov 115 E2 island
  Severnaya Zemlya, N Russia
Bol'shezemel'skaya Tundra 110 E3
  physical region NW Russia
Bol'shoy Lyakhovskiy, Ostrov 115 F2
  island NE Russia
Bolton 89 D5 prev. Bolton-le-Moors.
  NW England, United Kingdom
Bolton-le-Moors see Bolton
Bolu 116 B3 Bolu, NW Turkey (Türkiye)
Bolungarvík 83 E4 Vestfirdhir,
  NW Iceland
Bolyarovo 104 D3 prev. Pashkeni.
  Yambol, E Bulgaria
Bolzano 96 C1 Ger. Bozen; anc.
  Bauzanum. Trentino-Alto Adige,
  N Italy
Boma 77 B6 Bas-Congo, W Dem.
  Rep. Congo
Bombay see Mumbai
Bomu 76 D4 var. Mbomou, Mbomu,
  M'Bomu. river Central African
  Republic/Dem. Rep. Congo
Bonaire 55 F5 Dutch special municipality
  S West Indies
Bonanza 52 D2 Región Autónoma
  Atlántico Norte, NE Nicaragua
Bonaparte Archipelago 146 C2 island
  group Western Australia
Bon, Cap 102 D3 headland N Tunisia
Bonda 77 B6 Ogooué-Lolo, C Gabon
Bondoukou 75 E4 E Ivory Coast
Bône see Annaba, Algeria
Bone, Teluk 139 E4 bay Sulawesi,
  C Indonesia
Bongaigaon 135 G3 Assam, NE India
Bongo, Massif des 76 D4 var. Chaîne des
  Mongos. mountain range NE Central
  African Republic
Bongor 76 B3 Mayo-Kébbi, SW Chad
Bonifacio 91 E7 Corse, France,
  C Mediterranean Sea
**Bonifacio, Bocche de/Bonifacio,
  Bouches de** see Bonifacio, Strait of
Bonifacio, Strait of 96 A4 Fr. Bouches
  de Bonifacio, It. Bocche di Bonifacio.
  strait C Mediterranean Sea
Bonn 95 A5 Nordrhein-Westfalen,
  W Germany
Bononia see Vidin, Bulgaria
Bononia see Boulogne-sur-Mer, France
Boosaaso 72 E4 var. Bandar Kassim,
  Bender Qaasim, Bosaso, It. Bender
  Cassim. Bari, N Somalia
Boothia Felix see Boothia Peninsula
Boothia, Gulf of 37 F2 gulf Nunavut,
  NE Canada

Boothia Peninsula 37 F2 prev. Boothia
  Felix. peninsula Nunavut, NE Canada
Boppard 95 A5 Rheinland-Pfalz,
  W Germany
Boquete 53 E5 var. Bajo Boquete.
  Chiriquí, W Panama
Boquillas 50 D2 var. Boquillas del
  Carmen. Coahuila, NE Mexico
Boquillas del Carmen see Boquillas
Bor 100 E4 Serbia, E Serbia
Bor 73 B5 Jonglei, C South Sudan
Borås 85 B7 Västra Götaland, S Sweden
Borbetomagus see Worms
Borborema, Planalto da 56 E3 plateau
  NE Brazil
Bordeaux 91 B5 anc. Burdigala.
  Gironde, SW France
Bordj Omar Driss 71 E3 E Algeria
Borgå see Porvoo
Børgefjell 84 C4 mountain range
  C Norway
Borger 86 E2 Drenthe, NE Netherlands
Borger 49 E1 Texas, SW USA
Borgholm 85 C7 Kalmar, S Sweden
Borgo Maggiore 96 E1 NW San Marino
Borislav see Boryslav
Borisoglebsk 111 B6 Voronezhskaya
  Oblast', W Russia
Borisov see Barysaw
Borlänge 85 C6 Dalarna, C Sweden
Borne 86 E3 Overijssel, E Netherlands
Borneo 138 C4 island Brunei/Indonesia/
  Malaysia
Bornholm 85 B8 island E Denmark
Borohoro Shan 126 B2 mountain range
  NW China
Borongo see Black Volta
Boron'ki see Baron'ki
Borosjenő see Ineu
Borovan 104 C2 Vratsa, NW Bulgaria
Borovichi 110 B4 Novgorodskaya
  Oblast', W Russia
Borovo 100 C3 Vukovar-Srijem,
  NE Croatia
Borriana 93 F3 var. Burriana. Comunitat
  Valenciana, E Spain
Borşa 108 C3 Hung. Borsa. Maramureş,
  N Romania
Boryslav 108 B2 Pol. Borysław, Rus.
  Borislav. L'vivs'ka Oblast', NW Ukraine
Borysław see Boryslav
Bosanska Dubica 100 B3 var. Kozarska
  Dubica. Republika Srpska, NW Bosnia
  and Herzegovina
Bosanska Gradiška 100 B3 var.
  Gradiška. Republika Srpska, N Bosnia
  and Herzegovina
Bosanski Novi 100 B3 var. Novi Grad.
  Republika Srpska, NW Bosnia and
  Herzegovina
Bosanski Šamac 100 C3 var. Šamac.
  Republika Srpska, N Bosnia and
  Herzegovina
Bosaso see Boosaaso
Bösing see Pezinok
Boskovice 99 B5 Ger. Boskowitz.
  Jihomoravský Kraj, SE Czechia
  (Czech Republic)
Boskowitz see Boskovice
Bosna 100 C4 river N Bosnia and
  Herzegovina
Bosnia and Herzegovina 100 B3
  country SE Europe
Boso-hanto 131 D6 peninsula Honshū,
  S Japan
Bosphorus/Bosporus see İstanbul Boğazı
Bosporus Cimmerius see Kerch Strait
Bosporus Thracius see İstanbul Boğazı
Bossangoa 76 C4 Ouham, C Central
  African Republic
Bossembélé 76 C4 Ombella-Mpoko,
  C Central African Republic
Bossier City 42 A2 Louisiana, S USA
Bosten Hu 126 C3 var. Bagrax Hu. lake
  NW China
Boston 89 E6 prev. St.Botolph's Town.
  E England, United Kingdom
Boston 41 G3 state capital Massachusetts,
  NE USA
Boston Mountains 42 B1 mountain
  range Arkansas, C USA
Bostyn' see Bastyn'
Botany 148 E2 New South Wales,
  E Australia
Botany Bay 148 E2 inlet New South
  Wales, SE Australia
Boteti 78 C3 var. Botletle. river
  N Botswana
Bothnia, Gulf of 85 D5 Fin. Pohjanlahti,
  Swe. Bottniska Viken. gulf N Baltic Sea

Botletle see Boteti
Botoşani 108 C3 Hung. Botosány.
  Botoşani, NE Romania
Botosány see Botoşani
Botou 128 C4 prev. Bozhen. Hebei,
  E China
Botrange 87 D6 mountain E Belgium
Botswana 78 C3 off. Republic of
  Botswana. country S Africa
Botswana, Republic of see Botswana
Bottniska Viken see Bothnia, Gulf of
Bouar 76 B4 Nana-Mambéré, W Central
  African Republic
Bou Craa 70 B3 var. Bu Craa.
  NW Western Sahara
Bougainville Island 142 B3 island
  NE Papua New Guinea
Bougaroun, Cap 102 C3 headland
  NE Algeria
Bougouni 74 D4 Sikasso, SW Mali
Boujdour 70 A3 var. Bojador.
  W Western Sahara
Boulder 44 C4 Colorado, C USA
Boulder 44 B2 Montana, NW USA
Boulogne see Boulogne-sur-Mer
Boulogne-Billancourt 90 D1
  Île-de-France, N France Europe
Boulogne-sur-Mer 90 C2 var. Boulogne;
  anc. Bononia, Gesoriacum,
  Gessoriacum. Pas-de-Calais, N France
Boûmdeïd 74 C3 var. Boumdeït. Assaba,
  S Mauritania
Boumdeït see Boûmdeïd
Boundiali 74 D4 N Ivory Coast
Bountiful 44 B4 Utah, W USA
Bounty Basin see Bounty Trough
Bounty Islands 142 D5 island group
  S New Zealand
Bounty Trough 152 C5 var. Bounty
  Basin. trough S Pacific Ocean
Bourbonnais 90 C4 cultural region
  C France
Bourbon Vendée see la Roche-sur-Yon
Bourg see Bourg-en-Bresse
Bourgas see Burgas
Bourge-en-Bresse see Bourg-en-Bresse
Bourg-en-Bresse 91 D5 var. Bourg,
  Bourge-en-Bresse. Ain, E France
Bourges 90 C4 anc. Avaricum. Cher,
  C France
Bourgogne 90 C4 Eng. Burgundy.
  cultural region E France
Bourke 149 C5 New South Wales,
  SE Australia
Bournemouth 89 D7 S England,
  United Kingdom
Boutilimit 74 C3 Trarza, SW Mauritania
Bouvet Island 67 D7 Norwegian
  dependency S Atlantic Ocean
Bowen 148 D3 Queensland, NE Australia
Bowling Green 40 B5 Kentucky,
  S USA
Bowling Green 40 C3 Ohio, N USA
Boxmeer 86 D4 Noord-Brabant,
  SE Netherlands
Boyarka see Boiarka
Boychinovtsi 104 C2 prev. Lekhchevo.
  Montana, NW Bulgaria
Boysun 123 E3 Rus. Baysun.
  Surkhondaryo Viloyati, S Uzbekistan
Bozeman 44 B2 Montana, NW USA
Bozen see Bolzano
Bozhen see Botou
Bozüyük 116 B3 Bilecik,
  NW Turkey (Türkiye)
Brač 100 B4 var. Brach, It. Brazza; anc.
  Brattia. island S Croatia
Bracara Augusta see Braga
Brach see Brač
Brades 55 G3 de facto dependent territory
  capital, de jure capital , Plymouth,
  destroyed by volcano in 1995
  (Montserrat) W Montserrat
Bradford 89 D5 N England,
  United Kingdom
Brady 49 F3 Texas, SW USA
Braga 92 B2 anc. Bracara Augusta.
  Braga, NW Portugal
Bragança 92 C2 Eng. Braganza; anc. Julio
  Briga. Bragança, NE Portugal
Braganza see Bragança
Brahestad see Raahe
Brahmanbaria 135 G4 Chattogram,
  E Bangladesh
Brahmapur 135 F5 Odisha, E India
Brahmaputra 135 H3 var. Padma,
  Tsangpo, Ben. Jamuna, Chin. Yarlung
  Zangbo Jiang, Ind. Dihang, Dihang,
  Dihang, Siang. river S Asia
Brăila 108 D4 Brăila, E Romania

Braine-le-Comte *87 B6* Hainaut,
   SW Belgium
Brainerd *45 F2* Minnesota, N USA
Brak *see* Bīrāk
Bramaputra *see* Brahmaputra
Brampton *38 D5* Ontario, S Canada
Branco, Rio *56 C3* *river* N Brazil
Brandberg *78 A3* *mountain*
   NW Namibia
Brandenburg *94 C3* *var.* Brandenburg
   an der Havel. Brandenburg,
   NE Germany
Brandenburg an der Havel *see*
   Brandenburg
Brandon *37 F5* Manitoba, S Canada
Braniewo *98 D2* *Ger.* Braunsberg.
   Warmińsko-mazurskie, N Poland
Brasil *see* Brazil
Brasília *63 F3* *country capital* (Brazil)
   Distrito Federal, C Brazil
Brasil, República Federativa do *see*
   Brazil
Braşov *108 C4* *Ger.* Kronstadt, *Hung.*
   Brassó; *prev.* Oraşul Stalin. Braşov,
   C Romania
Brassó *see* Braşov
Bratislava *99 C6* *Ger.* Pressburg, *Hung.*
   Pozsony. *country capital* (Slovakia)
   Bratislavský Kraj, W Slovakia
Bratsk *115 E4* Irkutskaya Oblast',
   C Russia
Brattia *see* Brač
Braunsberg *see* Braniewo
Braunschweig *94 C4* *Eng./Fr.*
   Brunswick. Niedersachsen, N Germany
Brava *see* Baraawe
Brava, Costa *93 H2* *coastal region*
   NE Spain
Bravo del Norte, Río/Bravo, Río *see*
   Grande, Rio
Bravo, Río *50 C1* *river* Mexico/USA
   North America
Brawley *47 D8* California, W USA
Brazil *62 C2* *off.* Federative Republic of
   Brazil, *Port.* República Federativa do
   Brasil, *Sp.* Brasil; *prev.* United States of
   Brazil. *country* South America
Brazil Basin *67 C5* *var.* Brazilian Basin,
   Brazil'skaya Kotlovina. *undersea basin*
   W Atlantic Ocean
Brazil, Federative Republic of *see* Brazil
Brazilian Basin *see* Brazil Basin
Brazilian Highlands *see* Central, Planalto
Brazil'skaya Kotlovina *see* Brazil Basin
Brazil, United States of *see* Brazil
Brazos River *49 G3* *river* Texas, SW USA
Brazza *see* Brač
Brazzaville *77 B6* *country capital*
   (Congo) Capital District, S Congo
Brčko *100 C3* Republika Srpska,
   NE Bosnia and Herzegovina
Brecht *87 C5* Antwerpen, N Belgium
Brecon Beacons *see* Bannau Brycheiniog
Breda *86 C4* Noord-Brabant,
   S Netherlands
Bree *87 D5* Limburg, NE Belgium
Bregalnica *101 E6* *river*
   E North Macedonia
Bregenz *57 B7* *anc.* Brigantium.
   Vorarlberg, W Austria
Bregovo *104 B1* Vidin, NW Bulgaria
Bremen *94 B3* *Fr.* Brême. Bremen,
   NW Germany
Bremerhaven *94 B3* Bremen,
   NW Germany
Bremerton *46 B2* Washington, NW USA
Brenham *49 G3* Texas, SW USA
Brenner, Col du/Brennero, Passo del
   see Brenner Pass
Brenner Pass *96 C1* *var.* Brenner Sattel,
   Fr. Col du Brenner, *Ger.* Brennerpass,
   It. Passo del Brennero. *pass* Austria/
   Italy
Brennerpass *see* Brenner Pass
Brenner Sattel *see* Brenner Pass
Brescia *96 B2* *anc.* Brixia. Lombardia,
   N Italy
Breslau *see* Wrocław
Bressanone *96 C3* *Ger.* Brixen.
   Trentino-Alto Adige, N Italy
Brest *107 A6* *Pol.* Brześć nad Bugiem,
   Rus. Brest-Litovsk; *prev.* Brześć
   Litewski. Brestskaya Voblasts',
   SW Belarus
Brest *90 A3* Finistère, NW France
Brest-Litovsk *see* Brest
Bretagne *90 A3* *Eng.* Brittany, *Lat.*
   Britannia Minor. *cultural region*
   NW France
Brewster, Kap *see* Kangikajik

Brewton *42 C3* Alabama,
   S USA
Brezhnev *see* Naberezhnyye Chelny
Brezovo *104 D2* *prev.* Abrashlare.
   Plovdiv, C Bulgaria
Bria *76 D4* Haute-Kotto, C Central
   African Rep.
Briançon *91 D5* *anc.* Brigantio.
   Hautes-Alpes, SE France
Bricgstow *see* Bristol
Bridgeport *41 F3* Connecticut, NE USA
Bridgetown *55 G2* *country capital*
   (Barbados) SW Barbados
Bridlington *89 D5* E England,
   United Kingdom
Bridport *89 D7* S England,
   United Kingdom
Brieg *see* Brzeg
Brig *95 A7* *Fr.* Brigue, *It.* Briga. Valais,
   SW Switzerland
Briga *see* Brig
Brigantio *see* Briançon
Brigantium *see* Bregenz
Brigham City *44 B3* Utah, W USA
Brighton *89 E7* SE England,
   United Kingdom
Brighton *44 D4* Colorado, C USA
Brigue *see* Brig
Brindisi *97 E5* *anc.* Brundisium,
   Brundusium. Puglia, SE Italy
Briovera *see* St-Lô
Brisbane *149 E5* *state capital*
   Queensland, E Australia
Bristol *89 D7* *anc.* Bricgstow.
   SW England, United Kingdom
Bristol *41 F3* Connecticut, NE USA
Bristol *40 D5* Tennessee, S USA
Bristol Bay *36 B3* *bay* Alaska, USA
Bristol Channel *89 C7* *inlet* England/
   Wales, United Kingdom
Britain *80 C3* *var.* Great Britain. *island*
   United Kingdom
Britannia Minor *see* Bretagne
British Columbia *36 D4* *Fr.*
   Colombie-Britannique. *province*
   SW Canada
British Guiana *see* Guyana
British Honduras *see* Belize
British Indian Ocean Territory *141 B5*
   UK Overseas Territory C Indian Ocean
British North Borneo *see* Sabah
British Solomon Islands Protectorate
   see Solomon Islands
British Virgin Islands *55 F3* *var.* Virgin
   Islands. *UK Overseas Territory*
   E West Indies
Brittany *see* Bretagne
Briva Curretia *see* Brive-la-Gaillarde
Briva Isarae *see* Pontoise
Brive *see* Brive-la-Gaillarde
Brive-la-Gaillarde *91 C5* *prev.* Brive;
   anc. Briva Curretia. Corrèze, C France
Brixen *see* Bressanone
Brixia *see* Brescia
Brno *99 B5* *Ger.* Brünn. Jihomoravský
   Kraj, SE Czechia (Czech Republic)
Brocēni *106 B3* Saldus, SW Latvia
Brod/Bród *see* Slavonski Brod
Brodeur Peninsula *37 F2* *peninsula*
   Baffin Island, Nunavut, NE Canada
Brod na Savi *see* Slavonski Brod
Brodnica *98 C3* *Ger.* Buddenbrock.
   Kujawski-pomorskie, C Poland
Broek-in-Waterland *86 C3*
   Noord-Holland, C Netherlands
Broken Arrow *49 G1* Oklahoma, C USA
Broken Bay *148 E1* *bay* New South
   Wales, SE Australia
Broken Hill *149 B6* New South Wales,
   SE Australia
Broken Ridge *141 D6* *undersea plateau*
   S Indian Ocean
Bromberg *see* Bydgoszcz
Bromley *89 B8* United Kingdom
Brookhaven *42 B3* Mississippi, S USA
Brookings *45 F3* South Dakota, N USA
Brooks Range *36 D2* *mountain range*
   Alaska, USA
Brookton *147 B6* Western Australia
Broome *146 B3* Western Australia
Broomfield *44 D4* Colorado, C USA
Broucsella *see* Brussel/Bruxelles
Brovary *109 E2* Kyivska Oblast',
   N Ukraine
Brownfield *49 E2* Texas, SW USA
Brownsville *49 G5* Texas, SW USA
Brownwood *49 F3* Texas, SW USA
Brozha *107 D7* Mahilyowskaya Voblasts',
   E Belarus

Bruges *see* Brugge
Brugge *87 A5* *Fr.* Bruges.
   West-Vlaanderen, NW Belgium
Brummen *86 D3* Gelderland,
   E Netherlands
Brundisium/Brundusium *see* Brindisi
Brunei *138 D3* *off.* Brunei Darussalam,
   Mal. Negara Brunei Darussalam.
   country SE Asia
Brunei Darussalam *see* Brunei
Brunei Town *see* Bandar Seri Begawan
Brünn *see* Brno
Brunner, Lake *151 C5* *lake* South Island,
   New Zealand
Brunswick *43 E3* Georgia, SE USA
Brunswick *see* Braunschweig
Brusa *see* Bursa
Brus Laguna *52 D2* Gracias a Dios,
   E Honduras
Brussa *see* Bursa
Brussel *87 C6* *var.* Brussels, *Fr.* Bruxelles,
   Ger. Brüssel; *anc.* Broucsella. *country
   capital* (Belgium) Brussels, C Belgium
Brüssel/Brussels *see* Brussel/Bruxelles
Brüx *see* Most
Bruxelles *see* Brussel
Bryan *49 G3* Texas, SW USA
Bryansk *111 A5* Bryanskaya Oblast',
   W Russia
Brzeg *98 C4* *Ger.* Brieg; *anc.* Civitas Altae
   Ripae. Opolskie, S Poland
Brześć Litewski/Brześć nad Bugiem *see*
   Brest
Brzeżany *see* Berezhany
Bucaramanga *58 B2* Santander,
   N Colombia
Buchanan *74 C5* *prev.* Grand Bassa.
   SW Liberia
Buchanan, Lake *49 F3* *reservoir* Texas,
   SW USA
Bucharest *see* Bucureşti
Buckeye State *see* Ohio
Bu Craa *see* Bou Craa
Bucureşti *108 C5* *Eng.* Bucharest, *Ger.*
   Bukarest, *prev.* Altenburg; *anc.* Cetatea
   Dambovoiţei. *country capital* (Romania)
   Bucureşti, S Romania
Buda-Kashalyova *107 D7* *Rus.*
   Buda-Koshelevo. Homyel'skaya
   Voblasts', SE Belarus
Buda-Koshelëvo *see* Buda-Kashalyova
Budapest *99 C6* *off.* Budapest Főváros,
   Croatian Budimpešta. *country capital*
   (Hungary) Pest, N Hungary
Budapest Főváros *see* Budapest
Budaun *134 D3* Uttar Pradesh,
   N India
Buddenbrock *see* Brodnica
Budimpešta *see* Budapest
Budweis *see* České Budějovice
Budyšin *see* Bautzen
Buena Park *46 E2* California,
   W USA
Buenaventura *58 A3* Valle del Cauca,
   W Colombia
Buena Vista *61 G4* Santa Cruz, C Bolivia
Buena Vista *93 H5* S Gibraltar Europe
Buenos Aires *64 D4* *hist.* Santa Maria del
   Buen Aire. *country capital* (Argentina)
   Buenos Aires, E Argentina
Buenos Aires *53 E5* Puntarenas,
   SE Costa Rica
Buenos Aires, Lago *65 B6* *var.* Lago
   General Carrera. *lake* Argentina/Chile
Buffalo *41 E3* New York, NE USA
Buffalo Narrows *37 F4* Saskatchewan,
   C Canada
Buff Bay *54 B5* E Jamaica
Buftea *108 C5* Ilfov, S Romania
Bug *81 E3* *Bel.* Zakhodni Buh, *Eng.*
   Western Bug, *Rus.* Zapadnyy Bug, *Ukr.*
   Zakhidnyi Buh. *river* E Europe
Buga *58 B3* Valle del Cauca, W Colombia
Bughotu *see* Santa Isabel
Buguruslan *111 D6* Orenburgskaya
   Oblast', W Russia
Buitenzorg *see* Bogor
Bujalance *92 D4* Andalucía, S Spain
Bujanovac *101 E5* SE Serbia
Bujnurd *see* Bojnūrd
Bujumbura *73 B7* *prev.* Usumbura.
   country capital (Burundi - commercial)
   W Burundi
Bukavu *77 E6* *prev.* Costermansville.
   Sud-Kivu, E Dem. Rep. Congo
Bukhara *see* Buxoro
Bukoba *73 B6* Kagera, NW Tanzania
Bülach *95 B7* Zürich, NW Switzerland
Bulawayo *78 D3* Matabeleland North,
   SW Zimbabwe
Bulgan *127 E2* Bulgan, N Mongolia

Bulgaria *104 C2* *off.* Republic of Bulgaria,
   Bul. Republika Bulgaria; *prev.* People's
   Republic of Bulgaria. *country*
   SE Europe
Bulgaria, People's Republic of *see*
   Bulgaria
Bulgaria, Republic of *see* Bulgaria
Bulgaria, Republika *see* Bulgaria
Bullion State *see* Missouri
Bull Shoals Lake *42 B1* *reservoir*
   Arkansas/Missouri, C USA
Bulukumba *139 E4* *prev.* Boeloekoemba.
   Sulawesi, C Indonesia
Bumba *77 D5* Equateur, N Dem.
   Rep. Congo
Bunbury *147 A7* Western Australia
Bundaberg *148 E4* Queensland,
   E Australia
Bungo-suido *131 B7* *strait* SW Japan
Bunia *77 E5* Orientale, NE Dem.
   Rep. Congo
Bünyan *116 D3* Kayseri,
   C Turkey (Türkiye)
Buraida *see* Buraydah
Buraydah *120 B4* *var.* Buraida. Al
   Qaşim, N Saudi Arabia
Burdigala *see* Bordeaux
Burdur *116 B4* *var.* Buldur. Burdur,
   SW Turkey (Türkiye)
Burdur Gölü *116 B4* *salt lake*
   SW Turkey (Türkiye)
Burë *72 C4* Āmara, N Ethiopia
Burgas *104 E2* *var.* Bourgas. Burgas,
   E Bulgaria
Burgaski Zaliv *104 E2* *gulf* E Bulgaria
Burgos *92 D2* Castilla y León, N Spain
Burgundy *see* Bourgogne
Burhan Budai Shan *126 D4* *mountain
   range* C China
Buriram *137 D5* *var.* Buri Ram,
   Puriramya. Buri Ram, E Thailand
Buri Ram *see* Buriram
Burjassot *93 F3* Comunitat Valenciana,
   E Spain
Burketown *148 B3* Queensland,
   NE Australia
Burkina *see* Burkina Faso
Burkina Faso *75 E4* *off.* Burkina Faso;
   var. Burkina; *prev.* Upper Volta.
   country W Africa
Burley *46 D4* Idaho, NW USA
Burlington *45 G4* Iowa, C USA
Burlington *41 F2* Vermont, NE USA
Burma *see* Myanmar
Burnie *149 C8* Tasmania, SE Australia
Burns *46 C3* Oregon, NW USA
Burnside *37 F3* *river* Nunavut,
   NW Canada
Burnsville *45 F2* Minnesota,
   N USA
Burrel *101 D6* *var.* Burreli. Dibër,
   C Albania
Burreli *see* Burrel
Burriana *see* Borriana
Bursa *116 B3* *var.* Brussa, *prev.* Brusa;
   anc. Prusa. Brusa,
   NW Turkey (Türkiye)
Bûr Sa'îd *72 B1* *var.* Port Said. N Egypt
Burtnieks *106 C3* *var.* Burtnieks Ezers.
   lake N Latvia
Burtnieks Ezers *see* Burtnieks
Burundi *73 B7* *off.* Republic of Burundi;
   prev. Kingdom of Burundi, Urundi.
   country C Africa
Burundi, Kingdom of *see* Burundi
Burundi, Republic of *see* Burundi
Buru, Pulau *139 F4* *prev.* Boeroe. *island*
   E Indonesia
Busan *129 E4* *off.* Busan Gwang-yeoksi,
   prev. Pusan, *Jap.* Fusan.
   SE South Korea
Busan Gwang-yeoksi *see* Busan
Büsheher/Bushire *see* Bandar-e Büshehr
Busra *see* Al Başrah, Iraq
Busselton *147 A7* Western Australia
Busta *77 D5* Orientale, N Dem.
   Rep. Congo
Butembo *77 E5* Nord-Kivu, NE Dem.
   Rep. Congo
Butler *41 E4* Pennsylvania, NE USA
Buton, Pulau *139 E4* *var.* Pulau Butung;
   prev. Boetoeng. *island* C Indonesia
Bütow *see* Bytów
Butte *44 B2* Montana, NW USA
Butterworth *138 B3* Pinang, Peninsular
   Malaysia
Button Islands *39 E1* *island group*
   Nunavut, NE Canada

Butuan *139 F2 off.* Butuan City.
Mindanao, S Philippines
Butuan City *see* Butuan
Butung, Pulau *see* Buton, Pulau
Butuntum *see* Bitonto
Buulobarde *73 D5 var.* Buulo Berde.
Hiiraan, C Somalia
Buulo Berde *see* Buulobarde
Buur Gaabo *73 D6* Jubbada Hoose,
S Somalia
Buxoro *122 D2 var.* Bokhara, *Rus.*
Bukhara. Buxoro Viloyati,
C Uzbekistan
Buynaksk *111 B8* Respublika Dagestan,
SW Russia
Büyük Ağrı Dağı *117 F3 var.* Ağrı Dağı,
Küh-e Ārārat-e Bozorg, Masis, *Eng.*
Great Ararat, Mount Ararat. *mountain*
E Turkey (Türkiye)
Büyükmenderes Nehri *116 A4 river*
SW Turkey (Türkiye)
Buzău *108 C4* Buzău, SE Romania
Büzmeyin *see* Abadan
Buzuluk *111 D6* Orenburgskaya Oblast',
W Russia
Byahoml' *107 D5 Rus.* Begoml'.
Vitsyebskaya Voblasts', N Belarus
Byalynichy *107 D6 Rus.* Belynichi.
Mahilyowskaya Voblasts', E Belarus
Byan Tumen *see* Choybalsan
Byarezina *107 D6 prev.* Byerezino, *Rus.*
Berezina. *river* C Belarus
Bydgoszcz *98 C3 Ger.* Bromberg.
Kujawski-pomorskie, C Poland
Byelaruskaya Hrada *107 B6 Rus.*
Belorusskaya Gryada. *ridge* N Belarus
Byerezino *see* Byarezina
Byron Island *see* Nikunau
Bystrovka *see* Kemin
Bytča *99 C5* Žilinský Kraj, N Slovakia
Bytom *99 C5 Ger.* Beuthen. Śląskie,
S Poland
Bytów *98 C2 Ger.* Bütow. Pomorskie,
N Poland
Byuzmeyin *see* Abadan
Byval'ki *107 D8* Homyel'skaya Voblasts',
SE Belarus
Byzantium *see* Istanbul

C

Caála *78 B2 var.* Kaala, Robert Williams,
Port. Vila Robert Williams. Huambo,
C Angola
Caazapá *64 D3* Caazapá, S Paraguay
Caballo Reservoir *48 C3 reservoir* New
Mexico, SW USA
Cabanaquinta *92 D1 var.* Cabañaquinta.
Asturias, N Spain
Cabañaquinta *see* Cabanaquinta
Cabanatuan *139 E1 off.* Cabanatuan
City. Luzon, N Philippines
Cabanatuan City *see* Cabanatuan
Cabillonum *see* Chalon-sur-Saône
Cabimas *58 C1* Zulia, NW Venezuela
Cabinda *78 A1 var.* Kabinda. Cabinda,
NW Angola
Cabinda *78 A1 var.* Kabinda. *province*
NW Angola
Cabo Verde, Republic of *see* Cape Verde
Cahora Bassa, Albufeira de *78 D2 var.*
Lake Cabora Bassa. *reservoir*
NW Mozambique
Cabora Bassa, Lake *see* Cahora Bassa,
Albufeira de
Caborca *50 B1* Sonora, NW Mexico
Cabot Strait *39 G4 strait* E Canada
Cabo Verde, Ilhas do *see* Cape Verde
Cabras, Ilha das *76 E2 island*
S Sao Tome and Principe, Africa,
E Atlantic Ocean
Cabrera, Illa de *93 G3 island* E Spain
Cáceres *92 C3 Ar.* Qazris. Extremadura,
W Spain
Cachimbo, Serra do *63 E2 mountain
range* C Brazil
Caconda *78 B2* Huíla, C Angola
Cadca *99 C5 Hung.* Csaca. Žilinský Kraj,
N Slovakia
Cadillac *40 C2* Michigan, N USA
Cadiz *139 E2 off.* Cadiz City. Negros,
C Philippines
Cádiz *92 C5 anc.* Gades, Gadier, Gadir,
Gadire. Andalucía, SW Spain
Cadiz City *see* Cadiz
Cádiz, Golfo de *92 B5 Eng.* Gulf of
Cadiz. *gulf* Portugal/Spain
Cadiz, Gulf of *see* Cádiz, Golfo de
Cadurcum *see* Cahors

Caen *90 B3* Calvados, N France
Caene/Caenepolis *see* Qinā
Caerdydd *see* Cardiff
Caer Glou *see* Gloucester
Caer Gybi *see* Holyhead
Caerleon *see* Chester
Caer Luel *see* Carlisle
Caesaraugusta *see* Zaragoza
Caesarea Mazaca *see* Kayseri
Caesarobriga *see* Talavera de la Reina
Caesarodunum *see* Tours
Caesaromagus *see* Beauvais
Caesena *see* Cesena
Cafayate *64 C2* Salta, N Argentina
Cagayan de Oro *139 E2 off.* Cagayan de
Oro City. Mindanao, S Philippines
Cagayan de Oro City *see* Cagayan de Oro
Cagliari *97 A6 anc.* Caralis. Sardegna,
Italy, C Mediterranean Sea
Caguas *55 F3* E Puerto Rico
Cahors *91 C5 anc.* Cadurcum. Lot,
S France
Cahul *108 D4 Rus.* Kagul. S Moldova
Caicos Passage *54 D2 strait* The
Bahamas/Turks and Caicos Islands
Caiffa *see* Hefa
Cailungo *96 E1* N San Marino
Caiphas *see* Hefa
Cairns *148 D3* Queensland,
NE Australia
Cairo *see* Al Qāhirah
Cairo *40 B5* Illinois, N USA
Caisleán an Bharraigh *see* Castlebar
Cajamarca *60 B3 prev.* Caxamarca.
Cajamarca, NW Peru
Čakovec *100 B2 Ger.* Csakathurn, *Hung.*
Csáktornya; *prev. Ger.* Tschakathurn.
Medimurje, N Croatia
Calabar *75 G5* Cross River, S Nigeria
Calabozo *58 D2* Guárico, C Venezuela
Calafat *108 B5* Dolj, SW Romania
Calafate *see* El Calafate
Calahorra *93 E2* La Rioja, N Spain
Calais *90 C2* Pas-de-Calais, N France
Calais *41 H2* Maine, NE USA
Calais, Pas de *see* Dover, Strait of
Calama *64 B2* Antofagasta, N Chile
Calarasi *see* Călăraşi
Călăraşi *108 D4 var.* Călăraş, *Rus.*
Kalarash. C Moldova
Călăraşi *108 C5* Călăraşi, SE Romania
Calatayud *93 E2* Aragón, NE Spain
Calbayog *139 E2 off.* Calbayog City.
Samar, C Philippines
Calbayog City *see* Calbayog
Calcutta *see* Kolkāta
Caldas da Rainha *92 B3* Leiria,
W Portugal
Caldera *64 B3* Atacama, N Chile
Caldwell *46 C3* Idaho, NW USA
Caledonia *52 C1* Corozal, N Belize
Caleta Olivia *65 B6* Santa Cruz,
SE Argentina
Calgary *37 E5* Alberta, SW Canada
Cali *58 B3* Valle del Cauca, W Colombia
Calicut *see* Kozhikode
California *47 B7 off.* State of California,
also known as El Dorado, The Golden
State. *state* W USA
California, Golfo de *50 B2 Eng.* Gulf of
California; *prev.* Sea of Cortez. *gulf*
W Mexico
California, Gulf of *see* California,
Golfo de
Călimăneşti *108 B4* Vâlcea, SW Romania
Calisia *see* Kalisz
Callabonna, Lake *149 B5 lake* South
Australia
Callao *60 C4* Callao, W Peru
Callatis *see* Mangalia
Callosa de Segura *93 F4* Comunitat
Valenciana, E Spain
Calmar *see* Kalmar
Caloundra *149 E5* Queensland,
E Australia
Caltanissetta *97 C7* Sicilia, Italy,
C Mediterranean Sea
Caluula *72 E4* Bari, NE Somalia
Calvinia *78 C5* W South Africa
Camabatela *78 B1* Cuanza Norte,
NW Angola
Camacupa *78 B2 var.* General Machado,
Port. Vila General Machado. Bié,
C Angola
Camagüey *54 C2 prev.* Puerto Príncipe.
Camagüey, C Cuba
Camagüey, Archipiélago de *54 C2
island group* C Cuba
Camana *61 E4 var.* Camaná. Arequipa,
SW Peru

Camargue *91 D6 physical region*
SE France
Ca Mau *137 D6 var.* Quan Long. Minh
Hai, S Vietnam
Cambay, Gulf of *see* Khambhāt, Gulf of
Camberia *see* Chambéry
Cambodia *137 D5 off.* Kingdom of
Cambodia, *var.* Democratic
Kampuchea, Roat Kampuchea, *Khm.*
Kampuchea; *prev.* People's Democratic
Republic of Kampuchea. *country*
SE Asia
Cambodia, Kingdom of *see* Cambodia
Cambrai *90 C2 Flem.* Kambryk, *prev.*
Cambray; *anc.* Cameracum. Nord,
N France
Cambray *see* Cambrai
Cambrian Mountains *89 C6 mountain
range* C Wales, United Kingdom
Cambridge *54 A4* W Jamaica
Cambridge *150 D3* Waikato, North
Island, New Zealand
Cambridge *89 E6 Lat.* Cantabrigia.
E England, United Kingdom
Cambridge *41 F4* Maryland, NE USA
Cambridge *40 D4* Ohio, NE USA
Cambridge Bay *37 F3 var.* Ikaluktutiak.
Victoria Island, Nunavut, NW Canada
Camden *42 B2* Arkansas, C USA
Camellia State *see* Alabama
Cameracum *see* Cambrai
Cameroon *76 A4 off.* Republic of
Cameroon, *Fr.* Cameroun. *country*
W Africa
Cameroon, Republic of *see* Cameroon
Cameroun *see* Cameroon
Camocim *63 F2* Ceará, E Brazil
Camopi *59 H3* E French Guiana
Campamento *52 C2* Olancho,
C Honduras
Campania *97 D5 Eng.* Champagne.
region S Italy
Campbell, Cape *151 D5 headland* South
Island, New Zealand
Campbell Island *142 D5 var.* Motu
Ihupuku *island* S New Zealand
Campbell Plateau *142 D5 undersea
plateau* SW Pacific Ocean
Campbell River *36 C5* Vancouver
Island, British Columbia, SW Canada
Campeche *51 G4* Campeche, SE Mexico
Campeche, Bahía de *51 F4 Eng.* Bay of
Campeche. *bay* E Mexico
Campeche, Bay of *see* Campeche,
Bahía de
Câm Pha *136 E3* Quang Ninh,
N Vietnam
Câmpina *108 C4 prev.* Cîmpina.
Prahova, SE Romania
Campina Grande *63 G2* Paraíba, E Brazil
Campinas *63 F4* São Paulo, S Brazil
Campobasso *97 D5* Molise, C Italy
Campo de Criptana *see* Campo de Criptana
Campo de Criptana *93 E3 var.* Campo
Criptana. Castilla-La Mancha, C Spain
Campo dos Goytacazes *63 F4 var.*
Campos. Rio de Janeiro, SE Brazil
Campo Grande *63 E4 state capital* Mato
Grosso do Sul, SW Brazil
Campos *see* Campo dos Goytacazes
Câmpulung *108 B4 prev.*
Câmpulung-Muşcel, Cîmpulung.
Argeş, S Romania
Câmpulung-Muşcel *see* Câmpulung
Campus Stellae *see* Santiago
de Compostela
Cam Ranh *137 E6* Khanh Hoa,
S Vietnam
Canada *34 B4 country* N North America
Canada Basin *34 C2 undersea basin*
Arctic Ocean
Canadian River *49 E2 river* SW USA
Çanakkale *116 A2 var.* Dardanelli; *prev.*
Chanak, Kale Sultanie. Çanakkale,
W Turkey (Türkiye)
Cananea *50 B1* Sonora, NW Mexico
Canarreos, Archipiélago de los *54 B2
island group* W Cuba
Canary Islands *70 A2 Eng.* Canary
Islands. *island group* Spain,
NE Atlantic Ocean
Canary Islands *see* Canarias, Islas
Cañas *52 D4* Guanacaste, NW Costa Rica
Canaveral, Cape *43 E4 headland* Florida,
SE USA
Canberra *142 C4 country capital*
(Australia) Australian Capital Territory,
SE Australia
Cancún *51 H3* Quintana Roo, SE Mexico

Candia *see* Irákleio
Canea *see* Chaniá
Cangzhou *128 D4* Hebei, E China
Caniapiscau *39 E2 river* Québec,
E Canada
Caniapiscau, Réservoir de *38 D3
reservoir* Québec, C Canada
Canik Dağları *116 D2 mountain range*
N Turkey (Türkiye)
Canillo *91 A7* Canillo, C Andorra Europe
Çankırı *116 C3 var.* Chankiri; *anc.*
Gangra, Germanicopolis. Çankın,
N Turkey (Türkiye)
Cannanore *see* Kannur
Cannes *91 D6* Alpes-Maritimes,
SE France
Canoas *63 E5* Rio Grande do Sul, S Brazil
Canon City *44 C5* Colorado, C USA
Cantabria *92 D1 autonomous
community* N Spain
Cantábrica, Cordillera *92 C1 mountain
range* N Spain
Cantabrigia *see* Cambridge
Cantaura *59 E2* Anzoátegui,
NE Venezuela
Canterbury *89 E7 hist.* Cantwaraburh;
anc. Durovernum, *Lat.* Cantuaria.
SE England, United Kingdom
Canterbury Bight *151 C6 bight* South
Island, New Zealand
Canterbury Plains *151 C6 plain* South
Island, New Zealand
Cân Thơ *137 E6* Cân Thơ, S Vietnam
Canton *42 B2* Mississippi, S USA
Canton *40 D4* Ohio, N USA
Canton *see* Guangzhou
Canton Island *see* Kanton
Cantuaria/Cantwaraburh *see*
Canterbury
Canyon *49 E2* Texas, SW USA
Cao Băng *136 D3 var.* Caobang. Cao
Băng, N Vietnam
Caobang *see* Cao Băng
Cap-Breton, Île du *see* Cape
Breton Island
Cape Barren Island *149 C8 island*
Furneaux Group, Tasmania,
SE Australia
Cape Basin *69 B7 undersea basin*
S Atlantic Ocean
Cape Breton Island *39 G4 Fr.* Île du
Cap-Breton. *island* Nova Scotia,
SE Canada
Cape Charles *41 F5* Virginia, NE USA
Cape Coast *75 E5 prev.* Cape Coast Castle.
S Ghana
Cape Coast Castle *see* Cape Coast
Cape Girardeau *45 H5* Missouri,
C USA
Capelle aan den IJssel *86 C4*
Zuid-Holland, SW Netherlands
Cape Palmas *see* Harper
Cape Saint Jacques *see* Vung Tau
Cape Town *78 B5 var.* Ekapa, *Afr.*
Kaapstad, Kapstad. *country capital*
(South Africa-legislative capital)
Western Cape, SW South Africa
Cape Verde *74 A2 off.* Republic of Cabo
Verde, *Port.* Cabo Verde, Ilhas do Cabo
Verde. *country* E Atlantic Ocean
Cape Verde Basin *66 C4 undersea basin*
E Atlantic Ocean
Cape Verde Plain *66 C4 abyssal plain*
E Atlantic Ocean
Cape York Peninsula *148 C2 peninsula*
Queensland, N Australia
Cap-Haïtien *54 D3 var.* Le Cap. N Haiti
Capira *53 G5* Panamá, C Panama
Capitán Pablo Lagerenza *64 D1 var.*
Mayor Pablo Lagerenza. Chaco,
N Paraguay
Capodistria *see* Koper
Capri *97 C5 island* S Italy
Caprivi Concession *see* Caprivi Strip
Caprivi Strip *78 C3 var.* Caprivizipfel;
prev. Caprivi Concession. *cultural
region* NE Namibia
Caprivizipfel *see* Caprivi Strip
Cap Saint-Jacques *see* Vung Tau
Capuetá, Río *58 C5 var.* Río Japurá,
Yapurá. *river* Brazil/Colombia
Caquetá, Río *see* Japurá, Rio
CAR *see* Central African Republic
Caracal *108 B5* Olt, S Romania
Caracaraí *62 D1* Rondônia, W Brazil
Caracas *58 D1 country capital*
(Venezuela) Distrito Federal,
N Venezuela
Caralis *see* Cagliari
Caratasca, Laguna de *53 E2 lagoon*
NE Honduras
Carballiño *see* O Carballiño

Carbón, Laguna del *65 B7 physical feature* SE Argentina
Carbondale *40 B5* Illinois, N USA
Carbonia *97 A6 var.* Carbonia Centro. Sardegna, Italy, C Mediterranean Sea
Carbonia Centro *see* Carbonia
Carcaso *see* Carcassonne
Carcassonne *91 C6 anc.* Carcaso. Aude, S France
Cardamomes, Chaîne des *see* Krâvanh, Chuŏr Phnum
Cardamom Mountains *see* Krâvanh, Chuŏr Phnum
Cárdenas *54 B2* Matanzas, W Cuba
Cardiff *89 C7 Wel.* Caerdydd. *national capital* S Wales, United Kingdom
Cardigan Bay *89 C6 bay* W Wales, United Kingdom
Carei *108 B3 Ger.* Gross-Karol, Karol, *Hung.* Nagykároly; *prev.* Careii-Mari. Satu Mare, NW Romania
Careii-Mari *see* Carei
Carey, Lake *147 B6 lake* Western Australia
Cariaco *59 E1* Sucre, NE Venezuela
Caribbean Sea *54 C4 sea* W Atlantic Ocean
Caribrod *see* Dimitrovgrad
Carlisle *88 C4 anc.* Caer Luel, Luguvallium, Luguvallum. NW England, United Kingdom
Carlow *89 B6 Ir.* Ceatharlach. SE Ireland
Carlsbad *48 D3* New Mexico, SW USA
Carlsbad *see* Karlovy Vary
Carlsberg Ridge *140 B4 undersea ridge* S Arabian Sea
Carlsruhe *see* Karlsruhe
Carmana/Carmania *see* Kermān
Carmarthen *89 C6* SW Wales, United Kingdom
Carmaux *91 C6* Tarn, S France
Carmel *40 C4* Indiana, N USA
Carmelita *52 B1* Petén, N Guatemala
Carmen *51 G4 var.* Ciudad del Carmen. Campeche, SE Mexico
Carmona *92 C4* Andalucía, S Spain
Carmona *see* Uíge
Carnaro *see* Kvarner
Carnarvon *147 A5* Western Australia
Carnegie, Lake *147 B5 salt lake* Western Australia
Car Nicobar *133 F3 island* Nicobar Islands, India, NE Indian Ocean
Caroaço, Ilha *71 E1 island* N Sao Tome and Principe, Africa, E Atlantic Ocean
Carolina *63 F2* Maranhão, E Brazil
Caroline Island *see* Millennium Island
Caroline Islands *142 B2 island group* C Federated States of Micronesia
Carolopois *see* Châlons-en-Champagne
Caroní, Río *59 E2 river* E Venezuela
Caronium *see* A Coruña
Carora *58 C1* Lara, N Venezuela
Carpathian Mountains *81 E4 var.* Carpathians, *Cz./Pol.* Karpaty, *Ger.* Karpaten. *mountain range* E Europe
Carpathians *see* Carpathian Mountains
Carpathos/Carpathus *see* Kárpathos
Carpaţii Meridionali *108 B4 var.* Alpi Transilvaniei, Carpaţii Sudici, *Eng.* South Carpathians, Transylvanian Alps, *Ger.* Südkarpaten, Transsylvanische Alpen, *Hung.* Déli-Kárpátok, Erdélyi-Havasok. *mountain range* C Romania
Carpaţii Sudici *see* Carpaţii Meridionali
Carpentaria, Gulf of *148 B2 gulf* N Australia
Carpi *96 C2* Emilia-Romagna, N Italy
Carrara *96 B3* Toscana, C Italy
Carson City *47 C5 state capital* Nevada, W USA
Carson Sink *47 C5 salt flat* Nevada, W USA
Carstensz, Puntjak *see* Jaya, Puncak
Cartagena de los *see* Cartagena de los Indes. Bolívar, NW Colombia
Cartagena *93 F4 anc.* Carthago Nova. Murcia, SE Spain
Cartagena de los Indes *see* Cartagena
Cartago *53 E4* Cartago, C Costa Rica
Cartago *58 B3* Valle del Cauca, W Colombia
Carthage *45 F5* Missouri, C USA
Carthago Nova *see* Cartagena
Cartwright *39 F2* Newfoundland and Labrador, E Canada
Carúpano *59 E1* Sucre, NE Venezuela
Carusbur *see* Cherbourg
Caruthersville *45 H5* Missouri, C USA
Cary *43 F1* North Carolina, SE USA

Casablanca *70 C2 Ar.* Dar-el-Beida. NW Morocco
Casa Grande *48 B2* Arizona, SW USA
Cascade Range *46 B3 mountain range* Oregon/Washington, NW USA
Cascadia Basin *34 A4 undersea basin* NE Pacific Ocean
Cascais *92 B4* Lisboa, C Portugal
Caserta *97 D5* Campania, S Italy
Casey *154 D4 Australian research station* Antarctica
Čáslav *99 B5 Ger.* Tschaslau. Střední Čechy, C Czechia (Czech Republic)
Casper *44 C3* Wyoming, C USA
Caspian Depression *111 B7 Kaz.* Kaspiy Mangy Oypaty, *Rus.* Prikaspiyskaya Nizmennost'. *depression* Kazakhstan/Russia
Caspian Sea *114 A4 Az.* Xäzär Dänizi, *Kaz.* Kaspiy Tengizi, *Per.* Baḥr-e Khazar, *Darya-ye* Khazar, *Rus.* Kaspiyskoye More. *inland sea* Asia/Europe
Cassai *see* Kasai
Cassel *see* Kassel
Castamoni *see* Kastamonu
Casteggio *96 B2* Lombardia, N Italy
Castellón/Castellón de la Plana *see* Castelló de la Plana
Castelló de la Plana *93 F3 var.* Castellón, *Cat.* Castelló de la Plana. Comunitat Valenciana, E Spain
Castelnaudary *91 C6* Aude, S France
Castelo Branco *92 C3* Castelo Branco, C Portugal
Castelsarrasin *91 B6* Tarn-et-Garonne, S France
Castelvetrano *97 C7* Sicilia, Italy, C Mediterranean Sea
Castilla-La Mancha *93 E3 autonomous community* NE Spain
Castilla y León *92 C2 autonomous community* NW Spain
Castlebar *89 A5 Ir.* Caisleán an Bharraigh. W Ireland
Castleford *89 D5* N England, United Kingdom
Castle Harbour *42 B5 inlet* Bermuda, NW Atlantic Ocean
Castra Regina *see* Regensburg
Castricum *86 C3* Noord-Holland, N Netherlands
Castries *55 F1 country capital* (Saint Lucia) N Saint Lucia
Castro *65 B6* Los Lagos, W Chile
Castrovillari *97 D6* Calabria, SW Italy
Castuera *92 D3* Extremadura, W Spain
Caswell Sound *see* Taitetimu
Catacamas *52 D2* Olancho, C Honduras
Catacaos *60 B3* Piura, NW Peru
Catalan Bay *93 H4 bay* E Gibraltar, Mediterranean Sea
Cataluña *93 G2* N Spain
Catamarca *see* San Fernando del Valle de Catamarca
Catania *97 D7* Sicilia, Italy, C Mediterranean Sea
Catanzaro *97 D6* Calabria, SW Italy
Catarroja *93 F3* Comunitat Valenciana, E Spain
Cat Island *54 C1 island* C The Bahamas
Catskill Mountains *41 F3 mountain range* New York, NE USA
Cattaro *see* Kotor
Cauca, Río *58 B2 river* N Colombia
Caucasia *58 B2* Antioquia, NW Colombia
Caucasus *81 G4 Rus.* Kavkaz. *mountain range* Georgia/Russia
Caura, Río *59 E3 river* C Venezuela
Cavaia *see* Kavajë
Cavalla *74 D5 var.* Cavally, Cavally Fleuve. *river* Ivory Coast/Liberia
Cavally/Cavally Fleuve *see* Cavalla
Caviana de Fora, Ilha *63 E1 var.* Ilha Caviana. *island* N Brazil
Caviana, Ilha *see* Caviana de Fora, Ilha
Cawnpore *see* Kānpur
Caxias *63 F2* Maranhão, E Brazil
Caxito *78 B1* Bengo, NW Angola
Cayenne *59 H3 dependent territory/arrondissement capital* (French Guiana) NE French Guiana
Cayes *54 D3 var.* Les Cayes. SW Haiti
Cayman Brac *54 B3 island* E Cayman Islands
Cayman Islands *54 B3 UK Overseas Territory* W West Indies
Cayo *see* San Ignacio
Cay Sal *54 B2 islet* SW The Bahamas

Cazin *100 B3* Federacija Bosna I Hercegovina, NW Bosnia and Herzegovina
Ceadâr-Lunga *108 D4 var.* Ciadâr-Lunga, *Rus.* Chadyr-Lunga. S Moldova
Ceará *63 F2 off.* Estado do Ceará. *state/region* C Brazil
Ceará *see* Fortaleza
Ceará Abyssal Plain *see* Ceará Plain
Ceará, Estado do *see* Ceará
Ceará Plain *56 E3 var.* Ceara Abyssal Plain. *abyssal plain* W Atlantic Ocean
Ceatharlach *see* Carlow
Cébaco, Isla *53 F5 island* SW Panama
Cebu *139 E2 off.* Cebu City. Cebu, C Philippines
Cebu City *see* Cebu
Čechy *see* Bohemia
Cecina *96 B3* Toscana, C Italy
Cedar Falls *45 G3* Iowa, C USA
Cedar Lake *38 A2 lake* Manitoba, C Canada
Cedar Rapids *45 G3* Iowa, C USA
Ceduna *149 A6* South Australia
Cedros, Isla *50 A2 island* W Mexico
Cefalù *97 C7 anc.* Cephaloedium. Sicilia, Italy, C Mediterranean Sea
Celebes *see* Sulawesi
Celebes Sea *139 E3 Ind.* Laut Sulawesi. *sea* Indonesia/Philippines
Celje *95 E7 Ger.* Cilli. C Slovenia
Celldömölk *99 C6* Vas, W Hungary
Celle *94 B3 var.* Zelle. Niedersachsen, N Germany
Celovec *see* Klagenfurt
Celtic Sea *89 B7 Ir.* An Mhuir Cheilteach. *sea* SW British Isles
Celtic Shelf *80 B3 continental shelf* E Atlantic Ocean
Cenderawasih, Teluk *139 G4 var.* Teluk Irian, Teluk Sarera. *bay* W Pacific Ocean
Cenon *91 B5* Gironde, SW France
Centennial State *see* Colorado
Centrafricaine, République *see* Central African Republic
Central African Republic *76 C4 var.* République Centrafricaine, *abbrev.* CAR; *prev.* Ubangi-Shari, Oubangui-Chari, Territoire de l'Oubangui-Chari. *country* C Africa
Central, Cordillera *58 B3 mountain range* W Colombia
Cordillera Central *55 E3 mountain range* C Dominican Republic
Cordillera Central *53 F5 mountain range* C Panama
Central, Cordillera *139 E1 mountain range* Luzon, N Philippines
Central Group *see* Inner Islands
Central Indian Ridge *see* Mid-Indian Ridge
Central Makran Range *134 A3 mountain range* W Pakistan
Central Pacific Basin *142 D1 undersea basin* C Pacific Ocean
Central, Planalto *63 F3 var.* Brazilian Highlands. *mountain range* E Brazil
Central Provinces and Berar *see* Madhya Pradesh
Central Range *144 B3 mountain range* NW Papua New Guinea
Central Russian Upland *see* Srednerusskaya Vozvyshennost'
Central Siberian Plateau *114 D3 var.* Central Siberian Uplands, *Eng.* Central Siberian Plateau. *mountain range* N Russia
Central Siberian Plateau/Central Siberian Uplands *see* Srednesibirskoye Ploskogor'ye
Central, Sistema *92 D3 mountain range* C Spain
Central Valley *47 B6 valley* California, W USA
Centum Cellae *see* Civitavecchia
Ceos *see* Tziá
Cephaloedium *see* Cefalù
Ceram *see* Seram, Pulau
Ceram Sea *see* Laut Seram
Cerasus *see* Giresun
Cereté *58 B2* Córdoba, NW Colombia
Cergy-Pontoise *see* Pontoise
Cerignola *97 D5* Puglia, SE Italy
Çerkeş *116 C2* Çankin, N Turkey (Türkiye)
Cernăuți *see* Chernivtsi

Cernay *90 E4* Haut-Rhin, NE France
Cerro de Pasco *60 C3* Pasco, C Peru
Cervera *93 F2* Cataluña, NE Spain
Cervino, Monte *see* Matterhorn
Cesena *96 C3 anc.* Caesena. Emilia-Romagna, N Italy
Cēsis *106 D3 Ger.* Wenden. Cēsis, C Latvia
České Budějovice *99 B5 Ger.* Budweis. Jihočeský Kraj, S Czechia (Czech Republic)
Česko *see* Czechia (Czech Republic)
Český Krumlov *99 A5 var.* Böhmisch-Krumau, *Ger.* Krummau. Jihočeský Kraj, S Czechia Republic (Czechia)
Český Les *see* Bohemian Forest
Cetatea Albă *see* Bilhorod-Dnistrovskyi
Cetatea Damboviţei *see* Bucureşti
Cetinje *101 C5 It.* Cettigne. S Montenegro
Cette *see* Sète
Ceuta *70 C2 autonomous city of Spain* Spain, N Africa
Cévennes *91 C6 mountain range* S France
Ceyhan *116 D4* Adana, S Turkey (Türkiye)
Ceylanpınar *117 E4* Şanlıurfa, SE Turkey (Türkiye)
Ceylon *see* Sri Lanka
Ceylon Plain *124 B4 abyssal plain* N Indian Ocean
Ceyre to the Caribs *see* Marie-Galante
Chachapoyas *60 B2* Amazonas, NW Peru
Chachevichy *107 D6 Rus.* Chechevichi. Mahilyowskaya Voblasts', E Belarus
Chaco *see* Gran Chaco
Chad *76 C3 off.* Republic of Chad, *Fr.* Tchad. *country* C Africa
Chad, Lake *76 B3 Fr.* Lac Tchad. *lake* C Africa
Chad, Republic of *see* Chad
Chadron *44 D3* Nebraska, C USA
Chadyr-Lunga *see* Ceadir-Lunga
Chagai Hills *134 A2 var.* Chah Gay. *mountain range* Afghanistan/Pakistan
Chaghasarāy *see* Asadābād
Chagos-Laccadive Plateau *124 B4 undersea plateau* N Indian Ocean
Chagos Trench *141 C5 trench* N Indian Ocean
Chah Gay *see* Chagai Hills
Chaillu, Massif du *77 B6 mountain range* C Gabon
Chain Ridge *140 B4 undersea ridge* W Indian Ocean
Chajul *52 B2* Quiché, W Guatemala
Chakhānsūr *122 D5* Nīmrūz, SW Afghanistan
Chala *60 D4* Arequipa, SW Peru
Chalatenango *52 C3* Chalatenango, N El Salvador
Chalcidice *see* Chalkidikí
Chalcis *see* Chalkída
Chalki *105 E7 island* Dodekánisa, Greece, Aegean Sea
Chalkída *105 C5 var.* Halkida, *prev.* Khalkís; *anc.* Chalcis. Evvoia, E Greece
Chalkidikí *104 C4 var.* Khalkidhikí; *anc.* Chalcidice. *peninsula* NE Greece
Challans *90 B4* Vendée, NW France
Challapata *61 F4* Oruro, SW Bolivia
Challenger Deep *152 B3 trench* W Pacific Ocean
Challenger Fracture Zone *153 F4 tectonic feature* SE Pacific Ocean
Châlons-en-Champagne *90 D3 prev.* Châlons-sur-Marne, *hist.* Arcae Remorum; *anc.* Carolopois. Marne, NE France
Châlons-sur-Marne *see* Châlons-en-Champagne
Chalon-sur-Saône *90 D4 anc.* Cabillonum. Saône-et-Loire, C France
Cha Mai *see* Thung Song
Chaman *134 B2* Baluchistan, SW Pakistan
Chambéry *91 D5 anc.* Cambaria. Savoie, E France
Champagne *90 D3 cultural region* N France
Champagne *see* Campania
Champaign *40 B4* Illinois, N USA
Champasak *137 D5* Champasak, S Laos
Champlain, Lake *41 F2 lake* Canada/USA

Champotón 51 G4 Campeche, SE Mexico
Chanak see Çanakkale
Chañaral 64 B3 Atacama, N Chile
Chan-chiang/Chanchiang see Zhanjiang
Chandeleur Islands 42 C3 island group Louisiana, S USA
Chandigarh 134 D2 state capital Punjab, N India
Chandrapur 135 E5 Mahārāshtra, C India
Changan see Xi'an, Shaanxi, C China
Changane 79 E3 river S Mozambique
Changchun 128 D3 var. Ch'angch'un, Ch'ang-ch'un; prev. Hsinking. province capital Jilin, NE China
Chang Jiang 129 A5 Eng. Yangtze; var. Yangtze Kiang. river SW China
Changkiakow see Zhangjiakou
Chang, Ko 137 C6 island S Thailand
Changsha 129 C5 var. Ch'angsha, Ch'ang-sha. province capital Hunan, S China
Ch'angsha/Ch'ang-sha see Changsha
Changzhi 128 C4 Shanxi, C China
Chaniá 83 C8 var. Hania, Khaniá, Eng. Canea; anc. Cydonia. Kríti, Greece, E Mediterranean Sea
Chañi, Nevado de 64 B2 mountain NW Argentina
Chankiri see Çankırı
Channel Islands 89 C8 Fr. Iles Normandes. island group S English Channel
Channel Islands 47 B8 island group California, W USA
Channel-Port aux Basques 39 G4 Newfoundland and Labrador, SE Canada
Channel, The see English Channel
Channel Tunnel 90 C2 tunnel France/United Kingdom
Chantabun/Chantaburi see Chanthaburi
Chantada 92 C1 Galicia, NW Spain
Chanthaburi 137 C6 var. Chantabun, Chantaburi. Chantaburi, S Thailand
Chanute 45 F5 Kansas, C USA
Chaouèn see Chefchaouen
Chaoyang 128 D3 Liaoning, NE China
Chapala, Lago de 50 D4 lake C Mexico
Chapan, Gora 122 B3 mountain C Turkmenistan
Chapayevsk 111 C6 Samarskaya Oblast', W Russia
Chapra see Chhapra
Charcot Seamounts 80 B3 seamount range E Atlantic Ocean
Chardzhev see Türkmenabat
Chardzhou/Chardzhui see Türkmenabat
Charente 91 B5 cultural region W France
Charente 91 B5 river W France
Chari 76 B3 var. Shari. river Central African Republic/Chad
Chārikār 123 E4 Parvān, NE Afghanistan
Charity 59 F2 NW Guyana
Chärjew see Türkmenabat
Charkhlik/Charkhliq see Ruoqiang
Charleroi 87 C7 Hainaut, S Belgium
Charlesbourg 39 E4 Québec, SE Canada
Charles de Gaulle 90 E1 (Paris) Seine-et-Marne, N France
Charles Island 38 D1 island Nunavut, NE Canada
Charles Island see Santa María, Isla
Charleston 43 F2 South Carolina, SE USA
Charleston 40 D5 state capital West Virginia, NE USA
Charleville 149 D5 Queensland, E Australia
Charleville-Mézières 90 D3 Ardennes, N France
Charlie-Gibbs Fracture Zone 66 C2 tectonic feature N Atlantic Ocean
Charlotte 43 E1 North Carolina, SE USA
Charlotte Amalie 55 F3 prev. Saint Thomas. dependent territory capital (Virgin Islands (US)) Saint Thomas, N Virgin Islands (US)
Charlotte Harbor 43 E5 inlet Florida, SE USA
Charlottenhof see Aegviidu
Charlottesville 41 E5 Virginia, NE USA
Charlottetown 39 F4 province capital Prince Edward Island, Prince Edward Island, SE Canada

Charlotte Town see Roseau, Dominica
Charsk see Shar
Charters Towers 148 D3 Queensland, NE Australia
Chartres 90 C4 anc. Autricum, Civitas Carnutum. Eure-et-Loir, C France
Chashniki 107 D5 Vitsyebskaya Voblasts', N Belarus
Châteaubriant 90 B4 Loire-Atlantique, NW France
Châteaudun 90 C3 Eure-et-Loir, C France
Châteauroux 90 C4 prev. Indreville. Indre, C France
Château-Thierry 90 C3 Aisne, N France
Châtelet 87 C7 Hainaut, S Belgium
Châtellerault see Châtellerault
Châtellerault 90 B4 var. Châtelherault. Vienne, W France
Chatham Island see San Cristóbal, Isla
Chatham Island see Chatham Rise
Chatham Islands 143 E5 island group New Zealand, SW Pacific Ocean
Chatham Rise 142 D5 var. Chatham Island Rise. undersea rise S Pacific Ocean
Chatkal Range 123 F2 Rus. Chatkal'skiy Khrebet. mountain range Kyrgyzstan/Uzbekistan
Chatkal'skiy Khrebet see Chatkal Range
Chāttagām see Chittagong
Chattahoochee River 42 D3 river SE USA
Chattanooga 42 D1 Tennessee, S USA
Chattgram 135 G4 Ben. Chāttagām, prev. Chittagong. Chattogram, SE Bangladesh
Chatyr-Tash 123 G2 Naryn Oblusu/Narynskaya Oblast', C Kyrgyzstan
Châu Đốc 137 D6 var. Chauphu, Chau Phu. An Giang, S Vietnam
Chauk 136 A3 Magway, W Myanmar (Burma)
Chaumont 90 D4 prev. Chaumont-en-Bassigny. Haute-Marne, N France
Chaumont-en-Bassigny see Chaumont
Chau Phu see Châu Đốc
Chausy see Chavusy
Chaves 92 C2 anc. Aquae Flaviae. Vila Real, N Portugal
Chávez, Isla see Santa Cruz, Isla
Chavusy 107 E6 Rus. Chausy. Mahilyowskaya Voblasts', E Belarus
Chaykovskiy 111 D5 Permskaya Oblast', NW Russia
Cheb 99 A5 Ger. Eger. Karlovarský Kraj, W Czechia (Czech Republic)
Cheboksary 111 C5 Chuvashskaya Respublika, W Russia
Cheboygan 40 C2 Michigan, N USA
Chechaouèn see Chefchaouen
Chech, Erg 74 D1 desert Algeria/Mali
Chechevichi see Chachevichy
Che-chiang see Zhejiang
Cheduba Island see Manaung Island)
Chefchaouen 70 C2 var. Chaouèn, Chechaouèn, Sp. Xauen. N Morocco
Chefoo see Yantai
Cheju-do see Jejudo
Cheju Strait see Jeju Strait
Chekiang see Zhejiang
Cheleken see Hazar
Chelkar see Shalqar
Chełm 98 E4 Rus. Kholm. Lubelskie, SE Poland
Chełmno 98 C3 Ger. Culm, Kulm. Kujawski-pomorskie, C Poland
Chełmża 98 C3 Ger. Culmsee, Kulmsee. Kujawski-pomorskie, C Poland
Cheltenham 89 D6 C England, United Kingdom
Chelyabinsk 114 C3 Chelyabinskaya Oblast', C Russia
Chemnitz 94 D4 prev. Karl-Marx-Stadt. Sachsen, E Germany
Chemulpo see Incheon
Chenab 134 C2 river India/Pakistan
Chengchiatun see Liaoyuan
Ch'eng-chou/Chengchow see Zhengzhou
Chengde 128 D3 var. Jehol. Hebei, E China
Chengdu 129 B5 var. Chengtu, Ch'eng-tu. province capital Sichuan, C China
Chenghsien see Zhengzhou
Chengtu/Ch'eng-tu see Chengdu
Chennai 132 D2 prev. Madras. state capital Tamil Nādu, S India
Chenstokhov see Częstochowa

Chen Xian/Chenxian/Chen Xiang see Chenzhou
Chenzhou 129 C6 var. Chenxian, Chen Xian, Chen Xiang. Hunan, S China
Chepelare 104 C3 Smolyan, S Bulgaria
Chepén 60 B3 La Libertad, C Peru
Cher 90 C4 river C France
Cherbourg 90 B3 anc. Carusbur. Manche, N France
Cherepovets 110 B4 Vologodskaya Oblast', NW Russia
Chergui, Chott ech 70 D2 salt lake NW Algeria
Cherikaw see Cherykaw
Cherkassy see Cherkasy
Cherkasy 109 E2 Rus. Cherkassy. Cherkas'ka Oblast', C Ukraine
Cherkessk 111 B7 Karachayevo-Cherkesskaya Respublika, SW Russia
Chernigov see Chernihiv
Chernihiv 109 E1 Rus. Chernigov. Chernihivs'ka Oblast', NE Ukraine
Chernivtsi 108 C3 Ger. Czernowitz, Rom. Cernăuţi, Rus. Chernovtsy. Chernivets'ka Oblast', W Ukraine
Cherno More see Black Sea
Chernomorskoye see Chornomorske
Chernovtsy see Chernivtsi
Chernoye More see Black Sea
Chernyakhovsk 106 A4 Ger. Insterburg. Kaliningradskaya Oblast', W Russia
Cherry Hill 41 F4 New Jersey, NE USA
Cherski Range see Cherskogo, Khrebet
Cherskiy 115 G2 Respublika Sakha (Yakutiya), NE Russia
Cherskogo, Khrebet 115 F2 var. Cherski Range. mountain range NE Russia
Cherso see Cres
Cherven' see Chervyen'
Chervonograd see Chervonohrad
Chervonohrad 108 C2 Rus. Chervonograd. L'vivs'ka Oblast', NW Ukraine
Chervyen' 107 D6 Rus. Cherven'. Minskaya Voblasts', C Belarus
Cherykaw 107 E7 Rus. Cherikov. Mahilyowskaya Voblasts', E Belarus
Chesapeake Bay 41 F5 inlet NE USA
Chesha Bay see Chëshskaya Guba
Chëshskaya Guba 155 D5 var. Archangel Bay, Chesha Bay, Dvina Bay. bay NW Russia
Chester 89 C6 Wel. Caerlleon, hist. Legaceaster, Lat. Deva, Devana Castra. C England, United Kingdom
Chetumal 51 H4 var. Payo Obispo. Quintana Roo, SE Mexico
Cheviot Hills 88 D4 hill range England/Scotland, United Kingdom
Cheyenne 44 D4 state capital Wyoming, C USA
Cheyenne River 44 D3 river South Dakota/Wyoming, N USA
Chezdi-Oşorheiu see Târgu Secuiesc
Chhapra 135 F3 prev. Chapra. Bihār, N India
Chhattisgarh 135 E4 cultural region E India
Chiai/Chia-i see Chiayi
Chiang-hsi see Jiangxi
Chiang Mai 136 B4 var. Chiangmai, Chiengmai, Kiangmai. Chiang Mai, NW Thailand
Chiangmai see Chiang Mai
Chiang Rai 136 C3 var. Chianpai, Chienrai, Muang Chiang Rai. Chiang Rai, NW Thailand
Chiang-su see Jiangsu
Chianning/Chian-ning see Nanjing
Chianpai see Chiang Rai
Chianti 96 C3 cultural region C Italy
Chiapa see Chiapa de Corzo
Chiapa de Corzo 51 G5 var. Chiapa. Chiapas, SE Mexico
Chiayi 129 D6 var. Chiai, Chia-i, Kiayi, Jiayi, Jap. Kagi. C Taiwan
Chiba 130 B1 var. Tiba. Chiba, Honshū, S Japan
Chibougamau 38 D3 Québec, SE Canada
Chicago 40 B3 Illinois, N USA
Chi'i-ch'i-ha-erh see Qiqihar
Chickasha 49 G2 Oklahoma, C USA
Chiclayo 60 B3 Lambayeque, NW Peru
Chico 47 B5 California, W USA
Chico, Río 65 B7 river S Argentina
Chico, Río 65 B6 river S Argentina
Chicoutimi 39 E4 Québec, SE Canada
Chiengmai see Chiang Mai
Chienrai see Chiang Rai

Chiesanuova 96 D2 SW San Marino
Chieti 96 D4 var. Teate. Abruzzo, C Italy
Chifeng 127 G2 var. Ulanhad. Nei Mongol Zizhiqu, N China
Chigirin see Chyhyryn
Chih-fu see Yantai
Chihli see Hebei
Chihli, Gulf of see Bo Hai
Chihuahua 50 C2 Chihuahua, NW Mexico
Childress 49 F2 Texas, SW USA
Chile 64 B3 off. Republic of Chile. country SW South America
Chile Basin 57 A5 undersea basin E Pacific Ocean
Chile Chico 65 B6 Aisén, W Chile
Chile, Republic of see Chile
Chile Rise 57 A7 undersea rise SE Pacific Ocean
Chilia-Nouă see Kiliya
Chililabombwe 78 D2 Copperbelt, C Zambia
Chi-lin see Jilin
Chillán 65 B5 Bío Bío, C Chile
Chillicothe 40 D4 Ohio, N USA
Chill Mhantáin, Sléibhte see Wicklow Mountains
Chiloé, Isla de 65 A6 var. Isla Grande de Chiloé. island W Chile
Chilpancingo 51 E5 var. Chilpancingo de los Bravos. Guerrero, S Mexico
Chilpancingo de los Bravos see Chilpancingo
Chilung see Keelung
Chimán 53 G5 Panamá, E Panama
Chimbay see Chimboy
Chimborazo 60 A1 volcano C Ecuador
Chimbote 60 C3 Ancash, W Peru
Chimboy 122 D1 Rus. Chimbay. Qoraqalpog'iston Respublikasi, NW Uzbekistan
China 124 C2 off. People's Republic of China, Chin. Chung-hua Jen-min Kung-ho-kuo, Zhonghua Renmin Gongheguo; prev. Chinese Empire. country E Asia
Chi-nan/Chinan see Jinan
Chinandega 52 C3 Chinandega, NW Nicaragua
China, People's Republic of see China
China, Republic of see Taiwan
Chincha Alta 60 D4 Ica, SW Peru
Chin-chiang see Quanzhou
Chin-chou/Chinchow see Jinzhou
Chindwin see Chindwin
Chindwinn 136 B2 var. Chindwin. river N Myanmar (Burma)
Chinese Empire see China
Ch'ing Hai see Qinghai Hu, China
Chinghai see Qinghai
Chingola 78 D2 Copperbelt, C Zambia
Ching-Tao/Ch'ing-tao see Qingdao
Chinguetti 74 C2 var. Chinguetti. Adrar, C Mauritania
Chin Hills 136 A3 mountain range W Myanmar (Burma)
Chinhsien see Jinzhou
Chinnereth see Kinneret, Yam
Chinook Trough 113 H4 trough N Pacific Ocean
Chioggia 96 C2 anc. Fossa Claudia. Veneto, NE Italy
Chíos 105 D5 var. Hios, Khíos, It. Scio, Turk. Sakíz-Adasi. Chíos, E Greece
Chíos 105 D5 island E Greece
Chipata 78 D2 prev. Fort Jameson. Eastern, E Zambia
Chiquián 60 C3 Ancash, W Peru
Chiquimula 52 B2 Chiquimula, SE Guatemala
Chirāla 132 D1 Andhra Pradesh, E India
Chirchik see Chirchiq
Chirchiq 123 E2 Rus. Chirchik. Toshkent Viloyati, E Uzbekistan
Chiriquí Gulf 53 E5 Eng. Chiriquí Gulf. gulf SW Panama
Chiriquí Gulf see Chiriquí, Golfo de
Chiriquí, Laguna de 53 E5 lagoon NW Panama
Chiriquí, Volcán de see Barú, Volcán
Chirripó, Cerro see Chirripó Grande, Cerro
Chirripó Grande, Cerro 52 D4 var. Cerro Chirripó. mountain SE Costa Rica
Chisec 52 B2 Alta Verapaz, C Guatemala
Chisholm 45 F1 Minnesota, N USA
Chisimaio/Chisimayu see Kismaayo

Chişinău 108 D4 *Rus.* Kishinev. *country capital* (Moldova) C Moldova
Chita 115 F4 Chitinskaya Oblast', S Russia
Chitangwiza *see* Chitungwiza
Chitato 78 C1 Lunda Norte, NE Angola
Chitina 36 D3 Alaska, USA
Chitose 130 D2 *var.* Titose. Hokkaidō, NE Japan
Chitré 53 F5 Herrera, S Panama
Chittagong *see* Chattogram
Chitungwiza 78 D3 *prev.* Chitangwiza. Mashonaland East, NE Zimbabwe
Chkalov *see* Orenburg
Chlef 70 D2 *var.* Ech Cheliff, Ech Chleff; *prev.* Al-Asnam, El Asnam, Orléansville. NW Algeria
Chocolate Mountains 47 D8 *mountain range* California, W USA
Chodorów *see* Khodoriv
Chodzież 98 C3 Wielkopolskie, C Poland
Choele Choel 65 C5 Río Negro, C Argentina
Choiseul 144 C3 *var.* Lauru. *island* NW Solomon Islands
Chojnice 98 C2 *Ger.* Konitz. Pomorskie, N Poland
Ch'ok'ē 72 C4 *var.* Choke Mountains. *mountain range* NW Ethiopia
Choke Mountains *see* Ch'ok'ē
Cholet 90 B4 Maine-et-Loire, NW France
Choluteca 52 C3 Choluteca, S Honduras
Choluteca, Río 52 C3 *river* SW Honduras
Choma 78 D2 Southern, S Zambia
Chomutov 98 A4 *Ger.* Komotau. Ústecký Kraj, NW Czechia (Czech Republic)
Chona 113 E2 *river* C Russia
Chon Buri 137 C5 *prev.* Bang Pla Soi. Chon Buri, S Thailand
Chone 60 A1 Manabí, W Ecuador
Ch'ŏngjin 129 E1 NE North Korea
Chongqing 129 B5 *var.* Ch'ung-ching, Ch'ung-ch'ing, Chungking, Pahsien, Tchongking, Yuzhou. Chongqing, C China
Chongqing 129 B5 *province* C China
Chonnacht *see* Connaught
Chonos, Archipiélago de los 65 A6 *island group* S Chile
Chóra 105 D7 Kykládes, Greece, Aegean Sea
Chóra Sfakíon 105 C8 *var.* Sfakiá. Kríti, Greece, E Mediterranean Sea
Chorne More *see* Black Sea
Chornomorsk 109 E4 *prev.* Illichivsk. Odes'ka Oblast', SW Ukraine
Chornomorske 109 E4 *Rus.* Chernomorskoye. Respublika Krym, S Ukraine
Chorokh/Chorokhi *see* Çoruh Nehri
Chortkiv 108 C2 *Rus.* Chortkov. Ternopil's'ka Oblast', W Ukraine
Chortkov *see* Chortkiv
Chorzów 99 C5 *Ger.* Königshütte; *prev.* Królewska Huta. Śląskie, S Poland
Chóśebuz *see* Cottbus
Chōsen-kaikyō *see* Korea Strait
Chōshi 131 D5 *var.* Tyōsi. Chiba, Honshū, S Japan
Chosŏn-minjujuŭi-inmin-kanghwaguk *see* North Korea
Choszczno 98 B3 *Ger.* Arnswalde. Zachodnio-pomorskie, NW Poland
Chota Nagpur 135 E4 *plateau* N India
Choûm 74 C2 Adrar, C Mauritania
Choybalsan 127 F2 *prev.* Byan Tumen. Dornod, E Mongolia
Christchurch 151 C6 Canterbury, South Island, New Zealand
Christiana 54 B5 C Jamaica
Christiania *see* Oslo
Christiansand *see* Kristiansand
Christianshåb *see* Qasigiannguit
Christiansund *see* Kristiansund
Christmas Island 141 D5 *Australian external territory* E Indian Ocean
Christmas Island *see* Kiritimati
Christmas Ridge 143 E1 *undersea ridge* C Pacific Ocean
Chuan *see* Sichuan
Ch'uan-chou *see* Quanzhou
Chubek *see* Maskav
Chubut, Río 65 B6 *river* SE Argentina
Ch'u-chiang *see* Shaoguan
Chudoku-sanchi 131 B6 *mountain range* Honshū, SW Japan
Chuí *see* Chuy

Chukai *see* Cukai
Chukchi Plain 155 B2 *abyssal plain* Arctic Ocean
Chukchi Plateau 34 C2 *undersea plateau* Arctic Ocean
Chukchi Sea 34 B2 *Rus.* Chukotskoye More. *sea* Arctic Ocean
Chukotskoye More *see* Chukchi Sea
Chula Vista 47 C8 California, W USA
Chulucanas 60 B2 Piura, NW Peru
Chulym 114 D4 *river* C Russia
Chumphon 137 C6 *var.* Jumporn. Chumphon, SW Thailand
Chuncheon 129 E4 *prev.* Ch'unch'ŏn, *Jap.* Shunsen. N South Korea
Ch'unch'ŏn *see* Chuncheon
Ch'ung-ch'ing/Ch'ung-ching *see* Chongqing
Chung-hua Jen-min Kung-ho-kuo *see* China
Chungking *see* Chongqing
Chunya 115 E3 *river* C Russia
Chuquicamata 64 B2 Antofagasta, N Chile
Chuquisaca *see* Sucre
Chur 95 B7 *Fr.* Coire, *It.* Coira, *Rmsch.* Cuera, Quera; *anc.* Curia Rhaetorum. Graubünden, E Switzerland
Churchill 37 G4 Manitoba, C Canada
Churchill 38 B2 *river* Manitoba/ Saskatchewan, C Canada
Churchill 39 F2 *river* Newfoundland and Labrador, E Canada
Chuska Mountains 48 C1 *mountain range* Arizona/New Mexico, SW USA
Chusovoy 111 D5 Permskaya Oblast', NW Russia
Chust *see* Khust
Chuuk Islands 144 B2 *var.* Hogoley Islands; *prev.* Truk Islands. *island group* Caroline Islands, C Federated States of Micronesia
Chuy 64 E4 *var.* Chuí. Rocha, E Uruguay
Chyhyryn 109 E2 *Rus.* Chigirin. Cherkas'ka Oblast', C Ukraine
Chystiakove 109 H3 *prev.* Chystyakove, *Rus.* Torez. Donets'ka Oblast', SE Ukraine
Chystyakove *see* Chystiakove
Ciadîr-Lunga *see* Ceadir-Lunga
Cide 116 C2 Kastamonu, N Turkey (Türkiye)
Ciechanów 98 D3 *prev.* Zichenau. Mazowieckie, C Poland
Ciego de Ávila 54 C2 Ciego de Ávila, C Cuba
Ciénaga 58 B1 Magdalena, N Colombia
Cienfuegos 54 B2 Cienfuegos, C Cuba
Cieza 93 E4 Murcia, SE Spain
Cihanbeyli 116 C3 Konya, C Turkey (Türkiye)
Cikobia 145 E4 *prev.* Thikombia. *island* N Fiji
Cilacap 138 C5 *prev.* Tjilatjap. Jawa, C Indonesia
Cill Airne *see* Killarney
Cill Chainnigh *see* Kilkenny
Cilli *see* Celje
Cill Mhantáin *see* Wicklow
Cîmpina *see* Câmpina
Cîmpulung *see* Câmpulung
Cincinnati 40 C4 Ohio, N USA
Ciney 87 C7 Namur, SE Belgium
Cinto, Monte 91 E7 *mountain* Corse, France, C Mediterranean Sea
Cintra *see* Sintra
Cipolletti 65 B5 Río Negro, C Argentina
Cirebon 138 C4 *prev.* Tjirebon. Jawa, S Indonesia
Cirkvenica *see* Crikvenica
Cirò Marina 97 E6 Calabria, S Italy
Cirquenizza *see* Crikvenica
Cisnădie 108 B4 *Ger.* Heltau, *Hung.* Nagydisznód. Sibiu, SW Romania
Citharista *see* la Ciotat
Citlaltépetl *see* Orizaba, Volcán Pico de
Citrus Heights 47 B5 California, W USA
Ciudad Acuña *see* Villa Acuña
Ciudad Bolívar 59 E2 *prev.* Angostura. Bolívar, E Venezuela
Ciudad Camargo 50 D2 Chihuahua, N Mexico
Ciudad Cortés *see* Cortés
Ciudad Darío 52 D3 *var.* Dario. Matagalpa, W Nicaragua
Ciudad de Dolores Hidalgo *see* Dolores Hidalgo

Ciudad de Guatemala 52 B2 *Eng.* Guatemala City; *prev.* Santiago de los Caballeros. *country capital* (Guatemala) Guatemala, C Guatemala
Ciudad del Carmen *see* Carmen
Ciudad del Este 64 E2 *prev.* Ciudad Presidente Stroessner, Presidente Stroessner, Puerto Presidente Stroessner. Alto Paraná, SE Paraguay
Ciudad Delicias *see* Delicias
Ciudad de México *see* México
Ciudad de Panamá *see* Panamá
Ciudad Guayana 59 E2 *prev.* San Tomé de Guayana, Santo Tomé de Guayana. Bolívar, NE Venezuela
Ciudad Guzmán 50 D4 Jalisco, SW Mexico
Ciudad Hidalgo 51 G5 Chiapas, SE Mexico
Ciudad Juárez 50 C1 Chihuahua, N Mexico
Ciudad Lerdo 50 D3 Durango, C Mexico
Ciudad Madero 51 E3 *var.* Villa Cecilia. Tamaulipas, C Mexico
Ciudad Mante 51 E3 Tamaulipas, C Mexico
Ciudad Miguel Alemán 51 E2 C Mexico
Ciudad Obregón 50 B2 Sonora, NW Mexico
Ciudad Ojeda 58 C1 Zulia, NW Venezuela
Ciudad Porfirio Díaz *see* Piedras Negras
Ciudad Presidente Stroessner *see* Ciudad del Este
Ciudad Quesada *see* Quesada
Ciudad Real 92 D3 Castilla-La Mancha, C Spain
Ciudad-Rodrigo 92 C3 Castilla y León, N Spain
Ciudad Trujillo *see* Santo Domingo
Ciudad Valles 51 E3 San Luis Potosí, C Mexico
Ciudad Victoria 51 E3 Tamaulipas, C Mexico
Ciutadella 93 H3 *var.* Ciutadella de Menorca. Menorca, Spain, W Mediterranean Sea
Ciutadella de Menorca *see* Ciutadella
Civitanova Marche 96 D3 Marche, C Italy
Civitas Altae Ripae *see* Brzeg
Civitas Carnutum *see* Chartres
Civitas Eburovicum *see* Évreux
Civitavecchia 96 C4 *anc.* Centum Cellae, Trajani Portus. Lazio, C Italy
Claremore 49 G1 Oklahoma, C USA
Clarence 151 C5 Canterbury, South Island, New Zealand
Clarence 151 C5 *river* South Island, New Zealand
Clarence Town 54 D2 Long Island, C The Bahamas
Clarinda 45 F4 Iowa, C USA
Clarion Fracture Zone 153 E2 *tectonic feature* NE Pacific Ocean
Clarión, Isla 50 A5 *island* W Mexico
Clark Fork 44 A1 *river* Idaho/Montana, NW USA
Clark Hill Lake 43 E2 *var.* J.Storm Thurmond Reservoir. *reservoir* Georgia/South Carolina, SE USA
Clarksburg 40 D4 West Virginia, NE USA
Clarksdale 42 B2 Mississippi, S USA
Clarksville 42 C1 Tennessee, S USA
Clausentum *see* Southampton
Clayton 49 F1 New Mexico, SW USA
Clearwater 43 E4 Florida, SE USA
Clearwater Mountains 46 D2 *mountain range* Idaho, NW USA
Cleburne 49 G3 Texas, SW USA
Clermont 148 D4 Queensland, E Australia
Clermont-Ferrand 91 C5 Puy-de-Dôme, C France
Cleveland 40 D3 Ohio, N USA
Cleveland 42 D1 Tennessee, S USA
Clifton 48 C2 Arizona, SW USA
Clinton 42 B2 Mississippi, S USA
Clinton 49 F1 Oklahoma, C USA
Clipperton Fracture Zone 153 E3 *tectonic feature* E Pacific Ocean
Clipperton, Ile *see* Clipperton Island
Clipperton Island 153 F3 *var.* La Passion-Clipperton, Fr. Clipperton, Ile. *administered from France* E Pacific Ocean
Cloncurry 148 B3 Queensland, C Australia

Clonmel 89 B6 *Ir.* Cluain Meala. S Ireland
Cloppenburg 94 B3 Niedersachsen, NW Germany
Cloquet 45 G2 Minnesota, N USA
Cloud Peak 44 C3 *mountain* Wyoming, C USA
Clovis 49 E2 New Mexico, SW USA
Cluain Meala *see* Clonmel
Cluj *see* Cluj-Napoca
Cluj-Napoca 108 B3 *Ger.* Klausenburg, *Hung.* Kolozsvár; *prev.* Cluj. Cluj, NW Romania
Clutha 151 B7 *var.* Mata-Au. *river* South Island, New Zealand
Clyde 88 C4 *river* W Scotland, United Kingdom
Coari 62 D2 Amazonas, N Brazil
Coast Mountains 36 D4 *Fr.* Chaîne Côtière. *mountain range* Canada/USA
Coast Ranges 46 A4 *mountain range* W USA
Coats Island 37 G3 *island* Nunavut, NE Canada
Coats Land 154 B2 *physical region* Antarctica
Coatzacoalcos 51 G4 *var.* Quetzalcoalco; *prev.* Puerto México. Veracruz-Llave, E Mexico
Cobán 52 B2 Alta Verapaz, C Guatemala
Cobar 149 C6 New South Wales, SE Australia
Cobija 61 E3 Pando, NW Bolivia
Coblence/Coblenz *see* Koblenz
Coburg 95 C5 Bayern, SE Germany
Coca *see* Puerto Francisco de Orellana
Cocanada *see* Kākināda
Cochabamba 61 F4 *hist.* Oropeza. Cochabamba, C Bolivia
Cochin *see* Kochi
Cochinos, Bahía de 54 B2 *Eng.* Bay of Pigs. *bay* SE Cuba
Cochrane 38 C4 Ontario, S Canada
Cochrane 65 B7 Aisén, S Chile
Cocibolca *see* Nicaragua, Lago de
Cockade State *see* Maryland
Cockburn Town 55 E2 San Salvador, E The Bahamas
Cockpit Country, The 54 A4 *physical region* W Jamaica
Cocobeach 77 A5 Estuaire, NW Gabon
Coconino Plateau 48 B1 *plain* Arizona, SW USA
Coco, Río 53 E2 *var.* Río Wanki, Segovino Wangkí. *river* Honduras/ Nicaragua
Cocos Basin 124 C5 *undersea basin* E Indian Ocean
Cocos Island Ridge *see* Cocos Ridge
Cocos Islands 141 D5 *island group* E Indian Ocean
Cocos Ridge 35 C8 *var.* Cocos Island Ridge. *undersea ridge* E Pacific Ocean
Cod, Cape 41 G3 *headland* Massachusetts, NE USA
Codfish Island 151 A8 *var.* Whenua Hou. *island* SW New Zealand
Codlea 108 C4 *Ger.* Zeiden, *Hung.* Feketehalom. Braşov, C Romania
Cody 44 C2 Wyoming, C USA
Coeur d'Alene 46 C2 Idaho, NW USA
Coevorden 86 E2 Drenthe, NE Netherlands
Coffs Harbour 149 E6 New South Wales, SE Australia
Cognac 91 B5 *anc.* Compniacum. Charente, W France
Cohatin *see* Rupea
Coiba, Isla de 53 E5 *island* SW Panama
Coihaique 65 B6 *var.* Coyhaique. Aisén, S Chile
Coimbatore 132 C3 Tamil Nādu, S India
Coimbra 92 B3 *anc.* Conímbria, Conimbriga. Coimbra, W Portugal
Coín 92 D5 Andalucía, S Spain
Coira/Coire *see* Chur
Coirib, Loch *see* Corrib, Lough
Colby 45 E4 Kansas, C USA
Colchester 89 E6 Connecticut, NE USA
Coleman 49 F3 Texas, SW USA
Coleraine 88 B4 *Ir.* Cúil Raithin. N Northern Ireland, United Kingdom
Colesberg 78 C5 Northern Cape, C South Africa
Colima 50 D4 Colima, S Mexico
Coll 88 B3 *island* W Scotland, United Kingdom
College Station 49 G3 Texas, SW USA
Collie 147 A7 Western Australia
Collipo *see* Leiria

olmar *90 E4 Ger.* Kolmar. Haut-Rhin, NE France

oln *see* Köln

ologne *see* Köln

olomb-Béchar *see* Béchar

olombia *58 B3 off.* Republic of Colombia. *country* N South America

olombian Basin *56 A1 undersea basin* SW Caribbean Sea

olombia, Republic of *see* Colombia

olombie-Britannique *see* British Columbia

olombo *132 C4 commercial capital* (Sri Lanka) Western Province, W Sri Lanka

olón *53 G4 prev.* Aspinwall. Colón, C Panama

olón, Archipiélago de *see* Galápagos Islands

olón Ridge *35 B8 undersea ridge* E Pacific Ocean

olorado *44 C4 off.* State of Colorado, *also known as* Centennial State, Silver State. *state* C USA

olorado City *49 F3* Texas, SW USA

olorado Plateau *48 B1 plateau* W USA

olorado, Río *65 C5 river* E Argentina

olorado, Río *see* Colorado River

olorado River *35 B5 var.* Río Colorado. *river* Mexico/USA

olorado River *49 G4 river* Texas, SW USA

olorado Springs *44 D5* Colorado, C USA

olumbia *41 E4* Maryland, NE USA

olumbia *45 G4* Missouri, C USA

olumbia *43 E2 state capital* South Carolina, SE USA

olumbia *42 C1* Tennessee, S USA

olumbia River *46 B3 river* Canada/USA

olumbia Plateau *46 C3 plateau* Idaho/Oregon, NW USA

olumbus *42 D2* Georgia, SE USA

olumbus *40 C4* Indiana, N USA

olumbus *42 C2* Mississippi, S USA

olumbus *45 F4* Nebraska, C USA

olumbus *40 D4 state capital* Ohio, N USA

olville Channel *150 D2 channel* North Island, New Zealand

olville River *36 D2 river* Alaska, USA

omacchio *96 C3 var.* Commachio; *anc.* Comactium. Emilia-Romagna, N Italy

omactium *see* Comacchio

omalcalco *51 G4* Tabasco, SE Mexico

oma Pedrosa, Pic de *91 A7 mountain* NW Andorra

omarapa *61 F4* Santa Cruz, C Bolivia

omayagua *52 C2* Comayagua, W Honduras

omer See *see* Como, Lago di

omilla *see* Cumilla

omino *see* Kemmuna

omitán *51 G5 var.* Comitán de Domínguez. Chiapas, SE Mexico

omitán de Domínguez *see* Comitán

ommachio *see* Comacchio

ommissioner's Point *42 A5 headland* W Bermuda

ommunism Peak *123 F3 prev.* Qullai Kommunizm. *mountain* E Tajikistan

omo *96 B2 anc.* Comum. Lombardia, N Italy

omodoro Rivadavia *65 B6* Chubut, SE Argentina

omo, Lake *96 B2 var.* Lario, *Eng.* Lake Como, *Ger.* Comer See. *lake* N Italy

omo, Lake *see* Como, Lago di

omores, Union des *see* Comoros

omoros *79 F2 off.* Union of the Comoros, *Fr.* Union des Comoros. *country* W Indian Ocean

omoros, Union of the *see* Comoros

ompiègne *90 C3* Oise, N France

omplutum *see* Alcalá de Henares

ompniacum *see* Cognac

ompostela *see* Santiago de Compostela

omrat *108 D4 Rus.* Komrat. S Moldova

omum *see* Como

onakry *74 C4 country capital* (Guinea) SW Guinea

onca *see* Cuenca

oncarneau *90 A3* Finistère, NW France

oncepción *61 G3* Santa Cruz, E Bolivia

oncepción *65 B5* Bío Bío, C Chile

oncepción *64 D2 var.* Villa Concepción. Concepción, C Paraguay

oncepción *see* La Concepción

oncepción de la Vega *see* La Vega

onchos, Río *48 D4 river* NW Mexico

Conchos, Río *50 D2 river* C Mexico

Concord *41 G3 state capital* New Hampshire, NE USA

Concordia *64 D4* Entre Ríos, E Argentina

Concordia *45 E4* Kansas, C USA

Concordia *154 C4 French/Italian research station* Antarctica

Côn Đao *see* Côn Đao Son

Côn Đao Son *137 E7 var.* Côn Đao, Con Son. *island* S Vietnam

Condate *see* Rennes, Ille-et-Vilaine, France

Condate *see* St-Claude, Jura, France

Condega *52 D3* Estelí, NW Nicaragua

Condivincum *see* Nantes

Confluentes *see* Koblenz

Cộng Hoa Xa Hôi Chu Nghia Viêt Nam *see* Vietnam

Congo *77 D5 off.* Republic of the Congo, *Fr.* Moyen-Congo; *prev.* Middle Congo. *country* C Africa

Congo *77 C6 off.* Democratic Republic of Congo; *prev.* Zaire, Belgian Congo, Congo (Kinshasa). *country* C Africa

Congo *77 C6 var.* Kongo, *Fr.* Zaire. *river* C Africa

Congo Basin *77 C6 drainage basin* W Dem. Rep. Congo

Congo/Congo (Kinshasa) *see* Congo (Democratic Republic of)

Coni *see* Cuneo

Conimbria/Conimbriga *see* Coimbra

Conjeeveram *see* Kānchipuram

Connacht *see* Connaught

Connaught *89 A5 var.* Connacht, *Ir.* Chonnacht, Cúige. *province* W Ireland

Connecticut *41 F3 river* Canada/USA

Connecticut *41 G3 var.* Blue Law State, Constitution State, Land of Steady Habits, Nutmeg State. *state* NE USA

Conroe *49 G3* Texas, SW USA

Consentia *see* Cosenza

Consolación del Sur *54 A2* Pinar del Río, W Cuba

Con Son *see* Côn Đao Son

Constance *see* Konstanz

Constance, Lake *95 B7 Ger.* Bodensee. *lake* C Europe

Constanţa *108 D5 var.* Küstendje, *Eng.* Constanza, *Ger.* Konstanza, *Turk.* Küstence. Constanţa, SE Romania

Constantia *see* Coutances

Constantia *see* Konstanz

Constantine *71 E2 var.* Qacentina, *Ar.* Qoussantîna. NE Algeria

Constantinople *see* İstanbul

Constantiola *see* Olteniţa

Constanz *see* Konstanz

Constanza *see* Constanţa

Constitution State *see* Connecticut

Coo *see* Kos

Coober Pedy *149 A5* South Australia

Cookeville *42 D1* Tennessee, S USA

Cook Islands *145 F4 self-governing territory in free association with New Zealand* S Pacific Ocean

Cook, Mount *see* Aoraki

Cook Strait *151 D5 var.* Raukawa. *strait* New Zealand

Cooktown *148 D2* Queensland, NE Australia

Coolgardie *147 B6* Western Australia

Cooma *149 D7* New South Wales, SE Australia

Coomassie *see* Kumasi

Coon Rapids *45 F2* Minnesota, N USA

Cooper Creek *148 C4 var.* Barcoo, Cooper's Creek. *seasonal river* Queensland/South Australia, Australia

Cooper's Creek *see* Cooper Creek

Coos Bay *46 A3* Oregon, NW USA

Cootamundra *149 D6* New South Wales, SE Australia

Copacabana *61 E4* La Paz, W Bolivia

Copenhagen *see* København

Copiapó *64 B3* Atacama, N Chile

Copperas Cove *49 G3* Texas, SW USA

Coppermine *see* Kugluktuk

Copper State *see* Arizona

Coquilhatville *see* Mbandaka

Coquimbo *64 B3* Coquimbo, N Chile

Corabia *108 B5* Olt, S Romania

Coral Harbour *37 G3 var.* Salliq. Southampton Island, Nunavut, NE Canada

Coral Sea *142 B3 sea* SW Pacific Ocean

Coral Sea Islands *144 B4 Australian external territory* SW Pacific Ocean

Corantijn Rivier *see* Courantyne River

Corcovado, Golfo *65 B6 gulf* S Chile

Corcyra Nigra *see* Korčula

Cordele *42 D3* Georgia, SE USA

Córdoba *64 C3* Córdoba, C Argentina

Córdoba *51 F4* Veracruz-Llave, E Mexico

Córdoba *92 D4 var.* Cordoba, *Eng.* Cordova; *anc.* Corduba. Andalucía, SW Spain

Cordova *36 C3* Alaska, USA

Cordova/Cordoba *see* Córdoba

Corduba *see* Córdoba

Corentyne River *see* Courantyne River

Corfu *see* Kérkyra

Coria *92 C3* Extremadura, W Spain

Corinth *42 C1* Mississippi, S USA

Corinth *see* Kórinthos

Corinth, Gulf of/Corinthiacus Sinus *see* Korinthiakós Kólpos

Corinthus *see* Kórinthos

Corinto *52 C3* Chinandega, NW Nicaragua

Cork *89 A6 Ir.* Corcaigh. S Ireland

Çorlu *116 A2* Tekirdağ, NW Turkey (Türkiye)

Corner Brook *39 G3* Newfoundland, Newfoundland and Labrador, E Canada

Cornhusker State *see* Nebraska

Corn Islands *53 E3 var.* Corn Islands. *island group* SE Nicaragua

Corn Islands *see* Maíz, Islas del

Cornwallis Island *37 F2 Island* Nunavut, N Canada

Coro *58 C1 prev.* Santa Ana de Coro. Falcón, NW Venezuela

Corocoro *61 F4* La Paz, W Bolivia

Coromandel *150 D2* Waikato, North Island, New Zealand

Coromandel Coast *132 D2 coast* E India

Coromandel Peninsula *150 D2 peninsula* North Island, New Zealand

Coronado, Bahía de *52 D5 bay* S Costa Rica

Coronel Dorrego *65 C5* Buenos Aires, E Argentina

Coronel Oviedo *64 D2* Caaguazú, SE Paraguay

Corozal *52 C1* Corozal, N Belize

Corpus Christi *49 G4* Texas, SW USA

Corrales *48 D2* New Mexico, SW USA

Corrib, Lough *89 A5 Ir.* Loch Coirib. *lake* W Ireland

Corrientes *64 D3* Corrientes, NE Argentina

Corriza *see* Korçë

Corsica *91 E7 Eng.* Corsica. *island* France, C Mediterranean Sea

Corsica *see* Corse

Corsicana *49 G3* Texas, SW USA

Cortegana *92 C4* Andalucía, S Spain

Cortés *53 E5 var.* Ciudad Cortés. Puntarenas, SE Costa Rica

Cortez, Sea of *see* California, Golfo de

Cortina d'Ampezzo *96 C1* Veneto, NE Italy

Coruche *92 B3* Santarém, C Portugal

Çoruh Nehri *117 E3 Geor.* Chorokh, *Rus.* Chorokhi. *river* Georgia/Turkey (Türkiye)

Çorum *116 D3 var.* Chorum. Çorum, N Turkey (Türkiye)

Coruña *see* A Coruña

Corumbá *63 G3* Mato Grosso do Sul, SW Brazil

Corvallis *46 B3* Oregon, NW USA

Corvo *92 A5 var.* Ilha do Corvo. *island* Azores, Portugal, NE Atlantic Ocean

Corvo, Ilha do *see* Corvo

Cos *see* Kos

Cosenza *97 D6 anc.* Consentia. Calabria, SW Italy

Cosne-Cours-sur-Loire *90 C4* Nièvre, Bourgogne, C France Europe

Costa Mesa *46 D2* California, W USA

Costa Rica *53 E4 off.* Republic of Costa Rica. *country* Central America

Costa Rica, Republic of *see* Costa Rica

Costermansville *see* Bukavu

Cotagaita *61 F5* Potosí, S Bolivia

Côte d'Ivoire *see* Ivory Coast

Côte d'Ivoire, République de la *see* Ivory Coast

Côte d'Or *90 D4 cultural region* C France

Côte Française des Somalis *see* Djibouti

Côtière, Chaîne *see* Coast Mountains

Cotonou *75 F5 var.* Kotonu. *country capital* (Benin - seat of government) S Benin

Cotswold Hills *89 D6 var.* Cotswolds. *hill range* S England, United Kingdom

Cotswolds *see* Cotswold Hills

Cottbus *94 D4 Lus.* Chośebuz; *prev.* Kottbus. Brandenburg, E Germany

Cotton State, The *see* Alabama

Cotyora *see* Ordu

Couentrey *see* Coventry

Council Bluffs *45 F4* Iowa, C USA

Courantyne River *59 G4 var.* Corantijn Rivier, Corentyne River. *river* Guyana/Suriname

Courland Lagoon *106 A4 Ger.* Kurisches Haff, *Rus.* Kurskiy Zaliv. *lagoon* Lithuania/Russia

Courtrai *see* Kortrijk

Coutances *90 B3 anc.* Constantia. Manche, N France

Couvin *87 C7* Namur, S Belgium

Coventry *89 D6 anc.* Couentrey. C England, United Kingdom

Covilhã *92 C3* Castelo Branco, E Portugal

Cowan, Lake *147 B6 lake* Western Australia

Coxen Hole *see* Roatán

Coxin Hole *see* Roatán

Coyhaique *see* Coihaique

Coyote State, The *see* South Dakota

Cozhê *126 C5* Xizang Zizhiqu, W China

Cozumel, Isla *51 H3 island* SE Mexico

Cracovia/Cracow *see* Kraków

Cradock *see* Nxuba

Craig *44 C4* Colorado, C USA

Craiova *108 B5* Dolj, SW Romania

Cranbrook *37 E5* British Columbia, SW Canada

Crane *see* The Crane

Cranz *see* Zelenogradsk

Crawley *89 E7* SE England, United Kingdom

Cremona *96 B2* Lombardia, N Italy

Creole State *see* Louisiana

Cres *100 A3 It.* Cherso; *anc.* Crexa. *island* W Croatia

Crescent City *46 A4* California, W USA

Crescent Group *129 C7 island group* C Paracel Islands

Creston *45 F4* Iowa, C USA

Crestview *42 D3* Florida, SE USA

Crete *see* Kríti

Créteil *90 E2* Val-de-Marne, N France

Crete, Sea of/Creticum, Mare *see* Kritikó Pélagos

Creuse *90 B4 river* C France

Crewe *89 D6* C England, United Kingdom

Crexa *see* Cres

Crikvenica *100 A3 It.* Cirquenizza; *prev.* Cirkvenica, Crikvenica. Primorje-Gorski Kotar, NW Croatia

Crimea *see* Krym

Cristóbal *53 G4* Colón, C Panama

Cristóbal Colón, Pico *58 B1 mountain* N Colombia

Cristur/Cristuru Săcuiesc *see* Cristuru Secuiesc

Cristuru Secuiesc *108 C4 prev.* Cristur, Cristuru Săcuiesc, Sitaş Cristuru, *Ger.* Kreutz, *Hung.* Székelykeresztúr, Szitás-Keresztúr. Harghita, C Romania

Crjkvenica *see* Crikvenica

Crna Gora *see* Montenegro

Crna Reka *101 D6 river* S North Macedonia

Crni Drim *see* Black Drin

Croatia *100 B3 off.* Republic of Croatia, *Cro.* Hrvatska, Republika Hrvatska. *country* SE Europe

Croatia, Republic of *see* Croatia

Crocodile *see* Limpopo

Croia *see* Krujë

Croker Island *146 E2 island* Northern Territory, N Australia

Cromwell *151 B7* Otago, South Island, New Zealand

Crooked Island *54 D2 island* SE The Bahamas

Crooked Island Passage *54 C2 channel* SE The Bahamas

Crookston *45 F1* Minnesota, N USA

Crossen *see* Krosno Odrzańskie

Croton/Crotona *see* Crotone

Crotone *97 E6 var.* Cotrone; *anc.* Croton, Crotona. Calabria, SW Italy

Croydon *89 A8* SE England, United Kingdom

Crozet Basin *141 B6 undersea basin* S Indian Ocean

Crozet Islands *141 B7 island group* French Southern and Antarctic Lands

Crozet Plateau *141 B7 var.* Crozet
Plateaus. *undersea plateau*
SW Indian Ocean
Crozet Plateaus *see* Crozet Plateau
Crystal Brook *149 B6* South Australia
Csaca *see* Čadca
Csakathurn/Csáktornya *see* Čakovec
Csíkszereda *see* Miercurea-Ciuc
Csorna *99 C6* Győr-Moson-Sopron,
NW Hungary
Csurgó *99 C7* Somogy, SW Hungary
Cuando *78 C2 var.* Kwando. *river*
S Africa
Cuango *see* Kwango
Cuanza *78 B1 var.* Kwanza. *river*
C Angola
Cuauhtémoc *50 C2* Chihuahua,
N Mexico
Cuautla *51 E4* Morelos, S Mexico
Cuba *54 B2 off.* Republic of Cuba.
country W West Indies
Cubal *78 B2* Benguela, W Angola
Cubango *78 B2 var.* Kuvango, *Port.* Vila
Artur de Paiva, Vila da Ponte. Huíla,
SW Angola
Cubango *78 B2 var.* Kavango, Kavengo,
Kubango, Okavango, Okavanggo.
river S Africa
Cuba, Republic of *see* Cuba
Cúcuta *58 C2 var.* San José de Cúcuta.
Norte de Santander, N Colombia
Cuddapah *see* Kadapa
Cuenca *60 B2* Azuay, S Ecuador
Cuenca *93 E3 anc.* Conca. Castilla-La
Mancha, C Spain
Cuera *see* Chur
Cuernavaca *51 E4* Morelos, S Mexico
Cuiabá *63 E3 prev.* Cuyabá. *state capital*
Mato Grosso, SW Brazil
Cúige *see* Connaught
Cúige Laighean *see* Leinster
Cúige Mumhan *see* Munster
Cuijck *86 D4* Noord-Brabant,
SE Netherlands
Cúil Raithin *see* Coleraine
Cuito *78 B2 var.* Kuito; *Port.* Silva Porto.
Bié, C Angola
Cuito *78 B2 var.* Kwito. *river* SE Angola
Cukai *138 B3 var.* Chukai, Kemaman.
Terengganu, Peninsular Malaysia
Cularo *see* Grenoble
Culiacán *50 C3 var.* Culiacán Rosales,
Culiacán-Rosales. Sinaloa, C Mexico
Culiacán-Rosales/Culiacán Rosales *see*
Culiacán
Cullera *93 F3* Comunitat Valenciana,
E Spain
Cullman *42 C2* Alabama, S USA
Culm *see* Chełmno
Culmsee *see* Chełmża
Cumaná *59 E1* Sucre, NE Venezuela
Cumbal, Nevado de *58 A4 elevation*
S Colombia
Cumberland *41 E4* Maryland,
NE USA
Cumberland Plateau *42 D1 plateau*
E USA
Cumberland Sound *37 H3 inlet* Baffin
Island, Nunavut, NE Canada
Cumilla *135 G4 prev.* Comilla, *Ben.*
Kumillā. Chattogram, E Bangladesh
Cumpas *50 B2* Sonora, NW Mexico
Cuneo *96 A2 Fr.* Coni. Piemonte,
NW Italy
Cunnamulla *149 C5* Queensland,
E Australia
Ćuprija *100 E4* Serbia, E Serbia
Curaçao *55 E5 Dutch self-governing
territory* S West Indies
Curia Rhaetorum *see* Chur
Curicó *64 B4* Maule, C Chile
Curieta *see* Krk
Curitiba *63 E4 prev.* Curytiba. *state
capital* Paraná, S Brazil
Curtbunar *see* Tervel
Curtea de Argeş *108 C4 var.*
Curtea-de-Argeş. Argeş, S Romania
Curtici *108 A4 Ger.* Kurtitsch, *Hung.*
Kürtös. Arad, W Romania
Curtis Island *148 E4 island* Queensland,
SE Australia
Curytiba *see* Curitiba
Curzola *see* Korčula
Cusco *61 E4 var.* Cuzco. Cusco, C Peru
Cusset *91 C5* Allier, C France
Cutch, Gulf of *see* Kachchh, Gulf of
Cuttack *135 F4* Odisha, E India
Cuvier Plateau *141 E6 undersea plateau*
E Indian Ocean

Cuxhaven *94 B2* Niedersachsen,
NW Germany
Cuyabá *see* Cuiabá
Cuyuni, Río *see* Cuyuni River
Cuyuni River *59 F3 var.* Río Cuyuni.
river Guyana/Venezuela
Cuzco *see* Cusco
Cyclades *see* Kykládes
Cydonia *see* Chaniá
Cymru *see* Wales
Cyprus *102 C4 off.* Republic of Cyprus,
Gk. Kypriakí Dimokratía, Kýpros, *Turk.*
Kıbrıs, Kıbrıs Cumhuriyeti. *country*
E Mediterranean Sea
Cyprus, Republic of *see* Cyprus
Cyrenaica *71 H2 region* NE Libya
Cythnos *see* Kýthnos
Czech Republic *see* Czechia
Czechia *99 A5 off.* Czech Republic, *Cz.*
Česko. *country* C Europe
Czenstochau *see* Częstochowa
Czernowitz *see* Chernivtsi
Częstochowa *98 C4 Ger.* Czenstochau,
Tschenstochau, *Rus.* Chenstokhov.
Śląskie, S Poland
Człuchów *98 C3 Ger.* Schlochau.
Pomorskie, NW Poland

D

Dabajuro *58 C1* Falcón, NW Venezuela
Dabeiba *58 B2* Antioquia, NW Colombia
Dąbrowa Tarnowska *99 D5*
Małopolskie, S Poland
Dabryn' *107 C8 Rus.* Dobryn'.
Homyel'skaya Voblasts', SE Belarus
Dacca *see* Dhaka
Daegu *129 E4 off.* Daegu Gwang-yeoksi,
prev. Taegu, *Jap.* Taikyū.
SE South Korea
Daegu Gwang-yeoksi *see* Daegu
Dagana *74 B3* N Senegal
Dagda *106 D4* Krāslava, SE Latvia
Dagden *see* Hiiumaa
Dagenham *89 B8* United Kingdom
Dagö *see* Hiiumaa
Dagupan *139 E1 off.* Dagupan City.
Luzon, N Philippines
Dagupan City *see* Dagupan
Da Hinggan Ling *127 G1 Eng.* Great
Khingan Range. *mountain range*
NE China
Dahm, Ramlat *121 B6 desert* NW Yemen
Dahomey *see* Benin
Daihoku *see* Taibei
Daimiel *92 D3* Castilla-La Mancha,
C Spain
Daimoniá *105 B7* Pelopónnisos, S Greece
Dainan *see* Tainan
Daingin, Bá an *see* Dingle Bay
Dairen *see* Dalian
Daejeon *129 E4 off.* Daejeon
Gwang-yeoksi, *prev.* Taejŏn, *Jap.*
Taiden. C South Korea
Daejeon Gwang-yeoksi *see* Daejeon
Dakar *74 B3 country capital* (Senegal)
W Senegal
Dakhla *see* Ad Dakhla
Dakoro *75 G3* Maradi, S Niger
Đakovica *see* Gjakovë
Đakovo *100 C3 var.* Djakovo, *Hung.*
Diakovár. Osijek-Baranja, E Croatia
Dakshin *see* Deccan
Dalain Hob *126 D3 var.* Ejin Qi. Nei
Mongol Zizhiqu, N China
Dalai Nor *see* Hulun Nur
Dalaman *116 A4* Muğla,
SW Turkey (Türkiye)
Dalandzadgad *127 E3* Ömnögovĭ,
S Mongolia
Đa Lat *137 E6* Lâm Đồng, S Vietnam
Dalby *149 D5* Queensland, E Australia
Dale City *41 E4* Virginia, NE USA
Dalhart *49 E1* Texas, SW USA
Dali *129 A6 var.* Xiaguan. Yunnan,
SW China
Dalian *128 D4 var.* Dairen, Dalien, Jay
Dairen, Lüda, Ta-lien, *Rus.* Dalny.
Liaoning, NE China
Dalien *see* Dalian
Dallas *49 G2* Texas, SW USA
Dalmacija *100 B4 Eng.* Dalmatia, *Ger.*
Dalmatien, *It.* Dalmazia. *cultural region*
S Croatia
Dalmatia/Dalmatien/Dalmazia *see*
Dalmacija
Dalny *see* Dalian
Dalton *42 D1* Georgia, SE USA

Dálvvadis *see* Jokkmokk
Daly Waters *148 A2* Northern Territory,
N Australia
Damachava *107 A6 var.* Damachova,
Pol. Domaczewo, *Rus.* Domachёvo.
Brestskaya Voblasts', SW Belarus
Damachova *see* Damachava
Daman *134 C4* Damān and Diu, W India
Damara *76 C4* Ombella-Mpoko,
S Central African Republic
Damas *see* Dimashq
Damasco *see* Dimashq
Damascus *see* Dimashq
Damavand, Qolleh-ye *120 D3 mountain*
N Iran
Damietta *see* Dumyāṭ
Dammām *see* Ad Dammām
Damoûr *119 A5 var.* Ad Dāmūr.
W Lebanon
Dampier *146 A4* Western Australia
Dampier, Selat *139 F4 strait* Papua,
E Indonesia
Damqawt *121 D6 var.* Damqut.
E Yemen
Damqut *see* Damqawt
Damxung *126 C5 var.* Gongtang. Xizang
Zizhiqu, W China
Danakil Desert *72 D4 var.* Afar
Depression, Danakil Plain. *desert*
E Africa
Danakil Plain *see* Danakil Desert
Danané *74 D5* W Ivory Coast
Đà Nẵng *137 E5 prev.* Tourane. Quang
Nam-Đa Nẵng, C Vietnam
Danborg *see* Daneborg
Dandong *128 D3 var.* Tan-tung; *prev.*
An-tung. Liaoning, NE China
Daneborg *83 E3 var.* Danborg.
NE Greenland
Dänew *see* Galkynyş
Dangara *see* Danghara
Dangerous Archipelago *see*
Tuamotu, Îles
Danghara *123 E3 Rus.* Dangara.
SW Tajikistan
Danghe Nanshan *126 D3 mountain
range* W China
Dangrek, Chaîne des *see* Dangrek
Mountains
Dângrêk, Chuŏr Phnum *see* Dangrek
Mountains
Dangrek Mountains *137 D5 Khm.*
Chuŏr Phnum Dângrêk, *Thai.* Thiu
Khao Phanom Dong Rak, *Fr.* Chaîne
des Dangrek. *mountain range*
Cambodia/Thailand
Dangriga *52 C1 prev.* Stann Creek.
Stann Creek, E Belize
Danish West Indies *see* Virgin Islands
(US)
Danli *52 D2* El Paraíso, S Honduras
Danmark *see* Denmark
Danmarksstraedet *see* Denmark Strait
Dannenberg *94 C3* Niedersachsen,
N Germany
Dannevirke *150 D4* Manawatu-
Wanganui, North Island, New Zealand
Dantzig *see* Gdańsk
Danube *81 E4 Bul., Croatian, Serb.*
Dunav, *Cz., Slov.* Dunaj, *Ger.* Donau,
Hung. Duna, *Rom.* Dunărea, *Ukr.*
Dunai. *river* C Europe
Danubian Plain *see* Dunavska Ravnina
Danum *see* Doncaster
Danville *41 E5* Virginia, NE USA
Danxian/Dan Xian *see* Danzhou
Danzhou *129 C7 prev.* Danxian, Dan
Xian, Nada. Hainan, S China
Danzig *see* Gdańsk
Danziger Bucht *see* Gdańsk, Gulf of
Danzig, Gulf of *see* Gdańsk, Gulf of
Daqm *see* Duqm
Dar'ā *119 B5 var.* Der'a, *Fr.* Déraa.
Dar'ā, SW Syria
Darabani *108 C3* Botoşani,
NW Romania
Darazat-Kurgan *see* Daroot-Korgon
Dardanelles *116 A2 Eng.* Dardanelles.
strait NW Turkey (Türkiye)
Dardanelles *see* Çanakkale Boğazı
Dardanelli *see* Çanakkale
Dar-el-Beida *see* Casablanca
Dar es Salaam *73 C7* Dar es Salaam,
E Tanzania
Darfield *151 C6* Canterbury, South
Island, New Zealand
Darfur *72 A4 var.* Darfur Massif.
cultural region W Sudan
Darfur Massif *see* Darfur
Darhan *127 E2* Darhan Uul, N Mongolia

Darién, Golfo del *see* Darien, Gulf of
Darien, Gulf of *58 A2 Sp.* Golfo del
Darién. *gulf* S Caribbean Sea
Darien, Isthmus of *see* Panama, Istmo de
Darién, Serranía del *53 H5 mountain
range* Colombia/Panama
Darío *see* Ciudad Darío
Darioriguum *see* Vannes
Darjeeling *see* Dārjiling
Dārjiling *135 F3 prev.* Darjeeling. West
Bengal, NE India
Darling River *149 C6 river* New South
Wales, SE Australia
Darlington *89 D5* N England,
United Kingdom
Darmstadt *95 B5* Hessen, SW Germany
Darnah *71 G2 var.* Dérna. NE Libya
Darnley, Cape *154 D2 cape* Antarctica
Daroca *93 E2* Aragón, NE Spain
Daroot-Korgon *123 F3 var.*
Daraut-Kurgan. Osh Oblusu/Oshskay
Oblast', SW Kyrgyzstan
Dartford *89 B8* SE England,
United Kingdom
Dartmoor *89 C7 moorland* SW England
United Kingdom
Dartmouth *39 F4* Nova Scotia,
SE Canada
Darvaza *see* Derweze, Turkmenistan
Darwin *146 D2 prev.* Palmerston, Port
Darwin. *territory capital* Northern
Territory, N Australia
Darwin, Isla *60 A4 island* Galápagos
Islands, W Ecuador
Dashhowuz *see* Daşoguz
Dashkawka *107 D6 Rus.* Dashkovka.
Mahilyowskaya Voblasts', E Belarus
Dashkovka *see* Dashkawka
Daşoguz *122 C2 Rus.* Dashkhovuz,
Turkm. Dashhowuz; *prev.* Tashauz.
Daşoguz Welaýaty, N Turkmenistan
Đa, Sông *see* Black River
Datong *128 C3 var.* Tatung, Ta-t'ung.
Shanxi, C China
Daugava *see* Western Dvina
Daugavpils *106 D4 Ger.* Dünaburg; *prev.*
Rus. Dvinsk. Daugvapils, SE Latvia
Daung Kyun *137 B6 island*
S Myanmar (Burma)
Dauphiné *91 D5 cultural region* E France
Dāvangere *132 C2* Karnātaka, W India
Davao *139 F3 off.* Davao City.
Mindanao, S Philippines
Davao City *see* Davao
Davao Gulf *139 F3 gulf* Mindanao,
S Philippines
Davenport *45 G3* Iowa, C USA
David *53 E5* Chiriquí, W Panama
Davie Ridge *141 A5 undersea ridge*
W Indian Ocean
Davis *154 D3 Australian research station*
Antarctica
Davis Sea *154 D3 sea* Antarctica
Davis Strait *82 B3 strait* Baffin Bay/
Labrador Sea
Dawei *137 B5 var.* Tavoy, Htawei.
Tanintharyi, S Myanmar (Burma)
Dawlat Qatar *see* Qatar
Dax *91 B6 var.* Ax; *anc.* Aquae Augustae,
Aquae Tarbelicae. Landes, SW France
Dayr az Zawr *118 D3 var.* Deir ez Zor.
Dayr az Zawr, E Syria
Dayton *40 C4* Ohio, N USA
Daytona Beach *43 E4* Florida, SE USA
De Aar *78 C5* Northern Cape,
C South Africa
Dead Sea *119 B6 var.* Bahret Lut, Lacus
Asphaltites, *Ar.* Al Baḥr al Mayyit,
Baḥrat Lūṭ, *Heb.* Yam HaMelaḥ. *salt
lake* Israel/Jordan
Dē Afghānistān Islāmī Jumhūriyat *see*
Afghanistan
Deán Funes *64 C3* Córdoba, C Argentina
Death Valley *47 C7 valley* California,
W USA
Deatnu *84 D2 Fin.* Tenojoki, *Nor.* Tana.
river Finland/Norway
Debar *101 D6 Ger.* Dibra, *Turk.* Debre.
W North Macedonia
De Behagle *see* Laï
Dębica *99 D5* Podkarpackie, SE Poland
De Bilt *86 C3* Utrecht,
C Netherlands
De Bilt *86 C3 var.* De Bildt. Utrecht,
C Netherlands
Debno *98 B3* Zachodnio-pomorskie,
NW Poland
Debre *see* Debar
Debrecen *99 D6 Ger.* Debreczin, *Rom.*
Debreţin; *prev.* Debreczen.
Hajdú-Bihar, E Hungary

Debreczen/Debreczin *see* Debrecen
Debreţin *see* Debrecen
Decatur *42 C1* Alabama, S USA
Decatur *40 B4* Illinois, N USA
Deccan *134 D5 Hind.* Dakshin. *plateau*
　C India
Děčín *98 B4 Ger.* Tetschen. Ústecký
　Kraj, NW Czechia (Czech Republic)
Dedeagaç/Dedeagach *see* Alexandroúpoli
Dedemsvaart *86 E3* Overijssel,
　E Netherlands
Dee *88 C3 river* NE Scotland,
　United Kingdom
Deering *36 C2* Alaska, USA
Deés *see* Dej
Deggendorf *95 D6* Bayern, SE Germany
Değirmenlik *102 C5 Gk.* Kythréa.
　N Cyprus
Deh Bīd *see* Şafāshahr
Dehli *see* Delhi
Deh-e Shū *122 D5 prev.* Deh Shū, *var.*
　Deshu, Dishū. Helmand, S Afghanistan
Deh Shū *see* Deh-e Shū
Deinze *87 B5* Oost-Vlaanderen,
　NW Belgium
Deir ez Zor *see* Dayr az Zawr
Deirgeirt, Loch *see* Derg, Lough
Dej *108 B3 Hung.* Dés; *prev.* Deés. Cluj,
　NW Romania
De Jouwer *see* Joure
Dekéleia *see* Dhekelia Sovereign
　Base Area
Dékoa *76 C4* Kémo, C Central
　African Republic
De Land *43 E4* Florida, SE USA
Delano *47 C7* California, W USA
Delārām *see* Dilārām
Delaware *40 D4* Ohio, N USA
Delaware *41 F4 off.* State of Delaware,
　also known as Blue Hen State, Diamond
　State, First State. *state* NE USA
Delft *86 B4* Zuid-Holland,
　W Netherlands
Delfzijl *86 E1* Groningen,
　NE Netherlands
Delgo *72 B3* Northern, N Sudan
Delhi *134 D3 var.* Dehli, *Hind.* Dilli, *hist.*
　Shahjahanabad. *union territory capital*
　Delhi, N India
Delicias *50 D2 var.* Ciudad Delicias.
　Chihuahua, N Mexico
Déli-Kárpátok *see* Carpaţii Meridionali
Delmenhorst *94 B3* Niedersachsen,
　NW Germany
Del Rio *49 F4* Texas, SW USA
Deltona *43 E4* Florida, SE USA
Demba *77 D6* Kasai-Occidental,
　C Dem. Rep. Congo
Dembia *76 D4* Mbomou, SE Central
　African Rep.
Demchok *see* Dêmqog
Demerara Plain *56 C2 abyssal plain*
　W Atlantic Ocean
Deming *48 C3* New Mexico, SW USA
Demmin *94 C2* Mecklenburg-
　Vorpommern, NE Germany
Demopolis *42 C2* Alabama, S USA
Dêmqog *126 A5 var.* Demchok. *disputed*
　region China/India
Denali *36 C3 var.* Mount McKinley.
　mountain Alaska, USA
Denali Park *36 C3 prev.* McKinley Park,
　Alaska, USA
Denau *see* Denov
Dender *87 B6 Fr.* Dendre. *river*
　W Belgium
Dendre *see* Dender
Denekamp *86 E3* Overijssel,
　E Netherlands
Den Haag *see* 's-Gravenhage
Denham *147 A5* Western Australia
Den Ham *86 E3* Overijssel,
　E Netherlands
Den Helder *86 C2* Noord-Holland,
　NW Netherlands
Dénia *93 F4* Comunitat Valenciana,
　E Spain
Deniliquin *149 C7* New South Wales,
　SE Australia
Denison *45 F3* Iowa, C USA
Denison *49 G2* Texas, SW USA
Denizli *116 B4* Denizli,
　SW Turkey (Türkiye)
Denmark *85 A7 off.* Kingdom of
　Denmark, *Dan.* Danmark; *anc.* Hafnia.
　country N Europe
Denmark, Kingdom of *see* Denmark
Denmark Strait *82 D4 var.*
　Danmarksstraedet. *strait* Greenland/
　Iceland

Dennery *55 F1* E Saint Lucia
Denov *123 E3 Rus.* Denau.
　Surkhondaryo Viloyati, S Uzbekistan
Denpasar *138 D5 prev.* Paloe. Bali,
　C Indonesia
Denton *49 G2* Texas, SW USA
D'Entrecasteaux Islands *144 B3 island*
　group SE Papua New Guinea
Denver *44 D4* state capital Colorado,
　C USA
Der'a/Derá/Déraa *see* Dar'ā
Dera Ghazi Khan *134 C2 var.* Dera
　Ghazikhan. Punjab, C Pakistan
Dera Ghazikhan *see* Dera Ghazi Khan
Đeravica *101 D5 mountain* S Serbia
Derbent *111 B8* Respublika Dagestan,
　SW Russia
Derby *89 D6* C England,
　United Kingdom
Dereli *see* Gónnoi
Dergachi *see* Derhachi
Derg, Lough *89 A6 Ir.* Loch Deirgeirt.
　lake W Ireland
Derhachi *109 G2 Rus.* Dergachi.
　Kharkivs'ka Oblast', E Ukraine
De Ridder *42 A3* Louisiana, S USA
Dérna *see* Darnah
Derry *see* Londonderry
Dertosa *see* Tortosa
Derventa *100 B3* Republika Srpska,
　N Bosnia and Herzegovina
Derweze *122 C2 Rus.* Darvaza. Ahal
　Welaýaty, C Turkmenistan
Dés *see* Dej
Deschutes River *46 B3 river* Oregon,
　NW USA
Desē *72 C4 var.* Desse, *It.* Dessie. Āmara,
　N Ethiopia
Deseado, Río *65 B7 river* S Argentina
Desertas, Ilhas *70 A2 island group*
　Madeira, Portugal, NE Atlantic Ocean
Deshu *see* Deh-e Shū
Des Moines *45 F3* state capital Iowa,
　C USA
Desna *109 E2 river* Russia/Ukraine
Dessau *94 C4* Sachsen-Anhalt,
　E Germany
Desse *see* Desē
Dessie *see* Desē
Destêrro *see* Florianópolis
Detroit *40 D3* Michigan, N USA
Detroit Lakes *45 F2* Minnesota, N USA
Deurne *87 D5* Noord-Brabant,
　SE Netherlands
Deutschendorf *see* Poprad
Deutsch-Eylau *see* Iława
Deutsch Krone *see* Wałcz
Deutschland/Deutschland,
　Bundesrepublik *see* Germany
Deutsch-Südwestafrika *see* Namibia
Deva *108 B4 Ger.* Diemrich, *Hung.* Déva.
　Hunedoara, W Romania
Déva *see* Deva
Deva *see* Chester
Devana *see* Aberdeen
Devana Castra *see* Chester
Devđelija *see* Gevgelija
Deventer *86 D3* Overijssel,
　E Netherlands
Devils Lake *45 E1* North Dakota, N USA
Devoll *see* Devollit, Lumi i
Devollit, Lumi i *101 D6 var.* Devoll. *river*
　SE Albania
Devon Island *37 F2 prev.* North Devon
　Island. *island* Parry Islands, Nunavut,
　NE Canada
Devonport *149 C8* Tasmania,
　SE Australia
Devrek *116 C2* Zonguldak,
　N Turkey (Türkiye)
Dexter *45 H5* Missouri, C USA
Deynau *see* Galkynyş
Dezfūl *120 C3 var.* Dizful. Khūzestān,
　SW Iran
Dezhou *128 D4* Shandong, E China
Dhaka *135 G4 prev.* Dacca. *country*
　capital (Bangladesh) Dhaka,
　C Bangladesh
Dhanbād *135 F4* Jhārkhand, NE India
Dhekelia Sovereign Base Area *102 C5*
　Gk. Dekéleia. *UK Sovereign Base Area*
　SE Cyprus
Dhidhimótikhon *see* Didymóteicho
Dhíkti Ori *see* Díkti
Dhodhekánisos *see* Dodekánisa
Dhomokós *see* Domokós
Dhráma *see* Dráma
Dhrepanon, Akrotírio *see* Drépano,
　Akrotírio
Dhún na nGall, Bá *see* Donegal Bay

Dhuusa Marreeb *73 E5 var.* Dusa
　Marreb, *It.* Dusa Mareb. Galguduud,
　C Somalia
Diakovár *see* Đakovo
Diamantina, Chapada *63 F3 mountain*
　range E Brazil
Diamantina Fracture Zone *141 E6*
　tectonic feature E Indian Ocean
Diamond State *see* Delaware
Diaoyutai Lieyu/ Diaoyutai Qundao *see*
　Senkaku-shoto
Diarbekr *see* Diyarbakır
Dibio *see* Dijon
Dibra *see* Debar
Dibrugarh *135 H3* Assam, NE India
Dickinson *44 D2* North Dakota, N USA
Dicle *see* Tigris
Didimotiho *see* Didymóteicho
Didymóteicho *104 D3 var.*
　Dhidhimótikhon, Didimotiho.
　Anatolikí Makedonía kai Thráki,
　NE Greece
Diedenhofen *see* Thionville
Diekirch *87 D7* Diekirch, C Luxembourg
Diemrich *see* Deva
Điện Biên *136 D3 var.* Bien Bien, Dien
　Bien Phu. Lai Châu, N Vietnam
Dien Bien Phu *see* Điện Biên
Diepenbeek *87 D6* Limburg, NE Belgium
Diepholz *94 B3* Niedersachsen,
　NW Germany
Dieppe *90 C2* Seine-Maritime, N France
Dieren *86 D3* Gelderland, E Netherlands
Differdange *87 D8* Luxembourg,
　SW Luxembourg
Digne *91 D6 var.* Digne-les-Bains.
　Alpes-de-Haute-Provence, SE France
Digne-les-Bains *see* Digne
Digoel *see* Digul, Sungai
Digoin *90 C4* Saône-et-Loire, C France
Digul, Sungai *139 H5 prev.* Digoel. *river*
　Papua, E Indonesia
Dihang *see* Brahmaputra
Dijlah *see* Tigris
Dijon *90 D4 anc.* Dibio. Côte d'Or,
　C France
Dikhil *72 D4* SW Djibouti
Dikson *114 D2* Taymyrskiy
　(Dolgano-Nenetskiy) Avtonomnyy
　Okrug, N Russia
Dikti *105 D8 var.* Dhíkti Ori. *mountain*
　range Kríti, Greece,
　E Mediterranean Sea
Dilārām *122 D5 var.* Delārām. Nīmrūz,
　SW Afghanistan
Dili *139 F5 var.* Dilli, Dílly. *country*
　capital (East Timor) N East Timor
　(Timor-Leste)
Dilia *75 G3 var.* Dillia. *river* SW Niger
Di Linh *137 E6* Lâm Đồng, S Vietnam
Dilli *see* Dili, East Timor (Timor-Leste)
Dilli *see* Delhi, India
Dillia *see* Dilia
Dilling *72 B4 var.* Ad Dalanj. Southern
　Kordofan, C Sudan
Dillon *44 B2* Montana, NW USA
Dilly *see* Dili
Dilolo *77 D7* Katanga, S Dem.
　Rep. Congo
Dimashq *119 B5 var.* Ash Shām, Esh
　Sham, *Eng.* Damascus, *Fr.* Damas, *It.*
　Damasco. *country capital* (Syria)
　Dimashq, SW Syria
Dimitrovgrad *104 D3* Haskovo,
　S Bulgaria
Dimitrovgrad *111 C6 prev.* Caribrod.
　Serbia, SE Serbia
Dimitrovo *see* Pernik
Dimovo *104 B1* Vidin, NW Bulgaria
Dinajpur *135 F3* Rajshahi,
　NW Bangladesh
Dinan *90 B3* Côtes d'Armor, NW France
Dinant *87 C7* Namur, S Belgium
Dinar *116 B4* Afyon,
　SW Turkey (Türkiye)
Dinaric Alps *100 C4 var.* Dinara.
　mountain range Bosnia and
　Herzegovina/Croatia
Dindigul *132 C3* Tamil Nādu, SE India
Dingle Bay *89 A6 Ir.* Bá an Daingin. *bay*
　SW Ireland
Dinguiraye *74 C4* N Guinea
Diourbel *74 B3* W Senegal
Dirē Dawa *73 D5* Dirē Dawa, E Ethiopia
Dirk Hartog Island *147 A5 island*
　Western Australia
Dirschau *see* Tczew
Disappointment, Lake *146 C4 salt lake*
　Western Australia

Discovery Bay *54 B4* Middlesex,
　Jamaica, Greater Antilles, C Jamaica
　Caribbean Sea
Dishū *see* Deh-e Shū
Disko Bugt *see* Qeqertarsuup Tunua
Dispur *135 G3 state capital* Assam,
　NE India
Divinópolis *63 F4* Minas Gerais,
　SE Brazil
Divo *74 D5* S Ivory Coast
Divodurum Mediomatricum *see* Metz
Diyarbakır *117 E4 var.* Diarbekr; *anc.*
　Amida. Diyarbakır,
　SE Turkey (Türkiye)
Dizful *see* Dezfūl
Djailolo *see* Halmahera, Pulau
Djajapura *see* Jayapura
Djakarta *see* Jakarta
Djakovo *see* Đakovo
Djambala *77 B6* Plateaux, C Congo
Djambi *see* Jambi
Djambi *see* Hari, Batang
Djanet *71 E4 prev.* Fort Charlet.
　SE Algeria
Djéblé *see* Jablah
Djelfa *70 D2 var.* El Djelfa. N Algeria
Djéma *76 D4* Haut-Mbomou, E Central
　African Republic
Djember *see* Jember
Djérablous *see* Jarābulus
Djerba, Île de *71 F2 var.* Djerba, Jazīrat
　Jarbah. *island* E Tunisia
Djerba *see* Djerba, Île de
Djérem *76 B4 river* C Cameroon
Djevdjelija *see* Gevgelija
Djibouti *72 D4 var.* Jibuti. *country*
　capital (Djibouti) E Djibouti
Djibouti *72 D4 off.* Republic of Djibouti,
　var. Jibuti; *prev.* French Somaliland,
　French Territory of the Afars and Issas,
　Fr. Côte Française des Somalis,
　Territoire Française des Afars et des
　Issas. *country* E Africa
Djibouti, Republic of *see* Djibouti
Djokjakarta *see* Yogyakarta
Djourab, Erg du *76 C2 desert* N Chad
Djúpivogur *83 E5* Austurland,
　SE Iceland
Dmitriyevsk *see* Makiivka
Dnepr *see* Dnieper
Dneprodzerzhinskoye
　Vodokhranilishche *see*
　Dniprodzerzhynske Vodoskhovyshche
Dneprorudnoye *see* Dniprorudne
Dnester *see* Dniester
Dnieper *81 F4 Bel.* Dnyapro, *Rus.* Dnepr,
　Ukr. Dnipro. *river* E Europe
Dnieper Lowland *109 E2 Bel.*
　Prydnyaprowskaya Nizina, *Ukr.*
　Prydniprovska Nyzovyna. *lowlands*
　Belarus/Ukraine
Dniester *81 E4 Rom.* Nistru, *Rus.* Dnestr,
　Ukr. Dnister; *anc.* Tyras. *river*
　Moldova/Ukraine
Dnipro *see* Dnieper
Dnipro *109 F3 prev.* Dnipropetrovsk,
　Yekaterinoslav. Dnipropetrovska
　Oblast, E Ukraine
Dniprodzerzhynsk *see* Kamianske
Dniprodzerzhynske Vodoskhovyshche
　109 F3 Rus. Dneprodzerzhinskoye
　Vodokhranilishche. *reservoir*
　C Ukraine
Dnipropetrovsk *see* Dnipro
Dniprorudne *109 F3 Rus.*
　Dneprorudnoye. Zaporiz'ka Oblast',
　SE Ukraine
Dnister *see* Dniester
Dnyapro *see* Dnieper
Doba *76 C4* Logone-Oriental, S Chad
Döbeln *94 D4* Sachsen, E Germany
Doberai Peninsula *139 G4 Dut.*
　Vogelkop. *peninsula* Papua,
　E Indonesia
Doboj *100 C3* Republiks Srpska,
　N Bosnia and Herzegovina
Dobre Miasto *98 D2 Ger.* Guttstadt.
　Warmińsko-mazurskie, NE Poland
Dobrich *104 E1 Rom.* Bazargic; *prev.*
　Tolbukhin. Dobrich, NE Bulgaria
Dobrush *107 D7* Homyel'skaya Voblasts',
　SE Belarus
Dobryn' *see* Dabryn'
Dodecanese *see* Dodekánisa
Dodekánisa *105 D6 var.* Nóties
　Sporádes, *Eng.* Dodecanese; *prev.*
　Dhodhekánisos, Dodekanisos. *island*
　group SE Greece
Dodekanisos *see* Dodekánisa
Dodge City *45 E5* Kansas, C USA

Dodoma 73 C7 *country capital* (Tanzania) Dodoma, C Tanzania
Dogana 96 E1 NE San Marino Europe
Dogo 131 B6 *island* Oki-shotō, SW Japan
Dogondoutchi 75 F3 Dosso, SW Niger
Dogrular see Pravda
Doğubayazıt 117 F3 Ağrı, E Turkey (Türkiye)
Doğu Karadeniz Dağları 117 E3 *var.* Anadolu Dağları. *mountain range* NE Turkey (Türkiye)
Doha see Ad Dawḥah
Doire see Londonderry
Dokdo see Liancourt Rocks
Dokkum 86 D1 Fryslân, N Netherlands
Dokuchaevs'k see Dokuchayevsk.
Dokuchayevsk 109 G3 *prev.* Dokuchayevs'k, var. Dokuchayevsk. Donets'ka Oblast', SE Ukraine
Dokuchayevsk/Dokuchayevs'k see Dokuchayevsk
Doldrums Fracture Zone 66 C4 *fracture zone* W Atlantic Ocean
Dôle 90 D4 Jura, E France
Dolina see Dolyna
Dolinskaya see Dolynska
Dolisie 77 B6 *prev.* Loubomo. Niari, S Congo
Dolna Oryahovitsa 104 D2 *prev.* Polikrayshte. Veliko Tarnovo, N Bulgaria
Dolni Chiflik 104 E2 *prev.* Rudnik. Varna, E Bulgaria
Dolomites 96 C1 *var.* Dolomiti, *Eng.* Dolomites. *mountain range* NE Italy
Dolomites/Dolomiti see Dolomitiche, Alpi
Dolores 64 D4 Buenos Aires, E Argentina
Dolores 52 B1 Petén, N Guatemala
Dolores 64 D4 Soriano, SW Uruguay
Dolores Hidalgo 51 E4 *var.* Ciudad de Dolores Hidalgo. Guanajuato, C Mexico
Dolyna 108 B2 *Rus.* Dolina. Ivano-Frankivs'ka Oblast', W Ukraine
Dolynska 109 F3 *Rus.* Dolinskaya. Kirovohrads'ka Oblast', S Ukraine
Domachëvo/Domaczewo see Damachava
Dombås 85 B5 Oppland, S Norway
Domel Island see Letsôk-aw Kyun
Domesnes, Cape see Kolkasrags
Domeyko 64 B3 Atacama, N Chile
Dominica 55 H4 *off.* Commonwealth of Dominica. *country* E West Indies
Dominica Channel see Martinique Passage
Dominica, Commonwealth of see Dominica
Dominican Republic 55 E2 *country* C West Indies
Domokós 105 B5 *var.* Dhomokós. Stereá Ellás, C Greece
Don 111 B6 *var.* Duna, Tanais. *river* SW Russia
Donau see Danube
Donauwörth 95 C6 Bayern, S Germany
Don Benito 92 C3 Extremadura, W Spain
Doncaster 89 D5 *anc.* Danum. N England, United Kingdom
Dondo 78 B1 Cuanza Norte, NW Angola
Donegal 89 B5 *Ir.* Dún na nGall. Donegal, NW Ireland
Donegal Bay 89 A5 *Ir.* Bá Dhún na nGall. *bay* NW Ireland
Donets see Siverskyi Donets
Donetsk 109 G3 *prev.* Stalino. Donets'ka Oblast', E Ukraine
Dongfang 129 B7 *var.* Basuo. Hainan, S China
Dongguan 129 C6 Guangdong, S China
Đông Ha 136 E4 Quang Tri, C Vietnam
Dong Hai see East China Sea
Đông Hơi 136 D4 Quang Binh, C Vietnam
Dongliao see Liaoyuan
Dongola 72 B3 *var.* Donqola, Dunqulah. Northern, N Sudan
Dongou 77 C5 Likouala, NE Congo
Dong Rak, Thiu Khao Phanom see Dangrak Mountains
Dongting Hu 129 C5 *var.* Tung-t'ing Hu. *lake* S China
Donostia 93 E1 País Vasco, N Spain *see also* San Sebastián
Donqola see Dongola
Doolow 73 D5 Sumalê, E Ethiopia
Doornik see Tournai
Door Peninsula 40 C2 *peninsula* Wisconsin, N USA

Dooxo Nugaaleed 73 E5 *var.* Nogal Valley. *valley* E Somalia
Dordogne 91 B5 *cultural region* SW France
Dordogne 91 B5 *river* W France
Dordrecht 86 C4 *var.* Dordt, Dort. Zuid-Holland, SW Netherlands
Dordt see Dordrecht
Dorohoi 108 C3 Botoşani, NE Romania
Dorotea 84 C4 Västerbotten, N Sweden
Dorpat see Tartu
Dorre Island 147 A5 *island* Western Australia
Dort see Dordrecht
Dortmund 94 A4 Nordrhein-Westfalen, W Germany
Dos Hermanas 92 C4 Andalucía, S Spain
Dospad Dagh see Rhodope Mountains
Dospat 104 C3 Smolyan, S Bulgaria
Dothan 42 D3 Alabama, S USA
Dotnuva 106 B4 Kaunas, C Lithuania
Douai 90 C2 *prev.* Douay; *anc.* Duacum. Nord, N France
Douala 77 A5 *var.* Duala. Littoral, W Cameroon
Douay see Douai
Douglas 89 C5 *dependent territory capital* (Isle of Man) E Isle of Man
Douglas 48 C3 Arizona, SW USA
Douglas 44 D3 Wyoming, C USA
Douma see Dūmā
Douro see Duero
Douvres see Dover
Dover 89 E7 *Fr.* Douvres, *Lat.* Dubris Portus. SE England, United Kingdom
Dover 41 F4 *state capital* Delaware, NE USA
Dover, Strait of 90 C2 *var.* Straits of Dover, *Fr.* Pas de Calais. *strait* England, United Kingdom/France
Dover, Straits of see Dover, Strait of
Dovrefjell 85 B5 *plateau* S Norway
Downpatrick 89 B5 *Ir.* Dún Pádraig. SE Northern Ireland, United Kingdom
Dozen 131 B6 *island* Oki-shotō, SW Japan
Drāa, Hammada du see Dra, Hamada du
Drač/Draç see Durrës
Drachten 86 D2 Fryslân, N Netherlands
Drăgăşani 108 B5 Vâlcea, SW Romania
Dragoman 104 B2 Sofiya, W Bulgaria
Dra, Hamada du 70 C3 *var.* Hammada du Drâa, Haut Plateau du Dra. *plateau* W Algeria
Dra, Haut Plateau du see Dra, Hamada du
Drahichyn 107 B6 *Pol.* Drohiczyn Poleski, *Rus.* Drogichin. Brestskaya Voblasts', SW Belarus
Drakensberg 78 D5 *mountain range* Lesotho/South Africa
Drake Passage 57 B8 *passage* Atlantic Ocean/Pacific Ocean
Dralfa 104 D2 Tŭrgovishte, N Bulgaria
Dráma 104 C3 *var.* Dhráma. Anatolikí Makedonía kai Thráki, NE Greece
Dramburg see Drawsko Pomorskie
Drammen 85 B6 Buskerud, S Norway
Drau see Drava
Drava 100 C3 *var.* Drau, *Eng.* Drave, *Hung.* Dráva. *river* C Europe
Dráva/Drave see Drau/Drava
Drawsko Pomorskie 98 B3 *Ger.* Dramburg. Zachodnio-pomorskie, NW Poland
Drépano, Akrotírio 104 C4 *var.* Akrotirio Dhrepanon. *headland* N Greece
Drepanum see Trapani
Dresden 94 D4 Sachsen, E Germany
Drin see Drinit, Lumi i
Drina 100 C3 *river* Bosnia and Herzegovina/Serbia
Drinit, Lumi i 101 D5 *var.* Drin. *river* NW Albania
Drinit të Zi, Lumi i see Black Drin
Drissa see Drysa
Drobeta-Turnu Severin 108 B5 *prev.* Turnu Severin. Mehedinţi, SW Romania
Drogheda 89 B5 *Ir.* Droichead Átha. NE Ireland
Drogichin see Drahichyn
Drogobych see Drohobych
Drohiczyn Poleski see Drahichyn
Drohobych 108 B2 *Pol.* Drohobycz, *Rus.* Drogobych. L'vivs'ka Oblast', NW Ukraine
Drohobycz see Drohobych
Droichead Átha see Drogheda

Drôme 91 D5 *cultural region* E France
Dronning Maud Land 154 B2 *physical region* Antarctica
Drontheim see Trondheim
Drug see Durg
Druk-yul see Bhutan
Drummondville 39 E4 Québec, SE Canada
Druskieniki see Druskininkai
Druskininkai 107 B5 *Pol.* Druskieniki. Alytus, S Lithuania
Dryden 38 B3 Ontario, C Canada
Drysa 107 D5 *Rus.* Drissa. *river* N Belarus
Duacum see Douai
Duala see Douala
Dubai see Dubayy
Dubăsari 108 D3 *Rus.* Dubossary. NE Moldova
Dubawnt 37 F4 *river* Nunavut, NW Canada
Dubayy 120 D4 *Eng.* Dubai. Dubayy, NE United Arab Emirates
Dubbo 149 D6 New South Wales, SE Australia
Dublin 89 B5 *Ir.* Baile Átha Cliath; *anc.* Eblana. *country capital* (Ireland) Dublin, E Ireland
Dublin 43 E2 Georgia, SE USA
Dubno 108 C2 Rivnens'ka Oblast', NW Ukraine
Dubossary see Dubăsari
Dubris Portus see Dover
Dubrovnik 101 B5 *It.* Ragusa. Dubrovnik-Neretva, SE Croatia
Dubuque 45 G3 Iowa, C USA
Dudelange 87 D8 *var.* Forge du Sud, *Ger.* Dudelingen. Luxembourg, S Luxembourg
Dudelingen see Dudelange
Duero 92 D2 *Port.* Douro. *river* Portugal/Spain
Duesseldorf see Düsseldorf
Duffel 87 C5 Antwerpen, C Belgium
Dugi Otok 100 A4 *var.* Isola Grossa, *It.* Isola Lunga. *island* W Croatia
Duinekerke see Dunkerque
Duisburg 94 A4 *prev.* Duisburg-Hamborn. Nordrhein-Westfalen, W Germany
Duisburg-Hamborn see Duisburg
Duiven 86 D4 Gelderland, E Netherlands
Duk Faiwil 73 B5 Jonglei, C South Sudan
Dulan 126 D4 *var.* Qagan Us. Qinghai, C China
Dulce, Golfo 53 E5 *gulf* S Costa Rica
Dulce, Golfo see Izabal, Lago de
Dülmen 94 A4 Nordrhein-Westfalen, W Germany
Dulovo 104 E1 Silistra, NE Bulgaria
Duluth 45 G2 Minnesota, N USA
Dūmā 119 B5 *Fr.* Douma. Dimashq, SW Syria
Dumas 49 E1 Texas, SW USA
Dumfries 88 C4 S Scotland, United Kingdom
Dumont d'Urville 154 C4 *French research station* Antarctica
Dumyāṭ 72 B1 *Eng.* Damietta. N Egypt
Duna see Danube, C Europe
Düna see Western Dvina
Duna see Don
Dünaburg see Daugavpils
Dunai see Danube
Dunaj see Wien, Austria
Dunaj see Danube, C Europe
Dunapentele see Dunaújváros
Dunărea see Danube
Dunaújváros 99 C7 *prev.* Dunapentele, Sztálinváros. Fejér, C Hungary
Dunav see Danube
Dunavska Ravnina 104 C2 *Eng.* Danubian Plain. *lowlands* N Bulgaria
Duncan 49 G2 Oklahoma, C USA
Dundalk 89 B5 *Ir.* Dún Dealgan. Louth, NE Ireland
Dún Dealgan see Dundalk
Dundee 78 D4 KwaZulu/Natal, E South Africa
Dundee 88 C4 E Scotland, United Kingdom
Dunedin 151 B7 Otago, South Island, New Zealand
Dunfermline 88 C4 C Scotland, United Kingdom
Dungu 77 E5 Orientale, NE Dem. Rep. Congo
Dungun 138 B3 *var.* Kuala Dungun. Terengganu, Peninsular Malaysia

Dunholme see Durham
Dunkerque 90 C2 *Eng.* Dunkirk, *Flem.* Duinekerke; *prev.* Dunquerque. Nord, N France
Dunkirk see Dunkerque
Dún Laoghaire 89 B6 *Eng.* Dunleary; *prev.* Kingstown. E Ireland
Dunleary see Dún Laoghaire
Dún Pádraig see Downpatrick
Dunquerque see Dunkerque
Dunqulah see Dongola
Dupnitsa 104 C2 *prev.* Marek, Stanke Dimitrov. Kyustendil, W Bulgaria
Duqm 121 E5 *var.* Daqm. E Oman
Durance 91 D6 *river* SE France
Durango 50 D3 *var.* Victoria de Durango. Durango, W Mexico
Durango 44 C5 Colorado, C USA
Durankulak 104 E1 *Rom.* Răcari; *prev.* Blatnitsa, Duranulac. Dobrich, NE Bulgaria
Durant 49 G2 Oklahoma, C USA
Duranulac see Durankulak
Durazzo see Durrës
Durban 78 D4 *var.* Port Natal. KwaZulu/Natal, E South Africa
Durbe 106 B3 *Ger.* Durben. Liepāja, W Latvia
Durben see Durbe
Durg 135 E4 *prev.* Drug. Chhattisgarh, C India
Durham 89 D5 *hist.* Dunholme. N England, United Kingdom
Durham 43 F1 North Carolina, SE USA
Durocortorum see Reims
Durostorum see Silistra
Durovernum see Canterbury
Durrës 101 C6 *var.* Durrësi, Dursi, *It.* Durazzo, *Croatian* Drač, *Turk.* Draç. Durrës, W Albania
Durrësi see Durrës
Dursi see Durrës
Durūz, Jabal ad 119 C5 *mountain* SW Syria
D'Urville Island see Rangitoto ki te Tonga
Dusa Mareb/Dusa Marreb see Dhuusa Marreeb
Dushanbe 123 E3 *var.* Dyushambe; *prev.* Stalinabad, *Taj.* Stalinobod. *country capital* (Tajikistan) W Tajikistan
Düsseldorf 94 A4 *var.* Duesseldorf. Nordrhein-Westfalen, W Germany
Düstí 123 E3 *Rus.* Dusti. SW Tajikistan
Dutch East Indies see Indonesia
Dutch Guiana see Suriname
Dutch Harbor 36 B3 Unalaska Island, Alaska, USA
Dutch New Guinea see Papua
Duzdab see Zāhedān
Dvina Bay see Chëshskaya Guba
Dvinsk see Daugavpils
Dyanev see Galkynyş
Dyersburg 42 C1 Tennessee, S USA
Dyushambe see Dushanbe
Dza Chu see Mekong
Dzaudzhikau see Vladikavkaz
Dzerzhinsk 111 C5 Nizhegorodskaya Oblast', W Russia
Dzerzhinskiy see Nar'yan-Mar
Dzhalal-Abad see Jalal-Abad
Dzhambul see Taraz
Dzhankoi 109 F4 *prev.* Dzhankoy, Respublika Krym, S Ukraine
Dzhankoy see Dzhankoi
Dzharkurgan see Jarqo'rg'on
Dzhelandy see Jelondi
Dzhergalan see Jyrgalan
Dzhezkazgan see Zhezqazghan
Dzhizak see Jizzax
Dzhugdzhur, Khrebet 115 G3 *mountain range* E Russia
Dzhusaly see Zhosaly
Działdowo 98 D3 Warmińsko-Mazurskie, C Poland
Dzuunmod 127 E2 Töv, C Mongolia
Dzüün Soyonï Nuruu see Vostochnyy Sayan
Dzvina see Western Dvina

E

Eagle Pass 49 F4 Texas, SW USA
East Açores Fracture Zone see East Azores Fracture Zone
East Antarctica 154 C3 *var.* Greater Antarctica. *physical region* Antarctica
East Australian Basin see Tasman Basin

East Azores Fracture Zone *66 C3 var.*
  East Açores Fracture Zone. *tectonic*
  feature E Atlantic Ocean
Eastbourne *89 E7* SE England,
  United Kingdom
East Cape *150 E3 headland* North Island,
  New Zealand
East China Sea *125 E2 Chin.* Dong Hai.
  sea W Pacific Ocean
Easter Fracture Zone *153 G4 tectonic*
  feature E Pacific Ocean
Easter Island *see* Pascua, Isla de
Eastern Desert *see* Şaḥrā' ash Sharqīyah
Eastern Ghats *124 B3 mountain range*
  SE India
Eastern Sayans *see* Vostochnyy Sayan
Eastern Sierra Madre *see* Madre Oriental,
  Sierra
East Falkland *65 D8 var.* Isla Soledad.
  island E Falkland Islands
East Frisian Islands *94 A3 Eng.* East
  Frisian Islands. *island group*
  NW Germany
East Frisian Islands *see*
  Ostfriesische Inseln
East Grand Forks *45 E1* Minnesota,
  N USA
East Indiaman Ridge *141 D6 undersea*
  ridge E Indian Ocean
East Indies *152 A3 island group* SE Asia
East Kilbride *88 C4* S Scotland,
  United Kingdom
East Korea Bay *129 E3 bay*
  E North Korea
Eastleigh *89 D7* S England,
  United Kingdom
East London *78 D5 Afr.* Oos-Londen;
  prev. Emonti, Port Rex. Eastern Cape,
  S South Africa
Eastmain *38 D3 river* Québec, C Canada
East Mariana Basin *142 B1 undersea*
  basin W Pacific Ocean
East Novaya Zemlya Trough *112 C1*
  var. Novaya Zemlya Trough. *trough*
  W Kara Sea
East Pacific Rise *153 F4 undersea rise*
  E Pacific Ocean
East Pakistan *see* Bangladesh
East Saint Louis *40 B4* Illinois, N USA
East Scotia Basin *67 C7 undersea basin*
  SE Scotia Sea
East Sea *130 A4 var.* Sea of Japan, *Rus.*
  Yaponskoye More. *Sea*
  NW Pacific Ocean
East Siberian Sea *see* Vostochno-
  Sibirskoye More
East Timor *139 F5 var.* Timor Lorosa'e,
  Port. Timor Timur; *prev.* Portuguese
  Timor, Timor Timur. *country*
  S Indonesia
Eau Claire *40 A2* Wisconsin, N USA
Eau Claire, Lac à L' *see* St. Clair, Lake
Eauripik Rise *142 B2 undersea rise*
  W Pacific Ocean
Ebensee *95 D6* Oberösterreich, N Austria
Eberswalde-Finow *94 D3* Brandenburg,
  E Germany
Ebetsu *130 D2 var.* Ebetu. Hokkaidō,
  NE Japan
Ebetu *see* Ebetsu
Eblana *see* Dublin
Ebolowa *77 A5* Sud, S Cameroon
Ebon Atoll *144 D2 var.* Epoon. *atoll*
  Ralik Chain, S Marshall Islands
Ebora *see* Évora
Eboracum *see* York
Ebro *93 E2 river* NE Spain
Eburacum *see* York
Ebusus *see* Eivissa
Ecbatana *see* Hamadān
Ech Cheliff/Ech Chleff *see* Chlef
Echo Bay *37 E3* Northwest Territories,
  NW Canada
Echt *87 D5* Limburg, SE Netherlands
Écija *92 D4 anc.* Astigi. Andalucía,
  SW Spain
Eckengraf *see* Viesīte
Ecuador *60 B1 off.* Republic of Ecuador.
  country NW South America
Ecuador, Republic of *see* Ecuador
Ed Da'ein *72 A4* Southern Darfur,
  W Sudan
Ed Damazin *72 C4 var.* Ad Damazīn.
  Blue Nile, E Sudan
Ed Damer *72 C3 var.* Ad Dāmir, Ad
  Damar. River Nile, NE Sudan
Ed Debba *72 B3* Northern, N Sudan
Ede *86 D3* Gelderland, C Netherlands
Ede *75 F5* Osun, SW Nigeria
Edéa *77 A5* Littoral, SW Cameroon

Edessa *see* Şanlıurfa
Edfu *see* Idfū
Edgeøya *83 G2 island* S Svalbard
Edgware *89 A7* Harrow, SE England,
  United Kingdom
Edinburg *49 G5* Texas, SW USA
Edinburgh *88 C4 national capital*
  S Scotland, United Kingdom
Edingen *see* Enghien
Edirne *116 A2 Eng.* Adrianople; *anc.*
  Adrianopolis, Hadrianopolis. Edirne,
  NW Turkey (Türkiye)
Edmonds *46 B2* Washington,
  NW USA
Edmonton *37 E5 province capital*
  Alberta, SW Canada
Edmundston *39 F4* New Brunswick,
  SE Canada
Edna *49 G4* Texas, SW USA
Edolo *96 B1* Lombardia, N Italy
Edremit *116 A3* Balıkesir,
  NW Turkey (Türkiye)
Edward, Lake *77 E5 var.* Albert Edward
  Nyanza, Edward Nyanza, Lac Idi
  Amin, Lake Rutanzige. *lake* Uganda/
  Dem. Rep. Congo
Edward Nyanza *see* Edward, Lake
Edwards Plateau *49 F3 plain* Texas,
  SW USA
Edzo *53 E4 prev.* Rae-Edzo. Northwest
  Territories, NW Canada
Eeklo *87 B5 var.* Eekloo.
  Oost-Vlaanderen, NW Belgium
Eekloo *see* Eeklo
Eems *see* Ems
Eersel *87 C5* Noord-Brabant,
  S Netherlands
Eesti Vabariik *see* Estonia
Efate *144 D4 var.* Efate, *Fr.* Vaté; *prev.*
  Sandwich Island. *island* C Vanuatu
Efate *see* Efate
Effingham *40 B4* Illinois, N USA
Eforie-Sud *108 D5* Constanţa,
  E Romania
Egadi Is. *97 B7 island group* S Italy
Ege Denizi *see* Aegean Sea
Eger *99 D6 Ger.* Erlau. Heves,
  NE Hungary
Eger *see* Cheb, Czechia (Czech Republic)
Eger *see* Ohře
Éghezée *87 C6* Namur, C Belgium
Egina *see* Aígina
Egio *see* Aígio
Egmont *see* Taranaki, Mount
Egmont, Cape *150 C4 headland* North
  Island, New Zealand
Egoli *see* Johannesburg
Egypt *72 B2 off.* Arab Republic of Egypt,
  Ar. Jumhūrīyat Mişr al 'Arabīyah, Mişr,
  prev. United Arab Republic; *anc.*
  Aegyptus. *country* NE Africa
Eibar *93 E1* País Vasco, N Spain
Eibergen *86 E3* Gelderland,
  E Netherlands
Eidfjord *85 A5* Hordaland, S Norway
Eier-Berg *see* Suur Munamägi
Eifel *95 A5 plateau* W Germany
Eiger *95 B7 mountain* C Switzerland
Eigg *88 B3 island* W Scotland,
  United Kingdom
Eight Degree Channel *132 B3 channel*
  India/Maldives
Eighty Mile Beach *146 B4 beach*
  Western Australia
Eijsden *87 D6* Limburg, SE Netherlands
Eilat see Elat
Eindhoven *87 D5* Noord-Brabant,
  S Netherlands
Eipel *see* Ipel'
Éire *see* Ireland
Éireann, Muir *see* Irish Sea
Eisenhüttenstadt *94 D4* Brandenburg,
  E Germany
Eisenmarkt *see* Hunedoara
Eisenstadt *95 E6* Burgenland, E Austria
Eisleben *94 C4* Sachsen-Anhalt,
  C Germany
Eivissa *93 G3 var.* Iviza, *Cast.* Ibiza; *anc.*
  Ebusus. Ibiza, Spain,
  W Mediterranean Sea
Ejea de los Caballeros *93 E2* Aragón,
  NE Spain
Ejin Qi *see* Dalain Hob
Ekapa *see* Cape Town
Ekaterinodar *see* Krasnodar
Ekvyvatapskiy Khrebet *115 G1*
  mountain range NE Russia
El 'Alamein *see* Al 'Alamayn
El Asnam *see* Chlef

Elat *119 B8 var.* Eilat, Elath. Southern,
  S Israel
Elat, Gulf of *see* Aqaba, Gulf of
Elath *see* Elat, Israel
Elath *see* Al 'Aqabah, Jordan
Elâziğ *117 E3 var.* Elâzig, Elâziz. Elâziğ,
  E Turkey (Türkiye)
Elba *96 B4 island* Archipelago Toscano,
  C Italy
Elbasan *101 D6 var.* Elbasani. Elbasan,
  C Albania
Elbasani *see* Elbasan
Elbe *80 D3 Cz.* Labe. *river* Czechia
  (Czech Republic) /Germany
Elbert, Mount *44 C4 mountain*
  Colorado, C USA
Elbing *see* Elbląg
Elbląg *98 C2 var.* Elbląg, *Ger.* Elbing.
  Warmińsko-Mazurskie, NE Poland
El Boulaida/El Boulaïda *see* Blida
El'brus *111 A8 var.* Gora El'brus.
  mountain SW Russia
El'brus, Gora *see* El'brus
El Burgo de Osma *93 E2* Castilla y León,
  C Spain
Elburz Mountains *see* Alborz, Reshteh-ye
  Kūhhā-ye
El Cajon *47 C8* California, W USA
El Calafate *65 B7 var.* Calafate. Santa
  Cruz, S Argentina
El Callao *59 E2* Bolívar, E Venezuela
El Campo *49 G4* Texas, SW USA
El Carmen de Bolívar *58 B2* Bolívar,
  NW Colombia
El Cayo *see* San Ignacio
El Centro *47 D8* California, W USA
Elche *see* Elx
Elda *93 F4* Comunitat Valenciana,
  E Spain
El Djazaïr *see* Alger
El Djelfa *see* Djelfa
Eldorado *63 E4* Misiones, NE Argentina
El Dorado *42 B2* Arkansas, C USA
El Dorado *45 F5* Kansas, C USA
El Dorado *59 E2* Bolívar, E Venezuela
El Dorado *see* California
Eldoret *73 C6* Rift Valley, W Kenya
Elektrostal' *111 B5* Moskovskaya
  Oblast', W Russia
Elephant Butte Reservoir *48 C2*
  reservoir New Mexico, SW USA
Élesd *see* Aleşd
Eleuthera Island *54 C1 island*
  N The Bahamas
El Fasher *72 A4 var.* Al Fāshir. Northern
  Darfur, W Sudan
El Ferrol/El Ferrol del Caudillo *see*
  Ferrol
El Gedaref *see* Gedaref
El Geneina *72 A4 var.* Ajjinena,
  Al-Genain, Al Junaynah. Western
  Darfur, W Sudan
Elgin *88 C3* NE Scotland,
  United Kingdom
Elgin *40 B3* Illinois, N USA
El Giza *see* Al Jīzah
El Goléa *70 D3 var.* Al Golea. C Algeria
El Hank *74 D1 cliff* N Mauritania
El Haseke *see* Al Ḥasakah
Elimberrum *see* Auch
Eliocroca *see* Lorca
Élisabethville *see* Lubumbashi
Elista *111 B7* Respublika Kalmykiya,
  SW Russia
Elizabeth *149 B6* South Australia
Elizabeth City *43 G1* North Carolina,
  SE USA
Elizabethtown *40 C5* Kentucky, S USA
El-Jadida *70 C2 prev.* Mazagan.
  W Morocco
Elk *98 E2 Ger.* Lyck. Warmińsko-
  mazurskie, NE Poland
Elk City *49 F1* Oklahoma, C USA
El Khalil *see* Hebron
El Khârga *see* Al Khārjah
Elkhart *40 C3* Indiana, N USA
El Khartûm *see* Khartoum
Elk River *45 F2* Minnesota, N USA
El Kuneitra *see* Al Qunayţirah
Elláda *see* Greece
Elláda *see* Greece
Ellef Ringnes Island *37 E1 island*
  Nunavut, N Canada
Ellen, Mount *44 B5 mountain* Utah,
  W USA
Ellensburg *46 B2* Washington,
  NW USA

Ellesmere Island *37 F1 island* Queen
  Elizabeth Islands, Nunavut, N Canada
Ellesmere, Lake *151 C6 lake* South
  Island, New Zealand
Ellice Islands *see* Tuvalu
Elliniki Dimokratía *see* Greece
Elliston *149 A6* South Australia
Ellsworth Land *154 A3 physical region*
  Antarctica
El Mahbas *70 B3 var.* Mahbés.
  SW Western Sahara
El Mina *118 B4 var.* Al Mīnā'.
  N Lebanon
El Minya *see* Al Minyā
Elmira *41 E3* New York, NE USA
El Mreyyé *74 D2 desert* E Mauritania
Elmshorn *94 B3* Schleswig-Holstein,
  N Germany
El Muglad *72 B4* Western Kordofan,
  C Sudan
El Obeid *72 B4 var.* Al Obayyid, Al
  Ubayyiḏ. Northern Kordofan, C Sudan
El Ouâdi *see* El Oued
El Oued *71 E2 var.* Al Oued, El Ouâdi, El
  Wad. NE Algeria
Eloy *48 B2* Arizona, SW USA
El Paso *48 D3* Texas, SW USA
El Porvenir *53 G4* Kuna Yala, N Panama
El Progreso *52 C2* Yoro, NW Honduras
El Puerto de Santa María *92 C5*
  Andalucía, S Spain
El Qâhira *see* Al Qāhirah
El Quneitra *see* Al Qunayţirah
El Queseir *see* Al Quşayr
El Quweira *see* Al Quwayrah
El Rama *53 E3* Región Autónoma
  Atlántico Sur, SE Nicaragua
El Real *53 H5 var.* El Real de Santa María.
  Darién, SE Panama
El Real de Santa María *see* El Real
El Reno *49 F1* Oklahoma, C USA
El Salvador *52 B3 off.* Republic of El
  Salvador, *Sp.* República de El Salvador.
  country Central America
El Salvador, Republic of *see* El Salvador
El Salvador, Republic of *see* El Salvador
Elsass *see* Alsace
El Sáuz *50 C2* Chihuahua, N Mexico
El Serrat *91 A7* N Andorra Europe
Elst *86 D4* Gelderland, E Netherlands
El Sueco *50 C2* Chihuahua, N Mexico
El Suweida *see* As Suwaydā'
El Suweis *see* As Suways
Eltanin Fracture Zone *153 E5 tectonic*
  feature SE Pacific Ocean
El Tigre *59 E2* Anzoátegui, NE Venezuela
Elvas *92 C4* Portalegre, C Portugal
El Vendrell *93 G2* Cataluña, NE Spain
El Vigía *58 C2* Mérida, NW Venezuela
El Wad *see* El Oued
Elwell, Lake *44 B1 reservoir* Montana,
  NW USA
Elx *93 F4 var.* Elche; *anc.* Ilici, Lat. Illicis.
  Comunitat Valenciana, E Spain
Ely *47 D5* Nevada, W USA
El Yopal *see* Yopal
Emajõgi *106 D3 Ger.* Embach. *river*
  SE Estonia
Emämrüd *see* Shāhrūd
Emämshahr *see* Shāhrūd
Emba *114 B4 Kaz.* Embi. Aktyubinsk,
  W Kazakhstan
Embach *see* Emajõgi
Embi *see* Emba
Emden *94 A3* Niedersachsen,
  NW Germany
Emerald *148 D4* Queensland, E Australia
Emerald Isle *see* Montserrat
Emesa *see* Ḥimş
Emmaste *106 C2* Hiiumaa, W Estonia
Emmeloord *86 D2* Flevoland,
  N Netherlands
Emmen *86 E2* Drenthe, NE Netherlands
Emmendingen *95 A6* Baden-
  Württemberg, SW Germany
Emona *see* Ljubljana
Emonti *see* East London
Emory Peak *49 E4 mountain* Texas,
  SW USA
Empalme *50 B2* Sonora, NW Mexico
Emperor Seamounts *113 G3 seamount*
  range NW Pacific Ocean
Empire State of the South *see* Georgia
Emporia *45 F5* Kansas, C USA
Empty Quarter *see* Ar Rub 'al Khālī
Ems *94 A3 Dut.* Eems. *river*
  NW Germany
Enareträsk *see* Inarijärvi
Encamp *91 A8* Encamp, C Andorra
Encarnación *64 D3* Itapúa, S Paraguay

Encinitas *47 C8* California,
 W USA
Encs *99 D6* Borsod-Abaúj-Zemplén,
 NE Hungary
Endeavour Strait *148 C1 strait*
 Queensland, NE Australia
Enderbury Island *145 F3 atoll* Phoenix
 Islands, C Kiribati
Enderby Land *154 C2 physical region*
 Antarctica
Enderby Plain *154 D2 abyssal plain*
 S Indian Ocean
Endersdorf *see* Jędrzejów
Enewetak Atoll *144 C1 var.* Ānewetak,
 Eniwetok. *atoll* Ralik Chain,
 W Marshall Islands
Enfield *89 A7* United Kingdom
Engeten *see* Aiud
Enghien *89 B6 Dut.* Edingen. Hainaut,
 SW Belgium
England *89 D5 Lat.* Anglia. *cultural
 region* England, United Kingdom
Englewood *44 D4* Colorado,
 C USA
English Channel *89 D8 var.* The
 Channel, *Fr.* la Manche. *channel*
 NW Europe
Engure *106 C3* Tukums, W Latvia
Engures Ezers *106 B3 lake* NW Latvia
Enguri *117 F1 Rus.* Inguri. *river*
 NW Georgia
Enid *49 F1* Oklahoma, C USA
Enikale Strait *see* Kerch Strait
Eniwetok *see* Enewetak Atoll
En Nâqoûra *119 A5 var.* An Nāqūrah.
 SW Lebanon
En Nazira *see* Natzrat
Ennedi *76 D2 plateau* E Chad
Ennis *89 A6 Ir.* Inis. Clare, W Ireland
Ennis *49 G3* Texas, SW USA
Enniskillen *89 B5 var.* Inniskilling, *Ir.*
 Inis Ceithleann. SW Northern Ireland,
 United Kingdom
Enns *95 D6 river* C Austria
Enschede *86 E3* Overijssel, E Netherlands
Ensenada *50 A1* Baja California Norte,
 NW Mexico
Entebbe *73 B6* S Uganda
Entroncamento *92 B3* Santarém,
 C Portugal
Enugu *75 G5* Enugu, S Nigeria
Epanomi *104 B4* Kentrikí Makedonía,
 N Greece
Epéna *77 B5* Likouala, NE Congo
Eperies/Eperjes *see* Prešov
Épi *144 D4 var.* Épi. *island* C Vanuatu
Épi *see* Épi
Épinal *90 D4* Vosges, NE France
Epiphania *see* Ḩamāh
Epoon *see* Ebon Atoll
Epsom *89 A8* United Kingdom
Equality State *see* Wyoming
Equatorial Guinea *77 A5 off.* Equatorial
 Guinea, Republic of. *country* C Africa
Equatorial Guinea, Republic of *see*
 Equatorial Guinea
Erautini *see* Johannesburg
Erbil *see* Arbīl
Erciş *117 F3* Van, E Turkey (Türkiye)
Erdély *see* Transylvania
Erdélyi-Havasok *see* Carpaţii Meridionali
Erdenet *127 E2* Orhon, N Mongolia
Erdi *76 C2 plateau* NE Chad
Erdi Ma *76 D2 desert* NE Chad
Erebus, Mount *154 B4 volcano* Ross
 Island, Antarctica
Ereğli *116 C4* Konya, S Turkey (Türkiye)
Erenhot *127 F2 var.* Erlian. Nei Mongol
 Zizhiqu, NE China
Erfurt *94 C4* Thüringen, C Germany
Ergene Çayı *see* Ergene Irmaği
Ergene Irmaği *116 A2 var.* Ergene Çayı.
 river NW Turkey (Türkiye)
Ergun *127 F1 var.* Labudalin; *prev.* Ergun
 Youqi. Nei Mongol Zizhiqu, N China
Ergun He *see* Argun
Ergun Youqi *see* Ergun
Erie *40 D3* Pennsylvania, NE USA
Érié, Lac *see* Erie, Lake
Erie, Lake *40 D3 Fr.* Lac Érié. *lake*
 Canada/USA
Eritrea *72 C4 off.* State of Eritrea, Iertra.
 country E Africa
Eritrea, State of *see* Eritrea
Erivan *see* Yerevan
Erlangen *95 C5* Bayern, S Germany
Erlau *see* Eger
Erlian *see* Erenhot
Ermelo *86 D3* Gelderland, C Netherlands
Ermióni *105 C6* Pelopónnisos, S Greece

Ermoúpoli *105 D6 var.* Hermoupolis;
 prev. Ermoúpolis. Sýyros, Kykládes,
 Greece, Aegean Sea
Ermoúpolis *see* Ermoúpoli
Ernákulam *132 C3* Kerala, SW India
Erode *132 C2* Tamil Nādu, SE India
Erquelinnes *87 B7* Hainaut, S Belgium
Er-Rachidia *70 C2 var.* Ksar al Souk.
 E Morocco
Er Rahad *72 B4 var.* Ar Rahad. Northern
 Kordofan, C Sudan
Erromango *144 D4 island* S Vanuatu
Ertis *see* Irtysh, C Asia
Eryri *89 C6 Eng.* Snowdonia. *mountain
 range* N Wales, United Kingdom
Erzerum *see* Erzurum
Erzgebirge *95 C5 Cz.* Krušné Hory, *Eng.*
 Ore Mountains. *mountain range*
 Czechia (Czech Republic) /Germany
Erzincan *117 E3 var.* Erzinjan. Erzincan,
 E Turkey (Türkiye)
Erzinjan *see* Erzincan
Erzurum *117 E3 prev.* Erzerum.
 Erzurum, NE Turkey (Türkiye)
Esbjerg *85 A7* Ribe, W Denmark
Esbo *see* Espoo
Escaldes *91 A8* Escaldes Engordany,
 C Andorra Europe
Escanaba *40 C2* Michigan, N USA
Escaut *see* Scheldt
Esch-sur-Alzette *87 D8* Luxembourg,
 S Luxembourg
Esclaves, Grand Lac des *see* Great
 Slave Lake
Escondido *47 C8* California, W USA
Escuinapa *50 D3 var.* Escuinapa de
 Hidalgo. Sinaloa, C Mexico
Escuinapa de Hidalgo *see* Escuinapa
Escuintla *52 B2* Escuintla, S Guatemala
Escuintla *51 G5* Chiapas, SE Mexico
Esenguly *122 B3 Rus.* Gasan-Kuli.
 Balkan Welaýaty, W Turkmenistan
Eşfahān *120 C3 Eng.* Isfahan; *anc.*
 Aspadana. Eşfahān, C Iran
Esh Sharā *see* Ash Sharāh
Esil *see* Ishim, Kazakhstan/Russia
Eskimo Point *see* Arviat
Eskişehir *116 B3 var.* Eskishehr.
 Eskişehir, W Turkey (Türkiye)
Eskishehr *see* Eskişehir
Eslāmābād *120 C3 var.* Eslāmābād-e
 Gharb; *prev.* Harunabad, Shāhābād.
 Kermānshāhān, W Iran
Eslāmābād-e Gharb *see* Eslāmābād
Esmeraldas *60 A1* Esmeraldas,
 N Ecuador
Esna *see* Isnā
España *see* Spain
Espanola *48 D1* New Mexico, SW USA
Esperance *147 B7* Western Australia
Esperanza *50 B2* Sonora, NW Mexico
Esperanza *154 A2 Argentinian research
 station* Antarctica
Espinal *58 B3* Tolima, C Colombia
Espinhaço, Serra do *56 D4 mountain
 range* SE Brazil
Espírito Santo *63 F4 off.* Estado do
 Espírito Santo. *state/region* E Brazil
Espírito Santo, Estado do *see*
 Espírito Santo
Espiritu Santo *144 C4 var.* Santo. *island*
 W Vanuatu
Espoo *85 D6 Swe.* Esbo. Uusimaa,
 S Finland
Esquel *65 B6* Chubut, SW Argentina
Essaouira *70 B2 prev.* Mogador.
 W Morocco
Esseg *see* Osijek
Es Semara *see* Smara
Essen *87 C5* Antwerpen, N Belgium
Essen *94 A4 var.* Essen an der Ruhr.
 Nordrhein-Westfalen, W Germany
Essen an der Ruhr *see* Essen
Essequibo River *59 F3 river* C Guyana
Es Suweida *see* As Suwaydā'
Estacado, Llano *49 E2 plain* New
 Mexico/Texas, SW USA
Estados, Isla de los *65 C8 prev. Eng.*
 Staten Island. *island* S Argentina
Estância *63 G3* Sergipe, E Brazil
Estelí *52 D3* Estelí, NW Nicaragua
Estella *93 E1 Bas.* Lizarra. Navarra,
 N Spain
Estepona *92 D5* Andalucía, S Spain
Estevan *37 F5* Saskatchewan, S Canada
Estland *see* Estonia
Estonia *106 D2 off.* Republic of Estonia,
 Est. Eesti Vabariik, *Ger.* Estland, *Latv.*
 Igaunija; *prev.* Estonian SSR, *Rus.*
 Estonskaya SSR. *country* NE Europe
Estonian SSR *see* Estonia
Estonia, Republic of *see* Estonia

Estonskaya SSR *see* Estonia
Estrela, Serra da *92 C3 mountain range*
 C Portugal
Estremadura *see* Extremadura
Estremoz *92 C4* Évora, S Portugal
Eswatini *78 D4 off.* Kingdom of Eswatini;
 prev. Swaziland. *country* S Africa
Eswatini, Kingdom of *see* Eswatini
Eszék *see* Osijek
Esztergom *99 C6 Ger.* Gran; *anc.*
 Strigonium. Komárom-Esztergom,
 N Hungary
Étaile *87 D8* Luxembourg, SE Belgium
Etāwah *134 D3* Uttar Pradesh, N India
Ethiopia *73 C5 off.* Federal Democratic
 Republic of Ethiopia; *prev.* Abyssinia,
 People's Democratic Republic of
 Ethiopia. *country* E Africa
**Ethiopia, Federal Democratic Republic
 of** *see* Ethiopia
Ethiopian Highlands *73 C5 var.*
 Ethiopian Plateau. *plateau* N Ethiopia
Ethiopian Plateau *see*
 Ethiopian Highlands
**Ethiopia, People's Democratic Republic
 of** *see* Ethiopia
Etna, Monte *97 C7 Eng.* Mount Etna.
 volcano Sicilia, Italy,
 C Mediterranean Sea
Etna, Mount *see* Etna, Monte
Etosha Pan *78 B3 salt lake* N Namibia
Etoumbi *77 B5* Cuvette Ouest,
 NW Congo
Etsch *see* Adige
Et Tafila *see* Aţ Ţafīlah
Ettelbrück *87 D8* Diekirch,
 C Luxembourg
'Eua *145 E5 prev.* Middleburg Island.
 island Tongatapu Group, SE Tonga
Euboea *see* Évvoia
Eucla *147 D6* Western Australia
Euclid *40 D3* Ohio, N USA
Eufaula Lake *49 G1 var.* Eufaula
 Reservoir. *reservoir* Oklahoma,
 C USA
Eufaula Reservoir *see* Eufaula Lake
Eugene *46 B3* Oregon, NW USA
Eumolpias *see* Plovdiv
Eupen *87 D6* Liège, E Belgium
Euphrates *112 B4 Ar.* Al-Furāt, *Turk.*
 Firat Nehri. *river* SW Asia
Eureka *47 A5* California, W USA
Eureka *44 A1* Montana, NW USA
Europa Point *93 H5 headland*
 S Gibraltar
Europe *80 continent*
Eutin *94 C2* Schleswig-Holstein,
 N Germany
Euxine Sea *see* Black Sea
Evansdale *45 G3* Iowa, C USA
Evanston *40 B3* Illinois, N USA
Evanston *44 B4* Wyoming, C USA
Evansville *40 B5* Indiana, N USA
Eveleth *45 G1* Minnesota, N USA
Everard, Lake *149 A6 salt lake*
 South Australia
Everest, Mount *126 B5 Chin.*
 Qomolangma Feng, *Nep.* Sagarmāthā.
 mountain China/Nepal
Everett *46 B2* Washington, NW USA
Everglades, The *43 F5 wetland* Florida,
 SE USA
Evje *85 A6* Aust-Agder, S Norway
Evmolpia *see* Plovdiv
Évora *92 B4 anc.* Ebora, *Lat.* Liberalitas
 Julia. Évora, C Portugal
Évreux *90 C3 anc.* Civitas Eburovicum.
 Eure, N France
Évros *see* Maritsa
Évry *90 E2* Essonne, N France
Évvoia *105 C5 var.* Euboea, *Lat.* Euboea.
 island C Greece
Ewarton *54 B5* C Jamaica
Excelsior Springs *45 F4* Missouri,
 C USA
Exe *89 C7 river* SW England,
 United Kingdom
Exeter *89 C7 anc.* Isca Damnoniorum.
 SW England, United Kingdom
Exmoor *89 C7 moorland* SW England,
 United Kingdom
Exmouth *146 A4* Western Australia
Exmouth *89 C7* SW England,
 United Kingdom
Exmouth Gulf *146 A4 gulf*
 Western Australia
Exmouth Plateau *141 E5 undersea
 plateau* E Indian Ocean
Extremadura *92 C3 var.* Estremadura.
 autonomous community W Spain
Exuma Cays *54 C1 islets* C The Bahamas

Exuma Sound *54 C1 sound*
 C The Bahamas
Eyre Mountains *151 A7 mountain range*
 South Island, New Zealand
Eyre, Lake *149 A5 var.* Kati Thanda. *salt
 lake* South Australia
Eyre Peninsula *149 A6 peninsula*
 South Australia
Eyre South, Lake *149 A5 salt lake*
 South Australia
Ezo *see* Hokkaidō

F

Faadhippolhu Atoll *132 B4 var.*
 Fadiffolu, Lhaviyani Atoll. *atoll*
 N Maldives
Fabens *48 D3* Texas, SW USA
Fada *76 C2* Borkou-Ennedi-Tibesti,
 E Chad
Fada-Ngourma *75 E4* E Burkina
Fadiffolu *see* Faadhippolhu Atoll
Faenza *96 C3 anc.* Faventia.
 Emilia-Romagna, N Italy
Faeroe Islands *see* Faero Islands
Færøerne *see* Faroe Islands
Faetano *96 E2* E San Marino
Făgăraş *108 C4 Ger.* Fogarasch, *Hung.*
 Fogaras. Braşov, C Romania
Fagibina, Lake *see* Faguibine, Lac
Fagne *87 C7 hill range* S Belgium
Faguibine, Lac *75 E3 var.* Lake Fagibina.
 lake NW Mali
Fahlun *see* Falun
Fahraj *120 E4* Kermān, SE Iran
Faial *92 A5 var.* Ilha do Faial. *island*
 Azores, Portugal, NE Atlantic Ocean
Faial, Ilha do *see* Faial
Faifo *see* Hội An
Fairbanks *36 D3* Alaska, USA
Fairfield *47 B6* California, W USA
Fair Isle *88 D2 island* NE Scotland,
 United Kingdom
Fairlie *151 B6* Canterbury, South Island,
 New Zealand
Fairmont *45 F3* Minnesota, N USA
Faisalabad *134 C2 prev.* Lyallpur.
 Punjab, NE Pakistan
Faizābād *135 E3* Uttar Pradesh, N India
Faizābād *123 F3 var.* Feyzābād, Fyzabad.
 Badakhshān, NE Afghanistan
Fakaofo Atoll *145 F3 island* SE Tokelau
Falam *136 A3* Chin State,
 W Myanmar (Burma)
Falconara Marittima *96 C3* Marche,
 C Italy
Falkenau an der Eger *see* Sokolov
Falkland Islands *65 D7 var.* Falklands,
 Islas Malvinas. *UK Overseas Territory*
 SW Atlantic Ocean
Falkland Plateau *57 D7 var.* Argentine
 Rise. *undersea feature*
 SW Atlantic Ocean
Falklands *see* Falkland Islands
Falknov nad Ohří *see* Sokolov
Fallbrook *47 C8* California, W USA
Falmouth *54 A4* W Jamaica
Falmouth *89 C7* SW England,
 United Kingdom
Falster *85 B8 island* SE Denmark
Fălticeni *108 C3 Hung.* Falticsén.
 Suceava, NE Romania
Falticsén *see* Fălticeni
Falun *85 C6 var.* Fahlun. Kopparberg,
 C Sweden
Famagusta *see* Gazimağusa
Famagusta Bay *see* Gazimağusa Körfezi
Famenne *87 C7 physical region*
 SE Belgium
Fang *136 C3* Chiang Mai, NW Thailand
Fanning Island *see* Tabuaeran
Fanø *96 C3* island W Denmark
Farafangana *79 G4* Fianarantsoa,
 SE Madagascar
Farāh *122 D4 var.* Farah, Fararud.
 Farāh, W Afghanistan
Farāh Rōd *122 D4 river* W Afghanistan
Faranah *74 C4* Haute-Guinée, S Guinea
Fararud *see* Farāh
Farasan Islands *see* Farasān, Juzur
Farasān, Juzur *121 A6 var.* Farasan
 Islands. *island group* SW Saudi Arabia
Farewell, Cape *150 C4 headland* South
 Island, New Zealand
Farewell, Cape *see* Nunap Isua
Fargo *45 F2* North Dakota, N USA
Farg'ona *123 F2 Rus.* Fergana; *prev.*
 Novyy Margilan. Farg'ona Viloyati,
 E Uzbekistan

Faribault *45 F2* Minnesota, N USA
Farīdābād *134 D3* Haryāna, N India
Farkhor *123 E3 Rus.* Parkhar.
 SW Tajikistan
Farmington *45 G5* Missouri, C USA
Farmington *48 C1* New Mexico,
 SW USA
Faro *92 B5* Faro, S Portugal
Faroe-Iceland Ridge *80 C1 undersea
 ridge* NW Norwegian Sea
Faroe Islands *83 E5 var.* Faero Islands,
 Dan. Færøerne, *Far.* Føroyar.
 Self-governing territory of Denmark
 N Atlantic Ocean
Faroe-Shetland Trough *80 C2 trough*
 NE Atlantic Ocean
Farquhar Group *79 G2 island group*
 S Seychelles
Fars, Khalīj-e *see* Persian Gulf
Farvel, Kap *see* Nunap Isua
Fastiv *109 E2 Rus.* Fastov. Kyivska
 Oblast, N Ukraine
Fastov *see* Fastiv
Fauske *84 C3* Nordland, C Norway
Faventia *see* Faenza
Faxa Bay *see* Faxaflói
Faxaflói *82 C5 Eng.* Faxa Bay. *bay*
 W Iceland
Faya *76 C2 prev.* Faya-Largeau, Largeau.
 Borkou-Ennedi-Tibesti, N Chad
Faya-Largeau *see* Faya
Fayetteville *42 A1* Arkansas, C USA
Fayetteville *43 F1* North Carolina,
 SE USA
Fdérick *see* Fdérik
Fdérik *74 C2 var.* Fdérick, *Fr.* Fort
 Gouraud. Tiris Zemmour,
 NW Mauritania
Fear, Cape *43 F2 headland* Bald Head
 Island, North Carolina, SE USA
Fécamp *90 B3* Seine-Maritime, N France
Fédala *see* Mohammedia
Federal Capital Territory *see* Australian
 Capital Territory
Fehérgyarmat *99 E6* Szabolcs-Szatmár-
 Bereg, E Hungary
Fehértemplom *see* Bela Crkva
Fehmarn *94 C2 island* N Germany
Fehmarn Belt *94 C2 Dan.* Femern Bælt,
 Ger. Fehmarnbelt. *strait* Denmark /
 Germany
Fehmarnbelt Fehmarn Belt/Femer Bælt
Feijó *62 C2* Acre, W Brazil
Feilding *150 D4* Manawatu-Wanganui,
 North Island, New Zealand
Feira *see* Feira de Santana
Feira de Santana *63 G3 var.* Feira.
 Bahia, E Brazil
Feketehalom *see* Codlea
Felanitx *93 G3* Mallorca, Spain,
 W Mediterranean Sea
Felicitas Julia *see* Lisboa
Felidhu Atoll *132 B4 atoll* C Maldives
Felipe Carrillo Puerto *51 H4* Quintana
 Roo, SE Mexico
Felixstowe *89 E6* E England,
 United Kingdom
Fellin *see* Viljandi
Felsőbánya *see* Baia Sprie
Felsőmuzslya *see* Mužlja
Femunden *85 B5* lake S Norway
Fénérive *see* Fenoarivo Atsinanana
Fengcheng *128 D3 var.* Feng-cheng,
 Fenghwangcheng. Liaoning, NE China
Feng-cheng *see* Fengcheng
Fenghwangcheng *see* Fengcheng
Fengtien *see* Shenyang, China
Fengtien *see* Liaoning, China
Fenoarivo Atsinanana *79 G3 Fr.*
 Fénérive. Toamasina, E Madagascar
Fens, The *89 E6 wetland* E England,
 United Kingdom
Feodosia *see* Feodosiia
Feodosiia *109 F5 prev.* Feodosiya, *var.*
 Kefe, *It.* Kaffa; *anc.* Theodosia.
 Respublika Krym, S Ukraine
Feodosiya *see* Feodosiia
Ferdinand *see* Montana, Bulgaria
Ferdinandsberg *see* Oţelu Roşu
Féres *104 D3* Anatolikí Makedonía kai
 Thráki, NE Greece
Fergana *see* Farg'ona
Fergus Falls *45 F2* Minnesota, N USA
Ferizaj *101 D5 Serb.* Uroševac.
 C Kosovo
Ferkessédougou *74 D4* N Ivory Coast
Fermo *96 C4 anc.* Firmum Picenum.
 Marche, C Italy
Fernandina, Isla *60 A5 var.* Narborough
 Island. *island* Galápagos Islands,
 Ecuador, E Pacific Ocean

Fernando de Noronha *63 H2 island*
 E Brazil
Fernando Po/Fernando Póo *see* Bioco,
 Isla de
Ferrara *96 C2 anc.* Forum Alieni.
 Emilia-Romagna, N Italy
Ferreñafe *60 B3* Lambayeque, W Peru
Ferro *see* Hierro
Ferrol *92 B1 var.* El Ferrol; *prev.* El Ferrol
 del Caudillo. Galicia, NW Spain
Fertő *see* Neusiedler See
Ferwerd *see* Ferwert
Ferwert *86 D1 Dutch.* Ferwerd. Fryslân,
 N Netherlands
Fès *70 C2 Eng.* Fez. N Morocco
Feteşti *108 D5* Ialomiţa, SE Romania
Fethiye *116 B4* Muğla,
 SW Turkey (Türkiye)
Fetlar *88 D1 island* NE Scotland,
 United Kingdom
Feuilles, Rivière aux *38 D2 river*
 Québec, E Canada
Feyzābād *see* Faizābād
Fez *see* Fès
Fezzan *71 G4 region* S Libya
Fianarantsoa *79 F3* Fianarantsoa,
 C Madagascar
Fianga *76 B4* Mayo-Kébbi, SW Chad
Fier *101 C6 var.* Fieri. Fier, SW Albania
Fieri *see* Fier
Figeac *91 C5* Lot, S France
Figig *see* Figuig
Figueira da Foz *92 B3* Coimbra,
 W Portugal
Figueres *93 G2* Cataluña, E Spain
Figuig *70 D2 var.* Figig. E Morocco
Fiji *145 E5 off.* Republic of Fiji, *Fij.* Viti.
 country SW Pacific Ocean
Fiji, Republic of *see* Fiji
Filadelfia *52 D4* Guanacaste,
 W Costa Rica
Filiaşi *108 B5* Dolj, SW Romania
Filipstad *85 B6* Värmland, C Sweden
Finale Ligure *96 A3* Liguria, NW Italy
Finchley *89 A7* United Kingdom
Findlay *40 C4* Ohio, N USA
Finike *116 B4* Antalya,
 SW Turkey (Türkiye)
Finland *84 D4 off.* Republic of Finland,
 Fin. Suomen Tasavalta, Suomi. *country*
 N Europe
Finland, Gulf of *85 D6 Est.* Soome Laht,
 Fin. Suomenlahti, *Ger.* Finnischer
 Meerbusen, *Rus.* Finskiy Zaliv, *Swe.*
 Finska Viken. *gulf* E Baltic Sea
Finland, Republic of *see* Finland
Finnischer Meerbusen *see*
 Finland, Gulf of
Finnmarksvidda *84 D2 physical region*
 N Norway
Finska Viken/Finskiy Zaliv *see*
 Finland, Gulf of
Finsterwalde *94 D4* Brandenburg,
 E Germany
Fiordland *151 A7 physical region* South
 Island, New Zealand
Fiorina *96 E1* NE San Marino
Firat Nehri *see* Euphrates
Firenze *96 C3 Eng.* Florence; *anc.*
 Florentia. Toscana, C Italy
Firmum Picenum *see* Fermo
First State *see* Delaware
Fischbacher Alpen *95 E7 mountain
 range* E Austria
Fischhausen *see* Primorsk
Fish *78 B4 var.* Vis. *river* S Namibia
Fishguard *89 C6 Wel.* Abergwaun.
 SW Wales, United Kingdom
Fisterra, Cabo *92 B1 headland*
 NW Spain
Fitzroy Crossing *146 C3*
 Western Australia
Fitzroy River *146 C3 river*
 Western Australia
Fiume *see* Rijeka
Flagstaff *48 B2* Arizona, SW USA
Flanders *87 A6 Dut.* Vlaanderen, *Fr.*
 Flandre. *cultural region* Belgium/
 France
Flandre *see* Flanders
Flathead Lake *44 B1 lake* Montana,
 NW USA
Flat Island *129 C8 island*
 NE Spratly Islands
Flatts Village *42 B5 var.* The Flatts
 Village. C Bermuda
Flensburg *94 B2* Schleswig-Holstein,
 N Germany
Flessingue *see* Vlissingen
Flickertail State *see* North Dakota

Flinders Island *149 C8 island* Furneaux
 Group, Tasmania, SE Australia
Flinders Ranges *149 B6 mountain range*
 South Australia
Flinders River *148 C3 river* Queensland,
 NE Australia
Flin Flon *37 F5* Manitoba, C Canada
Flint *40 C3* Michigan, N USA
Flint Island *145 G4 island* Line Islands,
 E Kiribati
Floreana, Isla *see* Santa María, Isla
Florence *42 C1* Alabama, S USA
Florence *43 F2* South Carolina, SE USA
Florence *see* Firenze
Florencia *58 B4* Caquetá, S Colombia
Florentia *see* Firenze
Flores *52 B1* Petén, N Guatemala
Flores *139 E5 island* Nusa Tenggara,
 C Indonesia
Flores *92 A5 island* Azores, Portugal,
 NE Atlantic Ocean
Flores, Laut *see* Flores Sea
Flores Sea *138 D5 Ind.* Laut Flores. *sea*
 C Indonesia
Floriano *63 F2* Piauí, E Brazil
Florianópolis *63 F5 prev.* Destêrro. *state
 capital* Santa Catarina, S Brazil
Florida *64 D4* Florida, S Uruguay
Florida *43 E4 off.* State of Florida, *also
 known as* Peninsular State, Sunshine
 State. *state* SE USA
Florida Bay *43 E5 bay* Florida, SE USA
Florida Keys *43 E5 island group* Florida,
 SE USA
Florida, Straits of *54 B1 strait* Atlantic
 Ocean/Gulf of Mexico
Flórina *104 B4 var.* Phlórina. Dytikí
 Makedonía, N Greece
Florissant *45 G4* Missouri, C USA
Floúda, Akrotírio *105 D7 headland*
 Astypálaia, Kykládes, Greece,
 Aegean Sea
Flushing *see* Vlissingen
Flylân *see* Vlieland
Foča *100 C4 var.* Srbinje. SE Bosnia
 and Herzegovina
Focşani *108 C4* Vrancea, E Romania
Fogaras/Fogarasch *see* Făgăraş
Foggia *97 D5* Puglia, SE Italy
Fogo *74 A3 island* Ilhas de Sotavento,
 SW Cape Verde (Cabo Verde)
Foix *91 B6* Ariège, S France
Folégandros *105 C7 island* Kykládes,
 Greece, Aegean Sea
Foleyet *38 C4* Ontario, S Canada
Foligno *96 C4* Umbria, C Italy
Folkestone *89 E7* SE England,
 United Kingdom
Fond du Lac *40 B2* Wisconsin, N USA
Fonseca, Golfo de *see* Fonseca, Gulf of
Fonseca, Gulf of *52 C3 Sp.* Golfo de
 Fonseca. *gulf* C Central America
Fontainebleau *90 C3* Seine-et-Marne,
 N France
Fontenay-le-Comte *90 B4* Vendée,
 NW France
Fontvieille *91 B8* SW Monaco Europe
Fonyód *99 C7* Somogy, W Hungary
Foochow *see* Fuzhou
Forchheim *95 C5* Bayern, SE Germany
Forel, Mont *82 D4 mountain*
 SE Greenland
Forfar *88 C4* E Scotland,
 United Kingdom
Forge du Sud *see* Dudelange
Forlì *96 C3 anc.* Forum Livii.
 Emilia-Romagna, N Italy
Formentera *93 G4 anc.* Ophiusa, *Lat.*
 Frumentum. *island* Islas Baleares,
 Spain, W Mediterranean Sea
Formosa *64 D3* Formosa, NE Argentina
Formosa/Formo'sa *see* Taiwan
Formosa, Serra *63 E3 mountain range*
 C Brazil
Formosa Strait *see* Taiwan Strait
Føroyar *see* Faroe Islands
Forrest City *42 B1* Arkansas, C USA
Fort Albany *38 C3* Ontario, C Canada
Fortaleza *61 F2* Pando, N Bolivia
Fortaleza *63 G2 prev.* Ceará. *state capital*
 Ceará, NE Brazil
Fort-Archambault *see* Sarh
Fort-Bayard *see* Zhanjiang
Fort-Cappolani *see* Tidjikja
Fort Charlet *see* Djanet
Fort-Chimo *see* Kuujjuaq
Fort Collins *44 D4* Colorado, C USA
Fort-Crampel *see* Kaga Bandoro
Fort Davis *49 E3* Texas,
 SW USA

Fort-de-France *55 H4 prev.* Fort-Royal.
 dependent territory capital (Martinique)
 W Martinique
Fort Dodge *45 F3* Iowa, C USA
Fortescue River *146 A4 river*
 Western Australia
Fort-Foureau *see* Kousséri
Fort Frances *38 B4* Ontario, S Canada
Fort Good Hope *37 E3 var.* Rádeyilikóé.
 Northwest Territories, NW Canada
Fort Gouraud *see* Fdérik
Forth *88 C4 river* C Scotland,
 United Kingdom
Forth, Firth of *88 C4 estuary* E Scotland,
 United Kingdom
Fortín General Eugenio Garay *see*
 General Eugenio A. Garay
Fort Jameson *see* Chipata
Fort-Lamy *see* Ndjamena
Fort Lauderdale *43 F5* Florida, SE USA
Fort Liard *37 E4 var.* Liard. Northwest
 Territories, W Canada
Fort Madison *45 G4* Iowa, C USA
Fort McMurray *37 E4* Alberta, C Canada
Fort McPherson *36 D3 var.* McPherson.
 Northwest Territories, NW Canada
Fort Morgan *44 D4* Colorado, C USA
Fort Myers *43 E5* Florida, SE USA
Fort Nelson *37 E4* British Columbia,
 W Canada
Fort Peck Lake *44 C1 reservoir* Montana,
 NW USA
Fort Pierce *43 F4* Florida, SE USA
Fort Providence *37 E4 var.* Providence.
 Northwest Territories, W Canada
Fort-Repoux *see* Akjoujt
Fort Rosebery *see* Mansa
Fort Rousset *see* Owando
Fort-Royal *see* Fort-de-France
Fort St. John *37 E4* British Columbia,
 W Canada
Fort Scott *45 F5* Kansas, C USA
Fort Severn *38 C2* Ontario, C Canada
Fort-Shevchenko *114 A4* Mangistau,
 W Kazakhstan
Fort-Sibut *see* Sibut
Fort Simpson *37 E4 var.* Simpson.
 Northwest Territories, W Canada
Fort Smith *37 E4* Northwest Territories,
 W Canada
Fort Smith *42 B1* Arkansas, C USA
Fort Stockton *49 E3* Texas, SW USA
Fort-Trinquet *see* Bîr Mogreïn
Fort Vermilion *37 E4* Alberta,
 W Canada
Fort Victoria *see* Masvingo
Fort Walton Beach *42 C3* Florida,
 SE USA
Fort Wayne *40 C4* Indiana, N USA
Fort William *88 C3* N Scotland,
 United Kingdom
Fort Worth *49 G2* Texas, SW USA
Fort Yukon *36 D3* Alaska, USA
Forum Alieni *see* Ferrara
Forum Livii *see* Forlì
Fossa Claudia *see* Chioggia
Fougamou *77 A6* Ngounié, C Gabon
Fougères *90 B3* Ille-et-Vilaine,
 NW France
Fou-hsin *see* Fuxin
Foulwind, Cape *151 B5 headland* South
 Island, New Zealand
Foumban *76 A4* Ouest, NW Cameroon
Fou-shan *see* Fushun
Foveaux Strait *151 A8 strait*
 S New Zealand
Foxe Basin *37 G3 sea* Nunavut,
 N Canada
Fox Glacier *151 B6* West Coast, South
 Island, New Zealand
Fraga *93 F2* Aragón, NE Spain
Fram Basin *155 C3 var.* Amundsen
 Basin. *undersea basin* Arctic Ocean
France *90 B4 off.* French Republic, *It./Sp.*
 Francia; *prev.* Gaul, Gaule, *Lat.* Gallia.
 country W Europe
Franceville *77 B6 var.* Massoukou,
 Masuku. Haut-Ogooué, E Gabon
Francfort *see* Frankfurt am Main
Franche-Comté *90 D4 cultural region*
 E France
Francia *see* France
Francis Case, Lake *45 E3 reservoir* South
 Dakota, N USA
Francisco Escárcega *51 G4* Campeche,
 SE Mexico
Francistown *78 D3* North East,
 NE Botswana
Franconian Jura *see* Fränkische Alb
Frankenalb *see* Fränkische Alb

Frankenstein/Frankenstein in Schlesien *see* Ząbkowice Śląskie
Frankfort *40 C5 state capital* Kentucky, S USA
Frankfort on the Main *see* Frankfurt am Main
Frankfurt *see* Frankfurt am Main, Germany
Frankfurt *see* Słubice, Poland
Frankfurt am Main *95 B5 var.* Frankfurt, *Fr.* Francfort; *prev. Eng.* Frankfort on the Main. Hessen, SW Germany
Frankfurt an der Oder *94 D3* Brandenburg, E Germany
Fränkische Alb *95 C6 var.* Frankenalb, *Eng.* Franconian Jura. *mountain range* S Germany
Franklin *42 C1* Tennessee, S USA
Franklin D. Roosevelt Lake *46 C1 reservoir* Washington, NW USA
Franz Josef Land *114 D1 Eng.* Franz Josef Land. *island group* N Russia
Franz Josef Land *see* Frantsa-Iosifa, Zemlya
Fraserburgh *88 D3* NE Scotland, United Kingdom
Fraser Island *see* K'gari
Frauenbach *see* Baia Mare
Frauenburg *see* Saldus, Latvia
Fredericksburg *41 E5* Virginia, NE USA
Fredericton *39 F4 province capital* New Brunswick, SE Canada
Frederikshåb *see* Paamiut
Frederikshald *see* Halden
Fredrikstad *85 B6* Østfold, S Norway
Freeport *54 C1* Grand Bahama Island, N The Bahamas
Freeport *49 H4* Texas, SW USA
Free State *see* Maryland
Freetown *74 C4 country capital* (Sierra Leone) W Sierra Leone
Freiburg *see* Freiburg im Breisgau, Germany
Freiburg im Breisgau *95 A6 var.* Freiburg, *Fr.* Fribourg-en-Brisgau. Baden-Württemberg, SW Germany
Freiburg in Schlesien *see* Świebodzice
Fremantle *147 A6* Western Australia
Fremont *45 F4* Nebraska, C USA
French Guiana *59 H3 var.* Guiana, Guyane. *French overseas department* N South America
French Guinea *see* Guinea
French Polynesia *143 F4 Overseas Country of France* S Pacific Ocean
French Republic *see* France
French Somaliland *see* Djibouti
French Southern and Antarctic Lands *141 B7 Fr.* Terres Australes et Antarctiques Françaises. *French overseas territory* S Indian Ocean
French Sudan *see* Mali
French Territory of the Afars and Issas *see* Djibouti
French Togoland *see* Togo
Fresnillo *50 D3 var.* Fresnillo de González Echeverría. Zacatecas, C Mexico
Fresnillo de González Echeverría *see* Fresnillo
Fresno *47 C6* California, W USA
Frías *64 C3* Catamarca, N Argentina
Fribourg-en-Brisgau *see* Freiburg im Breisgau
Friedek-Mistek *see* Frýdek-Místek
Friedrichshafen *95 B7* Baden-Württemberg, S Germany
Friendly Islands *see* Tonga
Frisches Haff *see* Vistula Lagoon
Frobisher Bay *32 B3 inlet* Baffin Island, Nunavut, NE Canada
Frobisher Bay *see* Iqaluit
Frohavet *84 B4 sound* C Norway
Frome, Lake *149 B6 var.* Munda. *salt lake* South Australia
Frontera *51 G4* Tabasco, SE Mexico
Frontignan *91 C6* Hérault, S France
Frostviken *see* Kvarnbergsvattnet
Frøya *84 A4 island* W Norway
Frumentum *see* Formentera
Frunze *see* Bishkek
Frýdek-Místek *90 C5 Ger.* Friedek-Mistek. Moravskoslezský Kraj, E Czechia (Czech Republic)
Fu-chien *see* Fujian
Fu-chou *see* Fuzhou
Fuengirola *92 D5* Andalucía, S Spain
Fuerte Olimpo *64 D2 var.* Olimpo. Alto Paraguay, NE Paraguay

Fuerte, Río *48 C5 river* C Mexico
Fuerteventura *70 B3 island* Islas Canarias, Spain, NE Atlantic Ocean
Fuhkien *see* Fujian
Fu-hsin *see* Fuxin
Fuji *131 D6 var.* Huzi. Shizuoka, Honshū, S Japan
Fujian *129 D6 var.* Fu-chien, Fuhkien, Fukien, Min, Fujian Sheng. *province* SE China
Fujian Sheng *see* Fujian
Fuji, Mount/Fujiyama *see* Fuji-san
Fuji-san *131 C6 var.* Fujiyama, *Eng.* Mount Fuji. *mountain* Honshū, SE Japan
Fukang *126 C2* Xinjiang Uygur Zizhiqu, W China
Fukien *see* Fujian
Fukui *131 C6 var.* Hukui. Fukui, Honshū, SW Japan
Fukuoka *131 A7 var.* Hukuoka, *hist.* Najima. Fukuoka, Kyūshū, SW Japan
Fukushima *130 D4 var.* Hukusima. Fukushima, Honshū, C Japan
Fulda *95 B5* Hessen, C Germany
Funafuti Atoll *145 E3 atoll and capital* (Tuvalu) C Tuvalu
Funchal *70 A2* Madeira, Portugal, NE Atlantic Ocean
Fundy, Bay of *39 F5 bay* Canada/USA
Fünen *see* Fyn
Fünfkirchen *see* Pécs
Furnes *see* Veurne
Fürth *95 C5* Bayern, S Germany
Furukawa *130 D4 var.* Hukukawa, Ōsaki. Miyagi, Honshū, C Japan
Fusan *see* Busan
Fushë Kosovë *101 D5 Serb.* Kosovo Polje. C Kosovo
Fushun *128 D3 var.* Fou-shan, Fu-shun. Liaoning, NE China
Fu-shun *see* Fushun
Fusin *see* Fuxin
Füssen *95 C7* Bayern, S Germany
Futog *100 D3* Vojvodina, NW Serbia
Futuna, Île *145 E4 island* S Wallis and Futuna
Fuxin *128 D3 var.* Fou-hsin, Fu-hsin, Fusin. Liaoning, NE China
Fuzhou *129 D6 var.* Foochow, Fu-chou. *province capital* Fujian, SE China
Fyn *85 B8 Ger.* Fünen. *island* C Denmark
Fyzabad *see* Faiżābād

G

Gaafu Alifu Atoll *see* North Huvadhu Atoll
Gaalkacyo *73 E5 var.* Galka'yo, *It.* Galcaio. Mudug, C Somalia
Gabela *78 B2* Cuanza Sul, W Angola
Gaberones *see* Gaborone
Gabès *71 E2 var.* Qābis. E Tunisia
Gabès, Golfe de *71 F2 Ar.* Khalij Qābis. *gulf* E Tunisia
Gabon *77 B6 off.* Gabonese Republic. *country* C Africa
Gabonese Republic *see* Gabon
Gaborone *78 C4 prev.* Gaberones. *country capital* (Botswana) South East, SE Botswana
Gabrovo *104 D2* Gabrovo, N Bulgaria
Gadag *132 C1* Karnātaka, W India
Gades/Gadier/Gadir/Gadire *see* Cádiz
Gadsden *42 D2* Alabama, S USA
Gaeta *97 C5* Lazio, C Italy
Gaeta, Golfo di *97 C5 var.* Gulf of Gaeta. *gulf* C Italy
Gaeta, Gulf of *see* Gaeta, Golfo di
Gäfle *see* Gävle
Gafsa *71 E2 var.* Qafşah. W Tunisia
Gagnoa *74 D5* C Ivory Coast
Gagra *117 E1* NW Georgia
Gaillac *91 C6 var.* Gaillac-sur-Tarn. Tarn, S France
Gaillac-sur-Tarn *see* Gaillac
Gaillimh *see* Galway
Gaillimhe, Cuan na *see* Galway Bay
Gainesville *43 E3* Florida, SE USA
Gainesville *42 D2* Georgia, SE USA
Gainesville *49 G2* Texas, SW USA
Lake Gairdner *149 A6 salt lake* South Australia
Gaizina Kalns *see* Gaiziņkalns
Gaiziņkalns *106 C3 var.* Gaizina Kalns. *mountain* E Latvia
Galán, Cerro *64 B3 mountain* NW Argentina

Galanta *99 C6 Hung.* Galánta. Trnavský Kraj, W Slovakia
Galapagos Fracture Zone *153 E3 tectonic feature* E Pacific Ocean
Galápagos Islands *153 F3 var.* Islas de los Galápagos, *Sp.* Archipiélago de Colón, *Eng.* Galapagos Islands, Tortoise Islands. *island group* Ecuador, E Pacific Ocean
Galápagos Islands *see* Galápagos Islands
Galápagos, Islas de los *see* Galápagos Islands
Galapagos Rise *153 F3 undersea rise* E Pacific Ocean
Galashiels *88 C4* SE Scotland, United Kingdom
Galaţi *108 D4 Ger.* Galatz. Galaţi, E Romania
Galatz *see* Galaţi
Galcaio *see* Gaalkacyo
Galesburg *40 B3* Illinois, N USA
Galicia *92 B1 anc.* Gallaecia. *autonomous community* NW Spain
Galicia Bank *80 B4 undersea bank* E Atlantic Ocean
Galilee, Sea of *see* Kinneret, Yam
Galka'yo *see* Gaalkacyo
Galkynyş *122 D3 prev. Rus.* Deynau, Dyanev, *Turkm.* Dänew. Lebap Welaýaty, NE Turkmenistan
Gallaecia *see* Galicia
Galle *132 D4 prev.* Point de Galle. Southern Province, SW Sri Lanka
Gallego Rise *153 F3 undersea rise* E Pacific Ocean
Gallegos *see* Río Gallegos
Gallia *see* France
Gallipoli *97 E6* Puglia, SE Italy
Gällivare *84 C3 Smi.* Váhtjer. Norrbotten, N Sweden
Gallup *48 C1* New Mexico, SW USA
Galtat-Zemmour *70 B3* C Western Sahara
Galveston *49 H4* Texas, SW USA
Galway *89 A5 Ir.* Gaillimh. W Ireland
Galway Bay *89 A6 Ir.* Cuan na Gaillimhe. *bay* W Ireland
Gámas *see* Kaamanen
Gambell *36 C2* Saint Lawrence Island, Alaska, USA
Gambia *see* Gambia, The
Gambia *74 C3 Fr.* Gambie. *river* W Africa
Gambia, Republic of The *see* Gambia, The
Gambia, The *74 B3 off.* Republic of The Gambia *var.* Gambia. *country* W Africa
Gambie *see* Gambia, The
Gambier, Îles *143 G4 island group* E French Polynesia
Gamboma *77 B6* Plateaux, E Congo
Gamlakarleby *see* Kokkola
Gan *132 B5* Addu Atoll, C Maldives
Gan *see* Gansu, China
Gan *see* Jiangxi, China
Ganaane *see* Juba
Gäncä *117 G2 Rus.* Gyandzha; *prev.* Kirovabad, Yelisavetpol. W Azerbaijan
Gand *see* Gent
Gandajika *77 D7* Kasai-Oriental, S Dem. Rep. Congo
Gander *39 G3* Newfoundland and Labrador, SE Canada
Gāndhīdhām *134 C4* Gujarāt, W India
Gandia *93 F3 prev.* Gandía. Comunitat Valenciana, E Spain
Gandía *see* Gandia
Ganga *see* Ganges
Ganges *135 F3 var.* Ganga, *Ben.* Padma. *river* Bangladesh/India
Ganges Cone *see* Ganges Fan
Ganges Fan *140 D3 var.* Ganges Cone. *undersea fan* N Bay of Bengal
Ganges, Mouths of the *135 G4 delta* Bangladesh/India
Gangra *see* Çankırı
Gangtok *135 F3 state capital* Sikkim, N India
Gansos, Lago dos *see* Goose Lake
Gansu *128 B4 var.* Gan, Gansu Sheng, Kansu. *province* N China
Gansu Sheng *see* Gansu
Gantsevichi *see* Hantsavichy
Ganzhou *129 C6* Jiangxi, S China
Gao *75 E3* Gao, E Mali
Gaocheng *see* Litang
Gaoual *74 C4* N Guinea
Gaoxiong *see* Kaohsiung
Gap *91 D5 anc.* Vapincum. Hautes-Alpes, SE France

Gaplaňgyr Platosy *122 C4 Rus.* Plato Kaplangky. *ridge* Turkmenistan/ Uzbekistan
Gar *see* Gar Xincun
Garabil Belentligi *122 D3 Rus.* Vozvyshennost' Karabil'. *mountain range* S Turkmenistan
Garabogaz Aýlagy *see* Garabogaz Köli
Garabogaz Köli *122 B2 var.* Garabogaz Aýlagy *Rus.* Zaliv Kara-Bogaz-Gol. *bay* NW Turkmenistan
Garachiné *53 G5* Darién, SE Panama
Garagum *122 C3 var.* Garagumy, Qara Qum, *Eng.* Black Sand Desert, Kara Kum; *prev.* Peski Karakumy. *desert* C Turkmenistan
Garagum Canal *122 D3 var.* Kara Kum Canal, *Rus.* Karagumskiy Kanal, Karakumskiy Kanal. *canal* C Turkmenistan
Garagumy *see* Garagum
Gara Hitrino *104 D2 var.* Gara Khitrino. Shumen, NE Bulgaria
Gara Khitrino *see* Gara Hitrino
Gárasavon *see* Karesuando
Garda, Lago di *96 C2 var.* Benaco, *Eng.* Lake Garda, *Ger.* Gardasee. *lake* NE Italy
Garda, Lake *see* Garda, Lago di
Gardasee *see* Garda, Lago di
Garden City *45 E5* Kansas, C USA
Garden State, The *see* New Jersey
Gardēz *123 E4 prev.* Gardīz. E Afghanistan
Gardīz *see* Gardēz
Gardner Island *see* Nikumaroro
Garegegasnjárga *see* Karigasniemi
Gargždai *106 B3* Klaipėda, W Lithuania
Garissa *73 D6* Coast, S Kenya
Garland *49 G2* Texas, SW USA
Garoe *see* Garoowe
Garonne *91 B5 anc.* Garumna. *river* S France
Garoowe *73 E5 var.* Garoe. Nugaal, N Somalia
Garoua *76 B4 var.* Garua. Nord, N Cameroon
Garrygala *see* Magtymguly
Garry Lake *37 F3 lake* Nunavut, C Canada
Garsen *73 D6* Coast, S Kenya
Garua *see* Garoua
Garumna *see* Garonne
Garwolin *98 D4* Mazowieckie, E Poland
Gar Xincun *126 A4 prev.* Gar. Xizang Zizhiqu, W China
Gary *40 B3* Indiana, N USA
Garzón *58 B4* Huila, S Colombia
Gasan-Kuli *see* Esenguly
Gascogne *91 B6 Eng.* Gascony. *cultural region* S France
Gascony *see* Gascogne
Gascoyne River *147 A5 river* Western Australia
Gaspé *39 F3* Québec, SE Canada
Gaspé, Péninsule de *39 E4 var.* Péninsule de la Gaspésie. *peninsula* Québec, SE Canada
Gaspésie, Péninsule de la *see* Gaspé, Péninsule de
Gastonia *43 E1* North Carolina, SE USA
Gastoúni *105 B6* Dytikí Ellás, S Greece
Gatchina *110 B4* Leningradskaya Oblast', NW Russia
Gatineau *38 D4* Québec, SE Canada
Gatooma *see* Kadoma
Gatún, Lake *53 E4 reservoir* C Panama
Gauhāti *see* Guwāhāti
Gauja *106 D3 Ger.* Aa. *river* Estonia/ Latvia
Gaul/Gaule *see* France
Gauteng *see* Johannesburg, South Africa
Gävbandī *120 D4* Hormozgān, S Iran
Gávdos *105 C8 island* SE Greece
Gävle *85 C6 var.* Gäfle; *prev.* Gefle. Gävleborg, C Sweden
Gawler *149 B6* South Australia
Gaya *135 F3* Bihār, N India
Gayndah *149 E5* Queensland, E Australia
Gaz-Achak *see* Gazojak
Gazandzhyk/Gazanjyk *see* Bereket
Gaza Strip *119 A7 Ar.* Qita Ghazzah. *disputed region* SW Asia

Gaziantep 116 D4 *var.* Gazi Antep; *prev.*
Aintab, Antep. Gaziantep,
S Turkey (Türkiye)
Gazi Antep *see* Gaziantep
Gazimağusa 102 D5 *var.* Famagusta, *Gk.*
Ammóchostos. E Cyprus
Gazimağusa Körfezi 102 C5 *var.*
Famagusta Bay, *Gk.* Kólpos
Ammóchostos. *bay* E Cyprus
Gazli 122 D2 Buxoro Viloyati,
C Uzbekistan
Gazojak 122 D2 *Rus.* Gaz-Achak. Lebap
Welaýaty, NE Turkmenistan
Gbanga 74 D5 *var.* Gbarnga. N Liberia
Gbarnga *see* Gbanga
Gdańsk 98 C2 *Fr.* Dantzig, *Ger.* Danzig.
Pomorskie, N Poland
Gdańsk, Gulf of 98 C2 *var.* Gulf of
Danzig, *Ger.* Danziger Bucht, *Pol.*
Zakota Gdańska, *Rus.* Gdan'skaya
Bukhta. *gulf* N Europe
Gdan'skaya Bukhta *see*
Gdańsk, Gulf of
Gdańska, Zakota *see* Gdańsk, Gulf of
Gdingen *see* Gdynia
Gdynia 98 C2 *Ger.* Gdingen. Pomorskie,
N Poland
Gedaref 72 C4 *var.* Al Qaḍārif, El
Gedaref. Gedaref, E Sudan
Gediz 116 B3 Kütahya,
W Turkey (Türkiye)
Gediz Nehri 116 A3 *river*
W Turkey (Türkiye)
Geel 87 C5 *var.* Gheel. Antwerpen,
N Belgium
Geelong 149 C7 Victoria, SE Australia
Ge'e'mu *see* Golmud
Gefle *see* Gävle
Geilo 85 A5 Buskerud, S Norway
Gejiu 129 B6 *var.* Kochiu. Yunnan,
S China
Gëkdepe *see* Gökdepe
Gela 97 C7 *prev.* Terranova di Sicilia.
Sicilia, Italy, C Mediterranean Sea
Geldermalsen 86 C4 Gelderland,
C Netherlands
Geleen 87 D6 Limburg, SE Netherlands
Gelib *see* Jilib
Gellinsor 73 E5 Mudug, C Somalia
Gembloux 87 C6 Namur, Belgium
Gemena 77 C5 Equateur, NW Dem.
Rep. Congo
Gem of the Mountains *see* Idaho
Gemona del Friuli 96 D2 Friuli-Venezia
Giulia, NE Italy
Gem State *see* Idaho
Genalë Wenz *see* Juba
Genck *see* Genk
General Alvear 64 B4 Mendoza,
W Argentina
General Carrera, Lago *see* Buenos Aires,
Lago
General Eugenio A. Garay 64 C1 *var.*
Fortín General Eugenio Garay; *prev.*
Yrendagüé. Nueva Asunción,
NW Paraguay
General José F.Uriburu *see* Zárate
General Machado *see* Camacupa
General Santos 139 F3 *off.* General
Santos City. Mindanao, S Philippines
General Santos City *see* General Santos
Gênes *see* Genova
Geneva *see* Genève
Geneva, Lake 95 A7 *Fr.* Lac de Genève,
Lac Léman, *le* Léman, *Ger.* Genfer See.
lake France/Switzerland
Genève 95 A7 *Eng.* Geneva, *Ger.* Genf, *It.*
Ginevra. Genève, SW Switzerland
Genève, Lac de *see* Geneva, Lake
Genf *see* Genève
Genfer See *see* Geneva, Lake
Genichesk *see* Henichesk
Genk 87 D6 *var.* Genck. Limburg,
NE Belgium
Genney 86 D4 Limburg, SE Netherlands
Genoa *see* Genova
Genoa, Gulf of 96 A3 *Eng.* Gulf of
Genoa. *gulf* NW Italy
Genoa, Gulf of *see* Genova, Golfo di
Genova 102 D1 *Eng.* Genoa; *anc.* Genua,
Fr. Gênes. Liguria, NW Italy
Genovesa, Isla 60 B5 *var.* Tower Island.
island Galápagos Islands, Ecuador,
E Pacific Ocean
Gent 87 B5 *Eng.* Ghent, *Fr.* Gand.
Oost-Vlaanderen, NW Belgium
Genua *see* Genova
Geok-Tepe *see* Gökdepe
George 78 C5 Western Cape,
S South Africa

George 82 A4 *river* Newfoundland and
Labrador/Québec, E Canada
George, Lake 43 E3 *lake* Florida, SE USA
Georgenburg *see* Jurbarkas
Georges Bank 35 D5 *undersea bank*
W Atlantic Ocean
George Sound *see* Te Houhou
Georges River 148 D2 *river* New South
Wales, E Australia
Georgetown 59 F2 *country capital*
(Guyana) N Guyana
George Town 54 C2 Great Exuma Island,
C The Bahamas
George Town 54 B3 *var.* Georgetown.
dependent territory capital (Cayman
Islands) Grand Cayman, SW Cayman
Islands
George Town 138 B3 *var.* Penang,
Pinang. Pinang, Peninsular Malaysia
Georgetown 43 F2 South Carolina,
SE USA
Georgetown *see* George Town
George V Land 154 C4 *physical region*
Antarctica
Georgia 117 F2 *off.* Georgia, Geor.
Sakart'velo, *prev.* Gruzinskaya SSR, *Rus.*
Gruziya. *country* SW Asia
Georgia 42 D2 *off.* State of Georgia, *also
known as* Empire State of the South,
Peach State. *state* SE USA
Georgian Bay 40 D2 *lake bay* Ontario,
S Canada
Georgia, Strait of 46 A1 *strait* British
Columbia, W Canada
Georgi Dimitrov *see* Kostenets
Georgiu-Dezh *see* Liski
Gera 94 C4 Thüringen, E Germany
Geráki 105 B6 Pelopónnisos, S Greece
Geraldine 151 B6 Canterbury, South
Island, New Zealand
Geraldton 147 A6 Western Australia
Geral, Serra 57 D5 *mountain range*
S Brazil
Gerede 116 C2 Bolu, N Turkey (Türkiye)
Gereshk *see* Girishk
Gering 44 D3 Nebraska, C USA
German East Africa *see* Tanzania
Germanicopolis *see* Çankırı
Germanicum, Mare/German Ocean *see*
North Sea
German Southwest Africa *see* Namibia
Germany 94 B4 *off.* Federal Republic of
Germany, Bundesrepublik
Deutschland, *Ger.* Deutschland.
country N Europe
Germany, Federal Republic of *see*
Germany
Gerolimênas 105 B7 Pelopónnisos,
S Greece
Gerona *see* Girona
Gerpinnes 87 C7 Hainaut, S Belgium
Gerunda *see* Girona
Gerze 116 D2 Sinop, N Turkey (Türkiye)
Gesoriacum *see* Boulogne-sur-Mer
Gessoriacum *see* Boulogne-sur-Mer
Getafe 92 D3 Madrid, C Spain
Gevaş 117 F3 Van, SE Turkey (Türkiye)
Gevgeli *see* Gevgelija
Gevgelija 101 E6 *var.* Đevdelija,
Djevdjelija, *Turk.* Gevgeli. SE North
Macedonia
Ghaba *see* Al Ghābah
Ghana 75 E5 *off.* Republic of Ghana.
country W Africa
Ghanzi 78 C3 *var.* Khanzi. Ghanzi,
W Botswana
Gharandal 119 B7 Al'Aqabah,
SW Jordan
Gharbt, Jabal al *see* Liban, Jebel
Ghardaïa 70 D2 N Algeria
Gharvàn *see* Gharyàn
Gharyàn 71 F2 *var.* Gharvàn. NW Libya
Ghawdex *see* Gozo
Ghazni 123 E4 *var.* Ghazni. Ghaznī,
E Afghanistan
Ghazzah *see* Gaza
Gheel *see* Geel
Ghent *see* Gent
Gheorgheni 108 C4 *prev.* Gheorghieni,
Sîn-Miclăuş, *Ger.* Niklasmarkt, *Hung.*
Gyergyószentmiklós. Harghita,
C Romania
Gheorghieni *see* Gheorgheni
Ghôriân *see* Ghôriyân
Ghôriyân 122 D4 *var.* Ghôriân, Ghûriān.
Herāt, W Afghanistan
Ghūdara 123 F3 *var.* Gudara, *Rus.*
Kudara. SE Tajikistan
Ghurdaqah *see* Al Ghurdaqah
Ghûriān *see* Ghôriyân

Giamame *see* Jamaame
Giannitsá 104 B4 *var.* Yiannitsá.
Kentrikí Makedonía, N Greece
Gibraltar 93 G4 *UK Overseas Territory*
SW Europe
Gibraltar, Bay of 93 G5 *bay* Gibraltar/
Spain Europe Mediterranean Sea
Atlantic Ocean
**Gibraltar, Détroit de/Gibraltar,
Estrecho de** *see* Gibraltar, Strait of
Gibraltar, Strait of 92 C5 *Fr.* Détroit de
Gibraltar, *Sp.* Estrecho de Gibraltar.
strait Atlantic Ocean/Mediterranean Sea
Gibson Desert 147 B5 *desert*
Western Australia
Giedraičiai 107 C5 Utena, E Lithuania
Giessen 95 B5 Hessen, W Germany
Gifu 131 C6 *var.* Gihu. Gifu, Honshū,
SW Japan
Giganta, Sierra de la 50 B3 *mountain
range* NW Mexico
Gihu *see* Gifu
G'ijduvon 122 D2 *Rus.* Gizhduvon.
Buxoro Viloyati, C Uzbekistan
Gijón 92 D1 *var.* Xixón. Asturias,
NW Spain
Gila River 48 A2 *river* Arizona, SW USA
Gilbert Islands *see* Tungaru
Gilbert River 148 C3 *river* Queensland,
NE Australia
Gilf Kebir Plateau *see* Haḍabat al Jilf
al Kabir
Gillette 44 D3 Wyoming, C USA
Gilolo *see* Halmahera, Pulau
Gilroy 47 B6 California, W USA
Gimie, Mount 55 F1 *mountain*
C Saint Lucia
Gimma *see* Jīma
Ginevra *see* Genève
Gingin 147 A6 Western Australia
Giohar *see* Jawhar
Gipeswic *see* Ipswich
Girardot 58 B3 Cundinamarca,
C Colombia
Giresun 117 E2 *var.* Kerasunt; *anc.*
Cerasus, Pharnacia. Giresun,
NE Turkey (Türkiye)
Girgenti *see* Agrigento
Girin *see* Jilin
Girishk 122 D5 *var.* Gereshk. Helmand,
SW Afghanistan
Girne 102 C5 *Gk.* Kerýneia, Kyrenia.
N Cyprus
Giron *see* Kiruna
Girona 93 G2 *var.* Gerona; *anc.* Gerunda.
Cataluña, NE Spain
Gisborne 150 E3 Gisborne, North Island,
New Zealand
Gissar Range 123 E3 *var.* Hisar Range,
Rus. Gissarskiy Khrebet, *Taj.* Qatorkūhi
Hisor, *Uzb.* Hisor Tizmasi. *mountain
range* Tajikistan/Uzbekistan
Gissarskiy Khrebet *see* Gissar Range
Gitega 73 B7 *prev.* Kitega. *country
capital* (Burundi - political) W Burundi
Githio *see* Gýtheio
Giulianova 96 D4 Abruzzi, C Italy
Giumri *see* Gyumri
Giurgiu 108 C5 Giurgiu, S Romania
Giza *see* Al Jīzah
Gizeh *see* Al Jīzah
Gizhduvon *see* G'ijduvon
Giżycko 98 D2 *Ger.* Lötzen.
Warmińsko-Mazurskie, NE Poland
Gjakovë 101 D5 *Serb.* Đakovica.
W Kosovo
Gjilan 101 D5 *Serb.* Gnjilane. E Kosovo
Gjirokastër 101 C7 *var.* Gjirokastra;
prev. Gjinokastër, *Gk.* Argyrokastron,
It. Argirocastro. Gjirokastër, S Albania
Gjirokastra *see* Gjirokastër
Gjoa Haven 37 F3 *var.* Uqsuqtuuq. King
William Island, Nunavut, N Canada
Gjøvik 85 B5 Oppland, S Norway
Glace Bay 39 G4 Cape Breton Island,
Nova Scotia, SE Canada
Gladstone 148 E4 Queensland,
E Australia
Gláma 85 B5 *var.* Glommen. *river*
S Norway
Glasgow 88 C4 S Scotland,
United Kingdom
Glavinitsa 104 D1 *prev.* Pravda,
Dogrular. Silistra, NE Bulgaria
Glavn'a Morava *see* Velika Morava
Glazov 111 D5 Udmurtskaya Respublika,
NW Russia
Gleiwitz *see* Gliwice
Glendale 48 B2 Arizona, SW USA

Glendive 44 D2 Montana, NW USA
Glens Falls 41 F3 New York, NE USA
Glevum *see* Gloucester
Glina 100 B3 *var.* Banijska Palanka.
Sisak-Moslavina, NE Croatia
Glittertind 85 A5 *mountain* S Norway
Gliwice 99 C5 *Ger.* Gleiwitz. Śląskie,
S Poland
Globe 48 B2 Arizona, SW USA
Globino *see* Hlobyne
Glogau *see* Głogów
Głogów 98 B4 *Ger.* Glogau, Glogów.
Dolnośląskie, SW Poland
Glogow *see* Głogów
Glomma *see* Gláma
Glommen *see* Gláma
Gloucester 89 D6 *hist.* Caer Glou, *Lat.*
Glevum. C England, United Kingdom
Głowno 98 D4 Łódź, C Poland
Glubokoye *see* Hlybokaye
Glukhov *see* Hlukhiv
Gnesen *see* Gniezno
Gniezno 98 C3 *Ger.* Gnesen.
Weilkopolskie, C Poland
Gnjilane *see* Gjilan
Gobabis 78 B3 Omaheke, E Namibia
Gobi 126 D3 *desert* China/Mongolia
Gobō 131 C6 Wakayama, Honshū,
SW Japan
Godávari 124 B3 *var.* Godavari. *river*
C India
Godavari *see* Godávari
Godhavn *see* Qeqertarsuaq
Godhra 134 C4 Gujarāt, W India
Gôding *see* Hodonín
Godoy Cruz 64 B4 Mendoza,
W Argentina
Godthaab/Godthåb *see* Nuuk
Godwin Austen, Mount *see* K2
Goede Hoop, Kaap de *see* Good Hope,
Cape of
Goeie Hoop, Kaap die *see* Good Hope,
Cape of
Goeree 86 B4 *island* SW Netherlands
Goes 87 B5 Zeeland, SW Netherlands
Goettingen *see* Göttingen
Gogebic Range 40 B1 *hill range*
Michigan/Wisconsin, N USA
Goiânia 63 E3 *prev.* Goyania. *state
capital* Goiás, C Brazil
Goiás 63 E3 *off.* Estado de Goiás; *prev.*
Goiaz, Goyaz, *state/region* C Brazil
Goiás, Estado de *see* Goiás
Goiaz *see* Goiás
Goidhoo Atoll *see* Horsburgh Atoll
Gojôme 130 D4 Akita, Honshū,
NW Japan
Gökçeada 104 D4 *var.* Imroz Adasï, *Gk.*
Imbros. *island* NW Turkey (Türkiye)
Gökdepe 122 C3 *Rus.* Gëkdepe,
Geok-Tepe. Ahal Welaýaty,
C Turkmenistan
Göksun 116 D4 Kahramanmaraş,
C Turkey (Türkiye)
Gol 85 A5 Buskerud, S Norway
Golan Heights 119 B5 *Ar.* Al Jawlān,
Heb. HaGolan. *mountain range*
SW Syria
Golaya Pristan *see* Hola Prystan
Gołdap 98 E2 *Ger.* Goldap.
Warmińsko-Mazurskie, NE Poland
Gold Coast 149 E5 *cultural region*
Queensland, E Australia
Golden Bay 150 C4 *var.* Mohua. *bay*
South Island, New Zealand
Golden State, The *see* California
Goldingen *see* Kuldīga
Goldsboro 43 F1 North Carolina,
SE USA
Goleniów 98 B3 *Ger.* Gollnow.
Zachodnio-pomorskie, NW Poland
Gollnow *see* Goleniów
Golmo *see* Golmud
Golmud 126 D4 *var.* Ge'e'mu, Golmo,
Chin. Ko-erh-mu. Qinghai, C China
Golovanevsk *see* Holovanivsk
Golub-Dobrzyń 98 C3 Kujawski-
pomorskie, C Poland
Goma 77 E6 Nord-Kivu, NE Dem.
Rep. Congo
Gombi 75 H4 Adamawa, E Nigeria
Gombroon *see* Bandar-e'Abbās
Gomel' *see* Homyel'
Gomera 70 A3 *island* Islas Canarias,
Spain, NE Atlantic Ocean
Gómez Palacio 50 D3 Durango,
C Mexico
Gonaïves 54 D3 *var.* Les Gonaïves.
N Haiti
Gonâve, Île de la 54 D3 *island* C Haiti

Gondar *see* Gonder
Gonder 72 *C4* *var.* Gondar. Āmara, NW Ethiopia
Gondia 135 *E4* Mahārāshtra, C India
Gonggar 126 *C5* *var.* Gyixong. Xizang Zizhiqu, W China
Gongola 75 *G4* *river* E Nigeria
Gongtang *see* Damxung
Gonni/Gónnos *see* Gónnoi
Gónnoi 104 *B4* *var.* Gonni, Gónnos; *prev.* Dereli. Thessalía, C Greece
Good Hope, Cape of 78 *B5* *Afr.* Kaap de Goede Hoop, Kaap die Goeie Hoop. *headland* SW South Africa
Goodland 44 *D4* Kansas, C USA
Goondiwindi 149 *D5* Queensland, E Australia
Goor 86 *E3* Overijssel, E Netherlands
Goose Green 65 *D7* *var.* Prado del Ganso. East Falkland, Falkland Islands
Goose Lake 46 *B4* *var.* Lago dos Gansos. *lake* California/Oregon, W USA
Gopher State *see* Minnesota
Göppingen 95 *B6* Baden-Württemberg, SW Germany
Góra Kalwaria 114 *D4* Mazowieckie, C Poland
Gorakhpur 135 *E3* Uttar Pradesh, N India
Gorany *see* Harany
Goražde 100 *C4* Federacija Bosna I Hercegovina, SE Bosnia and Herzegovina
Gorbovichi *see* Harbavichy
Goré 76 *C4* Logone-Oriental, S Chad
Gorë 73 *C5* Oromīya, C Ethiopia
Gore 151 *B7* Southland, South Island, New Zealand
Gorgān 120 *D2* *var.* Astarabad, Astrabad, Gurgan, *prev.* Asterābād; *anc.* Hyrcania. Golestán, N Iran
Gori 117 *F2* C Georgia
Gorinchem 86 *C4* *var.* Gorkum. Zuid-Holland, C Netherlands
Goris 117 *G3* SE Armenia
Gorki *see* Horki
Gor'kiy *see* Nizhniy Novgorod
Gorkum *see* Gorinchem
Görlitz 94 *D4* Sachsen, E Germany
Görlitz *see* Zgorzelec
Gorlovka *see* Horlivka
Gorna Dzhumaya *see* Blagoevgrad
Gornja Mužlja *see* Mužlja
Gornji Milanovac 100 *C4* Serbia, C Serbia
Gorodets *see* Haradzyets
Gorodishche *see* Horodyshche
Gorodnya *see* Horodnia
Gorodok *see* Haradok
Gorodok/Gorodok Yagellonski *see* Horodok
Gorontalo 139 *E4* Sulawesi, C Indonesia
Gorontalo, Teluk *see* Tomini, Gulf of
Gorssel 86 *D3* Gelderland, E Netherlands
Goryn *see* Horyn'
Gorzów Wielkopolski 98 *B3* *Ger.* Landsberg, Landsberg an der Warthe. Lubuskie, W Poland
Gosford 149 *D6* New South Wales, SE Australia
Goshogawara 130 *D3* *var.* Gosyogawara. Aomori, Honshū, C Japan
Gospić 100 *A3* Lika-Senj, C Croatia
Gostivar 101 *D6* W Macedonia
Gosyogawara *see* Goshogawara
Göteborg 85 *B7* *Eng.* Gothenburg. Västra Götaland, S Sweden
Gotel Mountains 75 *G5* *mountain range* E Nigeria
Gotha 94 *C4* Thüringen, C Germany
Gothenburg *see* Göteborg
Gotland 85 *C7* *island* SE Sweden
Goto-retto 131 *A7* *island group* SW Japan
Gotska Sandön 106 *B1* *island* SE Sweden
Gōtsu 131 *B6* *var.* Gôtu. Shimane, Honshū, SW Japan
Göttingen 94 *B4* *var.* Goettingen. Niedersachsen, C Germany
Gottschee *see* Kočevje
Gottwaldov *see* Zlín
Gôtu *see* Gōtsu
Gouda 86 *C4* Zuid-Holland, C Netherlands
Gough Fracture Zone 67 *C6* *tectonic feature* S Atlantic Ocean
Gough Island 69 *B8* *UK Overseas territory,* S Atlantic Ocean
Gouin, Réservoir 38 *D3* *reservoir* Québec, SE Canada

Goulburn 149 *D6* New South Wales, SE Australia
Goundam 75 *E3* Tombouctou, NW Mali
Gouré 75 *G3* Zinder, SE Niger
GovernAla, Gora *see* Hoverla, Hora
Governador Valadares 63 *F4* Minas Gerais, SE Brazil
Govi Altayn Nuruu 127 *E3* *mountain range* S Mongolia
Goya 64 *D3* Corrientes, NE Argentina
Goyania *see* Goiânia
Goyaz *see* Goiás
Goz Beïda 76 *C3* Ouaddaï, SE Chad
Gozo 97 *C8* *var.* Ghawdex. *island* N Malta
Gqeberha 78 *C5* *prev.* Port Elizabeth. Eastern Cape, S South Africa
Graciosa 92 *A5* *var.* Ilha Graciosa. *island* Azores, Portugal, NE Atlantic Ocean
Graciosa, Ilha *see* Graciosa
Gradačac 100 *C3* Federacija Bosna I Hercegovina, N Bosnia and Herzegovina
Gradaús, Serra dos 63 *E3* *mountain range* C Brazil
Gradiška *see* Bosanska Gradiška
Grafton 149 *E5* New South Wales, SE Australia
Grafton 45 *E1* North Dakota, N USA
Graham Land 154 *A2* *physical region* Antarctica
Grajewo 98 *E3* Podlaskie, NE Poland
Grampian Mountains 88 *C3* *mountain range* C Scotland, United Kingdom
Gran *see* Esztergom, Hungary
Granada 52 *D3* Granada, SW Nicaragua
Granada 92 *D5* Andalucía, S Spain
Gran Canaria 70 *A3* *var.* Grand Canary. *island* Islas Canarias, Spain, NE Atlantic Ocean
Gran Chaco 64 *D2* *var.* Chaco. *lowland plain* South America
Grand Bahama Island 54 *B1* *island* N The Bahamas
Grand Banks of Newfoundland 34 *E4* *undersea basin* NW Atlantic Ocean
Grand Bassa *see* Buchanan
Grand Canary *see* Gran Canaria
Grand Canyon 48 *A1* *canyon* Arizona, SW USA
Grand Canyon State *see* Arizona
Grand Cayman 54 *B3* *island* SW Cayman Islands
Grand Duchy of Luxembourg *see* Luxembourg
Grande, Bahía 65 *B7* *bay* S Argentina
Grande-Comor *see* Ngazidja
Grande de Chiloé, Isla *see* Chiloé, Isla de
Grand Prairie 37 *E4* Alberta, W Canada
Grand Erg Occidental 70 *D3* *desert* W Algeria
Grand Erg Oriental 71 *E3* *desert* Algeria/Tunisia
Río Grande 51 *E2* *var.* Río Bravo, *Sp.* Río Bravo del Norte, Bravo del Norte. *river* Mexico/USA
Grande Terre 55 *G3* *island* E West Indies
Grand Falls 39 *G3* Newfoundland, Newfoundland and Labrador, SE Canada
Grand Forks 45 *E1* North Dakota, N USA
Grandichi *see* Hrandzichy
Grand Island 45 *E4* Nebraska, C USA
Grand Junction 44 *C4* Colorado, C USA
Grand Paradis *see* Gran Paradiso
Grand Rapids 40 *C3* Michigan, N USA
Grand Rapids 45 *F1* Minnesota, N USA
Grand-Saint-Bernard, Col du *see* Great Saint Bernard Pass
Grand-Santi 59 *G3* W French Guiana
Granite State *see* New Hampshire
Gran Lago *see* Nicaragua, Lago de
Gran Malvina *see* West Falkland
Gran Paradiso 96 *A2* *Fr.* Grand Paradis. *mountain* NW Italy
Gran San Bernardo, Passo di *see* Great Saint Bernard Pass
Gran Santiago *see* Santiago
Grants 48 *C2* New Mexico, SW USA
Grants Pass 46 *B4* Oregon, NW USA
Granville 90 *B3* Manche, N France
Gratianopolis *see* Grenoble
Gratz *see* Graz
Graudenz *see* Grudziądz
Graulhet 91 *C6* Tarn, S France
Grave 86 *D4* Noord-Brabant, SE Netherlands
Grayling 36 *C2* Alaska, USA

Graz 95 *E7* *prev.* Gratz. Steiermark, SE Austria
Great Abaco 54 *C1* *var.* Abaco Island. *island* N The Bahamas
Great Alföld *see* Great Hungarian Plain
Great Ararat *see* Büyük Ağrı Dağı
Great Australian Bight 147 *D7* *bight* S Australia
Great Barrier Island 150 *D2* *island* N New Zealand
Great Barrier Reef 148 *D2* *reef* Queensland, NE Australia
Great Basin 47 *C5* *basin* W USA
Great Bear Lake 37 *E3* *Fr.* Grand Lac de l'Ours. *lake* Northwest Territories, NW Canada
Great Belt 85 *B8* *var.* Store Bælt, *Eng.* Great Belt, Storebelt. *channel* Baltic Sea/Kattegat
Great Belt *see* Storebælt
Great Bend 45 *E5* Kansas, C USA
Great Britain *see* Britain
Great Dividing Range 148 *D4* *mountain range* NE Australia
Greater Antilles 54 *D3* *island group* West Indies
Greater Caucasus 117 *G2* *mountain range* Azerbaijan/Georgia/Russia Asia/Europe
Greater Sunda Islands 124 *D5* *var.* Sunda Islands. *island group* Indonesia
Great Exhibition Bay 150 *C1* *inlet* North Island, New Zealand
Great Exuma Island 54 *C2* *island* C The Bahamas
Great Falls 44 *B1* Montana, NW USA
Great Grimsby *see* Grimsby
Great Hungarian Plain 99 *C7* *var.* Great Alföld, Plain of Hungary, *Hung.* Alföld. *plain* SE Europe
Great Inagua 54 *D2* *var.* Inagua Islands. *island* S The Bahamas
Great Indian Desert *see* Thar Desert
Great Khingan Range *see* Da Hinggan Ling
Great Lake *see* Tônlé Sap
Great Lakes 35 *C5* *lakes* Ontario, Canada/USA
Great Lakes State *see* Michigan
Great Meteor Seamount *see* Great Meteor Tablemount
Great Meteor Tablemount 66 *B3* *var.* Great Meteor Seamount. *seamount* E Atlantic Ocean
Great Nicobar 133 *G3* *island* Nicobar Islands, India, NE Indian Ocean
Great Plain of China 125 *E2* *plain* E China
Great Plains 45 *E3* *var.* High Plains. *plains* Canada/USA
Great Rift Valley 73 *C5* *var.* Rift Valley. *depression* Asia/Africa
Great Ruaha 73 *C7* *river* S Tanzania
Great Saint Bernard Pass 96 *A1* *Fr.* Col du Grand-Saint-Bernard, *It.* Passo del Gran San Bernardo. *pass* Italy/Switzerland
Great Salt Desert *see* Kavīr, Dasht-e
Great Salt Lake 44 *A3* *salt lake* Utah, W USA
Great Salt Lake Desert 44 *A3* *plain* Utah, W USA
Great Sand Sea 71 *H3* *desert* Egypt/Libya
Great Sandy Desert 146 *C4* *desert* Western Australia
Great Sandy Desert *see* Ar Rub 'al Khālī
Great Sandy Island *see* K'gari
Great Slave Lake 37 *E4* *Fr.* Grand Lac des Esclaves. *lake* Northwest Territories, NW Canada
Great Socialist People's Libyan Arab Jamahiriya *see* Libya
Great Sound 42 *A5* *sound* Bermuda, NW Atlantic Ocean
Great Victoria Desert 147 *C5* *desert* South Australia/Western Australia
Great Wall of China 128 *C4* *ancient monument* N China Asia
Great Yarmouth 89 *E6* *var.* Yarmouth. E England, United Kingdom
Grebenka *see* Hrebinka
Gredos, Sierra de 92 *D3* *mountain range* W Spain
Greece 105 *A5* *off.* Hellenic Republic, *Gk.* Ellinikí Dímokratía, Elláda, Ellás; *anc.* Hellas. *country* SE Europe
Greeley 44 *D4* Colorado, C USA
Green Bay 40 *B2* Wisconsin, N USA
Green Bay 40 *B2* *lake bay* Michigan/Wisconsin, N USA

Greeneville 43 *E1* Tennessee, S USA
Greenland 82 *D3* *Dan.* Grønland, *Inuit* Kalaallit Nunaat. *Danish self-governing territory* NE North America
Greenland Sea 83 *F2* *sea* Arctic Ocean
Green Mountains 41 *G2* *mountain range* Vermont, NE USA
Green Mountain State *see* Vermont
Greenock 88 *C4* W Scotland, United Kingdom
Green River 44 *B3* Wyoming, C USA
Green River 40 *C5* *river* Kentucky, C USA
Green River 44 *B4* *river* Utah, W USA
Greensboro 43 *F1* North Carolina, SE USA
Greenville 42 *B2* Mississippi, S USA
Greenville 43 *F1* North Carolina, SE USA
Greenville 43 *E1* South Carolina, SE USA
Greenville 49 *G2* Texas, SW USA
Greenwich 89 *B8* United Kingdom
Greenwood 42 *B2* Mississippi, S USA
Greenwood 43 *E2* South Carolina, SE USA
Gregory Range 148 *C3* *mountain range* Queensland, E Australia
Greifenberg/Greifenberg in Pommern *see* Gryfice
Greifswald 94 *D2* Mecklenburg-Vorpommern, NE Germany
Grenada 42 *C2* Mississippi, S USA
Grenada 55 *G5* *country* SE West Indies
Grenadines, The 55 *H4* *island group* Grenada/St Vincent and the Grenadines
Grenoble 91 *D5* *anc.* Cularo, Gratianopolis. Isère, E France
Gresham 46 *B3* Oregon, NW USA
Grevená 104 *B4* Dytikí Makedonía, N Greece
Grevenmacher 87 *E8* Grevenmacher, E Luxembourg
Greymouth 151 *B5* West Coast, South Island, New Zealand
Grey Range 149 *C5* *mountain range* New South Wales/Queensland, E Australia
Greytown *see* San Juan del Norte
Griffin 42 *D2* Georgia, SE USA
Grimari 76 *C4* Ouaka, C Central African Republic
Grimsby 89 *E5* *prev.* Great Grimsby. E England, United Kingdom
Grobin *see* Grobiņa
Grobiņa 106 *B3* *Ger.* Grobin. Liepāja, W Latvia
Gródek Jagielloński *see* Horodok
Grodno *see* Hrodna
Grodzisk Wielkopolski 98 *B3* Wielkopolskie, C Poland
Groesbeek 86 *D4* Gelderland, SE Netherlands
Grójec 98 *D4* Mazowieckie, C Poland
Groningen 86 *E1* Groningen, NE Netherlands
Grønland *see* Greenland
Groote Eylandt 148 *B2* *island* Northern Territory, N Australia
Grootfontein 78 *B3* Otjozondjupa, N Namibia
Groot Karasberge 78 *B4* *mountain range* S Namibia
Gros Islet 55 *F1* N Saint Lucia
Grossa, Isola *see* Dugi Otok
Grossbetschkerek *see* Zrenjanin
Grosse Morava *see* Velika Morava
Grosser Sund *see* Suur Väin
Grosseto 96 *B4* Toscana, C Italy
Grossglockner 95 *C7* *mountain* W Austria
Grosskanizsa *see* Nagykanizsa
Gross-Karol *see* Carei
Grosskikinda *see* Kikinda
Grossmichel *see* Michalovce
Gross-Schlatten *see* Abrud
Grosswardein *see* Oradea
Groznyy 111 *B8* Chechenskaya Respublika, SW Russia
Grudovo *see* Sredets
Grudziądz 98 *C3* *Ger.* Graudenz. Kujawsko-pomorskie, C Poland
Grums 85 *B6* Värmland, C Sweden
Grünberg/Grünberg in Schlesien *see* Zielona Góra
Grüneberg *see* Zielona Góra
Gruzinskaya SSR/Gruziya *see* Georgia
Gryazi 111 *B6* Lipetskaya Oblast', W Russia
Gryfice 98 *B2* *Ger.* Greifenberg, Greifenberg in Pommern. Zachodnio-pomorskie, NW Poland

Guabito *53 E4* Bocas del Toro, NW Panama

Guadalajara *50 D4* Jalisco, C Mexico

Guadalajara *93 E3* Ar. Wad Al-Hajarah; *anc.* Arriaca. Castilla-La Mancha, C Spain

Guadalcanal *144 C3* island C Solomon Islands

Guadalqivir *92 D4* river W Spain

Guadalupe *50 D3* Zacatecas, C Mexico

Guadalupe Peak *48 D3* mountain Texas, SW USA

Guadalupe River *49 G4* river SW USA

Guadarrama, Sierra de *93 E2* mountain range C Spain

Guadeloupe *55 H3* French overseas department E West Indies

Guadiana *92 C4* river Portugal/Spain

Guadix *93 E4* Andalucía, S Spain

Guaimaca *52 C2* Francisco Morazán, C Honduras

Guajira, Península de la *58 B1* peninsula N Colombia

Gualaco *52 D2* Olancho, C Honduras

Gualán *52 B2* Zacapa, C Guatemala

Gualdicciolo *96 D1* NW San Marino E Argentina

Gualeguaychú *64 D4* Entre Ríos, E Argentina

Guam *144 B1* US unincorporated territory W Pacific Ocean

Guamúchil *50 C3* Sinaloa, C Mexico

Guanabacoa *54 B2* La Habana, W Cuba

Guanajuato *51 E4* Guanajuato, C Mexico

Guanare *58 C2* Portuguesa, N Venezuela

Guanare, Río *58 D2* river N Venezuela

Guangdong *129 C6* var. Guangdong Sheng, Kuang-tung, Kwangtung, Yue. province S China

Guangdong Sheng see Guangdong

Guangju see Gwangju

Guangxi see Guangxi Zhuang Zizhiqu

Guangxi Zhuangzu Zizhiqu *129 C6* var. Guangxi, Gui, Kuang-hsi, Kwangsi, Eng. Kwangsi Chuang Autonomous Region. autonomous region S China

Guangyuan *129 B5* var. Kuang-yuan, Kwangyuan. Sichuan, C China

Guangzhou *129 C6* var. Kuang-chou, Kwangchow, Eng. Canton. province capital Guangdong, SE China

Guantánamo *54 D3* Guantánamo, SE Cuba

Guantánamo, Bahía de *54 D3* Eng. Guantanamo Bay. US military base SE Cuba

Guantanamo Bay see Guantánamo, Bahía de

Guapore, Río *62 D3* var. Río Iténez. river Bolivia/Brazil

Guarda *92 C3* Guarda, N Portugal

Guarumal *53 F5* Veraguas, S Panama

Guasave *50 C3* Sinaloa, C Mexico

Guatemala *52 A2* off. Republic of Guatemala. country Central America

Guatemala Basin *35 B7* undersea basin E Pacific Ocean

Guatemala City see Ciudad de Guatemala

Guatemala, Republic of see Guatemala

Guaviare *56 B2* off. Comisaría Guaviare. province S Colombia

Guaviare, Comisaría see Guaviare

Guaviare, Río *58 D3* river E Colombia

Guayanas, Macizo de las see Guiana Highlands

Guayaquil *60 A2* var. Santiago de Guayaquil. Guayas, SW Ecuador

Guayaquil, Golfo de *60 A2* var. Gulf of Guayaquil. gulf SW Ecuador

Guayaquil, Gulf of see Guayaquil, Golfo de

Guaymas *50 B2* Sonora, NW Mexico

Gubadag *122 C2* Turkm. Tel'man; prev. Tel'mansk. Daşoguz Welaýaty, N Turkmenistan

Guben *94 D4* var. Wilhelm-Pieck-Stadt. Brandenburg, E Germany

Gudara see Ghūdara

Gudauta *117 E1* NW Georgia

Guéret *90 C4* Creuse, C France

Guernsey *89 D8* British Crown Dependency Channel Islands, NW Europe

Guerrero Negro *50 A2* Baja California Sur, NW Mexico

Gui see Guangxi Zhuangzu Zizhiqu

Guiana see French Guiana

Guiana Highlands *62 D1* var. Macizo de las Guayanas. mountain range N South America

Guiba see Juba

Guidder see Guider

Guider *76 B4* var. Guidder. Nord, N Cameroon

Guidimouni *75 G3* Zinder, S Niger

Guildford *89 D7* SE England, United Kingdom

Guilin *129 C6* var. Kuei-lin, Kweilin. Guangxi Zhuangzu Zizhiqu, S China

Guimarães *92 B2* var. Guimaráes. Braga, N Portugal

Guimaráes see Guimarães

Guinea *74 C4* off. Republic of Guinea, var. Guinée; prev. French Guinea, People's Revolutionary Republic of Guinea. country W Africa

Guinea Basin *69 A5* undersea basin E Atlantic Ocean

Guinea-Bissau *74 B4* off. Republic of Guinea-Bissau, Fr. Guinée-Bissau, Port. Guiné-Bissau; prev. Portuguese Guinea. country W Africa

Guinea-Bissau, Republic of see Guinea-Bissau

Guinea, Gulf of *68 B4* Fr. Golfe de Guinée. gulf E Atlantic Ocean

Guinea, People's Revolutionary Republic of see Guinea

Guinea, Republic of see Guinea

Guiné-Bissau see Guinea-Bissau

Guinée see Guinea

Guinée-Bissau see Guinea-Bissau

Guinée, Golfe de see Guinea, Gulf of

Güiria *59 E1* Sucre, NE Venezuela

Guiyang *129 B6* var. Kuei-Yang, Kuei-yang, Kueyang, Kweiyang; prev. Kweichu. province capital Guizhou, S China

Guizhou *129 B6* Guangdong, SE China

Gujarāt *134 C4* var. Gujerat. cultural region W India

Gujerat see Gujarāt

Gujranwala *134 D2* Punjab, NE Pakistan

Gujrat *134 D2* Punjab, E Pakistan

Gulbarga see Kalaburagi

Gulbene *106 D3* Ger. Alt-Schwanenburg. Gulbene, NE Latvia

Gulf of Liaotung see Liaodong Wan

Gulfport *42 C3* Mississippi, S USA

Gulf, The see Persian Gulf

Gulistan see Guliston

Guliston *123 E2* Rus. Gulistan. Sirdaryo Viloyati, E Uzbekistan

Gulja see Yining

Gulkana *36 D3* Alaska, USA

Gulu *73 B6* N Uganda

Gulyantsi *104 C1* Pleven, N Bulgaria

Guma see Pishan

Gumbinnen see Gusev

Gumpolds see Humpolec

Gümülcine/Gümüljina see Komotiní

Gümüşane see Gümüşhane

Gümüşhane *117 E3* var. Gümüşane, Gumushkhane. Gümüşhane, NE Turkey (Türkiye)

Gumushkhane see Gümüşhane

Güney Doğu Toroslar *117 E4* mountain range SE Turkey (Türkiye)

Gunnbjörns Fjeld *82 D4* var. Gunnbjörns Bjerge. mountain C Greenland

Gunnbjörns Bjerge see Gunnbjørn Fjeld

Gunnedah *149 D6* New South Wales, SE Australia

Gunnison *44 C5* Colorado, C USA

Gurbansoltan Eje see Andalyp

Gurbantünggüt Shamo *126 B2* desert W China

Gurgan see Gorgān

Guri, Embalse de *59 E2* reservoir E Venezuela

Gurkfeld see Krško

Gurktaler Alpen *95 D7* mountain range S Austria

Gürün *116 D3* Sivas, C Turkey (Türkiye)

Gur'yev see Atyraū

Gusau *75 G4* Zamfara, NW Nigeria

Gusev *106 B4* Ger. Gumbinnen. Kaliningradskaya Oblast', W Russia

Gustavus *36 D4* Alaska, USA

Güstrow *94 C3* Mecklenburg-Vorpommern, NE Germany

Guta/Gúta see Kolárovo

Gütersloh *94 B4* Nordrhein-Westfalen, W Germany

Gutta see Kolárovo

Guttstadt see Dobre Miasto

Guwāhāti *135 G3* prev. Gauhāti. Assam, NE India

Guyana *59 F3* off. Co-operative Republic of Guyana; prev. British Guiana. country N South America

Guyana, Co-operative Republic of see Guyana

Guyane see French Guiana

Guymon *49 F1* Oklahoma, C USA

Güzelyurt *102 C5* Gk. Mórfu, Morphou. W Cyprus

Gvardeysk *106 A4* Ger. Tapaiu. Kaliningradskaya Oblast', W Russia

Gwadar *134 A3* var. Gwadur. Baluchistan, SW Pakistan

Gwadur see Gwadar

Gwalior *134 D3* Madhya Pradesh, C India

Gwanda *78 D3* Matabeleland South, SW Zimbabwe

Gwangju *129 E4* off. Gwangju Gwang-yeoksi, prev. Kwangju, var. Guangju, Kwangchu, Jap. Kōshū. SW South Korea

Gwangju Gwang-yeoksi see Gwangju

Gwy see Wye

Gyandzha see Gäncä

Gyangzê *126 C5* Xizang Zizhiqu, W China

Gyaring Co *126 C5* lake W China

Gyêgu see Yushu

Gyergyószentmiklós see Gheorgheni

Gyixong see Gonggar

Gympie *149 E5* Queensland, E Australia

Gyomaendrőd *99 D7* Békés, SE Hungary

Gyöngyös *99 D6* Heves, NE Hungary

Győr *99 C6* Ger. Raab, Lat. Arrabona. Győr-Moson-Sopron, NW Hungary

Gýtheio *105 B6* var. Githio; prev. Ýthion. Pelopónnisos, S Greece

Gyulafehérvár see Alba Iulia

Gyumri *117 F2* var. Giumri, Rus. Kumayri; prev. Aleksandropol', Leninakan. W Armenia

Gyzyrlabat *122 C2* prev. Serdar, Rus. Kizyl-Arvat. Balkan Welaýaty, W Turkmenistan

H

Haabai see Ha'apai Group

Haacht *87 C6* Vlaams Brabant, C Belgium

Haaksbergen *86 E3* Overijssel, E Netherlands

Ha'apai Group *145 F4* var. Haabai. island group C Tonga

Haapsalu *106 D2* Ger. Hapsal. Läänemaa, W Estonia

Ha'Arava see 'Arabah, Wādī al

Haarlem *86 C3* prev. Harlem. Noord-Holland, W Netherlands

Haast *151 B6* West Coast, South Island, New Zealand

Hachijo-jima *131 D6* island Izu-shotō, SE Japan

Hachinohe *130 D3* Aomori, Honshū, C Japan

Hacqabul *117 H3* prev. Qazimämmäd, Rus. Kazi Magomed. SE Azerbaijan

Ḥaḍabat al Jilf al Kabīr *72 A2* var. Gilf Kebir Plateau. plateau SW Egypt

Hadama see Nazrēt

Hadejia *75 G4* Jigawa, N Nigeria

Hadejia *75 G3* river N Nigeria

Hadera *119 A6* var. Khadera; prev. Ḥadera. Haifa, C Israel

Ḥadera see Hadera

Hadhdhunmathi Atoll *132 A5* atoll S Maldives

Ḥadhramawt see Ḥaḍramawt

Ha Đông *136 D3* var. Hadong. Ha Tây, N Vietnam

Hadong see Ha Đông

Ḥaḍramawt *121 C6* Eng. Hadramaut. mountain range S Yemen

Hadrianopolis see Edirne

Haerbin/Haerhpin/Ha-erh-pin see Harbin

Hafnia see Denmark

Hafnia see København

Hafren see Severn

Hafun, Ras see Xaafuun, Raas

Hagåtña *144 B1* var. Agaña. dependent territory capital (Guam) NW Guam

Hagerstown *41 E4* Maryland, NE USA

Ha Giang *136 D3* Ha Giang, N Vietnam

Hagios Evstrátios see Ágios Efstrátios

HaGolan see Golan Heights

Hagondange *90 D3* Moselle, NE France

Haguenau *90 E3* Bas-Rhin, NE France

Haibowan see Wuhai

Haicheng *128 D3* Liaoning, NE China

Haida Gwaii *36 C5* prev. Queen Charlotte Islands, Fr. Îles de la Reine-Charlotte. island group British Columbia, SW Canada

Haidarabad see Hyderabad

Haifa see Hefa

Haifa, Bay of see Mifrats Hefa

Haifong see Hai Phong

Haikou *129 C7* var. Hai-k'ou, Hoihow, Fr. Hoï-Hao. province capital Hainan, S China

Hai-k'ou see Haikou

Ḥa'il *120 B4* var. Ḥayil, NW Saudi Arabia

Hailuoto *84 D4* Swe. Karlö. island W Finland

Hainan *129 B7* var. Hainan Sheng, Qiong. province S China

Hainan Dao *129 C7* island S China

Hainan Sheng see Hainan

Hainasch see Ainaži

Haines *36 D4* Alaska, USA

Hainichen *94 D4* Sachsen, E Germany

Hai Phong *136 D3* var. Haifong, Haiphong. N Vietnam

Haiphong see Hai Phong

Haisyn *108 D3* prev. Haysyn, Rus. Gaysin. Vinnyts'ka Oblast', C Ukraine

Haiti *54 D3* off. Republic of Haiti. country C West Indies

Haiti, Republic of see Haiti

Haiya *72 C3* Red Sea, NE Sudan

Hajdúhadház *99 D6* Hajdú-Bihar, E Hungary

Hajîn *118 E3* var. Abū Ḥardān, Hajîne. Dayr az Zawr, E Syria

Hajîne see Hajîn

Hajnówka *98 E3* Ger. Hermhausen. Podlaskie, NE Poland

Hakodate *130 D3* Hokkaidō, NE Japan

Hal see Halle

Ḥalab *118 B2* Eng. Aleppo, Fr. Alep; anc. Beroea. Ḥalab, NW Syria

Hala'ib Triangle *72 C3* region Egypt/Sudan

Ḥalāniyāt, Juzur al *121 D6* var. Jazā'ir Bin Ghalfan, Eng. Kuria Muria Islands. island group S Oman

Halberstadt *94 C4* Sachsen-Anhalt, C Germany

Halden *85 B6* prev. Fredrikshald. Østfold, S Norway

Halfmoon Bay *151 A8* var. Oban. Stewart Island, Southland, New Zealand

Haliacmon see Aliákmonas

Halifax *39 F4* province capital Nova Scotia, SE Canada

Halkida see Chalkída

Halle *87 B6* Fr. Hal. Vlaams Brabant, C Belgium

Halle *94 C4* var. Halle an der Saale. Sachsen-Anhalt, C Germany

Halle an der Saale see Halle

Halle-Neustadt *94 C4* Sachsen-Anhalt, C Germany

Halley *154 B2* UK research station Antarctica

Hall Islands *142 B2* island group C Federated States of Micronesia

Halls Creek *146 C3* Western Australia

Halmahera, Laut *139 F3* Eng. Halmahera Sea; sea E Indonesia

Halmahera, Pulau *139 F3* prev. Djailolo, Gilolo, Jailolo. island E Indonesia

Halmahera Sea see Halmahera, Laut

Halmstad *85 B7* Halland, S Sweden

Ha Long *136 E3* prev. Hông Gai; var. Hon Gai, Hongay. Quang Ninh, N Vietnam

Hälsingborg see Helsingborg

Hamada *131 B6* Shimane, Honshū, SW Japan

Hamadān *120 C3* anc. Ecbatana. Hamadān, W Iran

Ḥamāh *118 B3* var. Hama; anc. Epiphania, Bibl. Hamath. Ḥamāh, W Syria

Hamamatsu *131 D6* var. Hamamatu. Shizuoka, Honshū, S Japan

Hamamatu see Hamamatsu

Hamar *85 B5* prev. Storhammer. Hedmark, S Norway

Hamath see Ḥamāh

Hamburg *94 B3* Hamburg, N Germany

Hamd, Wādī al *120 A4* dry watercourse W Saudi Arabia

Hämeenlinna *85 D5* Swe. Tavastehus. Kanta-Häme, S Finland

HaMelah, Yam see Dead Sea

Hamersley Range *146 A4 mountain range* Western Australia
Hamhŭng *129 E3* C North Korea
Hami *126 C3 var.* Ha-mi, *Uygh.* Kumul, Qomul. Xinjiang Uygur Zizhiqu, NW China
Ha-mi *see* Hami
Hamilton *42 A5 dependent territory capital* (Bermuda) C Bermuda
Hamilton *38 D5* Ontario, S Canada
Hamilton *150 D3* Waikato, North Island, New Zealand
Hamilton *88 C4* S Scotland, United Kingdom
Hamilton *42 C2* Alabama, S USA
Hamim, Wadi al *71 G2 river* NE Libya
Hamïs Musait *see* Khamïs Mushayt
Hamm *94 B4 var.* Hamm in Westfalen. Nordrhein-Westfalen, W Germany
Ḥammāmāt, Khalīj al *see* Hammamet, Golfe de
Hammamet, Golfe de *102 D3 Ar.* Khalīj al Ḥammāmāt. *gulf* NE Tunisia
Ḥammēr, Hawr al *120 C3 lake* SE Iraq
Hamm in Westfalen *see* Hamm
Hampden *151 B7* Otago, South Island, New Zealand
Hampstead *89 A7* Maryland, USA
Ḥamrun *102 B5* C Malta
Hāmūn, Daryācheh-ye *see* Ṣāberī, Hāmūn-e/Sīstān, Daryācheh-ye
Hamwih *see* Southampton
Hâncești *see* Hîncești
Hancewicze *see* Hantsavichy
Handan *128 C4 var.* Han-tan. Hebei, E China
Haneda *130 A2* (Tōkyō) Tōkyō, Honshū, S Japan
HaNegev *119 A7 Eng.* Negev. *desert* S Israel
Hanford *47 C6* California, W USA
Hangayn Nuruu *126 D2 mountain range* C Mongolia
Hang-chou/Hangchow *see* Hangzhou
Hangō *see* Hanko
Hangzhou *129 D5 var.* Hang-chou, Hangchow. *province capital* Zhejiang, SE China
Hania *see* Chaniá
Hanka, Lake *see* Khanka, Lake
Hanko *85 D6 Swe.* Hangö. Uusimaa, SW Finland
Han-kou/Han-k'ou/Hankow *see* Wuhan
Hanmer Springs *151 C5* Canterbury, South Island, New Zealand
Hannibal *45 G4* Missouri, C USA
Hannover *94 B3 Eng.* Hanover. Niedersachsen, NW Germany
Hanöbukten *85 B7 bay* S Sweden
Ha Nội *136 D3 Eng.* Hanoi, *Fr.* Hanoï. *country capital* (Vietnam) N Vietnam
Hanover *see* Hannover
Han Shui *127 E4 river* C China
Han-tan *see* Handan
Hantsavichy *107 B6 Pol.* Hancewicze, *Rus.* Gantsevichi. Brestskaya Voblasts', SW Belarus
Hanyang *see* Wuhan
Hanzhong *129 B5* Shaanxi, C China
Hāora *135 F4 prev.* Howrah. West Bengal, NE India
Haparanda *84 D4* Norrbotten, N Sweden
Hapsal *see* Haapsalu
Haradok *107 E5 Rus.* Gorodok. Vitsyebskaya Voblasts', N Belarus
Haradzyets *107 B6 Rus.* Gorodets. Brestskaya Voblasts', SW Belarus
Haramachi *130 D4* Fukushima, Honshū, E Japan
Harany *107 D5 Rus.* Gorany. Vitsyebskaya Voblasts', N Belarus
Harare *78 D3 prev.* Salisbury. *country capital* (Zimbabwe) Mashonaland East, NE Zimbabwe
Harbavichy *107 E6 Rus.* Gorbovichi. Mahilyowskaya Voblasts', E Belarus
Harbin *129 E2 var.* Haerbin, Ha-erh-pin, Kharbin; *prev.* Haerhpin, Pingkiang, Pinkiang. *province capital* Heilongjiang, NE China
Hardangerfjorden *85 A6 fjord* S Norway
Hardangervidda *85 A6 plateau* S Norway
Hardenberg *86 E3* Overijssel, E Netherlands
Harelbeke *87 A6 var.* Harlebeke. West-Vlaanderen, W Belgium
Harem *see* Ḥārim
Haren *86 E2* Groningen, NE Netherlands

Härer *73 D5* E Ethiopia
Hargeisa *see* Hargeysa
Hargeysa *73 D5 var.* Hargeisa. Woqooyi Galbeed, NW Somalia
Hariana *see* Haryāna
Hari, Batang *138 B4 prev.* Djambi. *river* Sumatera, W Indonesia
Ḥārim *118 B2 var.* Harem. Idlib, W Syria
Harima-nada *131 B6 sea* S Japan
Harī Rōd *123 E4 var.* Harīrūd, Tedzhen, *Turkm.* Tejen. *river* Afghanistan/Iran
Harīrūd *see* Harī Rōd
Harlan *45 F3* Iowa, C USA
Harlebeke *see* Harelbeke
Harlem *see* Haarlem
Harlingen *86 D2 Fris.* Harns. Fryslân, N Netherlands
Harlingen *49 G5* Texas, SW USA
Harlow *89 E6* E England, United Kingdom
Harmanli *104 D3 prev.* Kharmanli. Haskovo, S Bulgaria
Harney Basin *46 B4 basin* Oregon, NW USA
Härnösand *85 C5 var.* Hernösand. Västernorrland, C Sweden
Harns *see* Harlingen
Harper *74 D5 var.* Cape Palmas. NE Liberia
Harricana *38 D3 river* Québec, SE Canada
Harris *88 B3 physical region* NW Scotland, United Kingdom
Harrisburg *41 E4 state capital* Pennsylvania, NE USA
Harrisonburg *41 E4* Virginia, NE USA
Harrison, Cape *39 F2 headland* Newfoundland and Labrador, E Canada
Harris Ridge *see* Lomonosov Ridge
Harrogate *89 D5* N England, United Kingdom
Hârşova *108 D5 prev.* Hîrşova. Constanţa, SE Romania
Harstad *84 C2* Troms, N Norway
Hartford *41 G3 state capital* Connecticut, NE USA
Hartlepool *89 D5* N England, United Kingdom
Harunabad *see* Eslāmābād
Har Us Gol *126 C2 lake* Hovd, W Mongolia
Har Us Nuur *126 C2 lake* NW Mongolia
Harwich *89 E6* E England, United Kingdom
Haryāna *134 D2 var.* Hariana. *cultural region* N India
Hashemite Kingdom of Jordan *see* Jordan
Haskovo *104 D3 var.* Khaskovo. Haskovo, S Bulgaria
Hasselt *87 C6* Limburg, NE Belgium
Hassetché *see* Al Ḥasakah
Hasta Colonia/Hasta Pompeia *see* Asti
Hastings *150 E4* Hawke's Bay, North Island, New Zealand
Hastings *89 E7* SE England, United Kingdom
Hastings *45 E4* Nebraska, C USA
Haţeg *108 B4 Ger.* Wallenthal, *Hung.* Hátszeg; *prev.* Hatzeg, Hötzing. Hunedoara, SW Romania
Hátszeg *see* Haţeg
Hattem *86 D3* Gelderland, E Netherlands
Hatteras, Cape *43 G1 headland* North Carolina, SE USA
Hatteras Plain *35 D6 abyssal plain* Atlantic Ocean
Hattiesburg *42 C3* Mississippi, S USA
Hatton Bank *see* Hatton Ridge
Hatton Ridge *80 B2 var.* Hatton Bank. *undersea ridge* N Atlantic Ocean
Hat Yai *137 C7 var.* Ban Hat Yai. Songkhla, SW Thailand
Hatzeg *see* Haţeg
Hatzfeld *see* Jimbolia
Haugesund *85 A6* Rogaland, S Norway
Haukeligrend *85 A6* Telemark, S Norway
Haukivesi *85 E5 lake* SE Finland
Hauraki Gulf *150 D2 var.* Tīkapa Moana. *gulf* North Island, N New Zealand
Hauroko, Lake *151 A7 lake* South Island, New Zealand
Haut Atlas *70 C2 Eng.* High Atlas. *mountain range* C Morocco
Hautes Fagnes *87 D6 Ger.* Hohes Venn. *mountain range* E Belgium
Hauts Plateaux *70 D2 plateau* Algeria/Morocco

Hauzenberg *95 D6* Bayern, SE Germany
Havana *35 D6* Illinois, N USA
Havana *see* La Habana
Havant *89 D7* S England, United Kingdom
Havelock *43 F1* North Carolina, SE USA
Havelock North *150 E4* Hawke's Bay, North Island, New Zealand
Haverfordwest *89 C6* SW Wales, United Kingdom
Havíŕov *99 C5* Moravskoslezský Kraj, E Czechia (Czech Republic)
Havre *44 C1* Montana, NW USA
Havre *see* le Havre
Havre-St-Pierre *39 F3* Québec, E Canada
Hawaii *47 A8 off.* State of Hawaii, *also known as* Aloha State, Paradise of the Pacific, *var.* Hawai'i. *state* USA, C Pacific Ocean
Hawai'i *47 B8 var.* Hawaii. *island* Hawaiian Islands, USA, C Pacific Ocean
Hawaiian Islands *152 D2 prev.* Sandwich Islands. *island group* Hawaii, USA
Hawaiian Ridge *152 H4 undersea ridge* N Pacific Ocean
Hawea, Lake *151 B6 lake* South Island, New Zealand
Hawera *150 D4* Taranaki, North Island, New Zealand
Hawick *88 C4* SE Scotland, United Kingdom
Hawke Bay *150 E4 bay* North Island, New Zealand
Hawkeye State *see* Iowa
Hawlêr *see* Arbīl
Hawthorne *47 C6* Nevada, W USA
Hay *149 C6* New South Wales, SE Australia
HaYarden *see* Jordan
Hayastan *see* Armenia
Hayastani Hanrapetut'yun *see* Armenia
Hayes *38 B2 river* Manitoba, C Canada
Hay River *37 E4* Northwest Territories, W Canada
Hays *45 E5* Kansas, C USA
Haysyn *see* Haisyn
Hazar *122 B2 prev. Rus.* Cheleken. Balkan Welaýaty, W Turkmenistan
Heard Island and McDonald Islands *141 B7 Australian external territory* S Indian Ocean
Hearst *38 C4* Ontario, S Canada
Heart of Dixie *see* Alabama
Heathrow *89 A8* (London) SE England, United Kingdom
Hebei *128 C4 var.* Hebei Sheng, Hopeh, Hopei, Ji; *prev.* Chihli. *province* E China
Hebei Sheng *see* Hebei
Hebron *119 A6 var.* Al Khalīl, El Khalil, *Heb.* Hevron; *anc.* Kiriath-Arba. S West Bank, SW Asia
Heemskerk *86 C3* Noord-Holland, W Netherlands
Heerde *86 D3* Gelderland, E Netherlands
Heerenveen *86 D2 Fris.* It Hearrenfean. Fryslân, N Netherlands
Heerhugowaard *86 C2* Noord-Holland, NW Netherlands
Heerlen *87 D6* Limburg, SE Netherlands
Heerwegen *see* Polkowice
Hefa *119 A5 var.* Haifa, *hist.* Caiffa, Caiphas; *anc.* Sycaminum. Haifa, N Israel
Hefa, Mifraz *see* Mifrats Hefa
Hefei *129 D5 var.* Hofei, *hist.* Luchow. *province capital* Anhui, E China
Hegang *129 E2* Heilongjiang, NE China
Hei *see* Heilongjiang
Heide *94 B2* Schleswig-Holstein, N Germany
Heidelberg *95 B5* Baden-Württemberg, SW Germany
Heidenheim *see* Heidenheim an der Brenz
Heidenheim an der Brenz *95 B6 var.* Heidenheim. Baden-Württemberg, S Germany
Hei-ho *see* Nagqu
Heilbronn *95 B6* Baden-Württemberg, SW Germany
Heiligenbeil *see* Mamonovo
Heilongjiang *128 D2 var.* Hei, Heilongjiang Sheng, Hei-lung-chiang, Heilungkiang. *province* NE China
Heilong Jiang *see* Amur
Heilongjiang Sheng *see* Heilongjiang

Heiloo *86 C3* Noord-Holland, NW Netherlands
Heilsberg *see* Lidzbark Warmiński
Hei-lung-chiang/Heilungkiang *see* Heilongjiang
Heimdal *85 B5* Sør-Trøndelag, S Norway
Heinaste *see* Ainaži
Hekimhan *116 D3* Malatya, C Turkey (Türkiye)
Helena *44 B2 state capital* Montana, NW USA
Helensville *150 D2* Auckland, North Island, New Zealand
Helgoland Bay *see* Helgoländer Bucht
Helgoländer Bucht *94 A2 var.* Helgoland Bay, Heligoland Bight. *bay* NW Germany
Heligoland Bight *see* Helgoländer Bucht
Heliopolis *see* Baalbek
Hellas *see* Greece
Hellenic Republic *see* Greece
Hellevoetsluis *86 B4* Zuid-Holland, SW Netherlands
Hellín *93 E4* Castilla-La Mancha, C Spain
Helmand, Darya-ye *see* Helmand Rōd
Helmand Rōd *122 D5 var.* Daryā-ye Helmand, Rūd-e Hirmand, Helmand Rūd. *river* Afghanistan/Iran
Helmand Rūd *see* Helmand Rōd
Helmantica *see* Salamanca
Helmond *87 D5* Noord-Brabant, S Netherlands
Helsingborg *85 B7 prev.* Hälsingborg. Skåne, S Sweden
Helsingfors *see* Helsinki
Helsinki *85 D6 Swe.* Helsingfors. *country capital* (Finland) Uusimaa, S Finland
Heltau *see* Cisnădie
Helvetia *see* Switzerland
Henan *129 C5 var.* Henan Sheng, Honan, Yu. *province* C China
Henderson *40 B5* Kentucky, S USA
Henderson *47 D7* Nevada, W USA
Henderson *49 H3* Texas, SW USA
Hendū Kosh *see* Hindu Kush
Hengchow *see* Hengyang
Hengduan Shan *123 A5 mountain range* SW China
Hengelo *86 E3* Overijssel, E Netherlands
Hengnan *see* Hengyang
Hengyang *129 C6 var.* Hengnan, Heng-yang; *prev.* Hengchow. Hunan, S China
Heng-yang *see* Hengyang
Henichesk *109 F4 Rus.* Genichesk. Khersons'ka Oblast', S Ukraine
Hennebont *90 A3* Morbihan, NW France
Henrique de Carvalho *see* Saurimo
Henzada *see* Hinthada
Herakleion *see* Irákleio
Herāt *122 D4 var.* Herat; *anc.* Aria. Herāt, W Afghanistan
Heredia *53 E4* Heredia, C Costa Rica
Hereford *49 E2* Texas, SW USA
Herford *94 B4* Nordrhein-Westfalen, NW Germany
Héristal *see* Herstal
Herk-de-Stad *87 C6* Limburg, NE Belgium
Herlen Gol/Herlen He *see* Kerulen
Hermannstadt *see* Sibiu
Hermansverk *85 A5* Sogn Og Fjordane, S Norway
Hermhausen *see* Hajnówka
Hermiston *46 C2* Oregon, NW USA
Hermon, Mount *119 B5 Ar.* Jabal ash Shaykh. *mountain* S Syria
Hermosillo *50 B2* Sonora, NW Mexico
Hermoupolis *see* Ermoúpoli
Hernösand *see* Härnösand
Herrera del Duque *92 D3* Extremadura, W Spain
Herselt *87 C5* Antwerpen, C Belgium
Herstal *87 D6 Fr.* Héristal. Liège, E Belgium
Herzogenbusch *see* 's-Hertogenbosch
Hesse *see* Hessen
Hessen *95 B5 Eng./Fr.* Hesse. *state* C Germany
Hevron *see* Hebron
Hewlêr *see* Arbīl
Heydebrech *see* Kędzierzyn-Kozle
Heydekrug *see* Šilutė
Heywood Islands *146 C3 island group* Western Australia
Hibbing *45 F1* Minnesota, N USA
Hibernia *see* Ireland
Hidalgo del Parral *50 C2 var.* Parral. Chihuahua, N Mexico

Hida-sanmyaku *131 C5 mountain range* Honshū, S Japan

Hierosolyma *see* Jerusalem

Hierro *70 A3 var.* Ferro. *island* Islas Canarias, Spain, NE Atlantic Ocean

High Atlas *see* Haut Atlas

High Plains *see* Great Plains

High Point *43 E1* North Carolina, SE USA

Hiiumaa *106 C2 Ger.* Dagden, *Swe.* Dagö. *island* W Estonia

Hikurangi *150 D2* Northland, North Island, New Zealand

Hildesheim *94 B4* Niedersachsen, N Germany

Hilla *see* Al Ḥillah

Hillaby, Mount *55 G1 mountain* N Barbados

Hill Bank *52 C1* Orange Walk, N Belize

Hillegom *86 C3* Zuid-Holland, W Netherlands

Hilo *47 B8* Hawaii, USA, C Pacific Ocean

Hilton Head Island *43 E2* South Carolina, SE USA

Hilversum *86 C3* Noord-Holland, C Netherlands

Himalaya/Himalaya Shan *see* Himalayas

Himalayas *135 E2 var.* Himalaya, *Chin.* Himalaya Shan. *mountain range* S Asia

Himeji *131 C6 var.* Himezi. Hyōgo, Honshū, SW Japan

Himezi *see* Himeji

Hims *118 B4 var.* Homs; *anc.* Emesa. C Syria

Hînceşti *108 D4 var.* Hânceşti; *prev.* Kotovsk. C Moldova

Hinchinbrook Island *148 D3 island* Queensland, NE Australia

Hinds *151 C6* Canterbury, South Island, New Zealand

Hindu Kush *123 F4 Per.* Hendū Kosh. *mountain range* Afghanistan/Pakistan

Hinesville *43 E3* Georgia, SE USA

Hinnøya *84 C3 Smi.* Iinnasuolu. *island* C Norway

Hinson Bay *42 A5 bay* W Bermuda, W Atlantic Ocean

Hinthada *136 B4 var.* Henzada. Ayeyarwady, SW Myanmar (Burma)

Híos *see* Chíos

Hîrlanli Baraji *116 C3 reservoir* C Turkey (Türkiye)

Hîrmand, Rūd-e *see* Helmand Rōd

Hirosaki *130 D3* Aomori, Honshū, C Japan

Hiroshima *131 B6 var.* Hirosima. Hiroshima, Honshū, SW Japan

Hirschberg/Hirschberg im Riesengebirge/Hirschberg in Schlesien *see* Jelenia Góra

Hirson *90 D3* Aisne, N France

Hîrşova *see* Hârşova

Hisar Range *see* Gissar Range

Hisor Tizmasi *see* Gissar Range

Hispalis *see* Sevilla

Hispana/Hispania *see* Spain

Hispaniola *56 B1 island* Dominion Republic/Haiti

Hitachi *131 D5 var.* Hitati. Ibaraki, Honshū, S Japan

Hitati *see* Hitachi

Hitra *84 A4 prev.* Hitteren. *island* S Norway

Hitteren *see* Hitra

Hjälmaren *85 C6 Eng.* Lake Hjalmar. *lake* C Sweden

Hjalmar, Lake *see* Hjälmaren

Hjørring *85 B7* Nordjylland, N Denmark

Hkakabo Razi *136 B1 mountain* Myanmar (Burma)/China

Hlobyne *109 F2 Rus.* Globino. Poltavs'ka Oblast', NE Ukraine

Hlukhiv *109 F1 Rus.* Glukhov. Sums'ka Oblast', NE Ukraine

Hlybokaye *107 D5 Rus.* Glubokoye. Vitsyebskaya Voblasts', N Belarus

Hoa Binh *136 D3* Hoa Binh, N Vietnam

Hoang Lien Son *136 D3 mountain range* N Vietnam

Hobart *149 C8 prev.* Hobarton, Hobart Town. *state capital* Tasmania, SE Australia

Hobarton/Hobart Town *see* Hobart

Hobbs *49 E3* New Mexico, SW USA

Hobro *85 A7* Nordjylland, N Denmark

Hô Chí Minh *137 E6 var.* Ho Chi Minh City; *prev.* Saigon. S Vietnam

Ho Chi Minh City *see* Hô Chí Minh

Hodeidah *see* Al Ḥudayda

Hódmezővásárhely *99 D7* Csongrád, SE Hungary

Hodna, Chott El *102 C4 var.* Chott el-Hodna, *Ar.* Shatt al-Hodna. *salt lake* N Algeria

Hodna, Chott el-/Hodna, Shatt al- *see* Hodna, Chott El

Hodonín *99 C5 Ger.* Göding. Jihomoravský Kraj, SE Czechia (Czech Republic)

Hoei/Hoey *see* Huy

Hof *95 C5* Bayern, SE Germany

Hofei *see* Hefei

Hōfu *131 B7* Yamaguchi, Honshū, SW Japan

Hofuf *see* Al Hufūf

Hogoley Islands *see* Chuuk Islands

Hohensalza *see* Inowrocław

Hohenstadt *see* Zábřeh

Hohes Venn *see* Hautes Fagnes

Hohe Tauern *95 C7 mountain range* W Austria

Hohhot *127 F3 var.* Huhehot, Huhuohaote, *Mong.* Kukukhoto; *prev.* Kweisui, Kwesui. Nei Mongol Zizhiqu, N China

Hôi An *137 E5 prev.* Faifo. Quang Nam-Ða Nāng, C Vietnam

Hoi-Hao/Hoihow *see* Haikou

Hokianga Harbour *150 C2 inlet* SE Tasman Sea

Hokitika *151 B5* West Coast, South Island, New Zealand

Hokkaidō *130 C2 prev.* Ezo, Yeso, Yezo. *island* NE Japan

Hola Prystan *109 E4 Rus.* Golaya Pristan. Khersons'ka Oblast', S Ukraine

Holbrook *48 B2* Arizona, SW USA

Holetown *55 G1 prev.* Jamestown. W Barbados

Holguín *54 C2* Holguín, SE Cuba

Hollabrunn *95 E6* Niederösterreich, NE Austria

Holland *see* Netherlands

Hollandia *see* Jayapura

Holly Springs *42 C1* Mississippi, S USA

Holman *see* Ulukhaktok

Holmsund *84 D4* Västerbotten, N Sweden

Holon *119 A6 var.* Kholon; *prev.* Ḥolon. Tel Aviv, C Israel

Ḥolon *see* Holon

Holovanivsk *109 E3 Rus.* Golovanevsk. Kirovohrads'ka Oblast', C Ukraine

Holstebro *85 A7* Ringkøbing, W Denmark

Holsteinborg/Holsteinsborg/ Holstenborg/Holstensborg *see* Sisimiut

Holyhead *89 C5 Wel.* Caer Gybi. NW Wales, United Kingdom

Hombori *75 E3* Mopti, S Mali

Homs *see* Al Khums, Libya

Homs *see* Ḥimş

Homyel' *107 D7 Rus.* Gomel'. Homyel'skaya Voblasts', SE Belarus

Honan *see* Luoyang, China

Honan *see* Henan, China

Hondo *49 F4* Texas, SW USA

Hondo *see* Honshū

Honduras *52 C2 off.* Republic of Honduras. *country* Central America

Honduras, Golfo de *see* Honduras, Gulf of

Honduras, Gulf of *52 C2 Sp.* Golfo de Honduras. *gulf* W Caribbean Sea

Honduras, Republic of *see* Honduras

Hønefoss *85 B6* Buskerud, S Norway

Honey Lake *47 B5 lake* California, W USA

Hon Gai *see* Ha Long

Hongay *see* Ha Long

Hông Gai *see* Ha Long

Hông Hà, Sông *see* Red River

Hong Kong *128 A1 Special administrative region* Hong Kong, S China

Hong Kong Island *128 B2 island* S China-Asia

Honiara *144 C3 country capital* (Solomon Islands) Guadalcanal, C Solomon Islands

Honjō *130 D4 var.* Honzyô, Yurihonjô. Akita, Honshū, C Japan

Honolulu *47 A8 state capital* O'ahu, Hawaii, USA, C Pacific Ocean

Honshu *131 E5 var.* Hondo, Honsyû. *island* SW Japan

Honsyû *see* Honshū

Honzyô *see* Honjō

Hoogeveen *86 E2* Drenthe, NE Netherlands

Hoogezand-Sappemeer *86 E2* Groningen, NE Netherlands

Hoorn *86 C2* Noord-Holland, NW Netherlands

Hoosier State *see* Indiana

Hopa *117 F2* Artvin, NE Turkey (Türkiye)

Hope *36 C3* British Columbia, SW Canada

Hopedale *39 F2* Newfoundland and Labrador, NE Canada

Hopeh/Hopei *see* Hebei

Hopkinsville *40 B5* Kentucky, S USA

Horasan *117 F3* Erzurum, NE Turkey (Türkiye)

Horizon Deep *152 D4 trench* W Pacific Ocean

Horki *107 E6 Rus.* Gorki. Mahilyowskaya Voblasts', E Belarus

Horlivka *109 G3 Rom.* Adâncata, *Rus.* Gorlovka. Donets'ka Oblast', E Ukraine

Hormoz, Tangeh-ye *see* Hormuz, Strait of

Hormuz, Strait of *120 D4 var.* Strait of Ormuz, *Per.* Tangeh-ye Hormoz. *strait* Iran/Oman

Horn, Cape *see* Hornos, Cabo de

Hornos, Cabo de *65 C8 Eng.* Cape Horn. *headland* S Chile

Hornsby *148 E1* New South Wales, SE Australia

Horodnia *109 E1 prev.* Horodnya, *Rus.* Gorodnya. Chernihivs'ka Oblast', NE Ukraine

Horodnya *see* Horodnia

Horodok *108 B2 Pol.* Gródek Jagielloński, *Rus.* Gorodok, Gorodok Yagellonski. L'vivs'ka Oblast', NW Ukraine

Horodyshche *109 E2 Rus.* Gorodishche. Cherkas'ka Oblast', C Ukraine

Horoshiri-dake *see* Horoshiri Dake

Horoshiri Dake *130 D2 var.* Horosiri Dake. *mountain* Hokkaidō, N Japan

Horosiri Dake *see* Horoshiri-dake

Horsburgh Atoll *132 A4 var.* Goidhoo Atoll. *atoll* N Maldives

Horseshoe Bay *42 A5 bay* W Bermuda W Atlantic Ocean

Horseshoe Seamounts *80 A4 seamount range* E Atlantic Ocean

Horsham *149 B7* Victoria, SE Australia

Horst *87 D5* Limburg, SE Netherlands

Horten *85 B6* Vestfold, S Norway

Horyn *108 D1 Rus.* Goryn. *river* NW Ukraine

Hosingen *87 D7* Diekirch, NE Luxembourg

Hospitalet *see* L'Hospitalet de Llobregat

Hotan *126 B4 var.* Khotan, *Chin.* Ho-t'ien. Xinjiang Uygur Zizhiqu, NW China

Ho-t'ien *see* Hotan

Hoting *84 C4* Jämtland, C Sweden

Hot Springs *42 B1* Arkansas, C USA

Hötzing *see* Hațeg

Houayxay *136 C3 var.* Ban Houayxay. Bokèo, N Laos

Houghton *40 B1* Michigan, N USA

Houilles *91 B5* Yvelines, Île-de-France, N France Europe

Houlton *41 H1* Maine, NE USA

Houma *42 B3* Louisiana, S USA

Houston *49 H4* Texas, SW USA

Hovd *126 D2 var.* Khovd, Kobdo; *prev.* Jirgalanta. Hovd, W Mongolia

Hove *89 E7* SE England, United Kingdom

Hoverla, Hora *108 C3 Rus.* Gora Goverla. *mountain* W Ukraine

Hovsgol, Lake *see* Hövsgöl Nuur

Hövsgöl Nuur *126 D1 var.* Lake Hovsgol. *lake* N Mongolia

Howa, Ouadi *see* Howar, Wādi

Howar, Wādi *72 A3 var.* Ouadi Howa. *river* Chad/Sudan

Howland Island *145 E2 US unincorporated territory* W Polynesia

Howrah *see* Haora

Hoy *88 C2 island* N Scotland, United Kingdom

Hoyerswerda *94 D4 Lus.* Wojerecy. Sachsen, E Germany

Hpa-An *136 B4 var.* Pa-an. Kayin State, S Myanmar (Burma)

Hpyu *see* Phyu

Hradec Králové *99 B5 Ger.* Königgrätz. Královéhradecký Kraj, N Czechia (Czech Republic)

Hrandzichy *107 B5 Rus.* Grandichi. Hrodzyenskaya Voblasts', W Belarus

Hranice *99 C5 Ger.* Mährisch-Weisskirchen. Olomoucký Kraj, E Czechia (Czech Republic)

Hrebinka *109 E2 Rus.* Grebenka. Poltavs'ka Oblast', NE Ukraine

Hrodna *107 B5 Pol.* Grodno. Hrodzyenskaya Voblasts', W Belarus

Hrvatska/ Republika Hrvatska *see* Croatia

Hsia-men *see* Xiamen

Hsiang-t'an *see* Xiangtan

Hsi Chiang *see* Xi Jiang

Hsing-K'ai Hu *see* Khanka, Lake

Hsi-ning/Hsining *see* Xining

Hsinking *see* Changchun

Hsin-yang *see* Xinyang

Hsu-chou *see* Xuzhou

Htawei *see* Dawei

Huacho *60 C4* Lima, W Peru

Hua Hin *see* Ban Hua Hin

Huaihua *129 C5* Hunan, S China

Huailai *128 C3 var.* Shacheng. Hebei, E China

Huainan *129 D5 var.* Huai-nan, Hwainan. Anhui, E China

Huai-nan *see* Huainan

Huajuapan *51 F5 var.* Huajuapan de León. Oaxaca, SE Mexico

Huajuapan de León *see* Huajuapan

Hualapai Peak *48 A2 mountain* Arizona, SW USA

Huallaga, Río *60 C3 river* N Peru

Huambo *78 B2 Port.* Nova Lisboa. Huambo, C Angola

Huancavelica *60 D4* Huancavelica, SW Peru

Huancayo *60 D3* Junín, C Peru

Huang Hai *see* Yellow Sea

Huang He *128 C4 var.* Yellow River. *river* C China

Huangshi *129 C5 var.* Huang-shih, Hwangshih. Hubei, C China

Huang-shih *see* Huangshi

Huanta *60 D4* Ayacucho, C Peru

Huánuco *60 C3* Huánuco, C Peru

Huanuni *61 F4* Oruro, W Bolivia

Huaral *60 C4* Lima, W Peru

Huarás *see* Huaraz

Huaraz *60 C3 var.* Huarás. Ancash, W Peru

Huarmey *60 C3* Ancash, W Peru

Huatabampo *50 C2* Sonora, NW Mexico

Hubballi *124 B2 prev.* Hubli. Karnātaka, SW India

Hubli *see* Hubballi

Huddersfield *89 D5* N England, United Kingdom

Hudiksvall *85 C5* Gävleborg, C Sweden

Hudson, Détroit d' *see* Hudson Strait

Hudson Bay *37 G4 bay* NE Canada

Hudson Strait *37 H3 Fr.* Détroit d'Hudson. *strait* Northwest Territories/Québec, NE Canada

Hudur *see* Xuddur

Huê *136 D4 var.* Tha Thiên-Huê, C Vietnam

Huehuetenango *52 A2* Huehuetenango, W Guatemala

Huelva *92 C4 anc.* Onuba. Andalucía, SW Spain

Huesca *93 F2 anc.* Osca. Aragón, NE Spain

Huéscar *93 E4* Andalucía, S Spain

Hughenden *148 C3* Queensland, NE Australia

Hugo *49 G2* Oklahoma, C USA

Huhehot/Huhuohaote *see* Hohhot

Huíla Plateau *78 B2 plateau* S Angola

Huixtla *51 G5* Chiapas, SE Mexico

Hulingol *127 G2 prev.* Huolin Gol. Nei Mongol Zizhiqu, N China

Hull *38 D4* Québec, SE Canada

Hull *see* Kingston upon Hull

Hull Island *see* Orona

Hulst *87 B5* Zeeland, SW Netherlands

Hulun *see* Hulun Buir

Hulun Buir *127 F1 var.* Hailar; *prev.* Hulun. Nei Mongol Zizhiqu, N China

Hu-lun Ch'ih *see* Hulun Nur

Hulun Nur *127 F1 var.* Hu-lun Ch'ih; *prev.* Dalai Nor. Lake N China

Humaitá *62 D2* Amazonas, N Brazil

Humboldt River *47 C5 river* W USA

Humphreys Peak *48 B1 mountain* Arizona, SW USA

Humpolec *99 B5 Ger.* Gumpolds, Humpoletz. Vysočina, C Czechia (Czech Republic)

Humpoletz *see* Humpolec

Hunan *129 C6 var.* Hunan Sheng,
Xiang. *province* S China
Hunan Sheng *see* Hunan
Hunedoara *108 B4 Ger.* Eisenmarkt,
Hung. Vajdahunyad. Hunedoara,
SW Romania
Hünfeld *95 B5* Hessen, C Germany
Hungarian People's Republic *see*
Hungary
Hungary *99 C6 off.* Hungary, *Ger.*
Ungarn, *Hung.* Magyarország, *Rom.*
Ungaria, *Croatian* Mađarska, *Ukr.*
Uhorshchyna; *prev.* Hungarian People's
Republic. *country* C Europe
Hungary, Plain of *see* Great
Hungarian Plain
Hunjiang *see* Baishan
Hunter Island *149 B8 island* Tasmania,
SE Australia
Huntington *40 D4* West Virginia,
NE USA
Huntington Beach *47 B8* California,
W USA
Huntly *150 D3* Waikato, North Island,
New Zealand
Huntsville *42 D1* Alabama, S USA
Huntsville *49 G3* Texas, SW USA
Huolin Gol *see* Hulingol
Hurghada *see* Al Ghardaqah
Huron *45 E3* South Dakota, N USA
Huron, Lake *40 D2 lake* Canada/USA
Hurukawa *see* Furukawa
Hurunui *151 C5 river* South Island,
New Zealand
Húsavík *83 E4* Norðurland Eystra,
NE Iceland
Husté *see* Khust
Husum *94 B2* Schleswig-Holstein,
N Germany
Huszt *see* Khust
Hutchinson *45 E5* Kansas, C USA
Hutchinson Island *43 F4 island* Florida,
SE USA
Huy *87 C6 Dut.* Hoei, Hoey. Liège,
E Belgium
Huzi *see* Fuji
Hvannadalshnúkur *83 E5 volcano*
S Iceland
Hvar *100 B4 It.* Lesina; *anc.* Pharus.
island S Croatia
Hwainan *see* Huainan
Hwange *78 D3 prev.* Wankie.
Matabeleland North, W Zimbabwe
Hwang-Hae *see* Yellow Sea
Hwangshih *see* Huangshi
Hyargas Nuur *126 C2 lake*
NW Mongolia
Hyderābād *134 D5 var.* Haidarabad.
state capital Telangana, C India
Hyderabad *134 B4 var.* Haidarabad.
Sindh, SE Pakistan
Hyères *91 D6* Var, SE France
Hyères, Îles d' *91 D6 island group*
S France
Hypanis *see* Kuban'
Hyrcania *see* Gorgān
Hyvinge *see* Hyvinkää
Hyvinkää *85 D5 Swe.* Hyvinge. Uusimaa,
S Finland

I

Iader *see* Zadar
Ialomiţa *108 C5 river* SE Romania
Iaşi *108 D3 Ger.* Jassy. Iaşi, NE Romania
Ibadan *75 F5* Oyo, SW Nigeria
Ibagué *58 B3* Tolima, C Colombia
Ibar *100 D4 Alb.* Ibër. *river* C Serbia
Ibarra *60 B1 var.* San Miguel de Ibarra.
Imbabura, N Ecuador
Ibër *see* Ibar
Iberia *see* Spain
Iberian Mountains *see* Ibérico, Sistema
Iberian Peninsula *80 B4 physical region*
Portugal/Spain
Iberian Plain *80 B4 abyssal plain*
E Atlantic Ocean
Ibérica, Cordillera *see* Ibérico, Sistema
Ibérico, Sistema *93 E2 var.* Cordillera
Ibérica, *Eng.* Iberian Mountains.
mountain range NE Spain
Ibiza *see* Eivissa
Ibo *see* Sassandra
Ica *60 D4* Ica, SW Peru
Icaria *see* Ikaría
Içá, Río *see* Putumayo, Río
İçel *see* Mersin
Iceland *83 E4 off.* Iceland, *Dan.* Island,
Icel. Ísland. *country* N Atlantic Ocean

Iceland Basin *80 B1 undersea basin*
N Atlantic Ocean
Icelandic Plateau *see* Iceland Plateau
Iceland Plateau *155 B6 var.* Icelandic
Plateau. *undersea plateau*
S Greenland Sea
Iceland, Republic of *see* Iceland
Iconium *see* Konya
Iculisma *see* Angoulême
Idabel *49 H2* Oklahoma, C USA
Idaho *46 D3 off.* State of Idaho, *also
known as* Gem of the Mountains, Gem
State. *state* NW USA
Idaho Falls *46 E3* Idaho, NW USA
Idensalmi *see* Iisalmi
Idfû *72 B2 var.* Edfu. SE Egypt
Idi Amin, Lac *see* Edward, Lake
Idini *74 B2* Trarza, W Mauritania
Idlib *118 B3* Idlib, NW Syria
Idre *85 B5* Dalarna, C Sweden
Iecava *106 C3* Bauska, S Latvia
Ieper *87 A6 Fr.* Ypres. West-Vlaanderen,
W Belgium
Ierápetra *105 D8* Kríti, Greece,
E Mediterranean Sea
Ierisós *see* Ierissós
Ierissós *104 C4 var.* Ierisós. Kentrikí
Makedonía, N Greece
Iertra *see* Ertra
Iferouâne *75 G2* Agadez, N Niger
Ifôghas, Adrar des *75 E2 var.* Adrar des
Iforas. *mountain range* NE Mali
Iforas, Adrar des *see* Ifôghas, Adrar des
Igarka *114 D3* Krasnoyarskiy Kray,
N Russia
Igaunija *see* Estonia
Iglau/Iglawa/Igława *see* Jihlava
Iglesias *97 A5* Sardegna, Italy,
C Mediterranean Sea
Igloolik *37 G2* Nunavut, N Canada
Igoumenítsa *104 A4* Ípeiros, W Greece
Iguaçu, Rio *63 E4 Sp.* Río Iguazú. *river*
Argentina/Brazil
Iguaçu, Saltos do *63 E4 Sp.* Cataratas del
Iguazú; *prev.* Victoria Falls. *waterfall*
Argentina/Brazil
Iguala *51 E4 var.* Iguala de la
Independencia. Guerrero, S Mexico
Iguala de la Independencia *see* Iguala
Iguazú, Cataratas del *see* Iguaçu,
Saltos do
Iguazú, Río *see* Iguaçu, Rio
Iguid, 'Erg *70 C3 var.* Iguidi, 'Erg
Iguidi, 'Erg *70 C3 var.* Erg Iguid. *desert*
Algeria/Mauritania
Ihavandhippolhu Atoll *132 A3 var.*
Ihavandiffulu Atoll. *atoll* N Maldives
Ihavandiffulu Atoll *see* Ihavandhippolhu
Atoll
Ihosy *79 F4* Fianarantsoa, S Madagascar
Ihupuku, Motu *see* Campbell Island
Iinnasuolu *see* Hinnøya
Iisalmi *84 E4 var.* Idensalmi.
Pohjois-Savo, C Finland
IJmuiden *86 C3* Noord-Holland,
W Netherlands
IJssel *86 D3 var.* Yssel. *river* Netherlands
IJsselmeer *86 C2 prev.* Zuider Zee. *lake*
N Netherlands
IJsselmuiden *86 D3* Overijssel,
E Netherlands
Ijzer *87 A6 river* W Belgium
Ikaahuk *see* Sachs Harbour
Ikaluktutiak *see* Cambridge Bay
Ikaría *105 D6 var.* Kariot, Nicaria,
Nikaria; *anc.* Icaria. *island* Dodekánisa,
Greece, Aegean Sea
Ikela *77 D6* Équateur, C Dem. Rep.
Congo
Iki *131 A7 island* SW Japan
Ilagan *139 E1* Luzon, N Philippines
Ilave *61 E4* Puno, S Peru
Iława *98 D3 Ger.* Deutsch-Eylau.
Warmińsko-Mazurskie, NE Poland
Ilebo *77 C6 prev.* Port-Francqui.
Kasai-Occidental, W Dem. Rep. Congo
Île-de-France *90 C3 cultural region*
N France
Ilemi Triangle *73 C6 disputed region*
Kenya/South Sudan
Ilerda *see* Lleida
Ilfracombe *89 C7* SW England, United
Kingdom
Ílhavo *92 B2* Aveiro, N Portugal
Iliamna Lake *36 C3 lake* Alaska, USA
Ilici *see* Elx
Iligan *139 E2 off.* Iligan City. Mindanao,
S Philippines
Iligan City *see* Iligan
Illapel *64 B4* Coquimbo, C Chile

Illichivsk *see* Chornomorsk
Illicis *see* Elx
Illinois *40 A4 off.* State of Illinois, *also
known as* Prairie State, Sucker State.
state C USA
Illinois River *40 B4 river* Illinois, N USA
Illurco *see* Lorca
Illuro *see* Mataró
Ilo *61 E4* Moquegua, SW Peru
Iloilo *139 E2 off.* Iloilo City. Panay
Island, C Philippines
Iloilo City *see* Iloilo
Ilorin *75 F4* Kwara, W Nigeria
Ilovlya *111 B6* Volgogradskaya Oblast',
SW Russia
İluh *see* Batman
Il'yaly *see* Andalyp
Imatra *85 E5* Etelä-Kariala, SE Finland
Imbros *see* Gökçeada
Imishli *see* İmişli
İmişli *117 H3 Rus.* Imishli. C Azerbaijan
Imola *96 C3* Emilia-Romagna, N Italy
Imperatriz *63 F2* Maranhão, NE Brazil
Imperia *96 A3* Liguria, NW Italy
Impfondo *77 C5* Likouala, NE Congo
Imphāl *135 H3 state capital* Manipur,
NE India
Imroz Adası *see* Gökçeada
Inagua Islands *see* Little Inagua
Inagua Islands *see* Great Inagua
Inarijärvi *84 D2 Smi.* Aanaarjávri, *Swe.*
Enareträsk. *lake* N Finland
Inău *see* Ineu
Inawashiro-ko *131 D5 var.* Inawasiro
Ko. *lake* Honshū, C Japan
Inawasiro Ko *see* Inawashiro-ko
İncesu *116 D3* Kayseri, Turkey (Türkiye)
Incheon *129 E4 off.* Incheon
Gwang-yeoksi, *prev.* Inch'ŏn, *Jap.*
Jinsen; *prev.* Chemulpo.
NW South Korea
Incheon-Gwang-yeoksi *see* Incheon
Inch'ŏn *see* Incheon
Incudine, Monte *91 E7 mountain* Corse,
France, C Mediterranean Sea
Indefatigable Island *see* Santa Cruz, Isla
Independence *45 F4* Missouri, C USA
Independence Fjord *83 E1 fjord*
N Greenland
Independence Island *see* Malden Island
Independence Mountains *46 C4
mountain range* Nevada, W USA
India *124 B3 off.* Republic of India, *var.*
Indian Union, Union of India, *Hind.*
Bhārat. *country* S Asia
India *see* Indija
Indiana *40 B4 off.* State of Indiana, *also
known as* Hoosier State. *state* N USA
Indianapolis *40 C4 state capital* Indiana,
N USA
Indian Church *52 C1* Orange Walk,
N Belize
Indian Desert *see* Thar Desert
Indianola *45 F4* Iowa, C USA
Indian Union *see* India
India, Republic of *see* India
India, Union of *see* India
Indigirka *115 F2 river* NE Russia
Indija *100 D3 Hung.* India; *prev.* Indjija.
Vojvodina, N Serbia
Indira Point *132 G3 headland* Andaman
and Nicobar Island, India,
NE Indian Ocean
Indjija *see* Indija
Indomed Fracture Zone *141 B6 tectonic
feature* SW Indian Ocean
Indonesia *138 B4 off.* Republic of
Indonesia, *Ind.* Republik Indonesia;
prev. Dutch East Indies, Netherlands
East Indies, United States of Indonesia.
country SE Asia
Indonesian Borneo *see* Kalimantan
Indonesia, Republic of *see* Indonesia
Indonesia, Republik *see* Indonesia
Indonesia, United States of *see*
Indonesia
Indore *134 D4* Madhya Pradesh,
C India
Indreville *see* Châteauroux
Indus *134 C2 Chin.* Yindu He; *prev.*
Yin-tu Ho. *river* S Asia
Indus Cone *see* Indus Fan
Indus Fan *112 C5 var.* Indus Cone.
undersea fan N Arabian Sea
Indus, Mouths of the *134 B4 delta*
S Pakistan
İnebolu *116 C2* Kastamonu,
N Turkey (Türkiye)
Ineu *108 A4 Hung.* Borosjenő; *prev.*
Inău. Arad, W Romania

Infiernillo, Presa del *51 E4 reservoir*
S Mexico
Inglewood *46 D2* California, W USA
Ingolstadt *95 C6* Bayern, S Germany
Ingulets *see* Inhulets
Inguri *see* Enguri
Inhambane *79 E4* Inhambane,
SE Mozambique
Inhulets *109 F3 Rus.* Ingulets.
Dnipropetrovska Oblast, E Ukraine
I-ning *see* Yining
Inis *see* Ennis
Inis Ceithleann *see* Enniskillen
Inn *95 C6 river* C Europe
Innaanganeq *82 C1 var.* Kap York.
headland NW Greenland
Inner Hebrides *88 B4 island group*
W Scotland, United Kingdom
Inner Islands *79 H1 var.* Central Group.
island group NE Seychelles
Innisfail *148 D3* Queensland,
NE Australia
Inniskilling *see* Enniskillen
Innsbruck *see* Innsbruck
Innsbruck *95 C7 var.* Innsbruch. Tirol,
W Austria
Inoucdjouac *see* Inukjuak
Inowraclaw *see* Inowrocław
Inowrocław *98 C3 Ger.* Hohensalza;
prev. Inowrazlaw. Kujawski-
pomorskie, C Poland
I-n-Sakane, 'Erg *75 E2 desert* N Mali
I-n-Salah *70 D3 var.* In Salah. C Algeria
Insterburg *see* Chernyakhovsk
Insula *see* Lille
Inta *110 E3* Respublika Komi, NW Russia
Interamna *see* Teramo
Interamna Nahars *see* Terni
International Falls *45 F1* Minnesota,
N USA
Inukjuak *38 D2 var.* Inoucdjouac; *prev.*
Port Harrison. Québec, NE Canada
Inuuvik *see* Inuvik
Inuvik *36 D3 var.* Inuuvik. Northwest
Territories, NW Canada
Invercargill *151 A7* Southland, South
Island, New Zealand
Inverness *88 C3* N Scotland,
United Kingdom
Investigator Ridge *141 D5 undersea
ridge* E Indian Ocean
Investigator Strait *149 B7 strait*
South Australia
Inyangani *78 D3 mountain*
NE Zimbabwe
Ioánnina *104 A4 var.* Janina, Yannina.
Ípeiros, W Greece
Iola *45 F5* Kansas, C USA
Ionia Basin *see* Ionian Basin
Ionian Basin *80 D5 var.* Ionia Basin.
undersea basin Ionian Sea,
C Mediterranean Sea
Ionian Islands *see* Iónia Nisiá
Ionian Sea *103 E3 Gk.* Iónio Pélagos, *It.*
Mar Ionio. *sea* C Mediterranean Sea
Iónia Nisiá *105 A5 Eng.* Ionian Islands.
island group W Greece
Ionio, Mar/Iónio Pélagos *see* Ionian Sea
Íos *105 D6 var.* Nio. *island* Kykládes,
Greece, Aegean Sea
Iouliída *105 C6 prev.* Ioulís, Kykládes,
Greece, Aegean Sea
Iowa *45 F3 off.* State of Iowa, *also known
as* Hawkeye State. *state* C USA
Iowa City *45 G3* Iowa, C USA
Iowa Falls *45 G3* Iowa, C USA
Ipel' *99 C6 var.* Ipoly, *Ger.* Eipel. *river*
Hungary/Slovakia
Ipiales *58 A4* Nariño, SW Colombia
Ipoh *138 B3* Perak, Peninsular Malaysia
Ipoly *see* Ipel'
Ippy *76 C4* Ouaka, C Central African
Republic
Ipswich *149 E5* Queensland, E Australia
Ipswich *89 E6 hist.* Gipeswic. E England,
United Kingdom
Iqaluit *37 H3 prev.* Frobisher Bay.
province capital Baffin Island, Nunavut,
NE Canada
Iquique *64 B1* Tarapacá, N Chile
Iquitos *60 C1* Loreto, N Peru
Irákleio *105 D7 var.* Herakleion, *Eng.*
Candia; *prev.* Iráklion. Kríti, Greece,
E Mediterranean Sea
Iráklion *see* Irákleio
Iran *120 D3 off.* Islamic Republic of Iran;
prev. Persia. *country* SW Asia
Iranian Plateau *120 D3 var.* Plateau of
Iran. *plateau* N Iran
Iran, Islamic Republic of *see* Iran

Iran, Plateau of see Iranian Plateau
Irapuato 51 E4 Guanajuato, C Mexico
Iraq 120 B3 off. Republic of Iraq, *Ar.*
'Irāq. *country* SW Asia
'Irāq see Iraq
Iraq, Republic of see Iraq
Irbid 119 B5 Irbid, N Jordan
Irbil see Arbil
Ireland 89 A5 off. Ireland, *Ir.* Éire.
country NW Europe
Ireland 80 C3 *Lat.* Hibernia. *island*
Ireland/United Kingdom
Irian see New Guinea
Irian Barat see Papua
Irian Jaya see Papua
Irian, Teluk see Cenderawasih, Teluk
Iringa 73 C7 Iringa, C Tanzania
Iriomote-jima 130 A4 *island*
Sakishima-shotō, SW Japan
Iriona 52 D2 Colón, NE Honduras
Irish Sea 89 C5 *Ir.* Muir Éireann. *sea*
C British Isles
Irkutsk 115 E4 Irkutskaya Oblast',
S Russia
Irminger Basin see Reykjanes Basin
Iroise 90 A3 *sea* NW France
Iron Mountain 40 B2 Michigan, N USA
Ironwood 40 B1 Michigan, N USA
Irrawaddy see Ayeyarwady
Irrawaddy, Mouths of the 137 A5 *delta*
SW Myanmar (Burma)
Irtish see Irtysh
Irtysh 114 C4 *var.* Irtish, *Kaz.* Ertis. *river*
C Asia
Irun 93 E1 *Cast.* Irún. País Vasco,
N Spain
Irún see Irun
Iruña see Pamplona
Isabela, Isla 60 A5 *var.* Albemarle Island.
island Galápagos Islands, Ecuador,
E Pacific Ocean
Isaccea 108 D4 Tulcea, E Romania
Isachsen 37 F1 Ellef Ringnes Island,
Nunavut, N Canada
Ísafjörður 83 E4 Vestfirðhir,
NW Iceland
Isbarta see Isparta
Isca Damnoniorum see Exeter
Ise 131 C6 Mie, Honshū, SW Japan
Iseghem see Izegem
Isère 91 D5 *river* E France
Isernia 97 D5 *var.* Aesernia. Molise,
C Italy
Ise-wan 131 C6 *bay* S Japan
Isfahan see Eşfahān
Isha Baydhabo see Baydhabo
Ishigaki-jima 130 A4 *island*
Sakishima-shotō, SW Japan
Ishikari-wan 130 C2 *bay* Hokkaidō,
NE Japan
Ishim 114 C4 Tyumenskaya Oblast',
C Russia
Ishim 114 C4 *Kaz.* Esil. *river*
Kazakhstan/Russia
Ishinomaki 130 D4 *var.* Isinomaki.
Miyagi, Honshū, C Japan
Ishkashim see Ishkoshim
Ishkoshim 123 F3 *Rus.* Ishkashim.
S Tajikistan
Isinomaki see Ishinomaki
Isiro 77 E5 Orientale, NE Dem.
Rep. Congo
Iskar 104 C2 *var.* Iskŭr. *river*
NW Bulgaria
İskenderun 116 D4 *Eng.* Alexandretta.
Hatay, S Turkey (Türkiye)
İskenderun Körfezi 118 A2 *Eng.* Gulf of
Alexandretta. *gulf* S Turkey (Türkiye)
Iskŭr see Iskar
Iskar, Yazovir 104 B2 *prev.* Yazovir
Stalin. *reservoir* W Bulgaria
Isla Cristina 92 C4 Andalucía, S Spain
Isla de León see San Fernando
Islamabad 134 C1 *country capital*
(Pakistan) Federal Capital Territory
Islamabad, NE Pakistan
Island/Ísland see Iceland
Islay 88 B4 *island* SW Scotland,
United Kingdom
Isle 91 B5 *river* W France
Isle of Man 89 B5 *British Crown*
Dependency NW Europe
Isles of Scilly 89 B8 *island group*
SW England, United Kingdom
Ismailia see Al Ismā'īliya
Ismâ'iliya see Al Ismā'īliya
Ismid see İzmit
Isna 72 B2 *var.* Esna. SE Egypt
Isoka 78 D1 Northern, NE Zambia
Isparta 116 B4 *var.* Isbarta. Isparta,
SW Turkey (Türkiye)

İspir 117 E3 Erzurum,
NE Turkey (Türkiye)
Israel 119 A7 off. State of Israel, *var.*
Medinat Israel, *Heb.* Yisrael, Yisra'el.
country SW Asia
Israel, State of see Israel
Issa see Vis
Issiq Köl see Ysyk-Köl
Issoire 91 C5 Puy-de-Dôme, C France
Issyk-Kul' see Balykchy
Issyk-Kul', Ozero see Ysyk-Köl
İstanbul 116 B2 *Bul.* Tsarigrad, *Eng.*
Istanbul, *prev.* Constantinople; *anc.*
Byzantium. Istanbul,
NW Turkey (Türkiye)
İstanbul Boğazi 116 B2 *var.* Bosporus
Thracius, *Eng.* Bosphorus, Bosporus,
Turk. Karadeniz Boğazi. *strait*
NW Turkey (Türkiye)
Istaravshan 123 E2 *prev.* Ŭroteppa, *Rus.*
Ura-Tyube. NW Tajikistan
Istarska Županija see Istra
Istra 100 A3 off. Istarska Županija.
province NW Croatia
Istra 100 A3 *Eng.* Istria, *Ger.* Istrien.
cultural region NW Croatia
Istria/Istrien see Istra
Itabuna 63 G3 Bahia, E Brazil
Itagüí 58 B3 Antioquia, W Colombia
Itaipú, Represa de 63 E4 *reservoir* Brazil/
Paraguay
Itaituba 62 B2 Pará, NE Brazil
Italia/Italiana, Republica/Italian
Republic, The see Italy
Italy 96 C3 off. The Italian Republic, *It.*
Italia, Repubblica Italiana. *country*
S Europe
Iténez, Río see Guaporé, Rio
Ithaca 41 E3 New York, NE USA
It Hearrenfean see Heerenveen
Itoigawa 131 C5 Niigata, Honshū,
C Japan
Itseqqortoormiit see Ittoqqortoormiit
Ittoqqortoormiit 83 E3 *var.*
Itseqqortoormiit, *Dan.* Scoresbysund,
Eng. Scoresby Sound. Tunu,
C Greenland
Iturup, Ostrov 130 E1 *island* Kuril'skiye
Ostrova, SE Russia
Itzehoe 94 B2 Schleswig-Holstein,
N Germany
Ivalo 84 D2 *Smi.* Avveel, Avvil. Lappi,
N Finland
Ivanava 107 B7 *Pol.* Janów, Janów
Poleski, *Rus.* Ivanovo. Brestskaya
Voblasts', SW Belarus
Ivangrad see Berane
Ivanhoe 149 C6 New South Wales,
SE Australia
Ivano-Frankivsk 108 C2 *Ger.* Stanislau,
Pol. Stanisławów, *Rus.*
Ivano-Frankovsk; *prev.* Stanislav.
Ivano-Frankivs'ka Oblast',
W Ukraine
Ivano-Frankovsk see Ivano-Frankivsk
Ivanovo 111 B5 Ivanovskaya Oblast',
W Russia
Ivanovo see Ivanava
Ivantsevichi/Ivatsevichi see Ivatsevichy
Ivatsevichy 107 B6 *Pol.* Iwacewicze, *Rus.*
Ivantsevichi, Ivatsevichi. Brestskaya
Voblasts', SW Belarus
Ivigtut see Ivittuut
Ivittuut 82 B4 *var.* Ivigtut. Sermersooq,
S Greenland
Iviza see Eivissa\Ibiza
Ivory Coast 74 D4 off. République de la
Côte d'Ivoire. *country* W Africa
Ivujivik 38 D1 Québec, NE Canada
Iwacewicze see Ivatsevichy
Iwaki 131 D5 Fukushima, Honshū,
N Japan
Iwakuni 131 B7 Yamaguchi, Honshū,
SW Japan
Iwanai 130 C2 Hokkaidō, NE Japan
Iwate 130 D3 Iwate, Honshū, N Japan
Ixtapa 51 E5 Guerrero, S Mexico
Ixtepec 51 F5 Oaxaca, SE Mexico
Iyo-nada 131 B7 *sea* S Japan
Izabal, Lago de 52 B2 *prev.* Golfo Dulce.
lake E Guatemala
Izad Khvāst 120 D3 Fārs, C Iran
I'zāz 118 B2 *var.* A'zāz. Ḩalab, NW Syria
Izegem 87 A6 *prev.* Iseghem.
West-Vlaanderen, W Belgium
Izhevsk 111 D5 *prev.* Ustinov.
Udmurtskaya Respublika, NW Russia
Izyaslav 108 C2 *prev.* Izyaslav.
Khmel'nyts'ka Oblast', W Ukraine

Izium 109 G2 *prev.* Izyum. Kharkivs'ka
Oblast', E Ukraine
Izmail see Izmayil
Izmayil 108 D4 *Rus.* Izmail. Odes'ka
Oblast', SW Ukraine
İzmir 116 A3 *prev.* Smyrna. İzmir,
W Turkey (Türkiye)
İzmit 116 B2 *var.* Ismid; *anc.* Astacus.
Kocaeli, NW Turkey (Türkiye)
İznik Gölü 116 B3 *lake*
NW Turkey (Türkiye)
Izu-hanto 131 D6 *peninsula* Honshū,
S Japan
Izu Shichito see Izu-shotō
Izu-shotō 131 C6 *var.* Izu Shichito.
island group S Japan
Izvor 104 B2 Pernik, W Bulgaria
Izyaslav see Iziaslav
Izyum see Izium

J

Jabal ash Shifa 120 A4 *desert* NW Saudi
Arabia
Jabalpur 135 E4 *prev.* Jubbulpore.
Madhya Pradesh, C India
Jabbúl, Sabkhat al 118 B2 *sabkha*
NW Syria
Jablah 118 A3 *var.* Jeble, *Fr.* Djéblé. Al
Lādhiqiyah, NW Syria
Jaca 93 F1 Aragón, NE Spain
Jacaltenango 52 A2 Huehuetenango,
W Guatemala
Jackson 42 B2 *state capital* Mississippi,
S USA
Jackson 45 H5 Missouri, C USA
Jackson 42 C1 Tennessee, S USA
Jackson Head 151 A6 *headland* South
Island, New Zealand
Jacksonville 43 E3 Florida, SE USA
Jacksonville 40 B4 Illinois, N USA
Jacksonville 43 F1 North Carolina,
SE USA
Jacksonville 49 G3 Texas, SW USA
Jacmel 54 D3 *var.* Jaquemel. S Haiti
Jacob see Nkayi
Jacobabad 134 B3 Sindh, SE Pakistan
Jadotville see Likasi
Jadransko More/Jadransko Morje see
Adriatic Sea
Jaén 60 B2 Cajamarca, N Peru
Jaén 92 D4 Andalucía, SW Spain
Jaffna 132 D3 Northern Province,
N Sri Lanka
Jagannath see Puri
Jagdalpur 135 E5 Chhattisgarh, C India
Jagdaqi 127 G1 Nei Mongol Zizhiqu,
N China
Jagodina 100 D4 *prev.* Svetozarevo.
Serbia, C Serbia
Jahra see Al Jahrā'
Jailolo see Halmahera, Pulau
Jaipur 134 D3 *prev.* Jeypore. *state capital*
Rājasthān, N India
Jaisalmer 134 C3 Rājasthān,
NW India
Jajce 100 B3 Federacija Bosna I
Hercegovina, W Bosnia and
Herzegovina
Jakarta 138 C5 *prev.* Djakarta, *Dut.*
Batavia. *country capital* (Indonesia)
Jawa, C Indonesia
Jakobstad 84 D4 *Fin.* Pietarsaari.
Österbotten, W Finland
Jakobstadt see Jēkabpils
Jalālābād 123 F4 *var.* Jalalabad,
Jelalabad. Nangarhār, E Afghanistan
Jalal-Abad 123 F2 *Rus.* Dzhalal-Abad.
Jalal-Abad Oblusu/Dzhalal-Abadskaya
Oblast', W Kyrgyzstan
Jalandhar 134 D2 *prev.* Jullundur.
Punjab, N India
Jalapa 52 D3 Nueva Segovia,
NW Nicaragua
Jalpa 50 D4 Zacatecas, C Mexico
Jālū 71 G3 *var.* Jūla. NE Libya
Jaluit Atoll 144 D2 *var.* Jālwōj. *atoll*
Ralik Chain, S Marshall Islands
Jālwōj see Jaluit Atoll
Jamaame 73 D6 *It.* Giamame; *prev.*
Margherita. Jubbada Hoose, S Somalia
Jamaica 54 A4 *country* W West Indies
Jamaica 56 A1 *island* W West Indies
Jamaica Channel 54 D3 *channel* Haiti/
Jamaica
Jamālpur 135 F3 Bihār, NE India
Jambi 138 B4 *var.* Telanaipura; *prev.*
Djambi. Sumatera, W Indonesia
Jamdena see Yamdena, Pulau

James Bay 38 C3 *bay* Ontario/Québec,
E Canada
James River 45 E2 *river* North Dakota/
South Dakota, N USA
James River 41 E5 *river* Virginia,
NE USA
Jamestown 41 E3 New York, NE USA
Jamestown 45 E2 North Dakota, N USA
Jamestown see Holetown
Jammu 134 D2 *prev.* Jummoo. *state*
capital Jammu and Kashmir, NW India
Jammu and Kashmir 134 D1 *disputed*
region India/Pakistan
Jāmnagar 134 C4 *prev.* Navanagar.
Gujarāt, W India
Jamshedpur 135 F4 Jhārkhand, NE India
Jamuna see Brahmaputra
Janaúba 63 F3 Minas Gerais, SE Brazil
Janesville 40 B3 Wisconsin, N USA
Jang Bogo 154 C4 *South Korean research*
station Antarctica
Janina see Ioánnina
Janin 119 E6 *var.* Jenin. N West Bank,
SW Asia
Janischken see Joniškis
Jankovac see Jánoshalma
Jan Mayen 83 F4 *constituent part of*
Norway. island N Atlantic Ocean
Jánoshalma 99 C7 *Croatian* Jankovac.
Bács-Kiskun, S Hungary
Janów see Ivanava, Belarus
Janów/Janów see Jonava, Lithuania
Janów Poleski see Ivanava
Japan 130 C4 *var.* Nippon, *Jap.* Nihon.
country E Asia
Japan, Sea of 130 A4 *var.* East Sea, *Rus.*
Yapanskoye More. *sea*
NW Pacific Ocean
Japan Trench 125 F1 *trench*
NW Pacific Ocean
Japen see Yapen, Pulau
Japiim 62 C2 *var.* Máncio Lima. Acre,
W Brazil
Japurá, Rio 62 C2 *var.* Río Caquetá,
Yapurá. *river* Brazil/Colombia
Japurá, Rio see Caquetá, Río
Jaqué 53 G5 Darién, SE Panama
Jaquemel see Jacmel
Jarablos see Jarābulus
Jarābulus 118 C2 *var.* Jarablos, Jerablus,
Fr. Djérablous. Ḩalab, N Syria
Jarbah, Jazirat see Djerba, Île de
Jardines de la Reina, Archipiélago de
los 54 B2 *island group* C Cuba
Jarid, Shaṭṭ al see Jerid, Chott el
Jarocin 98 C4 Wielkopolskie, C Poland
Jarosław 99 E5 *Ger.* Jaroslau, *Rus.*
Yaroslav. Podkarpackie, SE Poland
Jarqo'rg'on 123 E3 *Rus.* Dzharkurgan.
Surkhondaryo Viloyati, S Uzbekistan
Jarvis Island 145 G2 *US unincorporated*
territory C Pacific Ocean
Jashore 135 G4 *prev.* Jessore. Khulna,
W Bangladesh
Jasło 99 D5 Podkarpackie, SE Poland
Jastrzębie-Zdrój 99 C5 Śląskie, S Poland
Jataí 63 E3 Goiás, C Brazil
Jativa see Xàtiva
Jauf see Al Jawf
Jaunpiebalga 106 D3 Gulbene, NE Latvia
Jaunpur 135 E3 Uttar Pradesh, N India
Java 152 A3 North Dakota, N USA
Javalambre 93 E3 *mountain* E Spain
Java Sea 138 D4 *Ind.* Laut Jawa. *sea*
W Indonesia
Java Trench 124 D5 *var.* Sunda Trench.
trench E Indian Ocean
Jawa, Laut see Java Sea
Jawhar 73 D6 *var.* Jowhar, *It.* Giohar.
Shabeellaha Dhexe, S Somalia
Jaworów see Yavoriv
Jaya, Puncak 139 G4 *prev.* Puntjak
Carstensz, Puntjak Sukarno. *mountain*
Papua, E Indonesia
Jayapura 139 H4 *var.* Djajapura, *Dut.*
Hollandia; *prev.* Kotabaru,
Sukarnapura. Papua, E Indonesia
Jay Dairen see Dalian
Jayhawker State see Kansas
Jaz Murian, Hamun-e 120 E4 *lake*
SE Iran
Jebba 75 F4 Kwara, W Nigeria
Jebel, Bahr el see White Nile
Jeble see Jablah
Jeddah see Jiddah
Jędrzejów 98 D4 *Ger.* Endersdorf.
Świętokrzyskie, C Poland
Jefferson City 45 G5 *state capital*
Missouri, C USA

Jega *75 F4* Kebbi, NW Nigeria

Jehol *see* Chengde

Jejudo *129 E4 Jap.* Saishū; *prev.* Cheju-do, Quelpart. *island* S South Korea

Jeju Strait *129 E4 var.* Jeju-haehyŏp; *prev.* Cheju-Strait. *strait* S South Korea

Jeju-haehyŏp *see* Jeju Strait

Jēkabpils *106 D4 Ger.* Jakobstadt. Jēkabpils, S Latvia

Jelalabad *see* Jalālābād

Jelenia Góra *98 B4 Ger.* Hirschberg, Hirschberg im Riesengebirge, Hirschberg in Riesengebirge, Hirschberg in Schlesien. Dolnośląskie, SW Poland

Jelgava *106 C3 Ger.* Mitau. Jelgava, C Latvia

Jelondi *123 F3 prev.* Dzhelandy. SE Tajikistan

Jemappes *87 B6* Hainaut, S Belgium

Jember *138 D5 prev.* Djember. Jawa, C Indonesia

Jena *94 C4* Thüringen, C Germany

Jengish Chokusu *see* Tömür Feng

Jenin *see* Janīn

Jerablus *see* Jarābulus

Jerada *70 D2* NE Morocco

Jérémie *54 D3* SW Haiti

Jerez *see* Jeréz de la Frontera, Spain

Jerez de la Frontera *92 C4 var.* Jerez; *prev.* Xeres. Andalucía, SW Spain

Jerez de los Caballeros *92 C4* Extremadura, W Spain

Jericho *119 E7 var.* Arīḥā, Yerīḥo. C West Bank, SW Asia

Jerid, Chott el *71 E2 var.* Shaṭṭ al Jarīd. *salt lake* SW Tunisia

Jersey *89 D8 British Crown Dependency* Channel Islands, NW Europe

Jerusalem *103 H4 Ar.* Al Quds, Al Quds ash Sharīf, *Heb.* Yerushalayim; *anc.* Hierosolyma. *country capital - not recognized by all countries* (Israel), NE Israel

Jesenice *95 D7 Ger.* Assling. NW Slovenia

Jesselton *see* Kota Kinabalu

Jessore *see* Jashore

Jesús María *64 C3* Córdoba, C Argentina

Jeypore *see* Jaipur, Rājasthān, India

Jhānsi *134 D3* Uttar Pradesh, N India

Jhārkhand *135 F4 cultural region* NE India

Jhelum *134 C2* Punjab, NE Pakistan

Ji *see* Hebei, China

Ji *see* Jilin, China

Jiangmen *129 C6* Guangdong, S China

Jiangsu *128 D4 var.* Chiang-su, Jiangsu Sheng, Kiangsu, Su. *province* E China

Jiangsu *see* Nanjing

Jiangsu Sheng *see* Jiangsu

Jiangxi *129 C6 var.* Chiang-hsi, Gan, Jiangxi Sheng, Kiangsi. *province* S China

Jiangxi Sheng *see* Jiangxi

Jiaxing *129 D5* Zhejiang, SE China

Jiayi *see* Chiayi

Jibuti *see* Djibouti

Jiddah *121 A5 Eng.* Jeddah. Makkah, W Saudi Arabia

Jih-k'a-tse *see* Xigazê

Jihlava *99 B5 Ger.* Iglau, *Pol.* Iglawa. Vysocina, S Czechia (Czech Republic)

Jilib *73 D6 It.* Gelib. Jubbada Dhexe, S Somalia

Jilin *128 E3 var.* Chi-lin, Girin, Kirin; *prev.* Yungki, Yunki. Jilin, NE China

Jilin *128 E3 var.* Chi-lin, Girin, Ji, Jilin Sheng, Kirin. *province* NE China

Jilin Sheng *see* Jilin

Jilong *see* Keelung

Jīma *73 C5 var.* Jimma, It. Gimma. Oromīya, C Ethiopia

Jimbolia *108 A4 Ger.* Hatzfeld, *Hung.* Zsombolya. Timiș, W Romania

Jiménez *50 D2* Chihuahua, N Mexico

Jimma *see* Jīma

Jimsar *126 C3* Xinjiang Uygur Zizhiqu, NW China

Jin *see* Shanxi

Jin *see* Tianjin Shi

Jinan *128 C4 var.* Chinan, Chi-nan, Tsinan. *province capital* Shandong, E China

Jingdezhen *129 C5* Jiangxi, S China

Jinghong *129 A6 var.* Yunjinghong. Yunnan, SW China

Jinhua *129 D5* Zhejiang, SE China

Jining *127 F3* Shandong, E China

Jinja *73 C6* S Uganda

Jinotega *52 D3* Jinotega, NW Nicaragua

Jinotepe *52 D3* Carazo, SW Nicaragua

Jinsen *see* Incheon

Jinzhong *128 C4 var.* Yucī. Shanxi, C China

Jinzhou *128 D3 var.* Chin-chou, Chinchow; *prev.* Chinhsien. Liaoning, NE China

Jirgalanta *see* Hovd

Jisr ash Shadadi *see* Ash Shadādah

Jiu *108 B5 Ger.* Schil, Schyl, *Hung.* Zsil, Zsily. *river* S Romania

Jiujiang *129 C5* Jiangxi, S China

Jixi *128 E2* Heilongjiang, NE China

Jizan *see* Jīzān

Jīzān *121 B6 var.* Qīzān, Jizan. Jīzān, SW Saudi Arabia

Jizzax *123 E2 Rus.* Dzhizak. Jizzax Viloyati, C Uzbekistan

João Belo *see* Xai-Xai

João Pessoa *63 G2 prev.* Paraíba. *state capital* Paraíba, E Brazil

Joazeiro *see* Juazeiro

Job'urg *see* Johannesburg

Jo-ch'iang *see* Ruoqiang

Jodhpur *134 C3* Rājasthān, NW India

Joensuu *85 E5* Pohjois-Karjala, SE Finland

Jōetsu *131 C5 var.* Zyôetu. Niigata, Honshū, C Japan

Jogjakarta *see* Yogyakarta

Johannesburg *78 D4 var.* Egoli, Erautini, Gauteng, *abbrev.* Job'urg. Gauteng, NE South Africa

Johannisburg *see* Pisz

John Day River *46 C3 river* Oregon, NW USA

John o'Groats *88 C2* N Scotland, United Kingdom

Johnston Atoll *143 E1 US unincorporated territory* C Pacific Ocean

Johor Baharu *see* Johor Bahru

Johor Bahru *138 B3 var.* Johor Baharu, Johore Bahru. Johor, Peninsular Malaysia

Johore Bahru *see* Johor Bahru

Johor Strait *138 A1 strait* Johor, Peninsular Malaysia, Malaysia/ Singapore Asia Andaman Sea/South China Sea

Joinvile *see* Joinville

Joinville *63 E4 var.* Joinvile. Santa Catarina, S Brazil

Jokkmokk *84 C3 Smi.* Dálvvadis. Norrbotten, N Sweden

Jokyakarta *see* Yogyakarta

Joliet *40 B3* Illinois, N USA

Jonava *106 B4 Ger.* Janow, *Pol.* Janów. Kaunas, C Lithuania

Jonesboro *42 B1* Arkansas, C USA

Joniškis *106 C3 Ger.* Janischken. Šiauliai, N Lithuania

Jönköping *85 B7* Jönköping, S Sweden

Jonquière *39 E4* Québec, SE Canada

Joplin *45 F5* Missouri, C USA

Jordan *119 B6 off.* Hashemite Kingdom of Jordan, *Ar.* Al Mamlaka al Urdunīya al Hashemīyah, Al Urdunn; *prev.* Transjordan. *country* SW Asia

Jordan *119 B5 Ar.* Urdun, *Heb.* HaYarden. *river* SW Asia

Jorhat *135 H3* Assam, NE India

Jos *75 G4* Plateau, C Nigeria

Joseph Bonaparte Gulf *146 D2 gulf* N Australia

Jos Plateau *75 G4 plateau* C Nigeria

Jotunheimen *85 A5 mountain range* S Norway

Joûnié *118 A4 var.* Juniyah. W Lebanon

Joure *86 D2 Fris.* De Jouwer. Fryslân, N Netherlands

Joutseno *85 E5* Etelä-Kariala, SE Finland

Jowhar *see* Jawhar

J.Storm Thurmond Reservoir *see* Clark Hill Lake

Juan Aldama *50 D3* Zacatecas, C Mexico

Juan de Fuca, Strait of *46 A1 strait* Canada/USA

Juan Fernández, Islas *57 A6 Eng.* Juan Fernandez Islands. *island group* W Chile

Juan Fernandez Islands *see* Juan Fernández, Islas

Juazeiro *63 G2 prev.* Joazeiro. Bahia, E Brazil

Juazeiro do Norte *63 G2* Ceará, E Brazil

Juba *73 B5 var.* Jūbā. *country capital* (South Sudan) Bahr el Gabel, S South Sudan

Juba *73 D6 Amh.* Genalē Wenz, *It.* Guiba, *Som.* Ganaane, Webi Jubba. *river* Ethiopia/Somalia

Jubba, Webi *see* Juba

Jubbulpore *see* Jabalpur

Júcar *93 E3 var.* Jucar. *river* C Spain

Juchitán *51 F5 var.* Juchitán de Zaragosa. Oaxaca, SE Mexico

Juchitán de Zaragosa *see* Juchitán

Judayyidat Hamir *120 B3* Al Anbār, S Iraq

Judenburg *95 D7* Steiermark, C Austria

Juigalpa *52 D3* Chontales, S Nicaragua

Juiz de Fora *63 F4* Minas Gerais, SE Brazil

Jujuy *see* San Salvador de Jujuy

Jūlā *see* Jālū, Libya

Julia Beterrae *see* Béziers

Juliaca *61 E4* Puno, SE Peru

Juliana Top *59 G3 mountain* C Suriname

Julianehåb *see* Qaqortoq

Julio Briga *see* Bragança

Juliobriga *see* Logroño

Juliomagus *see* Angers

Jullundur *see* Jalandhar

Jumilla *93 E4* Murcia, SE Spain

Jummoo *see* Jammu

Jumna *see* Yamuna

Jumporn *see* Chumphon

Junction City *45 F4* Kansas, C USA

Juneau *36 D4 state capital* Alaska, USA

Junín *64 C4* Buenos Aires, E Argentina

Juniyah *see* Joûnié

Junkseylon *see* Phuket

Jur *73 B5 river* C Sudan

Jura *90 D4 cultural region* E France

Jura *95 A7 var.* Jura Mountains. *mountain range* France/Switzerland

Jura *88 B4 island* SW Scotland, United Kingdom

Jura Mountains *see* Jura

Jurbarkas *106 B4 Ger.* Georgenburg, Jurburg. Tauragė, W Lithuania

Jurburg *see* Jurbarkas

Jūrmala *106 C3* Rīga, C Latvia

Juruá, Rio *62 C2 var.* Río Yuruá. *river* Brazil/Peru

Juruena, Rio *62 D3 river* W Brazil

Jutiapa *52 B2* Jutiapa, S Guatemala

Juticalpa *52 D2* Olancho, C Honduras

Jutland *85 A7 Den.* Jylland. *peninsula* W Denmark

Juvavum *see* Salzburg

Juventud, Isla de la *54 A2 var.* Isla de Pinos, *Eng.* Isle of Youth; *prev.* The Isle of the Pines. *island* W Cuba

Južna Morava *101 E5 Ger.* Südliche Morava. *river* SE Serbia

Jwaneng *78 C4* Southern, S Botswana

Jylland *see* Jutland

Jyrgalan *123 G2 Rus.* Dzhergalan. Ysyk-Köl Oblusu/Issyk-Kul'skaya Oblast', NE Kyrgyzstan

Jyväskylä *85 D5* Keski-Suomi, C Finland

K

K2 *126 A4 Chin.* Qogir Feng, *Eng.* Mount Godwin Austen. *mountain* China/ Pakistan

Kaafu Atoll *see* Male' Atoll

Kaaimanston *59 G3* Sipaliwini, N Suriname

Kaakhka *see* Kaka

Kaala *see* Caála

Kaamanen *84 D2 Smi.* Gámas. Lappi, N Finland

Kaapstad *see* Cape Town

Kaaresuvanto *84 C3 Smi.* Gárasavon. Lappi, N Finland

Kabale *73 B6* SW Uganda

Kabinda *77 D7* Kasai-Oriental, SE Dem. Rep. Congo

Kabinda *see* Cabinda

Kābol *see* Kābul

Kabompo *78 C2 river* W Zambia

Kābul *123 E4 var.* Kābol, *Pash.* Kābul. *country capital* (Afghanistan) Kābul, E Afghanistan

Kabul *123 E4 var.* Daryā-ye Kābul. *river* Afghanistan/Pakistan

Kābul, Daryā-ye *see* Kabul

Kabwe *78 D2* Central, C Zambia

Kachchh, Gulf of *134 B4 var.* Gulf of Cutch, Gulf of Kutch. *gulf* W India

Kachchh, Rann of *134 B4 var.* Rann of Kachh, Rann of Kutch. *salt marsh* India/Pakistan

Kachh, Rann of *see* Kachchh, Rann of

Kadan Kyun *137 B5 prev.* King Island. *island* Myeik Archipelago, S Myanmar (Burma)

Kadapa *132 C2 prev.* Cuddapah. Andhra Pradesh, S India

Kadavu *145 E4 prev.* Kandavu. *island* S Fiji

Kadiivka *109 H3 prev.* Kadiyivka, *Rus.* Stakhanov. Luhans'ka Oblast', E Ukraine

Kadiyivka *see* Kadiivka

Kadoma *78 D3 prev.* Gatooma. Mashonaland West, C Zimbabwe

Kadugli *72 B4* Southern Kordofan, S Sudan

Kaduna *75 G4* Kaduna, C Nigeria

Kadzhi-Say *see* Kaji-Say

Kaédi *74 C3* Gorgol, S Mauritania

Kaffa *see* Feodosiia

Kafue *78 D2* Lusaka, SE Zambia

Kafue *78 C2 river* C Zambia

Kaga Bandoro *76 C4 prev.* Fort-Crampel. Nana-Grébizi, C Central African Republic

Kagan *see* Kogon

Kåghet *74 D1 var.* Karet. *physical region* N Mauritania

Kagi *see* Chiayi

Kagoshima *131 B8 var.* Kagosima. Kagoshima, Kyūshū, SW Japan

Kagoshima-wan *131 A8 bay* SW Japan

Kagosima *see* Kagoshima

Kagul *see* Cahul

Kharga *see* Al Khārjah

Kahmard, Darya-ye *123 E4 prev.* Darya-i-surkhab. *river* NE Afghanistan

Kahramanmaraş *116 D4 var.* Kahraman Maraş, Maraş, Marash. Kahramanmaraş, S Turkey (Türkiye)

Kaiapoi *151 C6* Canterbury, South Island, New Zealand

Kaifeng *128 C4* Henan, C China

Kai, Kepulauan *139 F4 prev.* Kei Islands. *island group* Maluku, SE Indonesia

Kaikohe *150 C2* Northland, North Island, New Zealand

Kaikoura *151 C5* Canterbury, South Island, New Zealand

Kaikoura Peninsula *151 C5 peninsula* South Island, New Zealand

Kainji Lake *see* Kainji Reservoir

Kainji Reservoir *75 F4 var.* Kainji Lake. *reservoir* W Nigeria

Kaipara Harbour *150 C2 harbour* North Island, New Zealand

Kairouan *71 E2 var.* Al Qayrawān. E Tunisia

Kaisaria *see* Kayseri

Kaiserslautern *95 A5* Rheinland-Pfalz, SW Germany

Kaišiadorys *107 B5* Kaunas, S Lithuania

Kaitaia *150 C2* Northland, North Island, New Zealand

Kajaani *84 E4 Swe.* Kajana. Kainuu, C Finland

Kajan *see* Kayan, Sungai

Kajana *see* Kajaani

Kaji-Say *123 G2 Rus.* Kadzhi-Say. Ysyk-Köl Oblusu/Issyk-Kul'skaya Oblast', NE Kyrgyzstan

Kaka *122 C3 var.* Kaakhka. Ahal Welaýaty, S Turkmenistan

Kake *36 D4* Kupreanof Island, Alaska, USA

Kakhovka *109 F4* Khersons'ka Oblast', S Ukraine

Kakhovs'ke Vodoskhovyshche *109 F4 Rus.* Kakhovskoye Vodokhranilishche. *reservoir* SE Ukraine

Kakhovskoye Vodokhranilishche *see* Kakhovs'ke Vodoskhovyshche

Kākināda *132 D1 prev.* Cocanada. Andhra Pradesh, E India

Kakshaal Range *see* Kokshaal-Too

Kaktovik *36 D2* Alaska, USA

Kalaallit Nunaat *see* Greenland

Kalaburagi *132 C1 prev.* Gulbarga. Karnātaka, C India

Kalahari Desert *78 B4 desert* Southern Africa

Kalaikhum *see* Qalaikhumb

Kalámai *see* Kalámata

Kalamariá *104 B4* Kentrikí Makedonía, N Greece

Kalamás *see* Thýamis

Kalámata *105 B6 prev.* Kalámai.
  Pelopónnisos, S Greece
Kalamazoo *44 C3* Michigan, N USA
Kalambaka *see* Kalampáka
Kálamos *105 C5* Attikí, C Greece
Kalampáka *104 B4 var.* Kalambaka.
  Thessalía, C Greece
Kalanchak *109 F4* Khersons'ka Oblast',
  S Ukraine
Kalarash *see* Călărași
Kalasin *136 D4 var.* Muang Kalasin.
  Kalasin, E Thailand
Kalat *134 B2 var.* Kelat, Khelat.
  Baluchistan, SW Pakistan
Kalāt *see* Qalāt
Kalbarri *147 A5* Western Australia
Kalecik *116 C3* Ankara,
  N Turkey (Türkiye)
Kalemie *77 E6 prev.* Albertville. Katanga,
  SE Dem. Rep. Congo
Kale Sultanie *see* Çanakkale
Kalgan *see* Zhangjiakou
Kalgoorlie *147 B6* Western Australia
Kalima *77 D6* Maniema, E Dem.
  Rep. Congo
Kalimantan *138 D4 Eng.* Indonesian
  Borneo. *geopolitical region* Borneo,
  C Indonesia
Kalinin *see* Tver'
Kaliningrad *106 A4* Kaliningradskaya
  Oblast', W Russia
Kaliningrad *see* Kaliningradskaya Oblast'
Kaliningradskaya Oblast' *106 B4 var.*
  Kaliningrad. *province and enclave*
  W Russia
Kalinkavichy *107 C7 Rus.* Kalinkovichi.
  Homyel'skaya Voblasts', SE Belarus
Kalinkovichi *see* Kalinkavichy
Kalisch/Kalish *see* Kalisz
Kalispell *44 B1* Montana, NW USA
Kalisz *98 C4 Ger.* Kalisch, *Rus.* Kalish;
  anc. Calisia. Wielkopolskie, C Poland
Kalix *84 D4* Norrbotten, N Sweden
Kalixälven *84 D3 river* N Sweden
Kallaste *106 E3 Ger.* Krasnogor.
  Tartumaa, SE Estonia
Kallavesi *85 E5 lake* SE Finland
Kalloní *105 D5* Lésvos, E Greece
Kalmar *85 C7 var.* Calmar. Kalmar,
  S Sweden
Kalmthout *87 C5* Antwerpen,
  N Belgium
Kalpáki *104 A4* Ípeiros, W Greece
Kalpeni Island *132 B3 island*
  Lakshadweep, India, N Indian Ocean
Kaltdorf *see* Pruszków
Kaluga *111 B5* Kaluzhskaya Oblast',
  W Russia
Kalush *108 C2 Pol.* Kałusz.
  Ivano-Frankivs'ka Oblast', W Ukraine
Kałusz *see* Kalush
Kalutara *132 D4* Western Province,
  SW Sri Lanka
Kalvarija *107 B5 Pol.* Kalwaria.
  Marijampolė, S Lithuania
Kalwaria *see* Kalvarija
Kalyān *134 C5* Mahārāshtra, W India
Kálymnos *105 D6 var.* Kálimnos. *island*
  Dodekánisa, Greece, Aegean Sea
Kama *110 D4 river* NW Russia
Kamarang *59 F3* W Guyana
Kambryk *see* Cambrai
Kamchatka *see* Kamchatka, Poluostrov
Kamchatka, Poluostrov *115 G3 Eng.*
  Kamchatka. *peninsula* E Russia
Kamenets-Podol'skiy *see*
  Kamianets-Podilskyi
Kamenka Dneprovskaya *see*
  Kamianka-Dniprovska
Kamenskoye *see* Dniprodzerzhyns'k
Kamensk-Shakhtinskiy *111 B6*
  Rostovskaya Oblast', SW Russia
Kamianets-Podilskyi *108 C3 prev.*
  Kam"yanets'-Podil's'kyy, *Rus.*
  Kamenets-Podol'skiy. Khmel'nyts'ka
  Oblast', W Ukraine
Kamianka-Dniprovska *109 F3 prev.*
  Kam"yanka-Dniprovs'ka, *Rus.*
  Kamenka Dneprovskaya. Zaporiz'ka
  Oblast', E Ukraine
Kamina *77 D7* Katanga, S Dem.
  Rep. Congo
Kamishli *see* Al Qāmishlī
Kamloops *37 E5* British Columbia,
  SW Canada
Kammu Seamount *152 C2 guyot*
  N Pacific Ocean

Kampala *73 B6 country capital* (Uganda)
  S Uganda
Kampong Cham *137 D6* Khmer.
  Kâmpóng Cham. Kampong Cham,
  C Cambodia
Kâmpóng Cham *see* Kampong Cham
Kampong Chhnang *137 D6* Khmer.
  Kâmpóng Chhnăng. Kampong
  Chhang, C Cambodia
Kâmpóng Chhnăng *see* Kampong
  Chhang
Kampong Speu *137 D6* Khmer.
  Kâmpóng Spœ. Kampong Speu,
  S Cambodia
Kâmpóng Spœ *see* Kampong Speu
Kampong Thom *137 D5* Khmer.
  Kâmpóng Thum, *prev.* Trâpeăng Vêng.
  Kampong Thom, C Cambodia
Kâmpóng Thum *see* Kampong Thom
Kâmpóng Trâbêk *137 D5 prev.* Phumĭ
  Kâmpóng Trâbêk, Phum Kompong
  Trabek. Kampong Thom, C Cambodia
Kampot *137 D6* Khmer. Kâmpôt.
  Kampot, SW Cambodia
Kâmpôt *see* Kampot
Kampuchea *see* Cambodia
Kampuchea, Democratic *see* Cambodia
**Kampuchea, People's Democratic
  Republic of** *see* Cambodia
Kam"yanets'-Podil's'kyy *see*
  Kamianets-Podilskyi
Kam"yanka-Dniprovs'ka *see*
  Kamianka-Dniprovska
Kam"yans'ke *see* Kamianske
Kamyshin *111 B6* Volgogradskaya Oblast',
  SW Russia
Kanaky *see* New Caledonia
Kananga *77 D6 prev.* Luluabourg.
  Kasai-Occidental, S Dem. Rep. Congo
Kanara *see* Karnātaka
Kanash *111 C5* Chuvashskaya
  Respublika, W Russia
Kanazawa *131 C5* Ishikawa, Honshū,
  SW Japan
Kanbe *136 B4* Yangon,
  SW Myanmar (Burma)
Kānchipuram *132 C2 prev.*
  Conjeeveram. Tamil Nādu, SE India
Kandahār *123 E5 Per.* Qandahār.
  Kandahār, S Afghanistan
Kandalakša *see* Kandalaksha
Kandalaksha *110 B2 var.* Kandalakša,
  Fin. Kantalahti. Murmanskaya Oblast',
  NW Russia
Kandangan *138 D4* Borneo, C Indonesia
Kandau *see* Kandava
Kandava *106 C3 Ger.* Kandau. Tukums,
  W Latvia
Kandavu *see* Kadavu
Kandi *75 F4* N Benin
Kandy *132 D3* Central Province,
  C Sri Lanka
Kane Fracture Zone *66 B4 fracture zone*
  NW Atlantic Ocean
Kāne'ohe *47 A8 var.* Kaneohe. O'ahu,
  Hawaii, USA, C Pacific Ocean
Kanestron, Akrotírio *see* Palioúri,
  Akrotírio
Kanëv *see* Kaniv
Kanevskoye Vodokhranilishche *see*
  Kanivske Vodoskhovyshche
Kangān *see* Bandar-e Kangān
Kangaroo Island *149 A7 island* South
  Australia
Kangertittivaq *83 E4 Dan.* Scoresby
  Sund. *fjord* E Greenland
Kangikajik *83 E4 var.* Kap Brewster.
  headland E Greenland
Kangiqliniq *37 G3 var.* Rankin Inlet.
  Nunavut, C Canada
Kaniv *109 E2 Rus.* Kanëv. Cherkas'ka
  Oblast', C Ukraine
Kanivske Vodoskhovyshche *109 E2
  prev.* Kaniys'ke Vodoskhovyshche. *Rus.*
  Kanevskoye Vodokhranilishche.
  reservoir C Ukraine
Kaniys'ke Vodoskhovyshche *see*
  Kanivske Vodoskhovyshche
Kanjiža *100 D2 Ger.* Altkanischa, *Hung.*
  Magyarkanizsa, Ókanizsa; *prev.* Stara
  Kanjiža. Vojvodina, N Serbia
Kankaanpää *85 D5* Satakunta,
  SW Finland
Kankakee *40 B3* Illinois, N USA
Kankan *74 D4* E Guinea
Kannur *132 B2 var.* Cannanore. Kerala,
  SW India
Kano *75 F4* Kano, N Nigeria
Kānpur *135 E3 Eng.* Cawnpore. Uttar
  Pradesh, N India

Kansas *45 F5 off.* State of Kansas, *also
  known as* Jayhawker State, Sunflower
  State. *state* C USA
Kansas City *45 F4* Kansas, C USA
Kansas City *45 F4* Missouri, C USA
Kansas River *45 F5 river* Kansas, C USA
Kansk *115 E4* Krasnoyarskiy Kray,
  S Russia
Kansu *see* Gansu
Kantalahti *see* Kandalaksha
Kántanos *105 C7* Kríti, Greece,
  E Mediterranean Sea
Kantemirovka *111 B6* Voronezhskaya
  Oblast', W Russia
Kantipur *see* Kathmandu
Kanton *145 F3 var.* Abariringa, Canton
  Island; *prev.* Mary Island. *atoll* Phoenix
  Islands, C Kiribati
Kanye *78 C4* Southern, SE Botswana
Kaohsiung *129 D6 var.* Gaoxiong, *Jap.*
  Takao, Takow. S Taiwan
Kaolack *74 B3 var.* Kaolak. W Senegal
Kaolak *see* Kaolack
Kaolan *see* Lanzhou
Kaoma *78 C2* Western, W Zambia
Kapelle *87 B5* Zeeland, SW Netherlands
Kapellen *87 C5* Antwerpen, N Belgium
Kapka, Massif du *76 C2 mountain range*
  E Chad
Kaplangky, Plato *see* Gaplaňgyr Platosy
Kapoeas *see* Kapuas, Sungai
Kapoeta *73 C5* E Equatoria, SE
  South Sudan
Kaposvár *99 C7* Somogy, SW Hungary
Kappeln *94 B2* Schleswig-Holstein,
  N Germany
Kaproncza *see* Koprivnica
Kapstad *see* Cape Town
Kapsukas *see* Marijampolė
Kaptsevichy *107 C7 Rus.* Koptsevichi.
  Homyel'skaya Voblasts', SE Belarus
Kapuas, Sungai *138 C4 prev.* Kapoeas.
  river Borneo, C Indonesia
Kapuskasing *38 C4* Ontario, S Canada
Kapyl' *107 C6 Rus.* Kopyl'. Minskaya
  Voblasts', C Belarus
Kara-Balta *123 F2* Chüy Oblusu/
  Chuyskaya Oblast', N Kyrgyzstan
Karabil', Vozvyshennost' *see* Garabil
  Belentligi
Kara-Bogaz-Gol, Zaliv *see* Garabogaz
  Köli
Karabük *116 C2* Karabük,
  NW Turkey (Türkiye)
Karachi *134 B3* Sindh, SE Pakistan
Karácsonkó *see* Piatra-Neamţ
Karadeniz *see* Black Sea
Karadeniz Boğazı *see* İstanbul Boğazı
Karaferiye *see* Véroia
Karagandy/Karaganda *see* Qaraghandy
Karaginskiy, Ostrov *115 H2 island*
  E Russia
Karagumskiy Kanal *see* Garagum Kanaly
Karak *see* Al Karak
Kara-Kala *see* Magtymguly
Karakax *see* Moyu
Karaklıssse *see* Ağrı
Karakol *123 G2 prev.* Przheval'sk.
  Ysyk-Köl Oblusu/Issyk-Kul'skaya
  Oblast', NE Kyrgyzstan
Karakol *123 G2 var.* Karakolka.
  Ysyk-Köl Oblusu/Issyk-Kul'skaya
  Oblast', NE Kyrgyzstan
Karakolka *see* Karakol
Karakoram Range *134 D1 mountain
  range* C Asia
Karaköse *see* Ağrı
Karakul' *see* Qarokŭl, Tajikistan
Kara Kum *see* Garagum
Kara Kum Canal/Karakumskiy Kanal
  see Garagum Kanaly
Karakumy, Peski *see* Garagum
Karamai *see* Karamay
Karaman *116 C4* Karaman,
  S Turkey (Türkiye)
Karamay *126 B2 var.* Karamai,
  Kelamayi; *prev. Chin.* K'o-la-ma-i.
  Xinjiang Uygur Zizhiqu, NW China
Karamea Bight *151 B5 gulf* South Island,
  New Zealand
Karapelit *104 E1 Rom.* Stejarul.
  Dobrich, NE Bulgaria
Kara-Say *123 G2* Ysyk-Köl Oblusu/
  Issyk-Kul'skaya Oblast', NE Kyrgyzstan
Karasburg *78 B4* Karas, S Namibia
Kara Sea *see* Karskoye More
Kara Strait *see* Karskiye Vorota, Proliv
Karatau *123 G4 Kaz.* Qarataŭ. Zhambyl,
  S Kazakhstan
Karavás *105 B7* Kýthira, S Greece

Karbalā' *120 B3 var.* Kerbala, Kerbela.
  Karbalā', S Iraq
Kardeljevo *see* Ploče
Kardhítsa *see* Kardítsa
Karditsa *105 B5 var.* Kardhítsa.
  Thessalía, C Greece
Kärdla *106 C2 Ger.* Kertel. Hiiumaa,
  W Estonia
Kardzhali *104 D3 var.* Kirdzhali,
  Kŭrdzhali. Kardzhali, S Bulgaria
Karet *see* Kāghet
Kargı *116 C2* Çorum,
  N Turkey (Türkiye)
Kargilik *see* Yecheng
Kariba *78 D2* Mashonaland West,
  N Zimbabwe
Kariba, Lake *78 D3* reservoir Zambia/
  Zimbabwe
Karibib *78 B3* Erongo, C Namibia
Kariega *78 C5 prev.* Uitenhage. Eastern
  Cape, S South Africa
Karies *see* Karyés
Karigasniemi *84 D2 Smi.*
  Garegasnjárga. Lappi, N Finland
Karimata, Selat *138 C4 strait*
  W Indonesia
Karīmnagar *134 D5* Telangana, C India
Karin *72 D4* Sahil, N Somalia
Kariot *see* Ikaría
Káristos *see* Kárystos
Karkinits'ka Zatoka *109 E4 Rus.*
  Karkinitskiy Zaliv. *gulf* S Ukraine
Karkinitskiy Zaliv *see* Karkinits'ka Zatoka
Karkük *see* Kirkük
Karleby *see* Kokkola
Karl-Marx-Stadt *see* Chemnitz
Karlő *see* Hailuoto
Karlovac *100 B3 Ger.* Karlstadt, *Hung.*
  Károlyváros. Karlovac, C Croatia
Karlovy Vary *99 A5 Ger.* Karlsbad; *prev.
  Eng.* Carlsbad. Karlovarský Kraj,
  W Czechia (Czech Republic)
Karlsbad *see* Karlovy Vary
Karlsburg *see* Alba Iulia
Karlskrona *85 C7* Blekinge, S Sweden
Karlsruhe *95 B6 var.* Carlsruhe.
  Baden-Württemberg, SW Germany
Karlstad *85 B6* Värmland, C Sweden
Karlstadt *see* Karlovac
Karnāl *134 D2* Haryāna, N India
Karnātaka *132 C1 var.* Kanara; *prev.*
  Maisur, Mysore. *cultural region*
  W India
Karnobat *104 D2* Burgas, E Bulgaria
Karnul *see* Kurnool
Karol *see* Carei
Károly-Fehérvár *see* Alba Iulia
Károlyváros *see* Karlovac
Karpaten *see* Carpathian Mountains
Kárpathos *105 E7* Kárpathos, SE Greece
Kárpathos *105 E7 It.* Scarpanto; *anc.*
  Carpathos, Carpathus. *island*
  SE Greece
Karpaty *see* Carpathian Mountains
Karpenísi *105 B5 prev.* Karpenísion.
  Stereá Ellás, C Greece
Karpenísion *see* Karpenísi
Karpilovka *see* Aktsyabrski
Kars *117 F2 var.* Qars. Kars,
  NE Turkey (Türkiye)
Karsau *see* Kārsava
Kārsava *106 D4 Ger.* Karsau; *prev. Rus.*
  Korsovka. Ludza, E Latvia
Karshi *see* Qarshi, Uzbekistan
Karskiye Vorota, Proliv *110 E2 Eng.*
  Kara Strait. *strait* N Russia
Karskoye More *114 D2 Eng.* Kara Sea.
  sea Arctic Ocean
Kárystos *105 C6 var.* Káristos. Évvoia,
  C Greece
Kasai *77 C6 var.* Cassai, Kassai. *river*
  Angola/Dem. Rep. Congo
Kasaji *77 D7* Katanga, S Dem.
  Rep. Congo
Kasama *78 D1* Northern, N Zambia
Kasan *see* Koson
Kāsaragod *132 B2* Kerala, SW India
Kaschau *see* Košice
Kāshān *120 C3* Eşfahān, C Iran
Kashgar *see* Kashi
Kashi *126 A3 Chin.* Kaxgar, K'o-shih,
  Uygh. Kaxgar. Xinjiang Uygur
  Zizhiqu, NW China
Kasi *see* Vārānasi
Kasongo *77 D6* Maniema, E Dem.
  Rep. Congo
Kasongo-Lunda *77 C7* Bandundu,
  SW Dem. Rep. Congo

Kásos 105 D7 *island* S Greece
Kaspiy Mangy Oypaty *see* Caspian Depression
Kaspiysk 111 B8 Respublika Dagestan, SW Russia
Kaspiyskoye More/Kaspiy Tengizi *see* Caspian Sea
Kassa *see* Košice
Kassai *see* Kasai
Kassala 72 C4 Kassala, E Sudan
Kassel 94 B4 *prev.* Cassel. Hessen, C Germany
Kasserine 71 E2 *var.* Al Qaşrayn. W Tunisia
Kastamonu 116 C2 *var.* Castamoni, Kastamuni. Kastamonu, N Turkey (Türkiye)
Kastamuni *see* Kastamonu
Kastaneá 104 B4 Kentrikí Makedonía, N Greece
Kastélli *see* Kíssamos
Kastoriá 104 B4 Dytikí Makedonía, N Greece
Kástro 105 C6 Sífnos, Kykládes, Greece, Aegean Sea
Kastsyukovichy 107 E7 *Rus.* Kostyukovichi. Mahilyowskaya Voblasts', E Belarus
Kastsyukowka 107 D7 *Rus.* Kostyukovka. Homyel'skaya Voblasts', SE Belarus
Kasulu 73 B7 Kigoma, W Tanzania
Kasumiga-ura 131 D5 *lake* Honshū, S Japan
Katahdin, Mount 41 G1 *mountain* Maine, NE USA
Katalla 36 C3 Alaska, USA
Katana *see* Qaṭanā
Katanning 147 B7 Western Australia
Katawaz *see* Zarghūn Shahr
Katchall Island 133 F3 *island* Nicobar Islands, India, NE Indian Ocean
Kateríni 104 B4 Kentrikí Makedonía, N Greece
Katha 136 B2 Sagaing, N Myanmar (Burma)
Katherine 148 A2 Northern Territory, N Australia
Kathmandu 124 C3 *prev.* Kantipur. *country capital* (Nepal) Central, C Nepal
Katikati 150 D3 Bay of Plenty, North Island, New Zealand
Katima Mulilo 78 C3 Caprivi, NE Namibia
Katiola 74 D4 C Ivory Coast
Kati Thanda *see* Eyre, Lake
Kā Tiritiri o Te Moana *see* Southern Alps
Káto Achaḯa 105 B5 *var.* Kato Ahaia, Káto Akhaía. Dytikí Ellás, S Greece
Kato Ahaia/Káto Akhaía *see* Káto Achaḯa
Katoúna 105 A5 Dytikí Ellás, C Greece
Katowice 99 C5 *Ger.* Kattowitz. Śląskie, S Poland
Katsina 75 G3 Katsina, N Nigeria
Kattakurgan *see* Kattaqo'rg'on
Kattaqo'rg'on 123 E2 *Rus.* Kattakurgan. Samarqand Viloyati, C Uzbekistan
Kattavía 105 E7 Ródos, Dodekánisa, Greece, Aegean Sea
Kattegat 85 B7 *Dan.* Kattegatt. *strait* N Europe
Kattegatt *see* Kattegat
Kattowitz *see* Katowice
Kaua'i 47 A7 *var.* Kauai. *island* Hawaiian Islands, Hawaii, USA, C Pacific Ocean
Kauai *see* Kaua'i
Kauen *see* Kaunas
Kaufbeuren 95 C6 Bayern, S Germany
Kaunas 106 B4 *Ger.* Kauen, *Pol.* Kowno; *prev. Rus.* Kovno. Kaunas, C Lithuania
Kavadar *see* Kavadarci
Kavadarci 101 E6 *Turk.* Kavadar. C Macedonia
Kavaja *see* Kavajë
Kavajë 101 C6 *It.* Cavaia, Kavaja. Tiranë, W Albania
Kavakli *see* Topolovgrad
Kavála 104 C3 *prev.* Kaválla. Anatolikí Makedonía kai Thráki, NE Greece
Kavali 132 D2 Andhra Pradesh, E India
Kaválla *see* Kavála
Kavango *see* Cubango/Okavango
Kavaratti Island 132 A3 *island* Lakshadweep, Lakshadweep, SW India Asia N Indian Ocean
Kavarna 104 E2 Dobrich, NE Bulgaria
Kavengo *see* Cubango/Okavango

Kavir, Dasht-e 120 D3 *var.* Great Salt Desert. *salt pan* N Iran
Kavkaz *see* Caucasus
Kawagoe 131 D5 Saitama, Honshū, S Japan
Kawasaki 130 A2 Kanagawa, Honshū, S Japan
Kawerau 150 E3 Bay of Plenty, North Island, New Zealand
Kaxgar *see* Kashi
Kaya 75 E4 C Burkina
Kayan 136 B4 Yangon, SW Burma (Myanmar)
Kayan, Sungai 138 D3 *prev.* Kajan. *river* Borneo, C Indonesia
Kayes 74 C3 Kayes, W Mali
Kayseri 116 D3 *var.* Kaisaria; *anc.* Caesarea Mazaca, Mazaca. Kayseri, C Turkey (Türkiye)
Kazach'ye 115 F2 Respublika Sakha (Yakutiya), NE Russia
Kazakh Soviet Socialist Republic *see* Kazakhstan
Kazakhskiy Melkosopchnik *see* Saryarqa
Kazakhstan 114 B4 *off.* Republic of Kazakhstan, *var.* Kazakstan, *Kaz.* Qazaqstan, Qazaqstan Respublikasy; *prev.* Kazakh Soviet Socialist Republic, *Rus.* Respublika Kazakhstan. *country* C Asia
Kazakhstan, Republic of *see* Kazakhstan
Kazakhstan, Respublika *see* Kazakhstan
Kazakh Steppe 114 C4 *var.* Kirghiz Steppe. *uplands* C Kazakhstan
Kazakstan *see* Kazakhstan
Kazan' 111 C5 Respublika Tatarstan, W Russia
Kazandzhik *see* Bereket
Kazanlak 104 D2 *var.* Kazanlŭk, Kazanlik. Stara Zagora, C Bulgaria
Kazanlik *see* Kazanlak
Kazanlŭk *see* Kazanlak
Kazatin *see* Koziatyn
Kazbegi *see* Kazbek
Kazbek 117 F1 *var.* Kazbegi, *Geor.* Mqinvartsveri. *mountain* N Georgia
Kazi Magomed *see* Hacıqabul
Kazvin *see* Qazvin
Kéa 105 C6 *var.* Tziá, Kéos; *anc.* Ceos. *island* Kykládes, Greece, Aegean Sea
Kea, Mauna 47 B8 *mountain* Hawaii, USA
Kéamu *see* Aneityum
Kearney 45 E4 Nebraska, C USA
Keban Baraji 117 E3 *reservoir* C Turkey (Türkiye)
Kebkabiya 72 A4 Northern Darfur, W Sudan
Kebnekaise 84 C3 *mountain* N Sweden
Kecskemét 99 D7 Bács-Kiskun, C Hungary
Kediri 138 D5 Jawa, C Indonesia
Kędzierzyn-Kozle 99 C5 *Ger.* Heydebrech. Opolskie, S Poland
Keelung 129 D6 *var.* Chilung, Jilong; *Jap.* Kirun, Kirun'; *prev. Sp.* Santissima Trinidad. N Taiwan
Keetmanshoop 78 B4 Karas, S Namibia
Kefallinía *see* Kefalloniá
Kefalloniá 105 A5 *var.* Kefallinía. *island* Iónia Nisiá, Greece, C Mediterranean Sea
Kefe *see* Feodosiia
Kegel *see* Keila
Kehl 95 A6 Baden-Württemberg, SW Germany
Kei Islands *see* Kai, Kepulauan
Keijō *see* Seoul
Keila 106 D2 *Ger.* Kegel. Harjumaa, NW Estonia
Keïta 75 F3 Tahoua, C Niger
Keith 149 B7 South Australia
Kēk-Art *see* Kök-Art
Kékes 99 C6 *mountain* N Hungary
Kelamayi *see* Karamay
Kelang *see* Klang
Kelat *see* Kälat
Kelifskiy Uzboy *see* Kelif Uzboýy
Kelif Uzboýy 122 D3 *Rus.* Kelifskiy Uzboy. *salt marsh* E Turkmenistan
Kelkit Çayı 117 E3 *river* N Turkey (Türkiye)
Kelmė 106 B4 Šiauliai, C Lithuania
Kélo 76 B4 Tandjilé, SW Chad
Kelowna 37 E5 British Columbia, SW Canada
Kelso 46 B2 Washington, NW USA
Keltsy *see* Kielce

Keluang 138 B3 *var.* Kluang. Johor, Peninsular Malaysia, Malaysia
Kem' 110 B3 Respublika Kareliya, NW Russia
Kemah 117 E3 Erzincan, E Turkey (Türkiye)
Kemaman *see* Cukai
Kemerovo 114 D4 *prev.* Shcheglovsk. Kemerovskaya Oblast', C Russia
Kemi 84 D4 Lappi, NW Finland
Kemijärvi 84 D3 *Swe.* Kemiträsk. Lappi, N Finland
Kemijoki 84 D3 *river* NW Finland
Kemin 123 G2 *prev.* Bystrovka. Chüy Oblusu/Chuyskaya Oblast', N Kyrgyzstan
Kemins Island *see* Nikumaroro
Kemiträsk *see* Kemijärvi
Kemmuna 102 A5 *var.* Comino. *island* C Malta
Kempele 84 D4 Pohjois-Pohjanmaa, C Finland
Kempten 95 B7 Bayern, S Germany
Kendal 89 D5 NW England, United Kingdom
Kendari 139 E4 Sulawesi, C Indonesia
Kenedy 49 G4 Texas, SW USA
Kenema 74 C4 SE Sierra Leone
Kēneurgench *see* Köneürgenç
Kenge 77 C6 Bandundu, SW Dem. Rep. Congo
Kengtung 136 C3 *prev.* Keng Tung. Shan State, E Myanmar (Burma)
Keng Tung *see* Kengtung
Kénitra 70 C2 *prev.* Port-Lyautey. NW Morocco
Kennett 45 H5 Missouri, C USA
Kennewick 46 C2 Washington, NW USA
Kenora 38 A3 Ontario, S Canada
Kenosha 40 B3 Wisconsin, N USA
Kentau 114 B5 Türkistan/Turkestan, S Kazakhstan
Kentucky 40 C5 *off.* Commonwealth of Kentucky, *also known as* Bluegrass State. *state* C USA
Kentucky Lake 40 B5 *reservoir* Kentucky/Tennessee, S USA
Kentung *see* Keng Tung
Kenya 73 C6 *off.* Republic of Kenya. *country* E Africa
Kenya, Mount *see* Kirinyaga
Kenya, Republic of *see* Kenya
Keokuk 45 G4 Iowa, C USA
Kéos *see* Kéa
Kępno 98 C4 Wielkopolskie, C Poland
Keppel Island *see* Niuatoputapu
Kerak *see* Al Karak
Kerala 132 C2 *cultural region* S India
Kerasunt *see* Giresun
Keratéa 105 C6 *var.* Keratea. Attikí, C Greece
Keratea *see* Keratéa
Kerbala/Kerbela *see* Karbalā'
Kerch 109 G5 *Rus.* Kerch'. Respublika Krym, SE Ukraine
Kerch' *see* Kerch
Kerchens'ka Protska/Kerchenskiy Proliv *see* Kerch Strait
Kerch Strait 109 G4 *var.* Bosporus Cimmerius, Enikale Strait, *Rus.* Kerchenskiy Proliv, *Ukr.* Kerchens'ka Protska. *strait* Black Sea/Sea of Azov
Keremitlik *see* Lyulyakovo
Kerguelen 141 C7 *island* C French Southern and Antarctic Lands
Kerguelen Plateau 141 C7 *undersea plateau* S Indian Ocean
Kerí 105 A6 Zákynthos, Iónia Nisiá, Greece, C Mediterranean Sea
Kerikeri 150 D2 Northland, North Island, New Zealand
Kerkenah, Îles de 71 F2 *var.* Kerkennah Islands, *Ar.* Juzur Qarqannah. *island group* E Tunisia
Kerkennah Islands *see* Kerkenah, Îles de
Kerki 122 D3 *prev.* Atamyrat. Lebap Welaýaty, E Turkmenistan
Kérkira *see* Kérkyra
Kerkrade 87 D6 Limburg, SE Netherlands
Kerkuk *see* Kirkūk
Kérkyra 104 A4 *var.* Kérkira, *Eng.* Corfu. Kérkyra, Iónia Nisiá, Greece, C Mediterranean Sea
Kermadec Islands 152 C4 *island group* New Zealand, SW Pacific Ocean
Kermadec Trench 143 E4 *trench* SW Pacific Ocean
Kermān 120 D3 *var.* Kirman; *anc.* Carmana. Kermān, C Iran

Kermānshāh 120 C3 *var.* Qahremānshahr; *prev.* Bākhtarān. Kermānshāhān, W Iran
Kerrville 49 F4 Texas, SW USA
Kertel *see* Kärdla
Kerulen 127 E2 *Chin.* Herlen He, *Mong.* Herlen Gol. *river* China/Mongolia
Keryneia *see* Girne
Kesennuma 130 D4 Miyagi, Honshū, C Japan
Keszthely 99 C7 Zala, SW Hungary
Ketchikan 36 D4 Revillagigedo Island, Alaska, USA
Kętrzyn 98 D2 *Ger.* Rastenburg. Warmińsko-Mazurskie, NE Poland
Kettering 89 D6 C England, United Kingdom
Kettering 40 C4 Ohio, N USA
Keupriya *see* Primorsko
Keuruu 85 D5 Keski-Suomi, C Finland
Keweenaw Peninsula 40 B1 *peninsula* Michigan, N USA
Key Largo 43 F5 Key Largo, Florida, SE USA
Keystone State *see* Pennsylvania
Key West 43 E5 Florida Keys, Florida, SE USA
Kezdivásárhely *see* Târgu Secuiesc
K'gari 148 E4 *prev.* Fraser Island, *var.* Great Sandy Island. *island* Queensland, E Australia
Khabarovsk 115 G4 Khabarovskiy Kray, SE Russia
Khachmas *see* Xaçmaz
Khadera *see* Hadera
Khairpur 134 B3 Sindh, SE Pakistan
Khalij as Suways 72 B2 *var.* Suez, Gulf of. *gulf* NE Egypt
Khalkidhikí *see* Chalkidikí
Khalkís *see* Chalkída
Khambhat, Gulf of 134 C4 *Eng.* Gulf of Cambay. *gulf* W India
Khamis Mushayt 121 B6 *var.* Hamis Musait. 'Asīr, SW Saudi Arabia
Khānābād 123 E3 Kunduz, NE Afghanistan
Khān al Baghdādī *see* Al Baghdādī
Khandwa 134 D4 Madhya Pradesh, C India
Khanh Hung *see* Soc Trăng
Khaniá *see* Chaniá
Khanka, Lake 129 E2 *var.* Hsing-K'ai Hu, Lake Hanka, *Chin.* Xingkai Hu, *Rus.* Ozero Khanka. *lake* China/Russia
Khanka, Ozero *see* Khanka, Lake
Khankendi *see* Xankändi
Khanthabouli 136 D4 *prev.* Savannakhét. Savannakhét, S Laos
Khanty-Mansiysk 114 C3 *prev.* Ostyako-Voguls'k. Khanty-Mansiyskiy Avtonomnyy Okrug-Yugra, C Russia
Khān Yūnis 119 A7 *var.* Khān Yunus. S Gaza Strip
Khān Yūnus *see* Khān Yūnis
Khanzi *see* Ghanzi
Kharagpur 135 F4 West Bengal, NE India
Kharbin *see* Harbin
Kharkiv 109 G2 *Rus.* Khar'kov. Kharkivs'ka Oblast', NE Ukraine
Khar'kov *see* Kharkiv
Kharmanli *see* Harmanli
Khartoum 72 B4 *var.* El Khartûm, Khartum. *country capital* (Sudan) Khartoum, C Sudan
Khartum *see* Khartoum
Khasavyurt 111 B8 Respublika Dagestan, SW Russia
Khash, Dasht-e *see* Khash Desert
Khash Desert 122 D5 *Eng.* Khash Desert. *desert* SW Afghanistan
Khashim Al Qirba/Khashm al Qirbah *see* Khashm el Girba
Khashm el Girba 72 C4 *var.* Khashim Al Qirba, Khashm al Qirbah. Kassala, E Sudan
Khaskovo *see* Haskovo
Khaybar, Kowtal-e *see* Khyber Pass
Khaydarken *see* Aydarken
Khazar, Baḩr-e/Khazar, Daryā-ye *see* Caspian Sea
Khelat *see* Kälat
Kherson 109 E4 Khersons'ka Oblast', S Ukraine
Kheta 115 E2 *river* N Russia
Khíos *see* Chíos
Khiva/Khiwa *see* Xiva
Khmel'nitskiy *see* Khmel 'nyts'kyy

Khmelnytskyi 108 C2 prev.
  Khmel'nyts'kyy, Rus. Khmel'nitskiy;
  prev. Proskurov. Khmel'nyts'ka
  Oblast', W Ukraine
Khmel'nyts'kyy see Khmelnytskyi
Khodasy 107 E6 Rus. Khodosy.
  Mahilyowskaya Voblasts', E Belarus
Khodoriv 108 C2 Pol. Chodorów, Rus.
  Khodorov. L'vivs'ka Oblast',
  NW Ukraine
Khodorov see Khodoriv
Khodosy see Khodasy
Khodzhent see Khujand
Khoi see Khvoy
Khojend see Khujand
Khokand see Qo'qon
Kholm see Khulm
Kholm see Chelm
Kholon see Holon
Khoms see Al Khums
Khong Sedone see Muang Khôngxédôn
Khon Kaen 136 D4 var. Muang Khon
  Kaen. Khon Kaen, E Thailand
Khor 115 G4 Khabarovsky Kray,
  SE Russia
Khorat see Nakhon Ratchasima
Khorog see Khorugh
Khorugh 123 F3 Rus. Khorog.
  S Tajikistan
Khöst 123 F4 prev. Khowst. Khöst,
  E Afghanistan
Khotan see Hotan
Khouribga 70 B2 C Morocco
Khovd see Hovd
Khowst see Khöst
Khoy see Khvoy
Khoyniki 107 D8 Homyel'skaya
  Voblasts', SE Belarus
Khrustalnyi 109 H3 prev. Khrustal'nyy,
  Krindachevka. Rus. Krasnyi Luch.
  Luhans'ka Oblast', E Ukraine
Khrustal'nyy see Khrustalnyi
Khudzhand see Khujand
Khujand 123 E2 var. Khodzhent,
  Khojend, Rus. Khudzhand; prev.
  Leninabad, Taj. Leninobod.
  N Tajikistan
Khulm 123 E3 var. Kholm, Tashqurghan.
  Balkh, N Afghanistan
Khulna 135 G4 Khulna, SW Bangladesh
Khums see Al Khums
Khust 108 B3 var. Husté, Cz. Chust,
  Hung. Huszt. Zakarpats'ka Oblast',
  W Ukraine
Khvoy 120 C2 var. Khoi, Khoy.
  Āžarbāyjān-e Bākhtarī, NW Iran
Khyber Pass 134 C1 var. Kowtal-e
  Khaybar. pass Afghanistan/Pakistan
Kiangmai see Chiang Mai
Kiang-ning see Nanjing
Kiangsi see Jiangxi
Kiangsu see Jiangsu
Kiáto 105 B6 prev. Kiáton. Pelopónnisos,
  S Greece
Kiáton see Kiáto
Kiayi see Chiayi
Kibangou 77 B6 Niari, SW Congo
Kibombo 77 D6 Maniema, E Dem.
  Rep. Congo
Kıbrıs/Kıbrıs Cumhuriyeti see Cyprus
Kičevo 101 D6 SW Macedonia
Kidderminster 89 D6 C England,
  United Kingdom
Kiel 94 B2 Schleswig-Holstein,
  N Germany
Kielce 98 D4 Rus. Keltsy. Świętokrzyskie,
  C Poland
Kieler Bucht 94 B2 bay N Germany
Kiev see Kyiv
Kiffa 74 C3 Assaba, S Mauritania
Kigali 73 B6 country capital (Rwanda)
  C Rwanda
Kigoma 73 B7 Kigoma, W Tanzania
Kihnu 106 C2 var. Kihnu Saar, Ger.
  Kühnö. island SW Estonia
Kihnu Saar see Kihnu
Kii-suido 131 C7 strait S Japan
Kikinda 100 D3 Ger. Grosskikinda,
  Hung. Nagykikinda; prev. Velika
  Kikinda. Vojvodina, N Serbia
Kikládhes see Kykládes
Kikwit 77 C6 Bandundu, W Dem.
  Rep. Congo
Kilien Mountains see Qilian Shan
Kilimane see Quelimane
Kilimanjaro 69 E5 region E Tanzania
Kilimanjaro 73 C7 var. Uhuru Peak.
  volcano NE Tanzania
Kilingi-Nõmme 106 D3 Ger. Kurkund.
  Pärnumaa, SW Estonia

Kilis 116 D4 Kilis, S Turkey (Türkiye)
Kiliya 108 D4 Rom. Chilia-Nouă.
  Odes'ka Oblast', SW Ukraine
Kilkenny 89 B6 Ir. Cill Chainnigh.
  Kilkenny, S Ireland
Kilkís 104 B3 Kentrikí Makedonía,
  N Greece
Killarney 89 A6 Ir. Cill Airne. Kerry,
  SW Ireland
Killeen 49 G3 Texas, SW USA
Kilmain see Quelimane
Kilmarnock 88 C4 W Scotland,
  United Kingdom
Kilwa see Kilwa Kivinje
Kilwa Kivinje 73 C7 var. Kilwa. Lindi,
  SE Tanzania
Kimberley 78 C4 Northern Cape,
  C South Africa
Kimberley Plateau 146 C3 plateau
  Western Australia
Kimch'aek 129 E3 prev. Sŏngjin.
  E North Korea
Kími see Kými
Kinabalu, Gunung 138 D3 mountain
  East Malaysia
Kindersley 37 F5 Saskatchewan,
  S Canada
Kindia 74 C4 Guinée-Maritime,
  SW Guinea
Kindley Field 42 A4 air base E Bermuda
Kindu 77 D6 prev. Kindu-Port-Empain.
  Maniema, C Dem. Rep. Congo
Kindu-Port-Empain see Kindu
Kineshma 111 C5 Ivanovskaya Oblast',
  W Russia
King Abdullah Economic City 121 A5
  W Saudi Arabia
King Charles Islands see Kong
  Karls Land
King Christian IX Land see Kong
  Christian IX Land
King Frederik VI Coast see Kong
  Frederik VI Kyst
King Frederik VIII Land see Kong
  Frederik VIII Land
King Island 149 B8 island Tasmania,
  SE Australia
King Island see Kadan Kyun
Kingissepp see Kuressaare
Kingman 48 A1 Arizona, SW USA
Kingman Reef 145 E2 US territory
  C Pacific Ocean
Kingsford Smith 148 E2 (Sydney) New
  South Wales, SE Australia
King's Lynn 89 E6 var. Bishop's Lynn,
  Kings Lynn, Lynn, Lynn Regis.
  E England, United Kingdom
Kings Lynn see King's Lynn
King Sound 146 B3 sound
  Western Australia
Kingsport 43 E1 Tennessee, S USA
Kingston 54 B5 country capital (Jamaica)
  E Jamaica
Kingston 38 D5 Ontario, SE Canada
Kingston 41 F3 New York, NE USA
Kingston upon Hull 89 D5 var. Hull.
  E England, United Kingdom
Kingston upon Thames 89 A8
  SE England, United Kingdom
Kingstown 55 H4 country capital (Saint
  Vincent and the Grenadines) Saint
  Vincent, Saint Vincent and
  the Grenadines
Kingstown see Dún Laoghaire
Kingsville 49 G5 Texas, SW USA
King William Island 37 F3 island
  Nunavut, N Canada
Kinneret, Yam 119 B5 var. Chinnereth,
  Sea of Bahr Tabariya, Sea of Galilee,
  Lake Tiberias, Ar. Bahrat Tabariya. lake
  N Israel
Kinrooi 87 D5 Limburg, NE Belgium
Kinshasa 77 B6 prev. Léopoldville.
  country capital (Dem. Rep. Congo)
  Kinshasa, W Dem. Rep. Congo
Kintyre 88 B4 peninsula W Scotland,
  United Kingdom
Kinyeti 73 B5 mountain S South Sudan
Kiparissía see Kyparissía
Kipili 73 B7 Rukwa, W Tanzania
Kipushi 77 D8 Katanga, SE Dem.
  Rep. Congo
Kirdzhali see Kardzhali
Kirghizia see Kyrgyzstan
Kirghiz Range 123 F2 Rus. Kirgizskiy
  Khrebet; prev. Alexander Range.
  mountain range Kazakhstan/
  Kyrgyzstan
Kirghiz SSR see Kyrgyzstan
Kirghiz Steppe see Kazakh Steppe

Kirgizskaya SSR see Kyrgyzstan
Kirgizskiy Khrebet see Kirghiz Range
Kiriath-Arba see Hebron
Kiribati 145 F2 off. Republic of Kiribati.
  country C Pacific Ocean
Kiribati, Republic of see Kiribati
Kırıkhan 116 D4 Hatay,
  S Turkey (Türkiye)
Kırıkkale 116 C3 province
  C Turkey (Türkiye)
Kirin see Jilin
Kirinyaga 73 C6 prev. Mount Kenya.
  volcano C Kenya
Kirishi 110 B4 var. Kirisi.
  Leningradskaya Oblast', NW Russia
Kirisi see Kirishi
Kiritimati 145 G2 prev. Christmas
  Island. atoll Line Islands, E Kiribati
Kirkenes 84 E2 Fin. Kirkkoniemi.
  Finnmark, N Norway
Kirk-Kilissa see Kırklareli
Kirkland Lake 38 D4 Ontario, S Canada
Kırklareli 116 A2 prev. Kirk-Kilissa.
  Kırklareli, NW Turkey (Türkiye)
Kirkpatrick, Mount 154 B3 mountain
  Antarctica
Kirksville 45 G4 Missouri, C USA
Kirkūk 120 B3 var. Karkūk, Kerkuk. At
  Ta'mīn, N Iraq
Kirkwall 88 C2 NE Scotland,
  United Kingdom
Kirkwood 45 G4 Missouri, C USA
Kir Moab/Kir of Moab see Al Karak
Kirov 111 C5 prev. Vyatka. Kirovskaya
  Oblast', NW Russia
Kirovabad see Gäncä
Kirovakan see Vanadzor
Kirovo-Chepetsk 111 D5 Kirovskaya
  Oblast', NW Russia
Kirovohrad see Kropyvnytskyi
Kirovo see Kropyvnytskyi
Kirthar Range 134 B3 mountain range
  S Pakistan
Kiruna 84 C3 Smi. Giron. Norrbotten,
  N Sweden
Kirun/Kirun' see Keelung
Kisalföld see Little Alföld
Kisangani 77 D5 prev. Stanleyville.
  Orientale, NE Dem. Rep. Congo
Kishinev see Chişinău
Kislovodsk 111 B7 Stavropol'skiy Kray,
  SW Russia
Kismaayo 73 D6 var. Chisimayu,
  Kismayu, It. Chisimaio. Jubbada
  Hoose, S Somalia
Kismayu see Kismaayo
Kíssamos 105 C7 prev. Kastélli. Kríti,
  Greece, E Mediterranean Sea
Kissidougou 74 C4 Guinée-Forestière,
  S Guinea
Kissimmee, Lake 43 E4 lake Florida,
  SE USA
Kistna see Krishna
Kisumu 73 C6 prev. Port Florence.
  Nyanza, W Kenya
Kisvárda 99 E6 Ger. Kleinwardein.
  Szabolcs-Szatmár-Bereg, E Hungary
Kita 74 D3 Kayes, W Mali
Kitab see Kitob
Kitakyūshū 131 A7 var. Kitakyūshū.
  Fukuoka, Kyūshū, SW Japan
Kitakyūsyū see Kitakyūshū
Kitami 130 D2 Hokkaidō, NE Japan
Kitchener 38 C5 Ontario, S Canada
Kitega see Gitega
Kíthnos see Kýthnos
Kitimat 36 D4 British Columbia,
  SW Canada
Kitinen 84 D3 river N Finland
Kitob 123 E3 Rus. Kitab. Qashqadaryo
  Viloyati, S Uzbekistan
Kitwe 78 D2 var. Kitwe-Nkana.
  Copperbelt, C Zambia
Kitwe-Nkana see Kitwe
Kitzbühler Alpen 95 C7 mountain
  range W Austria
Kivalina 36 C2 Alaska, USA
Kivalo 84 D3 ridge C Finland
Kivertsi 108 C1 Pol. Kiwerce, Rus.
  Kivertsy. Volyns'ka Oblast',
  NW Ukraine
Kivertsy see Kivertsi
Kivu, Lac see Kivu, Lake
Kivu, Lake 77 E6 Fr. Lac Kivu. lake
  Rwanda/Dem. Rep. Congo
Kiwerce see Kivertsi
Kiyev see Kyiv
Kiyevskoye Vodokhranilishche see
  Kyivske Vodoskhovyshche

Kızıl Irmak 116 C3 river
  C Turkey (Türkiye)
Kizil Kum see Kyzylkum Desert
Kizyl-Arvat see Gyzylarbat
Kjølen see Kölen
Kladno 99 A5 Středočeský,
  NW Czechia (Czech Republic)
Klagenfurt 95 D7 Slvn. Celovec.
  Kärnten, S Austria
Klaipėda 106 B3 Ger. Memel. Klaipeda,
  NW Lithuania
Klamath Falls 46 B4 Oregon, NW USA
Klamath Mountains 46 A4 mountain
  range California/Oregon, W USA
Klang 138 B3 var. Kelang; prev. Port
  Swettenham. Selangor, Peninsular
  Malaysia
Klarälven 85 B6 river Norway/Sweden
Klatovy 99 A5 Ger. Klattau. Plzeňský
  Kraj, W Czechia (Czech Republic)
Klattau see Klatovy
Klausenburg see Cluj-Napoca
Klazienaveen 86 E2 Drenthe,
  NE Netherlands
Kleines Ungarisches Tiefland see
  Little Alföld
Klein Karas 78 B4 Karas, S Namibia
Kleinwardein see Kisvárda
Kleisoúra 105 A5 Ípeiros, W Greece
Klerksdorp 78 D4 North-West,
  N South Africa
Klimavichy 107 E7 Rus. Klimovichi.
  Mahilyowskaya Voblasts', E Belarus
Klimovichi see Klimavichy
Klintsy 111 A5 Bryanskaya Oblast',
  W Russia
Klisura 104 C2 Plovdiv, C Bulgaria
Ključ 100 B3 Federacija Bosna I
  Hercegovina, NW Bosnia and
  Herzegovina
Klobuck 98 C4 Śląskie, S Poland
Klosters 95 B7 Graubünden,
  SE Switzerland
Kluang see Keluang
Kluczbork 98 C4 Ger. Kreuzburg,
  Kreuzburg in Oberschlesien. Opolskie,
  S Poland
Klyuchevskaya Sopka, Vulkan 115 H3
  volcano E Russia
Knin 100 B4 Šibenik-Knin, S Croatia
Knjaževac 100 E4 Serbia, E Serbia
Knokke-Heist 87 A5 West-Vlaanderen,
  NW Belgium
Knoxville 42 D1 Tennessee, S USA
Knud Rasmussen Land 82 D1 physical
  region N Greenland
Kobdo see Hovd
Kōbe 131 C6 Hyōgo, Honshū,
  SW Japan
København 85 B7 Eng. Copenhagen;
  anc. Hafnia. country capital (Denmark)
  Sjælland, København, E Denmark
Kobenni 74 D3 Hodh el Gharbi,
  S Mauritania
Koblenz 95 A5 prev. Coblenz, Fr.
  Coblence; anc. Confluentes.
  Rheinland-Pfalz, W Germany
Kobrin see Kobryn
Kobryn 107 A6 Rus. Kobrin. Brestskaya
  Voblasts', SW Belarus
Kobuleti 117 F2 prev. K'obulet'i.
  W Georgia
K'obulet'i see Kobuleti
Kočani 101 E6 NE Macedonia
Kočevje 95 D8 Ger. Gottschee.
  S Slovenia
Koch Bihār 135 G3 West Bengal,
  NE India
Kochi see Kochi
Kochi 132 C3 var. Cochin, Kochchi.
  Kerala SW India
Kōchi 131 B7 var. Kôti. Kōchi, Shikoku,
  SW Japan
Kochiu see Gejiu
Kodiak 36 C3 Kodiak Island, Alaska,
  USA
Kodiak Island 36 C3 island Alaska, USA
Koedoes see Kudus
Koeln see Köln
Koepang see Kupang
Ko-erh-mu see Golmud
Koetai see Mahakam, Sungai
Koetaradja see Bandaaceh
Kōfu 131 D5 var. Kôhu. Yamanashi,
  Honshū, S Japan
Kogarah 148 E2 New South Wales,
  E Australia
Kogon 122 D2 Rus. Kagan. Buxoro
  Viloyati, C Uzbekistan
Kôhalom see Rupea
Kohīma 135 H3 state capital Nāgāland,
  E India

213

Kohtla-Järve 106 E2 1da-Virumaa, NE Estonia

Kôhu see Kôfu

Kokand see Qo'qon

Kök-Art 123 G2 Rus. Këk-Art, prev. Alaykel', Alay-Kuu. Osh Oblusu/ Oshskaya Oblast', SW Kyrgyzstan

Kokchetav see Kökshetaü

Kokkola 84 D4 Swe. Karleby; prev. Swe. Gamlakarleby. Österbotten, W Finland

Koko 75 F4 Kebbi, W Nigeria

Kokomo 40 C4 Indiana, N USA

Koko Nor see Qinghai, China

Koko Nor see Qinghai Hu, China

Kokrines 36 C2 Alaska, USA

Koksoak 38 D2 river Québec, E Canada

Kokshaal-Too 123 G2 Rus. Kokshaal-Tau, Khrebet Kokshaltau, var. Kakshaal Range. mountain range China/Kyrgyzstan

Kokshaal-Tau see Kokshaal-Too

Kokshaltau, Khrebet see Kokshaal-Too

Kokshetau see Kökshetaü

Kökshetaü 114 C4 var. Kokshetau; prev. Kokchetav. Kokshetau, N Kazakhstan

Koksijde 87 A5 West-Vlaanderen, W Belgium

Kokstad 78 D5 KwaZulu/Natal, E South Africa

Kolaka 139 E4 Sulawesi, C Indonesia

K'o-la-ma-i see Karamay

Kola Peninsula see Kol'skiy Poluostrov

Kolari 84 D3 Lappi, NW Finland

Kolárovo 99 C6 Ger. Gutta; prev. Guta, Hung. Gúta. Nitriansky Kraj, SW Slovakia

Kolberg see Kołobrzeg

Kolda 74 C3 S Senegal

Kolding 85 A7 Vejle, C Denmark

Kölen 81 E1 Nor. Kjølen. mountain range Norway/Sweden

Kolguyev, Ostrov 110 C2 island NW Russia

Kolhapur 132 B1 Mahārāshtra, SW India

Kolhumadulu 132 A5 var. Thaa Atoll. atoll S Maldives

Kolín 99 B5 Ger. Kolin. Střední Čechy, C Czechia (Czech Republic)

Kolka 106 C2 Talsi, NW Latvia

Kolkasrags 106 C2 prev. Eng. Cape Domesnes. headland NW Latvia

Kolkāta 135 G4 prev. Calcutta. West Bengal, E India

Kollam 132 C4 var. Quilon. Kerala, SW India

Kolmar see Colmar

Köln 94 A4 var. Koeln, Eng./Fr. Cologne, prev. Cöln; anc. Colonia Agrippina, Oppidum Ubiorum. Nordrhein-Westfalen, W Germany

Kolo 98 C3 Wielkopolskie, C Poland

Kołobrzeg 98 B2 Ger. Kolberg. Zachodnie-pomorskie, NW Poland

Kolokani 74 D3 Koulikoro, W Mali

Kolomea see Kolomya

Kolomna 111 B5 Moskovskaya Oblast', W Russia

Kolomyya see Kolomya

Kolomya 108 C3 prev. Kolomyya, Ger. Kolomea. Ivano-Frankivs'ka Oblast', W Ukraine

Kolosjoki see Nikel'

Kolozsvár see Cluj-Napoca

Kolpa 100 A2 Ger. Kulpa, Croatian Kupa. river Croatia/Slovenia

Kolpino 110 B4 Leningradskaya Oblast', NW Russia

Kol'skiy Poluostrov 110 C2 Eng. Kola Peninsula. peninsula NW Russia

Kolwezi 77 D7 Katanga, S Dem. Rep. Congo

Kolyma 115 G2 river NE Russia

Komani 78 D5 prev. Queenstown. Eastern Cape, S South Africa

Komatsu 131 C5 var. Komatu. Ishikawa, Honshū, SW Japan

Komatu see Komatsu

Kommunizm, Qullai see Ismoili Somoní, Qullai

Komoé 75 E4 var. Komoé Fleuve. river E Ivory Coast

Komoé Fleuve see Komoé

Komotau see Chomutov

Komotiní 104 D3 var. Gümüljina, Turk. Gümülcine. Anatolikí Makedonía kai Thráki, NE Greece

Kompong Som see Sihanoukville

Komrat see Comrat

Komsomolets, Ostrov 115 E1 island Severnaya Zemlya, N Russia

Komsomol'sk-na-Amure 115 G4 Khabarovskiy Kray, SE Russia

Kondolovo 104 E3 Burgas, E Bulgaria

Kondopoga 110 B3 Respublika Kareliya, NW Russia

Kondoz see Kunduz

Köneürgenç 122 C2 var. Köneürgench, Rus. Këneurgench; prev. Kunya-Urgench. Daşoguz Welaýaty, N Turkmenistan

Kong Christian IX Land 82 D4 Eng. King Christian IX Land. physical region SE Greenland

Kong Frederik IX Land 82 C3 physical region SW Greenland

Kong Frederik VIII Land 83 E2 Eng. King Frederik VIII Land. physical region NE Greenland

Kong Frederik VI Kyst 82 C4 Eng. King Frederik VI Coast. physical region SE Greenland

Kong Karls Land 83 G2 Eng. King Charles Islands. island group SE Svalbard

Kongo see Congo (river)

Kongolo 77 D6 Katanga, E Dem. Rep. Congo

Kongor 73 B5 Jonglei, E South Sudan

Kong Oscar Fjord 83 E3 fjord E Greenland

Kongsberg 85 B6 Buskerud, S Norway

Kông, Tônlé 137 E5 var. Xê Kong. river Cambodia/Laos

Kong, Xê see Kông, Tônlé

Königgrätz see Hradec Králové

Königshütte see Chorzów

Konin 98 C3 Ger. Kuhnau. Weilkopolskie, C Poland

Koninkrijk der Nederlanden see Netherlands

Konispol 101 C7 var. Konispoli. Vlorë, S Albania

Konispoli see Konispol

Kónitsa 104 A4 Ípeiros, W Greece

Konitz see Chojnice

Konjic 100 C4 Federacija Bosna I Hercegovina, S Bosnia and Herzegovina

Konosha 110 C4 Arkhangel'skaya Oblast', NW Russia

Konotop 109 F1 Sums'ka Oblast', NE Ukraine

Konstantinovka see Kostiantynivka

Konstanz 95 B7 var. Constanz, Eng. Constance, hist. Kostnitz; anc. Constantia. Baden-Württemberg, S Germany

Konstanza see Constanţa

Konya 116 C4 var. Konieh, prev. Konia; anc. Iconium. Konya, C Turkey (Türkiye)

Kopaonik 101 D5 mountain range S Serbia

Kopar see Koper

Koper 95 D8 It. Capodistria; prev. Kopar. SW Slovenia

Köpetdag 122 C3 var. Kopet Dagh, Rus. Khrebet Kopetdag, Per. Kopeh Dāgh mountain range Iran/ Turkmenistan

Kopeh Dāgh/ Kopet Dagh see Köpetdag

Kopetdag, Khrebet see Köpetdag

Kopreinitz see Koprivnica

Koprivnica 100 B2 Ger. Kopreinitz, Hung. Kaproncza. Koprivnica-Križevci, N Croatia

Köprülü see Veles

Koptsevichi see Kaptsevichy

Kopyl' see Kapyl'

Korat see Nakhon Ratchasima

Korat Plateau 136 D4 plateau E Thailand

Korba 135 E4 Chhattīsgarh, C India

Korça see Korçë

Korçë 101 D6 var. Korça, Gk. Korytsa, It. Corriza; prev. Koritsa. Korçë, SE Albania

Korčula 100 B4 It. Curzola; anc. Corcyra Nigra. island S Croatia

Korea Bay 127 G3 bay China/ North Korea

Korea, Democratic People's Republic of see North Korea

Korea, Republic of see South Korea

Korea Strait 131 A7 Jap. Chōsen-kaikyō, Kor. Taehan-haehyŏp. channel Japan/ South Korea

Korhogo 74 D4 N Ivory Coast

Kórinthos 105 B6 anc. Corinthus Eng. Corinth. Pelopónnisos, S Greece

Korinthiakós Kólpos 105 B5 Eng. Gulf of Corinth; anc. Corinthiacus Sinus. gulf C Greece

Koritsa see Korçë

Kōriyama 131 D5 Fukushima, Honshū, C Japan

Korla 126 C3 Chin. K'u-erh-lo. Xinjiang Uygur Zizhiqu, NW China

Körmend 99 B7 Vas, W Hungary

Koróni 105 B6 Pelopónnisos, S Greece

Koror 144 A2 (Palau) Oreor, N Palau

Körös see Križevci

Korosten' 108 D1 Zhytomyrs'ka Oblast', NW Ukraine

Koro Toro 76 C2 Borkou-Ennedi-Tibesti, N Chad

Korsovka see Kārsava

Kortrijk 87 A6 Fr. Courtrai. West-Vlaanderen, W Belgium

Koryak Range 115 H2 var. Koryakskiy Khrebet, Eng. Koryak Range. mountain range NE Russia

Koryak Range see Koryakskoye Nagor'ye

Koryakskiy Khrebet see Koryakskoye Nagor'ye

Koryazhma 110 C4 Arkhangel'skaya Oblast', NW Russia

Korytsa see Korçë

Kos 105 E6 Kos, Dodekánisa, Greece, Aegean Sea

Kos 105 E6 It. Coo; anc. Cos. island Dodekánisa, Greece, Aegean Sea

Ko-saki 131 A7 headland Nagasaki, Tsushima, SW Japan

Kościan 98 B4 Ger. Kosten. Wielkopolskie, C Poland

Kościerzyna 98 C2 Pomorskie, NW Poland

Kosciusko, Mount see Kosciuszko, Mount

Kosciuszko, Mount 149 C7 prev. Mount Kosciusko. mountain New South Wales, SE Australia

K'o-shih see Kashi

Koshikijima-retto 131 A8 var. Kosikizima Rettô. island group SW Japan

Kôshū see Gwangju

Košice 99 D6 Ger. Kaschau, Hung. Kassa. Košický Kraj, E Slovakia

Kosikizima Rettô see Koshikijima-rettô

Köslin see Koszalin

Koson 123 E3 Rus. Kasan. Qashqadaryo Viloyati, S Uzbekistan

Kosovo 101 D5 prev. Autonomous Province of Kosovo and Metohija. country (not fully recognised) SE Europe

Kosovo and Metohija, Autonomous Province of see Kosovo

Kosovo Polje see Fushë Kosovë

Kosovska Mitrovica see Mitrovicë

Kosrae 144 C2 prev. Kusaie. island Caroline Islands, E Federated States of Micronesia

Kossou, Lac de 74 D5 lake C Ivory Coast

Kostanay see Qostanay

Kosten see Kościan

Kostenets 104 C2 prev. Georgi Dimitrov. Sofiya, W Bulgaria

Kostiantynivka 109 G3 prev. Kostyantynivka, Rus. Konstantinovka. Donets'ka Oblast', SE Ukraine

Kostnitz see Konstanz

Kostroma 110 B4 Kostromskaya Oblast', NW Russia

Kostyantynivka see Kostiantynivka

Kostyukovichi see Kastsyukovichy

Kostyukovka see Kastsyukowka

Koszalin 98 B2 Ger. Köslin. Zachodnio-pomorskie, NW Poland

Kota 134 D3 prev. Kotah. Rājasthān, N India

Kota Baharu see Kota Bharu

Kota Bahru see Kota Bharu

Kotabaru see Jayapura

Kota Bharu 138 B3 var. Kota Baharu, Kota Bahru. Kelantan, Peninsular Malaysia

Kotaboemi see Kotabumi

Kotabumi 138 B4 prev. Kotaboemi. Sumatera, W Indonesia

Kotah see Kota

Kota Kinabalu 138 D3 prev. Jesselton. Sabah, East Malaysia

Kotel'nyy, Ostrov 115 E2 island Novosibirskiye Ostrova, N Russia

Kotka 85 E5 Kymenlaakso, S Finland

Kotlas 110 C4 Arkhangel'skaya Oblast', NW Russia

Kotonu see Cotonou

Kotor 101 C5 It. Cattaro. SW Montenegro

Kotovsk see Podilsk

Kotovsk see Hînceşti

Kottbus see Cottbus

Kotto 76 D4 river Central African Republic/Dem. Rep. Congo

Kotuy 115 E2 river N Russia

Koudougou 75 E4 C Burkina

Koulamoutou 77 B6 Ogooué-Lolo, C Gabon

Koulikoro 74 D3 Koulikoro, SW Mali

Koumra 76 C4 Moyen-Chari, S Chad

Kourou 59 H3 N French Guiana

Kousseir see Al Quşayr

Kousséri 76 B3 prev. Fort-Foureau. Extrême-Nord, NE Cameroon

Koutiala 74 D4 Sikasso, S Mali

Kouvola 85 E5 Kymenlaakso, S Finland

Kovel 108 C1 Pol. Kowel. Volyns'ka Oblast', NW Ukraine

Kovno see Kaunas

Koweit see Kuwait

Kowel see Kovel

Kowloon 128 A2 Hong Kong, S China

Kowno see Kaunas

Kozáni 104 B4 Dytikí Makedonía, N Greece

Kozara 100 B3 mountain range NW Bosnia and Herzegovina

Kozarska Dubica see Bosanska Dubica

Kozhikode 132 C2 var. Calicut. Kerala, SW India

Koziatyn 108 D2 prev. Kozyatyn, Rus. Kazatin. Vinnyts'ka Oblast', C Ukraine

Kozu-shima 131 D6 island E Japan

Kozyatyn see Koziatyn

Kpalimé 75 E5 var. Palimé. SW Togo

Krâchéh see Kratie

Kragujevac 100 D4 Serbia, C Serbia

Krainburg see Kranj

Kra, Isthmus of 137 B6 isthmus Malaysia/Thailand

Krakau see Kraków

Kraków 99 D5 Eng. Cracow, Ger. Krakau; anc. Cracovia. Małopolskie, S Poland

Králánh 137 D5 Siěmréab, NW Cambodia

Kralendijk 55 E5 dependent territory capital (Bonaire) Lesser Antilles, S Caribbean Sea

Kraljevo 100 D4 prev. Rankovićevo. Serbia, C Serbia

Kramatorsk 109 G3 Donets'ka Oblast', SE Ukraine

Kramfors 85 C5 Västernorrland, C Sweden

Kranéa see Kraniá

Kraniá 104 B4 var. Kranéa. Dytikí Makedonía, N Greece

Kranj 95 D7 Ger. Krainburg. NW Slovenia

Kranz see Zelenogradsk

Krāslava 106 D4 Krāslava, SE Latvia

Krasnaye 107 C5 Rus. Krasnoye. Minskaya Voblasts', C Belarus

Krasnoarmeysk 111 C6 Saratovskaya Oblast', W Russia

Krasnodar 111 A7 prev. Ekaterinodar, Yekaterinodar. Krasnodarskiy Kray, SW Russia

Krasnodon see Sorokyne

Krasnogor see Kallaste

Krasnogvardeyskoye see Kurman

Krasnohvardiyske see Kurman

Krasnokamensk 115 F4 Chitinskaya Oblast', S Russia

Krasnokamsk 111 D5 Permskaya Oblast', W Russia

Krasnoperekopsk see Yany Kapu

Krasnostav see Krasnystaw

Krasnovodsk see Türkmenbaşy

Krasnovodskiy Zaliv/Krasnowodsk Aylagy see Türkmenbaşy Aylagy

Krasnoyarsk 114 D4 Krasnoyarskiy Kray, S Russia

Krasnoye see Krasnaye

Krasnystaw 98 E4 Rus. Krasnostav. Lubelskie, SE Poland

Krasnyy Luch see Khrustalnyi

Kratie 137 D6 Khmer. Krâchéh. Kratie, E Cambodia

Krâvanh, Chuŏr Phnum 137 C6 Eng. Cardamom Mountains, Fr. Chaîne des Cardamomes. mountain range W Cambodia

Krefeld *94 A4* Nordrhein-Westfalen, W Germany
Kreisstadt *see* Krosno Odrzańskie
Kremenchugskoye Vodokhranilishche/ Kremenchuk Reservoir *see* Kremenchutske Vodoskhovyshche
Kremenchuk *109 F2 Rus.* Kremenchug. Poltavs'ka Oblast', NE Ukraine
Kremenchuk Reservoir *see* Kremenchutske Vodoskhovyshche
Kremenchutske Vodoskhovyshche *109 F2. Eng.* Kremenchuk Reservoir, *Rus.* Kremenchugskoye Vodokhranilishche. *reservoir* C Ukraine
Kremenets *108 C2 Pol.* Krzemieniec. Ternopil's'ka Oblast', W Ukraine
Kremennaya *see* Kreminna
Kreminna *109 G2 Rus.* Kremennaya. Luhans'ka Oblast', E Ukraine
Kresna *see* Kresna
Kresna *104 C3 var.* Kresena. Blagoevgrad, SW Bulgaria
Kretikon Delagos *see* Kritikó Pélagos
Kretinga *106 B3 Ger.* Krottingen. Klaipėda, NW Lithuania
Kreutz *see* Cristuru Secuiesc
Kreuz *see* Križevci, Croatia
Kreuz *see* Risti, Estonia
Kreuzburg/Kreuzburg in Oberschlesien *see* Kluczbork
Krichëv *see* Krychaw
Krievija *see* Russia
Krindachevka *see* Khrustal'nyy
Krishna *132 C1 prev.* Kistna. *river* C India
Krishnagiri *132 C2* Tamil Nādu, SE India
Kristiania *see* Oslo
Kristiansand *85 A6 var.* Christiansand. Vest-Agder, S Norway
Kristianstad *85 B7* Skåne, S Sweden
Kristiansund *84 A4 var.* Christiansund. Møre og Romsdal, S Norway
Kríti *105 C7 Eng.* Crete. *island* Greece, Aegean Sea
Kritikó Pélagos *105 D7 var.* Kretikon Delagos, *Eng.* Sea of Crete; *anc.* Mare Creticum. *sea* Greece, Aegean Sea
Krivoy Rog *see* Kryvyy Rih
Križevci *100 B2 Ger.* Kreuz, *Hung.* Kőrös. Varaždin, NE Croatia
Krk *100 A3 It.* Veglia; *anc.* Curieta. *island* NW Croatia
Kroatien *see* Croatia
Krolevets *109 F1* Sums'ka Oblast', NE Ukraine
Królewska Huta *see* Chorzów
Kronach *95 C5* Bayern, E Germany
Kronstadt *see* Brașov
Kroonstad *78 D4* Free State, C South Africa
Kropotkin *111 A7* Krasnodarskiy Kray, SW Russia
Kropyvnytskyi *109 E3 prev.* Kropyvnyts'kyy, *Rus.* Kirovohrad; *prev.* Kirovo, Yelizavetgrad, Zinov'yevsk. Kirovohrads'ka Oblast', C Ukraine
Kropyvnyts'kyy *see* Kropyvnytskyi
Krosno *99 D5 Ger.* Krossen. Podkarpackie, SE Poland
Krosno Odrzańskie *98 B3 Ger.* Crossen, Kreisstadt. Lubuskie, W Poland
Krossen *see* Krosno
Krottingen *see* Kretinga
Krško *95 E8 Ger.* Gurkfeld; *prev.* Videm-Krško, *It.* Videm. E Slovenia
Krugloye *see* Kruhlaye
Kruhlaye *107 D6 Rus.* Krugloye. Mahilyowskaya Voblasts', E Belarus
Kruja *see* Krujë
Krujë *101 C6 var.* Kruja, *It.* Croia. Durrës, C Albania
Krummau *see* Český Krumlov
Krupki *107 D6* Minskaya Voblasts', C Belarus
Krušné Hory *see* Erzgebirge
Krychaw *107 E7 Rus.* Krichëv. Mahilyowskaya Voblasts', E Belarus
Krym *109 F5 prev.* Kryms'kyy Pivostriv, *Eng.* Crimea. (Ukrainian territory annexed by Russia since 2014). *peninsula* S Ukraine

Krymsky Hory *109 F5 mountain range* S Ukraine
Kryms'kyy Pivostriv *see* Krym
Krynica *99 D5 Ger.* Tannenhof. Małopolskie, S Poland
Kryve Ozero *109 E3* Odes'ka Oblast', SW Ukraine
Kryvyy Rih *109 F3 prev.* Kryvyy Rih, *Rus.* Krivoy Rog. Dnipropetrovska Oblast, SE Ukraine
Kryvyy Rih *see* Kryvyi Rih
Krzemieniec *see* Kremenets
Ksar al Kabir *see* Ksar-el-Kebir
Ksar al Soule *see* Er-Rachidia
Ksar-el-Kebir *70 C2 var.* Alcázar, Ksar al Kabir, Ksar-el-Kébir, *Ar.* Al-Kasr al-Kebir, Al-Qsar al-Kbir, *Sp.* Alcazarquivir. NW Morocco
Ksar-el-Kébir *see* Ksar-el-Kebir
Kuala Dungun *see* Dungun
Kuala Lumpur *138 B3 country capital* (Malaysia) Kuala Lumpur, Peninsular Malaysia
Kuala Terengganu *138 B3 var.* Kuala Trengganu. Terengganu, Peninsular Malaysia
Kualatungkal *138 B4* Sumatera, W Indonesia
Kuang-chou *see* Guangzhou
Kuang-hsi *see* Guangxi Zhuangzu Zizhiqu
Kuang-tung *see* Guangdong
Kuang-yuan *see* Guangyuan
Kuantan *138 B3* Pahang, Peninsular Malaysia
Kuba *see* Quba
Kuban' *109 G5 var.* Hypanis. *river* SW Russia
Kubango *see* Cubango/Okavango
Kuching *138 C3 prev.* Sarawak. Sarawak, East Malaysia
Kūchnay Darwēshān *122 D5 prev.* Kūchnay Darweyshān. Helmand, S Afghanistan
Kūchnay Darweyshān *see* Kūchnay Darwēshān
Kuçovë *see* Kuçovë
Kuçovë *101 C6 var.* Kuçova; *prev.* Qyteti Stalin. Berat, C Albania
Kudara *see* Ghūdara
Kudus *138 C5 prev.* Koedoes. Jawa, C Indonesia
Kuei-lin *see* Guilin
Kuei-Yang/Kuei-yang *see* Guiyang
K'u-erh-lo *see* Korla
Kueyang *see* Guiyang
Kugaaruk *37 G3 prev.* Pelly Bay. Nunavut, N Canada
Kugluktuk *53 E3 var.* Qurlurtuuq; *prev.* Coppermine. Nunavut, NW Canada
Kuhmo *84 E4* Kainuu, E Finland
Kuhnau *see* Konin
Kühnö *see* Kihnu
Kuibyshev *see* Kuybyshevskoye Vodokhranilishche
Kuito *see* Cuito
Kuji *130 D3 var.* Kuzi. Iwate, Honshū, C Japan
Kukës *101 D5 var.* Kukësi. Kukës, NE Albania
Kukësi *see* Kukës
Kukong *see* Shaoguan
Kukukhoto *see* Hohhot
Kula Kangri *135 G3 var.* Kulhakangri. *mountain* Bhutan/China
Kuldiga *106 B3 Ger.* Goldingen. Kuldiga, W Latvia
Kuldja *see* Yining
Kulhakangri *see* Kula Kangri
Kullorsuaq *82 D2 var.* Kuvdlorssuak. Avannaata, C Greenland
Kulm *see* Chełmno
Kulmsee *see* Chełmża
Kůlob *123 F3 Rus.* Kulyab. SW Tajikistan
Kulpa *see* Kolpa
Kulu *116 C3* Konya, W Turkey (Türkiye)
Kulunda Steppe *114 C4 Kaz.* Qulyndy Zhazyghy, *Rus.* Kulundinskaya Ravnina. *grassland* Kazakhstan/Russia
Kulundinskaya Ravnina *see* Kulunda Steppe
Kulyab *see* Kůlob
Kum *see* Qom
Kuma *111 B7 river* SW Russia
Kumamoto *131 A7* Kumamoto, Kyūshū, SW Japan
Kumanova *see* Kumanovo
Kumanovo *101 E5 Turk.* Kumanova. N Macedonia
Kumasi *75 E5 prev.* Coomassie. C Ghana

Kumayri *see* Gyumri
Kumba *77 A5* Sud-Ouest, W Cameroon
Kumertau *111 D6* Respublika Bashkortostan, W Russia
Kumilla *see* Cumilla
Kumo *75 G4* Gombe, E Nigeria
Kumon Range *136 B2 mountain range* N Myanmar (Burma)
Kumul *see* Hami
Kunashiri *see* Kunashir, Ostrov
Kunashir, Ostrov *130 E1 var.* Kunashiri. *island* Kuril'skiye Ostrova, SE Russia
Kunda *106 E2* Lääne-Virumaa, NE Estonia
Kunduz *123 E3 prev.* Kondoz. NE Afghanistan
Kunene *69 C6 var.* Kunene. *river* Angola/Namibia
Kunene *see* Cunene
Kungsbacka *85 B7* Halland, S Sweden
Kungur *111 D5* Permskaya Oblast', NW Russia
Kunlun Mountains *see* Kunlun Shan
Kunlun Shan *126 B4 Eng.* Kunlun Mountains. *mountain range* NW China
Kunming *129 B6 var.* K'un-ming; *prev.* Yunnan. *province capital* Yunnan, SW China
K'un-ming *see* Kunming
Kununurra *146 D3* Western Australia
Kunya-Urgench *see* Köneürgenç
Kuopio *85 E5* Pohjois-Savo, C Finland
Kupa *see* Kolpa
Kupang *139 E5 prev.* Koepang. Timor, C Indonesia
Kupiansk *109 G2 prev.* Kup"yans'k, *Rus.* Kupyansk. Kharkivs'ka Oblast', E Ukraine
Kup"yans'k *see* Kupiansk
Kupyansk *see* Kupiansk
Kür *see* Kura
Kura *117 H3 Az.* Kür, *Geor.* Mtkvari, *Turk.* Kura Nehri. *river* SW Asia
Kura Nehri *see* Kura
Kurashiki *131 B6 var.* Kurasiki. Okayama, Honshū, SW Japan
Kurasiki *see* Kurashiki
Kurdistan *117 F4 cultural region* SW Asia
Kürdzhali *see* Kardzhali
Kure *131 B7* Hiroshima, Honshū, SW Japan
Küre Dağları *116 C2 mountain range* N Turkey (Türkiye)
Kuressaare *106 C2 Ger.* Arensburg; *prev.* Kingissepp. Saaremaa, W Estonia
Kureyka *112 D2 river* N Russia
Kurgan-Tyube *see* Bokhtar
Kuria Muria Islands *see* Ḩalāniyāt, Juzur al
Kuril'skiye Ostrova *115 H4 Eng.* Kuril Islands. *island group* SE Russia
Kuril Islands *see* Kuril'skiye Ostrova
Kuril-Kamchatka Depression *see* Kuril-Kamchatka Trench
Kuril-Kamchatka Trench *113 F3 var.* Kuril-Kamchatka Depression. *trench* NW Pacific Ocean
Kuril'sk *130 E1 Jap.* Shana. Kuril'skiye Ostrova, Sakhalinskaya Oblats', SE Russia
Ku-ring-gai *148 E1* New South Wales, E Australia
Kurisches Haff *see* Courland Lagoon
Kurkund *see* Kilingi-Nõmme
Kurman *109 F4 prev.* Krasnohvardiiske, *Rus.* Krasnogvardeyskoye. Respublika Krym, S Ukraine
Kurnool *132 C1 var.* Karnul. Andhra Pradesh, S India
Kursk *111 A6* Kurskaya Oblast', W Russia
Kurskiy Zaliv *see* Courland Lagoon
Kuršumlija *101 D5* Serbia, S Serbia
Kurtbunar *see* Tervel
Kurtitsch/Kürtös *see* Curtici
Kuruktag *126 C3 mountain range* NW China
Kurume *131 A7* Fukuoka, Kyūshū, SW Japan
Kurupukari *59 F3* C Guyana
Kusaie *see* Kosrae
Kushiro *130 D2 var.* Kusiro. Hokkaidō, NE Japan
Kushka *see* Serhetabat
Kusiro *see* Kushiro
Kuskokwim Mountains *36 C3 mountain range* Alaska, USA
Kustanay *see* Qostanay
Küstence/Küstendje *see* Constanța

Kütahya *116 B3 prev.* Kutaia. Kütahya, W Turkey (Türkiye)
Kutai *see* Mahakam, Sungai
Kutaisi *117 F2 prev.* K'ut'aisi *117 F2* W Georgia W Georgia
K'ut'aisi *see* Kutaisi
Kůt al 'Amārah *see* Al Kūt
Kutaradja/Kutaraja *see* Bandaaceh
Kutch, Gulf of *see* Kachchh, Gulf of
Kutch, Rann of *see* Kachchh, Rann of
Kutina *100 B3* Sisak-Moslavina, NE Croatia
Kutno *98 C3* Łódzkie, C Poland
Kuujjuaq *39 E2 prev.* Fort-Chimo. Québec, E Canada
Kuusamo *84 E3* Pohjois-Pohjanmaa, E Finland
Kuvango *see* Cubango
Kuvdlorssuak *see* Kullorsuaq
Kuwait *120 C4 off.* State of Kuwait, *var.* Dawlat al Kuwait, Koweit, Kuweit. *country* SW Asia
Kuwait *see* Al Kuwayt
Kuwait City *see* Al Kuwayt
Kuwait, Dawlat al *see* Kuwait
Kuwait, State of *see* Kuwait
Kuwajleen *see* Kwajalein Atoll
Kuwayt *120 C3* Maysān, E Iraq
Kuybyshev *see* Samara
Kuybyshev Reservoir *see* Kuybyshevskoye Vodokhranilishche
Kuybyshevskoye Vodokhranilishche *111 C5 var.* Kuibyshev, *Eng.* Kuybyshev Reservoir. *reservoir* W Russia
Kuytun *126 B2* Xinjiang Uygur Zizhiqu, NW China
Kuzi *see* Kuji
Kuznetsk *111 B6* Penzenskaya Oblast', W Russia
Kuźnica *98 E2* Białystok, NE Poland
Kvaløya *84 C2 island* N Norway
Kvarnbergsvattnet *84 B4 var.* Frostviken. *lake* N Sweden
Kvarner *100 A3 var.* Carnaro, *It.* Quarnero. *gulf* W Croatia
Kvitøya *83 G1 island* N Svalbard
Kwajalein Atoll *144 C1 var.* Kuwajleen. *atoll* Ralik Chain, C Marshall Islands
Kwando *see* Cuando
Kwangchow *see* Guangzhou
Kwangchu *see* Gwangju
Kwangju *see* Gwangju
Kwango *77 C7 Port.* Cuango. *river* Angola/Dem. Rep. Congo
Kwangsi/Kwangsi Chuang Autonomous Region *see* Guangxi Zhuangzu Zizhiqu
Kwangtung *see* Guangdong
Kwangyuan *see* Guangyuan
Kwanza *see* Cuanza
Kweichu *see* Guiyang
Kweilin *see* Guilin
Kweisui *see* Hohhot
Kweiyang *see* Guiyang
Kwekwe *78 D3 prev.* Que Que. Midlands, C Zimbabwe
Kwesui *see* Hohhot
Kwidzyń *98 C2 Ger.* Marienwerder. Pomorskie, N Poland
Kwigillingok *36 C3* Alaska, USA
Kwilu *77 C6 river* W Dem. Rep. Congo
Kwito *see* Cuito
Kyabé *76 C4* Moyen-Chari, S Chad
Kyaikkami *137 B5 prev.* Amherst. Mon State, S Myanmar (Burma)
Kyaiklat *136 B4* Ayeyarwady, SW Myanmar (Burma)
Kyaikto *136 B4* Mon State, S Myanmar (Burma)
Kyakhta *115 F3* Respublika Buryatiya, S Russia
Kyauksauk *136 B3* Mandalay, C Myanmar (Burma)
Kyiv *109 E2 var.* Kyyiv, Kiev, *Rus.* Kiyev. *country capital* (Ukraine) Kyivska Oblast, N Ukraine
Kyiv Reservoir *see* Kyivske Vodoskhovyshche
Kyivske Vodoskhovyshche *109 E1 var.* Kyyivs'ke Vodoskhovyshche, *Eng.* Kyiv Reservoir, *Rus.* Kiyevskoye Vodokhranilishche. *reservoir* N Ukraine
Kyjov *99 C5 Ger.* Gaya. Jihomoravský Kraj, SE Czechia (Czech Republic)
Kykládes *105 D6 var.* Kikládhes, *Eng.* Cyclades. *island group* SE Greece

Kými 105 C5 prev. Kími. Évvoia,
C Greece
Kyŏngsŏng see Seoul
Kyōto 131 C6 Kyōto, Honshū, SW Japan
Kyparissía 105 B6 var. Kiparissís.
Pelopónnisos, S Greece
Kypriakí Dimokratía see Cyprus
Kýpros see Cyprus
Kyrá Panagía 105 C5 island Vóreies
Sporádes, Greece, Aegean Sea
Kyrenia see Girne
Kyrgyz Republic see Kyrgyzstan
Kyrgyzstan 123 F2 off. Kyrgyz Republic,
var. Kirghizia; prev. Kirgizskaya SSR,
Kirghiz SSR, Republic of Kyrgyzstan.
country C Asia
Kyrgyzstan, Republic of see Kyrgyzstan
Kythira 105 C7 var. Kíthira, It. Cerigo,
Lat. Cythera. island S Greece
Kýthnos 105 C6 Knýthnos, Kykládes,
Greece, Aegean Sea
Kythnos 105 C6 var. Kíthnos, Thermiá,
It. Termia; anc. Cythnos. island
Kykládes, Greece, Aegean Sea
Kythréa see Değirmenlik
Kyushu 131 B7 var. Kyûsyû. island
SW Japan
Kyushu-Palau Ridge 125 F3 var.
Kyusyu-Palau Ridge. undersea ridge
W Pacific Ocean
Kyustendil 104 B2 anc. Pautalia.
Kyustendil, W Bulgaria
Kyusyu-Palau Ridge see Kyushu-
Palau Ridge
Kyyiv see Kyiv
Kyyivs'ke Vodoskhovyshche see Kyivske
Vodoskhovyshche
Kyzyl 114 D4 Respublika Tyva, C Russia
Kyzyl Kum see Kyzylkum Desert
Kyzylkum, Peski see Kyzylkum Desert
Kyzylkum Desert 122 D2 var. Kizil Kum,
Kyzyl Kum, Qizil Qum, Uzb. Qizilqum,
Kaz. Qyzylqum, Rus. Peski Kyzylkum.
desert Kazakhstan/Uzbekistan
Kyzylrabot see Qizilrabot
Kyzyl-Suu 123 G2 prev. Pokrovka.
Ysyk-Köl Oblusu/Issyk-Kul'skaya
Oblast', NE Kyrgyzstan
Kzylorda see Qyzylorda
Kzyl-Orda see Qyzylorda

L

Laaland see Lolland
La Algaba 92 C4 Andalucía, S Spain
Laarne 87 B5 Oost-Vlaanderen,
NW Belgium
La Asunción 59 E1 Nueva Esparta,
NE Venezuela
Laatokka see Ladozhskoye, Ozero
Laâyoune 70 B3 var. Aaiún. country
capital (Western Sahara)
NW Western Sahara
La Banda Oriental see Uruguay
la Baule-Escoublac 90 A4
Loire-Atlantique, NW France
Labé 74 C4 NW Guinea
Labe see Elbe
Laborca see Laborec
Laborec 99 E5 Hung. Laborca. river
E Slovakia
Labrador 39 F2 cultural region
Newfoundland and Labrador,
SW Canada
Labrador Basin 34 E3 var. Labrador Sea
Basin. undersea basin Labrador Sea
Labrador Sea 82 A4 sea
NW Atlantic Ocean
Labrador Sea Basin see Labrador Basin
Labudalin see Ergun
Labutta 137 A5 Ayeyarwady,
SW Myanmar (Burma)
Laç 101 C6 var. Laci. Lezhë, C Albania
La Calera 64 B4 Valparaíso, C Chile
La Carolina 92 D4 Andalucía, S Spain
Laccadive Islands 132 A3 Eng. Laccadive
Islands. island group India,
N Indian Ocean
**Laccadive Islands/Laccadive Minicoy
and Amindivi Islands, the** see
Lakshadweep
La Ceiba 52 D2 Atlántida, N Honduras
Lachanás 104 B3 Kentrikí Makedonía,
N Greece
La Chaux-de-Fonds 95 A7 Neuchâtel,
W Switzerland
Lachlan River 149 C6 river New South
Wales, SE Australia

Laci see Laç
la Ciotat 91 D6 anc. Citharista.
Bouches-du-Rhône, SE France
Lacobriga see Lagos
La Concepción 53 E5 var. Concepción.
Chiriquí, W Panama
La Concepción 58 C1 Zulia,
NW Venezuela
La Condamine 91 C8 W Monaco
Laconia 41 G2 New Hampshire, NE USA
La Crosse 40 A2 Wisconsin, N USA
La Cruz 52 D4 Guanacaste,
NW Costa Rica
Ladoga, Lake see Ladozhskoye, Ozero
Ladozhskoye, Ozero 110 B3 Eng. Lake
Ladoga, Fin. Laatokka. lake NW Russia
Ladysmith 40 B2 Wisconsin, N USA
Lae 144 B3 Morobe, W Papua
New Guinea
La Esperanza 52 C2 Intibucá,
SW Honduras
Lafayette 40 C4 Indiana, N USA
Lafayette 42 B3 Louisiana, S USA
La Fé 54 A2 Pinar del Río, W Cuba
Lafia 75 G4 Nassarawa, C Nigeria
la Flèche 90 B4 Sarthe, NW France
Lagdo, Lac de 76 B4 lake N Cameroon
Laghouat 70 D2 N Algeria
Lagos 75 F5 Lagos, SW Nigeria
Lagos 92 B5 anc. Lacobriga. Faro,
S Portugal
Lagos de Moreno 51 E4 Jalisco,
SW Mexico
Lagouira 70 A4 SW Western Sahara
La Grande 46 C3 Oregon, NW USA
La Guaira 66 B4 Distrito Federal,
N Venezuela
Lagunas 61 B1 Tarapacá, N Chile
Lagunillas 61 G4 Santa Cruz, SE Bolivia
La Habana 54 B2 var. Havana. country
capital (Cuba) Ciudad de La Habana,
W Cuba
Lahat 138 B4 Sumatera, W Indonesia
La Haye see 's-Gravenhage
Laholm 85 B7 Halland, S Sweden
Lahore 134 D2 Punjab, NE Pakistan
Lahr 95 A6 Baden-Württemberg,
S Germany
Lahti 85 D5 Swe. Lahtis. Päijät-Häme,
S Finland
Lahtis see Lahti
Laï 76 B4 prev. Behagle, De Behagle.
Tandjilé, S Chad
Laibach see Ljubljana
Lai Châu 136 D3 Lai Châu, N Vietnam
Laila see Laylā
La Junta 44 D5 Colorado, C USA
Lake Charles 42 A3 Louisiana, S USA
Lake City 43 E3 Florida, SE USA
Lake District 89 C5 physical region
NW England, United Kingdom
Lake Havasu City 48 A2 Arizona,
SW USA
Lake Jackson 49 H4 Texas, SW USA
Lakeland 43 E4 Florida, SE USA
Lakeside 47 C8 California, W USA
Lake State see Michigan
Lakewood 44 D4 Colorado, C USA
Lakhnau see Lucknow
Lakonikós Kólpos 105 B7 gulf S Greece
Lakselv 84 D2 Smi. Leavdnja. Finnmark,
N Norway
la Laon see Laon
Lalibela 72 C4 Āmara, Ethiopia
La Libertad 52 B1 Petén, N Guatemala
La Ligua 64 B4 Valparaíso, C Chile
Lalín 92 C1 Galicia, NW Spain
Lalitpur 135 F3 Central, C Nepal
La Louvière 87 B6 Hainaut, S Belgium
la Maddalena 96 A4 Sardegna, Italy,
C Mediterranean Sea
Lamar 44 D5 Colorado, C USA
La Marmora, Punta 97 A5 mountain
Sardegna, Italy, C Mediterranean Sea
La Massana 91 A8 La Massana,
W Andorra Europe
Lambaréné 77 A6 Moyen-Ogooué,
W Gabon
Lamego 92 C2 Viseu, N Portugal
Lamesa 49 E3 Texas, SW USA
Lamezia Terme 97 D6 Calabria, SE Italy
Lamía 105 B5 Stereá Ellás, C Greece
Lamoni 45 F4 Iowa, C USA
Lampang 136 C4 var. Muang Lampang.
Lampang, NW Thailand
Lámpeia 105 B6 Dytikí Ellás, S Greece
Lanbi Kyun 137 B6 prev. Sullivan Island.
island Myeik Archipelago,
S Myanmar (Burma)

Lancang Jiang see Mekong
Lancaster 89 D5 NW England,
United Kingdom
Lancaster 47 C7 California, W USA
Lancaster 41 F4 Pennsylvania, NE USA
Lancaster Sound 37 F2 sound Nunavut,
N Canada
Lan-chou/Lan-chow/Lanchow see
Lanzhou
Landao see Lantau Island
Landen 87 C6 Vlaams Brabant,
C Belgium
Lander 44 C3 Wyoming, C USA
Landerneau 90 A3 Finistère, NW France
Landes 91 B5 cultural region SW France
Land of Enchantment see New Mexico
The Land of Opportunity see Arkansas
Land of Steady Habits see Connecticut
Land of the Midnight Sun see Alaska
Landsberg see Gorzów Wielkopolski,
Lubuskie, Poland
Landsberg an der Warthe see Gorzów
Wielkopolski
Land's End 89 B8 headland SW England,
United Kingdom
Landshut 95 C6 Bayern, SE Germany
Langar 123 E2 Rus. Lyangar. Navoiy
Viloyati, C Uzbekistan
Langfang 128 D4 Hebei, E China
Langkawi, Pulau 137 B7 island
Peninsular Malaysia
Langres 90 D4 Haute-Marne, N France
Langsa 138 A3 Sumatera, W Indonesia
Lang Shan 127 E3 mountain range
N China
Lang Sơn 136 D3 var. Langson. Lang
Sơn, N Vietnam
Langson see Lang Sơn
Lang Suan 137 B6 Chumphon,
SW Thailand
Languedoc 91 C6 cultural region S France
Länkäran 117 B3 Rus. Lenkoran'.
S Azerbaijan
Lansdowne Reef 129 C8 island
S Spratly Islands
Lansing 40 C3 state capital Michigan,
N USA
Lanta, Ko 137 B7 island S Thailand
Lantau Island 128 A2 Cant. Tai Yue
Shan, Chin. Landao. island Hong
Kong, S China
Lan-ts'ang Chiang see Mekong
Lantung, Gulf of see Liaodong Wan
Lanzarote 70 B3 island Islas Canarias,
Spain, NE Atlantic Ocean
Lanzhou 128 B4 var. Lan-chou,
Lanchow, Lan-chow; prev. Kaolan.
province capital Gansu, C China
Lao Cai 136 D3 Lao Cai, N Vietnam
Laodicea/Laodicea ad Mare see Al
Lādhiqīyah
Laoet see Laut, Pulau
Laojunmiao 128 A3 prev. Yumen.
Gansu, N China
Laon 90 D3 var. la Laon; anc. Laudunum.
Aisne, N France
Lao People's Democratic Republic see
Laos
La Orchila, Isla 58 D1 island
N Venezuela
La Oroya 60 C3 Junín, C Peru
Laos 136 D4 off. Lao People's Democratic
Republic. country SE Asia
La Palma 53 G5 Darién, SE Panama
La Palma 70 A3 island Islas Canarias,
Spain, NE Atlantic Ocean
La Passion-Clipperton see
Clipperton Island
La Paz 61 F4 var. La Paz de Ayacucho.
country capital (Bolivia-seat of
government) La Paz, W Bolivia
La Paz 50 B3 Baja California Sur,
NW Mexico
La Paz, Bahía de 50 B3 bay NW Mexico
La Paz de Ayacucho see La Paz
La Pérouse Strait 130 D1 Jap.
Sōya-saikyō, Rus. Proliv Laperuza.
strait Japan/Russia
Laperuza, Proliv see La Pérouse Strait
Lápithos see Lapta
Lapland see Sápmi
La Plata 64 D4 Buenos Aires, E Argentina
La Plata see Sucre
La Pola 92 D1 var. Pola de Lena.
Asturias, N Spain
Lappeenranta 85 E5 Swe. Villmanstrand.
Etelä-Karjala, SE Finland
Lappi see Sápmi
Lappo see Lapua
Lapta 102 C5 Gk. Lápithos. NW Cyprus

Laptev Sea see Laptevykh, More
Laptevykh, More 115 E2 Eng. Laptev
Sea. sea Arctic Ocean
Lapua 85 D5 Swe. Lappo.
Etelä-Pohjanmaa, W Finland
Lapurdum see Bayonne
Łapy 98 E3 Podlaskie, NE Poland
La Quiaca 64 C2 Jujuy, N Argentina
L'Aquila 96 C4 var. Aquila, Aquila degli
Abruzzi. Abruzzo, C Italy
Laracha 92 B1 Galicia, NW Spain
Laramie 44 C4 Wyoming, C USA
Laramie Mountains 44 C3 mountain
range Wyoming, C USA
Laredo 93 E1 Cantabria, N Spain
Laredo 49 F5 Texas, SW USA
La Réunion see Réunion
Largeau see Faya
Largo 43 E4 Florida, SE USA
Largo, Cayo 54 B2 island W Cuba
Lario see Como, Lago di
La Rioja 64 C3 La Rioja, NW Argentina
La Rioja 93 E2 autonomous community
N Spain
Lárisa 104 B4 var. Larissa. Thessalía,
C Greece
Larissa see Lárisa
Larkana 134 B3 var. Larkhana. Sindh,
SE Pakistan
Larkhana see Lārkāna
Larnaca see Lárnaka
Lárnaca 102 C5 var. Larnaca, Larnax.
SE Cyprus
Larnax see Lárnaka
La Rochelle 90 B4 anc. Rupella.
Charente-Maritime, W France
La Roche-sur-Yon 90 B4 prev. Bourbon
Vendée, Napoléon-Vendée. Vendée,
NW France
La Roda 93 E3 Castilla-La Mancha,
C Spain
La Romana 55 E3 E Dominican Republic
Larvotto 91 C8 N Monaco Europe
Lasa see Lhasa
Las Cabezas de San Juan 92 C5
Andalucía, S Spain
Las Cruces 48 D3 New Mexico,
SW USA
La See d'Urgel see La Seu d'Urgell
La Serena 64 B3 Coquimbo, C Chile
La Seu d'Urgell 93 G1 prev. La See
d'Urgel, Seo de Urgel. Cataluña,
NE Spain
La Seyne-sur-Mer 91 D6 Var, SE France
Lashio 136 B3 Shan State,
E Myanmar (Burma)
Lashkar Gāh 122 D5 var. Lash-Kar-Gar'.
Helmand, S Afghanistan
Lash-Kar-Gar' see Lashkar Gāh
La Sila 97 D6 mountain range SW Italy
La Sirena 52 D3 Región Autónoma
Atlántico Sur, E Nicaragua
Łask 98 C4 Łódzkie, C Poland
Las Lomitas 64 D2 Formosa,
N Argentina
La Solana 93 E4 Castilla-La Mancha,
C Spain
Las Palmas 70 A3 var. Las Palmas de
Gran Canaria. Gran Canaria, Islas
Canarias, Spain, NE Atlantic Ocean
Las Palmas de Gran Canaria see
Las Palmas
La Spezia 96 B3 Liguria, NW Italy
Lassa see Lhasa
Las Tablas 53 F5 Los Santos, S Panama
Last Frontier, The see Alaska
Las Tunas 54 C2 var. Victoria de las
Tunas. Las Tunas, E Cuba
La Suisse see Switzerland
Las Vegas 47 D7 Nevada, W USA
Latacunga 60 B1 Cotopaxi, C Ecuador
Latakia see Al Lādhiqīyah
La Teste 91 B5 Gironde, SW France
Latina 97 C5 prev. Littoria. Lazio, C Italy
La Tortuga, Isla 59 E1 var. Isla Tortuga.
island N Venezuela
La Tuque 39 E4 Québec, SE Canada
Latvia 106 C3 off. Republic of Latvia, Ger.
Lettland, Latv. Latvija, Latvijas
Republika; prev. Latvian SSR, Rus.
Latviyskaya SSR. country NE Europe
**Latvian SSR/Latvija/Latvijas
Republika/Latviyskaya SSR** see Latvia
Latvia, Republic of see Latvia
Laudus see St-Lô
Lauenburg/Lauenburg in Pommern see
Lębork
Lau Group 145 E4 island group E Fiji
Lauis see Lugano

Launceston 149 C8 Tasmania,
 SE Australia
La Unión 52 C2 Olancho, C Honduras
La Unión 93 F4 Murcia, SE Spain
Laurel 42 C3 Mississippi, S USA
Laurel 44 C2 Montana, NW USA
Laurentian Highlands see
 Laurentian Mountains
Laurentian Mountains 39 E3 var.
 Laurentian Highlands, Fr. Les
 Laurentides. plateau Newfoundland
 and Labrador/Québec, Canada
Laurentides, Les see
 Laurentian Mountains
Lauria 97 D6 Basilicata, S Italy
Laurinburg 43 F1 North Carolina,
 SE USA
Lauru see Choiseul
Lausanne 95 A7 It. Losanna. Vaud,
 SW Switzerland
Laut, Pulau 138 D4 prev. Laoet. island
 Borneo, C Indonesia
Laval 38 D4 Québec, SE Canada
Laval 90 B3 Mayenne, NW France
La Vall d'Uixó 93 F3 var. Vall D'Uxó.
 Comunitat Valenciana, E Spain
La Vega 55 E3 var. Concepción de la
 Vega. C Dominican Republic
La Vila Joiosa 93 F4 var. Villajoyosa.
 Comunitat Valenciana, E Spain
Lávrio 105 C6 prev. Lávrion. Attikí,
 C Greece
Lávrion see Lávrio
Lawrence 41 G3 Massachusetts, NE USA
Lawrenceburg 42 C1 Tennessee, S USA
Lawton 49 F2 Oklahoma, C USA
La Yarada 61 E4 Tacna, SW Peru
Laylā 121 C5 var. Laila. Ar Riyāḍ,
 C Saudi Arabia
Lazarev Sea 154 B1 sea Antarctica
Lázaro Cárdenas 51 E5 Michoacán,
 SW Mexico
Leal see Lihula
Leamhcán see Lucan
Leamington 38 C5 Ontario, S Canada
Leavdnja see Lakselv
Lebak 139 E3 Mindanao, S Philippines
Lebanese Republic see Lebanon
Lebanon 45 G5 Missouri, C USA
Lebanon 41 G2 New Hampshire,
 NE USA
Lebanon 46 B3 Oregon, NW USA
Lebanon 118 A4 off. Lebanese Republic,
 Ar. Lubnān, Fr. Liban. country SW Asia
Lebanon, Mount see Liban, Jebel
Lebap 122 D2 Lebapskiy Velayat,
 NE Turkmenistan
Lebedin see Lebedyn
Lebedyn 109 F2 Rus. Lebedin. Sums'ka
 Oblast', NE Ukraine
Łębork 98 C2 var. Lębórk, Ger.
 Lauenburg, Lauenburg in Pommern.
 Pomorskie, N Poland
Lebrija 92 C5 Andalucía, S Spain
Lebu 65 A5 Bío Bío, C Chile
le Cannet 91 D6 Alpes-Maritimes,
 SE France
Le Cap see Cap-Haïtien
Lecce 97 E6 Puglia, SE Italy
Lechainá 105 B6 var. Lehena, Lekhainá.
 Dytikí Ellás, S Greece
Ledo Salinarius see Lons-le-Saunier
Leduc 37 E5 Alberta, SW Canada
Leech Lake 45 F2 lake Minnesota, N USA
Leeds 89 D5 N England, United Kingdom
Leek 86 E2 Groningen, NE Netherlands
Leer 94 A3 Niedersachsen, NW Germany
Leeuwarden 86 D1 Fris. Ljouwert.
 Fryslân, N Netherlands
Leeuwin, Cape 142 A5 headland
 Western Australia
Leeward Islands 55 G3 island group
 E West Indies
Leeward Islands see Sotavento, Ilhas de
Lefkáda 105 A5 prev. Levkás. Lefkáda,
 Iónia Nisiá, Greece,
 C Mediterranean Sea
Lefkáda 105 A5 It. Santa Maura, prev.
 Levkás; anc. Leucas. island Iónia Nisiá,
 Greece, C Mediterranean Sea
Lefká Óri 105 C7 mountain range Kríti,
 Greece, E Mediterranean Sea
Lefkímmi 105 A5 var. Levkímmi.
 Kérkyra, Iónia Nisiá, Greece,
 C Mediterranean Sea
Lefkosía/Lefkoşa see Nicosia
Legaceaster see Chester
Legaspi see Legazpi City
Legazpi City 139 E2 var. Legaspi.
 C Philippines

Leghorn see Livorno
Legnica 98 B4 Ger. Liegnitz.
 Dolnośląskie, SW Poland
le Havre 90 B3 Eng. Havre; prev. le
 Havre-de-Grâce. Seine-Maritime,
 N France
le Havre-de-Grâce see le Havre
Lehena see Lechainá
Leicester 89 D6 Lat. Batae Coritanorum.
 C England, United Kingdom
Leiden 86 B3 prev. Leyden; anc.
 Lugdunum Batavorum. Zuid-Holland,
 W Netherlands
Leie 90 D2 Fr. Lys. river Belgium/France
Leinster 89 B6 Ir. Cúige Laighean.
 cultural region E Ireland
Leipsic see Leipzig
Leipsoí 105 E6 island Dodekánisa,
 Greece, Aegean Sea
Leipzig 94 C4 Pol. Lipsk, hist. Leipsic;
 anc. Lipsia. Sachsen, E Germany
Leiria 92 B3 anc. Collipo. Leiria,
 C Portugal
Leirvik 85 A6 Hordaland, S Norway
Lek 86 C4 river SW Netherlands
Lekhainá see Lechainá
Lekhchevo see Boychinovtsi
Leksand 85 C5 Dalarna, C Sweden
Lel'chitsy see Lyel'chytsy
le Léman see Geneva, Lake
Lelystad 86 D3 Flevoland, C Netherlands
Léman, Lac see Geneva, Lake
le Mans 90 B3 Sarthe, NW France
Lemberg see Lviv
Lemesós 102 C5 var. Limassol.
 SW Cyprus
Lemhi Range 46 D3 mountain range
 Idaho, C USA North America
Lemnos see Límnos
Lemovices see Limoges
Lena 115 F3 river NE Russia
Lena Tablemount 141 B7 seamount
 S Indian Ocean
Lengshuitan see Yongzhou
Leninabad see Khujand
Leninakan see Gyumri
Lenine see Yedy Kuiu
Leningrad see Sankt-Peterburg
Lenino see Yedy Kuiu
Leninobod see Khujand
Leninogor see Ridder/Ridder
Leninogorsk see Ridder/Ridder
Leninpol' see Bakay-Ata
Lenin-Turkmenski see Türkmenabat
Lenkoran' see Länkäran
Lenti 99 B7 Zala, SW Hungary
Lentia see Linz
Leoben 95 E7 Steiermark, C Austria
León 51 E4 var. León de los Aldamas.
 Guanajuato, C Mexico
León 52 D3 León, NW Nicaragua
León 92 D1 Castilla y León, NW Spain
León de los Aldamas see León
Leonídi see Leonídio
Leonídio 105 B6 var. Leonídi.
 Pelopónnisos, S Greece
Léopold II, Lac see Mai-Ndombe, Lac
Léopoldville see Kinshasa
Lepe 92 C4 Andalucía, S Spain
Lepel' see Lyepyel'
le Portel 90 C2 Pas-de-Calais, N France
Le Puglie see Puglia
le Puy 91 C5 prev. le Puy-en-Velay, hist.
 Anicium, Podium Anicensis.
 Haute-Loire, C France
le Puy-en-Velay see le Puy
Léré 76 B4 Mayo-Kébbi, SW Chad
Lérida see Lleida
Lerma 92 D2 Castilla y León,
 N Spain
Leros 105 D6 island Dodekánisa, Greece,
 Aegean Sea
Lerwick 88 D1 NE Scotland,
 United Kingdom
Lesbos see Lésvos
Les Cayes see Cayes
Les Gonaïves see Gonaïves
Leshan 129 B5 Sichuan, C China
les Herbiers 90 B4 Vendée, NW France
Lesh/Leshi see Lezhë
Lesina see Hvar
Leskovac 101 E5 Serbia, SE Serbia
Lesnoy 114 C3 Sverdlovskaya Oblast',
 C Russia
Lesotho 78 D4 off. Kingdom of Lesotho;
 prev. Basutoland. country S Africa
Lesotho, Kingdom of see Lesotho
les Sables-d'Olonne 90 B4 Vendée,
 NW France
Lesser Antarctica see West Antarctica

Lesser Antilles 55 G4 island group
 E West Indies
Lesser Caucasus 117 F2 Rus. Malyy
 Kavkaz. mountain range SW Asia
Lesser Khingan Range see Xiao
 Hinggan Ling
Lesser Sunda Islands 139 E5 Eng. Lesser
 Sunda Islands. island group
 C Indonesia
Lesser Sunda Islands see Nusa Tenggara
Lésvos 116 A3 anc. Lesbos. island
 E Greece
Leszno 98 B4 Ger. Lissa. Wielkopolskie,
 C Poland
Lethbridge 37 E5 Alberta, SW Canada
Lethem 59 F3 S Guyana
Leti, Kepulauan 139 F5 island group
 E Indonesia
Letpadan 136 B4 Bago, SW Myanmar
 (Burma)
Letsôk-aw Kyun 137 B6 var. Letsutan
 Island; prev. Domel Island. island
 Myeik Archipelago,
 S Myanmar (Burma)
Letsutan Island see Letsôk-aw Kyun
Lettland see Latvia
Lëtzebuerg see Luxembourg
Leucas see Lefkáda
Leuven 87 C6 Fr. Louvain, Ger. Löwen.
 Vlaams Brabant, C Belgium
Leuze see Leuze-en-Hainaut
Leuze-en-Hainaut 87 B6 var. Leuze.
 Hainaut, SW Belgium
Léva see Levice
Levanger 84 B4 Nord-Trøndelag,
 C Norway
Levelland 49 E2 Texas, SW USA
Leverkusen 94 A4 Nordrhein-Westfalen,
 W Germany
Levice 99 C6 Ger. Lewentz, Hung. Léva,
 Lewenz. Nitriansky Kraj, SW Slovakia
Levin 150 D4 Manawatu-Wanganui,
 North Island, New Zealand
Levkás see Lefkáda
Levkímmi see Lefkímmi
Lewis, Isle of 88 B2 island NW Scotland,
 United Kingdom
Lewis Range 44 B1 mountain range
 Montana, NW USA
Lewiston 46 C2 Idaho, NW USA
Lewiston 41 G2 Maine, NE USA
Lewistown 44 C1 Montana,
 NW USA
Lexington 40 C5 Kentucky, S USA
Lexington 45 E4 Nebraska, C USA
Leyden see Leiden
Leyte 139 F2 island C Philippines
Leżajsk 99 E5 Podkarpackie, SE Poland
Lezha see Lezhë
Lezhë 101 C6 var. Lezha; prev. Lesh,
 Leshi. Lezhë, NW Albania
Lhasa 126 C5 var. La-sa, Lassa. Xizang
 Zizhiqu, W China
Lhaviyani Atoll see Faadhippolhu Atoll
Lhazê 126 C5 var. Quxar. Xizang
 Zizhiqu, China E Asia
L'Hospitalet de Llobregat 93 G2 var.
 Hospitalet. Cataluña, NE Spain
Liancourt Rocks 131 A5 Jap. Takeshima,
 Kor. Dokdo. disputed island group
 Japan/South Korea
Lianyungang 128 D4 var. Xinpu.
 Jiangsu, E China
Liao see Liaoning
Liaodong Wan 127 G3 Eng. Gulf of
 Lantung, Gulf of Liaotung. gulf
 NE China
Liao He 125 E1 river NE China
Liaoning 128 D3 var. Liao, Liaoning
 Sheng, Shengking, hist. Fengtien,
 Shenking. province NE China
Liaoning Sheng see Liaoning
Liaoyuan 125 D3 var. Dongliao,
 Shuang-liao, Jap. Chengchiatun. Jilin,
 NE China
Liard see Fort Liard
Liban see Lebanon
Liban, Jebel 118 B4 Ar. Jabal al Gharbî,
 Jabal Lubnān, Eng. Mount Lebanon.
 mountain range C Lebanon
Libau see Liepāja
Libby 44 A1 Montana, NW USA
Liberal 45 E5 Kansas, C USA
Liberalitas Julia see Évora
Liberec 98 B4 Ger. Reichenberg.
 Liberecký Kraj, N Czechia (Czech
 Republic)
Liberia 52 D4 Guanacaste,
 NW Costa Rica

Liberia 74 C5 off. Republic of Liberia.
 country W Africa
Liberia, Republic of see Liberia
Libian Desert see Libyan Desert
Libīyah, Aş Şahrā' al see Libyan Desert
Libīyā, Dawlat see Libya
Libourne 91 B5 Gironde, SW France
Libreville 77 A5 country capital (Gabon)
 Estuaire, NW Gabon
Libya 71 F3 off. State of Libya, Ar. Dawlat
 Libīyā; prev. Libyan Arab Republic,
 Great Socialist People's Libyan Arab
 Jamahiriya. country N Africa
Libya, State of see Libya
Libyan Arab Republic see Libya
Libyan Desert 71 H4 var. Libian Desert,
 Ar. Aş Şahrā' al Libīyah. desert N Africa
Libyan Plateau 103 F4 var. Aḑ Diffah.
 plateau Egypt/Libya
Lichtenfels 95 C5 Bayern, SE Germany
Lichtenvoorde 86 E4 Gelderland,
 E Netherlands
Lichuan 129 C5 Hubei, C China
Lida 107 B5 Hrodzyenskaya Voblasts',
 W Belarus
Lidhorikíon see Lidoríki
Lidköping 85 B6 Västra Götaland,
 S Sweden
Lidoríki see Lidoríki
Lidoríki 105 B5 prev. Lidhorikíon,
 Lidokhorikíon. Stereá Ellás, C Greece
Lidzbark Warmiński 98 D2 Ger.
 Heilsberg. Olsztyn, N Poland
Liechtenstein 94 D1 off. Principality of
 Liechtenstein. country C Europe
Liechtenstein, Principality of see
 Liechtenstein
Liège 87 D6 Dut. Luik, Ger. Lüttich.
 Liège, E Belgium
Liegnitz see Legnica
Lienz 95 D7 Tirol, W Austria
Liepāja 106 B3 Ger. Libau. Liepāja,
 W Latvia
Lietuva see Lithuania
Lievenhof see Līvāni
Liezen 95 D7 Steiermark, C Austria
Liffey 89 B6 river Ireland
Lifou 146 D5 island Îles Loyauté,
 E New Caledonia
Ligure, Appennino 96 A2 Eng. Ligurian
 Mountains. mountain range NW Italy
Ligure, Mar see Ligurian Sea
Ligurian Mountains see Ligure,
 Appennino
Ligurian Sea 96 A3 Fr. Mer Ligurienne,
 It. Mar Ligure. sea N Mediterranean Sea
Ligurienne, Mer see Ligurian Sea
Lîhu'e 47 A7 var. Lihue. Kaua'i, Hawaii,
 USA
Lihue see Lîhu'e
Lihula 106 D2 Ger. Leal. Läänemaa,
 W Estonia
Liivi Laht see Riga, Gulf of
Likasi 77 D7 prev. Jadotville. Shaba,
 SE Dem. Rep. Congo
Liknes 85 A6 Vest-Agder, S Norway
Lille 90 C2 var. l'Isle, Dut. Rijssel, Flem.
 Ryssel, prev. Lisle; anc. Insula. Nord,
 N France
Lillehammer 85 B5 Oppland, S Norway
Lillestrøm 85 B6 Akershus, S Norway
Lilongwe 79 E2 country capital (Malawi)
 Central, W Malawi
Lilybaeum see Marsala
Lima 60 C4 country capital (Peru) Lima,
 W Peru
Limanowa 99 D5 Małopolskie, S Poland
Limassol see Lemesós
Limerick 89 A6 Ir. Luimneach.
 Limerick, SW Ireland
Limin Vathéos see Sámos
Límnos 103 E4 anc. Lemnos. island
 E Greece
Limoges 91 C5 anc. Augustoritum
 Lemovicensium, Lemovices.
 Haute-Vienne, C France
Limón 53 E4 var. Puerto Limón. Limón,
 E Costa Rica
Limón 52 D2 Colón, NE Honduras
Limonum see Poitiers
Limousin 91 C5 cultural region C France
Limoux 91 C6 Aude, S France
Limpopo 78 D3 var. Crocodile. river
 S Africa
Linares 64 A4 Maule, C Chile
Linares 51 E3 Nuevo León, NE Mexico
Linares 92 D4 Andalucía, S Spain
Lincoln 89 D5 anc. Lindum, Lindum
 Colonia. E England, United Kingdom

INDEX – GAZETTEER: **Lincoln – L'uban'**

Lincoln *41 H2* Maine, NE USA
Lincoln *45 F4 state capital* Nebraska, C USA
Lincoln Sea *34 D2 sea* Arctic Ocean
Linden *59 F3* E Guyana
Líndhos *see* Líndos
Lindi *73 D8* Lindi, SE Tanzania
Líndos *105 E7 var.* Líndhos. Ródos, Dodekánisa, Greece, Aegean Sea
Lindum/Lindum Colonia *see* Lincoln
Line Islands *145 G3 island group* E Kiribati
Lingeh *see* Bandar-e Lengeh
Lingen *94 A3 var.* Lingen an der Ems. Niedersachsen, NW Germany
Lingen an der Ems *see* Lingen
Lingga, Kepulauan *138 B4 island group* W Indonesia
Linköping *85 C6* Östergötland, S Sweden
Linz *95 D6 anc.* Lentia. Oberösterreich, N Austria
Lion, Golfe du *91 C7 Eng.* Gulf of Lion, Gulf of Lions; *anc.* Sinus Gallicus. *gulf* S France
Lion, Gulf of/Lions, Gulf of *see* Lion, Golfe du
Liozno *see* Lyozna
Lipari *97 D6 island* Isole Eolie, S Italy
Lipari Islands/Lipari, Isole *see* Eolie, Isole
Lipetsk *111 B5* Lipetskaya Oblast', W Russia
Lipno *98 C3* Kujawsko-pomorskie, C Poland
Lipova *108 A4 Hung.* Lippa. Arad, W Romania
Lipovets *see* Lypovets
Lippa *see* Lipova
Lipsia/Lipsk *see* Leipzig
Lira *73 B6* N Uganda
Lisala *77 C5* Equateur, N Dem. Rep. Congo
Lisboa *92 B4 Eng.* Lisbon; *anc.* Felicitas Julia, Olisipo. *country capital* (Portugal) Lisboa, W Portugal
Lisbon *see* Lisboa
Lisichansk *see* Lysychansk
Lisieux *90 B3 anc.* Noviomagus. Calvados, N France
Liski *111 B6 prev.* Georgiu-Dezh. Voronezhskaya Oblast', W Russia
Lisle/l'Isle *see* Lille
Lismore *149 E5* New South Wales, SE Australia
Lissa *see* Vis, Croatia
Lissa *see* Leszno, Poland
Lisse *86 C3* Zuid-Holland, W Netherlands
Litang *129 A5 var.* Gaocheng. Sichuan, C China
Līṭani, Nahr el *119 B5 var.* Nahr al Litant. *river* C Lebanon
Litant, Nahr al *see* Līṭani, Nahr el
Litauen *see* Lithuania
Lithgow *149 D6* New South Wales, SE Australia
Lithuania *106 B4 off.* Republic of Lithuania, *Ger.* Litauen, *Lith.* Lietuva, *Pol.* Litwa, *Rus.* Litva; *prev.* Lithuanian SSR, *Rus.* Litovskaya SSR. *country* NE Europe
Lithuanian SSR *see* Lithuania
Lithuania, Republic of *see* Lithuania
Litóchoro *104 B4 var.* Litohoro, Litókhoron. Kentrikí Makedonía, N Greece
Litohoro/Litókhoron *see* Litóchoro
Litovskaya SSR *see* Lithuania
Little Alföld *99 C6 Ger.* Kleines Ungarisches Tiefland, *Hung.* Kisalföld, *Slvk.* Podunajská Rovina. *plain* Hungary/Slovakia
Little Andaman *133 F2 island* Andaman Islands, India, NE Indian Ocean
Little Barrier Island *see* Te Hauturu-o-Toi
Little Bay *93 H5 bay* Alboran Sea, Mediterranean Sea
Little Cayman *54 B3 island* E Cayman Islands
Little Falls *45 F2* Minnesota, N USA
Littlefield *49 E2* Texas, SW USA
Little Inagua *54 D2 var.* Inagua Islands. *island* S The Bahamas
Little Minch, The *88 B3 strait* NW Scotland, United Kingdom
Little Missouri River *44 D2 river* NW USA
Little Nicobar *133 G3 island* Nicobar Islands, India, NE Indian Ocean
Little Rhody *see* Rhode Island
Little Rock *42 B1 state capital* Arkansas, C USA
Little Saint Bernard Pass *91 D5 Fr.* Col du Petit St-Bernard, *It.* Colle del Piccolo San Bernardo. *pass* France/Italy
Little Sound *42 A5 bay* Bermuda, NW Atlantic Ocean
Littleton *44 D4* Colorado, C USA
Littoria *see* Latina
Litva/Litwa *see* Lithuania
Liubotyn *109 G2 prev.* Lyubotyn, *Rus.* Lyubotin. Kharkivs'ka Oblast', E Ukraine
Liu-chou/Liuchow *see* Liuzhou
Liuzhou *129 C6 var.* Liu-chou, Liuchow. Guangxi Zhuangzu Zizhiqu, S China
Livanátai *see* Livanátes
Livanátes *105 B5 prev.* Livanátai. Stereá Ellás, C Greece
Līvāni *106 D4 Ger.* Lievenhof. Preiļi, SE Latvia
Liverpool *39 F5* Nova Scotia, SE Canada
Liverpool *89 C5* NW England, United Kingdom
Livingston *44 B2* Montana, NW USA
Livingston *49 H3* Texas, SW USA
Livingstone *78 C3 var.* Maramba. Southern, S Zambia
Livingstone Mountains *151 A7 mountain range* South Island, New Zealand
Livno *100 B4* Federicija Bosna I Hercegovina, SW Bosnia and Herzegovina
Livojoki *84 D4 river* C Finland
Livonia *40 D3* Michigan, N USA
Livorno *96 B3 Eng.* Leghorn. Toscana, C Italy
Lixian Jiang *see* Black River
Lixoúri *105 A5 prev.* Lixoúrion. Kefallinía, Iónia Nisiá, Greece, C Mediterranean Sea
Lixoúrion *see* Lixoúri
Lizarra *see* Estella
Ljouwert *see* Leeuwarden
Ljubelj *see* Loibl Pass
Ljubljana *95 D7 Ger.* Laibach, *It.* Lubiana; *anc.* Aemona, Emona. *country capital* (Slovenia) C Slovenia
Ljungby *85 B7* Kronoberg, S Sweden
Ljusdal *85 C5* Gävleborg, C Sweden
Ljusnan *85 C5 river* C Sweden
Llanelli *89 C6 prev.* Llanelly. SW Wales, United Kingdom
Llanelly *see* Llanelli
Llanes *92 D1* Asturias, N Spain
Llanos *58 D2 physical region* Colombia/Venezuela
Lleida *93 F2 Cast.* Lérida; *anc.* Ilerda. Cataluña, NE Spain
Llucmajor *93 G3* Mallorca, Spain, W Mediterranean Sea
Loaita Island *129 C8 island* W Spratly Islands
Loanda *see* Luanda
Lobamba *78 D4 country capital* (Eswatini- royal and legislative) NW Eswatini
Lobatse *78 C4 var.* Lobatsi. Kgatleng, SE Botswana
Lobatsi *see* Lobatse
Löbau *94 D4* Sachsen, E Germany
Lobito *78 B2* Benguela, W Angola
Lob Nor *see* Lop Nur
Lobositz *see* Lovosice
Loburi *see* Lop Buri
Locarno *95 B8 Ger.* Luggarus. Ticino, S Switzerland
Lochem *86 E3* Gelderland, E Netherlands
Lockport *41 E3* New York, NE USA
Lodja *77 D6* Kasai-Oriental, C Dem. Rep. Congo
Lodwar *73 C6* Rift Valley, NW Kenya
Łódź *98 D4 Rus.* Lodz. Łódź, C Poland
Loei *136 C4 var.* Loey, Muang Loei. Loei, C Thailand
Loey *see* Loei
Lofoten *84 B3 var.* Lofoten Islands. *island group* C Norway
Lofoten Islands *see* Lofoten
Logan *44 B3* Utah, W USA
Logan, Mount *36 D3 mountain* Yukon, W Canada
Logroño *93 E1 anc.* Vareia, *Lat.* Juliobriga. La Rioja, N Spain
Loibl Pass *95 D7 Ger.* Loiblpass, *Slvn.* Ljubelj. *pass* Austria/Slovenia
Loiblpass *see* Loibl Pass
Loikaw *136 B4* Kayah State, C Myanmar (Burma)
Loire *90 B4 var.* Liger. *river* C France
Loja *60 B2* Loja, S Ecuador
Lokitaung *73 C5* Rift Valley, NW Kenya
Lokoja *75 G4* Kogi, C Nigeria
Loksa *106 E2 Ger.* Loxa. Harjumaa, NW Estonia
Lolland *85 B8 prev.* Laaland. *island* S Denmark
Lom *104 C1 prev.* Lom-Palanka. Montana, NW Bulgaria
Lomami *77 D6 river* C Dem. Rep. Congo
Lomas *60 D4* Arequipa, SW Peru
Lomas de Zamora *64 D4* Buenos Aires, E Argentina
Lombamba *78 D4 country capital* (Eswatini - royal and legislative) NW Eswatini
Lombardia *96 B2 Eng.* Lombardy. *region* N Italy
Lombardy *see* Lombardia
Lombok, Pulau *138 D5 island* Nusa Tenggara, C Indonesia
Lomé *75 F5 country capital* (Togo) S Togo
Lomela *77 D6* Kasai-Oriental, C Dem. Rep. Congo
Lommel *87 C5* Limburg, N Belgium
Lomond, Loch *88 B4 lake* C Scotland, United Kingdom
Lomonosov Ridge *155 B3 var.* Harris Ridge, *Rus.* Khrebet Homonsova. *undersea ridge* Arctic Ocean
Lomonsova, Khrebet *see* Lomonosov Ridge
Lom-Palanka *see* Lom
Lompoc *47 B7* California, W USA
Lom Sak *136 C4 var.* Muang Lom Sak. Phetchabun, C Thailand
Łomża *98 D3 Rus.* Lomzha. Podlaskie, NE Poland
Lomzha *see* Łomża
Loncoche *65 B5* Araucanía, C Chile
Londinium *see* London
London *89 E7 anc.* Augusta, *Lat.* Londinium. *country capital* (United Kingdom) SE England, United Kingdom
London *38 C5* Ontario, S Canada
London *40 C5* Kentucky, S USA
Londonderry *88 B4 var.* Derry, *Ir.* Doire. NW Northern Ireland, United Kingdom
Londonderry, Cape *146 C2 cape* Western Australia
Londrina *63 E4* Paraná, S Brazil
Lone Star State *see* Texas
Long Bay *43 F2 bay* W Jamaica
Long Beach *47 C7* California, W USA
Longford *89 B5 Ir.* An Longfort. Longford, C Ireland
Long Island *54 D2 island* C The Bahamas
Long Island *41 G4 island* New York, NE USA
Longlac *38 C3* Ontario, S Canada
Longmont *44 D4* Colorado, C USA
Longreach *148 C4* Queensland, E Australia
Long Strait *115 G1 Eng.* Long Strait. *strait* NE Russia
Long Strait *see* Longa, Proliv
Longview *49 H3* Texas, SW USA
Longview *46 B2* Washington, NW USA
Long Xuyên *137 D6 var.* Longxuyen. An Giang, S Vietnam
Longxuyen *see* Long Xuyên
Longyan *129 D6* Fujian, SE China
Longyearbyen *83 G2 dependent territory capital* (Svalbard) Spitsbergen, W Svalbard
Lons-le-Saunier *90 D4 anc.* Ledo Salinarius. Jura, E France
Lop Buri *137 C5 var.* Loburi. Lop Buri, C Thailand
Lop Nor *see* Lop Nur
Lop Nur *126 C3 var.* Lob Nor, Lop Nor, Lo-pu Po. *seasonal lake* NW China
Loppersum *86 E1* Groningen, NE Netherlands
Lo-pu Po *see* Lop Nur
Lorca *93 E4 Ar.* Lurka; *anc.* Eliocroca, *Lat.* Illurco. Murcia, S Spain
Lord Howe Island *142 C4 island* E Australia
Lord Howe Rise *142 C4 undersea rise* SW Pacific Ocean
Loreto *50 B3* Baja California Sur, NW Mexico
Lorient *90 A3 prev.* l'Orient. Morbihan, NW France
l'Orient *see* Lorient
Lorn, Firth of *88 B4 inlet* W Scotland, United Kingdom
Lörrach *95 A7* Baden-Württemberg, S Germany
Lorraine *90 D3 cultural region* NE France
Los Alamos *48 C1* New Mexico, SW USA
Los Amates *52 B2* Izabal, E Guatemala
Los Ángeles *65 B5* Bío Bío, C Chile
Los Angeles *47 C7* California, W USA
Losanna *see* Lausanne
Lošinj *100 A3 Ger.* Lussin, *It.* Lussino. *island* W Croatia
Loslau *see* Wodzisław Śląski
Los Mochis *50 C3* Sinaloa, C Mexico
Losonc/Losontz *see* Lučenec
Los Roques, Islas *58 D1 island group* N Venezuela
Lot *91 B5 cultural region* S France
Lot *91 B5 river* S France
Lotagipi Swamp *73 C5 wetland* Kenya/South Sudan
Lötzen *see* Giżycko
Loualaba *see* Lualaba
Louangnamtha *136 C3 var.* Luong Nam Tha. Louang Namtha, N Laos
Louangphabang *124 D3 var.* Louangphrabang, Luang Prabang. Louangphabang, N Laos
Louangphrabang *see* Louangphabang
Loubomo *see* Dolisie
Loudéac *90 A3* Côtes d'Armor, NW France
Loudi *129 C5* Hunan, S China
Louga *74 B3* NW Senegal
Louisiade Archipelago *144 B4 island group* SE Papua New Guinea
Louisiana *42 A2 off.* State of Louisiana, *also known as* Creole State, Pelican State. *state* S USA
Louisville *40 C5* Kentucky, S USA
Louisville Ridge *143 E4 undersea ridge* S Pacific Ocean
Loup River *45 E4 river* Nebraska, C USA
Lourdes *91 B6* Hautes-Pyrénées, S France
Lourenço Marques *see* Maputo
Louth *89 E5* E England, United Kingdom
Loutrá *104 C4* Kentrikí Makedonía, N Greece
Louvain *see* Leuven
Louvain-la Neuve *87 C6* Walloon Brabant, C Belgium
Louviers *90 C3* Eure, N France
Lovech *104 C2* Lovech, N Bulgaria
Loveland *44 D4* Colorado, C USA
Lovosice *98 A4 Ger.* Lobositz. Ústecký Kraj, NW Czechia (Czech Republic)
Lóvua *78 C1* Moxico, E Angola
Lowell *41 G3* Massachusetts, NE USA
Löwen *see* Leuven
Lower California *see* Baja California
Lower Hutt *151 D5* Wellington, North Island, New Zealand
Lower Lough Erne *89 A5 lake* SW Northern Ireland, United Kingdom
Lower Red Lake *45 F1 lake* Minnesota, N USA
Lower Rhine *see* Neder Rijn
Lower Tunguska *see* Nizhnyaya Tunguska
Lowestoft *89 E6* E England, United Kingdom
Loxa *see* Loksa
Lo-yang *see* Luoyang
Loyauté, Îles *144 D5 island group* S New Caledonia
Loyev *see* Loyew
Loyew *107 D8 Rus.* Loyev. Homyel'skaya Voblasts', SE Belarus
Loznica *100 C3* Serbia, W Serbia
Lu *see* Shandong, China
Lualaba *77 D6 Fr.* Loualaba. *river* SE Dem. Rep. Congo
Luanda *78 A1 var.* Loanda, *Port.* São Paulo de Loanda. *country capital* (Angola) Luanda, NW Angola
Luang Prabang *see* Louangphabang
Luang, Thale *137 C7 lagoon* S Thailand
Luangua, Rio *see* Luangwa
Luangwa *73 B8 var.* Aruângua, Rio Luangua. *river* Mozambique/Zambia
Luanshya *78 D2* Copperbelt, C Zambia
Luarca *92 C1* Asturias, N Spain
Lubaczów *99 E5 var.* Lúbaczów. Podkarpackie, SE Poland
Lubnān *see* Lebanon
L'uban' *98 B4* SW Poland

Lubānas Ezers see Lubāns
Lubango 78 B2 Port. Sá da Bandeira. Huíla, SW Angola
Lubāns 106 D4 var. Lubānas Ezers. lake E Latvia
Lubao 77 D6 Kasai-Oriental, C Dem. Rep. Congo
Lübben 94 D4 Brandenburg, E Germany
Lübbenau 94 D4 Brandenburg, E Germany
Lubbock 49 E2 Texas, SW USA
Lübeck 94 C2 Schleswig-Holstein, N Germany
Lubelska, Wyżyna 98 E4 plateau SE Poland
Lüben see Lubin
Lubiana see Ljubljana
Lubin 98 B4 Ger. Lüben. Dolnośląskie, SW Poland
Lublin 98 E4 Rus. Lyublin. Lubelskie, E Poland
Lubliniec 98 C4 Śląskie, S Poland
Lubnān, Jabal see Liban, Jebel
Lubny 109 F2 Poltavs'ka Oblast', NE Ukraine
Lubsko 98 B4 Ger. Sommerfeld. Lubuskie, W Poland
Lubumbashi 77 E8 prev. Élisabethville. Shaba, SE Dem. Rep. Congo
Lubutu 77 D6 Maniema, E Dem. Rep. Congo
Luca see Lucca
Lucan 89 B5 Ir. Leamhcán. Dublin, E Ireland
Lucanian Mountains see Lucano, Appennino
Lucano, Appennino 97 D5 Eng. Lucanian Mountains. mountain range S Italy
Lucapa 78 C1 var. Lukapa. Lunda Norte, NE Angola
Lucca 96 B3 anc. Luca. Toscana, C Italy
Lucea 54 A4 W Jamaica
Lucena 139 E1 off. Lucena City. Luzon, N Philippines
Lucena 92 D4 Andalucía, S Spain
Lucena City see Lucena
Lučenec 99 D6 Ger. Losontz, Hung. Losonc. Banskobystrický Kraj, C Slovakia
Lucentum see Alicante
Lucerna/Lucerne see Luzern
Luchow see Hefei
Łuck see Lutsk
Lucknow 135 E3 var. Lakhnau. state capital Uttar Pradesh, N India
Lüda see Dalian
Luda Kamchia 104 D2 river E Bulgaria
Ludasch see Luduş
Lüderitz 78 B4 prev. Angra Pequena. Karas, SW Namibia
Ludhiäna 134 D2 Punjab, N India
Ludington 40 C2 Michigan, N USA
Ludsan see Ludza
Luduş 108 B4 Ger. Ludasch, Hung. Marosludas. Mureş, C Romania
Ludvika 85 C6 Dalarna, C Sweden
Ludwigsburg 95 B6 Baden-Württemberg, SW Germany
Ludwigsfelde 94 D3 Brandenburg, NE Germany
Ludwigshafen 95 B5 var. Ludwigshafen am Rhein. Rheinland-Pfalz, W Germany
Ludwigshafen am Rhein see Ludwigshafen
Ludwigslust 94 C3 Mecklenburg-Vorpommern, N Germany
Ludza 106 D4 Ger. Ludsan. Ludza, E Latvia
Luebo 77 C6 Kasai-Occidental, SW Dem. Rep. Congo
Luena 78 C2 var. Lwena, Port. Luso. Moxico, E Angola
Lufira 77 E7 river SE Dem. Rep. Congo
Lufkin 49 H3 Texas, SW USA
Luga 110 A4 Leningradskaya Oblast', NW Russia
Lugano 95 B8 Ger. Lauis. Ticino, S Switzerland
Lugdunum see Lyon
Lugdunum Batavorum see Leiden
Lugenda, Rio 79 E2 river N Mozambique
Luggarus see Locarno
Lugh Ganana see Luuq
Lugo 92 C1 anc. Lugus Augusti. Galicia, NW Spain
Lugoj 108 A4 Ger. Lugosch, Hung. Lugos. Timiş, W Romania
Lugos/Lugosch see Lugoj

Lugus Augusti see Lugo
Luguvallium/Luguvallum see Carlisle
Luhansk 109 H3 Rus. Voroshilovgrad. Luhans'ka Oblast', E Ukraine
Luimneach see Limerick
Lukapa see Lucapa
Lukenie 77 C6 river C Dem. Rep. Congo
Lukovit 104 C2 Lovech, N Bulgaria
Łuków 98 E4 Ger. Bogendorf. Lubelskie, E Poland
Lukuga 77 D7 river SE Dem. Rep. Congo
Luleå 84 D4 Norrbotten, N Sweden
Luleälven 84 C3 river N Sweden
Lulonga 77 C5 river NW Dem. Rep. Congo
Lulua 77 D7 river S Dem. Rep. Congo
Luluabourg see Kananga
Lumber State see Maine
Lumbo 79 F2 Nampula, NE Mozambique
Lumsden 151 A7 Southland, South Island, New Zealand
Lund 85 B7 Skåne, S Sweden
Lüneburg 94 C3 Niedersachsen, N Germany
Lunga, Isola see Dugi Otok
Lungkiang see Qiqihar
Lungué-Bungo 78 C2 var. Lungwebungu. river Angola/Zambia
Lungwebungu see Lungué-Bungo
Luninets see Luninyets
Łuniniec see Luninyets
Luninyets 107 B7 Pol. Łuniniec, Rus. Luninets. Brestskaya Voblasts', SW Belarus
Lunteren 86 D4 Gelderland, C Netherlands
Luong Nam Tha see Louangnamtha
Luoyang 128 C4 var. Honan, Lo-yang. Henan, C China
Lupatia see Altamura
Lúrio 79 F2 Nampula, NE Mozambique
Lúrio, Rio 79 E2 river NE Mozambique
Lurka see Lorca
Lusaka 78 D2 country capital (Zambia) Lusaka, SE Zambia
Lushnja see Lushnjë
Lushnjë 101 C6 var. Lushnja. Fier, C Albania
Luso see Luena
Lussin/Lussino see Lošinj
Lüt, Baḥrat/Lut, Bahret see Dead Sea
Lut, Dasht-e 120 D3 var. Kavīr-e Lūt. desert E Iran
Lutetia/Lutetia Parisiorum see Paris
Lüt, Kavīr-e see Lūt, Dasht-e
Luton 89 D6 E England, United Kingdom
Łutselk'e 37 F4 prev. Snowdrift. Northwest Territories, W Canada
Lutsk 108 C1 Pol. Łuck. Volyns'ka Oblast', NW Ukraine
Lüttich see Liège
Lützow-Holm Bay 154 C2 var. Lützow Holmbukta. bay Antarctica
Lutzow Holmbukta see Lützow-Holm Bay
Luuq 73 D6 It. Lugh Ganana. Gedo, SW Somalia
Luvua 77 D7 river SE Dem. Rep. Congo
Luwego 73 C8 river S Tanzania
Luxembourg 87 D8 country capital (Luxembourg) Luxembourg, S Luxembourg
Luxembourg 87 D8 off. Grand Duchy of Luxembourg, var. Lëtzebuerg, Luxemburg. country NW Europe
Luxemburg see Luxembourg
Luxor see El Uqsur
Luza 110 C4 Kirovskaya Oblast', NW Russia
Luz, Costa de la 92 C5 coastal region SW Spain
Luzern 95 B7 Fr. Lucerne, It. Lucerna. Luzern, C Switzerland
Luzon 139 E1 island N Philippines
Luzon Strait 125 E3 strait Philippines/Taiwan
Lviv 108 B2 Ger. Lemberg, Pol. Lwów, Rus. L'vov. L'vivs'ka Oblast', W Ukraine
L'vov see Lviv
Lwena see Luena
Lwów see Lviv
Lyakhavichy 107 B6 Rus. Lyakhovichi. Brestskaya Voblasts', SW Belarus
Lyakhovichi see Lyakhavichy
Lyallpur see Faisalābād
Lyangar see Langar
Lyck see Ełk
Lycksele 84 C4 Västerbotten, N Sweden
Lycopolis see Asyūṭ

Lyel'chytsy 107 C7 Rus. Lel'chitsy. Homyel'skaya Voblasts', SE Belarus
Lyepyel' 107 D5 Rus. Lepel'. Vitsyebskaya Voblasts', N Belarus
Lyme Bay 89 C7 bay S England, United Kingdom
Lynchburg 41 E5 Virginia, NE USA
Lynn see King's Lynn
Lynn Regis see King's Lynn
Lynn Lake 37 F4 Manitoba, C Canada
Lyon 91 D5 Eng. Lyons; anc. Lugdunum. Rhône, E France
Lyons see Lyon
Lyozna 107 E6 Rus. Liozno. Vitsyebskaya Voblasts', NE Belarus
Lypovets 108 D2 Rus. Lipovets. Vinnyts'ka Oblast', C Ukraine
Lys see Leie
Lysychansk 109 H3 Rus. Lisichansk. Luhans'ka Oblast', E Ukraine
Lyttelton 151 C6 South Island, New Zealand
Lyublin see Lublin
Lyubotin see Liubotyn
Lyubotyn see Liubotyn
Lyulyakovo 104 E2 prev. Keremitlik. Burgas, E Bulgaria
Lyusina 107 B6 Rus. Lyusino. Brestskaya Voblasts', SW Belarus
Lyusino see Lyusina

M

Maale 132 B4 var. Male'. country capital (Maldives) Male' Atoll, C Maldive
Ma'ān 119 B7 Ma'ān, SW Jordan
Maardu 106 D2 Ger. Maart. Harjumaa, NW Estonia
Ma'aret-en-Nu'man see Ma'arrat an Nu'mān
Ma'arrat an Nu'mān 118 B3 var. Ma'aret-en-Nu'man, Fr. Maarret enn Naamâne. Idlib, NW Syria
Maarret enn Naamâne see Ma'arrat an Nu'mān
Maart see Maardu
Maas see Meuse
Maaseik 87 D5 prev. Maeseyck. Limburg, NE Belgium
Maastricht 87 D6 var. Maestricht; anc. Traiectum ad Mosam, Traiectum Tungorum. Limburg, SE Netherlands
Macao 129 C6 Port. Macau. Special administrative region Guangdong, SE China
Macapá 63 E1 state capital Amapá, N Brazil
Macarsca see Makarska
Macassar see Makassar
Macāu see Makó, Hungary
Macau see Macao
MacCluer Gulf see Berau, Teluk
Macdonnell Ranges 146 D4 mountain range Northern Territory, C Australia
Macedonia see North Macedonia
Maceió 63 G3 state capital Alagoas, E Brazil
Machachi 60 B1 Pichincha, C Ecuador
Machala 60 B2 El Oro, SW Ecuador
Machanga 79 E3 Sofala, E Mozambique
Machilipatnam 132 D1 var. Bandar Masulipatnam. Andhra Pradesh, E India
Machiques 58 C2 Zulia, NW Venezuela
Macías Nguema Biyogo see Bioco, Isla de
Măcin 108 D5 Tulcea, SE Romania
Mackay 148 D4 Queensland, NE Australia
Mackay, Lake 146 C4 salt lake Northern Territory/Western Australia
Mackenzie 37 E3 river Northwest Territories, NW Canada
Mackenzie Bay 154 D3 bay Antarctica
Mackenzie Mountains 36 D3 mountain range Northwest Territories, NW Canada
Macleod, Lake 146 A4 lake Western Australia
Macomb 40 A4 Illinois, N USA
Macomer 97 A5 Sardegna, Italy, C Mediterranean Sea
Mâcon 91 D5 anc. Matisco, Matisco Ædourum. Saône-et-Loire, C France
Macon 42 D2 Georgia, SE USA
Macon 45 G4 Missouri, C USA
Macquarie Ridge 154 C5 undersea ridge SW Pacific Ocean
Macuspana 51 G4 Tabasco, SE Mexico

Ma'dabā 119 B6 var. Mādabā, Madeba; anc. Medeba. Ma'dabā, NW Jordan
Mādabā see Ma'dabā
Madagascar 79 F3 off. Republic of Madagascar, Malg. Madagasikara; prev. Democratic Republic of Madagascar, Malagasy Republic. country W Indian Ocean
Madagascar 79 F3 island W Indian Ocean
Madagascar Basin 69 E7 undersea basin W Indian Ocean
Madagascar, Democratic Republic of see Madagascar
Madagascar Plateau 69 E7 var. Madagascar Ridge, Madagascar Rise, Rus. Madagaskarskiy Khrebet. undersea plateau W Indian Ocean
Madagascar Rise/Madagascar Ridge see Madagascar Plateau
Madagasikara see Madagascar
Madagaskarskiy Khrebet see Madagascar Plateau
Madang 144 B3 Madang, N Papua New Guinea
Madanīyīn see Médenine
Madarska see Hungary
Made 86 C4 Noord-Brabant, S Netherlands
Madeba see Ma'dabā
Madeira 70 A2 var. Ilha da Madeira. island Madeira, Portugal, NE Atlantic Ocean
Madeira, Ilha de see Madeira
Madeira Plain 66 C3 abyssal plain E Atlantic Ocean
Madeira, Rio 62 D2 var. Río Madera. river Bolivia/Brazil
Madeleine, Îles de la 39 F4 Eng. Magdalen Islands. island group Québec, E Canada
Madera 47 B6 California, W USA
Madera, Río see Madeira, Rio
Madhya Pradesh 135 E4 prev. Central Provinces and Berar. cultural region C India
Madīnat al Kuwayt see Al Kuwayt
Madīnat ath Thawrah see Ath Thawrah
Madioen see Madiun
Madison 45 F3 South Dakota, N USA
Madison 40 B3 state capital Wisconsin, N USA
Madiun 138 D5 prev. Madioen. Jawa, C Indonesia
Madoera see Madura, Pulau
Madona 106 D4 Ger. Modohn. Madona, E Latvia
Madras see Chennai
Madras see Tamil Nādu
Madre de Dios, Río 61 E3 river Bolivia/Peru
Madre del Sur, Sierra 51 E5 mountain range S Mexico
Madre, Laguna 51 F3 lagoon NE Mexico
Madre, Laguna 49 G5 lagoon Texas, SW USA
Madre Occidental, Sierra 50 C3 var. Western Sierra Madre. mountain range C Mexico
Madre Oriental, Sierra 51 E3 var. Eastern Sierra Madre. mountain range C Mexico
Madre, Sierra 52 B2 var. Sierra de Soconusco. mountain range Guatemala/Mexico
Madrid 92 D3 country capital (Spain) Madrid, C Spain
Madura see Madurai
Madurai 132 C3 prev. Madura, Mathurai. Tamil Nādu, S India
Madura, Pulau 138 D5 prev. Madoera. island C Indonesia
Maebashi 131 D5 var. Maebasi, Mayebashi. Gunma, Honshū, S Japan
Maebasi see Maebashi
Mae Nam Khong see Mekong
Mae Nam Nan 136 C4 river NW Thailand
Mae Nam Yom 136 C4 river W Thailand
Maeseyck see Maaseik
Maestricht see Maastricht
Maéwo 144 D4 prev. Aurora. island C Vanuatu
Mafia 73 D7 island E Tanzania
Mafraq/Muḥāfaẓat al Mafraq see Al Mafraq
Magadan 115 G3 Magadanskaya Oblast', E Russia
Magallanes see Punta Arenas

Magallanes, Estrecho de *see* Magellan, Strait of
Magangué *58 B2* Bolívar, N Colombia
Magdalena *61 B3* Beni, N Bolivia
Magdalena *50 B1* Sonora, NW Mexico
Magdalena, Isla *50 B3* *island* NW Mexico
Magdalena, Río *58 B2* *river* C Colombia
Magdalen Islands *see* Madeleine, Îles de la
Magdeburg *94 C4* Sachsen-Anhalt, C Germany
Magelang *138 C5* Jawa, C Indonesia
Magellan, Strait of *65 B8 Sp.* Estrecho de Magallanes. *strait* Argentina/Chile
Magerøy *see* Magerøya
Magerøya *84 D1 var.* Magerøy, *Smi.* Máhkarávju. *island* N Norway
Maggiore, Lake *see* Maggiore, Lake
Maggiore, Lake *96 B1 It.* Lago Maggiore. *lake* Italy/Switzerland
Maglaj *100 C3* Federacija Bosna I Hercegovina, N Bosnia and Herzegovina
Maglie *97 E6* Puglia, SE Italy
Magna *44 B4* Utah, W USA
Magnesia *see* Manisa
Magnitogorsk *114 B4* Chelyabinskaya Oblast', C Russia
Magnolia State *see* Mississippi
Magta' Lahjar *74 C3 var.* Magta Lahjar, Magta' Lahjar, Magtá Lahjar. Brakna, SW Mauritania
Magtymguly *122 C3 prev.* Garrygala; *Rus.* Kara-gala. W Turkmenistan
Magway *136 A3 var.* Magwe. Magway, W Myanmar (Burma)
Magyar-Becse *see* Bečej
Magyarkanizsa *see* Kanjiža
Magyarország *see* Hungary
Mahajanga *79 F2 var.* Majunga. Mahajanga, NW Madagascar
Mahakam, Sungai *138 D4 var.* Koetai, Kutai. *river* Borneo, C Indonesia
Mahalapye *78 D3 var.* Mahalatswe. Central, SE Botswana
Mahalatswe *see* Mahalapye
Mahān *120 D3* Kermān, E Iran
Mahanādi *135 F4 river* E India
Mahārāshtra *134 D5 cultural region* W India
Mahbés *see* El Mahbas
Mahbūbnagar *134 D5* Telangana, C India
Mahdia *71 F2 var.* Al Mahdīyah, Mehdia. NE Tunisia
Mahé *79 H1 island* Inner Islands, NE Seychelles
Mahia Peninsula *150 E4 peninsula* North Island, New Zealand
Mahilyow *107 D6 Rus.* Mogilëv. Mahilyowskaya Voblasts', E Belarus
Máhkarávju *see* Magerøya
Mahmūd-e ʿErāqī *see* Maḩmūd-e Rāqī
Maḩmūd-e Rāqī *123 E4 var.* Mahmūd-e ʿErāqī. Kāpīsā, NE Afghanistan
Mahón *see* Maó
Mähren *see* Moravia
Mährisch-Weisskirchen *see* Hranice
Maicao *58 C1* La Guajira, N Colombia
Mai Ceu/Mai Chio *see* Maych'ew
Maïdān Shahr *123 E4 prev.* Meydān Shahr. Vardak, E Afghanistan
Maidstone *89 E7* SE England, United Kingdom
Maiduguri *75 H4* Borno, NE Nigeria
Mailand *see* Milano
Maīmanah *122 D3 var.* Meymaneh, Maymana. Fāryāb, NW Afghanistan
Main *95 B5 river* C Germany
Mai-Ndombe, Lac *77 C6 prev.* Lac Léopold II. *lake* W Dem. Rep. Congo
Maine *41 G2 off.* State of Maine, *also known as* Lumber State, Pine Tree State. *state* NE USA
Maine *90 B3 cultural region* NW France
Maine, Gulf of *41 H2 gulf* NE USA
Mainland *88 C2 island* N Scotland, United Kingdom
Mainland *88 D1 island* NE Scotland, United Kingdom
Mainz *95 B5 Fr.* Mayence. Rheinland-Pfalz, SW Germany
Maio *74 A3 var.* Mayo. *island* Ilhas de Sotavento, SE Cape Verde (Cabo Verde)
Maisūr *see* Karnātaka, India
Maisur *see* Karnātaka, India
Maitri *154 C2 Indian research station* Antarctica

Maizhokunggar *126 C5* Xizang Zizhiqu, W China
Majorca *see* Mallorca
Mājro *see* Majuro Atoll
Majunga *see* Mahajanga
Majuro Atoll *144 D2 var.* Mājro. *atoll and capital* (Marshall Islands) Ratak Chain, SE Marshall Islands
Makale *see* Mek'elē
Makarov Basin *155 B3 undersea basin* Arctic Ocean
Makarska *100 B4 It.* Macarsca. Split-Dalmacija, SE Croatia
Makasar *see* Makassar
Makasar, Selat *see* Makassar Straits
Makassar *139 E4 var.* Macassar, Makasar; *prev.* Ujungpandang. Sulawesi, C Indonesia
Makassar Straits *138 D4 Ind.* Makassar Selat. *strait* C Indonesia
Makay *79 F3 var.* Massif du Makay. *mountain range* SW Madagascar
Makay, Massif du *see* Makay
Makeni *74 C4* C Sierra Leone
Makeyevka *see* Makiivka
Makhachkala *114 A4 prev.* Petrovsk-Port. Respublika Dagestan, SW Russia
Makiivka *109 G3 prev.* Makiyivka, *Rus.* Makeyevka; *prev.* Dmitriyevsk. Donets'ka Oblast', E Ukraine
Makin *144 D2 prev.* Pitt Island. *atoll* Tungaru, W Kiribati
Makira *see* San Cristobal
Makiyivka *see* Makiivka
Makkah *121 A5 Eng.* Mecca. Makkah, W Saudi Arabia
Makkovik *39 F2* Newfoundland and Labrador, NE Canada
Makó *99 D7 Rom.* Macău. Csongrád, SE Hungary
Makoua *77 B5* Cuvette, C Congo
Makran Coast *120 E4 coastal region* SE Iran
Makrany *107 A6 Rus.* Mokrany. Brestskaya Voblasts', SW Belarus
Mākū *120 B2* Āžarbāyjān-e Gharbī, NW Iran
Makurdi *75 G4* Benue, C Nigeria
Mala *see* Malaita, Solomon Islands
Malabār Coast *132 B3 coast* SW India
Malabo *77 A5 prev.* Santa Isabel. *country capital* (Equatorial Guinea) Isla de Bioco, NW Equatorial Guinea
Malaca *see* Málaga
Malacca, Strait of *138 B3 Ind.* Selat Malaka. *strait* Indonesia/Malaysia
Malacka *see* Malacky
Malacky *99 C6 Hung.* Malacka. Bratislavský Kraj, W Slovakia
Maladzyechna *107 C5 Pol.* Molodeczno, *Rus.* Molodechno. Minskaya Voblasts', C Belarus
Málaga *92 D5 anc.* Malaca. Andalucía, S Spain
Malagarasi River *73 B7 river* W Tanzania Africa
Malagasy Republic *see* Madagascar
Malaita *144 C3 var.* Mala. *island* N Solomon Islands
Malakal *73 B5* Upper Nile, NE South Sudan
Malakula *see* Malekula
Malang *138 D5* Jawa, C Indonesia
Malange *see* Malanje
Malanje *78 B1 var.* Malange. Malanje, NW Angola
Mälaren *85 C6 lake* C Sweden
Malatya *117 E4 anc.* Melitene. Malatya, SE Turkey (Türkiye)
Mala Vyska *109 F3 Rus.* Malaya Viska. Kirovohrads'ka Oblast', S Ukraine
Malawi *79 E1 off.* Republic of Malawi; *prev.* Nyasaland, Nyasaland Protectorate. *country* S Africa
Malawi, Lake *see* Nyasa, Lake
Malawi, Republic of *see* Malawi
Malaya Viska *see* Mala Vyska
Malay Peninsula *124 D4 peninsula* Malaysia/Thailand
Malaysia *138 B3 off.* Malaysia, *var.* Federation of Malaysia; *prev.* the separate territories of Federation of Malaya, Sarawak and Sabah (North Borneo) and Singapore. *country* SE Asia
Malaysia, Federation of *see* Malaysia
Malbork *98 C2 Ger.* Marienburg, Marienburg in Westpreussen. Pomorskie, N Poland

Malchin *94 C3* Mecklenburg-Vorpommern, N Germany
Malden *45 H5* Missouri, C USA
Malden Island *145 G3 prev.* Independence Island. *atoll* E Kiribati
Maldives *132 A4 off.* Republic of Maldives. *country* N Indian Ocean
Maldives, Republic of *see* Maldives
Maldive *see* Maale
Male' Atoll *132 B4 var.* Kaafu Atoll. *atoll* C Maldives
Malekula *144 D4 var.* Malakula; *prev.* Mallicolo. *island* W Vanuatu
Malesína *105 C5* Stereá Ellás, E Greece
Malheur Lake *46 C3 lake* Oregon, NW USA
Mali *75 E3 off.* Republic of Mali, *Fr.* République du Mali; *prev.* French Sudan, Sudanese Republic. *country* W Africa
Malik, Wadi al *see* Milk, Wadi el
Mali Kyun *137 B5 var.* Tavoy Island. *island* Myeik Archipelago, S Myanmar (Burma)
Malin *see* Malyn
Malindi *73 D7* Coast, SE Kenya
Malines *see* Mechelen
Mali, Republic of *see* Mali
Mali, République du *see* Mali
Malkiye *see* Al Mālikīyah
Malko Tarnovo *104 E3 var.* Malko Türnovo. Burgas, E Bulgaria
Malko Türnovo *see* Malko Tarnovo
Mallaig *88 B3* N Scotland, United Kingdom
Mallawi *72 B2 var.* Mallawi. C Egypt
Mallawi *see* Mallawi
Mallicolo *see* Malekula
Mallorca *93 G3 Eng.* Majorca; *anc.* Baleares Major. *island* Islas Baleares, Spain, W Mediterranean Sea
Malmberget *84 C3 Lapp.* Malmivaara. Norrbotten, N Sweden
Malmédy *87 D6* Liège, E Belgium
Malmivaara *see* Malmberget
Malmö *85 B7* Skåne, S Sweden
Maloelap *see* Maloelap Atoll
Maloelap Atoll *144 D1 var.* Maḷoeḷap. *atoll* E Marshall Islands
Małopolska, Wyżyna *98 D4 plateau* S Poland
Malozemel'skaya Tundra *110 D3 physical region* NW Russia
Malta *106 A4* Rēzekne, SE Latvia
Malta *44 C1* Montana, NW USA
Malta *97 C8 off.* Republic of Malta. *country* C Mediterranean Sea
Malta *97 C8 island* Malta, C Mediterranean Sea
Malta, Canale di *see* Malta Channel
Malta Channel *97 C8 It.* Canale di Malta. *strait* Italy/Malta
Malta, Republic of *see* Malta
Maluku *139 F4 Dut.* Molukken, *Eng.* Moluccas; *prev.* Spice Islands. *island group* E Indonesia
Maluku, Laut *see* Molucca Sea
Malung *85 B6* Dalarna, C Sweden
Malventum *see* Benevento
Malvina, Isla Gran *see* West Falkland
Malvinas, Islas *see* Falkland Islands
Malyn *108 D2 Rus.* Malin. Zhytomyrs'ka Oblast', N Ukraine
Malyy Kavkaz *see* Lesser Caucasus
Mamberamo, Sungai *139 H4 river* Papua, E Indonesia
Mambij *see* Manbij
Mamonovo *106 A4 Ger.* Heiligenbeil. Kaliningradskaya Oblast', W Russia
Mamoré, Rio *61 F3 river* Bolivia/Brazil
Mamou *74 C4* W Guinea
Mamoudzou *79 F2 dependent territory capital* (Mayotte) C Mayotte
Mamuno *78 C3* Ghanzi, W Botswana
Manacor *93 G3* Mallorca, Spain, W Mediterranean Sea
Manado *139 F3 prev.* Menado. Sulawesi, C Indonesia
Managua *52 D3 country capital* (Nicaragua) Managua, W Nicaragua
Managua, Lago de *52 C3 var.* Xolotlán. *lake* W Nicaragua
Manakara *79 G4* Fianarantsoa, SE Madagascar
Manama *see* Al Manāmah
Mananjary *79 G3* Fianarantsoa, SE Madagascar
Manáos *see* Manaus
Manapouri, Lake *151 A7 lake* South Island, New Zealand

Manar *see* Mannar
Manas, Gora *123 E2 mountain* Kazakhstan/Kyrgyzstan
Manaung Island *136 A4 prev.* Cheduba Island. *island* W Myanmar (Burma)
Manaus *62 D2 prev.* Manáos. *state capital* Amazonas, NW Brazil
Manavgat *116 B4* Antalya, SW Turkey (Türkiye)
Manawatāwhi *150 C1 var.* Three Kings Islands. *island group* N New Zealand
Manbij *118 C2 var.* Mambij, *Fr.* Membidj. Ḩalab, N Syria
Manchester *89 D5 Lat.* Mancunium. NW England, United Kingdom
Manchester *41 G3* New Hampshire, NE USA
Man-chou-li *see* Manzhouli
Manchurian Plain *125 E1 plain* NE China
Máncio Lima *see* Japiim
Mancunium *see* Manchester
Mand *see* Mand, Rūd-e
Mandalay *136 B3* Mandalay, C Myanmar (Burma)
Mandan *45 E2* North Dakota, N USA
Mandeville *54 B5* C Jamaica
Mándra *105 C6* Attikí, C Greece
Mand, Rud-e *120 D4 var.* Mand. *river* S Iran
Mandurah *147 A6* Western Australia
Manduria *97 E5* Puglia, SE Italy
Mandya *132 C2* Karnātaka, C India
Manfredonia *97 D5* Puglia, SE Italy
Manfredonia *77 C6* Bandundu, W Dem. Rep. Congo
Mangaia *145 G5 island group* S Cook Islands
Mangalia *108 D5 anc.* Callatis. Constanța, SE Romania
Mangalmé *76 C3* Guéra, SE Chad
Mangalore *see* Mangalūru
Mangalūru *132 B2 prev.* Mangalore. Karnātaka, W India
Mangaung *see* Bloemfontein
Mango *75 E4 var.* Sansanné-Mango. N Togo
Mangoky *79 F3 river* W Madagascar
Manhattan *45 F4* Kansas, C USA
Manicouagan, Réservoir *38 D2 lake* Québec, E Canada
Manihiki *145 G4 atoll* N Cook Islands
Manihiki Plateau *143 E3 undersea plateau* C Pacific Ocean
Maniitsoq *82 C3 var.* Manîtsoq, *Dan.* Sukkertoppen. Qeqqata, S Greenland
Manila *139 E1 off.* City of Manila; *Fil.* Maynila. *country capital* (Philippines) Luzon, N Philippines
Manila, City of *see* Manila
Manisa *116 A3 var.* Manissa, *prev.* Saruhan; *anc.* Magnesia. Manisa, W Turkey (Türkiye)
Manissa *see* Manisa
Manitoba *37 F5 province* S Canada
Manitoba, Lake *37 F5 lake* Manitoba, S Canada
Manîtsoq *see* Maniitsoq
Manizales *58 B3* Caldas, W Colombia
Manjimup *147 A7* Western Australia
Mankato *45 F3* Minnesota, N USA
Manlleu *93 G2* Cataluña, NE Spain
Manly *148 E1* Iowa, C USA
Manmād *134 C5* Mahārāshtra, W India
Mannar *132 C3 var.* Manar. Northern Province, NW Sri Lanka
Mannar, Gulf of *132 C3 gulf* India/ Sri Lanka
Mannheim *95 B5* Baden-Württemberg, SW Germany
Manokwari *139 G4* New Guinea, E Indonesia
Manono *77 E7* Shaba, SE Dem. Rep. Congo
Manosque *91 D6* Alpes-de-Haute-Provence, SE France
Manra *145 F3 prev.* Sydney Island. *atoll* Phoenix Islands, C Kiribati
Mansa *78 D2 prev.* Fort Rosebery. Luapula, N Zambia
Mansel Island *37 G3 island* Nunavut, NE Canada
Mansfield *40 D4* Ohio, N USA
Manta *60 A2* Manabí, W Ecuador
Manteca *47 B6* California, W USA
Mantoue *see* Mantova
Mantova *96 B2 Eng.* Mantua, *Fr.* Mantoue. Lombardia, NW Italy

Mantua see Mantova
Manuae 145 G4 *island* S Cook Islands
Manukau see Manurewa
Manurewa 150 D3 *var.* Manukau.
Auckland, North Island, New Zealand
Manzanares 93 E3 Castilla-La Mancha,
C Spain
Manzanillo 54 C3 Granma, E Cuba
Manzanillo 50 D4 Colima, SW Mexico
Manzhouli 127 F1 *var.* Man-chou-li. Nei
Mongol Zizhiqu, N China
Mao 76 B3 Kanem, W Chad
Maó 93 H3 *Cast.* Mahón, *Eng.* Port
Mahon; *anc.* Portus Magonis. Menorca,
Spain, W Mediterranean Sea
Maoke, Pegunungan 139 H4 *Dut.*
Sneeuw-gebergte, *Eng.* Snow
Mountains. *mountain range* Papua,
E Indonesia
Maoming 129 C6 Guangdong, S China
Mapmaker Seamounts 125 H2
seamount range N Pacific Ocean
Maputo 78 D4 *prev.* Lourenço Marques.
country capital (Mozambique) Maputo,
S Mozambique
Maraba 63 F2 Pará, NE Brazil
Maracaibo 58 C1 Zulia, NW Venezuela
Maracaibo, Golf of see Venezuela,
Golfo de
Maracaibo, Lago de 58 C2 *var.* Lake
Maracaibo. *inlet* NW Venezuela
Maracaibo, Lake see Maracaibo, Lago de
Maracay 58 D2 Aragua, N Venezuela
Marada see Marādah
Marādah 71 G3 *var.* Marada. N Libya
Maradi 75 G3 Maradi, S Niger
Maragha see Marāgheh
Marāgheh 120 C2 *var.* Maragha.
Āzarbāyjān-e Khāvarī, NW Iran
Marajó, Baía de 63 F1 *bay* N Brazil
Marajó, Ilha de 63 E1 *island* N Brazil
Marakesh see Marrakech
Maramba see Livingstone
Maranhão 63 F2 *off.* Estado do
Maranhão. *state/region* E Brazil
Maranhão, Estado do see Maranhão
Marañón, Río 60 B2 *river* N Peru
Marathon 34 C4 Ontario, S Canada
Marathón see Marathónas
Marathónas 105 C5 *prev.* Marathón.
Attikí, C Greece
Marbella 92 D5 Andalucía, S Spain
Marble Bar 146 B4 Western Australia
Marburg see Marburg an der Lahn,
Germany
Marburg see Maribor, Slovenia
Marburg an der Lahn 94 B4 *hist.*
Marburg. Hessen, W Germany
March see Morava
Marche 96 C3 *Eng.* Marches. *region*
C Italy
Marche 91 C5 *cultural region* C France
Marche-en-Famenne 87 C7
Luxembourg, SE Belgium
Marchena, Isla 60 B5 *var.* Bindloe
Island. *island* Galápagos Islands,
Ecuador, E Pacific Ocean
Marches see Marche
Mar Chiquita, Laguna 64 C3 *lake*
C Argentina
Marcounda see Markounda
Mardan 134 C1 North-West Frontier
Province, N Pakistan
Mar del Plata 65 D5 Buenos Aires,
E Argentina
Mardin 117 E4 Mardin,
SE Turkey (Türkiye)
Maré 144 D5 *island* Îles Loyauté,
E New Caledonia
Marea Neagrǎ see Black Sea
Mareeba 148 D3 Queensland,
NE Australia
Marek see Dupnitsa
Marganets see Marhanets
Margarita, Isla de 59 E1 *island*
N Venezuela
Margate 89 E7 *prev.* Mergate.
SE England, United Kingdom
Margherita see Jamaame
Margherita, Lake 73 C5 *Eng.* Lake
Margherita, *It.* Abbaia. *lake*
SW Ethiopia
Margherita, Lake see Ābaya Hāyk'
Marghita 108 B3 Hung. Margitta. Bihor,
NW Romania
Margitta see Marghita
Mārgō, Dasht-e 122 D5 *desert* SW
Afghanistan
Marhanets 109 F3 *Rus.* Marganets.
Dnipropetrovska Oblast, E Ukraine

María Cleofas, Isla 50 C4 *island*
C Mexico
Maria Island 149 C8 *island* Tasmania,
SE Australia
María Madre, Isla 50 C4 *island*
C Mexico
María Magdalena, Isla 50 C4 *island*
C Mexico
Mariana Trench 125 G4 *trench*
W Pacific Ocean
Mariánské Lázně 99 A5 *Ger.* Marienbad.
Karlovarský Kraj, W Czechia (Czech
Republic)
Marías, Islas 50 C4 *island group*
C Mexico
Maria-Theresiopel see Subotica
Maribor 95 E7 *Ger.* Marburg.
NE Slovenia
Marica see Maritsa
Maridi 73 B5 W Equatoria,
S South Sudan
Marie Byrd Land 154 A3 *physical region*
Antarctica
Marie-Galante 55 G4 *var.* Ceyre to the
Caribs. *island* SE Guadeloupe
Marienbad see Mariánské Lázně
Marienburg see Alūksne, Latvia
Marienburg see Malbork, Poland
Marienburg in Westpreussen see
Malbork
Marienhausen see Viļaka
Mariental 78 B4 Hardap, SW Namibia
Marienwerder see Kwidzyń
Mariestad 85 B6 Västra Götaland,
S Sweden
Marietta 42 D2 Georgia, SE USA
Marijampolė 106 B4 *prev.* Kapsukas.
Marijampolė, S Lithuania
Marília 63 E4 São Paulo, S Brazil
Marín 92 B1 Galicia, NW Spain
Mar'ina Gorka see Mar"ina Horka
Mar"ina Horka 107 C6 *Rus.* Mar'ina
Gorka. Minskaya Voblasts', C Belarus
Maringá 63 E4 Paraná, S Brazil
Marion 45 G3 Iowa, C USA
Marion 40 D4 Ohio, N USA
Marion, Lake 43 E2 *reservoir* South
Carolina, SE USA
Mariscal Estigarribia 64 D2 Boquerón,
NW Paraguay
Maritsa 104 D3 *var.* Marica, *Gk.* Évros,
Turk. Meriç; *anc.* Hebrus. *river*
SW Europe
Maritzburg see Pietermaritzburg
Mariupol 109 G4 *prev.* Zhdanov.
Donets'ka Oblast', SE Ukraine
Marka 73 D6 *var.* Merca. Shabeellaha
Hoose, S Somalia
Markham, Mount 154 B4 *mountain*
Antarctica
Markounda 76 C4 *var.* Marcounda.
Ouham, NW Central African Republic
Marktredwitz 95 C5 Bayern, E Germany
Marlborough 148 D4 Queensland,
E Australia
Marlborough see Port Laoise
Marly Island see Kanton
Marmanda see Marmande
Marmande 91 B5 *anc.* Marmanda.
Lot-et-Garonne, SW France
Marmara, Sea of 116 A2 *Eng.* Sea of
Marmara. *sea* NW Turkey (Türkiye)
Marmara, Sea of see Marmara Denizi
Marmaris 116 A4 Muğla,
SW Turkey (Türkiye)
Marne 90 C3 *cultural region* N France
Marne 90 D3 *river* N France
Maro 76 C4 Moyen-Chari, S Chad
Maroantsetra 79 G2 Toamasina,
NE Madagascar
Maromokotro 79 G2 *mountain*
N Madagascar
Maroni 59 G3 *Dut.* Marowijne. *river*
French Guiana/Suriname
Marosheviz see Topliţa
Marosludas see Luduş
Marosvásárhely see Târgu Mureş
Marotiri 143 F4 *var.* Îlots de Bass,
Morotiri. *island group* Îles Australes,
SW French Polynesia
Maroua 76 B3 Extrême-Nord,
N Cameroon
Marowijne see Maroni
Marquesas Fracture Zone 153 E3
fracture zone E Pacific Ocean
Marquette 40 B1 Michigan, N USA
Marrakech 70 C2 *var.* Marakesh, *Eng.*
Marrakesh; *prev.* Morocco.
W Morocco
Marrakesh see Marrakech
Marrawah 149 C8 Tasmania,
SE Australia
Marree 149 B5 South Australia

Marsá al Burayqah 71 G3 *var.* Al
Burayqah. N Libya
Marsabit 73 C6 Eastern, N Kenya
Marsala 97 B7 *anc.* Lilybaeum. Sicilia,
Italy, C Mediterranean Sea
Marsberg 94 B4 Nordrhein-Westfalen,
W Germany
Marseille 91 D6 *Eng.* Marseilles; *anc.*
Massilia. Bouches-du-Rhône,
SE France
Marseilles see Marseille
Marshall 45 F2 Minnesota, N USA
Marshall 49 H2 Texas, SW USA
Marshall Islands 144 C1 *off.* Republic of
the Marshall Islands. *country*
W Pacific Ocean
Marshall Islands, Republic of the see
Marshall Islands
Marshall Seamounts 125 H3 *seamount
range* SW Pacific Ocean
Marsh Harbour 54 C1 Great Abaco,
W The Bahamas
Martaban see Mottama
Martha's Vineyard 41 G3 *island*
Massachusetts, NE USA
Martigues 91 D6 Bouches-du-Rhône,
SE France
Martin 99 C5 *Ger.* Sankt Martin, *Hung.*
Turócszentmárton; *prev.* Turčiansky
Svätý Martin. Žilinský Kraj, N Slovakia
Martinique 55 G4 *French overseas
department* E West Indies
Martinique Channel see Martinique
Passage
Martinique Passage 55 G4 *var.*
Dominica Channel, Martinique
Channel. *channel* Dominica/
Martinique
Marton 150 D4 Manawatu-Wanganui,
North Island, New Zealand
Martos 92 D4 Andalucía, S Spain
Marungu 77 E7 *mountain range*
SE Dem. Rep. Congo
Mary 122 D3 *prev.* Merv. Mary
Welaýaty, S Turkmenistan
Maryborough 149 D4 Queensland,
E Australia
Maryborough see Port Laoise
Mary Island see Kanton
Maryland 41 E5 *off.* State of Maryland,
also known as America in Miniature,
Cockade State, Free State, Old Line
State. *state* NE USA
Maryland, State of see Maryland
Maryville 45 F4 Missouri, C USA
Maryville 42 D1 Tennessee, S USA
Masai Steppe 73 C7 *grassland*
NW Tanzania
Masaka 73 B6 SW Uganda
Masalli 117 H3 *Rus.* Masally.
S Azerbaijan
Masally see Masalli
Masasi 73 C8 Mtwara, SE Tanzania
Masawa/Massawa see Mits'iwa
Masaya 52 D3 Masaya, W Nicaragua
Mascarene Basin 141 B5 *undersea basin*
W Indian Ocean
Mascarene Islands 79 H4 *island group*
W Indian Ocean
Mascarene Plain 141 B5 *abyssal plain*
W Indian Ocean
Mascarene Plateau 141 B5 *undersea
plateau* W Indian Ocean
Maseru 78 D4 *country capital* (Lesotho)
W Lesotho
Mashhad 120 E2 *var.* Meshed.
Khorāsān-Razavī, NE Iran
Masindi 73 B6 W Uganda
Masira see Maşīrah, Jazīrat
Masira, Gulf of see Maşīrah, Khalīj
Maşīrah, Jazīrat 121 E5 *var.* Masira.
island E Oman
Maşīrah, Khalīj 121 E5 *var.* Gulf of
Masira. *bay* E Oman
Masis see Büyük Ağrı Dağı
Maskat see Masqaţ
Maskav 123 E3 *Rus.* Moskovskiy; *prev.*
Chubek, Moskva. SW Tajikistan
Mason City 45 F3 Iowa, C USA
Masqaţ 121 E5 *var.* Maskat, *Eng.* Muscat.
country capital (Oman) NE Oman
Massa 96 B3 Toscana, C Italy
Massachusetts 41 G3 *off.*
Commonwealth of Massachusetts, *also
known as* Bay State, Old Bay State, Old
Colony State. *state* NE USA
Massenya 76 B3 Chari-Baguirmi,
SW Chad
Massif Central 91 C5 *plateau* C France
Massilia see Marseille

Massoukou see Franceville
Mastanli see Momchilgrad
Masterton 151 D5 Wellington, North
Island, New Zealand
Masty 107 B5 *Rus.* Mosty.
Hrodzyenskaya Voblasts', W Belarus
Masuda 131 B6 Shimane, Honshū,
SW Japan
Masuku see Franceville
Masvingo 78 D3 *prev.* Fort Victoria,
Nyanda, Victoria. Masvingo,
SE Zimbabwe
Maşyāf 118 B3 *Fr.* Misiaf. Ḥamāh,
C Syria
Mata-Au see Clutha
Matadi 77 B6 Bas-Congo, W Dem.
Rep. Congo
Matagalpa 52 D3 Matagalpa,
C Nicaragua
Matale 132 D3 Central Province,
C Sri Lanka
Matam 74 C3 NE Senegal
Matamata 150 D3 Waikato, North
Island, New Zealand
Matamoros 50 D3 Coahuila, NE Mexico
Matamoros 51 E2 Tamaulipas,
C Mexico
Matane 39 E4 Québec, SE Canada
Matanzas 54 B2 Matanzas, NW Cuba
Matara 132 D4 Southern Province,
S Sri Lanka
Mataram 138 D5 Pulau Lombok,
C Indonesia
Mataró 93 G2 *anc.* Illuro. Cataluña,
E Spain
Mataura 151 B7 Southland, South Island,
New Zealand
Mataura 151 B7 *river* South Island,
New Zealand
Mata Uta see Matā'utu
Matā'utu 145 E4 *var.* Mata Uta.
dependent territory capital (Wallis and
Futuna) Île Uvea, Wallis and Futuna
Matera 97 E5 Basilicata, S Italy
Mathura see Madurai
Matianus see Orūmīyeh, Daryācheh-ye
Matías Romero 51 F5 Oaxaca,
SE Mexico
Matisco/Matisco Ædourum see Mâcon
Mato Grosso 63 E3 *off.* Estado de Mato
Grosso; *prev.* Matto Grosso. *state/
region* W Brazil
Mato Grosso do Sul 63 E4 *off.* Estado de
Mato Grosso do Sul. *state/region*
S Brazil
Mato Grosso do Sul, Estado de see
Mato Grosso do Sul
Mato Grosso, Estado de see Mato Grosso
Mato Grosso, Planalto de 56 C4 *plateau*
C Brazil
Matosinhos 92 B2 *prev.* Matozinhos.
Porto, NW Portugal
Matozinhos see Matosinhos
Matsue 131 B6 *var.* Matsuye, Matue.
Shimane, Honshū, SW Japan
Matsumoto 131 C5 *var.* Matumoto.
Nagano, Honshū, S Japan
Matsuyama 131 B7 *var.* Matuyama.
Ehime, Shikoku, SW Japan
Matsuye see Matsue
Matterhorn 95 A8 *It.* Monte Cervino.
mountain Italy/Switzerland
Matthews Ridge 59 F2 N Guyana
Matthew Town 54 D2 Great Inagua,
S The Bahamas
Matto Grosso see Mato Grosso
Matucana 60 C4 Lima, W Peru
Matue see Matsue
Matumoto see Matsumoto
Maturín 59 E2 Monagas, NE Venezuela
Matuyama see Matsuyama
Mau 135 E3 *var.* Maunāth Bhanjan.
Uttar Pradesh, N India
Maui 47 B8 *island* Hawaii, USA,
C Pacific Ocean
Maun 78 C3 North-West, C Botswana
Maunāth Bhanjan see Mau
Mauren 94 E1 NE Liechtenstein Europe
Maurice see Mauritius
Mauritania 74 C2 *off.* Islamic Republic
of Mauritania, *Ar.* Mūrītāniyah.
country W Africa
Mauritania, Islamic Republic of see
Mauritania
Mauritius 79 H3 *off.* Republic of
Mauritius, *Fr.* Maurice. *country*
W Indian Ocean
Mauritius 141 B5 *island* W Indian Ocean
Mauritius, Republic of see Mauritius
Mawlamyaing see Mawlamyine

Mawlamyine 136 B4 var. Mawlamyaing, Moulmein. Mon State, S Myanmar (Burma)
Mawson 154 D2 Australian research station Antarctica
Mayadin see Al Mayādīn
Mayaguana 54 D2 island SE The Bahamas
Mayaguana Passage 54 D2 passage SE The Bahamas
Mayagüez 55 F3 W Puerto Rico
Mayamey 120 D2 Semnān, N Iran
Maya Mountains 52 B2 Sp. Montañas Mayas. mountain range Belize/ Guatemala
Mayas, Montañas see Maya Mountains
Maych'ew 72 C4 var. Mai Chio, It. Mai Ceu. Tigray, N Ethiopia
Mayebashi see Maebashi
Mayence see Mainz
Mayfield 151 B6 Canterbury, South Island, New Zealand
Maykop 111 A7 Respublika Adygeya, SW Russia
Maymana see Maīmanah
Maymyo see Pyin-Oo-Lwin
Maynila see Manila
Mayo see Maio
Mayor Island 150 D3 island NE New Zealand
Mayor Pablo Lagerenza see Capitán Pablo Lagerenza
Mayotte 79 F2 French overseas department E Africa
May Pen 54 B5 C Jamaica
Mayyit, Al Bahr al see Dead Sea
Mazabuka 78 D2 Southern, S Zambia
Mazaca see Kayseri
Mazagan see El-Jadida
Mazār-e Sharīf 123 E3 var. Mazâr-i Sharif. Balkh, N Afghanistan
Mazār-i Sharīf see Mazār-e Sharīf
Mazatlán 50 C3 Sinaloa, C Mexico
Mažeikiai 106 B3 Telšiai, NW Lithuania
Mazirbe 106 C2 Talsi, NW Latvia
Mazra'a see Al Mazra'ah
Mazury 98 D3 physical region NE Poland
Mazyr 107 C7 Rus. Mozyr'. Homyel'skaya Voblasts', SE Belarus
Mbabane 78 D4 country capital (Eswatini - administrative) NW Eswatini
Mbacké see Mbaké
Mbaïki 77 C5 var. M'Baiki. Lobaye, SW Central African Republic
M'Baiki see Mbaïki
Mbaké 74 B3 var. Mbacké. W Senegal
Mbala 78 D1 prev. Abercorn. Northern, NE Zambia
Mbale 73 C6 E Uganda
Mbandaka 77 C5 prev. Coquilhatville. Equateur, NW Dem. Rep. Congo
M'Banza Congo 78 B1 var. Mbanza Congo; prev. São Salvador, São Salvador do Congo. Dem. Rep. Congo, NW Angola
Mbanza-Ngungu 77 B6 Bas-Congo, W Dem. Rep. Congo
Mbarara 73 B6 SW Uganda
Mbé 76 B4 Nord, N Cameroon
Mbeya 73 C7 Mbeya, SW Tanzania
Mbomou/M'Bomu/Mbomu see Bomu
Mbour 74 B3 W Senegal
Mbuji-Mayi 77 D7 prev. Bakwanga. Kasai-Oriental, S Dem. Rep. Congo
McAlester 49 G2 Oklahoma, C USA
McAllen 49 G5 Texas, SW USA
McCamey 49 E3 Texas, SW USA
M'Clintock Channel 37 F2 channel Nunavut, N Canada
McComb 42 B3 Mississippi, S USA
McCook 45 E4 Nebraska, C USA
McKean Island 145 E3 island Phoenix Islands, C Kiribati
McKinley, Mount see Denali
McKinley Park see Denali Park
McMinnville 46 B3 Oregon, NW USA
McMurdo 154 B4 US research station Antarctica
McPherson 45 E5 Kansas, C USA
McPherson see Fort McPherson
Mdantsane 78 D5 Eastern Cape, SE South Africa
Mead, Lake 47 D6 reservoir Arizona/ Nevada, W USA
Mecca see Makkah
Mechelen 87 C5 Eng. Mechlin, Fr. Malines. Antwerpen, C Belgium
Mechlin see Mechelen

Mecklenburger Bucht 94 C2 bay N Germany
Mecsek 99 C7 mountain range SW Hungary
Medan 138 B3 Sumatera, E Indonesia
Medeba see Ma'dabā
Medellín 58 B3 Antioquia, NW Colombia
Médenine 71 F2 var. Madanīyīn. SE Tunisia
Medeshamstede see Peterborough
Medford 46 B4 Oregon, NW USA
Medgidia 108 D5 Constanţa, SE Romania
Medgyes see Mediaş
Mediaş 108 B4 Ger. Mediasch, Hung. Medgyes. Sibiu, C Romania
Mediasch see Mediaş
Medicine Hat 37 F5 Alberta, SW Canada
Medina see Al Madīnah
Medinaceli 93 E2 Castilla y León, N Spain
Medina del Campo 92 D2 Castilla y León, N Spain
Medinat Israel see Israel
Mediolanum see Saintes, France
Mediolanum see Milano, Italy
Mediomatrica see Metz
Mediterranean Sea 102 D3 Fr. Mer Méditerranée. sea Africa/Asia/Europe
Méditerranée, Mer see Mediterranean Sea
Médoc 91 B5 cultural region SW France
Medvezh'yegorsk 110 B3 Respublika Kareliya, NW Russia
Meekatharra 147 B5 Western Australia
Meemu Atoll see Mulakatholhu
Meerssen 87 D6 var. Mersen. Limburg, SE Netherlands
Meerut 134 D2 Uttar Pradesh, N India
Megáli Préspa, Límni see Prespa, Lake
Meghálaya 113 G3 cultural region NE India
Mehdia see Mahdia
Meheso see Mī'ēso
Me Hka see Nmai Hka
Mehriz 120 D3 Yazd, C Iran
Mehtar Lām 123 F4 var. Mehtarlām, Meterlam, Methariam, Metharlam, Laghmān, E Afghanistan
Mehtarlām see Mehtar Lām
Meiktila 136 B3 Mandalay, C Myanmar (Burma)
Méjico see Mexico
Mejillones 64 B2 Antofagasta, N Chile
Mek'elē 72 C4 var. Makale. Tigray, N Ethiopia
Mékhé 74 B3 NW Senegal
Mekong 124 D3 var. Lan-ts'ang Chiang, Khm. Mékôngk, Chin. Lancang Jiang, Lao. Mènam Khong, Th. Mae Nam Khong, Tib. Dza Chu, Vtn. Sông Tiên Giang. river SE Asia
Mékôngk see Mekong
Mekong, Mouths of the 137 E6 delta S Vietnam
Melaka 138 B3 var. Malacca. Melaka, Peninsular Malaysia
Melaka, Selat see Malacca, Strait of
Melanesia 144 D3 island group W Pacific Ocean
Melanesian Basin 142 C2 undersea basin W Pacific Ocean
Melbourne 149 C7 state capital Victoria, SE Australia
Melbourne 43 E4 Florida, SE USA
Meleda see Mljet
Melghir, Chott 71 E2 var. Chott Melrhir. salt lake E Algeria
Melilla 80 B5 anc. Rusaddir, Russadir. Melilla, Spain, N Africa
Melilla 70 D2 autonomous city of Spain Spain, N Africa
Melita 37 F5 Manitoba, S Canada
Melita see Mljet
Melitene see Malatya
Melitopol 109 F4 Zaporiz'ka Oblast', SE Ukraine
Melle 87 B5 Oost-Vlaanderen, NW Belgium
Mellerud 85 B6 Västra Götaland, S Sweden
Mellieha 102 B5 E Malta
Mellizo Sur, Cerro 65 A7 mountain S Chile
Melo 64 E4 Cerro Largo, NE Uruguay
Melodunum see Melun
Melrhir, Chott see Melghir, Chott
Melsungen 94 B4 Hessen, C Germany

Melun 90 C3 anc. Melodunum. Seine-et-Marne, N France
Melville Bay/Melville Bugt see Qimusseriarsuaq
Melville Island 146 D2 island Northern Territory, N Australia
Melville Island 37 E2 island Parry Islands, Northwest Territories, NW Canada
Melville, Lake 39 F2 lake Newfoundland and Labrador, E Canada
Melville Peninsula 37 G3 peninsula Nunavut, NE Canada
Melville Sound see Viscount Melville Sound
Membidj see Manbij
Memel see Neman, NE Europe
Memel see Klaipėda, Lithuania
Memmingen 95 B6 Bayern, S Germany
Memphis 42 C1 Tennessee, S USA
Menaam see Menaldum
Menado see Manado
Mènaka 75 F3 Goa, E Mali
Menaldum 86 D1 Fris. Menaam. Fryslân, N Netherlands
Mènam Khong see Mekong
Mendaña Fracture Zone 153 F4 fracture zone E Pacific Ocean
Mende 91 C5 anc. Mimatum. Lozère, S France
Mendeleyev Ridge 155 B2 undersea ridge Arctic Ocean
Mendocino Fracture Zone 152 D2 fracture zone NE Pacific Ocean
Mendoza 64 B4 Mendoza, W Argentina
Menemen 116 A3 İzmir, W Turkey (Türkiye)
Menengiyn Tal 127 F2 plain E Mongolia
Menongue 78 B2 var. Vila Serpa Pinto, Port. Serpa Pinto. Cuando Cubango, C Angola
Menorca 93 H3 Eng. Minorca; anc. Balearis Minor. island Islas Baleares, Spain, W Mediterranean Sea
Mentawai, Kepulauan 138 A4 island group W Indonesia
Meppel 86 D2 Drenthe, NE Netherlands
Meran see Merano
Merano 96 C1 Ger. Meran. Trentino-Alto Adige, N Italy
Merca see Marka
Mercedes 64 D3 Corrientes, NE Argentina
Mercedes 64 D4 Soriano, SW Uruguay
Meredith, Lake 49 E1 reservoir Texas, SW USA
Merefa 109 G2 Kharkivs'ka Oblast', E Ukraine
Mergate see Margate
Mergui see Myeik
Mergui Archipelago see Myeik Archipelago
Mérida 51 H3 Yucatán, SW Mexico
Mérida 92 C4 anc. Augusta Emerita. Extremadura, W Spain
Mérida 58 C2 Mérida, W Venezuela
Meridian 42 C2 Mississippi, S USA
Mérignac 91 B5 Gironde, SW France
Merin, Laguna see Mirim Lagoon
Merkulovichi see Myerkulavichy
Merowe 72 B3 Northern, N Sudan
Merredin 147 B6 Western Australia
Mersen see Meerssen
Mersey 89 C5 river NW England, United Kingdom
Mersin 116 C4 var. İçel. İçel, S Turkey (Türkiye)
Mērsrags 106 C3 Talsi, NW Latvia
Meru 73 C6 Eastern, C Kenya
Merv see Mary
Merzifon 116 D2 Amasya, N Turkey (Türkiye)
Merzig 95 A5 Saarland, SW Germany
Mesa 48 B2 Arizona, SW USA
Meseritz see Międzyrzecz
Meshed see Mashhad
Mesopotamia 57 C5 var. Mesopotamia Argentina. physical region NE Argentina
Mesopotamia Argentina see Mesopotamia
Messalo, Rio 79 E2 var. Mualo. river NE Mozambique
Messana/Messene see Messina
Messina see Musina
Messina, Strait of see Messina, Stretto di
Messina, Stretto di 97 D7 Eng. Strait of Messina. strait SW Italy
Messíni 105 B6 Pelopónnisos, S Greece
Mesta see Néstos

Mestghanem see Mostaganem
Mestia 117 F1 var. Mestiya. N Georgia
Mestiya see Mestia
Mestre 96 C2 Veneto, NE Italy
Metairie 42 B3 Louisiana, S USA
Metán 64 C2 Salta, N Argentina
Metapán 52 B2 Santa Ana, NW El Salvador
Meta, Río 58 D3 river Colombia/ Venezuela
Meterlam see Mehtar Lām
Methariam/Metharlam see Mehtar Lām
Metis see Metz
Metković 100 B4 Dubrovnik-Neretva, SE Croatia
Métsovo 104 B4 prev. Métsovon. Ípeiros, C Greece
Métsovon see Métsovo
Metz 90 D3 anc. Divodurum Mediomatricum, Mediomatrica, Metis. Moselle, NE France
Meulaboh 138 A3 Sumatera, W Indonesia
Meuse 87 C6 Dut. Maas. river W Europe
Mexcala, Río see Balsas, Río
Mexicali 50 A1 Baja California Norte, NW Mexico
Mexicanos, Estados Unidos see Mexico
Mexico 45 G4 Missouri, C USA
Mexico 50 C3 off. United Mexican States, var. Méjico, México, Sp. Estados Unidos Mexicanos. country N Central America
México see Mexico
Mexico City see México, Ciudad de
México, Ciudad de 51 E4 var. Mexico City. country capital (Mexico) México, C Mexico
México, Golfo de see Mexico, Gulf of
Mexico, Gulf of 51 F2 Sp. Golfo de México. gulf W Atlantic Ocean
Meyadine see Al Mayādīn
Meydān Shahr see Maidān Shahr
Meymaneh see Maīmanah
Mezen' 110 D3 river NW Russia
Mezőtúr 99 D7 Jász-Nagykun-Szolnok, E Hungary
Mgarr 102 A5 Gozo, N Malta
Miadzioł Nowy see Myadzyel
Miahuatlán 51 F5 var. Miahuatlán de Porfirio Díaz. Oaxaca, SE Mexico
Miahuatlán de Porfirio Díaz see Miahuatlán
Miami 43 F5 Florida, SE USA
Miami 49 G1 Oklahoma, C USA
Miami Beach 43 F5 Florida, SE USA
Miāneh 120 C2 var. Miyāneh. Āzārbāyjān-e Sharqī, NW Iran
Mianyang 129 B5 Sichuan, C China
Miastko 98 C2 Ger. Rummelsburg in Pommern. Pomorskie, N Poland
Mi Chai see Nong Khai
Michalovce 99 E5 Ger. Grossmichel, Hung. Nagymihály. Košický Kraj, E Slovakia
Michigan 40 C1 off. State of Michigan, also known as Great Lakes State, Lake State, Wolverine State. state N USA
Michigan, Lake 40 C2 lake N USA
Michurin see Tsarevo
Michurinsk 111 B5 Tambovskaya Oblast', W Russia
Micoud 55 F2 SE Saint Lucia
Micronesia, Federated States of 144 B1 country W Pacific Ocean
Micronesia 144 C1 island group W Pacific Ocean
Mid-Atlantic Cordillera see Mid-Atlantic Ridge
Mid-Atlantic Ridge 66 C3 var. Mid-Atlantic Cordillera, Mid-Atlantic Rise, Mid-Atlantic Swell. undersea ridge Atlantic Ocean
Mid-Atlantic Rise/Mid-Atlantic Swell see Mid-Atlantic Ridge
Middelburg 87 B5 Zeeland, SW Netherlands
Middelharnis 86 B4 Zuid-Holland, SW Netherlands
Middelkerke 87 A5 West-Vlaanderen, W Belgium
Middle America Trench 35 B7 trench E Pacific Ocean
Middle Andaman 133 F2 island Andaman Islands, India, NE Indian Ocean
Middle Atlas 70 C2 Eng. Middle Atlas. mountain range N Morocco
Middle Atlas see Moyen Atlas
Middleburg Island see 'Eua

Middle Congo *see* Congo (Republic of)
Middlesboro *40 C5* Kentucky, S USA
Middlesbrough *89 D5* N England, United Kingdom
Middletown *41 F4* New Jersey, NE USA
Middletown *41 F3* New York, NE USA
Mid-Indian Basin *141 C5 undersea basin* N Indian Ocean
Mid-Indian Ridge *141 C5 var.* Central Indian Ridge. *undersea ridge* C Indian Ocean
Midland *38 D5* Ontario, S Canada
Midland *40 C3* Michigan, N USA
Midland *49 E3* Texas, SW USA
Mid-Pacific Mountains *152 C2 var.* Mid-Pacific Seamounts. *seamount range* NW Pacific Ocean
Mid-Pacific Seamounts *see* Mid-Pacific Mountains
Midway Islands *152 D2 US unincorporated territory* C Pacific Ocean
Miechów *99 D5* Małopolskie, S Poland
Międzyrzec Podlaski *98 E3* Lubelskie, E Poland
Międzyrzecz *98 B3 Ger.* Meseritz. Lubuskie, W Poland
Mielec *99 D5* Podkarpackie, SE Poland
Miercurea-Ciuc *108 C4 Ger.* Szeklerburg, *Hung.* Csíkszereda. Harghita, C Romania
Mieres del Camín *92 D1 var.* Mieres del Camino. Asturias, NW Spain
Mieres del Camino *see* Mieres del Camín
Mi'eso *73 D5 var.* Meheso, Miesso. Oromīya, C Ethiopia
Miesso *see* Mi'eso
Mifrats Hefa *119 A5 Eng.* Bay of Haifa; *prev.* Mifraz Hefa. *bay* N Israel
Miguel Asua *50 D3 var.* Miguel Auza. Zacatecas, C Mexico
Miguel Auza *see* Miguel Asua
Mijdrecht *86 C3* Utrecht, C Netherlands
Mikashevichi *107 C7 Pol.* Mikaszewicze, *Rus.* Mikashevichi. Brestskaya Voblasts', SW Belarus
Mikaszewicze *see* Mikashevichi
Mikhaylovgrad *see* Montana
Mikhaylovka *111 B6* Volgogradskaya Oblast', SW Russia
Mikonos *see* Mýkonos
Mikre *104 C2* Lovech, N Bulgaria
Mikun' *110 D4* Respublika Komi, NW Russia
Mikuni-sanmyaku *131 D5 mountain range* Honshū, N Japan Asia
Mikura-jima *131 D6 island* E Japan
Milagro *60 B2* Guayas, SW Ecuador
Milan *see* Milano
Milange *79 E2* Zambézia, NE Mozambique
Milano *96 B2 Eng.* Milan, *Ger.* Mailand; *anc.* Mediolanum. Lombardia, N Italy
Milas *116 A4* Muğla, SW Turkey (Türkiye)
Milashavichy *107 C7 Rus.* Milashevichi. Homyel'skaya Voblasts', SE Belarus
Milashevichi *see* Milashavichy
Mildura *149 C6* Victoria, SE Australia
Mile *see* Mili Atoll
Miles *149 D5* Queensland, E Australia
Miles City *44 C2* Montana, NW USA
Milford *see* Milford Haven
Milk River *37 E5* Alberta, SW Canada
Milk River *44 C1 river* Montana, NW USA
Milk, Wadi el *88 B4 var.* Wadi al Malik. *river* C Sudan
Milledgeville *43 E2* Georgia, SE USA
Mille Lacs Lake *45 F2 lake* Minnesota, N USA
Millennium Island *143 F3 prev.* Caroline Island, Thornton Island. *atoll* Line Islands, E Kiribati
Millerovo *111 B6* Rostovskaya Oblast', SW Russia
Mílos *105 C2 island* Kykládes, Greece, Aegean Sea

Milton *151 B7* Otago, South Island, New Zealand
Milton Keynes *89 D6* SE England, United Kingdom
Milwaukee *40 B3* Wisconsin, N USA
Mimatum *see* Mende
Min *see* Fujian
Mīnā' as Sulṭān Qābūs *140 B3* NE Oman
Minas Gerais *63 F3 off.* Estado de Minas Gerais. *state/region* E Brazil
Minas Gerais, Estado de *see* Minas Gerais
Minatitlán *51 F4* Veracruz-Llave, E Mexico
Minbu *136 A3* Magway, W Myanmar (Burma)
Minch, The *88 B3 var.* North Minch. *strait* NW Scotland, United Kingdom
Mindanao *139 F2 island* S Philippines
Mindanao Sea *see* Bohol Sea
Mindelheim *95 C6* Bayern, S Germany
Mindello *see* Mindelo
Mindelo *74 A2 var.* Mindello; *prev.* Porto Grande. São Vicente, N Cape Verde (Cabo Verde)
Minden *94 B4 anc.* Minthun. Nordrhein-Westfalen, NW Germany
Mindoro *139 E2 island* N Philippines
Mindoro Strait *139 E2 strait* W Philippines
Mineral Wells *49 F2* Texas, SW USA
Mingäçevir *117 G2 Rus.* Mingechaur, Mingechevir. C Azerbaijan
Mingechaur/Mingechevir *see* Mingäçevir
Mingora *see* Saidu Sharif
Minho *92 B2 former province* N Portugal
Minho *92 B2 Sp.* Miño. *river* Portugal/ Spain
Minho, Rio *see* Miño
Minicoy Island *132 B3 island* SW India
Minius *see* Miño
Minna *75 G4* Niger, C Nigeria
Minneapolis *45 F2* Minnesota, N USA
Minnesota *45 F2 off.* State of Minnesota, *also known as* Gopher State, New England of the West, North Star State. *state* N USA
Miño *92 B2 var.* Mino, Minius, *Port.* Rio Minho. *river* Portugal/Spain
Miño *see* Minho, Rio
Minorca *see* Menorca
Minot *45 E1* North Dakota, N USA
Minsk *107 C6 country capital* (Belarus) Minskaya Voblasts', C Belarus
Minskaya Wzvyshsha *107 C6 mountain range* C Belarus
Mińsk Mazowiecki *98 D3 var.* Nowo-Minsk. Mazowieckie, C Poland
Minthun *see* Minden
Minto, Lac *38 D2 lake* Québec, C Canada
Minya *see* Al Minyā
Miraflores *50 C3* Baja California Sur, NW Mexico
Miranda de Ebro *93 E1* La Rioja, N Spain
Mirgorod *see* Myrhorod
Miri *138 D3* Sarawak, East Malaysia
Mirim Lagoon *63 E5 var.* Lake Mirim, *Sp.* Laguna Merín. *lagoon* Brazil/ Uruguay
Mirim, Lake *see* Mirim Lagoon
Mírina *see* Mýrina
Mīrjāveh *120 E4* Sīstān va Balūchestān, SE Iran
Mirny *154 C3 Russian research station* Antarctica
Mirnyy *115 F3* Respublika Sakha (Yakutiya), NE Russia
Mirpur Khas *134 B3* Sindh, SE Pakistan
Mirtóo Pélagos *see* Myrtóo Pélagos
Misiaf *see* Maşyāf
Miskito Coast *see* Mosquito Coast
Miskitos, Cayos *53 E2 island group* NE Nicaragua
Miskolc *99 D6* Borsod-Abaúj-Zemplén, NE Hungary
Misool, Pulau *139 F4 island* Maluku, E Indonesia
Mişr *see* Egypt
Mişrātah *71 F2 var.* Misurata. NW Libya
Mission *49 G5* Texas, SW USA
Mississippi *42 B2 off.* State of Mississippi, *also known as* Bayou State, Magnolia State. *state* SE USA
Mississippi Delta *42 B4 delta* Louisiana, S USA
Mississippi River *35 C6 river* C USA
Missoula *44 B1* Montana, NW USA

Missouri *45 F5 off.* State of Missouri, *also known as* Bullion State, Show Me State. *state* C USA
Missouri River *45 E3 river* C USA
Mistassini, Lac *38 D3* lake Québec, SE Canada
Mistelbach an der Zaya *95 E6* Niederösterreich, NE Austria
Misti, Volcán *61 E4 volcano* S Peru
Misurata *see* Mişrātah
Mitau *see* Jelgava
Mitchell *149 D5* Queensland, E Australia
Mitchell *45 E3* South Dakota, N USA
Mitchell, Mount *43 E1 mountain* North Carolina, SE USA
Mitchell River *148 C2 river* Queensland, NE Australia
Mi Tho *see* My Tho
Mitilíni *see* Mytilíni
Mito *131 D5* Ibaraki, Honshū, S Japan
Mitrovica *see* Mitrovicë
Mitrovica/Mitrowitz *see* Sremska Mitrovica, Serbia
Mitrovicë *101 D5 Serb.* Mitrovica, *prev.* Kosovska Mitrovica, Titova Mitrovica. N Kosovo
Mits'iwa *72 C4 var.* Masawa, Massawa. E Eritrea
Mitspe Ramon *119 A7 prev.* Mizpe Ramon. Southern, S Israel
Mittelstadt *see* Baia Sprie
Mitú *58 C4* Vaupés, SE Colombia
Mitumba, Chaine des/Mitumba Range *see* Mitumba, Monts
Mitumba Monts *77 E7 var.* Chaîne des Mitumba, Mitumba Range. *mountain range* E Dem. Rep. Congo
Miueru Wantipa, Lake *73 E7 lake* N Zambia
Miyake-jima *131 D6 island* Sakishima-shotō, SW Japan
Miyako *130 D4* Iwate, Honshū, C Japan
Miyakonojō *131 B8 var.* Miyakonzyô. Miyazaki, Kyūshū, SW Japan
Miyakonzyô *see* Miyakonojō
Miyāneh *see* Miāneh
Miyazaki *131 B8* Miyazaki, Kyūshū, SW Japan
Mizia *104 C1 var.* Miziya. Vratsa, NW Bulgaria
Mizil *108 C5* Prahova, SE Romania
Miziya *see* Mizia
Mizpe Ramon *see* Mitspe Ramon
Mjøsa *85 B6 var.* Mjøsen. *lake* S Norway
Mjøsen *see* Mjøsa
Mladenovac *100 D4* Serbia, C Serbia
Mława *98 D3* Mazowieckie, C Poland
Mljet *101 B5 It.* Meleda; *anc.* Melita. *island* S Croatia
Mmabatho *78 C4* North-West, N South Africa
Moab *44 B5* Utah, W USA
Moa Island *148 C1 island* Queensland, NE Australia
Moanda *77 B6 var.* Mouanda. Haut-Ogooué, SE Gabon
Moba *77 E7* Katanga, E Dem. Rep. Congo
Mobay *see* Montego Bay
Mobaye *77 C5* Basse-Kotto, S Central African Republic
Moberly *45 G4* Missouri, C USA
Mobile *42 C3* Alabama, S USA
Mobutu Sese Seko, Lac *see* Albert, Lake
Moçâmedes *see* Namibe
Mochudi *78 C4* Kgatleng, SE Botswana
Mocímboa da Praia *79 F2 var.* Vila de Mocímboa da Praia. Cabo Delgado, N Mozambique
Môco *78 B2 var.* Morro de Môco. *mountain* W Angola
Mocoa *58 A4* Putumayo, SW Colombia
Môco, Morro de *see* Môco
Mocuba *79 E3* Zambézia, NE Mozambique
Modena *96 B3 anc.* Mutina. Emilia-Romagna, N Italy
Modesto *47 B6* California, W USA
Modica *97 C7 anc.* Motyca. Sicilia, Italy, C Mediterranean Sea
Modimolle *78 D4 prev.* Nylstroom. Limpopo, NE South Africa
Modohn *see* Madona
Modriča *100 C3* Republika Srpska, N Bosnia and Herzegovina
Moe *149 C7* Victoria, SE Australia
Møen *see* Møn, Denmark
Moero, Lac *see* Mweru, Lake
Moeskroen *see* Mouscron
Mogadiscio/Mogadishu *see* Muqdisho
Mogador *see* Essaouira

Mogilëv *see* Mahilyow
Mogilev-Podol'skiy *see* Mohyliv-Podilskyi
Mogilno *98 C3* Kujawsko-pomorskie, C Poland
Moḩammadābād-e Rīgān *120 E4* Kermān, SE Iran
Mohammedia *70 C2 prev.* Fédala. NW Morocco
Mohave, Lake *47 D7 reservoir* Arizona/ Nevada, W USA
Mohawk River *41 F3 river* New York, NE USA
Mohéli *see* Mwali
Mohns Ridge *83 F3 undersea ridge* Greenland Sea/Norwegian Sea
Moho *61 E4* Puno, SE Peru
Mohoro *73 C7* Pwani, E Tanzania
Mohua *see* Golden Bay
Mohyliv-Podilskyi *108 D3 prev.* Mohyliv-Podol'skiy, *Rus.* Mogilev-Podol'skiy. Vinnyts'ka Oblast', C Ukrain
Mohyliv-Podil's'kyy *see* Mohyliv-Podilskyi
Moili *see* Mwali
Mo i Rana *84 C3* Nordland, C Norway
Mõisaküla *106 D3 Ger.* Moiseküll. Viljandimaa, S Estonia
Moiseküll *see* Mõisaküla
Moissac *91 B6* Tarn-et-Garonne, S France
Mojácar *93 E5* Andalucía, S Spain
Mojave Desert *47 D7 plain* California, W USA
Mokrany *see* Makrany
Moktama *see* Mottama
Mol *87 C5 prev.* Moll. Antwerpen, N Belgium
Moldavia *see* Moldova
Moldavian SSR/Moldavskaya SSR *see* Moldova
Molde *85 A5* Møre og Romsdal, S Norway
Moldo-Too, Khrebet *see* Moldo-Too, Khrebet
Moldo-Too, Khrebet *123 G2 prev.* Khrebet Moldotau. *mountain range* C Kyrgyzstan
Moldova *108 D3 off.* Republic of Moldova, *var.* Moldavia, *prev.* Moldavian SSR, *Rus.* Moldavskaya SSR. *country* SE Europe
Moldova Nouă *108 A4 Ger.* Neumoldowa, *Hung.* Újmoldova. Caraş-Severin, SW Romania
Moldova, Republic of *see* Moldova
Moldoveanul *see* Vârful Moldoveanu
Molfetta *97 E5* Puglia, SE Italy
Moll *see* Mol
Mollendo *61 E4* Arequipa, SW Peru
Mölndal *85 B7* Västra Götaland, S Sweden
Molochansk *109 G4* Zaporiz'ka Oblast', SE Ukraine
Molodechno/Molodeczno *see* Maladzyechna
Moloka'i *47 B8 var.* Molokai. *island* Hawaiian Islands, Hawaii, USA
Molokai Fracture Zone *153 E2 tectonic feature* NE Pacific Ocean
Molopo *78 C4 seasonal river* Botswana/ South Africa
Mólos *105 B5* Stereá Ellás, C Greece
Molotov *see* Severodvinsk, Arkhangel'skaya Oblast', Russia
Molotov *see* Perm', Permskaya Oblast', Russia
Moluccas *see* Maluku
Molucca Sea *139 F4 Ind.* Laut Maluku. *sea* E Indonesia
Molukken *see* Maluku
Mombasa *73 D7* Coast, SE Kenya
Mombetsu *see* Monbetsu
Momchilgrad *104 D3 prev.* Mastanli. Kardzhali, S Bulgaria
Møn *85 B8 prev.* Møen. *island* SE Denmark
Mona, Canal de la *see* Mona Passage
Monaco *91 C8 var.* Monaco-Ville; *anc.* Monoecus. *country capital* (Monaco) S Monaco
Monaco *91 E6 off.* Principality of Monaco. *country* W Europe
Monaco *see* München
Monaco, Port de *91 C8 bay* S Monaco W Mediterranean Sea
Monaco, Principality of *see* Monaco
Monaco-Ville *see* Monaco

Monahans *49 E3* Texas, SW USA
Mona, Isla *55 E3 island* W Puerto Rico
Mona Passage *55 E3 Sp.* Canal de la
    Mona. *channel* Dominican Republic/
    Puerto Rico
Monastir *see* Bitola
Monbetsu *130 D2 var.* Mombetsu,
    Monbetu. Hokkaidō, NE Japan
Monbetu *see* Monbetsu
Moncalieri *96 A2* Piemonte, NW Italy
Monchegorsk *110 C2* Murmanskaya
    Oblast', NW Russia
Monclova *50 D2* Coahuila, NE Mexico
Moncton *39 F4* New Brunswick,
    SE Canada
Mondovì *96 A2* Piemonte, NW Italy
Monfalcone *96 D2* Friuli-Venezia Giulia,
    NE Italy
Monforte de Lemos *92 C1* Galicia,
    NW Spain
Mongo *76 C3* Guéra, C Chad
Mongolia *126 C2 Mong.* Mongol Uls.
    country E Asia
Mongolia, Plateau of *124 D1 plateau*
    E Mongolia
Mongol Uls *see* Mongolia
Mongora *see* Saidu Sharif
Mongos, Chaîne des *see* Bongo,
    Massif des
Mongu *78 C2* Western, W Zambia
Monkchester *see* Newcastle upon Tyne
Monkey Bay *79 E2* Southern, SE Malawi
Monkey River *see* Monkey River Town
Monkey River Town *52 C2 var.* Monkey
    River. Toledo, SE Belize
Monoecus *see* Monaco
Mono Lake *47 C6* lake California,
    W USA
Monostor *see* Beli Manastir
Monóvar *see* Monòver
Monòver *93 F4 var.* Monóvar.
    Comunitat Valenciana, E Spain
Monroe *42 B2* Louisiana, S USA
Monrovia *74 C5 country capital* (Liberia)
    W Liberia
Mons *87 B6 Dut.* Bergen. Hainaut,
    S Belgium
Monselice *96 C2* Veneto, NE Italy
Montana *104 C2 prev.* Ferdinand,
    Mikhaylovgrad. Montana, NW Bulgaria
Montana *44 B1 off.* State of Montana,
    also known as Mountain State,
    Treasure State. *state* NW USA
Montargis *90 C4* Loiret, C France
Montauban *91 B6* Tarn-et-Garonne,
    S France
Montbéliard *90 D4* Doubs, E France
Mont Cenis, Col du *91 D5 pass* E France
Mont-de-Marsan *91 B6* Landes,
    SW France
Monteagudo *61 G4* Chuquisaca,
    S Bolivia
Montecarlo *61 C8* Misiones,
    NE Argentina
Monte Caseros *64 D3* Corrientes,
    NE Argentina
Monte Cristi *54 D3 var.* San Fernando
    de Monte Cristi. NW Dominican
    Republic
Monte Croce Carnico, Passo di *see*
    Plöcken Pass
Montegiardino *96 E2* SE San Marino
Montego Bay *54 A4 var.* Mobay.
    W Jamaica
Montélimar *91 D5 anc.* Acunum Acusio,
    Montilium Adhemari. Drôme, E France
Montemorelos *51 E3* Nuevo León,
    NE Mexico
Montenegro *101 C5 Serb.* Crna Gora.
    country SW Europe
Monte Patria *64 B3* Coquimbo, N Chile
Monterey *47 B6* California, W USA
Monterey *see* Monterrey
Monterey Bay *47 A6 bay* California,
    W USA
Montería *58 B2* Córdoba, NW Colombia
Montero *61 G4* Santa Cruz, C Bolivia
Monterrey *51 E3 var.* Monterey. Nuevo
    León, NE Mexico
Montes Claros *63 F3* Minas Gerais,
    SE Brazil
Montevideo *64 D4 country capital*
    (Uruguay) Montevideo, S Uruguay
Montevideo *45 F2* Minnesota, N USA
Montgenèvre, Col de *91 D5 pass* France/
    Italy
Montgomery *42 D2 state capital*
    Alabama, S USA
Montgomery *see* Sahiwal
Monthey *95 A7* Valais, SW Switzerland
Montilium Adhemari *see* Montélimar
Montluçon *90 C4* Allier, C France

Montoro *92 D4* Andalucía, S Spain
Montpelier *41 G2 state capital* Vermont,
    NE USA
Montpellier *91 C6* Hérault, S France
Montréal *39 E4 Eng.* Montreal. Québec,
    SE Canada
Montrose *88 D3* E Scotland, United
    Kingdom
Montrose *44 C5* Colorado, C USA
Montserrat *55 G3 var.* Emerald Isle. *UK*
    Overseas Territory E West Indies
Monywa *136 B3* Sagaing, C Myanmar
    (Burma)
Monza *96 B2* Lombardia, N Italy
Monze *78 D2* Southern, S Zambia
Monzón *93 F2* Aragón, NE Spain
Moonie *149 D5* Queensland, E Australia
Moon-Sund *see* Väinameri
Moora *147 A6* Western Australia
Moore *49 G1* Oklahoma, C USA
Moore, Lake *147 B6 lake* Western
    Australia
Moorhead *45 F2* Minnesota, N USA
Moose *38 C3 river* Ontario, S Canada
Moosehead Lake *41 G1 lake* Maine,
    NE USA
Moosonee *38 C3* Ontario, SE Canada
Mopti *75 E3* Mopti, C Mali
Moquegua *61 E4* Moquegua, SE Peru
Mora *85 C5* Dalarna, C Sweden
Morales *52 C2* Izabal, E Guatemala
Morant Bay *54 B5* E Jamaica
Moratalla *93 E4* Murcia, SE Spain
Morava *99 C5 var.* March. *river*
    C Europe
Morava *see* Moravia, Czechia
    (Czech Republic)
Morava *see* Velika Morava, Serbia
Moravia *99 B5 Cz.* Morava, Ger. Mähren.
    cultural region E Czechia
    (Czech Republic)
Moray Firth *88 C3 inlet* N Scotland,
    United Kingdom
Morea *see* Pelopónnisos
Moreau River *44 D2 river* South Dakota,
    N USA
Moree *149 D5* New South Wales,
    SE Australia
Morelia *51 E4* Michoacán, S Mexico
Morena, Sierra *92 C4 mountain range*
    S Spain
Moreni *108 C5* Dâmbovița, S Romania
Mórfou *see* Güzelyurt
Morgan City *42 B3* Louisiana, S USA
Morghab, Darya-ye *122 D3 Rus.*
    Murgab, Murghab, *Turkm.* Murgap,
    Murgap Deryasy. *river* Afghanistan/
    Turkmenistan
Morioka *130 D4* Iwate, Honshū, C Japan
Morlaix *90 A3* Finistère, NW France
Mormon State *see* Utah
Mornington Abyssal Plain *67 A7*
    abyssal plain SE Pacific Ocean
Mornington Island *148 B2 island*
    Wellesley Islands, Queensland,
    N Australia
Morocco *70 B3 off.* Kingdom of
    Morocco, *Ar.* Al Maghrib. *country*
    N Africa
Morocco *see* Marrakech
Morocco, Kingdom of *see* Morocco
Morogoro *73 C7* Morogoro, E Tanzania
Moro Gulf *139 E3 gulf* S Philippines
Morón *54 C2* Ciego de Ávila, C Cuba
Mörön *126 D2* Hövsgöl, N Mongolia
Morondava *79 F3* Toliara,
    W Madagascar
Moroni *79 F2 country capital* (Comoros)
    Grande Comore, NW Comoros
Morotai, Pulau *139 F3 island* Maluku,
    E Indonesia
Morotiri *see* Marotiri
Morphou *see* Güzelyurt
Morrinsville *150 D3* Waikato, North
    Island, New Zealand
Morris *45 F2* Minnesota, N USA
Morris Jesup, Kap *83 E1 headland*
    N Greenland
Morvan *90 D4 physical region* C France
Moscow *46 C2* Idaho, NW USA
Moscow *see* Moskva
Mosel *95 E5 Fr.* Moselle. *river* W Europe
Mosel *see* Moselle
Moselle *87 E8 Ger.* Mosel. *river*
    W Europe
Moselle *see* Mosel
Mosgiel *151 B7* Otago, South Island,
    New Zealand
Moshi *73 C7* Kilimanjaro, NE Tanzania
Mosjøen *84 B4* Nordland, C Norway

Moskovskiy *see* Maskav
Moskva *111 B5 Eng.* Moscow. *country*
    capital (Russia) Gorod Moskva,
    W Russia
Moskva *see* Maskav
Moson and Magyaróvár *see*
    Mosonmagyaróvár
Mosonmagyaróvár *99 C6 Ger.*
    Wieselburg-Ungarisch-Altenburg; *prev.*
    Moson and Magyaróvár, *Ger.*
    Wieselburg and Ungarisch-Altenburg.
    Győr-Moson-Sopron, NW Hungary
Mosquito Coast *53 E3 var.* Miskito
    Coast, *Eng.* Mosquito Coast. *coastal*
    region E Nicaragua
Mosquito Coast *see* La Mosquitia
Mosquito Gulf *53 F4 Eng.* Mosquito
    Gulf. *gulf* N Panama
Mosquito Gulf *see* Mosquitos,
    Golfo de los
Moss *85 B6* Østfold, S Norway
Mossâmedes *see* Namibe
Mosselbaai *78 C5 var.* Mosselbai, *Eng.*
    Mossel Bay. Western Cape,
    SW South Africa
Mosselbai/Mossel Bay *see* Mosselbaai
Mossendjo *77 B6* Niari, SW Congo
Mossoró *63 G2* Rio Grande do Norte,
    NE Brazil
Most *98 A4 Ger.* Brüx. Ústecký Kraj,
    NW Czechia (Czech Republic)
Mosta *102 B5 var.* Musta. C Malta
Mostaganem *70 D2 var.* Mestghanem.
    NW Algeria
Mostar *100 C4* Federacija Bosna I
    Hercegovina, S Bosnia and Herzegovina
Mosty *see* Masty
Mosul *see* Al Mawşil
Mota del Cuervo *93 E3* Castilla-La
    Mancha, C Spain
Motagua, Río *52 B2 river* Guatemala/
    Honduras
Mother of Presidents/Mother of States
    see Virginia
Motril *92 D5* Andalucía, S Spain
Motru *108 B4* Gorj, SW Romania
Mottama *136 B4 prev.* Martaban; *var.*
    Moktama. Mon State,
    S Myanmar (Burma)
Motueka *151 C5* Tasman, South Island,
    New Zealand
Motul *51 H3 var.* Motul de Felipe
    Carrillo Puerto. Yucatán, SE Mexico
Motul de Felipe Carrillo Puerto *see*
    Motul
Motyca *see* Modica
Mouanda *see* Moanda
Mouhoun *see* Black Volta
Mouila *77 A6* Ngounié, C Gabon
Moukden *see* Shenyang
Mould Bay *37 E2* Prince Patrick Island,
    Northwest Territories, N Canada
Moulins *90 C4* Allier, C France
Moulmein *see* Mawlamyine
Moundou *76 B4* Logone-Occidental,
    SW Chad
Moŭng Roessei *137 D5* Battambang,
    W Cambodia
Moun Hou *see* Black Volta
Mountain Home *42 B1* Arkansas, C USA
Mountain State *see* Montana
Mountain State *see* West Virginia
Mount Cook *see* Aoraki (mountain)
Mount Cook *see* Aoraki
    (populated place)
Mount Gambier *149 B7* South Australia
Mount Isa *148 B3* Queensland,
    C Australia
Mount Magnet *147 B5*
    Western Australia
Mount Pleasant *45 G4* Iowa, C USA
Mount Pleasant *40 C3* Michigan, N USA
Mount Vernon *40 B5* Illinois, N USA
Mount Vernon *46 B1* Washington,
    NW USA
Mourdi, Dépression du *76 C2 desert*
    lowland Chad/Sudan
Mouscron *87 A6 Dut.* Moeskroen.
    Hainaut, W Belgium
Mouse River *see* Souris River
Moussoro *76 B3* Kanem, W Chad
Moyen-Congo *see* Congo (Republic of)
Mo'ynoq *122 C1 Rus.* Muynak.
    Qoraqalpog'iston Respublikasi,
    NW Uzbekistan
Moyobamba *60 B2* San Martín, NW Peru
Moyu *126 B3 var.* Karakax. Xinjiang
    Uygur Zizhiqu, NW China

Moyynkum, Peski *123 F1 Kaz.*
    Moyynqum. *desert* S Kazakhstan
Moyynqum *see* Moyynkum, Peski
Mozambika, Lakandranon' i *see*
    Mozambique Channel
Mozambique *79 E3 off.* Republic of
    Mozambique; *prev.* People's Republic of
    Mozambique, Portuguese East Africa.
    country S Africa
Mozambique Basin *see* Natal Basin
Mozambique, Canal de *see* Mozambique
    Channel
Mozambique Channel *79 E3 Fr.* Canal
    de Mozambique, *Mal.* Lakandranon' i
    Mozambika. *strait* W Indian Ocean
Mozambique, People's Republic of *see*
    Mozambique
Mozambique Plateau *69 D7 var.*
    Mozambique Rise. *undersea plateau*
    SW Indian Ocean
Mozambique, Republic of *see*
    Mozambique
Mozambique Rise *see* Mozambique
    Plateau
Mozyr' *see* Mazyr
Mpama *77 B6 river* C Congo
Mpika *78 D2* Northern, NE Zambia
Mqinvartsveri *see* Kazbek
Mragowo *98 D2 Ger.* Sensburg.
    Warmińsko-Mazurskie, NE Poland
Mthatha *78 D5 prev.* Umtata. Eastern
    Cape, SE South Africa
Mtkvari *see* Kura
Mtwara *73 D8* Mtwara, SE Tanzania
Mualo *see* Messalo, Rio
Muang Chiang Rai *see* Chiang Rai
Muang Kalasin *see* Kalasin
Muang Khammouan *see* Thakhèk
Muang Không *137 D5* Champasak,
    S Laos
Muang Khôngxédôn *137 D5 var.* Khong
    Sedone. Salavan, S Laos
Muang Khon Kaen *see* Khon Kaen
Muang Lampang *see* Lampang
Muang Loei *see* Loei
Muang Lom Sak *see* Lom Sak
Muang Nakhon Sawan *see*
    Nakhon Sawan
Muang Namo *136 C3* Oudômxai, N Laos
Muang Nan *see* Nan
Muang Phalan *136 D4 var.* Muang
    Phalane. Savannakhét, S Laos
Muang Phalane *see* Muang Phalan
Muang Phayao *see* Phayao
Muang Phitsanulok *see* Phitsanulok
Muang Phrae *see* Phrae
Muang Roi Et *see* Roi Et
Muang Sakon Nakhon *see*
    Sakon Nakhon
Muang Samut Prakan *see* Samut Prakan
Muang Sing *136 C3* Louang Namtha,
    N Laos
Muang Ubon *see* Ubon Ratchathani
Muang Xaignabouri *see* Xaignabouli
Muar *138 B3 var.* Bandar Maharani.
    Johor, Peninsular Malaysia
Mucojo *79 F2* Cabo Delgado,
    N Mozambique
Mudanjiang *129 E3 var.* Mu-tan-chiang.
    Heilongjiang, NE China
Mudon *137 B5* Mon State,
    S Myanmar (Burma)
Muenchen *see* München
Muenster *see* Münster
Mufulira *78 D2* Copperbelt, C Zambia
Mughla *see* Muğla
Muğla *116 A4 var.* Mughla. Muğla,
    SW Turkey (Türkiye)
Muḥ, Sabkhat al *118 C3 lake* C Syria
Muhu *see* Väin see
Muisne *60 A1* Esmeraldas, NW Ecuador
Mukacheve *see* Mukachevo
Mukachevo *108 B3 prev.* Mukacheve,
    Hung. Munkács. Zakarpats'ka Oblast',
    W Ukraine
Mukalla *see* Al Mukallā
Mukden *see* Shenyang
Mula *93 E4* Murcia, SE Spain
Mulakatholhu *132 B4 var.* Meemu Atoll,
    Mulaku Atoll. *atoll* C Maldives
Mulaku Atoll *see* Mulakatholhu
Muleshoe *49 E2* Texas, SW USA
Mulhacén *93 E5 var.* Cerro de Mulhacén.
    mountain S Spain
Mulhacén, Cerro de *see* Mulhacén
Mülhausen *see* Mulhouse
Mülheim *95 A6 var.* Mulheim an der
    Ruhr. Nordrhein-Westfalen,
    W Germany
Mülheim an der Ruhr *see* Mülheim

Mulhouse 90 E4 *Ger.* Mülhausen.
Haut-Rhin, NE France
Müller-gerbergte *see* Muller,
Pegunungan
Muller, Pegunungan 138 D4 *Dut.*
Müller-gerbergte. *mountain range*
Borneo, C Indonesia
Mull, Isle of 88 B4 *island* W Scotland,
United Kingdom
Mulongo 77 D7 Katanga, SE Dem.
Rep. Congo
Multan 134 C2 Punjab, E Pakistan
Mumbai 134 C5 *prev.* Bombay. *state
capital* Mahārāshtra, W India
Munamägi *see* Suur Munamägi
Münchberg 95 C5 Bayern, E Germany
München 95 C6 *var.* Muenchen, *Eng.*
Munich, *It.* Monaco. Bayern,
SE Germany
Muncie 40 C4 Indiana, N USA
Munda *see* Frome, Lake
Mungbere 77 E5 Orientale, NE Dem.
Rep. Congo
Mu Nggawa *see* Rennell
Munich *see* München
Munkács *see* Mukacheve
Münster 94 A4 *var.* Muenster, Münster
in Westfalen. Nordrhein-Westfalen,
W Germany
Munster 89 A6 *Ir.* Cúige Mumhan.
cultural region S Ireland
Münster in Westfalen *see* Münster
Muong Xiang Ngeun 136 C4 *var.* Xieng
Ngeun. Louangphabang, N Laos
Muonio 84 D3 Lappi, N Finland
Muonioälv/Muoniojoki *see* Muonionjoki
Muonionjoki 84 D3 *var.* Muoniojoki,
Swe. Muonioälv. *river* Finland/Sweden
Muqāṭ 119 C5 Al Mafraq, E Jordan
Muqdisho 73 D6 *Eng.* Mogadishu, *It.*
Mogadiscio. *country capital* (Somalia)
Banaadir, S Somalia
Mur 95 E7 *Croatian* Mura. *river*
C Europe
Mura *see* Mur
Muradiye 117 F3 Van,
E Turkey (Türkiye)
Murapara *see* Murupara
Murata 96 E2 S San Marino
Murchison River 147 A5 *river*
Western Australia
Murcia 93 F4 Murcia, SE Spain
Murcia 93 E4 *autonomous community*
SE Spain
Mureş 108 A4 *river* Hungary/Romania
Murfreesboro 42 D1 Tennessee,
S USA
Murgab *see* Morghāb, Daryā-ye
Murgap 122 D3 Mary Welaýaty,
S Turkmenistan
Murgap *see* Morghāb, Daryā-ye
Murgap Deryasy *see* Morghāb, Daryā-ye
Murghab *see* Morghāb, Daryā-ye
Murghob 123 F3 *Rus.* Murgab.
SE Tajikistan
Murgon 149 E5 Queensland, E Australia
Müritäniyah *see* Mauritania
Müritz 94 C3 *var.* Müritzee. *lake*
NE Germany
Müritzee *see* Müritz
Murmansk 110 C2 Murmanskaya
Oblast', NW Russia
Murmashi 110 C2 Murmanskaya
Oblast', NW Russia
Murom 111 B5 Vladimirskaya Oblast',
W Russia
Muroran 130 D3 Hokkaidō, NE Japan
Muros 92 B1 Galicia, NW Spain
Murray Fracture Zone 153 E2 *fracture
zone* NE Pacific Ocean
Murray Range *see* Murray Ridge
Murray Ridge 112 C5 *var.* Murray
Range. *undersea ridge* N Arabian Sea
Murray River 149 B6 *river* SE Australia
Murrumbidgee River 149 C6 *river* New
South Wales, SE Australia
Murska Sobota 95 E7 *Ger.* Olsnitz.
NE Slovenia
Murupara 150 E3 *var.* Murapara. Bay of
Plenty, North Island, New Zealand
Murviedro *see* Sagunt
Murwāra 135 E4 Madhya Pradesh,
N India
Murwillumbah 149 E5 New South
Wales, SE Australia
Muruq, Edeyin *see* Murzuq, Idhān
Murzuq, Idhan 71 F4 *var.* Edeyin
Murzuq. *desert* SW Libya
Mürzzuschlag 95 E7 Steiermark,
E Austria

Muş 117 F3 *var.* Mush. Muş,
E Turkey (Türkiye)
Musa, Gebel 72 C2 *var.* Gebel Mûsa.
mountain NE Egypt
Mûsa, Gebel *see* Mûsá, Jabal
Musala 104 B3 *mountain* W Bulgaria
Muscat *see* Masqaṭ
Muscat and Oman *see* Oman
Muscatine 45 G3 Iowa, C USA
Musgrave Ranges 147 D5 *mountain
range* South Australia
Musina 97 D7 *var.* Messana, Messene;
anc. Zancle. Sicilia, Italy,
C Mediterranean Sea
Musina 78 D3 *prev.* Messina. Limpopo,
NE South Africa
Muskegon 40 C3 Michigan, N USA
Muskogean *see* Tallahassee
Muskogee 49 G1 Oklahoma, C USA
Musoma 73 C6 Mara, N Tanzania
Musta *see* Mosta
Mustafa-Pasha *see* Svilengrad
Musters, Lago 65 B6 *lake* S Argentina
Muswellbrook 149 D6 New South
Wales, SE Australia
Mut 116 C4 İçel, S Turkey (Türkiye)
Mu-tan-chiang *see* Mudanjiang
Mutare 78 D3 *var.* Mutari; *prev.* Umtali.
Manicaland, E Zimbabwe
Mutari *see* Mutare
Mutina *see* Modena
Mutsu-wan 130 D3 *bay* N Japan
Muttonbird Islands *see* Titi
Mu Us Shadi 127 E3 *var.* Ordos Desert;
prev. Mu Us Shamo. *desert* N China
Mu Us Shamo *see* Mu Us Shadi
Muy Muy 52 D3 Matagalpa, C Nicaragua
Muynak *see* Mo'ynoq
Mužlja 100 D3 *Hung.* Felsőmuzslya;
prev. Gornja Mužlja. Vojvodina,
N Serbia
Mwali 79 F2 *var.* Moili, *Fr.* Mohéli.
island S Comoros
Mwanza 73 B6 Mwanza, NW Tanzania
Mweka 77 C6 Kasai-Occidental,
C Dem. Rep. Congo
Mwene-Ditu 77 D7 Kasai-Oriental,
S Dem. Rep. Congo
Mweru, Lake 77 D7 *var.* Lac Moero. *lake*
Dem. Rep. Congo/Zambia
Myadel' *see* Myadzyel
Myadzyel 107 C5 *Pol.* Miadzioł Nowy,
Rus. Myadel'. Minskaya Voblasts',
N Belarus
Myanaung 136 B4 Ayeyarwady,
SW Myanmar (Burma)
Myanmar 136 A3 *off.* Republic of the
Union of Myanmar; *prev.* Union of
Myanmar, *var.* Burma. *country* SE Asia
Myanmar, Union of *see* Myanmar
Myanmar, Republic of the Union of *see*
Myanmar
Myaungmya 136 A4 Ayeyarwady,
SW Myanmar (Burma)
Myaydo *see* Aunglan
Myeik 137 B6 *var.* Mergui. Tanintharyi,
S Myanmar (Burma)
Myeik Archipelago 137 B6 *prev.* Mergui
Archipelago. *island group* S Myanmar
(Burma)
Myerkulavichy 107 D7 *Rus.*
Merkulovichi. Homyel'skaya Voblasts',
SE Belarus
Myingyan 136 B3 Mandalay,
C Myanmar (Burma)
Myitkyina 136 B2 Kachin State,
N Myanmar (Burma)
Mykolayiv 109 E4 *prev.* Mykolayiv, *Rus.*
Nikolayev. Mykolayivs'ka Oblast',
S Ukraine
Mykolayiv *see* Mykolaiv
Mykonos 105 D6 *var.* Míkonos. *island*
Kykládes, Greece, Aegean Sea
Myrhorod 109 F2 *Rus.* Mirgorod.
Poltavs'ka Oblast', NE Ukraine
Mýrina 104 D4 *var.* Mírina. Límnos,
SE Greece
Myrtle Beach 43 F2 South Carolina,
SE USA
Myrtoan Sea *see* Myrtóo Pélagos
Myrtóo Pélagos 105 C6 *var.* Mirtóo
Pélagos, *Eng.* Myrtoan Sea; *anc.*
Myrtoum Mare. *sea* S Greece
Mýrtos 105 D8 Kríti, Greece,
E Mediterranean Sea
Myrtoum Mare *see* Myrtóo Pélagos
Myślibórz 98 B3 Zachodnio-pomorskie,
NW Poland
Mysore *see* Mysūru
Mysore *see* Karnātaka

Mysūru 132 C2 *prev.* Mysore, *var.*
Maisur. Karnātaka, W India
My Tho 137 E6 *var.* Mi Tho. Tiên Giang,
S Vietnam
Mytilene *see* Mytilíni
Mytilíni 105 D5 *var.* Mitilíni; *anc.*
Mytilene. Lésvos, E Greece
Mzuzu 79 E2 Northern, N Malawi

N

Naberezhnyye Chelny 111 D5 *prev.*
Brezhnev. Respublika Tatarstan,
W Russia
Nablus 119 A6 *var.* Nablus, *Heb.*
Shekhem; *anc.* Neapolis, *Bibl.* Shechem.
N West Bank, NW Asia
Nābulus *see* Nablus
Nacala 79 F2 Nampula, NE Mozambique
Na-Ch'ii *see* Nagqu
Nada *see* Danzhou
Nadi 145 E4 *prev.* Nandi. Viti Levu,
W Fiji
Nadur 102 A5 Gozo, N Malta
Nadvirna 108 C3 *Pol.* Nadwórna, *Rus.*
Nadvornaya. Ivano-Frankivs'ka
Oblast', W Ukraine
Nadvoitsy 110 B3 Respublika Kareliya,
NW Russia
Nadvornaya/Nadwórna *see* Nadvirna
Nadym 114 C3 Yamalo-Nenetskiy
Avtonomnyy Okrug, N Russia
Náfpaktos 105 B5 *var.* Návpaktos.
Dytikí Ellás, C Greece
Náfplio 105 B6 *prev.* Návplion.
Pelopónnisos, S Greece
Naga 139 E2 *off.* Naga City; *prev.* Nueva
Caceres. Luzon, N Philippines
Naga City *see* Naga
Nagano 131 C5 Nagano, Honshū,
S Japan
Nagaoka 131 D5 Niigata, Honshū,
C Japan
Nagara Pathom *see* Nakhon Pathom
Nagara Sridharmaraj *see* Nakhon Si
Thammarat
Nagara Svarga *see* Nakhon Sawan
Nagasaki 131 A7 Nagasaki, Kyūshū,
SW Japan
Nagato 131 A7 Yamaguchi, Honshū,
SW Japan
Nāgercoil 132 C3 Tamil Nādu, SE India
Nagoya 131 C6 Aichi, Honshū,
SW Japan
Nāgpur 134 D4 Mahārāshtra, C India
Nagqu 126 C5 Chin. Na-ch'ii; *prev.*
Hei-ho. Xizang Zizhiqu, W China
Nagybánya *see* Baia Mare
Nagybecskerek *see* Zrenjanin
Nagydisznód *see* Cisnădie
Nagyenyed *see* Aiud
Nagykálló 99 E6 Szabolcs-Szatmár-
Bereg, E Hungary
Nagykanizsa 99 C7 *Ger.* Grosskanizsa.
Zala, SW Hungary
Nagykároly *see* Carei
Nagykikinda *see* Kikinda
Nagykőrös 99 D7 Pest, C Hungary
Nagymihály *see* Michalovce
Nagysurány *see* Šurany
Nagyszalonta *see* Salonta
Nagyszeben *see* Sibiu
Nagyszentmiklós *see* Sânnicolau Mare
Nagyszőllős *see* Vynohradiv
Nagyszombat *see* Trnava
Nagytapolcsány *see* Topol'čany
Nagyvárad *see* Oradea
Naha 130 A3 Okinawa, Okinawa,
SW Japan
Nahariya 119 A5 *prev.* Nahariyya.
Northern, N Israel
Nahariyya *see* Nahariya
Nahuel Huapí, Lago 65 B5 *lake*
W Argentina
Nain 39 F2 Newfoundland and Labrador,
NE Canada
Nā'īn 120 D3 Eşfahān, C Iran
Nairobi 69 E5 *country capital* (Kenya)
Nairobi Area, S Kenya
Nairobi 73 C6 Nairobi Area, S Kenya
Naissus *see* Niš
Najaf *see* An Najaf
Najima *see* Fukuoka
Najin *see* Rajin
Najrān 121 B6 *var.* Abā as Su'ūd. Najrān,
S Saudi Arabia
Nakambé *see* White Volta
Nakamura 131 B7 *var.* Shimanto. Kōchi,
Shikoku, SW Japan

Nakatsugawa 131 C6 *var.* Nakatugawa.
Gifu, Honshū, SW Japan
Nakatugawa *see* Nakatsugawa
Nakhichevan' *see* Naxçıvan
Nakhodka 115 G5 Primorskiy Kray,
SE Russia
Nakhon Pathom 137 C5 *var.* Nagara
Pathom, Nakorn Pathom. Nakhon
Pathom, W Thailand
Nakhon Ratchasima 137 C5 *var.*
Khorat, Korat. Nakhon Ratchasima,
E Thailand
Nakhon Sawan 137 C5 *var.* Muang
Nakhon Sawan, Nagara Svarga.
Nakhon Sawan, W Thailand
Nakhon Si Thammarat 137 C7 *var.*
Nagara Sridharmaraj, Nakhon
Sithamnaraj. Nakhon Si Thammarat,
SW Thailand
Nakhon Sithamnaraj *see* Nakhon Si
Thammarat
Nakorn Pathom *see* Nakhon Pathom
Nakuru 73 C6 Rift Valley, SW Kenya
Nal'chik 111 B8 Kabardino-Balkarskaya
Respublika, SW Russia
Nälüt 71 F2 NW Libya
Namakan Lake 40 A1 *lake* Canada/USA
Namangan 123 F2 Namangan Viloyati,
E Uzbekistan
Nambala 78 D2 Central, C Zambia
Nam Co 126 C5 *lake* W China
Nam Dinh 136 D3 Nam Ha, N Vietnam
Namib Desert 78 B3 *desert* W Namibia
Namibe 78 A2 *Port.* Moçâmedes,
Mossâmedes. Namibe, SW Angola
Namibia 78 B3 *off.* Republic of Namibia,
var. South West Africa, *Afr.*
Suidwes-Afrika, *Ger.* Deutsch-
Südwestafrika; *prev.* German Southwest
Africa, South-West Africa. *country*
S Africa
Namibia, Republic of *see* Namibia
Namnetes *see* Nantes
Namo *see* Namu Atoll
Nam Ou 136 C3 *river* N Laos
Nampa 46 D3 Idaho, NW USA
Nampula 79 E2 Nampula,
NE Mozambique
Namsos 84 B4 Nord-Trøndelag,
C Norway
Nam Tha 136 C4 *river* N Laos
Namu Atoll 144 D2 *var.* Namo. *atoll*
Ralik Chain, C Marshall Islands
Namur 87 C6 *Dut.* Namen. Namur,
SE Belgium
Namyit Island 129 C8 *island*
S Spratly Islands
Nan 136 C4 *var.* Muang Nan. Nan,
NW Thailand
Nanaimo 36 D5 Vancouver Island,
British Columbia, SW Canada
Nanchang 129 C5 *var.* Nan-ch'ang,
Nanch'ang-hsien. *province capital*
Jiangxi, S China
Nan-ch'ang *see* Nanchang
Nanch'ang-hsien *see* Nanchang
Nan-ching *see* Nanjing
Nancy 90 D3 Meurthe-et-Moselle,
NE France
Nandaime 52 D3 Granada,
SW Nicaragua
Nānded 134 D5 Mahārāshtra, C India
Nandi *see* Nadi
Nándorhgy *see* Oţelu Roşu
Nandyāl 132 C1 Andhra Pradesh,
E India
Naniwa *see* Ōsaka
Nanjing 129 D5 *var.* Nan-ching,
Nanking; *prev.* Chianning, Chian-ning,
Kiang-ning, Jiangsu. *province capital*
Jiangsu, E China
Nanking *see* Nanjing
Nanning 129 B6 *var.* Nan-ning; *prev.*
Yung-ning. Guangxi Zhuangzu
Zizhiqu, S China
Nan-ning *see* Nanning
Nanortalik 82 C5 Kujalleq, S Greenland
Nanpan Jiang 136 D2 *river* S China
Nanping 129 D6 *var.* Nan-p'ing; *prev.*
Yenping. Fujian, SE China
Nan-p'ing *see* Nanping
Nansei-shotō 130 A2 *Eng.* Ryukyu
Islands. *island group* SW Japan
Nansei Syotö *see* Ryukyu Trench
Nansen Basin 155 C4 *undersea basin*
Arctic Ocean
Nansen Cordillera 155 B3 *var.* Arctic
Mid Oceanic Ridge, Nansen Ridge.
seamount range Arctic Ocean
Nansen Ridge *see* Nansen Cordillera

Nanterre 90 D1 Hauts-de-Seine, N France
Nantes 90 B4 Bret. Naoned; anc. Condivincum, Namnetes. Loire-Atlantique, NW France
Nantucket Island 41 G3 island Massachusetts, NE USA
Nanumaga 145 E3 var. Nanumanga. atoll NW Tuvalu
Nanumanga see Nanumaga
Nanumea Atoll 145 E3 atoll NW Tuvalu
Nanyang 129 C5 var. Nan-yang. Henan, C China
Nan-yang see Nanyang
Naoned see Nantes
Napa 47 B6 California, W USA
Napier 150 E4 Hawke's Bay, North Island, New Zealand
Naples 43 E5 Florida, SE USA
Naples see Napoli
Napo 56 A3 province NE Ecuador
Napoléon-Vendée see la Roche-sur-Yon
Napoli 97 C5 Eng. Naples, Ger. Neapel; anc. Neapolis. Campania, S Italy
Napo, Río 60 C1 river Ecuador/Peru
Naracoorte 149 B7 South Australia
Naradhivas see Narathiwat
Narathiwat 137 C7 var. Naradhivas. Narathiwat, SW Thailand
Narbada see Narmada
Narbo Martius see Narbonne
Narbonne 91 C6 anc. Narbo Martius. Aude, S France
Narborough Island see Fernandina, Isla
Nares Abyssal Plain see Nares Plain
Nares Plain 35 E6 var. Nares Abyssal Plain. abyssal plain NW Atlantic Ocean
Nares Stræde see Nares Strait
Nares Strait 82 D1 Dan. Nares Stræde. strait Canada/Greenland
Narew 98 E3 river E Poland
Narmada 124 B3 var. Narbada. river C India
Narova see Narva
Narovlya see Narowlya
Narowlya 107 C8 Rus. Narovlya. Homyel'skaya Voblasts', SE Belarus
Närpes 85 D5 Fin. Närpiö. Österbotten, W Finland
Närpiö see Närpes
Narrabri 149 D6 New South Wales, SE Australia
Narrogin 147 B6 Western Australia
Narva 106 E2 Ida-Virumaa, NE Estonia
Narva 106 E2 prev. Narova. river Estonia/Russia
Narva Bay 106 E2 Est. Narva Laht, Ger. Narwa-Bucht, Rus. Narvskiy Zaliv. bay Estonia/Russia
Narva Laht see Narva Bay
Narva Reservoir 106 E2 Est. Narva Veehoidla, Rus. Narvskoye Vodokhranilishche. reservoir Estonia/Russia
Narva Veehoidla see Narva Reservoir
Narvik 84 C3 Nordland, C Norway
Narvskiy Zaliv see Narva Bay
Narvskoye Vodokhranilishche see Narva Reservoir
Narwa-Bucht see Narva Bay
Nar'yan-Mar 110 D2 prev. Beloshchel'ye, Dzerzhinskiy. Nenetskiy Avtonomnyy Okrug, NW Russia
Naryn 123 G2 Naryn Oblasu/Narynskaya Oblast', C Kyrgyzstan
Nassau 54 C1 country capital (The Bahamas) New Providence, N The Bahamas
Năsăud 108 B3 Ger. Nussdorf, Hung. Naszód. Bistrița-Năsăud, N Romania
Nase see Naze
Nāshik 134 C5 prev. Nāsik. Mahārāshtra, W India
Nashua 41 G3 New Hampshire, NE USA
Nashville 42 C1 state capital Tennessee, S USA
Näsijärvi 85 D5 lake SW Finland
Nāsik see Nāshik
Nasir, Buhayrat/Nâşir,Buḥeiret see Nasser, Lake
Nāşiri see Ahvāz
Nasiriya see An Nāşiriyah
Nasser, Lake 72 B3 var. Buhayrat Nasir, Buḥayrat Nâşir, Buḥeiret Nâşir. lake Egypt/Sudan
Naszód see Năsăud
Nata 78 C3 Central, NE Botswana
Natal 63 G2 state capital Rio Grande do Norte, E Brazil

Natal Basin 141 A6 var. Mozambique Basin. undersea basin W Indian Ocean
Natanya see Netanya
Natchez 42 B3 Mississippi, S USA
Natchitoches 42 A2 Louisiana, S USA
Nathanya see Netanya
Natitingou 75 F4 NW Benin
Natsrat see Natzrat
Natuna Islands see Natuna, Kepulauan
Natuna, Kepulauan 124 D4 var. Natuna Islands. island group W Indonesia
Naturaliste Plateau 141 E6 undersea plateau E Indian Ocean
Natzrat 119 A5 var. Natsrat, Ar. En Nazira, Eng. Nazareth; prev. Nazerat. Northern, N Israel
Naugard see Nowogard
Naujaat 37 G3 prev. Repulse Bay, Nunavut, N Canada
Naujamiestis 106 C4 Panevėžys, C Lithuania
Nauru 144 D2 off. Republic of Nauru; prev. Pleasant Island. country W Pacific Ocean
Nauru, Republic of see Nauru
Nauta 60 C2 Loreto, N Peru
Navahrudak 107 C6 Pol. Nowogródek, Rus. Novogrudok. Hrodzyenskaya Voblasts', W Belarus
Navanagar see Jāmnagar
Navapolatsk 107 D5 Rus. Novopolotsk. Vitsyebskaya Voblasts', N Belarus
Navarra 93 E2 Eng./Fr. Navarre. autonomous community N Spain
Navarre see Navarra
Navassa Island 54 C3 US unincorporated territory C West Indies
Navoi see Navoiy
Navoiy 123 E2 Rus. Navoi. Navoiy Viloyati, C Uzbekistan
Navojoa 50 C2 Sonora, NW Mexico
Navolat see Navolato
Navolato 50 C3 var. Navolat. Sinaloa, C Mexico
Návpaktos see Náfpaktos
Návplion see Náfplio
Nawabashah see Nawabshah
Nawabshah 134 B3 var. Nawabashah. Sindh, S Pakistan
Naxçıvan 117 G3 Rus. Nakhichevan'. SW Azerbaijan
Náxos 105 D6 var. Naxos. Náxos, Kykládes, Greece, Aegean Sea
Náxos 105 D6 island Kykládes, Greece, Aegean Sea
Nayoro 130 D2 Hokkaidō, NE Japan
Nay Pyi Taw 136 B4 country capital Myanmar (Burma)) Mandalay, C Myanmar (Burma)
Nazareth see Natzrat
Nazca 60 D4 Ica, S Peru
Nazca Ridge 57 A5 undersea ridge E Pacific Ocean
Naze 130 B3 var. Nase. Kagoshima, Amami-ōshima, SW Japan
Nazerat see Natzrat
Nazilli 116 A4 Aydın, SW Turkey (Türkiye)
Nazrēt 73 C5 var. Adama, Hadama. Oromīya, C Ethiopia
N'Dalatando 78 B1 Port. Salazar, Vila Salazar. Cuanza Norte, NW Angola
Ndélé 76 C4 Bamingui-Bangoran, N Central African Republic
Ndendé 77 B6 Ngounié, S Gabon
Ndindi 77 A6 Nyanga, S Gabon
N'djamena 76 B3 var. Ndjamena; prev. Fort-Lamy. country capital (Chad) Chari-Baguirmi, W Chad
Ndjamena see N'Djaména
Ndjolé 77 A5 Moyen-Ogooué, W Gabon
Ndola 78 D2 Copperbelt, C Zambia
Ndzouani see Anjouan
Neagh, Lough 89 B5 lake E Northern Ireland, United Kingdom
Néa Moudaniá 104 C4 var. Néa Moudhaniá. Kentrikí Makedonía, N Greece
Neapel see Napoli
Neápoli 104 B4 prev. Neápolis. Dytikí Makedonía, N Greece
Neápoli 105 D8 Kríti, Greece, E Mediterranean Sea
Neápoli 105 C7 Pelopónnisos, S Greece
Neápolis see Neápoli, Greece
Neapolis see Napoli, Italy
Neapolis see Nablus, West Bank
Near Islands 36 A2 island group Aleutian Islands, Alaska, USA

Néa Zíchni 104 C3 var. Néa Zíkhni; prev. Néa Zíkhna. Kentrikí Makedonía, NE Greece
Néa Zíkhni/Néa Zíkhni see Néa Zíchni
Nebaj 52 B2 Quiché, W Guatemala
Neblina, Pico da 62 C1 mountain NW Brazil
Nebraska 44 D4 off. State of Nebraska, also known as Blackwater State, Cornhusker State, Tree Planters State. state C USA
Nebraska City 45 F4 Nebraska, C USA
Neches River 49 H3 river Texas, SW USA
Neckar 95 B6 river SW Germany
Necochea 65 D5 Buenos Aires, E Argentina
Nederland see Netherlands
Neder Rijn 86 D4 Eng. Lower Rhine. river C Netherlands
Nederweert 87 D5 Limburg, SE Netherlands
Neede 86 E3 Gelderland, E Netherlands
Neerpelt 87 D5 Limburg, NE Belgium
Neftekamsk 111 D5 Respublika Bashkortostan, W Russia
Neftezavodsk see Seýdi
Negara Brunei Darussalam see Brunei
Negělě 73 D5 var. Negelli, It. Neghelli. Oromīya, C Ethiopia
Negelli see Negělě
Negev see HaNegev
Neghelli see Negělě
Negomane 79 E2 var. Negomano. Cabo Delgado, N Mozambique
Negomano see Negomane
Negombo 132 C3 Western Province, SW Sri Lanka
Negotin 100 E4 Serbia, E Serbia
Negra, Punto 60 A3 headland NW Peru
Negrești see Negrești-Oaş
Negrești-Oaş 108 B3 Hung. Avasfelsőfalu; prev. Negrești. Satu Mare, NE Romania
Negro, Río 65 C5 river E Argentina
Negro, Río 62 D1 river N South America
Negro, Río 64 D4 river Brazil/Uruguay
Negros 139 E2 island C Philippines
Nehbandān 120 E3 Khorāsān, E Iran
Neijiang 129 B5 Sichuan, C China
Neiva 58 B3 Huila, S Colombia
Nellore 132 D2 Andhra Pradesh, E India
Nelson 151 C5 Nelson, South Island, New Zealand
Nelson 37 G4 river Manitoba, C Canada
Néma 74 D3 Hodh ech Chargui, SE Mauritania
Neman 106 B4 Ger. Ragnit. Kaliningradskaya Oblast', W Russia
Neman 106 A4 Bel. Nyoman, Ger. Memel, Lith. Nemunas, Pol. Niemen. river NE Europe
Nemausus see Nîmes
Neméa 105 B6 Pelopónnisos, S Greece
Nemetocenna see Arras
Nemours 90 C3 Seine-et-Marne, N France
Nemunas see Neman
Nemuro 130 E2 Hokkaidō, NE Japan
Neochóri 105 B5 Dytikí Ellás, C Greece
Nepal 135 E3 off. Nepal. country S Asia
Nepal see Nepal
Nereta 106 C4 Aizkraukle, S Latvia
Neretva 100 C4 river Bosnia and Herzegovina/Croatia
Neris 107 C5 Bel. Viliya, Pol. Wilia; prev. Pol. Wilja. river Belarus/Lithuania
Neris see Viliya
Nerva 92 C4 Andalucía, S Spain
Neryungri 115 F3 Respublika Sakha (Yakutiya), NE Russia
Neskaupstaður 83 E5 Austurland, E Iceland
Ness, Loch 88 C3 lake N Scotland, United Kingdom
Nesterov see Zhovkva
Néstos 104 C3 Bul. Mesta, Turk. Kara Su. river Bulgaria/Greece
Nesvizh see Nyasvizh
Netanya 119 A6 var. Natanya, Nathanya. Central, C Israel
Netherlands 86 C3 off. Kingdom of the Netherlands, var. Holland, Dut. Koninkrijk der Nederlanden, Nederland. country NW Europe
Netherlands East Indies see Indonesia
Netherlands Guiana see Suriname
Netherlands, Kingdom of the see Netherlands

Netherlands New Guinea see Papua
Nettilling Lake 37 G3 lake Baffin Island, Nunavut, N Canada
Netze see Noteć
Neu Amerika see Puławy
Neubrandenburg 94 D3 Mecklenburg-Vorpommern, NE Germany
Neuchâtel 95 A7 Ger. Neuenburg. Neuchâtel, W Switzerland
Neuchâtel, Lac de 95 A7 Ger. Neuenburger See. lake W Switzerland
Neuenburg see Neuchâtel
Neuenburger See see Neuchâtel, Lac de
Neufchâteau 87 D8 Luxembourg, SE Belgium
Neugradiška see Nova Gradiška
Neuhof see Zgierz
Neukuhren see Pionerskiy
Neumarkt see Târgu Secuiesc, Covasna, Romania
Neumarkt see Târgu Mureş
Neumayer III 154 A2 German research station Antarctica
Neumoldowa see Moldova Nouă
Neumünster 94 B2 Schleswig-Holstein, N Germany
Neunkirchen 95 A5 Saarland, SW Germany
Neuquén 65 B5 Neuquén, SE Argentina
Neuruppin 94 C3 Brandenburg, NE Germany
Neusalz an der Oder see Nowa Sól
Neu Sandec see Nowy Sącz
Neusatz see Novi Sad
Neusiedler See 95 E6 Hung. Fertő. lake Austria/Hungary
Neusohl see Banská Bystrica
Neustadt see Baia Mare, Maramureş, Romania
Neustadt an der Haardt see Neustadt an der Weinstrasse
Neustadt an der Weinstrasse 95 B5 prev. Neustadt an der Haardt, hist. Niewenstat; anc. Nova Civitas. Rheinland-Pfalz, SW Germany
Neustadtl see Novo mesto
Neustettin see Szczecinek
Neustrelitz 94 D3 Mecklenburg-Vorpommern, NE Germany
Neutra see Nitra
Neu-Ulm 95 B6 Bayern, S Germany
Neuwied 95 A5 Rheinland-Pfalz, W Germany
Neuzen see Terneuzen
Nevada 47 C5 off. State of Nevada, also known as Battle Born State, Sagebrush State, Silver State. state W USA
Nevada, Sierra 92 D5 mountain range S Spain
Nevada, Sierra 47 C6 mountain range W USA
Nevers 90 C4 anc. Noviodunum. Nièvre, C France
Neves 76 E2 São Tomé, S Sao Tome and Principe, Africa
Nevinnomyssk 111 B7 Stavropol'skiy Kray, SW Russia
Nevşehir 116 C3 var. Nevsehir. Nevşehir, C Turkey (Türkiye)
Newala 73 C8 Mtwara, SE Tanzania
New Albany 40 C5 Indiana, N USA
New Amsterdam 59 G3 E Guyana
Newark 41 F4 New Jersey, NE USA
New Bedford 41 G3 Massachusetts, NE USA
Newberg 46 B3 Oregon, NW USA
New Bern 43 F1 North Carolina, SE USA
New Braunfels 49 G4 Texas, SW USA
Newbridge 89 B6 Ir. An Droichead Nua. Kildare, C Ireland
New Britain 144 B3 island E Papua New Guinea
New Brunswick 39 E4 Fr. Nouveau-Brunswick. province SE Canada
New Caledonia 144 D4 var. Kanaky, Fr. Nouvelle-Calédonie. French self-governing territory of special status SW Pacific Ocean
New Caledonia 144 C5 island SW Pacific Ocean
New Caledonia Basin 142 C4 undersea basin W Pacific Ocean
Newcastle 149 D6 New South Wales, SE Australia
Newcastle see Newcastle upon Tyne
Newcastle upon Tyne 88 D4 var. Newcastle, hist. Monkchester, Lat. Pons Aelii. NE England, United Kingdom

New Delhi 134 D3 *country capital* (India) Delhi, N India
New England of the West *see* Minnesota
Newfoundland 39 G3 *Fr.* Terre-Neuve. *island* Newfoundland and Labrador, SE Canada
Newfoundland and Labrador 39 F2 *Fr.* Terre Neuve. *province* E Canada
Newfoundland Basin 66 B3 *undersea feature* NW Atlantic Ocean
New Georgia Islands 144 C3 *island group* NW Solomon Islands
New Glasgow 39 F4 Nova Scotia, SE Canada
New Goa *see* Panaji
New Guinea 144 A3 *Dut.* Nieuw Guinea, *Ind.* Irian. *island* Indonesia/Papua New Guinea
New Hampshire 41 F2 *off.* State of New Hampshire, *also known as* Granite State. *state* NE USA
New Haven 41 G3 Connecticut, NE USA
New Hebrides *see* Vanuatu
New Iberia 42 B3 Louisiana, S USA
New Ireland 144 C3 *island* NE Papua New Guinea
New Jersey 41 F4 *off.* State of New Jersey, *also known as* The Garden State. *state* NE USA
Newman 146 B4 Western Australia
Newmarket 89 E6 E England, United Kingdom
New Mexico 48 C2 *off.* State of New Mexico, *also known as* Land of Enchantment, Sunshine State. *state* SW USA
New Orleans 42 B3 Louisiana, S USA
New Plymouth 150 C4 Taranaki, North Island, New Zealand
Newport 89 C7 SE Wales, United Kingdom
Newport 40 C4 Kentucky, S USA
Newport 41 G2 Vermont, NE USA
Newport News 41 F5 Virginia, NE USA
New Providence 54 C1 *island* N The Bahamas
Newquay 89 C7 SW England, United Kingdom
Newry 89 B5 *Ir.* An tÍúr. SE Northern Ireland, United Kingdom
New Sarum *see* Salisbury
New Siberian Islands *see* Novosibirskiye Ostrova
New South Wales 149 C6 *state* SE Australia
Newton 45 G3 Iowa, C USA
Newton 45 F5 Kansas, C USA
Newtownabbey 89 B5 *Ir.* Baile na Mainistreach. E Northern Ireland, United Kingdom
New Ulm 45 F2 Minnesota, N USA
New York 41 F3 New York, NE USA
New York 41 F3 *state* NE USA
New Zealand 150 A4 *var.* Aotearoa. *country* SW Pacific Ocean
Neyveli 132 C2 Tamil Nādu, SE India
Nezhin *see* Nizhyn
Ngangze Co 126 B5 *lake* W China
Ngaoundéré 76 B4 *var.* N'Gaoundéré. Adamaoua, N Cameroon
N'Gaoundéré *see* Ngaoundéré
Ngazidja 79 F2 *Fr.* Grande-Comore. *island* NW Comoros
Ngerulmud 78 D4 *country capital* (Palau) NW Pacific Ocean
N'Giva 78 B3 *var.* Ondjiva, *Port.* Vila Pereira de Eça. Cunene, S Angola
Ngo 77 B6 Plateaux, SE Congo
Ngoko 77 B5 *river* Cameroon/Congo
Ngourti 75 H3 Diffa, E Niger
Nguigmi 75 H3 *var.* N'Guigmi. Diffa, SE Niger
N'Guigmi *see* Nguigmi
N'Gunza *see* Sumbe
Nguru 75 G3 Yobe, NE Nigeria
Nha Trang 137 E6 Khanh Hoa, S Vietnam
Niagara Falls 38 D5 Ontario, S Canada
Niagara Falls 41 F3 New York, NE USA
Niagara Falls 40 D3 *waterfall* Canada/ USA
Niamey 75 F3 *country capital* (Niger) Niamey, SW Niger
Niangay, Lac 75 E3 *lake* E Mali
Nia-Nia 77 E5 Orientale, NE Dem. Rep. Congo
Nias, Pulau 138 A3 *island* W Indonesia
Nicaea *see* Nice
Nicaragua 52 D3 *off.* Republic of Nicaragua. *country* Central America

Nicaragua, Lago de 52 D4 *var.* Cocibolca, Gran Lago, *Eng.* Lake Nicaragua. *lake* S Nicaragua
Nicaragua, Lake *see* Nicaragua, Lago de
Nicaragua, Republic of *see* Nicaragua
Nicaria *see* Ikaría
Nice 91 D6 *It.* Nizza; *anc.* Nicaea. Alpes-Maritimes, SE France
Nicephorium *see* Ar Raqqah
Nicholas II Land *see* Severnaya Zemlya
Nicholls Town 54 C1 Andros Island, NW The Bahamas
Nicobar Islands 124 B4 *island group* India, E Indian Ocean
Nicosia 102 C5 *Gk.* Lefkosía, *Turk.* Lefkoşa. *country capital* (Cyprus) C Cyprus
Nicoya 52 D4 Guanacaste, W Costa Rica
Nicoya, Golfo de 52 D5 *gulf* W Costa Rica
Nicoya, Península de 52 D4 *peninsula* NW Costa Rica
Nida 106 A3 *Ger.* Nidden. Klaipėda, SW Lithuania
Nidaros *see* Trondheim
Nidden *see* Nida
Nidzica 98 D3 *Ger.* Niedenburg. Warmińsko-Mazurskie, NE Poland
Niedenburg *see* Nidzica
Niedere Tauern 99 A6 *mountain range* C Austria
Niemen *see* Neman
Nieśwież *see* Nyasvizh
Nieuw Amsterdam 59 G3 Commewijne, NE Suriname
Nieuw-Bergen 86 D4 Limburg, SE Netherlands
Nieuwegein 86 C4 Utrecht, C Netherlands
Nieuw Guinea *see* New Guinea
Nieuw Nickerie 59 G3 Nickerie, NW Suriname
Niewenstat *see* Neustadt an der Weinstrasse
Niğde 116 C4 Niğde, C Turkey (Türkiye)
Niger 75 F4 *off.* Republic of Niger. *country* W Africa
Niger 75 F4 *river* W Africa
Nigeria 75 F4 *off.* Federal Republic of Nigeria. *country* W Africa
Nigeria, Federal Republic of *see* Nigeria
Niger, Mouths of the 75 F5 *delta* S Nigeria
Niger, Republic of *see* Niger
Nihon *see* Japan
Niigata 131 D5 Niigata, Honshū, C Japan
Niihama 131 B7 Ehime, Shikoku, SW Japan
Ni'ihau 47 A7 *var.* Niihau. *island* Hawaii, USA, C Pacific Ocean
Nii-jima 131 D6 *island* E Japan
Nijkerk 86 D3 Gelderland, C Netherlands
Nijlen 87 C5 Antwerpen, N Belgium
Nijmegen 86 D4 *Ger.* Nimwegen; *anc.* Noviomagus. Gelderland, SE Netherlands
Nikaria *see* Ikaría
Nikel' 110 C2 *Finn.* Kolosjoki. Murmanskaya Oblast', NW Russia
Nikiniki 139 E5 Timor, S Indonesia
Niklasmarkt *see* Gheorgheni
Nikolaikaupunki *see* Vaasa
Nikolayev *see* Mykolayiv
Nikol'sk *see* Ussuriysk
Nikol'sk-Ussuriyskiy *see* Ussuriysk
Nikopol 109 F3 Dnipropetrovska Oblast', SE Ukraine
Nikšić 101 C5 C Montenegro
Nikumaroro 145 E3 ; *prev.* Gardner Island. *atoll* Phoenix Islands, C Kiribati
Nikunau 145 E3 *var.* Nukunau; *prev.* Byron Island. *atoll* Tungaru, W Kiribati
Nile 72 B2 *former province* NW Uganda
Nile 68 D3 *Ar.* Nahr an Nīl. *river* N Africa
Nile Delta 72 B1 *delta* N Egypt
Nīl, Nahr an *see* Nile
Nîmes 91 C6 *anc.* Nemausus, Nismes. Gard, S France
Nimwegen *see* Nijmegen
Nine Degree Channel 132 B3 *channel* India/Maldives
Ninetyeast Ridge 141 D5 *undersea feature* E Indian Ocean
Ninety Mile Beach *see* Te-Oneroa-a-Tōhē
Ningbo 128 D5 *var.* Ning-po, Yin-hsien; *prev.* Ninghsien. Zhejiang, SE China
Ning-hsia *see* Ningxia
Ninghsien *see* Ningbo

Ning-po *see* Ningbo
Ningsia/Ningsia Hui/Ningsia Hui Autonomous Region *see* Ningxia
Ningxia 128 B4 *off.* Ningxia Huizu Zizhiqu, *var.* Ning-hsia, Ningsia, *Eng.* Ningsia Hui, Ningsia Hui Autonomous Region. *autonomous region* N China
Ningxia Huizu Zizhiqu *see* Ningxia
Nio *see* Íos
Niobrara River 45 E3 *river* Nebraska/ Wyoming, C USA
Nioro 74 D3 *var.* Nioro du Sahel. Kayes, W Mali
Nioro du Sahel *see* Nioro
Niort 90 B4 Deux-Sèvres, W France
Nipigon 38 B4 Ontario, S Canada
Nipigon, Lake 38 B3 *lake* Ontario, S Canada
Nippon *see* Japan
Niš 101 E5 *Eng.* Nish, *Ger.* Nisch; *anc.* Naissus. Serbia, SE Serbia
Nişab 120 B4 Al Ḩudūd ash Shamālīyah, N Saudi Arabia
Nisch/Nish *see* Niš
Nisibin *see* Nusaybin
Nisiros *see* Nísyros
Nisko 98 E4 Podkarpackie, SE Poland
Nismes *see* Nîmes
Nistru *see* Dniester
Nísyros 105 E7 *var.* Nisiros. *island* Dodekánisa, Greece, Aegean Sea
Nitra 99 C6 *Ger.* Neutra, *Hung.* Nyitra. Nitriansky Kraj, SW Slovakia
Nitra 99 C6 *Ger.* Neutra, *Hung.* Nyitra. *river* W Slovakia
Niuatobutabu *see* Niuatoputapu
Niuatoputapu 145 E4 *var.* Niuatobutabu; *prev.* Keppel Island. *island* N Tonga
Niue 145 F4 *self-governing territory in free association with New Zealand* S Pacific Ocean
Niulakita 145 E3 *var.* Nurakita. *atoll* S Tuvalu
Niutao 145 E3 *atoll* NW Tuvalu
Nivernais 90 C4 *cultural region* C France
Nizāmābād 134 D5 Telangana, C India
Nizhnegorskiy *see* Nyzhnohirskyi
Nizhnekamsk 111 C5 Respublika Tatarstan, W Russia
Nizhnevartovsk 114 D3 Khanty-Mansiyskiy Avtonomnyy Okrug-Yugra, C Russia
Nizhniy Novgorod 111 C5 *prev.* Gor'kiy. Nizhegorodskaya Oblast', W Russia
Nizhniy Odes 110 D4 Respublika Komi, NW Russia
Nizhnyaya Tunguska 115 E3 *Eng.* Lower Tunguska. *river* N Russia
Nizhyn 109 E1 *Rus.* Nezhin. Chernihivs'ka Oblast', NE Ukraine
Nizza *see* Nice
Njombe 73 C8 Iringa, S Tanzania
Nkayi 77 B6 *prev.* Jacob. Bouenza, S Congo
Nkongsamba 76 A4 *var.* N'Kongsamba. Littoral, W Cameroon
N'Kongsamba *see* Nkongsamba
Nmai Hka 136 C2 *var.* Me Hka. *river* N Myanmar (Burma)
Nobeoka 131 B7 Miyazaki, Kyūshū, SW Japan
Noboribetsu 130 D3 *var.* Noboribetu. Hokkaidō, NE Japan
Noboribetu *see* Noboribetsu
Nogales 50 B1 Sonora, NW Mexico
Nogales 48 B3 Arizona, SW USA
Nogal Valley *see* Dooxo Nugaaleed
Noire, Rivi` *see* Black River
Nokia 85 D5 Pirkanmaa, W Finland
Nokou 76 B3 Kanem, W Chad
Nola 77 B5 Sangha-Mbaéré, SW Central African Republic
Nolinsk 111 C5 Kirovskaya Oblast', NW Russia
Nongkaya *see* Nong Khai
Nong Khai 136 C4 *var.* Mi Chai, Nongkaya. Nong Khai, E Thailand
Nonouti 144 D2 *prev.* Sydenham Island. *atoll* Tungaru, W Kiribati
Noord-Beveland 86 B4 *var.* North Beveland. *island* SW Netherlands
Noordwijk aan Zee 86 C3 Zuid-Holland, W Netherlands
Noordzee *see* North Sea
Nora 85 C6 Örebro, C Sweden
Norak 123 E3 *Rus.* Nurek. W Tajikistan
Nordaustlandet 83 G1 *island* NE Svalbard

Norden 94 A3 Niedersachsen, NW Germany
Norderstedt 94 B3 Schleswig-Holstein, N Germany
Nordfriesische Inseln *see* North Frisian Islands
Nordhausen 94 C4 Thüringen, C Germany
Nordhorn 94 A3 Niedersachsen, NW Germany
Nord, Mer du *see* North Sea
Nord-Ouest, Territoires du *see* Northwest Territories
Nordsee/Nordsjøen/Nordsøen *see* North Sea
Norfolk 45 E3 Nebraska, C USA
Norfolk 41 F5 Virginia, NE USA
Norfolk Island 142 D4 *Australian self-governing territory* SW Pacific Ocean
Norfolk Ridge 142 D4 *undersea feature* W Pacific Ocean
Norge *see* Norway
Norias 49 G5 Texas, SW USA
Noril'sk 114 D3 Taymyrskiy (Dolgano-Nenetskiy) Avtonomnyy Okrug, N Russia
Norman 49 G1 Oklahoma, C USA
Normandes, Îles *see* Channel Islands
Normandie 90 B3 *Eng.* Normandy. *cultural region* N France
Normandy *see* Normandie
Normanton 148 C3 Queensland, NE Australia
Norrköping 85 C6 Östergötland, S Sweden
Norrtälje 85 C6 Stockholm, C Sweden
Norseman 147 B6 Western Australia
Norske Havet *see* Norwegian Sea
North Albanian Alps 101 C5 *Alb.* Bjeshkët e Namuna, *Croatian* Prokletije. *mountain range* SE Europe
Northallerton 89 D5 N England, United Kingdom
Northam 147 A6 Western Australia
Northampton 89 D6 C England, United Kingdom
North Andaman 133 F2 *island* Andaman Islands, India, NE Indian Ocean
North Australian Basin 141 E5 *Fr.* Bassin Nord de l' Australie. *undersea feature* E Indian Ocean
North Bay 38 D4 Ontario, S Canada
North Beveland *see* Noord-Beveland
North Borneo *see* Sabah
North Cape 150 C1 *headland* North Island, New Zealand
North Cape 84 D1 *Eng.* North Cape. *headland* N Norway
North Cape *see* Nordkapp
North Carolina 43 E1 *off.* State of North Carolina, *also known as* Old North State, Tar Heel State, Turpentine State. *state* SE USA
North Channel 40 D2 *lake channel* Canada/USA
North Charleston 43 F2 South Carolina, SE USA
North Dakota 44 D2 *off.* State of North Dakota, *also known as* Flickertail State, Peace Garden State, Sioux State. *state* N USA
North Devon Island *see* Devon Island
North East Frontier Agency/North East Frontier Agency of Assam *see* Arunāchal Pradesh
Northeast Providence Channel 54 C1 *channel* N The Bahamas
Northeim 94 B4 Niedersachsen, C Germany
Northern Cook Islands 145 F4 *island group* N Cook Islands
Northern Dvina *see* Severnaya Dvina
Northern Ireland 88 B4 *var.* The Six Counties. *cultural region* Northern Ireland, United Kingdom
Northern Mariana Islands 142 B1 *US commonwealth territory* W Pacific Ocean
Northern Rhodesia *see* Zambia
Northern Sporades *see* Vóreíes Sporádes
Northern Territory 144 A3 *territory* N Australia
North European Plain 81 E3 *plain* N Europe
Northfield 45 F2 Minnesota, N USA
North Fiji Basin 142 D3 *undersea feature* N Coral Sea

North Frisian Islands *94 B2 var.*
Nordfriesische Inseln. *island group*
N Germany
North Huvadhu Atoll *132 B5 var.* Gaafu
Alifu Atoll. *atoll* S Maldives
North Island *150 B2 var.* Te Ika-a-Māui.
island N New Zealand
North Korea *129 E3 off.* Democratic
People's Republic of Korea, *Kor.*
Chosŏn-minjujuŭi-inmin-kanghwaguk.
country E Asia
North Little Rock *42 B1* Arkansas,
C USA
North Macedonia *101 D6 off.* Republic
of North Macedonia, *Mac.* Severna
Makedonija, *prev.* Macedonia. *country*
SE Europe
North Macedonia, Republic of *see*
North Macedonia
North Minch *see* Minch, The
North Mole *93 G4 harbour wall*
NW Gibraltar, Europe
North Platte *45 E4* Nebraska, C USA
North Platte River *44 D4 river* C USA
North Pole *155 B2 pole* Arctic Ocean
North Saskatchewan *37 E5 river*
Alberta/Saskatchewan, S Canada
North Sea *80 D3 Dan.* Nordsøen, *Dut.*
Noordzee, *Fr.* Mer du Nord, *Ger.*
Nordsee, *Nor.* Nordsjøen; *prev.* German
Ocean, *Lat.* Mare Germanicum. *sea*
NW Europe
North Siberian Lowland *115 E2 var.*
North Siberian Plain, *Eng.* North
Siberian Lowland. *lowlands* N Russia
North Siberian Lowland/North
Siberian Plain *see* Severo-Sibirskaya
Nizmennost'
North Star State *see* Minnesota
North Taranaki Bight *150 C3 gulf* North
Island, New Zealand
North Uist *88 B3 island* NW Scotland,
United Kingdom
Northwest Atlantic Mid-Ocean Canyon
34 E4 undersea feature
N Atlantic Ocean
North West Highlands *88 C3 mountain*
range N Scotland, United Kingdom
Northwest Pacific Basin *113 G4*
undersea feature NW Pacific Ocean
Northwest Territories *37 E3 Fr.*
Territoires du Nord-Ouest. *territory*
NW Canada
Northwind Plain *155 B2 undersea*
feature Arctic Ocean
Norton Sound *36 C2 inlet* Alaska, USA
Norway *85 A5 off.* Kingdom of Norway,
Nor. Norge. *country* N Europe
Norway, Kingdom of *see* Norway
Norwegian Basin *83 F4 undersea feature*
NW Norwegian Sea
Norwegian Sea *83 F4 var.* Norske Havet.
sea NE Atlantic Ocean
Norwich *89 E6* E England,
United Kingdom
Nösen *see* Bistriţa
Noshiro *130 D4 var.* Nosiro; *prev.*
Noshirominato. Akita, Honshū,
C Japan
Noshirominato/Nosiro *see* Noshiro
Nosivka *109 E1 Rus.* Nosovka.
Chernihivs'ka Oblast', NE Ukraine
Nosop *78 C4 river* E Namibia
Nosovka *see* Nosivka
Noşratābād *120 E3* Sīstān va
Balūchestān, E Iran
Noteć *98 C3 Ger.* Netze. *river*
NW Poland
Nóties Sporádes *see* Dodekánisa
Nottingham *89 D6* C England,
United Kingdom
Nouâdhibou *74 B2 prev.* Port-Étienne.
Dakhlet Nouâdhibou, W Mauritania
Nouakchott *74 B2 country capital*
(Mauritania) Nouakchott District,
SW Mauritania
Nouméa *144 C5 dependent territory*
capital (New Caledonia) Province Sud,
S New Caledonia
Nouveau-Brunswick *see* New Brunswick
Nouvelle-Calédonie *see* New Caledonia
Nouvelle Écosse *see* Nova Scotia
Nova Civitas *see* Neustadt an der
Weinstrasse
Nova Gorica *95 D8* W Slovenia
Nova Gradiška *100 C3 Ger.* Neugradisk,
Hung. Újgradiska. Brod-Posavina,
NE Croatia
Nova Iguaçu *63 F4* Rio de Janeiro,
SE Brazil

Nova Kakhovka *109 F4* S Ukraine
Nova Lisboa *see* Huambo
Novara *96 B2 anc.* Novaria. Piemonte,
NW Italy
Novaria *see* Novara
Nova Scotia *39 F4 Fr.* Nouvelle Écosse.
province SE Canada
Nova Scotia *35 E5 physical region*
SE Canada
Novaya Sibir', Ostrov *115 F1 island*
Novosibirskiye Ostrova,
NE Russia
Novaya Zemlya *110 D1 island group*
N Russia
Novaya Zemlya Trough *see* East Novaya
Zemlya Trough
Novgorod *see* Velikiy Novgorod
Novi Grad *see* Bosanski Novi
Novi Iskar *104 C2 var.* Novi Iksŭr.
Sofia-Grad, W Bulgaria
Novi Iskŭr *see* Novi Iksar
Noviodunum *see* Nevers, Nièvre, France
Noviomagus *see* Lisieux, Calvados,
France
Noviomagus *see* Nijmegen, Netherlands
Novi Pazar *101 D5 Turk.* Yenipazar.
Serbia, S Serbia
Novi Sad *100 D3 Ger.* Neusatz, *Hung.*
Újvidék. Vojvodina, N Serbia
Novoazovs'k *109 G4 Donets'ka Oblast'*,
E Ukraine
Novocheboksarsk *111 C5* Chuvashskaya
Respublika, W Russia
Novocherkassk *111 B7* Rostovskaya
Oblast', SW Russia
Novodvinsk *110 C3* Arkhangel'skaya
Oblast', NW Russia
Novohrad-Volynskiy *see*
Novohrad-Volyns'kyi
Novogrudok *see* Navahrudak
Novohrad-Volynskyi *108 D2 prev.*
Novohrad-Volyns'kyy, Rus.
Novograd-Volynskiy. Zhytomyrs'ka
Oblast', N Ukraine
Novohrad-Volyns'kyy *see*
Novohrad-Volynskyi
Novokazalinsk *see* Ayteke Bi
Novokuznetsk *114 D4 prev.* Stalinsk.
Kemerovskaya Oblast', S Russia
Novolazarevskaya *154 C2 Russian*
research station Antarctica
Novo mesto *95 E8 Ger.* Rudolfswert;
prev. Ger. Neustadtl. SE Slovenia
Novomoskovsk *111 B5* Tul'skaya
Oblast', W Russia
Novomoskovsk *109 F3*
Dnipropetrovska Oblast, E Ukraine
Novopolotsk *see* Navapolatsk
Novoradomsk *see* Radomsko
Novo Redondo *see* Sumbe
Novorossiysk *111 A7* Krasnodarskiy
Kray, SW Russia
Novoshakhtinsk *111 B6* Rostovskaya
Oblast', SW Russia
Novosibirsk *114 D4* Novosibirskaya
Oblast', C Russia
Novosibirskiye Ostrova *115 F1 Eng.*
New Siberian Islands. *island group*
N Russia
Novotroitsk *111 D6* Orenburgskaya
Oblast', W Russia
Novotroitskoye *see* Novotroyitske
Novotroyitske *109 F4 prev.*
Novotroyits'ke, *Rus.* Novotroitskoye.
Khersons'ka Oblast', S Ukraine
Novotroyits'ke *see* Novotroyitske
Novo-Urgench *see* Urganch
Novovolynsk *108 C1* Volyns'ka Oblast',
NW Ukraine
Novy Dvor *107 B6 Rus.* Novyy Dvor.
Hrodzyenskaya Voblasts', W Belarus
Novyy Bug *see* Novyy Buh
Novyy Buh *109 E3 Rus.* Novyy Bug.
Mykolayivs'ka Oblast', S Ukraine
Novy Dvor *see* Novy Dvor
Novyy Margilan *see* Farg'ona
Novyy Uzen' *see* Zhangaözen
Nowa Sól *98 B4 var.* Nowasól, *Ger.*
Neusalz an der Oder. Lubuskie,
W Poland
Nowasól *see* Nowa Sól
Nowogard *98 B2 var.* Nowógard, *Ger.*
Naugard. Zachodnio-pomorskie,
NW Poland
Nowogródek *see* Navahrudak
Nowo-Minsk *see* Mińsk Mazowiecki
Nowy Dwór Mazowiecki *98 D3*
Mazowieckie, C Poland
Nowy Sącz *99 D5 Ger.* Neu Sandec.
Małopolskie, S Poland

Nowy Tomyśl *98 B3 var.* Nowy Tomysl.
Wielkopolskie, C Poland
Nowy Tomysl *see* Nowy Tomyśl
Noyon *90 C3* Oise, N France
Nsanje *79 E3* Southern, S Malawi
Nsawam *75 E5* SE Ghana
Ntomba, Lac *77 C6 var.* Lac Tumba.
lake NW Dem. Rep. Congo
Nubian Desert *72 B3 desert* NE Sudan
Nu Chiang *see* Salween
Nueva Caceres *see* Naga
Nueva Gerona *54 B2* Isla de la Juventud,
S Cuba
Nueva Rosita *50 D2* Coahuila,
NE Mexico
Nuevitas *54 C2* Camagüey, E Cuba
Nuevo, Bajo *53 G1 island* NW Colombia
South America
Nuevo Casas Grandes *50 C1* Chihuahua,
N Mexico
Nuevo, Golfo *65 C6 gulf* S Argentina
Nuevo Laredo *51 E2* Tamaulipas,
NE Mexico
Nui Atoll *145 E3 atoll* W Tuvalu
Nu Jiang *see* Salween
Nûk *see* Nuuk
Nukha *see* Şäki
Nuku'alofa *145 E5 country capital*
(Tonga) Tongatapu, S Tonga
Nukufetau Atoll *145 E3 atoll* C Tuvalu
Nukulaelae Atoll *145 E3 var.* Nukulailai.
atoll E Tuvalu
Nukulailai *see* Nukulaelae Atoll
Nukunau *see* Nikunau
Nukunonu Atoll *145 E3 island*
C Tokelau
Nukus *122 C2* Qoraqalpog'iston
Respublikasi, W Uzbekistan
Nullarbor Plain *147 C6 plateau* South
Australia/Western Australia
Nunap Isua *82 C5 var.*
Uummannarsuaq, *Dan.* Kap Farvel,
Eng. Cape Farewell. *cape* S Greenland
Nunavut *37 F3 territory* N Canada
Nuneaton *89 D6* C England,
United Kingdom
Nunivak Island *36 B2 island* Alaska,
USA
Nunspeet *86 D3* Gelderland,
E Netherlands
Nuoro *97 A5* Sardegna, Italy,
C Mediterranean Sea
Nuqui *58 A3* Chocó, W Colombia
Nurakita *see* Niulakita
Nurata *see* Nurota
Nurek *see* Norak
Nuremberg *see* Nürnberg
Nurmes *84 E4* Pohjois-Karjala, E Finland
Nürnberg *95 C5 Eng.* Nuremberg.
Bayern, S Germany
Nurota *123 E2 Rus.* Nurata. Navoiy
Viloyati, C Uzbekistan
Nur-Sultan *see* Astana
Nusaybin *117 F4 var.* Nisibin. Manisa,
SE Turkey (Türkiye)
Nussdorf *see* Năsăud
Nutmeg State *see* Connecticut
Nuuk *82 C4 var.* Nûk, *Dan.* Godthaab,
Godthåb. *dependent territory capital*
(Greenland) Sermersooq,
SW Greenland
Nxua *78 C5 prev.* Cradock. Eastern
Cape, S South Africa
Nyagan' *111 C4* Khanty-Mansiyskiy
Avtonomnyy Okrug-Yugra, N Russia
Nyainqentanglha Shan *126 C5*
mountain range W China
Nyala *72 A4* Southern Darfur, W Sudan
Nyamapanda *78 D3* Mashonaland East,
NE Zimbabwe
Nyamtumbo *73 C8* Ruvuma, S Tanzania
Nyanda *see* Masvingo
Nyandoma *110 C4* Arkhangel'skaya
Oblast', NW Russia
Nyantakara *73 B7* Kagera, NW Tanzania
Nyasa, Lake *79 E2 var.* Lake Malawi;
prev. Lago Nyassa. *lake* E Africa
Nyasaland/Nyasaland Protectorate *see*
Malawi
Nyassa, Lago *see* Nyasa, Lake
Nyasvizh *107 C6 Pol.* Nieśwież, *Rus.*
Nesvizh. Minskaya Voblasts', C Belarus
Nyaunglebin *136 B4* Bago,
SW Myanmar (Burma)
Nyeri *73 C6* Central, C Kenya
Nyíregyháza *99 D6* Szabolcs-Szatmár-
Bereg, NE Hungary
Nyitra *see* Nitra
Nykøbing *85 B8* Storstrøm, SE Denmark
Nyköping *85 C6* Södermanland,
S Sweden

Nylstroom *see* Modimolle
Nyngan *149 D6* New South Wales,
SE Australia
Nyoman *see* Neman
Nyurba *115 F3* Respublika Sakha
(Yakutiya), NE Russia
Nyzhnohirskyi *109 F4 prev.*
Nyzhn'ohirs'kyy, *Rus.* Nizhnegorskiy.
Respublika Krym, S Ukraine
Nyzhn'ohirs'kyy *see* Nyzhnohirskyi
NZ *see* New Zealand
Nzega *73 C7* Tabora, C Tanzania
Nzérékoré *74 D4* SE Guinea
Nzwani *see* Anjouan

O

Oa'hu *47 A7 var.* Oahu. *island* Hawaiian
Islands, Hawaii, USA
Oak Harbor *46 B1* Washington,
NW USA
Oakland *47 B6* California, W USA
Oamaru *151 B7* Otago, South Island,
New Zealand
Oaxaca *51 F5 var.* Oaxaca de Juárez;
prev. Antequera. Oaxaca, SE Mexico
Oaxaca de Juárez *see* Oaxaca
Ob' *112 C2 river* C Russia
Obal' *107 D5 Rus.* Obol'. Vitsyebskaya
Voblasts', N Belarus
Oban *88 C4* W Scotland,
United Kingdom
Oban *see* Halfmoon Bay
Obando *see* Puerto Inírida
Obdorsk *see* Salekhard
Óbecse *see* Bečej
Obeliai *106 C4* Panevėžys, NE Lithuania
Oberhollabrunn *see* Tulln
Ob', Gulf of *see* Obskaya Guba
Obidovichi *see* Abidavichy
Obihiro *130 D2* Hokkaidō, NE Japan
Obo *76 D4* Haut-Mbomou, E Central
African Republic
Obock *72 D4* E Djibouti
Obol' *see* Obal'
Oborniki *98 C3* Wielkolpolskie,
W Poland
Obrovo *see* Abrova
Obskaya Guba *114 D3 Eng.* Gulf of Ob.
gulf N Russia
Ob' Tablemount *141 B7 undersea*
feature S Indian Ocean
Ocala *43 E4* Florida, SE USA
Ocaña *58 B2* Norte de Santander,
N Colombia
Ocaña *92 D3* Castilla-La Mancha,
C Spain
O Carballiño *92 C1 Cast.* Carballiño.
Galicia, NW Spain
Occidental, Cordillera *58 B2 mountain*
range W South America
Occidental, Cordillera *61 E4 mountain*
range Bolivia/Chile
Occupied Palestinian Territories *see*
West Bank/Gaza Strip
Ocean Falls *36 D5* British Columbia,
SW Canada
Ocean Island *see* Banaba
Oceanside *47 C8* California, W USA
Ocean State *see* Rhode Island
Ochakiv *109 E4 Rus.* Ochakov.
Mykolayivs'ka Oblast', S Ukraine
Ochakov *see* Ochakiv
Ochamchire *117 E2 prev.* Och'amch'ire.
W Georgia
Och'amch'ire *see* Ochamchire
Ocho Rios *54 B4* C Jamaica
Ochrida *see* Ohrid
Ochrida, Lake *see* Ohrid, Lake
Ocotal *52 D3* Nueva Segovia,
NW Nicaragua
Ocozocuautla *51 G5* Chiapas, SE Mexico
October Revolution Island *see*
Oktyabr'skoy Revolyutsii, Ostrov
Ocú *53 F5* Herrera, S Panama
Ōdate *130 D3* Akita, Honshū, C Japan
Oddur *see* Xuddur
Ödemiş *116 A4* İzmir,
SW Turkey (Türkiye)
Ödenburg *see* Sopron
Odenpäh *see* Otepää
Odense *85 B7* Fyn, C Denmark
Oder *98 B3 Cz./Pol.* Odra. *river*
C Europe
Oderhaff *see* Szczeciński, Zalew
Odesa *109 E4 Rus.* Odessa. Odes'ka
Oblast', SW Ukraine
Odessa *49 E3* Texas, SW USA
Odessa *see* Odesa

Odessus see Varna
Odienné 74 D4 NW Ivory Coast
Odisha 135 F4 prev. Orissa. cultural region NE India
Ôdôngk 137 D6 Kampong Speu, S Cambodia
Odoorn 86 E2 Drenthe, NE Netherlands
Odra see Oder
Oesel see Saaremaa
Of 117 E2 Trabzon, NE Turkey (Türkiye)
Ofanto 97 D5 river S Italy
Offenbach 95 B5 var. Offenbach am Main. Hessen, W Germany
Offenbach am Main see Offenbach
Offenburg 95 B6 Baden-Württemberg, SW Germany
Ogaaden see Ogaden
Ogaden 73 D5 Som. Ogaadeen. plateau Ethiopia/Somalia
Ōgaki 131 C6 Gifu, Honshū, SW Japan
Ogallala 44 D4 Nebraska, C USA
Ogbomosho 75 F4 var. Ogmoboso. Oyo, W Nigeria
Ogden 44 B4 Utah, W USA
Ogdensburg 41 F2 New York, NE USA
Ogmoboso see Ogbomosho
Ogulin 100 A3 Karlovac, NW Croatia
Ohio 40 C4 off. State of Ohio, also known as Buckeye State. state N USA
Ohio River 40 C4 river N USA
Ohlau see Oława
Ohri see Ohrid
Ohrid 101 D6 Turk. Ochrida, Ohrī. SW North Macedonia
Ohrid, Lake 101 D6 var. Lake Ochrida, Alb. Liqeni i Ohrit, Mac. Ohridsko Ezero. lake Albania/North Macedonia
Ohridsko Ezero/Ohrit, Liqeni i see Ohrid, Lake
Ohura 150 D3 Manawatu-Wanganui, North Island, New Zealand
Oirschot 87 C5 Noord-Brabant, S Netherlands
Oise 90 C3 river N France
Oistins 55 G2 S Barbados
Ōita 131 B6 Ōita, Kyūshū, SW Japan
Ojinaga 50 D2 Chihuahua, N Mexico
Ojos del Salado, Cerro 64 B3 mountain W Argentina
Okaihau 150 C2 Northland, North Island, New Zealand
Okanizsa see Kanjiža
Okara 134 C2 Punjab, E Pakistan
Okavanggo see Cubango/Okavango
Okavango see Cubango
Okavango Delta 78 C3 wetland N Botswana
Okayama 131 B6 Okayama, Honshū, SW Japan
Okazaki 131 C6 Aichi, Honshū, C Japan
Okeechobee, Lake 43 E4 lake Florida, SE USA
Okefenokee Swamp 43 E3 wetland Georgia, SE USA
Okhotsk 115 G3 Khabarovskiy Kray, E Russia
Okhotsk, Sea of 113 F3 sea NW Pacific Ocean
Okhtyrka 109 F2 Rus. Akhtyrka. Sums'ka Oblast', NE Ukraine
Oki-guntō see Oki-shotō
Okinawa 130 A3 island SW Japan
Okinawa-shoto 130 A3 island group Nansei-shotō, SW Japan Asia
Oki-shoto 131 B6 var. Oki-guntō. island group SW Japan
Oklahoma 49 F2 off. State of Oklahoma, also known as The Sooner State. state C USA
Oklahoma City 49 G1 state capital Oklahoma, C USA
Okmulgee 49 G1 Oklahoma, C USA
Oko, Wadi 72 C3 river NE Sudan
Oktyabr'skiy 111 D6 Volgogradskaya Oblast', SW Russia
Oktyabr'skiy see Aktsyabrski
Oktyabr'skoy Revolyutsii, Ostrov 115 E2 Eng. October Revolution Island. island Severnaya Zemlya, N Russia
Okulovka 110 B4 var. Okulovka. Novgorodskaya Oblast', W Russia
Okulovka see Okulovka
Okushiri-to 130 C3 var. Okusiri Tô. island NE Japan
Okusiri Tô see Okushiri-tô
Oláh-Toplicza see Toplița
Öland 85 C7 island S Sweden
Olavarría 65 D5 Buenos Aires, E Argentina

Oława 98 C4 Ger. Ohlau. Dolnośląskie, SW Poland
Olbia 135 F4 prev. Terranova Pausania. Sardegna, Italy, C Mediterranean Sea
Old Bay State/Old Colony State see Massachusetts
Old Dominion see Virginia
Oldebroek 86 D3 Gelderland, E Netherlands
Oldenburg 94 B3 Niedersachsen, NW Germany
Oldenburg 94 C2 var. Oldenburg in Holstein. Schleswig-Holstein, N Germany
Oldenburg in Holstein see Oldenburg
Oldenzaal 86 E3 Overijssel, E Netherlands
Old Harbour 54 B5 C Jamaica
Old Line State see Maryland
Old North State see North Carolina
Olëkma 115 F4 river C Russia
Olëkminsk 115 F3 Respublika Sakha (Yakutiya), NE Russia
Oleksandriia 109 F3 prev. Oleksandriya, Rus. Aleksandriya. Kirovohrads'ka Oblast', C Ukraine
Oleksandrivka 109 E3 Rus. Aleksandrovka. Kirovohrads'ka Oblast', C Ukraine
Oleksandriya see Oleksandriia
Olenegorsk 110 C2 Murmanskaya Oblast', NW Russia
Olenëk 115 E3 Respublika Sakha (Yakutiya), NE Russia
Olenëk 115 E3 river NE Russia
Oléron, Île d' 91 A5 island W France
Oleshky 109 E4 Rus. Tsiurupynsk. Khersons'ka Oblast', S Ukraine
Olevsk 108 D1 Zhytomyrs'ka Oblast', N Ukraine
Ölgiy 126 C2 Bayan-Ölgiy, W Mongolia
Olhão 92 B5 Faro, S Portugal
Olifa 78 B3 Kunene, NW Namibia
Ólimbos see Ólympos
Olimpo see Fuerte Olimpo
Olisipo see Lisboa
Olita see Alytus
Oliva 93 F4 Comunitat Valenciana, E Spain
Olivet 90 C4 Loiret, C France
Olmaliq 123 E2 Rus. Almalyk. Toshkent Viloyati, E Uzbekistan
Olmütz see Olomouc
Olomouc 99 C5 Ger. Olmütz, Pol. Ołomuniec. Olomoucký Kraj, E Czechia (Czech Republic)
Ołomuniec see Olomouc
Olonets 110 B3 Respublika Kareliya, NW Russia
Olovyannaya 115 F4 Chitinskaya Oblast', S Russia
Olpe 94 B4 Nordrhein-Westfalen, W Germany
Olsnitz see Murska Sobota
Olsztyn 98 D2 Ger. Allenstein. Warmińsko-Mazurskie, N Poland
Olt 108 B5 var. Oltul, Ger. Alt. river S Romania
Oltenița 108 C5 prev. Eng. Oltenitsa; anc. Constantiola. Călărași, SE Romania
Oltenitsa see Oltenița
Oltul see Olt
Olvera 92 D5 Andalucía, S Spain
Olviopol see Pervomaisk
Olympia 46 B2 state capital Washington, NW USA
Olympic Mountains 46 A2 mountain range Washington, NW USA
Olympus, Mount 46 A4 var. Ólimbos, Eng. Mount Olympus. mountain N Greece
Omagh 89 B5 Ir. An Ómaigh. W Northern Ireland, United Kingdom
Omaha 45 F4 Nebraska, C USA
Oman 121 D6 off. Sultanate of Oman, Ar. Salṭanat 'Umān; prev. Muscat and Oman. country SW Asia
Oman, Gulf of 120 E4 Ar. Khalīj 'Umān. gulf N Arabian Sea
Oman, Sultanate of see Oman
Omboué 77 A6 Ogooué-Maritime, W Gabon
Omdurman 72 B4 var. Umm Durmān. Khartoum, C Sudan
Ometepe, Isla de 52 D4 island S Nicaragua
Ommen 86 E3 Overijssel, E Netherlands
Omsk 114 C4 Omskaya Oblast', C Russia
Ōmuta 131 A7 Fukuoka, Kyūshū, SW Japan

Onda 93 F3 Comunitat Valenciana, E Spain
Ondjiva see N'Giva
Öndörhaan 127 E2 var. Undur Khan; prev. Tsetsen Khan. Hentiy, E Mongolia
Onega 110 C3 Arkhangel'skaya Oblast', NW Russia
Onega 110 B4 river NW Russia
Onega, lake see Onezhskoye Ozero
Onex 95 A7 Genève, SW Switzerland
Onezhskoye Ozero 110 B4 Eng. Lake Onega. lake NW Russia
Ongole 132 D1 Andhra Pradesh, E India
Onitsha 75 G5 Anambra, S Nigeria
Onon Gol 127 E2 river N Mongolia
Onslow 146 A4 Western Australia
Onslow Bay 43 F1 bay North Carolina, E USA
Ontario 38 B3 province S Canada
Ontario, Lake 41 E3 lake Canada/USA
Onteniente see Ontinyent
Ontinyent 93 F4 var. Onteniente. Comunitat Valenciana, E Spain
Ontong Java Rise 125 H4 undersea feature W Pacific Ocean
Onuba see Huelva
Oodeypore see Udaipur
Oos-Londen see East London
Oostakker 87 B5 Oost-Vlaanderen, NW Belgium
Oostburg 87 B5 Zeeland, SW Netherlands
Oostende 87 A5 Eng. Ostend, Fr. Ostende. West-Vlaanderen, NW Belgium
Oosterbeek 86 D4 Gelderland, SE Netherlands
Oosterhout 86 C4 Noord-Brabant, S Netherlands
Opatija 100 A2 It. Abbazia. Primorje-Gorski Kotar, NW Croatia
Opava 99 C5 Ger. Troppau. Moravskoslezský Kraj, E Czechia (Czech Republic)
Opazova see Stara Pazova
Opelika 42 D3 Alabama, S USA
Opelousas 42 B3 Louisiana, S USA
Ophiusa see Formentera
Opmeer 86 C2 Noord-Holland, NW Netherlands
Opochka 110 A4 Pskovskaya Oblast', W Russia
Opole 98 C4 Ger. Oppeln. Opolskie, S Poland
Oporto see Porto
Opotiki 150 E3 Bay of Plenty, North Island, New Zealand
Oppeln see Opole
Oppidum Ubiorum see Köln
Oqtosh 123 E2 Rus. Aktash. Samarqand Viloyati, C Uzbekistan
Oradea 108 B3 prev. Oradea Mare, Ger. Grosswardein, Hung. Nagyvárad. Bihor, NW Romania
Oradea Mare see Oradea
Orahovac see Rahovec
Oral 114 B3 Kaz. Zapadnyy Kazakhstan, NW Kazakhstan
Oran 70 D2 var. Ouahran, Wahran. NW Algeria
Orange 149 D6 New South Wales, SE Australia
Orange 91 D6 anc. Arausio. Vaucluse, SE France
Orangeburg 43 E2 South Carolina, SE USA
Orange Cone see Orange Fan
Orange Fan 69 C7 var. Orange Cone. undersea feature SW Indian Ocean
Orange Mouth/Orangemund see Oranjemund
Orange River 78 B4 Afr. Oranjerivier. river S Africa
Orange Walk 52 C1 Orange Walk, N Belize
Oranienburg 94 D3 Brandenburg, NE Germany
Oranjemund 78 B4 var. Orangemund; prev. Orange Mouth. Karas, SW Namibia
Oranjerivier see Orange River
Oranjestad 55 E5 dependent territory capital (Aruba) Lesser Antilles, S Caribbean Sea
Orantes see Orontes
Orany see Varėna
Orașul Stalin see Brașov
Oravicabánya see Oravița

Oravița 108 A4 Ger. Orawitza, Hung. Oravicabánya. Caraș-Severin, SW Romania
Orawitza see Oravița
Orbetello 96 B4 Toscana, C Italy
Orcadas 154 A1 Argentinian research station South Orkney Islands, Antarctica
Orchard Homes 44 B1 Montana, NW USA
Ordino 91 A8 Ordino, NW Andorra Europe
Ordos Desert see Mu Us Shadi
Ordu 116 D2 anc. Cotyora. Ordu, N Turkey (Türkiye)
Ordzhonikidze see Pokrov, Ukraine
Ordzhonikidze see Vladikavkaz, Russia
Ordzhonikidze see Yenakiieve
Orealla 59 G3 E Guyana
Örebro 85 C6 Örebro, C Sweden
Oregon 46 B3 off. State of Oregon, also known as Beaver State, Sunset State, Valentine State, Webfoot State. state NW USA
Oregon City 46 B3 Oregon, NW USA
Oregon, State of see Oregon
Orekhov see Orikhiv
Orël 111 B5 Orlovskaya Oblast', W Russia
Orem 44 B4 Utah, W USA
Ore Mountains see Erzgebirge/ Krušné Hory
Orenburg 111 D6 prev. Chkalov. Orenburgskaya Oblast', W Russia
Orense see Ourense
Orestiáda 104 D3 prev. Orestiás. Anatoliki Makedonía kai Thráki, NE Greece
Orestiás see Orestiáda
Organ Peak 48 D3 mountain New Mexico, SW USA
Orgeyev see Orhei
Orhei 108 D3 var. Orheiu, Rus. Orgeyev. N Moldova
Orheiu see Orhei
Oriental, Cordillera 60 D3 mountain range Bolivia/Peru
Oriental, Cordillera 58 B3 mountain range C Colombia
Orihuela 93 F4 Comunitat Valenciana, E Spain
Orikhiv 109 G3 Rus. Orekhov. Zaporiz'ka Oblast', SE Ukraine
Orinoco, Río 59 E2 river Colombia/ Venezuela
Orinoquía 58 C3 region NE Colombia
Orissa see Odisha
Orissaare 106 C2 Ger. Orissaar. Saaremaa, W Estonia
Oristano 97 A5 Sardegna, Italy, C Mediterranean Sea
Orito 58 A4 Putumayo, SW Colombia
Orizaba, Volcán Pico de 35 C7 var. Citlaltépetl. mountain S Mexico
Orkney Islands see Orkney Islands
Orkney Islands 88 C2 var. Orkney, Orkneys. island group N Scotland, United Kingdom
Orkneys see Orkney Islands
Orlando 43 E4 Florida, SE USA
Orléanais 90 C4 cultural region C France
Orléans 90 C4 anc. Aurelianum. Loiret, C France
Orléansville see Chlef
Orly 90 E2 (Paris) Essonne, N France
Orlya 107 B6 Hrodzyenskaya Voblasts', W Belarus
Ormsö see Vormsi
Ormuz, Strait of see Hormuz, Strait of
Örnsköldsvik 85 C5 Västernorrland, C Sweden
Orolaunum see Arlon
Orol Dengizi see Aral Sea
Oromocto 39 F4 New Brunswick, SE Canada
Orona 145 F3 prev. Hull Island. atoll Phoenix Islands, C Kiribati
Orontes 116 D5 var. Orantes, Nahr El Aassi, Nahr Al 'Āṣī, Asi Nehri. river Lebanon/Syria/Turkey
Oropeza see Cochabamba
Oroseirá Rodópis see Rhodope Mountains
Orpington 89 B8 United Kingdom
Orschowa see Orșova
Orsha 107 E6 Vitsyebskaya Voblasts', NE Belarus
Orsk 114 B4 Orenburgskaya Oblast', W Russia

Orşova 108 A4 Ger. Orschowa, Hung. Orsova. SW Romania

Ortelsburg see Szczytno

Orthez 91 B6 Pyrénées-Atlantiques, SW France

Ortona 96 D4 Abruzzo, C Italy

Oruba see Aruba

Orümïyeh, Daryächeh-ye 121 C2 var. Matianus, Sha Hi, Urumi Yeh, Eng. Lake Urmia; prev. Daryächeh-ye Reza'īyeh. lake NW Iran

Oruro 61 F4 Oruro, W Bolivia

Oryokko see Yalu

Oss 86 D4 Noord-Brabant, S Netherlands

Ōsaka 131 C6 hist. Naniwa. Ōsaka, Honshū, SW Japan

Ōsaki see Furukawa

Osa, Península de 53 E5 peninsula S Costa Rica

Osborn Plateau 141 D5 undersea feature E Indian Ocean

Osca see Huesca

Ösel see Saaremaa

Osh 123 F2 Osh Shaary/Gorod Osh, SW Kyrgyzstan

Oshawa 38 D5 Ontario, SE Canada

Oshikango 78 B3 Ohangwena, N Namibia

O-shima 131 D6 island S Japan

Oshkosh 40 B2 Wisconsin, N USA

Oshmyany see Ashmyany

Osiek see Osijek

Osijek 100 C3 prev. Osiek, Osjek, Ger. Esseg, Hung. Eszék. Osijek-Baranja, E Croatia

Osipenko see Berdiansk

Osipovichi see Asipovichy

Osjek see Osijek

Oskaloosa 45 G4 Iowa, C USA

Oskarshamn 85 C7 Kalmar, S Sweden

Öskemen 114 D5 var. Ust'-Kamenogorsk. Vostochnyy Kazakhstan, E Kazakhstan

Oskol see Oskil

Oskil 109 G2 Rus. Oskol. river Russia/Ukraine

Oslo 85 B6 prev. Christiania, Kristiania. country capital (Norway) Oslo, S Norway

Osmaniye 116 D4 Osmaniye, S Turkey (Türkiye)

Osnabrück 94 A3 Niedersachsen, NW Germany

Osogov Mountains 104 B3 var. Osogovske Planine, Osogovski Planina, Mac. Osogovski Planini. mountain range Bulgaria/North Macedonia

Osogovske Planine/Osogovski Planina/Osogovski Planini see Osogov Mountains

Oşorhei see Târgu Mureş

Osorno 65 B5 Los Lagos, C Chile

Ossa, Serra d' 92 C4 mountain range SE Portugal

Ossora 115 H2 Koryakskiy Avtonomnyy Okrug, E Russia

Ostee see Baltic Sea

Ostend/Ostende see Oostende

Oster 109 E1 Chernihivs'ka Oblast', N Ukraine

Östermyra see Seinäjoki

Osterode/Osterode in Ostpreussen see Ostróda

Österreich see Austria

Östersund 85 C5 Jämtland, C Sweden

Ostia Aterni see Pescara

Ostiglia 96 C2 Lombardia, N Italy

Ostrava 99 C5 Moravskoslezský Kraj, E Czechia (Czech Republic)

Ostróda 98 D3 Ger. Osterode, Osterode in Ostpreussen. Warmińsko-Mazurskie, NE Poland

Ostrołęka 98 D3 Ger. Wiesenhof, Rus. Ostrolenka. Mazowieckie, C Poland

Ostrolenka see Ostrołęka

Ostrov 110 A4 Latv. Austrava. Pskovskaya Oblast', W Russia

Ostrów see Ostrów Wielkopolski

Ostrovnoy 110 C2 Murmanskaya Oblast', NW Russia

Ostrowiec see Ostrowiec Świętokrzyski

Ostrowiec Świętokrzyski 98 D4 var. Ostrowiec, Rus. Ostrovets. Świętokrzyskie, C Poland

Ostrów Mazowiecka 98 D3 var. Ostrów Mazowiecki. Mazowieckie, NE Poland

Ostrów Mazowiecki see Ostrów Mazowiecka

Ostrowo see Ostrów Wielkopolski

Ostrów Wielkopolski 98 C4 var. Ostrów, Ger. Ostrowo. Wielkopolskie, C Poland

Ostyako-Voguls'k see Khanty-Mansiysk

Osum see Osumit, Lumi i

Osumi-shotō 131 A8 island group Kagoshima, Nansei-shotō, SW Japan Asia East China Sea Pacific Ocean

Osumit, Lumi i 101 D7 var. Osum. river SE Albania

Osuna 92 D4 Andalucía, S Spain

Oswego 41 F2 New York, NE USA

Otago Peninsula 151 B7 peninsula South Island, New Zealand

Otaki 150 D4 Wellington, North Island, New Zealand

Otaru 130 C2 Hokkaidō, NE Japan

Otavalo 60 B1 Imbabura, N Ecuador

Otavi 78 B3 Otjozondjupa, N Namibia

Oţelu Roşu 108 B4 Ger. Ferdinandsberg, Hung. Nándorhgy. Caras-Severin, SW Romania

Otepää 106 D3 Ger. Odenpäh. Valgamaa, SE Estonia

Oti 75 E4 river N Togo

Otira 151 C6 West Coast, South Island, New Zealand

Otjiwarongo 78 B3 Otjozondjupa, N Namibia

Otorohanga 150 D3 Waikato, North Island, New Zealand

Otranto, Canale d' see Otranto, Strait of

Otranto, Strait of 101 C6 It. Canale d'Otranto. strait Albania/Italy

Otrokovice 99 C5 Ger. Otrokowitz. Zlínský Kraj, E Czechia (Czech Republic)

Otrokowitz see Otrokovice

Ōtsu 131 C6 var. Ōtu. Shiga, Honshū, SW Japan

Ottawa 38 D5 country capital (Canada) Ontario, SE Canada

Ottawa 40 B3 Illinois, N USA

Ottawa 45 F5 Kansas, C USA

Ottawa 41 E2 Fr. Outaouais. river Ontario/Québec, SE Canada

Ottawa Islands 38 C1 island group Nunavut, C Canada

Ottignies 87 C6 Wallon Brabant, C Belgium

Ottumwa 45 G4 Iowa, C USA

Ōtu see Ōtsu

Ouachita Mountains 42 A1 mountain range Arkansas/Oklahoma, C USA

Ouachita River 42 B2 river Arkansas/Louisiana, C USA

Ouagadougou 75 E4 var. Wagadugu. country capital (Burkina) C Burkina

Ouahigouya 75 E3 NW Burkina

Ouahran see Oran

Oualâta 74 D3 var. Oualata. Hodh ech Chargui, SE Mauritania

Ouanary 59 H3 E French Guiana

Ouanda Djallé 76 D4 Vakaga, NE Central African Republic

Ouarâne 74 D2 desert C Mauritania

Ouargla 71 E2 var. Wargla. NE Algeria

Ouarzazate 70 C3 S Morocco

Oubangui see Ubangi

Oubangui-Chari see Central African Republic

Oubangui-Chari, Territoire de l' see Central African Republic

Oudjda see Oujda

Ouessant, Île d' 90 A3 Eng. Ushant. island NW France

Ouésso 77 B5 Sangha, NW Congo

Oujda 70 D2 Ar. Oudjda, Ujda. NE Morocco

Oujeft 74 C2 Adrar, C Mauritania

Oulu 84 D4 Swe. Uleåborg. Pohjois-Pohjanmaa, C Finland

Oulujärvi 84 D4 Swe. Uleträsk. lake C Finland

Oulujoki 84 D4 Swe. Uleälv. river C Finland

Ounasjoki 84 D3 river N Finland

Ounianga Kébir 76 C2 Borkou-Ennedi-Tibesti, N Chad

Oup see Auob

Oupeye 87 D6 Liège, E Belgium

Our 87 D6 river NW Europe

Ourense 92 C1 Cast. Orense, Lat. Aurium. Galicia, NW Spain

Ourique 92 B4 Beja, S Portugal

Ours, Grand Lac de l' see Great Bear Lake

Ourthe 87 D7 river E Belgium

Ouse 89 D5 river N England, United Kingdom

Outaouais see Ottawa

Outer Hebrides 88 B3 var. Western Isles. island group NW Scotland, United Kingdom

Outer Islands 79 G1 island group SW Seychelles Africa W Indian Ocean

Outes 92 B1 Galicia, NW Spain

Ouvéa 144 D5 island Îles Loyauté, NE New Caledonia

Ouyen 149 C6 Victoria, SE Australia

Ovalle 64 B3 Coquimbo, N Chile

Ovar 92 B2 Aveiro, N Portugal

Overflakkee 86 B4 island SW Netherlands

Overijse 87 C6 Vlaams Brabant, C Belgium

Oviedo 92 C1 var. Uviéu, anc. Asturias. Asturias, NW Spain

Ovilava see Wels

Ovruch 108 D1 Zhytomyrs'ka Oblast', N Ukraine

Owando 77 B5 prev. Fort Rousset. Cuvette, C Congo

Owase 131 C6 Mie, Honshū, SW Japan

Owatonna 45 F3 Minnesota, N USA

Owen Fracture Zone 140 B4 tectonic feature W Arabian Sea

Owen, Mount 151 C5 mountain South Island, New Zealand

Owensboro 40 B5 Kentucky, S USA

Owen Stanley Range 144 B3 mountain range S Papua New Guinea

Owerri 75 G5 Imo, S Nigeria

Owo 75 F5 Ondo, SW Nigeria

Owyhee River 46 C4 river Idaho/Oregon, NW USA

Oxford 151 C6 Canterbury, South Island, New Zealand

Oxford 89 D6 Lat. Oxonia. S England, United Kingdom

Oxkutzcab 51 H4 Yucatán, SE Mexico

Oxnard 47 B7 California, W USA

Oxonia see Oxford

Oxus see Amu Darya

Oyama 131 D5 Tochigi, Honshū, S Japan

Oyem 77 B5 Woleu-Ntem, N Gabon

Oyo 77 B6 Cuvette, C Congo

Oyo 75 F4 Oyo, W Nigeria

Ozark 42 D3 Alabama, S USA

Ozark Plateau 42 G5 plain Arkansas/Missouri, C USA

Ozarks, Lake of the 45 F5 reservoir Missouri, C USA

Ozbourn Seamount 152 D4 undersea feature W Pacific Ocean

Ózd 99 D6 Borsod-Abaúj-Zemplén, NE Hungary

Ozieri 97 A5 Sardegna, Italy, C Mediterranean Sea

P

Paamiut 82 B4 var. Pâmiut, Dan. Frederikshåb. S Greenland

Pa-an see Hpa-An

Pabianice 98 C4 Łódzskie, C Poland

Pabna 135 G4 Rajshahi, NW Bangladesh

Pacaraima, Sierra/Pacaraim, Serra see Pakaraima Mountains

Pachuca 51 E4 var. Pachuca de Soto. Hidalgo, C Mexico

Pachuca de Soto see Pachuca

Pacific-Antarctic Ridge 154 B5 undersea feature S Pacific Ocean

Pacific Ocean 152 D3 ocean

Padalung see Phatthalung

Padang 138 B4 Sumatera, W Indonesia

Paderborn 94 B4 Nordrhein-Westfalen, NW Germany

Padma see Brahmaputra

Padma see Ganges

Padova 96 C2 Eng. Padua; anc. Patavium. Veneto, NE Italy

Padre Island 49 G5 island Texas, SW USA

Padua see Padova

Paducah 40 B5 Kentucky, S USA

Paeroa 150 D3 Waikato, North Island, New Zealand

Páfos 102 C5 var. Paphos. W Cyprus

Pag 100 A3 It. Pago. island Zadar, C Croatia

Page 48 B1 Arizona, SW USA

Pago see Pag

Pago Pago 145 F4 dependent territory capital (American Samoa) Tutuila, W American Samoa

Pahiatua 150 D4 Manawatu-Wanganui, North Island, New Zealand

Pahsien see Chongqing

Paide 106 D2 Ger. Weissenstein. Järvamaa, N Estonia

Paihia 150 D2 Northland, North Island, New Zealand

Päijänne 85 D5 lake S Finland

Paine, Cerro 65 A7 mountain S Chile

Painted Desert 48 B1 desert Arizona, SW USA

Paisance see Piacenza

Paisley 88 C4 W Scotland, United Kingdom

País Vasco 93 E1 cultural region N Spain

Paita 60 B3 Piura, NW Peru

Pakanbaru see Pekanbaru

Pakaraima Mountains 59 E3 var. Serra Pacaraim, Sierra Pacaraima. mountain range N South America

Pakistan 134 A2 off. Islamic Republic of Pakistan, var. Islami Jamhuriya e Pakistan. country S Asia

Pakistan, Islamic Republic of see Pakistan

Pakistan, Islami Jamhuriya e see Pakistan

Paknam see Samut Prakan

Pakokku 136 A3 Magway, C Myanmar (Burma)

Pak Phanang 137 C7 var. Ban Pak Phanang. Nakhon Si Thammarat, SW Thailand

Pakruojis 106 C4 Šiauliai, N Lithuania

Paks 99 C7 Tolna, S Hungary

Paksé see Pakxé

Pakxé 137 D5 var. Paksé. Champasak, S Laos

Palafrugell 93 G2 Cataluña, NE Spain

Palagruža 101 B5 It. Pelagosa. island SW Croatia

Palaiá Epídavros 105 C6 Pelopónnisos, S Greece

Palaiseau 90 D2 Essonne, N France

Palamós 93 G2 Cataluña, NE Spain

Palamuse 106 E2 Ger. Sankt-Bartholomäi. Jõgevamaa, E Estonia

Palanka see Bačka Palanka

Pālanpur 134 C4 Gujarāt, W India

Palantia see Palencia

Palapye 78 D3 Central, SE Botswana

Palau 144 A2 var. Belau. country W Pacific Ocean

Palawan 139 E2 island W Philippines

Palawan Passage 138 D2 passage W Philippines

Paldiski 106 D2 prev. Baltiski, Eng. Baltic Port, Ger. Baltischport. Harjumaa, NW Estonia

Palembang 138 B4 Sumatera, W Indonesia

Palencia 92 D2 anc. Palantia, Pallantia. Castilla y León, NW Spain

Palerme see Palermo

Palermo 97 C7 Fr. Palerme; anc. Panhormus, Panormus. Sicilia, Italy, C Mediterranean Sea

Palestine, State of see West Bank; Gaza Strip

Pāli 134 C3 Rājasthān, N India

Palikir 144 C2 country capital (Federated States of Micronesia) Pohnpei, E Federated States of Micronesia

Palimé see Kpalimé

Palioúri, Akrotírio 104 C4 var. Akrotírio Kanestron. headland N Greece

Palk Strait 132 C3 strait India/Sri Lanka

Pallantia see Palencia

Palliser, Cape 151 D5 headland North Island, New Zealand

Palma 93 G3 var. Palma de Mallorca. Mallorca, Spain, W Mediterranean Sea

Palma del Río 92 D4 Andalucía, S Spain

Palma de Mallorca see Palma

Palmar Sur 53 E5 Puntarenas, SE Costa Rica

Palma Soriano 54 C3 Santiago de Cuba, E Cuba

Palm Beach 148 E1 New South Wales, E Australia

Palmer 154 A2 US research station Antarctica

Palmer Land 154 A3 physical region Antarctica

Palmerston 145 F4 island S Cook Islands

Palmerston see Darwin

Palmerston North 150 D4 Manawatu-Wanganui, North Island, New Zealand

Palmetto State, The see South Carolina

Palmi 97 D7 Calabria, SW Italy

Palmira *58 B3* Valle del Cauca,
W Colombia
Palm Springs *47 D7* California, W USA
Palmyra *see* Tadmur
Palmyra Atoll *145 G2* US incorporated
territory C Pacific Ocean
Palo Alto *47 B6* California, W USA
Paloe *see* Denpasar, Bali, C Indonesia
Paloe *see* Palu
Palu *139 E4* prev. Paloe. Sulawesi,
C Indonesia
Pamiers *91 B6* Ariège, S France
Pamir *123 F3* var. Daryā-ye Pāmīr, Taj.
Dar"yoi Pomir. river Afghanistan/
Tajikistan
Pāmīr, Daryā-ye *see* Pamir
Pāmīr/Pāmīr, Daryā-ye *see* Pamirs
Pamirs *123 F3* Pash. Daryā-ye Pāmīr,
Rus. Pamir. mountain range C Asia
Pāmiut *see* Paamiut
Pamlico Sound *43 G1* sound North
Carolina, SE USA
Pampa *49 E1* Texas, SW USA
Pampa Aullagas, Lago *see* Poopó, Lago
Pampas *64 C4* plain C Argentina
Pampeluna *see* Pamplona
Pamplona *58 C2* Norte de Santander,
N Colombia
Pamplona *93 E1* Basq. Iruña, prev.
Pampeluna; anc. Pompaelo. Navarra,
N Spain
Panaji *132 B1* var. Panaim, Panjim, New
Goa. state capital Goa, W India
Panamá *see* Panamá, Ciudad de
Panama *53 G5* off. Republic of Panama.
country Central America
Panama Basin *35 C8* undersea feature
E Pacific Ocean
Panama Canal *53 F4* canal E Panama
Panama City *42 D3* Florida, SE USA
Panama City *see* Panamá
Panamá, Ciudad de *53 G4* var. Panamá,
Eng. Panama City. country capital
(Panama) Panamá, C Panama
Panamá, Golfo de *53 G5* var. Gulf of
Panama. gulf S Panama
Panama, Gulf of *see* Panamá, Golfo de
Panama, Isthmus of *see* Panama,
Istmo de
Panama, Istmo de *53 G4* Eng. Isthmus
of Panama; prev. Isthmus of Darien.
isthmus E Panama
Panama, Republic of *see* Panama
Panay Island *139 E2* island C Philippines
Pančevo *100 D3* Ger. Pantschowa, Hung.
Pancsova. Vojvodina, N Serbia
Pancsova *see* Pančevo
Paneas *see* Bāniyās
Panevežys *106 C4* Panevežys,
C Lithuania
Pangim *see* Panaji
Pangkalpinang *138 C4* Pulau Bangka,
W Indonesia
Pang-Nga *see* Phang-Nga
Panhormus *see* Palermo
Panjim *see* Panaji
Panopolis *see* Akhmīm
Pánormos *105 C7* Kríti, Greece,
E Mediterranean Sea
Panormus *see* Palermo
Pantanal *63 E3* var. Pantanalmato-
Grossense. swamp SW Brazil
Pantanalmato-Grossense *see* Pantanal
Pantelleria, Isola di *97 B7* island
SW Italy
Pantschowa *see* Pančevo
Pánuco *51 E3* Veracruz-Llave, E Mexico
Pao-chi/Paoki *see* Baoji
Paola *102 B5* E Malta
Pao-shan *see* Baoshan
Pao-t'ou/Paotow *see* Baotou
Papagayo, Golfo de *52 C4* gulf
NW Costa Rica
Papakura *150 D3* Auckland, North
Island, New Zealand
Papantla *51 F4* var. Papantla de Olarte.
Veracruz-Llave, E Mexico
Papantla de Olarte *see* Papantla
Papeete *145 H4* dependent territory
capital (French Polynesia) Tahiti,
W French Polynesia
Paphos *see* Páfos
Papilé *106 B3* Šiauliai, NW Lithuania
Papillion *45 F4* Nebraska,
C USA
Papua *139 H4* var. Irian Barat, West
Irian, West New Guinea, West Papua;
prev. Dutch New Guinea, Irian Jaya,
Netherlands New Guinea. province
E Indonesia

Papua and New Guinea, Territory of *see*
Papua New Guinea
Papua, Gulf of *144 B3* gulf S Papua
New Guinea
Papua New Guinea *144 B3* off.
Independent State of Papua New
Guinea; prev. Territory of Papua and
New Guinea. country NW Melanesia
**Papua New Guinea, Independent State
of** *see* Papua New Guinea
Papuk *100 C3* mountain range
NE Croatia
Pará *63 E2* off. Estado do Pará. state/
region NE Brazil
Pará *see* Belém
Paracel Islands *125 E3* disputed territory
SE Asia
Paraćin *100 D4* Serbia, C Serbia
Paradise of the Pacific *see* Hawaii
Pará, Estado do *see* Pará
Paragua, Río *59 E3* river SE Venezuela
Paraguay *64 C2* country
C South America
Paraguay *64 D2* var. Río Paraguay. river
C South America
Paraguay, Río *see* Paraguay
Paraíba/Parahyba *see* Paraíba
Paraíba *63 G2* off. Estado da Paraíba;
prev. Parahiba, Parahyba. state/region
E Brazil
Paraíba *see* João Pessoa
Paraíba, Estado da *see* Paraíba
Parakou *75 F4* C Benin
Paramaribo *59 G3* country capital
(Suriname) Paramaribo, N Suriname
Paramushir, Ostrov *115 H3* island
SE Russia
Paraná *63 E4* Entre Ríos, E Argentina
Paraná *63 E5* off. Estado do Paraná.
state/region S Brazil
Paraná *57 C5* var. Alto Paraná. river
C South America
Paraná, Estado do *see* Paraná
Paranésti *104 C3* var. Paranéstio.
Anatolikí Makedonía kai Thráki,
NE Greece
Paranéstio *see* Paranésti
Paraparaumu *151 D5* Wellington, North
Island, New Zealand
Parchim *94 C3* Mecklenburg-
Vorpommern, N Germany
Parczew *98 E4* Lubelskie, E Poland
Pardubice *99 B5* Ger. Pardubitz.
Pardubický Kraj, C Czechia (Czech
Republic)
Pardubitz *see* Pardubice
Parechcha *107 B5* Pol. Porzecze, Rus.
Porech'ye. Hrodzyenskaya Voblasts',
W Belarus
Parecis, Chapada dos *62 D3* var. Serra
dos Parecis. mountain range W Brazil
Parecis, Serra dos *see* Parecis,
Chapada dos
Parenzo *see* Poreč
Prepare *139 E4* Sulawesi, C Indonesia
Párga *105 A5* Ípeiros, W Greece
Paria, Golfo de *see* Paria, Gulf of
Paria, Gulf of *59 E1* var. Golfo de Paria.
gulf Trinidad and Tobago/Venezuela
Parika *59 F2* NE Guyana
Parikia *105* Kykládes, Greece,
Aegean Sea
Paris *90 D1* anc. Lutetia, Lutetia
Parisiorum, Parisii. country capital
(France) Paris, N France
Paris *49 G2* Texas, SW USA
Parisii *see* Paris
Parkersburg *40 D4* West Virginia,
NE USA
Parkes *149 D6* New South Wales,
SE Australia
Parkhar *see* Farkhor
Parma *96 B2* Emilia-Romagna, N Italy
Parnahyba *see* Parnaíba
Parnaíba *63 F2* var. Parnahyba. Piauí,
E Brazil
Pärnu *106 D2* Ger. Pernau, Latv.
Pērnava; prev. Rus. Pernov. Pärnumaa,
SW Estonia
Pärnu *106 D2* var. Pärnu Jõgi, Ger.
Pernau. river SW Estonia
Pärnu-Jaagupi *106 D2* Ger. Sankt-
Jakobi. Pärnumaa, SW Estonia
Pärnu Jõgi *see* Pärnu
Pärnu Laht *106 D2* Ger. Pernauer Bucht.
bay SW Estonia
Paroikiá *see* Páros
Paropamisus Range *see* Safēd Kōh,
Silsilah-ye
Páros *105 D6* var. Paroikiá, Greece

Páros *105 D6* island Kykládes, Greece,
Aegean Sea
Parral *64 B4* Maule, C Chile
Parral *see* Hidalgo del Parral
Parramatta *148 D1* New South Wales,
SE Australia
Parras *50 D3* var. Parras de la Fuente.
Coahuila, NE Mexico
Parras de la Fuente *see* Parras
Parsons *45 F5* Kansas, C USA
Pasadena *47 C7* California, W USA
Pasadena *49 H4* Texas, SW USA
Pașcani *108 C3* Hung. Páskán. Iași,
NE Romania
Pasco *46 C2* Washington, NW USA
Pascua, Isla de *153 F4* var. Rapa Nui,
Easter Island. island E Pacific Ocean
Pasewalk *94 D3* Mecklenburg-
Vorpommern, NE Germany
Pashkeni *see* Bolyarovo
Pasinler *117 F3* Erzurum,
NE Turkey (Türkiye)
Páskán *see* Pașcani
Pasłęk *98 D2* Ger. Preußisch Holland.
Warmińsko-Mazurskie, NE Poland
Pasni *134 A3* Baluchistan, SW Pakistan
Paso de Indios *65 B6* Chubut,
S Argentina
Passarowitz *see* Požarevac
Passau *95 D6* Bayern, SE Germany
Passo Fundo *63 E5* Rio Grande do Sul,
S Brazil
Pastavy *107 C5* Pol. Postawy, Rus.
Postovy. Vitsyebskaya Voblasts',
NW Belarus
Pastaza, Río *60 B2* river Ecuador/Peru
Pasto *58 A4* Nariño, SW Colombia
Pasvalys *106 C4* Panevežys, N Lithuania
Patagonia *57 B7* physical region
Argentina/Chile
Patalung *see* Phatthalung
Patani *see* Pattani
Patavium *see* Padova
Patea *150 D4* Taranaki, North Island,
New Zealand
Paterson *41 F3* New Jersey, NE USA
Pathein *136 A4* var. Bassein.
Ayeyarwady, SW Myanmar (Burma)
Pátmos *105 D6* island Dodekánisa,
Greece, Aegean Sea
Patna *135 F3* var. Azimabad. state capital
Bihār, N India
Patnos *117 F3* Ağrı, E Turkey (Türkiye)
Patos, Lagoa dos *63 E5* lagoon S Brazil
Pátra *105 B5* Eng. Patras; prev. Pátrai.
Dytikí Ellás, S Greece
Pátrai/Patras *see* Pátra
Pattani *137 C7* var. Patani. Pattani,
SW Thailand
Pattaya *137 C5* Chon Buri, S Thailand
Patuca, Río *52 D2* river E Honduras
Pau *91 B6* Pyrénées-Atlantiques,
SW France
Paulatuk *37 E3* Northwest Territories,
NW Canada
Paungde *136 B4* Bago,
C Myanmar (Burma)
Pautalia *see* Kyustendil
Pavia *96 B2* anc. Ticinum. Lombardia,
N Italy
Pāvilosta *106 B3* Liepāja, W Latvia
Pavlikeni *104 D2* Veliko Tarnovo,
N Bulgaria
Pavlodar *114 C4* Pavlodar,
NE Kazakhstan
Pavlograd *see* Pavlohrad
Pavlohrad *109 G3* Rus. Pavlograd.
Dnipropetrovska Oblast, E Ukraine
Pawai, Pulau *138 A2* island
SW Singapore Asia
Pawn *136 B3* river C Myanmar (Burma)
Pax Augusta *see* Badajoz
Pax Julia *see* Beja
Paxoí *105 A5* island Iónia Nisiá, Greece,
C Mediterranean Sea
Payo Obispo *see* Chetumal
Paysandú *64 D4* Paysandú, W Uruguay
Pazar *117 E2* Rize, NE Turkey (Türkiye)
Pazardzhik *104 C3* prev. Tatar
Pazardzhik. Pazardzhik, SW Bulgaria
Peace Garden State *see* North Dakota
Peach State *see* Georgia
Pearl Islands *53 G5* Eng. Pearl Islands.
island group SE Panama
Pearl Islands *see* Perlas, Archipiélago de
las
Pearl Lagoon *see* Perlas, Laguna de
Pearl River *42 B3* river Louisiana/
Mississippi, S USA
Pearsall *49 F4* Texas, SW USA

Peawanuk *38 C2* Ontario, S Canada
Peč *see* Pejë
Pechora *110 D3* Respublika Komi,
NW Russia
Pechora *110 D3* river NW Russia
Pechorskoye More *110 D2* Eng. Pechora
Sea. sea NW Russia
Pecos *49 E3* Texas, SW USA
Pecos River *49 E3* river New Mexico/
Texas, SW USA
Pécs *99 C7* Ger. Fünfkirchen, Lat.
Sopianae. Baranya, SW Hungary
Pedra Lume *74 A3* Sal,
NE Cape Verde (Cabo Verde)
Pedro Cays *54 C3* island group Greater
Antilles, S Jamaica North America
N Caribbean Sea W Atlantic Ocean
Pedro Juan Caballero *64 D2* Amambay,
E Paraguay
Peer *87 D5* Limburg, NE Belgium
Pegasus Bay *151 C6* bay South Island,
New Zealand
Pegu *see* Bago
Pehuajó *64 C4* Buenos Aires, E Argentina
Pei-ching *see* Beijing/Beijing Shi
Peine *94 B3* Niedersachsen, C Germany
Pei-p'ing *see* Beijing/Beijing Shi
Peipsi Järv/Peipus-See *see* Peipus, Lake
Peipus, Lake *106 E3* Est. Peipsi Järv, Ger.
Peipus-See, Rus. Chudskoye Ozero.
lake Estonia/Russia
Peiraiás *105 C6* prev. Piraiévs, Eng.
Piraeus. Attikí, C Greece
Pejë *101 D5* Serb. Peć. W Kosovo
Pek *see* Phônsaven
Pekalongan *138 C4* Jawa, C Indonesia
Pekanbaru *138 B3* var. Pakanbaru.
Sumatera, W Indonesia
Pekin *40 B4* Illinois, N USA
Peking *see* Beijing/Beijing Shi
Pelagie *97 B8* island group SW Italy
Pelican State *see* Louisiana
Pelly Bay *see* Kugaaruk
Pélmonostor *see* Beli Manastir
Peloponnese *105 B6* var. Morea, Eng.
Peloponnese; anc. Peloponnesus.
peninsula S Greece
Peloponnese/Peloponnesus *see*
Pelopónnisos
Pematangsiantar *138 B3* Sumatera,
W Indonesia
Pemba *73 D7* island E Tanzania
Pembroke *38 D4* Ontario, SE Canada
Penang *see* Pinang, Pulau, Peninsular
Malaysia
Penang *see* George Town
Penas, Golfo de *65 A7* gulf S Chile
Penderma *see* Bandırma
Pendleton *46 C3* Oregon, NW USA
Pend Oreille, Lake *46 D2* lake Idaho,
NW USA
Peneius *see* Pineiós
Peng-pu *see* Bengbu
Penibético, Sistema *see* Béticos, Sistemas
Peniche *92 B3* Leiria, W Portugal
Peninsular State *see* Florida
Penninae, Alpes/Pennine, Alpi *see*
Pennine Alps
Pennine Alps *95 A8* Fr. Alpes Pennines,
It. Alpi Pennine, Lat. Alpes Penninae.
mountain range Italy/Switzerland
Pennine Chain *see* Pennines
Pennines *89 D5* var. Pennine Chain.
mountain range N England,
United Kingdom
Pennines, Alpes *see* Pennine Alps
Pennsylvania *41 E4* off. Commonwealth
of Pennsylvania, also known as
Keystone State. state NE USA
Penobscot River *41 G2* river Maine,
NE USA
Penong *149 A6* South Australia
Penonomé *53 F5* Coclé, C Panama
Penrhyn *145 G3* atoll N Cook Islands
Penrhyn Basin *143 F3* undersea feature
C Pacific Ocean
Penrith *148 D1* New South Wales,
SE Australia
Penrith *89 D5* NW England,
United Kingdom
Pensacola *42 C3* Florida, SE USA
Pentecost *144 D4* Fr. Pentecôte. island
C Vanuatu
Pentecôte *see* Pentecost
Penza *111 C6* Penzenskaya Oblast',
W Russia

Penzance *89 C7* SW England, United Kingdom
Peoria *40 B4* Illinois, N USA
Perchtoldsdorf *95 E6* Niederösterreich, NE Austria
Percival Lakes *146 C4 lakes* Western Australia
Perdido, Monte *93 F1 mountain* NE Spain
Perece Vela Basin *see* West Mariana Basin
Pereira *58 B3* Risaralda, W Colombia
Peremyshl *see* Przemyśl
Pergamino *64 C4* Buenos Aires, E Argentina
Périgueux *91 C5 anc.* Vesuna. Dordogne, SW France
Perito Moreno *65 B6* Santa Cruz, S Argentina
Perlas, Laguna de *53 E3 Eng.* Pearl Lagoon. *lagoon* E Nicaragua
Perleberg *94 C3* Brandenburg, N Germany
Perlepe *see* Prilep
Perm' *114 C3 prev.* Molotov. Permskaya Oblast', NW Russia
Pernambuco *63 G2 off.* Estado de Pernambuco. *state/region* E Brazil
Pernambuco *see* Recife
Pernambuco Abyssal Plain *see* Pernambuco Plain
Pernambuco, Estado de *see* Pernambuco
Pernambuco Plain *67 C5 var.* Pernambuco Abyssal Plain. *undersea feature* E Atlantic Ocean
Pernau *see* Pärnu
Pernauer Bucht *see* Pärnu Laht
Pērnava *see* Pärnu
Pernik *104 B2 prev.* Dimitrovo. Pernik, W Bulgaria
Pernov *see* Pärnu
Perote *51 F4* Veracruz-Llave, E Mexico
Pérouse *see* Perugia
Perovsk *see* Qyzylorda
Perpignan *91 C6* Pyrénées-Orientales, S France
Perryton *49 F1* Texas, SW USA
Perryville *45 H5* Missouri, C USA
Persia *see* Iran
Persian Gulf *120 C4 var.* Gulf, The, *Ar.* Khalij al 'Arabī, *Per.* Khalīj-e Fars. *gulf* SW Asia
Perth *147 A6 state capital* Western Australia
Perth *88 C4* C Scotland, United Kingdom
Perth Basin *141 E6 undersea feature* SE Indian Ocean
Peru *60 C3 off.* Republic of Peru. *country* W South America
Peru *see* Beru
Peru Basin *67 A5 undersea feature* E Pacific Ocean
Peru-Chile Trench *56 A4 undersea feature* E Pacific Ocean
Perugia *96 C3 anc.* Perusia. *Fr.* Pérouse; *anc.* Perusia. Umbria, C Italy
Perugia, Lake of *see* Trasimeno, Lago
Peru, Republic of *see* Peru
Perusia *see* Perugia
Péruwelz *87 B6* Hainaut, SW Belgium
Pervomaisk *see* Pervomaisk
Pervomais'k *109 E3 prev.* Pervomays'k, Olviopol. Mykolayivs'ka Oblast', S Ukraine
Pervomays'k *see* Pervomaisk
Pervyy Kuril'skiy Proliv *115 H3 strait* E Russia
Pesaro *96 C3 anc.* Pisaurum. Marche, C Italy
Pescara *96 D4 anc.* Aternum, Ostia Aterni. Abruzzo, C Italy
Peshawar *134 C1* North-West Frontier Province, N Pakistan
Peshkopi *101 C6 var.* Peshkopia, Peshkopija. Dibër, NE Albania
Peshkopia/Peshkopija *see* Peshkopi
Pessac *91 B5* Gironde, SW France
Petach-Tikva *see* Petah Tikva
Petah Tikva *119 A6 var.* Petach-Tikva, Petah Tiqwa, Petakh Tikva; *prev.* Petah Tiqwa. Tel Aviv, C Israel
Petah Tiqwa *see* Petah Tikva
Petakh Tikva/Petah Tiqwa *see* Petah Tikva
Pétange *87 D8* Luxembourg, SW Luxembourg
Petchaburi *see* Phetchaburi
Peterborough *149 B6* South Australia
Peterborough *38 D5* Ontario, SE Canada

Peterborough *89 E6 prev.* Medeshamstede. E England, United Kingdom
Peterhead *88 D3* NE Scotland, United Kingdom
Peter I Øy *154 A3 Norwegian dependency* Antarctica
Petermann Bjerg *83 E3 mountain* C Greenland
Petersburg *41 E5* Virginia, NE USA
Peters Mine *59 F3 var.* Peter's Mine. N Guyana
Petit St-Bernard, Col du *see* Little Saint Bernard Pass
Peto *51 H4* Yucatán, SE Mexico
Petoskey *40 C2* Michigan, N USA
Petra *see* Wādī Mūsā
Petrich *104 C3* Blagoevgrad, SW Bulgaria
Petrikau *see* Piotrków Trybunalski
Petrikov *see* Pyetrykaw
Petrinja *100 B3* Sisak-Moslavina, C Croatia
Petroaleksandrovsk *see* To'rtkok'l
Petrodvorets *110 A4 Fin.* Pietarhovi. Leningradskaya Oblast', NW Russia
Petrograd *see* Sankt-Peterburg
Petrokov *see* Piotrków Trybunalski
Petropavl *see* Petropavlovsk
Petropavlovsk *114 C4 Kaz.* Petropavl. Severnyy Kazakhstan, N Kazakhstan
Petropavlovsk-Kamchatskiy *115 H3* Kamchatskaya Oblast', E Russia
Petroşani *108 B4 var.* Petroşeni, *Ger.* Petroschen, *Hung.* Petrozsény. Hunedoara, W Romania
Petroschen/Petroşeni *see* Petroşani
Petroskoi *see* Petrozavodsk
Petrovgrad *see* Zrenjanin
Petrovsk-Port *see* Makhachkala
Petrozavodsk *114 B2 Fin.* Petroskoi. Respublika Kareliya, NW Russia
Petrozsény *see* Petroşani
Pettau *see* Ptuj
Pevek *115 G1* Chukotskiy Avtonomnyy Okrug, NE Russia
Pezinok *99 C6 Ger.* Bösing, *Hung.* Bazin. Bratislavský Kraj, W Slovakia
Pforzheim *95 B6* Baden-Württemberg, SW Germany
Pfungstadt *95 B5* Hessen, W Germany
Phangan, Ko *137 C6 island* SW Thailand
Phang-Nga *137 B7 var.* Pang-Nga, Phangnga. Phangnga, SW Thailand
Phangnga *see* Phang-Nga
Phan Rang/Phanrang *see* Phan Rang-Thap Cham
Phan Rang-Thap Cham *137 E6 var.* Phanrang, Phan Rang, Phan Rang Thap Cham. Ninh Thuận, S Vietnam
Phan Thiết *137 E6* Binh Thuận, S Vietnam
Pharnacia *see* Giresun
Pharus *see* Hvar
Phatthalung *137 C7 var.* Padalung, Patalung. Phatthalung, SW Thailand
Phayao *136 C4 var.* Muang Phayao. Phayao, NW Thailand
Phenix City *42 D2* Alabama, S USA
Phet Buri *see* Phetchaburi
Phetchaburi *137 C5 var.* Bejraburi, Petchaburi, Phet Buri. Phetchaburi, SW Thailand
Philadelphia *41 F4* Pennsylvania, NE USA
Philadelphia *see* 'Ammān
Philippine Basin *125 F3 undersea feature* W Pacific Ocean
Philippine Islands *139 E1 island group* W Pacific Ocean
Philippines *139 E1 off.* Republic of the Philippines. *country* SE Asia
Philippine Sea *125 F3 sea* W Pacific Ocean
Philippines, Republic of the *see* Philippines
Philippine Trench *142 A1 undersea feature* W Philippine Sea
Philippopolis *see* Plovdiv
Phitsanulok *136 C4 var.* Bisnulok, Muang Phitsanulok, Pitsanulok. Phitsanulok, C Thailand
Phlórina *see* Flórina
Phnom Penh *137 D6 Khmer.* Phnum Pénh. *country capital* (Cambodia) Phnom Penh, S Cambodia
Phnum Pénh *see* Phnom Penh
Phoenix *48 B2 state capital* Arizona, SW USA
Phoenix Islands *145 E3 island group* C Kiribati

Phôngsali *136 C3 var.* Phong Saly. Phôngsali, N Laos
Phong Saly *see* Phôngsali
Phônsavan *136 D4 var.* Xieng Khouang; *prev.* Pèk, Xiangkhoang. Xiangkhoang, N Laos
Phrae *136 C4 var.* Muang Phrae, Prae. Phrae, NW Thailand
Phra Nakhon Si Ayutthaya *see* Ayutthaya
Phra Thong, Ko *137 B6 island* SW Thailand
Phuket *137 B7 var.* Bhuket, Puket, *Mal.* Ujung Salang; *prev.* Junkseylon, Salang. Phuket, SW Thailand
Phuket, Ko *137 B7 island* SW Thailand
Phumĭ Kâmpóng Trâbêk *see* Kâmpóng Trâbêk
Phumĭ Sâmraông *see* Samraong
Phum Kompong Trabek *see* Kâmpóng Trâbêk
Phum Samrong *see* Samraong
Phu Vinh *see* Tra Vinh
Phyu *136 B4 var.* Hpyu, Pyu. Bago, C Myanmar (Burma)
Piacenza *96 B2 Fr.* Paisance; *anc.* Placentia. Emilia-Romagna, N Italy
Piatra-Neamţ *108 C4 Hung.* Karácsonkő. Neamţ, NE Romania
Piatykhatky *109 F3 prev.* P"yatykhatky, *Rus.* Pyatikhatki. Dnipropetrovska Oblast', E Ukraine
Piauhy *see* Piauí
Piauí *63 F2 off.* Estado do Piauí; *prev.* Piauhy. *state/region* E Brazil
Piauí, Estado do *see* Piauí
Picardie *90 C3 Eng.* Picardy. *cultural region* N France
Picardy *see* Picardie
Piccolo San Bernardo, Colle di *see* Little Saint Bernard Pass
Pichilemu *64 B4* Libertador, C Chile
Pico *92 A5 var.* Ilha do Pico. *island* Azores, Portugal, NE Atlantic Ocean
Pico, Ilha do *see* Pico
Picos *63 F2* Piauí, E Brazil
Picton *151 C5* Marlborough, South Island, New Zealand
Piedmont *see* Piemonte
Piedras Negras *51 E2 var.* Ciudad Porfirio Díaz. Coahuila, NE Mexico
Pielavesi *84 D4 lake* C Finland
Pielinen *84 E4 var.* Pielisjärvi. *lake* E Finland
Pielisjärvi *see* Pielinen
Piemonte *96 A2 Eng.* Piedmont. *region* NW Italy
Pierre *45 E3 state capital* South Dakota, N USA
Piešťany *99 C6 Ger.* Pistyan, *Hung.* Pöstyén. Tranavský Kraj, W Slovakia
Pietarhovi *see* Petrodvorets
Pietari *see* Sankt-Peterburg
Pietarsaari *see* Jakobstad
Pietermaritzburg *78 C5 var.* Maritzburg. KwaZulu/Natal, E South Africa
Pietersburg *see* Polokwane
Pigs, Bay of *see* Cochinos, Bahía de
Pihkva Järv *see* Pskov, Lake
Pijijiapán *51 G5* Chiapas, SE Mexico
Pikes Peak *44 C5 mountain* Colorado, C USA
Pikeville *40 D5* Kentucky, S USA
Pikinni *see* Bikini Atoll
Piła *98 B3 Ger.* Schneidemühl. Wielkopolskie, C Poland
Pilar *64 D3 var.* Villa del Pilar. Ñeembucú, S Paraguay
Pilcomayo, Río *57 C5 river* C South America
Pilos *see* Pýlos
Pilsen *see* Plzeň
Pilzno *see* Plzeň
Pinang *see* Pinang, Pulau, Peninsular Malaysia
Pinang *see* George Town
Pinang, Pulau *138 B3 var.* Penang, Pinang; *prev.* Prince of Wales Island. *island* Peninsular Malaysia
Pinar del Río *54 A2* Pinar del Río, W Cuba
Píndhos/Píndhos Óros *see* Píndos
Píndos *104 A4 var.* Píndhos Óros, *Eng.* Pindus Mountains; *prev.* Píndhos. *mountain range* C Greece
Pindus Mountains *see* Píndos
Pine Bluff *42 B2* Arkansas, C USA
Pine Creek *146 D2* Northern Territory, N Australia

Pinega *110 C3 river* NW Russia
Pineiós *104 B4 var.* Piniós; *anc.* Peneius. *river* C Greece
Pineland *49 H3* Texas, SW USA
Pínes, Akrotírio *104 C4 var.* Akrotírio Pínnes. *headland* N Greece
Pines, The Isle of the *see* Juventud, Isla de la
Pine Tree State *see* Maine
Pingdingshan *128 C4* Henan, C China
Pingkiang *see* Harbin
Ping, Mae Nam *136 B4 river* W Thailand
Piniós *see* Pineiós
Pinkiang *see* Harbin
Pínnes, Akrotírio *see* Pínes, Akrotírio
Pinos, Isla de *see* Juventud, Isla de la
Pinotepa Nacional *51 F5 var.* Santiago Pinotepa Nacional. Oaxaca, SE Mexico
Pinsk *107 B6 Pol.* Pińsk. Brestskaya Voblasts', SW Belarus
Pinta, Isla *60 A5 var.* Abingdon. *island* Galápagos Islands, Ecuador, E Pacific Ocean
Piombino *96 B3* Toscana, C Italy
Pioneer Mountains *46 D3 mountain range* Montana, N USA North America
Pionerskiy *106 A4 Ger.* Neukuhren. Kaliningradskaya Oblast', W Russia
Piopiotahi *see* Milford Sound
Piotrków Trybunalski *98 D4 Ger.* Petrikau, *Rus.* Petrokov. Łodzkie, C Poland
Piraeus/Piraiévs *see* Peiraías
Pírgos *see* Pýrgos
Pirineos *see* Pyrenees
Piripiri *63 F2* Piauí, E Brazil
Pirna *94 D4* Sachsen, E Germany
Pirot *101 E5* Serbia, SE Serbia
Piryatin *see* Pyriatyn
Pisa *96 B3 var.* Pisae. Toscana, C Italy
Pisae *see* Pisa
Pisaurum *see* Pesaro
Pisco *60 D4* Ica, SW Peru
Písek *99 A5* Budějovický Kraj, S Czechia (Czech Republic)
Pishan *126 A3 var.* Guma. Xinjiang Uygur Zizhiqu, NW China
Pishpek *see* Bishkek
Pistoia *96 B3 anc.* Pistoria, Pistoriæ. Toscana, C Italy
Pistoria/Pistoriæ *see* Pistoia
Pistyan *see* Piešťany
Pisz *98 D3 Ger.* Johannisburg. Warmińsko-Mazurskie, NE Poland
Pita *74 C4* NW Guinea
Pitalito *58 B4* Huila, S Colombia
Pitcairn Group of Islands *see* Pitcairn, Henderson, Ducie & Oeno Islands
Pitcairn Island *143 G4 island* Pitcairn, Henderson, Ducie & Oeno Islands
Pitcairn, Henderson, Ducie & Oeno Islands *143 G4 var.* Pitcairn Group of Islands. *UK Overseas Territory* C Pacific Ocean
Piteå *84 D4* Norrbotten, N Sweden
Piteşti *108 B5* Argeş, S Romania
Pitsanulok *see* Phitsanulok
Pitt Island *see* Makin
Pittsburg *45 F5* Kansas, C USA
Pittsburgh *41 E4* Pennsylvania, NE USA
Pittsfield *41 F3* Massachusetts, NE USA
Piura *60 B2* Piura, NW Peru
Pivdennyi Buh *109 E3 prev.* Pivdennyy Buh, *Rus.* Yuzhnyy Bug. *river* S Ukraine
Pivdennyy Buh *see* Pivdennyi Buh
Placentia *see* Piacenza
Placetas *54 B2* Villa Clara, C Cuba
Plainview *49 E2* Texas, SW USA
Pláka *105 C7* Kykládes, Greece, Aegean Sea
Planeta Rica *58 B2* Córdoba, NW Colombia
Planken *72 C1* Liechtenstein Europe
Plano *49 G2* Texas, SW USA
Plasencia *92 C3* Extremadura, W Spain
Plate, River *see* Plata, Río de la
Plate, River *64 D4 var.* River Plate. *estuary* Argentina/Uruguay
Plate, River *see* Plata, Río de la
Platinum *36 C3* Alaska, USA
Plattensee *see* Balaton
Platte River *45 E4 river* Nebraska, C USA
Plattsburgh *41 F2* New York, NE USA
Plauen *95 C5 var.* Plauen im Vogtland. Sachsen, E Germany
Plauen im Vogtland *see* Plauen
Pļaviņas *106 D4 Ger.* Stockmannshof. S Latvia
Plây Cu *137 E5 var.* Pleiku. Gia Lai, C Vietnam
Pleasant Island *see* Nauru

Pleiku *see* Plây Cu
Plenty, Bay of *150 E3 bay* North Island, New Zealand
Plérin *90 A3* Côtes d'Armor, NW France
Plesetsk *110 C3* Arkhangel'skaya Oblast', NW Russia
Pleshchenitsy *see* Plyeshchanitsy
Pleskau *see* Pskov
Pleskauer See *see* Pskov, Lake
Pleskava *see* Pskov
Pleszew *98 C4* Wielkopolskie, C Poland
Pleven *104 C2 prev.* Plevna. Pleven, N Bulgaria
Plevlja/Plevlje *see* Pljevlja
Plevna *see* Pleven
Pljevlja *100 C4 prev.* Plevlja, Plevlje. N Montenegro
Plocce *see* Ploče
Ploče *100 B4 It.* Plocce; *prev.* Kardeljevo. Dubrovnik-Neretva, SE Croatia
Płock *98 D3 Ger.* Plozk. Mazowieckie, C Poland
Plöcken Pass *95 C7 Ger.* Plöckenpass, *It.* Passo di Monte Croce Carnico. *pass* SW Austria
Plöckenpass *see* Plöcken Pass
Ploești *see* Ploiești
Ploiești *108 C5 prev.* Ploești. Prahova, SE Romania
Plomári *105 D5 prev.* Plomárion. Lésvos, E Greece
Plomárion *see* Plomári
Płońsk *98 D3* Mazowieckie, C Poland
Plovdiv *104 C3 prev.* Eumolpias; *anc.* Evmolpia, Philippopolis, *Lat.* Trimontium. Plovdiv, C Bulgaria
Plozk *see* Płock
Plungé *106 B3* Telšiai, W Lithuania
Plyeshchanitsy *107 D5 Rus.* Pleshchenitsy. Minskaya Voblasts', N Belarus
Plymouth *89 C7* SW England, United Kingdom
Plzeň *99 A5 Ger.* Pilsen, *Pol.* Pilzno. Plzeňský Kraj, W Czechia (Czech Republic)
Po *80 D4 river* N Italy
Pobedy, Pik *see* Tömür Feng
Po, Bocche del *see* Po, Foci del
Pocahontas *42 B1* Arkansas, C USA
Pocatello *46 E4* Idaho, NW USA
Pochinok *111 A5* Smolenskaya Oblast', W Russia
Pocking *95 D6* Bayern, SE Germany
Poděbrady *99 B5 Ger.* Podiebrad. Středočeský Kraj, C Czechia (Czech Republic)
Podgorica *101 C5 prev.* Titograd. *country capital* (Montenegro) S Montenegro
Podiebrad *see* Poděbrady
Podilsk *108 D3 prev.* Kotovsk. Odes'ka Oblast', SW Ukraine
Podilska Vysochyna *108 D3 prev.* Podil's'ka Vysochina. *plateau* W Ukraine
Podil's'ka Vysochina *see* Podilska Vysochyna
Podium Anicensis *see* le Puy
Podol'sk *111 B5* Moskovskaya Oblast', W Russia
Podravska Slatina *see* Slatina
Podujevë *101 D5 Serb.* Podujevo. N Kosovo
Podujevo *see* Podujevë
Podunajská Rovina *see* Little Alföld
Poetovio *see* Ptuj
Pogradec *101 D6 var.* Pogradeci. Korçë, SE Albania
Pogradeci *see* Pogradec
Pohjanlahti *see* Bothnia, Gulf of
Pohnpei *144 C2 prev.* Ponape Ascension Island. *island* E Federated States of Micronesia
Poinciers *see* Poitiers
Poinsett, Cape *154 D4 headland* Antarctica
Point de Galle *see* Galle
Pointe-à-Pitre *55 G3* Grande Terre, C Guadeloupe
Pointe-Noire *77 B6* Kouilou, S Congo
Point Lay *36 C2* Alaska, USA
Poitiers *90 B4 prev.* Poictiers; *anc.* Limonum. Vienne, W France
Poitou *90 B4 cultural region* W France
Pokharā *135 E3* Western, C Nepal
Pokrov *109 F3 Rus.* Ordzhonikidze. Dnipropetrovska Oblast, E Ukraine
Pokrovka *see* Kyzyl-Suu

Pokrovske *109 G3 Rus.* Pokrovskoye. Dnipropetrovska Oblast, E Ukraine
Pokrovskoye *see* Pokrovske
Pola *see* Pula
Pola de Lena *see* La Pola
Poland *98 B4 off.* Republic of Poland, *var.* Polish Republic, *Pol.* Polska, Rzeczpospolita Polska; *prev. Pol.* Polska Rzeczpospolita Ludowa, The Polish People's Republic. *country* C Europe
Poland, Republic of *see* Poland
Polatlı *116 C3* Ankara, C Turkey (Türkiye)
Polatsk *107 D5 Rus.* Polotsk. Vitsyebskaya Voblasts', N Belarus
Pol-e Khomrī *see* Pul-e Khumrī
Polevoy *see* Pólis
Polikastro/Polikastron *see* Polýkastro
Polikrayshte *see* Dolna Oryahovitsa
Pólis *102 C5 var.* Poli. W Cyprus
Polish People's Republic, The *see* Poland
Polish Republic *see* Poland
Polkowice *98 B4 Ger.* Heerwegen. Dolnośląskie, W Poland
Pollença *93 G3* Mallorca, Spain, W Mediterranean Sea
Pologi *see* Polohy
Polohy *109 G3 Rus.* Pologi. Zaporiz'ka Oblast', SE Ukraine
Polokwane *78 D4 prev.* Pietersburg. Limpopo, NE South Africa
Polonne *108 D2 Rus.* Polonnoye. Khmel'nyts'ka Oblast', NW Ukraine
Polonnoye *see* Polonne
Polotsk *see* Polatsk
Polska/Polska, Rzeczpospolita/Polska Rzeczpospolita Ludowa *see* Poland
Polski Trambesh *104 D2 prev.* Polsko Kosovo, *var.* Polski Trümbesh. Ruse, N Bulgari
Polski Trümbesh *see* Polski Trambesh
Polsko Kosovo *see* Polski Trambesh
Poltava *109 F2* Poltavs'ka Oblast', NE Ukraine
Poltoratsk *see* Aşgabat
Põlva *106 E3 Ger.* Põlwe. Põlvamaa, SE Estonia
Põlwe *see* Põlva
Polyarnyy *110 C2* Murmanskaya Oblast', NW Russia
Polýkastro *104 B3 var.* Polikastro; *prev.* Polílkastron. Kentrikí Makedonía, N Greece
Polynesia *143 F4 island group* C Pacific Ocean
Pomeranian Bay *94 D2 Ger.* Pommersche Bucht, *Pol.* Zatoka Pomorska. *bay* Germany/Poland
Pomir, Dar"yoi *see* Pamir/Pāmīr, Daryā-ye
Pommersche Bucht *see* Pomeranian Bay
Pomorska, Zatoka *see* Pomeranian Bay
Pomorskiy Proliv *110 D2 strait* NW Russia
Po, Mouth of the *96 C2 var.* Bocche del Po. *river* NE Italy
Pompaelo *see* Pamplona
Pompano Beach *43 F5* Florida, SE USA
Ponape Ascension Island *see* Pohnpei
Ponca City *49 F1* Oklahoma, C USA
Ponce *55 F3* C Puerto Rico
Pondicherry *see* Puducherry
Pondichéry *see* Puducherry
Ponferrada *92 C1* Castilla y León, NW Spain
Poniatowa *98 E4* Lubelskie, E Poland
Pons Aelii *see* Newcastle upon Tyne
Pons Vetus *see* Pontevedra
Ponta Delgada *92 B5* São Miguel, Azores, Portugal, NE Atlantic Ocean
Ponta Grossa *63 E4* Paraná, S Brazil
Pontarlier *90 D4* Doubs, E France
Ponteareas *92 B2* Galicia, NW Spain
Ponte da Barca *92 B2* Viana do Castelo, N Portugal
Pontevedra *92 B1 anc.* Pons Vetus. Galicia, NW Spain
Pontiac *40 D3* Michigan, N USA
Pontianak *138 C4* Borneo, C Indonesia
Pontisarae *see* Pontoise
Pontivy *90 A3* Morbihan, NW France
Pontoise *90 C3 anc.* Briva Isarae, Cergy-Pontoise, Pontisarae. Val-d'Oise, N France
Ponziane Island *97 C5 island* C Italy
Poole *89 D7* S England, United Kingdom
Poona *see* Pune
Poopó, Lago *61 F4 var.* Lago Pampa Aullagas. *lake* W Bolivia

Popayán *58 B4* Cauca, SW Colombia
Poperinge *87 A6* West-Vlaanderen, W Belgium
Poplar Bluff *45 G5* Missouri, C USA
Popocatépetl *51 E4 volcano* S Mexico
Popper *see* Poprad
Poprad *99 D5 Ger.* Deutschendorf, *Hung.* Poprád. Prešovský Kraj, E Slovakia
Poprád *99 D5 Ger.* Popper, *Hung.* Poprád. *river* Poland/Slovakia
Porbandar *134 B4* Gujarāt, W India
Porcupine Plain *80 B3 undersea feature* E Atlantic Ocean
Pordenone *96 C2 anc.* Portenau. Friuli-Venezia Giulia, NE Italy
Poreč *100 A2 It.* Parenzo. Istra, NW Croatia
Porech'ye *see* Parechcha
Pori *85 D5 Swe.* Björneborg. Satakunta, SW Finland
Porirua *151 D5* Wellington, North Island, New Zealand
Porkhov *110 A4* Pskovskaya Oblast', W Russia
Porlamar *59 E1* Nueva Esparta, NE Venezuela
Póros *105 C6* Póros, S Greece
Póros *105 A5* Kefallinía, Iónia Nisiá, Greece, C Mediterranean Sea
Pors *see* Porsangenfjorden
Porsangenfjorden *84 D2 Smi.* Pors. *fjord* N Norway
Porsgrunn *85 B6* Telemark, S Norway
Portachuelo *61 G4* Santa Cruz, C Bolivia
Portadown *89 B5 Ir.* Port An Dúnáin.
Portalegre *92 C3 anc.* Ammaia, Amoea. Portalegre, E Portugal
Port Alexander *36 C4* Baranof Island, Alaska, USA
Port Alfred *78 D5* Eastern Cape, S South Africa
Port Amelia *see* Pemba
Port An Dúnáin *see* Portadown
Port Angeles *46 B1* Washington, NW USA
Port Antonio *54 B5* NE Jamaica
Port Arthur *49 H4* Texas, SW USA
Port Augusta *149 B6* South Australia
Port-au-Prince *54 D3 prev.* Pôtoprens. *country capital* (Haiti) C Haiti
Port Blair *133 F2* Andaman and Nicobar Islands, SE India
Port Charlotte *43 E4* Florida, SE USA
Port Darwin *see* Darwin
Port d'Envalira *91 B8* E Andorra Europe
Port Douglas *148 D3* Queensland, NE Australia
Port Elizabeth *see* Gqeberha
Portenau *see* Pordenone
Porterville *47 C7* California, W USA
Port-Étienne *see* Nouâdhibou
Port Florence *see* Kisumu
Port-Francqui *see* Ilebo
Port-Gentil *77 A6* Ogooué-Maritime, W Gabon
Port Harcourt *75 G5* Rivers, S Nigeria
Port Hardy *36 D5* Vancouver Island, British Columbia, SW Canada
Port Harrison *see* Inukjuak
Port Hedland *146 B4* Western Australia
Port Huron *40 D3* Michigan, N USA
Portimão *92 B4 var.* Vila Nova de Portimão. Faro, S Portugal
Port Jackson *148 E1 harbour* New South Wales, E Australia
Portland *149 B7* Victoria, SE Australia
Portland *41 G2* Maine, NE USA
Portland *46 B3* Oregon, NW USA
Portland *49 G4* Texas, SW USA
Portland Bight *54 B5 bay* S Jamaica
Portlaoighise *see* Port Laoise
Port Laoise *89 B6 var.* Portlaoise, *Ir.* Portlaoighise; *prev.* Maryborough. C Ireland
Portlaoise *see* Port Laoise
Port Lavaca *49 G4* Texas, SW USA
Port Lincoln *149 A6* South Australia
Port Louis *79 H3 country capital* (Mauritius) NW Mauritius
Port-Lyautey *see* Kenitra
Port Macquarie *149 E6* New South Wales, SE Australia
Port Mahon *see* Mahón
Portmore *54 B5* C Jamaica
Port Moresby *144 B3 country capital* (Papua New Guinea) Central/National Capital District, SW Papua New Guinea
Port Natal *see* Durban

Porto *92 B2 Eng.* Oporto; *anc.* Portus Cale. Porto, NW Portugal
Porto Alegre *63 E5 var.* Pôrto Alegre. *state capital* Rio Grande do Sul, S Brazil
Porto Alegre *76 E2* São Tomé, S Sao Tome and Principe, Africa
Port Alexandre *see* Tombua
Porto Amélia *see* Pemba
Porto Bello *see* Portobelo
Portobelo *53 G4 var.* Porto Bello, Puerto Bello. Colón, N Panama
Port O'Connor *49 G4* Texas, SW USA
Porto Edda *see* Sarandë
Portoferraio *96 B4* Toscana, C Italy
Port of Spain *55 H5 country capital* (Trinidad and Tobago) Trinidad, Trinidad and Tobago
Porto Grande *see* Mindelo
Portogruaro *96 C2* Veneto, NE Italy
Porto-Novo *75 F5 country capital* (Benin - official) S Benin
Porto Rico *see* Puerto Rico
Porto Santo *70 A2 var.* Ilha do Porto Santo. *island* Madeira, Portugal, NE Atlantic Ocean
Porto Santo, Ilha do *see* Porto Santo
Porto Torres *97 A5* Sardegna, Italy, C Mediterranean Sea
Porto Velho *62 D2 var.* Velho. *state capital* Rondônia, W Brazil
Portoviejo *60 A2 var.* Puertoviejo. Manabí, W Ecuador
Port Pirie *149 B6* South Australia
Port Rex *see* East London
Port Said *see* Bûr Sa'îd
Portsmouth *89 D7* S England, United Kingdom
Portsmouth *41 G3* New Hampshire, NE USA
Portsmouth *40 D4* Ohio, N USA
Portsmouth *41 F5* Virginia, NE USA
Port Stanley *see* Stanley
Port Sudan *72 C3* Red Sea, NE Sudan
Port Swettenham *see* Klang/ Pelabuhan Klang
Port Talbot *89 C7* S Wales, United Kingdom
Portugal *92 B3 off.* Portuguese Republic. *country* SW Europe
Portuguese East Africa *see* Mozambique
Portuguese Guinea *see* Guinea-Bissau
Portuguese Republic *see* Portugal
Portuguese Timor *see* East Timor
Portuguese West Africa *see* Angola
Portus Cale *see* Porto
Portus Magnus *see* Almería
Portus Magonis *see* Mahón
Port-Vila *144 D4 var.* Vila. *country capital* (Vanuatu) Éfaté, C Vanuatu
Porvenir *61 E3* Pando, NW Bolivia
Porvenir *65 B8* Magallanes, S Chile
Porvoo *85 E6 Swe.* Borgå. Uusimaa, S Finland
Porzecze *see* Parechcha
Posadas *64 D3* Misiones, NE Argentina
Posad-Pokrovske *109 E4* Khersonska Oblast, S Ukraine
Poschega *see* Požega
Posen *see* Poznań
Posnania *see* Poznań
Postovy/Postawy *see* Pastavy
Posterholt *87 D5* Limburg, SE Netherlands
Postojna *95 D8 Ger.* Adelsberg, *It.* Postumia. SW Slovenia
Postumia *see* Postojna
Pöstyén *see* Piešťany
Potamós *105 C7* Antikýthira, S Greece
Potentia *see* Potenza
Potenza *97 D5 anc.* Potentia. Basilicata, S Italy
Poti *117 F2 prev.* P'ot'i. W Georgia
P'ot'i *see* Poti
Potiskum *75 G4* Yobe, NE Nigeria
Potomac River *41 E5 river* NE USA
Pòtoprens *see* Port-au-Prince
Potosí *61 F4* Potosí, S Bolivia
Potsdam *94 D3* Brandenburg, NE Germany
Potwar Plateau *134 C2 plateau* NE Pakistan
Poulterillsville *see* Pursat
Po, Valle del *see* Po Valley
Po Valley *96 C2* var. Valle del Po. *valley* N Italy
Považská Bystrica *99 C5 Ger.* Waagbistritz, *Hung.* Vágbeszterce. Trenčiansky Kraj, W Slovakia
Poverty Bay *see* Tūranganui-a-Kiwa

Póvoa de Varzim *92 B2* Porto, NW Portugal
Powder River *44 D2 river* Montana/ Wyoming, NW USA
Powell *44 C2* Wyoming, C USA
Powell, Lake *44 B5 lake* Utah, W USA
Požarevac *100 D4 Ger.* Passarowitz. Serbia, NE Serbia
Poza Rica *51 F4 var.* Poza Rica de Hidalgo. Veracruz-Llave, E Mexico
Poza Rica de Hidalgo *see* Poza Rica
Požega *100 D4 prev.* Slavonska Požega, *Ger.* Poschega, *Hung.* Pozsega. Požega-Slavonija, NE Croatia
Požega *100 D4* Serbia
Poznań *98 C3 Ger.* Posen, Posnania. Wielkopolskie, C Poland
Pozoblanco *92 D4* Andalucía, S Spain
Pozsega *see* Požega
Pozsony *see* Bratislava
Pozzallo *97 C8* Sicilia, Italy, C Mediterranean Sea
Prachatice *99 A5 Ger.* Prachatitz. Jihočeský Kraj, S Czechia (Czech Republic)
Prachatitz *see* Prachatice
Prado del Ganso *see* Goose Green
Prae *see* Phrae
Prag/Praga/Prague *see* Praha
Praha *99 A5 Eng.* Prague, *Ger.* Prag, *Pol.* Praga. *country capital* (Czechia (Czech Republic)) Středočeský Kraj, NW Czechia (Czech Republic)
Praia *74 A3 country capital* (Cape Verde) Santiago, S Cape Verde (Cabo Verde)
Prairie State *see* Illinois
Prathet Thai *see* Thailand
Prato *96 B3* Toscana, C Italy
Pratt *45 E5* Kansas, C USA
Prattville *42 D2* Alabama, S USA
Pravda *see* Glavinitsa
Pravia *92 C1* Asturias, N Spain
Prayagraj *133 E3 prev.* Allahābād. Uttar Pradesh, N India
Preăh Seihânŭ *see* Sihanoukville
Preny *see* Prienai
Prenzlau *94 D3* Brandenburg, NE Germany
Prerau *see* Přerov
Přerov *99 C5 Ger.* Prerau. Olomoucký Kraj, E Czechia (Czech Republic)
Preschau *see* Prešov
Prescott *48 B2* Arizona, SW USA
Prešovo *101 D5* Serbia, SE Serbia
Presidente Epitácio *63 E4* São Paulo, S Brazil
Presidente Stroessner *see* Ciudad del Este
Prešov *99 D5 var.* Preschau, *Ger.* Eperies, *Hung.* Eperjes. Prešovský Kraj, E Slovakia
Prespa, Lake *101 D6 Alb.* Liqeni i Prespës, *Gk.* Límni Megáli Préspa, Límni Prespa, *Mac.* Prespansko Ezero, *Serb.* Prespansko Jezero. *lake* SE Europe
Prespa, Limni/Prespansko Ezero/ Prespansko Jezero/Prespës, Liqen i *see* Prespa, Lake
Presque Isle *41 H1* Maine, NE USA
Pressburg *see* Bratislava
Preston *89 D5* NW England, United Kingdom
Prestwick *88 C4* W Scotland, United Kingdom
Pretoria *78 D4 var.* Epitoli. *country capital* (South Africa-administrative capital) Gauteng, NE South Africa
Preussisch Eylau *see* Bagrationovsk
Preußisch Holland *see* Pasłęk
Preussisch-Stargard *see* Starogard Gdański
Préveza *105 A5* Ípeiros, W Greece
Pribilof Islands *36 A3 island group* Alaska, USA
Priboj *100 C4* Serbia, W Serbia
Price *44 B4* Utah, W USA
Prichard *42 C3* Alabama, S USA
Priekulė *106 B3 Ger.* Prökuls. Klaipėda, W Lithuania
Prienai *107 B5 Pol.* Preny. Kaunas, S Lithuania
Prieska *78 C4* Northern Cape, C South Africa
Prijedor *100 B3* Republika Srpska, NW Bosnia and Herzegovina
Prijepolje *100 D4* Serbia, W Serbia
Prikaspiyskaya Nizmennost' *see* Caspian Depression

Prilep *101 D6 Turk.* Perlepe. S North Macedonia
Priluki *see* Pryluky
Primorsk *106 A4 Ger.* Fischhausen. Kaliningradskaya Oblast', W Russia
Primorsko *104 E2 prev.* Keupriya. Burgas, E Bulgaria
Primorskoye *see* Prymorsk
Prince Albert *37 F5* Saskatchewan, S Canada
Prince Edward Island *39 F4 Fr.* Île-du Prince-Édouard. *province* SE Canada
Prince Edward Islands *69 E8 island group* S South Africa
Prince George *37 E5* British Columbia, SW Canada
Prince of Wales Island *148 B1 island* Queensland, E Australia
Prince of Wales Island *37 F2 island* Queen Elizabeth Islands, Nunavut, NW Canada
Prince of Wales Island *see* Pinang, Pulau
Prince Patrick Island *37 E2 island* Parry Islands, Northwest Territories, NW Canada
Prince Rupert *36 D4* British Columbia, SW Canada
Prince's Island *see* Príncipe
Princess Charlotte Bay *148 C2 bay* Queensland, NE Australia
Princess Elizabeth Land *154 C3 physical region* Antarctica
Príncipe *77 A5 var.* Príncipe Island, *Eng.* Prince's Island. *island* N Sao Tome and Principe
Príncipe Island *see* Príncipe
Prinzapolka *53 E3* Región Autónoma Atlántico Norte, NE Nicaragua
Pripet *see* Pripyat
Pripet Marshes *see* Pripyat Marshes
Pripyat *107 C7 var.* Pripet, *Bel.* Prypyats', *Ukr.* Prypiat. *river* Belarus/Ukraine
Pripyat Marshes *107 B7 bel.* Prypyatskiya baloty, *Ukr.* Prypiatski bolota. *wetland* Belarus/Ukraine
Prishtinë *101 D5 Eng.* Pristina, *Serb.* Priština. *country capital* (Kosovo) C Kosovo
Pristina/Priština *see* Prishtinë
Privas *91 D5* Ardèche, E France
Privolzhskaya Vozvyshennost' *81 G3 var.* Volga Uplands. *mountain range* W Russia
Prizren *101 D5* S Kosovo
Probolinggo *138 D5* Jawa, C Indonesia
Probstberg *see* Wyszków
Progreso *51 H3* Yucatán, SE Mexico
Prokhladnyy *111 B8* Kabardino-Balkarskaya Respublika, SW Russia
Prokletije *see* North Albanian Alps
Prökuls *see* Priekulė
Prokuplje *101 D5* Serbia, SE Serbia
Prome *see* Pyay
Promyshlennyy *110 E3* Respublika Komi, NW Russia
Prościejów *see* Prostějov
Proskurov *see* Khmelnytskyi
Prossnitz *see* Prostějov
Prostějov *99 C5 Ger.* Prossnitz, *Pol.* Prościejów. Olomoucký Kraj, E Czechia (Czech Republic)
Provence *91 D6 cultural region* SE France
Providence *41 G3 state capital* Rhode Island, NE USA
Providence *see* Fort Providence
Providencia, Isla de *53 F3 island* NW Colombia, Caribbean Sea
Provideniya *155 B1* Chukotskiy Avtonomnyy Okrug, NE Russia
Provo *44 B4* Utah, W USA
Prudhoe Bay *36 D2* Alaska, USA
Prusa *see* Bursa
Pruszków *98 D3 Ger.* Kaltdorf. Mazowieckie, C Poland
Prut *108 D2 var.* Pruth. *river* E Europe
Pruth *see* Prut
Pružana *see* Pruzhany
Pruzhany *107 B6 Pol.* Pružana. Brestskaya Voblasts', SW Belarus
Prychornomor'ska Nyzovyna *see* Black Sea Lowland
Prydniprovska Nyzovyna/ Prydnyaprowskaya Nizina *see* Dnieper Lowland
Prydz Bay *154 D3 bay* Antarctica
Pryluky *109 E2 Rus.* Priluki. Chernihivs'ka Oblast', NE Ukraine
Prymorsk *109 G4 prev.* Primorskoye. Zaporiz'ka Oblast', SE Ukraine

Prypiatski bolota *see* Pripyat Marshes
Pryp"yat'/Prypyats' *see* Pripyat
Prypyatskiya baloty *see* Pripyat Marshes
Przemyśl *99 E5 Rus.* Peremyshl. Podkarpackie, C Poland
Przheval'sk *see* Karakol
Psará *105 D5 island* E Greece
Psel *109 F2 Rus.* Psël. *river* Russia/Ukraine
Psël *see* Psel
Pskov *114 B2 Ger.* Pleskau, *Latv.* Pleskava. Pskovskaya Oblast', W Russia
Pskov, Lake *106 E3 Est.* Pihkva Järv, *Ger.* Pleskauer See, *Rus.* Pskovskoye Ozero. *lake* Estonia/Russia
Pskovskoye Ozero *see* Pskov, Lake
Ptich' *see* Ptsich
Ptsich *107 C7 Rus.* Ptich'. Homyel'skaya Voblasts', SE Belarus
Ptsich *107 C7 Rus.* Ptich'. *river* SE Belarus
Ptuj *95 E7 Ger.* Pettau; *anc.* Poetovio. NE Slovenia
Pucallpa *60 C3* Ucayali, C Peru
Puck *98 C2* Pomorskie, N Poland
Pudasjärvi *84 D4* Pohjois-Pohjanmaa, C Finland
Puduchcheri *see* Puducherry
Puducherry *132 C2 prev.* Pondicherry; *var.* Puduchcheri, *Fr.* Pondichéry. Pondicherry, SE India
Puebla *51 F4 var.* Puebla de Zaragoza. Puebla, S Mexico
Puebla de Zaragoza *see* Puebla
Pueblo *44 D5* Colorado, C USA
Puerto Acosta *61 E4* La Paz, W Bolivia
Puerto Aisén *65 B6* Aisén, S Chile
Puerto Ángel *51 F5* Oaxaca, SE Mexico
Puerto Argentino *see* Stanley
Puerto Ayacucho *58 D3* Amazonas, SW Venezuela
Puerto Baquerizo Moreno *60 B5 var.* Baquerizo Moreno. Galápagos Islands, Ecuador, E Pacific Ocean
Puerto Barrios *52 C2* Izabal, E Guatemala
Puerto Bello *see* Portobelo
Puerto Berrío *58 B2* Antioquia, C Colombia
Puerto Cabello *58 D1* Carabobo, N Venezuela
Puerto Cabezas *53 E2 var.* Bilwi. Región Autónoma Atlántico Norte, NE Nicaragua
Puerto Carreño *58 D3* Vichada, E Colombia
Puerto Cortés *52 C2* Cortés, NW Honduras
Puerto Cumarebo *58 C1* Falcón, N Venezuela
Puerto Deseado *65 C7* Santa Cruz, SE Argentina
Puerto Escondido *51 F5* Oaxaca, SE Mexico
Puerto Francisco de Orellana *60 B1 var.* Coca. NE Ecuador
Puerto Gallegos *see* Río Gallegos
Puerto Inírida *58 D3 var.* Obando. Guainía, E Colombia
Puerto La Cruz *59 E1* Anzoátegui, NE Venezuela
Puerto Lempira *53 E2* Gracias a Dios, E Honduras
Puerto Limón *see* Limón
Puertollano *92 D4* Castilla-La Mancha, C Spain
Puerto López *58 C1* La Guajira, N Colombia
Puerto Maldonado *61 E3* Madre de Dios, E Peru
Puerto México *see* Coatzacoalcos
Puerto Montt *65 B5* Los Lagos, C Chile
Puerto Natales *65 B7* Magallanes, S Chile
Puerto Obaldía *53 H5* Kuna Yala, NE Panama
Puerto Plata *55 E3 var.* San Felipe de Puerto Plata. N Dominican Republic
Puerto Presidente Stroessner *see* Ciudad del Este
Puerto Princesa *139 E2 off.* Puerto Princesa City. Palawan, W Philippines
Puerto Princesa City *see* Puerto Princesa
Puerto Príncipe *see* Camagüey
Puerto Rico *55 F3 off.* Commonwealth of Puerto Rico; *prev.* Porto Rico. *US commonwealth territory* C West Indies
Puerto Rico *56 B1 island* C West Indies
Puerto Rico, Commonwealth of *see* Puerto Rico

Puerto Rico Trench *56 B1 trench* NE Caribbean Sea
Puerto San José *see* San José
Puerto San Julián *65 B7 var.* San Julián. Santa Cruz, SE Argentina
Puerto Suárez *61 H4* Santa Cruz, E Bolivia
Puerto Vallarta *50 D4* Jalisco, SW Mexico
Puerto Varas *65 B5* Los Lagos, C Chile
Puerto Viejo *53 E4* Heredia, NE Costa Rica
Puertoviejo *see* Portoviejo
Puget Sound *46 B1 sound* Washington, NW USA
Puglia *97 E5 var.* Le Puglie, *Eng.* Apulia. *region* SE Italy
Puhiwaero *see* South West Cape
Pukaki, Lake *151 B6 lake* South Island, New Zealand
Pukekohe *150 D3* Auckland, North Island, New Zealand
Puket *see* Phuket
Pukhavichy *107 C6 Rus.* Pukhovichi. Minskaya Voblasts', C Belarus
Pukhovichi *see* Pukhavichy
Pula *100 A3 It.* Pola; *prev.* Pulj. Istra, NW Croatia
Pulaski *40 D5* Virginia, NE USA
Puławy *98 D4 Ger.* Neu Amerika. Lubelskie, E Poland
Pu-le-Khumri *123 E4 prev.* Pol-e Khomrī, *var.* Pul-i-Khumri. Baghlān, NE Afghanistan
Pul-i-Khumri *see* Pu-le Khumri
Pulj *see* Pula
Pullman *46 C2* Washington, NW USA
Pułtusk *98 D3* Mazowieckie, C Poland
Puná, Isla *60 A2 island* SW Ecuador
Pune *134 C5 prev.* Poona. Mahārāshtra, W India
Punjab *134 C2 prev.* West Punjab, Western Punjab. *province* E Pakistan
Puno *61 E4* Puno, SE Peru
Punta Alta *65 C5* Buenos Aires, E Argentina
Punta Arenas *65 B8 prev.* Magallanes. Magallanes, S Chile
Punta Gorda *52 C2* Toledo, SE Belize
Punta Gorda *53 E4* Región Autónoma Atlántico Sur, SE Nicaragua
Puntarenas *52 D4* Puntarenas, W Costa Rica
Punto Fijo *58 C1* Falcón, N Venezuela
Pupuya, Nevado *61 E4 mountain* W Bolivia
Puri *135 F5 var.* Jagannath. Odisha, E India
Puriramya *see* Buriram
Purmerend *86 C3* Noord-Holland, C Netherlands
Pursat *137 D6 Khmer.* Poŭthĭsăt. Pursat, W Cambodia
Purus, Rio *62 C2 var.* Río Purús. *river* Brazil/Peru
Pusan *see* Busan
Pushkino *see* Biläsuvar
Püspökladány *99 D6* Hajdú-Bihar, E Hungary
Putorana, Gory/Putorana Mountains *see* Putorana, Plato
Putorana Mountains *115 E3 var.* Gory Putorana, *Eng.* Putorana Mountains. *mountain range* N Russia
Putrajaya *138 B3* Kuala Lumpur, Peninsular Malaysia
Puttalam *132 C3* North Western Province, W Sri Lanka
Puttgarden *94 C2* Schleswig-Holstein, N Germany
Putumayo, Río *58 B5 var.* Içá, Rio. *river* NW South America
Puurmani *106 D2 Ger.* Talkhof. Jõgevamaa, E Estonia
Pyatigorsk *111 B7* Stavropol'skiy Kray, SW Russia
Pyatikhatki *see* Piatykhatky
Pʺyatykhatky *see* Piatykhatky
Pyay *136 B4 var.* Prome, Pye. Bago, C Myanmar (Burma)
Pye *see* Pyay
Pyetrykaw *107 C7 Rus.* Petrikov. Homyel'skaya Voblasts', SE Belarus
Pyin-Oo-Lwin *136 B3 var.* Maymyo. Mandalay, C Myanmar (Burma)
Pýlos *105 B6 var.* Pilos. Pelopónnisos, S Greece

P'yŏngyang *129 E3 var.*
P'yŏngyang-jikhalsi, *Eng.* Pyongyang.
country capital (North Korea)
SW North Korea
P'yŏngyang-jikhalsi *see* P'yŏngyang
Pyramid Lake *47 C5 lake* Nevada,
W USA
Pyrenaei Montes *see* Pyrenees
Pyrenees *102 B2 Fr.* Pyrénées, *Sp.*
Pirineos; *anc.* Pyrenaei Montes.
mountain range SW Europe
Pýrgos *105 B6 var.* Pírgos. Dytikí Ellás,
S Greece
Pyriatyn *109 E2 prev.* Pyryatyn, *Rus.*
Piryatin. Poltavs'ka Oblast',
NE Ukraine
Pyritz *see* Pyrzyce
Pyryatyn *see* Pyriatyn
Pyrzyce *98 B3 Ger.* Pyritz.
Zachodnio-pomorskie, NW Poland
Pyu *see* Phyu
Pyuntaza *136 B4* Bago,
SW Myanmar (Burma)

Qā' al Jafr *119 C7 lake* S Jordan
Qaanaaq *82 D1 var.* Qânâq, *Dan.* Thule.
Avannaata, N Greenland
Qabātiyah *119 E6* N West Bank, SW Asia
Qābis *see* Gabès
Qābis, Khalīj *see* Gabès, Golfe de
Qacentina *see* Constantine
Qafşah *see* Gafsa
Qagan Us *see* Dulan
Qahremānshahr *see* Kermānshāh
Qaidam Pendi *126 C4 basin* C China
Qal'aikhum *see* Qalaikhumb
Qalaikhumb *123 F3 prev.* Qal'aikhum,
Rus. Kalaikhum. S Tajikistan
Qalāt *123 D4 Pash.* Kalāt. Zābol,
S Afghanistan
Qal'at Bīshah *121 B5* 'Asīr,
SW Saudi Arabia
Qalqīliya *see* Qalqīlyah
Qalqīlyah *119 D6 var.* Qalqiliya. Central,
W West Bank, SW Asia
Qamdo *126 D5* Xizang Zizhiqu, W China
Qamishly *see* Al Qāmishlī
Qânâq *see* Qaanaaq
Qaqortoq *82 C4 Dan.* Julianehåb.
Kujalleq, S Greenland
Qaraghandy *114 C4 var.* Karagandy;
prev. Karaganda. Karagandy,
C Kazakhstan
Qara Qum *see* Garagum
Qarataū *see* Karatau, Zhambyl,
Kazakhstan
Qarkilik *see* Ruoqiang
Qarokŭl *123 F3 Rus.* Karakul'.
E Tajikistan
Qaryannah, Juzur *see* Kerkenah, Îles de
Qars *see* Kars
Qarshi *123 E3 Rus.* Karshi; *prev.*
Bek-Budi. Qashqadaryo Viloyati,
S Uzbekistan
Qasigianguit *see* Qasigiannguit
Qasigiannguit *82 C3 var.* Qasigianguit,
Dan. Christianshåb. Qeqertalika,
C Greenland
Qaşr al Farāfirah *72 B2 var.* Qasr
Farâfra. W Egypt
Qasr Farâfra *see* Qaşr al Farāfirah
Qaţanā *119 B5 var.* Katana. Dimashq,
S Syria
Qatar *120 C4 off.* State of Qatar, *Ar.*
Dawlat Qatar. *country* SW Asia
Qatar, State of *see* Qatar
Qatorŭhii Hisor *see* Gissar Range
Qattara Depression *see* Qaţţārah,
Munkhafad al
Qaţţāra, Monkhafad el *see* Qaţţārah,
Munkhafad al
Qaţţārah, Munkhafad al *72 A1 var.*
Monkhafad el Qaţţâra, *Eng.* Qattara
Depression. *desert* NW Egypt
Qausuittuq *see* Resolute
Qazaqstan/Qazaqstan Respublikasy *see*
Kazakhstan
Qazimämmäd *see* Hacıqabul
Qazris *see* Cáceres
Qazvīn *120 C2 var.* Kazvin. Qazvīn,
N Iran
Qena *see* Qinā
Qeqertarsuaq *see* Qeqertarsuaq
Qeqertarsuaq *82 C3 var.* Qeqertarsuaq,
Dan. Godhavn. Qeqertalik,
S Greenland
Qeqertarsuaq *82 C3 island* W Greenland

Qeqertarsuup Tunua *82 C3 Dan.* Disko
Bugt. *inlet* W Greenland
Qerveh *see* Qorveh
Qeshm *120 D4 var.* Jazīreh-ye Qeshm,
Qeshm Island. *island* S Iran
Qeshm Island/Qeshm, Jazīreh-ye *see*
Qeshm
Qilian Shan *126 D3 var.* Kilien
Mountains. *mountain range* N China
Qimusseriarsuaq *82 C2 Dan.* Melville
Bugt, *Eng.* Melville Bay. *bay*
NW Greenland
Qinā *72 B2 var.* Qena; *anc.* Caene,
Caenepolis. E Egypt
Qing *see* Qinghai
Qingdao *128 D4 var.* Ching-Tao,
Ch'ing-tao, Tsingtao, Tsintao, *Ger.*
Tsingtau. Shandong, E China
Qinghai *126 C4 var.* Chinghai, Koko
Nor, Qing, Qinghai Sheng, Tsinghai.
province C China
Qinghai Hu *126 D4 var.* Ch'ing Hai,
Tsing Hai, *Mong.* Koko Nor. *lake*
C China
Qinghai Sheng *see* Qinghai
Qingzang Gaoyuan *126 B4 var.* Xizang
Gaoyuan, *Eng.* Plateau of Tibet. *plateau*
W China
Qinhuangdao *128 D3* Hebei, E China
Qinzhou *129 B6* Guangxi Zhuangzu
Zizhiqu, S China
Qiong *see* Hainan
Qiqihar *128 D2 var.* Ch'i-ch'i-ha-erh,
Tsitsihar; *prev.* Lungkiang.
Heilongjiang, NE China
Qira *126 B4* Xinjiang Uygur Zizhiqu,
NW China
Qita Ghazzah *see* Gaza Strip
Qitai *126 C3* Xinjiang Uygur Zizhiqu,
NW China
Qīzān *see* Jīzān
Qizil Orda *see* Qyzylorda
Qizil Qum/Qizilqum *see*
Kyzylkum Desert
Qizilrabot *123 G3 Rus.* Kyzylrabot.
SE Tajikistan
Qogir Feng *see* K2
Qom *120 C3 var.* Kum, Qum. Qom,
N Iran
Qomolangma Feng *see* Everest, Mount
Qomul *see* Hami
Qo'qon *123 F2 var.* Khokand, *Rus.*
Kokand. Farg'ona Viloyati,
E Uzbekistan
Qorveh *120 C3 var.* Qerveh, Qurveh.
Kordestān, W Iran
Qostanay *114 C4 var.* Kostanay,
Kustanay. Kostanay, N Kazakhstan
Qoubaïyât *118 B4 var.* Al Qubayyāt.
N Lebanon
Qoussantina *see* Constantine
Quang Ngai *137 E5 var.* Quangngai,
Quang Nghia. Quang Ngai, C Vietnam
Quangngai *see* Quang Ngai
Quang Nghia *see* Quang Ngai
Quan Long *see* Ca Mau
Quanzhou *129 D6 var.* Ch'uan-chou,
Tsinkiang; *prev.* Chin-chiang. Fujian,
SE China
Quanzhou *129 C6* Guangxi Zhuangzu
Zizhiqu, S China
Qu'Appelle *37 F5 river* Saskatchewan,
S Canada
Quarles, Pegunungan *139 E4 mountain
range* Sulawesi, C Indonesia
Quarnero *see* Kvarner
Quartu Sant' Elena *97 A6* Sardegna,
Italy, C Mediterranean Sea
Quba *117 H2 Rus.* Kuba. N Azerbaijan
Québec *39 E4 var.* Quebec. *province*
capital Québec, SE Canada
Québec *39 E3 var.* Quebec. *province*
SE Canada
Queen Charlotte Islands *see*
Haida Gwaii
Queen Charlotte Sound *36 C5 sea area*
British Columbia, W Canada
Queen Elizabeth Islands *37 E1 Fr.* Îles
de la Reine-Élisabeth. *island group*
Nunavut, N Canada
Queensland *148 B4 state* N Australia
Queenstown *151 B7* Otago, South Island,
New Zealand
Queenstown *see* Komani
Quelimane *79 E3 var.* Kilimane,
Kilmain, Quilimane. Zambézia,
NE Mozambique
Quelpart *see* Jejudo
Quepos *53 E4* Puntarenas, S Costa Rica

Que Que *see* Kwekwe
Quera *see* Chur
Querétaro *51 E4* Querétaro de Arteaga,
C Mexico
Quesada *53 E4 var.* Ciudad Quesada,
San Carlos. Alajuela, N Costa Rica
Quetta *134 B2* Baluchistan, SW Pakistan
Quetzalcoalco *see* Coatzacoalcos
Quezaltenango *52 A2 var.*
Quetzaltenango. Quezaltenango,
W Guatemala
Quibdó *58 A3* Chocó, W Colombia
Quilimane *see* Quelimane
Quillabamba *60 D3* Cusco, C Peru
Quilon *see* Kollam
Quimper *90 A3 anc.* Quimper Corentin.
Finistère, NW France
Quimper Corentin *see* Quimper
Quimperlé *90 A3* Finistère, NW France
Quincy *40 A4* Illinois, N USA
Qui Nhon/Quinhon *see* Quy Nhơn
Quissico *79 E4* Inhambane,
S Mozambique
Quito *60 B1 country capital* (Ecuador)
Pichincha, N Ecuador
Qulyndy Zhazyghy *see* Kulunda Steppe
Qum *see* Qom
Qurein *see* Al Kuwayt
Qūrghonteppa *see* Bokhtar
Qurlurtuuq *see* Kugluktuk
Qurveh *see* Qorveh
Quşayr *see* Al Quşayr
Quxar *see* Lhazê
Quy Nhơn *137 E5 var.* Quinhon, Qui
Nhon. Binh Định, C Vietnam
Qyteti Stalin *see* Kuçovë
Qyzylorda *114 B5 var.* Kzylorda,
Kzyl-Orda, Qizil Orda; *prev.* Perovsk.
Kyzylorda, S Kazakhstan
Qyzylqum *see* Kyzylkum Desert

R

Raab *100 B1 Hung.* Rába. *river*
Austria/Hungary
Raab *see* Rába
Raab *see* Győr
Raahe *84 D4 Swe.* Brahestad.
Pohjois-Pohjanmaa, W Finland
Raalte *86 D3* Overijssel, E Netherlands
Raamsdonksveer *86 C4* Noord-Brabant,
S Netherlands
Raasiku *106 D2 Ger.* Rasik. Harjumaa,
NW Estonia
Rába *99 B7 Ger.* Raab. *river*
Austria/Hungary
Rába *see* Raab
Rabat *70 C2 var.* al Dar al Baida. *country
capital* (Morocco) NW Morocco
Rabat *102 B5* W Malta
Rabat *see* Victoria
Rabbah Ammon/Rabbath Ammon *see*
'Ammān
Rabinal *52 B2* Baja Verapaz,
C Guatemala
Rabka *99 D5* Małopolskie, S Poland
Rabnita *see* Rîbniţa
Rabyānah, Ramlat *71 G4 var.* Rebiana
Sand Sea, Şaḩrā' Rabyānah. *desert*
SE Libya
Rabyānah, Şaḩrā' *see* Rabyānah, Ramlat
Răcari *see* Durankulak
Race, Cape *39 H3 headland*
Newfoundland, Newfoundland and
Labrador, E Canada
Rach Gia *137 D6* Kiên Giang, S Vietnam
Rach Gia, Vinh *137 D6 bay* S Vietnam
Racine *40 B3* Wisconsin, N USA
Rácz-Becse *see* Bečej
Rădăuți *108 C3 Ger.* Radautz, *Hung.*
Rádóc. Suceava, N Romania
Radautz *see* Rădăuți
Rădeyilikóé *see* Fort Good Hope
Rádóc *see* Rădăuți
Radom *98 D4* Mazowieckie, C Poland
Radomsko *98 D4 Rus.* Novoradomsk.
Łódzkie, C Poland
Radomyshl *108 D2* Zhytomyrs'ka
Oblast', N Ukraine
Radoviš *101 E6 prev.* Radovište.
E North Macedonia
Radovište *see* Radoviš
Radviliškis *106 B4* Šiauliai, N Lithuania
Radzyń Podlaski *98 E4* Lubelskie,
E Poland
Rae-Edzo *see* Edzo
Raetihi *150 D4* Manawatu-Wanganui,
North Island, New Zealand
Rafa *see* Rafah

Rafaela *64 C3* Santa Fe, E Argentina
Rafah *119 A7 var.* Rafa, Rafaḩ, *Heb.*
Rafiaḩ, Raphiah. SW Gaza Strip
Rafḩah *120 B4* Al Ḥudūd ash
Shamālīyah, N Saudi Arabia
Rafiaḩ *see* Rafah
Raga *73 A5* W Bahr el Ghazal,
S South Sudan
Ragged Island Range *54 C2 island group*
S The Bahamas
Ragnit *see* Neman
Ragusa *97 C7* Sicilia, Italy,
C Mediterranean Sea
Ragusa *see* Dubrovnik
Rahachow *107 D7 Rus.* Rogachëv.
Homyel'skaya Voblasts', SE Belarus
Rahaeng *see* Tak
Raḥaṭ, Ḥarrat *121 B5 lava flow*
W Saudi Arabia
Rahimyar Khan *134 C3* Punjab,
SE Pakistan
Rahovec *101 D5 Serb.* Orahovac.
W Kosovo
Raiatea *145 G4 island* Îles Sous le Vent,
W French Polynesia
Rāichūr *132 C1* Karnātaka, C India
Raidestos *see* Tekirdağ
Rainier, Mount *34 A4 volcano*
Washington, NW USA
Rainy Lake *38 A4 lake* Canada/USA
Raipur *135 E4* Chhattisgarh, C India
Rājahmundry *135 E5* Andhra Pradesh,
E India
Rajang *see* Rajang, Batang
Rajang, Batang *138 D3 var.* Rajang.
river East Malaysia
Rājapālaiyam *132 C3* Tamil Nādu,
SE India
Rajasthān *134 C3 cultural region*
NW India
Rajin *129 E3 prev.* Najin. NE North Korea
Rājkot *134 C4* Gujarāt, W India
Rāj Nāndgaon *135 E4* Chhattisgarh,
C India
Rajshahi *135 G3 prev.* Rampur Boalia.
Rajshahi, N Bangladesh
Rakahanga *145 F3 atoll* N Cook Islands
Rakaia *151 B6 river* South Island,
New Zealand
Rakiura *see* Stewart Island
Rakka *see* Ar Raqqah
Rakke *106 E2 Ger.* Wesenberg.
Lääne-Virumaa, NE Estonia
Rakvere *106 E2 Ger.* Wesenberg.
Lääne-Virumaa, N Estonia
Raleigh *43 F1 state capital* North
Carolina, SE USA
Ralik Chain *144 D1 island group* Ralik
Chain, W Marshall Islands
Ramadi *see* Ar Ramādī
Ramallah *119 E7* C West Bank, SW Asia
Râmnicul-Sărat *see* Râmnicu Sărat
Râmnicu Sărat *108 C4 prev.*
Râmnicul-Sărat, *Hung.* Râmnicu-Sărat.
Buzău, E Romania
Râmnicu Vâlcea *108 B4 prev.* Rîmnicu
Vîlcea. Vâlcea, C Romania
Rampur Boalia *see* Rajshahi
Ramree Island *136 A4 island*
W Myanmar (Burma)
Ramtha *see* Ar Ramthā
Rancagua *64 B4* Libertador, C Chile
Rānchi *135 F4* Jhārkhand, N India
Randers *85 B7* Århus, C Denmark
Ránes *see* Ringvassøya
Rangiora *151 C6* Canterbury, South
Island, New Zealand
Rangitikei *150 D4 river* North Island,
New Zealand
Rangitoto ki te Tonga *150 C4 var.*
D'Urville Island. *island* C New Zealand
Rangoon *see* Yangon
Rangpur *135 G3* Rajshahi, N Bangladesh
Rankin Inlet *see* Kangiqliniq
Rankovićevo *see* Kraljevo
Ranong *137 B6* Ranong, SW Thailand
Rapa Nui *see* Pascua, Isla de
Raphiah *see* Rafah
Rapid City *44 D3* South Dakota, N USA
Räpina *106 E3 Ger.* Rappin. Põlvamaa,
SE Estonia
Rapla *106 D2 Ger.* Rappel. Raplamaa,
NW Estonia
Rappel *see* Rapla
Rappin *see* Räpina
Rarotonga *145 G5 island* S Cook Islands,
C Pacific Ocean
Ras al'Ain *see* Ra's al 'Ayn
Ra's al 'Ayn *118 D1 var.* Ras al'Ain. Al
Ḥasakah, N Syria

INDEX – GAZETTEER: Ra's an Naqb – Rochester

Ra's an Naqb 119 B7 Ma'ān, S Jordan
Raseiniai 106 B4 Kaunas, C Lithuania
Rasht 120 C2 var. Resht. Gīlān, NW Iran
Rasik see Raasiku
Râşnov 108 C4 prev. Rîşno, Rozsnyó, Hung. Barcarozsnyó. Braşov, C Romania
Rastenburg see Kętrzyn
Ratak Chain 144 D1 island group Ratak Chain, E Marshall Islands
Ratän 85 C5 Jämtland, C Sweden
Rat Buri see Ratchaburi
Ratchaburi 137 C5 var. Rat Buri. Ratchaburi, W Thailand
Ratisbon/Ratisbona/Ratisbonne see Regensburg
Rat Islands 36 A2 island group Aleutian Islands, Alaska, USA
Ratlām 134 D4 prev. Rutlam. Madhya Pradesh, C India
Ratnapura 132 D4 Sabaragamuwa Province, S Sri Lanka
Raton 48 D1 New Mexico, SW USA
Rättvik 85 C5 Dalarna, C Sweden
Raudhatain see Ar Rawḍatayn
Raufarhöfn 83 E4 Nordhurland Eystra, NE Iceland
Raukawa see Cook Strait
Raukumara Range 150 E3 mountain range North Island, New Zealand
Rāulakela see Rāurkela
Rauma 85 D5 Swe. Raumo. Satakunta, SW Finland
Raumo see Rauma
Rāurkela 135 F4 var. Rāulakela, Rourkela. Odisha, E India
Ravenna 96 C3 Emilia-Romagna, N Italy
Ravi 134 C2 river India/Pakistan
Rawalpindi 134 C1 Punjab, NE Pakistan
Rawa Mazowiecka 98 D4 Łódzkie, C Poland
Rawicz 98 C4 Ger. Rawitsch. Wielkopolskie, C Poland
Rawitsch see Rawicz
Rawlins 44 C3 Wyoming, C USA
Rawson 65 C6 Chubut, SE Argentina
Rayak 118 B4 var. Rayaq, Riyāq. E Lebanon
Rayaq see Rayak
Rayong 137 C5 Rayong, S Thailand
Razazah, Buhayrat ar 120 B3 var. Baḥr al Milḥ. lake C Iraq
Razdolnoye see Rozdolne
Razelm, Lacul see Razim, Lacul
Razgrad 104 D2 Razgrad, N Bulgaria
Razim, Lacul 108 D5 prev. Lacul Razelm. lagoon NW Black Sea
Reading 89 D7 S England, United Kingdom
Reading 41 F4 Pennsylvania, NE USA
Realicó 64 C4 La Pampa, C Argentina
Reăng Kesei 137 D5 Battambang, W Cambodia
Greater Antarctica see East Antarctica
Rebecca, Lake 147 C6 lake Western Australia
Rebiana Sand Sea see Rabyānah, Ramlat
Rebun-to 130 C2 island NE Japan
Rechitsa see Rechytsa
Rechytsa 107 D7 Rus. Rechitsa. Brestskaya Voblasts', SW Belarus
Recife 63 G2 prev. Pernambuco. state capital Pernambuco, E Brazil
Recklinghausen 94 A4 Nordrhein-Westfalen, W Germany
Recogne 87 C7 Luxembourg, SE Belgium
Reconquista 64 D3 Santa Fe, C Argentina
Red Deer 37 E5 Alberta, SW Canada
Redding 47 B5 California, W USA
Redon 90 B4 Ille-et-Vilaine, NW France
Red River 45 E1 river Canada/USA
Red River 136 C2 var. Yuan, Chin. Yuan Jiang, Vtn. Sông Hông Hà. river China/Vietnam
Red River 35 C6 river S USA
Red River 42 B3 river Louisiana, S USA
Red Sea 72 C3 var. Sinus Arabicus. sea Africa/Asia
Red Wing 45 G2 Minnesota, N USA
Reefton 151 C5 West Coast, South Island, New Zealand
Reese River 47 C5 river Nevada, W USA
Refahiye 117 E3 Erzincan, C Turkey (Türkiye)
Regensburg 95 C6 Eng. Ratisbon, Fr. Ratisbonne, hist. Ratisbona; anc. Castra Regina, Reginum. Bayern, SE Germany
Regenstauf 95 C6 Bayern, SE Germany
Rēgestān see Rēgistān
Reggane 70 D3 C Algeria
Reggio see Reggio nell'Emilia
Reggio Calabria see Reggio di Calabria
Reggio di Calabria 97 D7 var. Reggio Calabria, Gk. Rhegion; anc. Regium, Rhegium. Calabria, SW Italy
Reggio Emilia see Reggio nell'Emilia
Reggio nell'Emilia 96 B2 var. Reggio Emilia, abbrev. Reggio; anc. Regium Lepidum. Emilia-Romagna, N Italy
Reghin 108 C4 Ger. Sächsisch-Reen, Hung. Szászrégen; prev. Reghinul Săsesc, Ger. Sächsisch-Regen. Mureş, C Romania
Reghinul Săsesc see Reghin
Regina 37 F5 province capital Saskatchewan, S Canada
Reginum see Regensburg
Rēgistān 122 D5 var. Rīgestān, Rēgestān. desert region S Afghanistan
Regium see Reggio di Calabria
Regium Lepidum see Reggio nell'Emilia
Rehoboth 78 B3 Hardap, C Namibia
Rehovot 119 A6 ; prev. Reḥovot. Central, C Israel
Reḥovot see Rehovot
Reichenau see Bogatynia, Poland
Reichenberg see Liberec
Reid 147 D6 Western Australia
Reikjavik see Reykjavík
Ré, Île de 90 A4 island W France
Reims 90 D3 Eng. Rheims; anc. Durocortorum, Remi. Marne, N France
Reindeer Lake 37 F4 lake Manitoba/Saskatchewan, C Canada
Reine-Charlotte, Îles de la see Haida Gwaii
Reine-Élisabeth, Îles de la see Queen Elizabeth Islands
Reinga, Cape 150 C1 var. Te Rerenga Wairua. headland North Island, New Zealand
Reinosa 92 D1 Cantabria, N Spain
Reka see Rijeka
Rekhovot see Rehovot
Reliance 37 F4 Northwest Territories, C Canada
Remi see Reims
Rendina see Rentína
Rendsburg 94 B2 Schleswig-Holstein, N Germany
Rengat 138 B4 Sumatera, W Indonesia
Reni 108 D4 Odes'ka Oblast', SW Ukraine
Rennell 144 C4 var. Mu Nggava. island S Solomon Islands
Rennes 90 B3 Bret. Roazon; anc. Condate. Ille-et-Vilaine, NW France
Reno 47 C5 Nevada, W USA
Renqiu 128 C4 Hebei, E China
Rentína 105 B5 var. Rendina. Thessalía, C Greece
Reps see Rupea
Repulse Bay see Naujaat
Reschitza see Reşiţa
Resht see Rasht
Resicabánya see Reşiţa
Resistencia 64 D3 Chaco, NE Argentina
Reşiţa 108 A4 Ger. Reschitza, Hung. Resicabánya. Caraş-Severin, W Romania
Resolute 37 F2 Inuit Qausuittuq. Cornwallis Island, Nunavut, N Canada
Resolution Island 39 E1 island Nunavut, NE Canada
Resolution Island 151 A7 island SW New Zealand
Réunion 79 H4 off. La Réunion. French overseas department W Indian Ocean
Réunion 141 B5 island W Indian Ocean
Reus 93 F2 Cataluña, E Spain
Reutlingen 95 B6 Baden-Württemberg, S Germany
Reuver 87 D5 Limburg, SE Netherlands
Reval/Revel see Tallinn
Revillagigedo Island 50 B5 island Alexander Archipelago, Alaska, USA
Rexburg 46 E3 Idaho, NW USA
Reyes 61 F3 Beni, NW Bolivia
Rey, Isla del 53 G5 island Archipiélago de las Perlas, SE Panama
Reykjanes Basin 82 C5 var. Irminger Basin. undersea basin N Atlantic Ocean
Reykjanes Ridge 80 A1 undersea ridge N Atlantic Ocean
Reykjavík 83 E5 var. Reikjavik. country capital (Iceland) Höfudhborgarsvaedhi, W Iceland
Reynosa 51 E2 Tamaulipas, C Mexico
Reẕā'īyeh, Daryācheh-ye see Orūmīyeh, Daryācheh-ye
Rezé 90 A4 Loire-Atlantique, NW France
Rēzekne 106 D4 Ger. Rositten; prev. Rus. Rezhitsa. Rēzekne, SE Latvia
Rezhitsa see Rēzekne
Rezovo 104 E3 Turk. Rezve. Burgas, E Bulgaria
Rezve see Rezovo
Rhaedestus see Tekirdağ
Rhegion/Rhegium see Reggio di Calabria
Rheims see Reims
Rhein see Rhine
Rheine 94 A3 var. Rheine in Westfalen. Nordrhein-Westfalen, NW Germany
Rheine in Westfalen see Rheine
Rheinisches Schiefergebirge 95 A5 var. Rhine State Uplands, Eng. Rhenish Slate Mountains. mountain range W Germany
Rhenish Slate Mountains see Rheinisches Schiefergebirge
Rhin see Rhine
Rhine 80 D4 Dut. Rijn, Fr. Rhin, Ger. Rhein. river W Europe
Rhinelander 40 B2 Wisconsin, N USA
Rhine State Uplands see Rheinisches Schiefergebirge
Rho 96 B2 Lombardia, N Italy
Rhode Island 41 G3 off. State of Rhode Island and Providence Plantations, also known as Little Rhody, Ocean State. state NE USA
Rhodes 105 E7 var. Ródhos, Eng. Rhodes, It. Rodi; anc. Rhodos. island Dodekánisa, Greece, Aegean Sea
Rhodes see Ródos
Rhodesia see Zimbabwe
Rhodope Mountains 104 C3 var. Rodópi Óri, Bul. Rodopi Planina, Rodopi, Gk. Oroseirá Rodópis, Turk. Dospad Dagh. mountain range Bulgaria/Greece
Rhône 80 C4 river France/Switzerland
Rhum 88 B3 var. Rum. island W Scotland, United Kingdom
Ribble 89 D5 river NW England, United Kingdom
Ribeira see Santa Uxía de Ribeira
Ribeirão Preto 63 F4 São Paulo, S Brazil
Riberalta 61 F2 Beni, N Bolivia
Rîbniţa 108 D3 var. Râbniţa, Rus. Rybnitsa. NE Moldova
Rice Lake 40 A2 Wisconsin, N USA
Richard Toll 74 B3 N Senegal
Richfield 44 B4 Utah, W USA
Richland 46 C2 Washington, NW USA
Richmond 151 C5 Tasman, South Island, New Zealand
Richmond 40 C5 Kentucky, S USA
Richmond 41 E5 state capital Virginia, NE USA
Richmond Range 151 C5 mountain range South Island, New Zealand
Ricobayo, Embalse de 92 C2 reservoir NW Spain
Ricomagus see Riom
Rīdder/Ridder 114 D4 prev. Lenīnogor, Leninogorsk. Vostochnyy Kazakhstan, E Kazakhstan
Ridgecrest 47 C7 California, W USA
Ried see Ried im Innkreis
Ried im Innkreis 95 D6 var. Ried. Oberösterreich, NW Austria
Riemst 87 D6 Limburg, NE Belgium
Riesa 94 D4 Sachsen, E Germany
Rift Valley see Great Rift Valley
Rīga 106 C3 Eng. Riga. country capital (Latvia) Rīga, C Latvia
Rigaer Bucht see Riga, Gulf of
Riga, Gulf of 106 C3 Est. Liivi Laht, Ger. Rigaer Bucht, Latv. Rīgas Jūras Līcis, Rus. Rizhskiy Zaliv; prev. Est. Riia Laht. gulf Estonia/Latvia
Rīgas Jūras Līcis see Riga, Gulf of
Rīgestān see Rēgistān
Riia Laht see Riga, Gulf of
Riihimäki 85 D5 Kanta-Häme, S Finland
Rijeka 100 A2 Ger. Sankt Veit am Flaum, It. Fiume, Slvn. Reka; anc. Tarsatica. Primorje-Gorski Kotar, NW Croatia
Rijn see Rhine
Rijssel see Lille
Rijssen 86 E3 Overijssel, E Netherlands
Rimah, Wadi ar 120 B4 var. Wādī ar Rummah. dry watercourse C Saudi Arabia
Rimini 96 C3 anc. Ariminum. Emilia-Romagna, N Italy
Rîmnicu-Sărat see Râmnicu Sărat
Rîmnicu Vîlcea see Râmnicu Vâlcea
Rimouski 39 E4 Québec, SE Canada
Ringebu 85 B5 Oppland, S Norway
Ringen see Rõngu
Ringkøbing Fjord 85 A7 fjord W Denmark
Ringvassøya 84 C2 Smi. Ráneš. island N Norway
Rio see Rio de Janeiro
Riobamba 60 B1 Chimborazo, C Ecuador
Rio Branco 56 B3 state capital Acre, W Brazil
Rio Branco, Território de see Roraima
Río Bravo 51 E2 Tamaulipas, C Mexico
Río Cuarto 64 C4 Córdoba, C Argentina
Rio de Janeiro 63 F4 var. Rio. state capital Rio de Janeiro, SE Brazil
Río Gallegos 65 B7 var. Gallegos, Puerto Gallegos. Santa Cruz, S Argentina
Rio Grande 63 E5 var. São Pedro do Rio Grande do Sul. Rio Grande do Sul, S Brazil
Río Grande 50 D3 Zacatecas, C Mexico
Rio Grande do Norte 63 G2 off. Estado do Rio Grande do Norte. state/region E Brazil
Rio Grande do Norte, Estado do see Rio Grande do Norte
Rio Grande do Sul 63 E5 off. Estado do Rio Grande do Sul. state/region S Brazil
Rio Grande do Sul, Estado do see Rio Grande do Sul
Rio Grande Plateau see Rio Grande Rise
Rio Grande Rise 57 E6 var. Rio Grande Plateau. undersea plateau SW Atlantic Ocean
Ríohacha 58 B1 La Guajira, N Colombia
Río Lagartos 51 H3 Yucatán, SE Mexico
Riom 91 C5 anc. Ricomagus. Puy-de-Dôme, C France
Río Verde 51 E4 var. Rioverde. San Luis Potosí, C Mexico
Rioverde see Río Verde
Ripoll 93 G2 Cataluña, NE Spain
Rishiri-to 130 C2 var. Risiri Tô. island NE Japan
Risiri Tô see Rishiri-tō
Rîşno see Râşnov
Risti 106 D2 Ger. Kreuz. Läänemaa, W Estonia
Rivas 52 D4 Rivas, SW Nicaragua
Rivera 64 D3 Rivera, NE Uruguay
River Falls 40 A2 Wisconsin, N USA
Riverside 47 C7 California, W USA
Riverton 151 A7 var. Aparima. Southland, South Island, New Zealand
Riverton 44 C3 Wyoming, C USA
Rivière-du-Loup 39 E4 Québec, SE Canada
Rivne 108 C2 Pol. Równe, Rus. Rovno. Rivnens'ka Oblast', NW Ukraine
Rivoli 96 A2 Piemonte, NW Italy
Riyadh/Riyāḍ, Minṭaqat ar see Ar Riyāḍ
Riyāq see Rayak
Rize 117 E2 Rize, NE Turkey (Türkiye)
Rizhao 128 D4 Shandong, E China
Rizhskiy Zaliv see Riga, Gulf of
Rkîz 74 C3 Trarza, W Mauritania
Road Town 55 F3 dependent territory capital (British Virgin Islands) Tortola, C British Virgin Islands
Roanne 91 C5 anc. Rodunma. Loire, E France
Roanoke 41 E5 Virginia, NE USA
Roanoke River 43 F1 river North Carolina/Virginia, SE USA
Roatán 52 C2 var. Coxen Hole, Coxin Hole. Islas de la Bahía, N Honduras
Roat Kampuchea see Cambodia
Roazon see Rennes
Robbie Ridge 143 E3 undersea ridge W Pacific Ocean
Robert Williams see Caála
Robinson Range 147 B5 mountain range Western Australia
Robson, Mount 37 E5 mountain British Columbia, SW Canada
Robstown 49 G4 Texas, SW USA
Roca Partida, Isla 50 B5 island W Mexico
Rocas, Atol das 63 G2 island E Brazil
Rochefort 87 C7 Namur, SE Belgium
Rochefort 90 B4 var. Rochefort sur Mer. Charente-Maritime, W France
Rochefort sur Mer see Rochefort
Rochester 45 G3 Minnesota, N USA
Rochester 41 G2 New Hampshire, NE USA

Rocheuses, Montagnes/Rockies *see* Rocky Mountains
Rockall Bank *80 B2* undersea bank N Atlantic Ocean
Rockall Trough *80 B2* trough N Atlantic Ocean
Rockdale *148 E2* Texas, SW USA
Rockford *40 B3* Illinois, N USA
Rockhampton *148 D4* Queensland, E Australia
Rock Hill *43 E1* South Carolina, SE USA
Rockingham *147 A6* Western Australia
Rock Island *40 B3* Illinois, N USA
Rock Sound *54 C1* Eleuthera Island, C The Bahamas
Rock Springs *44 C3* Wyoming, C USA
Rockstone *59 F3* C Guyana
Rocky Mount *43 F1* North Carolina, SE USA
Rocky Mountains *34 B4* var. Rockies, *Fr.* Montagnes Rocheuses. *mountain range* Canada/USA
Roden *86 E2* Drenthe, NE Netherlands
Rodez *91 C5* anc. Segodunum. Aveyron, S France
Ródhos/Ródi *see* Ródos
Rodópi/Rodópi Óri *see* Rhodope Mountains
Rodopi Planina *see* Rhodope Mountains
Rodosto *see* Tekirdağ
Rodunma *see* Roanne
Roermond *87 D5* Limburg, SE Netherlands
Roeselare *87 A6* Fr. Roulers; prev. Rousselaere. West-Vlaanderen, W Belgium
Rogachev *see* Rahachow
Rogatica *100 C4* Republika Srpska, SE Bosnia and Herzegovina
Rogers *42 A1* Arkansas, C USA
Roger Simpson Island *see* Abemama
Roi Ed *see* Roi Et
Roi Et *137 D5* var. Muang Roi Et, Roi Ed. Roi Et, E Thailand
Roja *106 C2* Talsi, NW Latvia
Rokiškis *106 C4* Panevėžys, NE Lithuania
Rokycany *99 A5* Ger. Rokytzan. Plzeňský Kraj, W Czechia (Czech Republic)
Rokytzan *see* Rokycany
Rôlas, Ilha das *76 E2* island S Sao Tome and Principe, Africa, E Atlantic Ocean
Rolla *45 G5* Missouri, C USA
Röm *see* Rømø
Roma *96 C4* Eng. Rome. *country capital* (Italy) Lazio, C Italy
Roma *149 D5* Queensland, E Australia
Roman *104 C2* Vratsa, NW Bulgaria
Roman *108 C4* Hung. Románvásár. Neamţ, NE Romania
Romania *108 B4* Bul. Rumŭniya, Ger. Rumänien, Hung. Románia, Rom. România, Croatian Rumunjska, Ukr. Rumuniya, prev. Republica Socialistă România, Roumania, Rumania, Socialist Republic of Romania, prev. Rom. Romínia. *country* SE Europe
România, Republica Socialistă *see* Romania
Romania, Socialist Republic of *see* Romania
Románvásár *see* Roman
Rome *42 D2* Georgia, SE USA
Rome *see* Roma
Romînia *see* Romania
Romny *109 F2* Sums'ka Oblast', NE Ukraine
Rømø *85 A7* Ger. Röm. island SW Denmark
Roncador, Serra do *56 D4* mountain range C Brazil
Ronda *92 D5* Andalucía, S Spain
Rondônia *62 D3* off. Estado de Rondônia; prev. Território de Rondônia. *state/region* W Brazil
Rondônia, Estado de *see* Rondônia
Rondônia, Território de *see* Rondônia
Rondonópolis *63 E3* Mato Grosso, W Brazil
Rongelap Atoll *144 D1* var. Rönlap. atoll Ralik Chain, NW Marshall Islands
Röngu *106 D3* Ger. Ringen. Tartumaa, SE Estonia
Rönlap *see* Rongelap Atoll
Rønne *85 B8* Bornholm, E Denmark
Ronne Ice Shelf *154 A3* ice shelf Antarctica
Roosendaal *87 C5* Noord-Brabant, S Netherlands
Roosevelt Island *154 B4* island Antarctica

Roraima *62 D1* off. Estado de Roraima; prev. Território de Rio Branco, Território de Roraima. state/region N Brazil
Roraima, Estado de *see* Roraima
Roraima, Mount *59 E3* mountain N South America
Røros *85 B5* Sør-Trøndelag, S Norway
Ross *151 B6* West Coast, South Island, New Zealand
Rosa, Lake *54 D2* lake Great Inagua, S The Bahamas
Rosario *64 D4* Santa Fe, C Argentina
Rosario *64 D2* San Pedro, C Paraguay
Rosario *see* Rosarito
Rosarito *50 A1* var. Rosario. Baja California Norte, NW Mexico
Roscianum *see* Rossano
Roscommon *40 C2* Michigan, N USA
Roseau *55 G4* prev. Charlotte Town. *country capital* (Dominica) SW Dominica
Roseburg *46 B4* Oregon, NW USA
Rosenau *see* Rožňava
Rosenberg *49 G4* Texas, SW USA
Rosenberg *see* Ružomberok, Slovakia
Rosengarten *94 B3* Niedersachsen, N Germany
Rosenheim *95 C6* Bayern, S Germany
Rosia *93 H5* W Gibraltar Europe
Rosia Bay *93 H5* bay SW Gibraltar Europe W Mediterranean Sea Atlantic Ocean
Roşiori de Vede *108 B5* Teleorman, S Romania
Rositten *see* Rēzekne
Roslavl' *111 A5* Smolenskaya Oblast', W Russia
Rosmalen *86 C4* Noord-Brabant, S Netherlands
Rossano *97 E6* anc. Roscianum. Calabria, SW Italy
Ross Ice Shelf *154 B4* ice shelf Antarctica
Rossiyskaya Federatsiya *see* Russia
Rosso *74 B3* Trarza, SW Mauritania
Rossosh' *111 B6* Voronezhskaya Oblast', W Russia
Ross Sea *154 B4* sea Antarctica
Rostak *see* Ar Rustāq
Rostock *94 C2* Mecklenburg-Vorpommern, NE Germany
Rostov-na-Donu *111 B7* var. Rostov, Eng. Rostov-on-Don. Rostovskaya Oblast', SW Russia
Rostov-on-Don *see* Rostov-na-Donu
Roswell *48 D2* New Mexico, SW USA
Rota *144 B1* island S Northern Mariana Islands
Rotcher Island *see* Tamana
Rothera *154 A2* UK research station Antarctica
Rotomagus *see* Rouen
Rotorua *150 D3* Bay of Plenty, North Island, New Zealand
Rotorua, Lake *150 D3* lake North Island, New Zealand
Rotterdam *86 C4* Zuid-Holland, SW Netherlands
Rottweil *95 B6* Baden-Württemberg, S Germany
Rotuma *145 E4* island NW Fiji Oceania, S Pacific Ocean
Roubaix *90 C2* Nord, N France
Rouen *90 C3* anc. Rotomagus. Seine-Maritime, N France
Roulers *see* Roeselare
Roumania *see* Romania
Round Rock *49 G3* Texas, SW USA
Rourkela *see* Räurkela
Rousselaere *see* Roeselare
Roussillon *91 C6* cultural region S France
Rouyn-Noranda *38 D4* Québec, SE Canada
Rovaniemi *84 D3* Lappi, N Finland
Rovigno *see* Rovinj
Rovigo *96 C2* Veneto, NE Italy
Rovinj *100 A3* It. Rovigno. Istra, NW Croatia
Rovno *see* Rivne
Rovuma, Rio *79 F2* var. Ruvuma. river Mozambique/Tanzania
Rovuma, Rio *see* Ruvuma
Równe *see* Rivne
Roxas City *139 E2* Panay Island, C Philippines
Royale, Isle *40 B1* island Michigan, N USA

Royan *91 B5* Charente-Maritime, W France
Rozdolne *109 F4* Rus. Razdolnoye. Respublika Krym, S Ukraine
Rožňava *99 D6* Ger. Rosenau, Hung. Rozsnyó. Košický Kraj, E Slovakia
Rózsahegy *see* Ružomberok
Rozsnyó *see* Rožňava
Ruanda *see* Rwanda
Ruapehu, Mount *150 D4* volcano North Island, New Zealand
Ruapuke Island *151 B8* island SW New Zealand
Ruatoria *150 E3* Gisborne, North Island, New Zealand
Ruawai *150 D2* Northland, North Island, New Zealand
Rubezhnoye *see* Rubizhne
Rubizhne *109 H3* Rus. Rubezhnoye. Luhans'ka Oblast', E Ukraine
Ruby Mountains *47 D5* mountain range Nevada, W USA
Rucava *106 B3* Liepāja, SW Latvia
Rudensk *see* Rudzyensk
Rūdiškės *107 B5* Vilnius, S Lithuania
Rudnik *see* Dolni Chiflik
Rudny *see* Rudnyy
Rudnyy *114 C4* var. Rudny. Kostanay, N Kazakhstan
Rudolf, Lake *see* Turkana, Lake
Rudolfswert *see* Novo mesto
Rudzyensk *107 C6* Rus. Rudensk. Minskaya Voblasts', C Belarus
Rufiji *73 C7* river E Tanzania
Rufino *64 C4* Santa Fe, C Argentina
Rugāji *106 D4* Balvi, E Latvia
Rügen *94 D2* headland NE Germany
Ruggell *94 E1* N Liechtenstein Europe
Ruhja *see* Rūjiena
Ruhnu *106 C3* Ruhnu Saar, Swe. Runö. island SW Estonia
Ruhnu Saar *see* Ruhnu
Rujen *see* Rūjiena
Rūjiena *106 D3* Est. Ruhja, Ger. Rujen. Valmiera, N Latvia
Rukwa, Lake *73 B7* lake SE Tanzania
Rum *see* Rhum
Ruma *100 D3* Vojvodina, N Serbia
Rumadiya *see* Ar Ramādī
Rumänia/Rumänien *see* Romania
Rumbek *73 B5* El Buhayrat, C South Sudan
Rumburk *see* Rimavská Sobota
Rumia *98 C2* Pomorskie, N Poland
Rummah, Wādī ar *see* Rimah, Wādī ar
Rummelsburg in Pommern *see* Miastko
Rumuniya/Rumūniya/Rumunjska *see* Romania
Runanga *151 B5* West Coast, South Island, New Zealand
Rundu *78 C3* var. Runtu. Okavango, NE Namibia
Runö *see* Ruhnu
Runtu *see* Rundu
Ruoqiang *126 C3* var. Jo-ch'iang, Uygh. Charkhlik, Charkhliq, Qarkilik. Xinjiang Uygur Zizhiqu, NW China
Rupea *108 C4* Ger. Reps, Hung. Kőhalom; prev. Cohalm. Braşov, C Romania
Rupel *87 B5* river N Belgium
Rupella *see* la Rochelle
Rupert, Rivière de *38 D3* river Québec, C Canada
Rusaddir *see* Melilla
Ruse *104 D1* var. Ruschuk, Rustchuk, Turk. Rusçuk. Ruse, N Bulgaria
Russadir *see* Melilla
Russellville *42 A1* Arkansas, C USA
Russia *112 D2* off. Russian Federation, Latv. Krievija, Rus. Rossiyskaya Federatsiya. *country* Asia/Europe
Russian America *see* Alaska
Russian Federation *see* Russia
Rustaq *see* Ar Rustāq
Rustavi *117 G2* prev. Rust'avi. SE Georgia
Rust'avi *see* Rustavi
Rustchuk *see* Ruse
Ruston *42 B2* Louisiana, S USA
Rutanzige, Lake *see* Edward, Lake
Rutba *see* Ar Ruţbah
Rutlam *see* Ratlām
Rutland *41 F2* Vermont, NE USA
Rutög *126 A4* var. Rutög, Rutok. Xizang Zizhiqu, W China
Rutok *see* Rutog
Ruvuma *see* Rovuma, Rio

Ruwenzori *77 E5* mountain range Dem. Rep. Congo/Uganda
Ruzhany *107 B6* Brestskaya Voblasts', SW Belarus
Ružomberok *99 C5* Ger. Rosenberg, Hung. Rózsahegy. Žilinský Kraj, N Slovakia
Rwanda *73 B6* off. Republic of Rwanda; prev. Ruanda. *country* C Africa
Rwanda, Republic of *see* Rwanda
Ryazan' *111 B5* Ryazanskaya Oblast', W Russia
Rybach'ye *see* Balykchy
Rybinsk *110 B4* prev. Andropov. Yaroslavskaya Oblast', W Russia
Rybnik *99 C5* Śląskie, S Poland
Rybnitsa *see* Rîbniţa
Ryde *148 E1* United Kingdom
Ryki *98 D4* Lubelskie, E Poland
Rykovo *see* Yenakiieve
Rypin *98 C3* Kujawsko-pomorskie, C Poland
Ryssel *see* Lille
Rysy *99 C5* mountain S Poland
Ryukyu Islands *see* Nansei-shotō
Ryukyu Trench *125 F3* var. Nansei Syotō Trench. trench S East China Sea
Rzeszów *99 E5* Podkarpackie, SE Poland
Rzhev *110 B4* Tverskaya Oblast', W Russia

S

Saale *94 C4* river C Germany
Saalfeld *95 C5* var. Saalfeld an der Saale. Thüringen, C Germany
Saalfeld an der Saale *see* Saalfeld
Saarbrücken *95 A6* Fr. Sarrebruck. Saarland, SW Germany
Sääre *106 C2* var. Sjar. Saaremaa, W Estonia
Saaremaa *106 C2* Ger. Oesel, Ösel; prev. Saare. island W Estonia
Saariselkä *84 D2* Smi. Suoločielgi. Lappi, N Finland
Sab' Ābār *118 C4* var. Sab'a Biyar, Sa'b Abar, Sab' Biyar, Sab'i Byar, Sab'i Abar. Ḩimş, C Syria
Sab'a Biyar *see* Sab' Ābār
Sabac *100 C4* Serbia, W Serbia
Sabadell *93 G2* Cataluña, E Spain
Sabah *138 D3* prev. British North Borneo, North Borneo. state East Malaysia
Sabanalarga *58 B1* Atlántico, N Colombia
Sabaneta *58 C1* Falcón, N Venezuela
Sabaria *see* Szombathely
Sab'atayn, Ramlat as *121 C6* desert C Yemen
Sabaya *61 F4* Oruro, S Bolivia
Sab' Bi'ār *see* Sab' Ābār
Saberi, Hamun-e *122 C5* var. Daryācheh-ye Hāmun, Daryācheh-ye Sīstān. lake Afghanistan/Iran
Sabḩā *71 F3* C Libya
Sabi *see* Save
Sabinas *51 E2* Coahuila, NE Mexico
Sabinas Hidalgo *51 E2* Nuevo León, NE Mexico
Sabine River *49 H3* river Louisiana/Texas, SW USA
Sabkha *see* As Sabkhah
Sable, Cape *43 E5* headland Florida, SE USA
Sable Island *39 G4* island Nova Scotia, SE Canada
Şabyā *121 B6* Jīzān, SW Saudi Arabia
Sabzawar *see* Sabzevār
Sabzevār *120 D2* var. Sabzawar. Khorāsān-Razavī, NE Iran
Sachsen *94 D4* Eng. Saxony, Fr. Saxe. state E Germany
Sachs Harbour *37 E2* var. Ikaahuk. Banks Island, Northwest Territories, N Canada
Sächsisch-Reen/Sächsisch-Regen *see* Reghin
Sacramento *47 B5* state capital California, W USA
Sacramento Mountains *48 D2* mountain range New Mexico, SW USA
Sacramento River *47 B5* river California, W USA
Sacramento Valley *47 B5* valley California, W USA
Sá da Bandeira *see* Lubango
Şa'dah *121 B6* NW Yemen

Sado 131 C5 *var.* Sadoga-shima. *island* C Japan
Sadoga-shima *see* Sado
Saena Julia *see* Siena
Safad *see* Tsefat
Şafāqis *see* Sfax
Şafāshahr 120 D3 *var.* Deh Bīd. Färs, C Iran
Safed *see* Tsefat
Safēd Kōh, Silsilah-ye 122 D4 *prev.* Selseleh-ye Safīd Kūh, Eng. Paropamisus Range. *mountain range* W Afghanistan
Säffle 85 B6 Värmland, C Sweden
Safford 48 C3 Arizona, SW USA
Safi 70 B2 W Morocco
Safīd Kuh, Selseleh-ye *see* Safēd Kōh, Silsilah-ye
Sagaing 136 B3 Sagaing, C Myanmar (Burma)
Sagami-nada 131 D6 *inlet* SW Japan
Sagan *see* Žagań
Sāgar 134 D4 *prev.* Saugor. Madhya Pradesh, C India
Sagarmāthā *see* Everest, Mount
Sagebrush State *see* Nevada
Saghez *see* Saqqez
Saginaw 40 C3 Michigan, N USA
Saginaw Bay 40 D2 *lake bay* Michigan, N USA
Sagua la Grande 54 B2 Villa Clara, C Cuba
Sagunto/Saguntum *see* Sagunt
Sagunt 93 F3 *var.* Sagunto, Ar. Murviedro; *anc.* Saguntum. Comunitat Valenciana, E Spain
Sahara 68 B3 *desert* Libya/Algeria
Sahara el Gharbîya *see* Şahrā' al Gharbīyah
Saharan Atlas *see* Atlas Saharien
Sahel 74 D3 *physical region* C Africa
Sāhīlīyah, Jībal as 118 B3 *mountain range* NW Syria
Sahiwal 134 C2 *prev.* Montgomery. Punjab, E Pakistan
Şahrā' al Gharbīyah 72 B2 *var.* Sahara el Gharbîya, Eng. Western Desert. *desert* C Egypt
Şahrā' ash Sharqīyah 103 H5 Eng. Arabian Desert, Eastern Desert. *desert* E Egypt
Saïda 119 A5 *var.* Şaydā, Sayida; *anc.* Sidon. W Lebanon
Sa'īdābād *see* Sīrjān
Saidpur 135 G3 *var.* Syedpur. Rajshahi, NW Bangladesh
Saidu Sharif 134 C1 *var.* Mingora, Mongora. North-West Frontier Province, N Pakistan
Saigon *see* Hô Chi Minh
Saimaa 85 E5 *lake* SE Finland
St Albans 89 E6 *anc.* Verulamium. E England, United Kingdom
Saint Albans 40 D5 West Virginia, NE USA
St Andrews 88 C4 E Scotland, United Kingdom
Saint Anna Trough *see* Svyataya Anna Trough
St. Ann's Bay 54 B4 C Jamaica
St. Anthony 39 G3 Newfoundland and Labrador, SE Canada
Saint Augustine 43 E3 Florida, SE USA
St Austell 89 C7 SW England, United Kingdom
Saint Croix 55 F3 *island* S Virgin Islands (US)
Saint Croix River 40 A2 *river* Minnesota/Wisconsin, N USA
St David's Island 42 B5 *island* E Bermuda
St-Denis 79 G4 *dependent territory capital* (Réunion) NW Réunion
St-Dié 90 E4 Vosges, NE France
St-Egrève 91 D5 Isère, E France
Sainte Marie, Cap *see* Vohimena, Tanjona
Saintes 91 B5 *anc.* Mediolanum. Charente-Maritime, W France
St-Étienne 91 D5 Loire, E France
St-Flour 91 C5 Cantal, C France
St-Gall/Saint Gall/St. Gallen *see* Sankt Gallen
St-Gaudens 91 B6 Haute-Garonne, S France
Saint George 149 D5 Queensland, E Australia
St George 42 B4 N Bermuda
Saint George 44 A5 Utah, W USA
St. George's 55 G5 *country capital* (Grenada) SW Grenada
St-Georges 39 E4 Québec, SE Canada
St-Georges 59 H3 E French Guiana
Saint George's Channel 89 B6 *channel* Ireland/Wales, United Kingdom
St George's Island 42 B4 *island* E Bermuda
St Helena *see* St Helena, Ascension and Tristan da Cunha
St Helena, Ascension & Tristan da Cunha 47 A6 *UK Overseas territory* C Atlantic Ocean
St Helier 89 D8 *dependent territory capital* (Jersey) S Jersey, Channel Islands
St.Iago de la Vega *see* Spanish Town
Saint Ignace 40 C2 Michigan, N USA
St-Jean, Lac 39 E4 *lake* Québec, SE Canada
Saint Joe River 46 D2 *river* Idaho, NW USA North America
St. John 39 F4 New Brunswick, SE Canada
Saint-John *see* Saint John
Saint John 41 H1 Fr. Saint-John. *river* Canada/USA
St John's 55 G3 *country capital* (Antigua and Barbuda) Antigua, Antigua and Barbuda
St. John's 39 H3 *province capital* Newfoundland and Labrador, E Canada
Saint Joseph 45 F4 Missouri, C USA
St Julian's *see* San Ġiljan
St Kilda 88 A3 *island* NW Scotland, United Kingdom
Saint Kitts and Nevis 55 F3 *off.* Federation of Saint Christopher and Nevis, *var.* Saint Christopher-Nevis. *country* E West Indies
St-Laurent *see* St-Laurent-du-Maroni
St-Laurent-du-Maroni 59 H3 *var.* St-Laurent. NW French Guiana
St-Laurent, Fleuve *see* St. Lawrence
St. Lawrence 39 E4 Fr. Fleuve St-Laurent. *river* Canada/USA
St. Lawrence, Gulf of 39 F3 *gulf* NW Atlantic Ocean
Saint Lawrence Island 36 B2 *island* Alaska, USA
St-Lô 90 B3 *anc.* Briovera, Laudus. Manche, N France
Saint-Louis 90 E4 Haut-Rhin, NE France
Saint Louis 74 B3 NW Senegal
Saint Louis 45 G4 Missouri, C USA
Saint Lucia 55 E1 *country* SE West Indies
Saint Lucia Channel 55 H4 *channel* Martinique/Saint Lucia
St-Malo 90 B3 Ille-et-Vilaine, NW France
St-Malo, Golfe de 90 A3 *gulf* NW France
Saint Martin *see* Sint Maarten
St Martin 55 G3 *French overseas collectivity*, E West Indies
St.Matthew's Island *see* Zadetkyi Kyun
St.Matthias Group 144 B3 *island group* NE Papua New Guinea
St. Moritz 95 B7 Ger. Sankt Moritz, Rmsch. San Murezzan. Graubünden, SE Switzerland
St-Nazaire 90 A4 Loire-Atlantique, NW France
Saint Nicholas *see* São Nicolau
Saint-Nicolas *see* Sint-Niklaas
St-Omer 90 C2 Pas-de-Calais, N France
Saint Paul 45 F2 *state capital* Minnesota, N USA
St-Paul, Île 141 C6 *var.* St.Paul Island. *island* Île St-Paul, NE French Southern and Antarctic Lands Antarctica Indian Ocean
St.Paul Island *see* St-Paul, Île
St Peter Port 89 D8 *dependent territory capital* (Guernsey) C Guernsey, Channel Islands
Saint Petersburg 43 E4 Florida, SE USA
Saint Petersburg *see* Sankt-Peterburg
St-Pierre and Miquelon 39 G4 Fr. Îles St-Pierre et Miquelon. *French overseas collectivity* NE North America
St-Quentin 90 C3 Aisne, N France
Saint Thomas *see* São Tomé, Sao Tome and Principe
Saint Thomas *see* Charlotte Amalie, Virgin Islands (US)
Saint Vincent 55 G4 *island* N Saint Vincent and the Grenadines
Saint Vincent *see* São Vicente
Saint Vincent and the Grenadines 55 H4 *country* SE West Indies
Saint Vincent, Cape *see* São Vicente, Cabo de
Saint Vincent Passage 55 H4 *passage* Saint Lucia/Saint Vincent and the Grenadines
Saint Yves *see* Setúbal
Saipan 142 B1 *island/country capital* (Northern Mariana Islands) S Northern Mariana Islands
Saishū *see* Jejudo
Sajama, Nevado 61 F4 *mountain* W Bolivia
Sajószentpéter 99 D6 Borsod-Abaúj-Zemplén, NE Hungary
Sakākah 120 B4 Al Jawf, NW Saudi Arabia
Sakakawea, Lake 44 D1 *reservoir* North Dakota, N USA
Sakartvelo *see* Georgia
Sakata 130 D4 Yamagata, Honshū, C Japan
Sakhalin 115 G4 *var.* Sakhalin. *island* SE Russia
Sakhalin *see* Sakhalin, Ostrov
Sakhon Nakhon *see* Sakon Nakhon
Şäki 117 G2 Rus. Sheki; *prev.* Nukha. NW Azerbaijan
Saki *see* Saky
Sakishima-shoto 130 A3 *var.* Sakisima Syotō. *island group* SW Japan
Sakisima Syotô *see* Sakishima-shotō
Sakiz *see* Saqqez
Sakiz-Adasi *see* Chíos
Sakon Nakhon 136 D4 *var.* Muang Sakon Nakhon, Sakhon Nakhon. Sakon Nakhon, E Thailand
Saky 109 F5 Rus. Saki. Respublika Krym, S Ukraine
Sal 74 A3 *island* Ilhas de Barlavento, NE Cape Verde (Cabo Verde)
Sala 85 C6 Västmanland, C Sweden
Salacgriva 106 C3 Est. Salatsi. Limbaži, N Latvia
Sala Consilina 97 D5 Campania, S Italy
Salado, Río 62 D5 *river* E Argentina
Salado, Río 64 C3 *river* C Argentina
Şalālah 121 D6 SW Oman
Salamá 52 B2 Baja Verapaz, C Guatemala
Salamanca 64 B4 Coquimbo, C Chile
Salamanca 92 D2 *anc.* Helmantica, Salmantica. Castilla y León, NW Spain
Salamīyah 118 B3 *var.* As Salamīyah. Ḥamāh, W Syria
Salang *see* Phuket
Salantai 106 B3 Klaipėda, NW Lithuania
Salatsi *see* Salacgriva
Salavan 137 D5 *var.* Saravan, Saravane. Salavan, S Laos
Salavat 111 D6 Respublika Bashkortostan, W Russia
Sala y Gomez 153 F4 *island* Chile, E Pacific Ocean
Sala y Gomez Fracture Zone *see* Sala y Gomez Ridge
Sala y Gomez Ridge 153 G4 *var.* Sala y Gomez Fracture Zone. *fracture zone* SE Pacific Ocean
Salazar *see* N'Dalatando
Šalčininkai 107 C5 Vilnius, SE Lithuania
Saldus 106 B3 Ger. Frauenburg. Saldus, W Latvia
Sale 149 C7 Victoria, SE Australia
Salé 70 C2 NW Morocco
Salekhard 114 D3 *prev.* Obdorsk. Yamalo-Nenetskiy Avtonomnyy Okrug, N Russia
Salem 132 C2 Tamil Nādu, SE India
Salem 46 B3 *state capital* Oregon, NW USA
Salerno 97 D5 *anc.* Salernum. Campania, S Italy
Salerno, Gulf of 97 C5 Eng. Gulf of Salerno. *gulf* S Italy
Salerno, Gulf of *see* Salerno, Golfo di
Salernum *see* Salerno
Salfit 119 E6 C West Bank, SW Asia
Salihorsk 107 C7 Rus. Soligorsk. Minskaya Voblasts', S Belarus
Salima 79 E2 Central, C Malawi
Salina 45 E5 Kansas, C USA
Salina Cruz 51 F5 Oaxaca, SE Mexico
Salinas 60 A2 Guayas, W Ecuador
Salinas 47 B6 California, W USA
Salisbury 89 D7 *var.* New Sarum. S England, United Kingdom
Salisbury *see* Harare
Sällan *see* Sørøya
Sallaq *see* Coral Harbour
Sallyana *see* Salyān
Salmantica *see* Salamanca
Salmon River 46 D3 *river* Idaho, NW USA
Salmon River Mountains 46 D3 *mountain range* Idaho, NW USA
Salo 85 D6 Länsi-Suomi, SW Finland
Salon-de-Provence 91 D6 Bouches-du-Rhône, SE France
Salonica/Salonika *see* Thessaloníki
Salonta 108 A3 Hung. Nagyszalonta. Bihor, W Romania
Sal'sk 111 B7 Rostovskaya Oblast', SW Russia
Salt *see* As Salt
Salta 64 C2 Salta, NW Argentina
Saltash 89 C7 SW England, United Kingdom
Saltillo 51 E3 Coahuila, NE Mexico
Salt Lake City 44 B4 *state capital* Utah, W USA
Salto 64 D4 Salto, N Uruguay
Salton Sea 47 D8 *lake* California, W USA
Salvador 63 G3 *prev.* São Salvador. *state capital* Bahia, E Brazil
Salween 124 E2 Bur. Thanlwin, Chin. Nu Chiang, Nu Jiang. *river* SE Asia
Şalyān 135 E3 *var.* Sallyana. Mid Western, W Nepal
Salzburg 95 D6 *anc.* Juvavum. Salzburg, N Austria
Salzgitter 94 C4 *prev.* Watenstedt-Salzgitter. Niedersachsen, C Germany
Salzwedel 94 C3 Sachsen-Anhalt, N Germany
Šamac *see* Bosanski Šamac
Samakhixai *see* Attapu
Samalayuca 50 C1 Chihuahua, N Mexico
Samar 139 F2 *island* C Philippines
Samara 114 B3 *prev.* Kuybyshev. Samarskaya Oblast', W Russia
Samarang *see* Semarang
Samarinda 138 D4 Borneo, C Indonesia
Samarkand *see* Samarqand
Samarkandski/Samarkandskoye *see* Temirtau
Samarobriva *see* Amiens
Samarqand 123 E2 Rus. Samarkand. Samarqand Viloyati, C Uzbekistan
Samawa *see* As Samāwah
Şamaxı 117 H2 E Azerbaijan
Sambalpur 135 F4 Odisha, E India
Sambava 79 G2 Antsiranana, NE Madagascar
Sambor 108 B2 Rus. Sambor. L'vivs'ka Oblast', W Ukraine
Sambor *see* Sambir
Sambre 90 D2 *river* Belgium/France
Samfya 78 D2 Luapula, N Zambia
Saminatal 94 E2 *valley* Austria/Liechtenstein, Europe
Samnān *see* Semnān
Sam Neua *see* Xam Nua
Samoa 145 E4 *off.* Independent State of Western Samoa, *var.* Sāmoa; *prev.* Western Samoa. *country* W Polynesia
Sāmoa *see* Samoa
Samoa Basin 143 E3 *undersea basin* W Pacific Ocean
Sambor 100 A2 Zagreb, N Croatia
Sámos 105 E6 *prev.* Limín Vathéos. Sámos, Dodekánisa, Greece, Aegean Sea
Sámos 105 D6 *island* Dodekánisa, Greece, Aegean Sea
Samothrace *see* Samothráki
Samothráki 104 D4 Samothráki, NE Greece
Samothraki 104 C4 *anc.* Samothrace. *island* NE Greece
Sampit 138 C4 Borneo, C Indonesia
Samraong 137 D5 *prev.* Phumĭ Sâmraông, Phum Samrong. Oddar Meanchey, NW Cambodia
Samsun 116 D2 *anc.* Amisus. Samsun, N Turkey (Türkiye)

Samt'redia *117 F2* W Georgia
Samui, Ko *137 C6 island* SW Thailand
Samut Prakan *137 C5 var.* Muang Samut
  Prakan, Paknam. Samut Prakan,
  C Thailand
San *74 D3* Ségou, C Mali
San *99 E5 river* SE Poland
Şan'ā' *121 B6 Eng.* Sanaa. *country capital*
  (Yemen) W Yemen
Sana *100 B3 river* NW Bosnia and
  Herzegovina
Sanaa *see* Şan'ā'
Sanae *154 B2 South African research
  station* Antarctica
Sanaga *77 B5 river* C Cameroon
San Ambrosio, Isla *57 A5 Eng.* San
  Ambrosio Island. *island* W Chile
San Ambrosio Island *see*
  San Ambrosio, Isla
Sanandaj *120 C3 prev.* Sinneh.
  Kordestän, W Iran
San Andrés, Isla de *53 F3 island*
  NW Colombia, Caribbean Sea
San Andrés Tuxtla *51 F4 var.* Tuxtla.
  Veracruz-Llave, E Mexico
San Angelo *49 F3* Texas, SW USA
San Antonio *52 B2* Toledo, S Belize
San Antonio *64 B4* Valparaíso, C Chile
San Antonio *49 F4* Texas, SW USA
San Antonio Oeste *65 C5* Río Negro,
  E Argentina
San Antonio River *49 G4 river* Texas,
  SW USA
Sanāw *121 C6 var.* Sanaw. NE Yemen
San Benedicto, Isla *50 B4 island*
  W Mexico
San Benito *52 B1* Petén, N Guatemala
San Benito *49 G5* Texas, SW USA
San Bernardino *47 C7* California,
  W USA
San Blas *50 C3* Sinaloa, C Mexico
San Blas, Cape *42 D3 headland* Florida,
  SE USA
San Blas, Cordillera de *53 G4 mountain
  range* NE Panama
San Carlos *52 D4* Río San Juan,
  S Nicaragua
San Carlos *48 B2* Arizona, SW USA
San Carlos *see* Quesada, Costa Rica
San Carlos de Ancud *see* Ancud
San Carlos de Bariloche *65 B5* Río
  Negro, SW Argentina
San Carlos del Zulia *58 C2* Zulia,
  W Venezuela
San Clemente Island *47 B8 island*
  Channel Islands, California, W USA
San Cristóbal *58 C2* Táchira,
  W Venezuela
San Cristóbal *144 C4 var.* Makira.
  island SE Solomon Islands
San Cristóbal *see* San Cristóbal
  de Las Casas
San Cristóbal de Las Casas *51 G5 var.*
  San Cristóbal. Chiapas, SE Mexico
San Cristóbal, Isla *60 B5 var.* Chatham
  Island. *island* Galápagos Islands,
  Ecuador, E Pacific Ocean
Sancti Spíritus *54 B2* Sancti Spíritus,
  C Cuba
Sandakan *138 D3* Sabah, East Malaysia
Sandalwood Island *see* Sumba, Pulau
Sandanski *104 C3 prev.* Sveti Vrach.
  Blagoevgrad, SW Bulgaria
Sanday *88 D2 island* NE Scotland,
  United Kingdom
Sanders *48 C2* Arizona, SW USA
Sand Hills *44 D3 mountain range*
  Nebraska, C USA
San Diego *47 C8* California, W USA
Sandnes *85 A6* Rogaland, S Norway
Sandomierz *98 D4 Rus.* Sandomir.
  Świętokrzyskie, C Poland
Sandomir *see* Sandomierz
Sandoway *see* Thandwe
Sandpoint *46 C1* Idaho, NW USA
Sand Springs *49 G1* Oklahoma,
  C USA
Sandusky *40 D3* Ohio, N USA
Sandvika *85 A6* Akershus, S Norway
Sandviken *85 C6* Gävleborg, C Sweden
Sandwich Island *see* Efate
Sandwich Islands *see* Hawaiian Islands
Sandy Bay *93 H5* Saskatchewan,
  C Canada
Sandy City *44 B4* Utah, W USA
Sandy Lake *38 B3 lake* Ontario,
  C Canada
San Esteban *52 D2* Olancho, C Honduras
San Eugenio/San Eugenio del Cuareim
  see Artigas

San Felipe *58 D1* Yaracuy,
  NW Venezuela
San Felipe de Puerto Plata *see* Puerto
  Plata
San Félix, Isla *57 A5 Eng.* San Felix
  Island. *island* W Chile
San Felix Island *see* San Félix, Isla
San Fernando *55 H5* Trinidad, Trinidad
  and Tobago
San Fernando *46 D1* California, W USA
San Fernando *58 D2 var.* San Fernando
  de Apure. Apure, C Venezuela
San Fernando de Apure *see* San
  Fernando
San Fernando del Valle de Catamarca
  64 C3 var. Catamarca. Catamarca,
  NW Argentina
San Fernando de Monte Cristi *see*
  Monte Cristi
San Francisco *47 B6* California, W USA
San Francisco del Oro *50 C2*
  Chihuahua, N Mexico
San Francisco de Macorís *55 E3*
  C Dominican Republic
San Fructuoso *see* Tacuarembó
San Gabriel *60 B1* Carchi, N Ecuador
San Gabriel Mountains *46 E1 mountain
  range* California, USA
Sangihe, Kepulauan *see* Sangir,
  Kepulauan
San Giljan *102 B5 var.* St Julian's.
  N Malta
Sangir, Kepulauan *139 F3 var.*
  Kepulauan Sangihe. *island group*
  N Indonesia
Sāngli *132 B1* Mahārāshtra, W India
Sangmélima *77 B5* Sud, S Cameroon
Sangre de Cristo Mountains *48 D1
  mountain range* Colorado/New Mexico,
  C USA
San Ignacio *52 B1 prev.* Cayo, El Cayo.
  Cayo, W Belize
San Ignacio *61 F3* Beni, N Bolivia
San Ignacio *50 B2* Baja California Sur,
  NW Mexico
San Joaquin Valley *47 B7 valley*
  California, W USA
San Jorge, Golfo *65 C6 var.* Gulf of San
  Jorge. *gulf* S Argentina
San Jorge, Gulf of *see* San Jorge, Golfo
San José *53 E4 country capital* (Costa
  Rica) San José, C Costa Rica
San José *61 G3 var.* San José de
  Chiquitos. Santa Cruz, E Bolivia
San José *52 B3 var.* Puerto San José.
  Escuintla, S Guatemala
San Jose *47 B6* California, W USA
San José *see* San José del Guaviare,
  Colombia
San José de Chiquitos *see* San José
San José de Cúcuta *see* Cúcuta
San José del Guaviare *58 C4 var.* San
  José. Guaviare, S Colombia
San Juan *64 B4* San Juan, W Argentina
San Juan *55 F3 dependent territory
  capital* (Puerto Rico) NE Puerto Rico
San Juan *see* San Juan de los Morros
San Juan Bautista *64 D3* Misiones,
  S Paraguay
San Juan Bautista Tuxtepec *see* Tuxtepec
San Juan de Alicante *see* Sant Joan
  d'Alacant
San Juan del Norte *53 E4 var.* Greytown.
  Río San Juan, SE Nicaragua
San Juan de los Morros *58 D2 var.* San
  Juan. Guárico, N Venezuela
San Juanito, Isla *50 C4 island* C Mexico
San Juan Mountains *48 D1 mountain
  range* Colorado, C USA
San Juan, Río *53 E4 river* Costa Rica/
  Nicaragua
San Juan River *48 C1 river* Colorado/
  Utah, W USA
San Julián *see* Puerto San Julián
Sankt-Bartholomäi *see* Palamuse
Sankt Gallen *95 B7 var.* St. Gallen, *Eng.*
  Saint Gall, *Fr.* St-Gall. Sankt Gallen,
  NE Switzerland
Sankt-Georgen *see* Sfântu Gheorghe
Sankt-Jakobi *see* Pärnu-Jaagupi,
  Pärnumaa, Estonia
Sankt Martin *see* Martin
Sankt Moritz *see* St. Moritz
Sankt-Peterburg *110 B4 prev.*
  Leningrad, Petrograd, *Eng.* Saint
  Petersburg, *Fin.* Pietari.
  Leningradskaya Oblast', NW Russia

Sankt Pölten *95 E6* Niederösterreich,
  N Austria
Sankt Veit am Flaum *see* Rijeka
Sankuru *77 D6 river* C Dem. Rep. Congo
Şanlıurfa *117 E4 prev.* Sanli Urfa, Urfa;
  anc. Edessa. Şanlıurfa,
  S Turkey (Türkiye)
Sanli Urfa *see* Şanlıurfa
San Lorenzo *61 G5* Tarija, S Bolivia
San Lorenzo *60 A1* Esmeraldas,
  N Ecuador
San Lorenzo, Isla *60 C4 island* W Peru
Sanlúcar de Barrameda *92 C5*
  Andalucía, S Spain
San Luis *64 C4* San Luis, C Argentina
San Luis *52 B2* Petén, N Guatemala
San Luis *see* San Luis Río Colorado
San Luis Obispo *47 B7* California,
  W USA
San Luis Potosí *51 E3* San Luis Potosí,
  C Mexico
San Luis Río Colorado *50 A1 var.* San
  Luis Río Colorado. Sonora,
  NW Mexico
San Marcos *52 A2* San Marcos,
  W Guatemala
San Marcos *49 G4* Texas, SW USA
San Marcos de Arica *see* Arica
San Marino *96 E1 country capital* (San
  Marino) C San Marino
San Marino *96 D1 off.* Republic of San
  Marino. *country* S Europe
San Marino, Republic of *see* San Marino
San Martín *154 A2 Argentinean research
  station* Antarctica
San Mateo *59 E2* Anzoátegui,
  NE Venezuela
San Matías *61 H3* Santa Cruz, E Bolivia
San Matías, Gulf of *65 C5 var.* Gulf
  of San Matías. *gulf* E Argentina
San Matías, Gulf of *see* San Matías, Golfo
Sanmenxia *128 C4 var.* Shan Xian.
  Henan, C China
San Miguel *52 C3* San Miguel,
  SE El Salvador
San Miguel *50 D2* Coahuila, N Mexico
San Miguel de Ibarra *see* Ibarra
San Miguel de Tucumán *64 C3 var.*
  Tucumán. Tucumán, N Argentina
San Miguelito *53 F4* Panama, C Panama
San Miguel, Río *61 G3 river* E Bolivia
San Murezzan *see* St. Moritz
Sannär *see* Sennar
Sânnicolau-Mare *see* Sânnicolau Mare
Sânnicolau Mare *108 A4 var.*
  Sânnicolau-Mare, *Hung.*
  Nagyszentmiklós; *prev.* Sânmiclăuş
  Mare, Sinnicolau Mare. Timiş,
  W Romania
Sanok *99 E5* Podkarpackie, SE Poland
San Pablo *61 F5* Potosí, S Bolivia
San Pedro *52 C1* Corozal, NE Belize
San-Pédro *74 D5* S Ivory Coast
San Pedro *50 D3 var.* San Pedro de las
  Colonias. Coahuila, NE Mexico
San Pedro de la Cueva *50 C2* Sonora,
  NW Mexico
San Pedro de las Colonias *see* San Pedro
San Pedro de Lloc *60 B3* La Libertad,
  NW Peru
San Pedro Mártir, Sierra *50 A1
  mountain range* NW Mexico
San Pedro Sula *52 C2* Cortés,
  NW Honduras
San Rafael *64 B4* Mendoza, W Argentina
San Rafael Mountains *47 C7 mountain
  range* California, W USA
San Ramón de la Nueva Orán *64 C2*
  Salta, N Argentina
San Remo *96 A3* Liguria, NW Italy
San Salvador *52 B3 country capital* (El
  Salvador) San Salvador, SW El Salvador
San Salvador *54 D2 prev.* Watlings
  Island. *island* E The Bahamas
San Salvador de Jujuy *64 C2 var.* Jujuy.
  Jujuy, N Argentina
San Salvador, Isla *60 A4 island* Ecuador
Sansanné-Mango *see* Mango
San Sebastián *93 E1* País Vasco, N Spain
  see also Donostia
Sansepolcro *96 C3* Toscana, C Italy
San Severo *97 D5* Puglia, SE Italy
Santa Ana *61 F3* Beni, N Bolivia
Santa Ana *52 B3* Santa Ana,
  NW El Salvador
Santa Ana *46 D2* California, W USA
Santa Ana de Coro *see* Coro
Santa Ana Mountains *46 E2 mountain
  range* California, W USA

Santa Barbara *50 C2* Chihuahua,
  N Mexico
Santa Barbara *47 C7* California, W USA
Santa Comba. Galicia, NW Spain
Santa Catalina de Armada *92 B1 var.*
  Santa Comba. Galicia, NW Spain
Santa Catalina Island *47 B8 island*
  Channel Islands, California, W USA
Santa Catarina *63 E5 off.* Estado de
  Santa Catarina. *state/region* S Brazil
Santa Catarina, Estado de *see*
  Santa Catarina
Santa Clara *54 B2* Villa Clara, C Cuba
Santa Clarita *46 D1* California, USA
Santa Comba *see* Santa Cataliña
  de Armada
Santa Cruz *76 E2* São Tomé, S Sao Tome
  and Principe, Africa
Santa Cruz *47 B6* California, W USA
Santa Cruz *61 G4 department* E Bolivia
Santa Cruz Barillas *see* Barillas
Santa Cruz del Quiché *52 B2* Quiché,
  W Guatemala
Santa Cruz de Tenerife *70 A3* Tenerife,
  Islas Canarias, Spain,
  NE Atlantic Ocean
Santa Cruz, Isla *60 B5 var.* Indefatigable
  Island, Isla Chávez. *island* Galápagos
  Islands, Ecuador, E Pacific Ocean
Santa Cruz Islands *144 D3 island group*
  E Solomon Islands
Santa Cruz, Río *65 B7 river* S Argentina
Santa Elena *52 B1* Cayo, W Belize
Santa Fe *64 C4* Santa Fe, C Argentina
Santa Fe *48 D1 state capital* New Mexico,
  SW USA
Santa Fe *see* Bogotá
Santa Fe de Bogotá *see* Bogotá
Santa Genoveva *50 B3 mountain*
  NW Mexico
Santa Isabel *144 C3 var.* Bughotu. *island*
  N Solomon Islands
Santa Isabel *see* Malabo
Santa Lucia Range *47 B7 mountain
  range* California, W USA
Santa Margarita, Isla *50 B3 island*
  NW Mexico
Santa Maria *63 E5* Rio Grande do Sul,
  S Brazil
Santa Maria *47 B7* California, W USA
Santa Maria *92 A5 island* Azores,
  Portugal, NE Atlantic Ocean
Santa Maria del Buen Aire *see*
  Buenos Aires
Santa María, Isla *60 A5 var.* Isla
  Floreana, Charles Island. *island*
  Galápagos Islands, Ecuador,
  E Pacific Ocean
Santa Marta *58 B1* Magdalena,
  N Colombia
Santa Maura *see* Lefkáda
Santa Monica *46 D1* California, W USA
Santana *76 E2* São Tomé, C Sao Tome
  and Principe, Africa
Santander *92 D1* Cantabria, N Spain
Santarém *63 E2* Pará, N Brazil
Santarém *92 B3 anc.* Scalabis. Santarém,
  W Portugal
Santa Rosa *64 C4* La Pampa, C Argentina
Santa Rosa *see* Santa Rosa de Copán
Santa Rosa de Copán *52 C2 var.* Santa
  Rosa. Copán, W Honduras
Santa Rosa Island *47 B8 island*
  California, W USA
Santa Uxía de Ribeira *92 B1 var.* Ribeira.
  NW Spain
Sant Carles de la Ràpida *see* Sant Carles
  de la Ràpita
Sant Carles de la Ràpita *93 F3 var.* Sant
  Carles de la Ràpida. Cataluña, NE Spain
Santiago *64 B4 var.* Gran Santiago.
  country capital (Chile) Santiago,
  C Chile
Santiago *55 E3 var.* Santiago de los
  Caballeros. N Dominican Republic
Santiago *53 F5* Veraguas, S Panama
Santiago *see* Santiago de Compostela
Santiago *74 A3 var.* São Tiago. *island*
  Ilhas de Sotavento, S Cape Verde
  (Cabo Verde)
Santiago de Cuba *see* Santiago de Cuba
Santiago de Compostela *92 B1 var.*
  Santiago, *Eng.* Compostella; *anc.*
  Campus Stellae. Galicia, NW Spain
Santiago de Cuba *54 C3 var.* Santiago.
  Santiago de Cuba, E Cuba
Santiago de Guayaquil *see* Guayaquil
Santiago del Estero *64 C3* Santiago del
  Estero, C Argentina
Santiago de los Caballeros *see* Santiago,
  Dominican Republic

Santiago de los Caballeros *see* Ciudad de Guatemala, Guatemala
Santiago Pinotepa Nacional *see* Pinotepa Nacional
Santiago, Río *60 B2 river* N Peru
Santi Quaranta *see* Sarandë
Santissima Trinidad *see* Keelung
San Joan d'Alacant *93 F4 Cast.* San Juan de Alicante. Comunitat Valenciana, E Spain
Sant Julià de Lòria *91 A8* Sant Julià de Lòria, SW Andorra Europe
Santo *see* Espíritu Santo
Santo Antão *74 A2 island* Ilhas de Barlavento, N Cape Verde (Cabo Verde)
Santo António *76 E1* Príncipe, N Sao Tome and Principe, Africa
Santo Domingo *55 D3 prev.* Ciudad Trujillo. *country capital* (Dominican Republic) SE Dominican Republic
Santo Domingo de los Colorados *60 B1* Pichincha, NW Ecuador
Santo Domingo Tehuantepec *see* Tehuantepec
Santorini *see* Thíra
San Tomé de Guayana *see* Ciudad Guayana
Santos *63 F4* São Paulo, S Brazil
Santos Plateau *57 D5 undersea plateau* SW Atlantic Ocean
Santo Tomé *64 D3* Corrientes, NE Argentina
Santo Tomé de Guayana *see* Ciudad Guayana
San Valentín, Cerro *65 A6 mountain* S Chile
San Vicente *52 C3* San Vicente, C El Salvador
São Francisco, Rio *63 F3 river* E Brazil
Sao Hill *73 C7* Iringa, S Tanzania
São João da Madeira *92 B2* Aveiro, N Portugal
São Jorge *92 A5 island* Azores, Portugal, NE Atlantic Ocean
São Luís *63 F2 state capital* Maranhão, NE Brazil
São Mandol *see* São Manuel, Rio
São Manuel, Rio *63 E3 var.* São Mandol, Teles Pirés. *river* C Brazil
São Marcos, Baía de *63 F1 bay* N Brazil
São Miguel *92 A5 island* Azores, Portugal, NE Atlantic Ocean
Saona, Isla *55 E3 island* SE Dominican Republic
Saône *90 D5 river* E France
São Nicolau *74 A3 Eng.* Saint Nicholas. *island* Ilhas de Barlavento, N Cape Verde (Cabo Verde)
São Paulo *63 E4 state capital* São Paulo, S Brazil
São Paulo *63 E4 off.* Estado de São Paulo. *state/region* S Brazil
São Paulo de Loanda *see* Luanda
São Paulo, Estado de *see* São Paulo
São Pedro do Rio Grande do Sul *see* Rio Grande
São Roque, Cabo de *63 G2 headland* E Brazil
São Salvador *see* Salvador, Brazil
São Salvador/São Salvador do Congo *see* M'Banza Congo, Angola
São Tiago *see* Santiago
São Tomé *77 A5 country capital* (Sao Tome and Principe) São Tomé, S Sao Tome and Principe
São Tomé *76 E2 Eng.* Saint Thomas. *island* S Sao Tome and Principe
Sao Tome and Principe *76 D1 off.* Democratic Republic of Sao Tome and Principe. *country* E Atlantic Ocean
Sao Tome and Principe, Democratic Republic of *see* Sao Tome and Principe
São Tomé, Pico de *76 D2 mountain* São Tomé, C Sao Tome and Principe, Africa
São Vicente *74 A3 Eng.* Saint Vincent. *island* Ilhas de Barlavento, N Cape Verde (Cabo Verde)
São Vicente, Cabo de *92 B5 Eng.* Cape Saint Vincent, *Port.* Cabode São Vicente. *cape* S Portugal
São Vicente, Cabo de *see* São Vicente, Cabo de
Sápai *see* Sápes
Sapele *75 F5* Delta, S Nigeria
Sápes *104 D3 var.* Sápai. Anatolikí Makedonía kai Thráki, NE Greece
Sapir *119 B7 prev.* Sappir. Southern, S Israel
Sápmi *84 D3 Fin.* Lappi, *var.* Lapland. *cultural region* N Europe
Sa Pobla *93 G3* Mallorca, Spain, W Mediterranean Sea
Sappir *see* Sapir
Sapporo *130 D2* Hokkaidō, NE Japan
Sapri *97 D6* Campania, S Italy
Sapulpa *49 G1* Oklahoma, C USA
Saqqez *120 C2 var.* Saghez, Sakiz, Saqqiz. Kordestān, NW Iran
Saqqiz *see* Saqqez
Sara Buri *137 C5 var.* Saraburi. Saraburi, C Thailand
Saraburi *see* Sara Buri
Saragossa *see* Zaragoza
Saragt *see* Sarahs
Saraguro *60 B2* Loja, S Ecuador
Sarahs *122 D3 var.* Saragt, *Rus.* Serakhs. Ahal Welaýaty, S Turkmenistan
Sarajevo *100 C4 country capital* (Bosnia and Herzegovina) Federacija Bosna I Hercegovina, SE Bosnia and Herzegovina
Sarakhs *122 E2* Khorāsān-Razavī, NE Iran
Saraktash *111 D6* Orenburgskaya Oblast', W Russia
Saran' *114 C4 Kaz.* Saran. Karagandy, C Kazakhstan
Saranda *see* Sarandë
Sarandë *101 C7 var.* Saranda, *It.* Porto Edda; *prev.* Santi Quaranta. Vlorë, S Albania
Saransk *111 C5* Respublika Mordoviya, W Russia
Sarasota *43 E4* Florida, SE USA
Saratov *114 B3* Saratovskaya Oblast', W Russia
Saravan/Saravane *see* Salavan
Sarawak *138 D3 state* East Malaysia
Sarawak *see* Kuching
Sarcelles *90 D1* Val-d'Oise, Île-de-France, N France Europe
Sardegna *97 A5 Eng.* Sardinia. *island* Italy, C Mediterranean Sea
Sardinia *see* Sardegna
Sarera, Teluk *see* Cenderawasih, Teluk
Sargasso Sea *66 B4 sea* W Atlantic Ocean
Sargodha *134 C2* Punjab, NE Pakistan
Sarh *76 C4 prev.* Fort-Archambault. Moyen-Chari, S Chad
Sārī *120 D2 var.* Sari, Sāri. Māzandarān, N Iran
Saría *105 E7 island* SE Greece
Sarıkamış *117 F3* Kars, NE Turkey (Türkiye)
Sarikol Range *123 G3 Rus.* Sarykol'skiy Khrebet. *mountain range* China/ Tajikistan
Sark *89 D8 Fr.* Sercq. *island* Channel Islands
Şarkışla *116 D3* Sivas, C Turkey (Türkiye)
Sarmiento *65 B6* Chubut, S Argentina
Sarnia *38 C5* Ontario, S Canada
Sarny *108 C1* Rivnens'ka Oblast', NW Ukraine
Sarochyna *107 D5 Rus.* Sorochino. Vitsyebskaya Voblasts', N Belarus
Sarov *111 C5 prev.* Sarova. Respublika Mordoviya, SW Russia
Sarova *see* Sarov
Sarpsborg *85 B6* Østfold, S Norway
Sarrebruck *see* Saarbrücken
Sartène *91 E7* Corse, France, C Mediterranean Sea
Sarthe *90 B4 river* N France
Sárti *104 C4* Kentrikí Makedonía, N Greece
Saruhan *see* Manisa
Saryarqa *114 C4 var.* Kazakhskiy Melkosopochnik. *uplands* C Kazakhstan
Sarykol'skiy Khrebet *see* Sarikol Range
Sary-Tash *123 F2* Osh Oblusu/Oshskaya Oblast', SW Kyrgyzstan
Saryyesik-Atyrau, Peski *123 G1 desert* E Kazakhstan
Sasebo *131 A7* Nagasaki, Kyūshū, SW Japan
Saskatchewan *37 F5 province* SW Canada
Saskatchewan *37 F5 river* Manitoba/ Saskatchewan, C Canada
Saskatoon *37 F5* Saskatchewan, S Canada
Sasovo *111 B5* Ryazanskaya Oblast', W Russia
Sassandra *74 D5* S Ivory Coast
Sassandra *74 D5 var.* Ibo, Sassandra Fleuve. *river* S Ivory Coast

Sassandra Fleuve *see* Sassandra
Sassari *97 A5* Sardegna, Italy, C Mediterranean Sea
Sassenheim *86 C3* Zuid-Holland, W Netherlands
Sassnitz *94 D2* Mecklenburg-Vorpommern, NE Germany
Sathmar *see* Satu Mare
Sátoraljaújhely *99 D6* Borsod-Abaúj-Zemplén, NE Hungary
Satpura Range *134 D4 mountain range* C India
Satsuma-Sendai *131 A8* Kagoshima, Kyūshū, SW Japan
Satsunan-shoto *130 A3 island group* Nansei-shotō, SW Japan Asia
Sattanen *84 D3* Lappi, NE Finland
Satu Mare *108 B3 Ger.* Sathmar, *Hung.* Szatmárnémeti. Satu Mare, NW Romania
Sau *see* Sava
Saudi Arabia *121 B5 off.* Kingdom of Saudi Arabia, *Al 'Arabīyah as Su'ūdīyah, Ar.* Al Mamlakah al 'Arabīyah as Su'ūdīyah. *country* SW Asia
Saudi Arabia, Kingdom of *see* Saudi Arabia
Sauer *see* Sûre
Saugor *see* Sāgar
Saulkrasti *106 C3* Rīga, C Latvia
Sault Sainte Marie *38 C4* Michigan, N USA
Sault Sainte Marie *40 C1* Michigan, N USA
Sault Ste. Marie *38 C4* Ontario, S Canada
Saumur *90 B4* Maine-et-Loire, NW France
Saurimo *78 C1 Port.* Henrique de Carvalho, Vila Henrique de Carvalho. Lunda Sul, NE Angola
Sava *107 E6* Mahilyowskaya Voblasts', E Belarus
Sava *100 B3 Eng.* Save, *Ger.* Sau, *Hung.* Száva. *river* SE Europe
Savá *52 D2* Colón, N Honduras
Savai'i *145 E4 island* NW Samoa
Savannah *43 E2* Georgia, SE USA
Savannah River *43 E2 river* Georgia/ South Carolina, SE USA
Savannakhét *see* Khanthabouli
Savanna-La-Mar *54 A5* W Jamaica
Savaria *see* Szombathely**Save** *see* Sava
Save, Rio *79 E3 var.* Sabi. *river* Mozambique/Zimbabwe
Saverne *90 E3 var.* Zabern; *anc.* Tres Tabernae. Bas-Rhin, NE France
Savigliano *96 A2* Piemonte, NW Italy
Savigsivik *see* Savissivik
Savinski *see* Savinskiy
Savinskiy *110 C3 var.* Savinski. Arkhangel'skaya Oblast', NW Russia
Savissivik *82 D1 var.* Savigsivik. Avannaata, N Greenland
Savoie *91 D5 cultural region* E France
Savona *96 A2* Liguria, NW Italy
Savu Sea *139 E5 Ind.* Laut Sawu. *sea* S Indonesia
Sawakin *see* Suakin
Sawdirī *see* Sodiri
Sawhāj *see* Sūhāj
Şawqirah *121 D6 var.* Suqrah. S Oman
Sawu, Laut *see* Savu Sea
Saxe *see* Sachsen
Saxony *see* Sachsen
Sayaboury *see* Xaignabouli
Sayanskiy Khrebet *112 D3 mountain range* S Russia
Saýat *122 D3 Rus.* Sayat. Lebap Welaýaty, E Turkmenistan
Sayaxché *52 B2* Petén, N Guatemala
Şaydā/Sayida *see* Saïda
Sayhūt *121 D6* E Yemen
Saynshand *127 E2* Dornogovĭ, SE Mongolia
Sayre *41 E3* Pennsylvania, NE USA
Say'ūn *121 C6 var.* Saywūn. C Yemen
Saywūn *see* Say'ūn
Scalabis *see* Santarém
Scandinavia *66 D2 geophysical region* NW Europe
Scarborough *89 D5* N England, United Kingdom
Scarpanto *see* Kárpathos
Scebeli *see* Shebeli
Schaan *94 E1* W Liechtenstein Europe
Schaerbeek *87 C6* Brussels, C Belgium
Schaffhausen *95 B7 Fr.* Schaffhouse. Schaffhausen, N Switzerland

Schaffhouse *see* Schaffhausen
Schagen *86 C2* Noord-Holland, NW Netherlands
Schaulen *see* Šiauliai
Schebschi Mountains *see* Shebshi Mountains
Scheessel *94 B3* Niedersachsen, NW Germany
Schefferville *39 E2* Québec, E Canada
Schelde *see* Scheldt
Scheldt *87 B5 Dut.* Schelde, *Fr.* Escaut. *river* W Europe
Schell Creek Range *47 D5 mountain range* Nevada, W USA
Schenectady *41 F3* New York, NE USA
Schertz *49 G4* Texas, SW USA
Schiermonnikoog *86 D1 Fris.* Skiermûntseach. *island* Waddeneilanden, N Netherlands
Schijndel *86 D4* Noord-Brabant, S Netherlands
Schil *see* Jiu
Schiltigheim *90 E3* Bas-Rhin, NE France
Schivelbein *see* Świdwin
Schleswig *94 B2* Schleswig-Holstein, N Germany
Schleswig-Holstein *94 B2 state* N Germany
Schlettstadt *see* Sélestat
Schlochau *see* Człuchów
Schneekoppe *see* Sněžka
Schneidemühl *see* Piła
Schoden *see* Skuodas
Schönebeck *94 C4* Sachsen-Anhalt, C Germany
Schönlanke *see* Trzcianka
Schooten *see* Schoten
Schoten *87 C5 var.* Schooten. Antwerpen, N Belgium
Schouwen *86 B4 island* SW Netherlands
Schwabenalb *see* Schwäbische Alb
Schwäbische Alb *95 B6 var.* Schwabenalb, *Eng.* Swabian Jura. *mountain range* S Germany
Schwandorf *95 C5* Bayern, SE Germany
Schwaz *95 C7* Tirol, W Austria
Schweidnitz *see* Świdnica
Schweinfurt *95 B5* Bayern, SE Germany
Schweiz *see* Switzerland
Schwerin *94 C3* Mecklenburg-Vorpommern, N Germany
Schwertberg *see* Świecie
Schwiebus *see* Świebodzin
Schwyz *95 B7 var.* Schwiz. Schwyz, C Switzerland
Schyl *see* Jiu
Scio *see* Chíos
Scoresby Sound/Scoresbysund *see* Ittoqqortoormiit
Scoresby Sund *see* Kangertittivaq
Scotia Sea *57 C8 sea* SW Atlantic Ocean
Scotland *88 C3 cultural region* Scotland, United Kingdom
Scott Base *154 B4 NZ research station* Antarctica
Scott Island *154 B5 island* Antarctica
Scottsbluff *44 D3* Nebraska, C USA
Scottsboro *42 D1* Alabama, S USA
Scottsdale *48 B2* Arizona, SW USA
Scranton *41 F3* Pennsylvania, NE USA
Scrobesbyrig' *see* Shrewsbury
Scupi *see* Skopje
Scutari *see* Shkodër
Scutari, Lake *101 C5 Alb.* Liqeni i Shkodrës, *Croatian* Skadarsko Jezero. *lake* Albania/Montenegro
Scyros *see* Skýros
Searcy *42 B1* Arkansas, C USA
Seattle *46 B2* Washington, NW USA
Sébaco *52 D3* Matagalpa, W Nicaragua
Sebaste/Sebastia *see* Sivas
Sebastián Vizcaíno, Bahía *50 A2 bay* NW Mexico
Sebastopol *see* Sevastopol
Sebenico *see* Šibenik
Sechura, Bahía de *60 A3 bay* NW Peru
Secunderābād *134 D5 var.* Sikandarabad. Telangana, C India
Sedan *90 D3* Ardennes, N France
Seddon *151 D5* Marlborough, South Island, New Zealand
Seddonville *151 C5* West Coast, South Island, New Zealand
Sédhiou *74 B3* SW Senegal
Sedlez *see* Siedlce
Sedona *48 B2* Arizona, SW USA
Sedunum *see* Sion
Seeland *see* Sjælland
Seenu Atoll *see* Addu Atoll
Seesen *94 B4* Niedersachsen, C Germany

Segestica *see* Sisak
Segezha *110 B3* Respublika Kareliya,
NW Russia
Seghedin *see* Szeged
Segna *see* Senj
Segodunum *see* Rodez
Ségou *74 D3 var.* Segu. Ségou, C Mali
Segovia *92 D2* Castilla y León,
C Spain
Segoviao Wangkí *see* Coco, Río
Segu *see* Ségou
Séguédine *75 H2* Agadez, NE Niger
Seguin *49 G4* Texas, SW USA
Segura *93 E4 river* S Spain
Seinäjoki *85 D5 Swe.* Östermyra.
Etelä-Pohjanmaa, W Finland
Seine *90 D1 river* N France
Seine, Baie de la *90 B3 bay* N France
Sejong *128 E4 var.* Sejong City.
C South Korea
Sejong City *see* Sejong
Sekondi *see* Sekondi-Takoradi
Sekondi-Takoradi *75 E5 var.* Sekondi.
S Ghana
Selânik *see* Thessaloníki
Selenga *127 E1 Mong.* Selenge Mörön.
river Mongolia/Russia
Selenge Mörön *see* Selenga
Sélestat *90 E4 Ger.* Schlettstadt.
Bas-Rhin, NE France
Seleucia *see* Silifke
Selfoss *83 E5* Sudhurland, SW Iceland
Sélibabi *74 C3 var.* Sélibaby. Guidimaka,
S Mauritania
Sélibaby *see* Sélibabi
Selma *47 C6* California, W USA
Selway River *46 D2 river* Idaho,
NW USA
Selwyn Range *148 B3 mountain range*
Queensland, C Australia
Selzaete *see* Zelzate
Semarang *138 C5 var.* Samarang. Jawa,
C Indonesia
Sembé *77 B5* Sangha, NW Congo
Semendria *see* Smederevo
Semey *114 D4 prev.* Semipalatinsk.
Vostochnyy Kazakhstan, E Kazakhstan
Semechevo *see* Syemyezhava
Seminole *49 E3* Texas, SW USA
Seminole, Lake *42 D3 reservoir* Florida/
Georgia, SE USA
Semipalatinsk *see* Semey
Semnān *120 D3 var.* Samnān. Semnān,
N Iran
Semois *87 C8 river* SE Belgium
Sendai *130 D4* Miyagi, Honshū, C Japan
Sendai-wan *130 D4 bay* E Japan
Senec *90 C6 Ger.* Wartberg, *Hung.* Szenc;
prev. Szempcz. Bratislavský Kraj,
W Slovakia
Senegal *50 B3 off.* Republic of Senegal,
Fr. Sénégal. *country* W Africa
Senegal *74 C3 Fr.* Sénégal. *river* W Africa
Senegal, Republic of *see* Senegal
Senftenberg *94 D4* Brandenburg,
E Germany
Senia *see* Senj
Senica *99 C6 Ger.* Senitz, *Hung.* Szenice.
Trnavský Kraj, W Slovakia
Senića *see* Senica
Senitz *see* Senica
Senj *100 A3 Ger.* Zengg, *It.* Segna; *anc.*
Senia. Lika-Senj, NW Croatia
Senja *84 C2 prev.* Senjen. *island*
N Norway
Senjen *see* Senja
Senkaku Islands *see* Senkaku-shoto
Senkaku-shoto *130 A3 var.* Senkaku
Islands, *Chin.* Diaoyutai Lieyu,
Diaoyutai Qundao. *disputed island
group* SW Japan
Senlis *90 C3* Oise, N France
Sennar *72 C4 var.* Sannār. Sinnar,
C Sudan
Senones *see* Sens
Sens *90 C3 anc.* Agendicum, Senones.
Yonne, C France
Sensburg *see* Mrągowo
Sên, Stœng *137 D5 river* C Cambodia
Senta *100 D3 Hung.* Zenta. Vojvodina,
N Serbia
Seo de Urgel *see* La Seo d'Urgel
Seoul *129 E4 off.* Seoul Teukbyeolsi,
prev. Sŏul, *Jap.* Keijō; *prev.* Kyŏngsŏng.
country capital (South Korea)
NW South Korea
Seoul Teukbyeolsi *see* Seoul
Şepşi-Sângeorz/Sepsiszentgyörgy *see*
Sfântu Gheorghe
Sept-Îles *39 E3* Québec, SE Canada

Seraing *87 D6* Liège, E Belgium
Serakhs *see* Sarahs
Seram, Laut *139 F4 Eng.* Ceram Sea. *sea*
E Indonesia
Pulau Seram *139 F4 var.* Serang, *Eng.*
Ceram. *island* Maluku, E Indonesia
Serang *138 C5* Jawa, C Indonesia
Serang *see* Seram, Pulau
Serasan, Selat *138 C3 strait* Indonesia/
Malaysia
Serbia *100 D4 off.* Republic of Serbia;
prev. Federal Republic of Serbia,
Yugoslavia, *Serb.* Srbija, Republika
Srbija. *country* SE Europe
Serbia, Federal Republic of *see* Serbia
Serbia, Republic of *see* Serbia
Sercq *see* Sark
Serdar *see* Gyzyrlabat
Serdica *see* Sofia
Serdobol' *see* Sortavala
Serenje *78 D2* Central, E Zambia
Seres *see* Sérres
Seret/Sereth *see* Siret
Serhetabat *122 D4 prev. Rus.* Gushgy,
Kushka. Mary Welaýaty,
S Turkmenistan
Sérifos *105 C6 anc.* Seriphos. *island*
Kykládes, Greece, Aegean Sea
Seriphos *see* Sérifos
Serov *114 C3* Sverdlovskaya Oblast',
C Russia
Serowe *78 D3* Central, SE Botswana
Serpa Pinto *see* Menongue
Serpent's Mouth, The *59 F2 Sp.* Boca de
la Serpiente. *strait* Trinidad and
Tobago/Venezuela
Serpiente, Boca de la *see* Serpent's
Mouth, The
Serpukhov *111 B5* Moskovskaya Oblast',
W Russia
Sérrai *see* Sérres
Serrana, Cayo de *53 F2 island group*
NW Colombia South America
Serranilla, Cayo de *53 F2 island group*
NW Colombia South America
Caribbean Sea
Serravalle *96 E1* N San Marino
Sérres *104 C3 var.* Seres; *prev.* Sérrai.
Kentrikí Makedonía, NE Greece
Sesdlets *see* Siedlce
Sesto San Giovanni *96 B2* Lombardia,
N Italy
Sesvete *100 B2* Zagreb, N Croatia
Setabis *see* Xàtiva
Sète *91 C6 prev.* Cette. Hérault, S France
Setesdal *85 A6 valley* S Norway
Sétif *71 E2 var.* Stif. N Algeria
Setté Cama *77 A6* Ogooué-Maritime,
SW Gabon
Setúbal *92 B4 Eng.* Saint Ubes, Saint
Yves. Setúbal, W Portugal
Setúbal, Baía de *92 B4 bay* W Portugal
Seul, Lac *38 B3 lake* Ontario, S Canada
Sevan *117 G2* C Armenia
Sevana Lich *117 G3 Eng.* Lake Sevan,
Rus. Ozero Sevan. *lake* E Armenia
Sevan, Lake/Sevan, Ozero *see*
Sevana Lich
Sevastopol *109 F5 Eng.* Sebastopol.
Misto Sevastopol, S Ukraine
Severn *38 B2 river* Ontario, S Canada
Severn *89 D6 Wel.* Hafren. *river*
England/Wales, United Kingdom
Severnaya Dvina *110 C4 var.* Northern
Dvina. *river* NW Russia
Severnaya Zemlya *115 E2 var.* Nicholas
II Land. *island group* N Russia
Severnyy *110 E3* Respublika Komi,
NW Russia
Severodonetsk *see* Sievierodonetsk
Severodvinsk *110 C3 prev.* Molotov,
Sudostroy. Arkhangel'skaya Oblast',
NW Russia
Severomorsk *110 C2* Murmanskaya
Oblast', NW Russia
Seversk *114 D4* Tomskaya Oblast',
C Russia
Sevier Lake *44 A4 lake* Utah, W USA
Sevilla *92 C4 Eng.* Seville; *anc.* Hispalis.
Andalucía, SW Spain
Seville *see* Sevilla
Sevlievo *104 D2* Gabrovo, N Bulgaria
Sevluš/Sevlyush *see* Vynohradiv
Seward's Folly *see* Alaska
Seychelles *79 G1 off.* Republic of
Seychelles. *country* W Indian Ocean
Seychelles, Republic of *see* Seychelles
Seyðisfjörður *83 E5* Austurland,
E Iceland

Seýdi *122 D2 Rus.* Seydi; *prev.*
Neftezavodsk. Lebap Welaýaty,
E Turkmenistan
Seyhan *see* Adana
Sfakiá *see* Chóra Sfakíon
Sfântu Gheorghe *108 C4 Ger.*
Sankt-Georgen, *Hung.*
Sepsiszentgyörgy; *prev.* Şepşi-Sângeorz,
Sfîntu Gheorghe. Covasna, C Romania
Sfax *71 F2 Ar.* Şafāqis. E Tunisia
Sfîntu Gheorghe *see* Sfântu Gheorghe
's-Gravenhage *86 B4 var.* Den Haag,
Eng. The Hague, *Fr.* La Haye. *country
capital* (Netherlands-seat of
government) Zuid-Holland,
W Netherlands
's-Gravenzande *86 B4* Zuid-Holland,
W Netherlands
Shaan/Shaanxi Sheng *see* Shaanxi
Shaanxi *129 B5 var.* Shaan, Shaanxi
Sheng,
Shan-hsi, Shenshi, Shensi. *province*
C China
Shabani *see* Zvishavane
Shabeelle, Webi *see* Shebeli
Shache *126 A3 var.* Yarkant. Xinjiang
Uygur Zizhiqu, NW China
Shacheng *see* Huailai
Shackleton Ice Shelf *154 D3 ice shelf*
Antarctica
Shaddādi *see* Ash Shadādah
Shāhābād *see* Eslāmābād
Sha Hi *see* Orūmīyeh, Daryācheh-ye
Shahjahanabad *see* Delhi
Shahr-e Kord *120 C3 var.* Shahr Kord.
Chahār MaIJal va Bakhtīārī, C Iran
Shahr Kord *see* Shahr-e Kord
Shāhrūd *120 D2 prev.* Emāmrūd,
Emāmshahr. Semnān, N Iran
Shalkar *see* Shalqar
Shalqar *114 B4 var.* Shalkar, Chelkar.
Aktyubinsk, W Kazakhstan
Shām, Bādiyat ash *see* Syrian Desert
Shana *see* Kuril'sk
Shandi *see* Shendi
Shandong *128 D4 var.* Lu, Shandong
Sheng, Shantung. *province* E China
Shandong Sheng *see* Shandong
Shanghai *129 D5 var.* Shang-hai.
Shanghai Shi, E China
Shangrao *129 D5* Jiangxi, S China
Shan-hsi *see* Shaanxi, China
Shan-hsi *see* Shanxi, China
Shannon *89 A6 Ir.* An tSionainn. *river*
W Ireland
Shan Plateau *136 B3 plateau*
E Myanmar (Burma)
Shansi *see* Shanxi
Shantar Islands *see* Shantarskiye Ostrova
Shantarskiye Ostrova *115 G3 Eng.*
Shantar Islands. *island group*
E Russia
Shantou *129 D6 var.* Shan-t'ou, Swatow.
Guangdong, S China
Shan-t'ou *see* Shantou
Shantung *see* Shandong
Shanxi *128 C4 var.* Jin, Shan-hsi,
Shansi, Shanxi Sheng. *province*
C China
Shan Xian *see* Sanmenxia
Shanxi Sheng *see* Shanxi
Shaoguan *129 C6 var.* Shao-kuan, *Cant.*
Kukong; *prev.* Ch'u-chiang.
Guangdong, S China
Shao-kuan *see* Shaoguan
Shaqrā' *120 B4 Ar.* Riyāḑ, C Saudi Arabia
Shaqrā *see* Shuqrah
Shar *114 D5 var.* Charsk. Vostochnyy
Kazakhstan, E Kazakhstan
Shari *130 D2* Hokkaidō, NE Japan
Shari *see* Chari
Sharjah *see* Ash Shāriqah
Shark Bay *147 A5 bay* Western Australia
Sharqi, Al Jabal ash/Sharqi, Jebel esh
see Anti-Lebanon
Shashe *78 D3 var.* Shashi. *river*
Botswana/Zimbabwe
Shashi *see* Shashe
Shatskiy Rise *125 G1 undersea rise*
N Pacific Ocean
Shawnee *49 G1* Oklahoma, C USA
Shaykh, Jabal ash *see* Hermon, Mount
Shchadryn *107 D7 Rus.* Shchedrin.
Homyel'skaya Voblasts', SE Belarus
Shchedrin *see* Shchadryn
Shcheglovsk *see* Kemerovo
Shchëkino *111 B5* Tul'skaya Oblast',
W Russia
Shchors *see* Snovsk
Shchuchin *see* Shchuchyn

Shchuchinsk *114 C4 prev.* Shchuchye.
Akmola, N Kazakhstan
Shchuchye *see* Shchuchinsk
Shchuchyn *107 B5 Pol.* Szczuczyn
Nowogródzki, *Rus.* Shchuchin.
Hrodzyenskaya Voblasts', W Belarus
Shebekino *111 A6* Belgorodskaya
Oblast', W Russia
Shebelē Wenz, Wabē *see* Shebeli
Shebeli *73 D5 Amh.* Wabē Shebelē Wenz,
It. Scebeli, *Som.* Webi Shabeelle. *river*
Ethiopia/Somalia
Sheberghan *see* Shiberghān
Sheboygan *40 B2* Wisconsin, N USA
Shebshi Mountains *76 A4 var.*
Schebschi Mountains. *mountain range*
E Nigeria
Shechem *see* Nablus
Shedadi *see* Ash Shadādah
Sheffield *89 D5* N England,
United Kingdom
Shekhem *see* Nablus
Sheki *see* Şäki
Shelby *44 B1* Montana, NW USA
Sheldon *45 F3* Iowa, C USA
Shelekhov Gulf *see* Shelikhova, Zaliv
Shelikhova, Zaliv *115 G2 Eng.*
Shelekhov Gulf. *gulf* E Russia
Shendi *72 C4 var.* Shandī. River Nile,
NE Sudan
Shengking *see* Liaoning
Shenking *see* Liaoning
Shenshi/Shensi *see* Shaanxi
Shenyang *128 D3 Chin.* Shen-yang, *Eng.*
Moukden, Mukden; *prev.* Fengtien.
province capital Liaoning, NE China
Shen-yang *see* Shenyang
Shepetivka *108 D2 Rus.* Shepetovka.
Khmel'nyts'ka Oblast', NW Ukraine
Shepetovka *see* Shepetivka
Shepparton *149 C7* Victoria,
SE Australia
Sherbrooke *39 E4* Québec, SE Canada
Shereik *72 C3* River Nile, N Sudan
Sheridan *44 C2* Wyoming, C USA
Sherman *49 G2* Texas, SW USA
's-Hertogenbosch *86 C4 Fr.*
Bois-le-Duc, *Ger.* Herzogenbusch.
Noord-Brabant, S Netherlands
Shetland Islands *88 D1 island group*
NE Scotland, United Kingdom
Shevchenko *see* Aqtaū
Shiberghan/Shiberghan *see* Shibirghān
Shibirghān *123 E3 var.* Sheberghān,
Shiberghan, Shiberghan. Jowzjān,
N Afghanistan
Shibetsu *130 D2 var.* Sibetu. Hokkaidō,
NE Japan
Shibh Jazirat Sīnā' *72 C2 var.* Sinai
Peninsula, *Ar.* Shibh Jazirat Sīnā', Sīnā'.
physical region NE Egypt
Shibushi-wan *131 B8 bay* SW Japan
Shigatse *see* Xigazê
Shih-chia-chuang/Shihmen *see*
Shijiazhuang
Shihezi *126 C2* Xinjiang Uygur Zizhiqu,
NW China
Shijiazhuang *128 C4 var.*
Shih-chia-chuang; *prev.* Shihmen.
province capital Hebei, E China
Shikarpur *134 B3* Sindh, S Pakistan
Shikoku *131 C7 var.* Sikoku. *island*
SW Japan
Shikoku Basin *125 F2 var.* Sikoku Basin.
undersea basin N Philippine Sea
Shikotan, Ostrov *130 E2 Jap.*
Shikotan-tō. *island* NE Russia
Shikotan-tō *see* Shikotan, Ostrov
Shilabo *73 D5* Sumalē, E Ethiopia
Shiliguri *135 F3 prev.* Siliguri. West
Bengal, NE India
Shilka *115 F4 river* S Russia
Shillong *135 G3 state capital* Meghālaya,
NE India
Shimanto *see* Nakamura
Shimbir Berris *see* Shimbiris
Shimbiris *72 E4 var.* Shimbir Berris.
mountain N Somalia
Shimoga *see* Shivamogga
Shimonoseki *131 A7 var.* Simonoseki,
hist. Akamagaseki, Bakan. Yamaguchi,
Honshū, SW Japan
Shinano-gawa *131 C5 var.* Sinano Gawa.
river Honshū, C Japan
Shindand *122 D4 prev.* Shīndand. Herāt,
W Afghanistan
Shīndand *see* Shindand
Shingū *131 C6 var.* Singū. Wakayama,
Honshū, SW Japan

241

Shinjō *130 D4 var.* Sinzyô. Yamagata,
Honshū, C Japan
Shinyanga *73 C7* Shinyanga,
NW Tanzania
Shiprock *48 C1* New Mexico, SW USA
Shīrāz *120 D4 var.* Shīrāz. Fārs, S Iran
Shishchitsy *see* Shyshchytsy
Shivamogga *132 C2 prev.* Shimoga.
Karnātaka, W India
Shivpuri *134 D3* Madhya Pradesh,
C India
Shizugawa *130 D4* Miyagi, Honshū,
NE Japan
Shizuoka *131 D6 var.* Sizuoka. Shizuoka,
Honshū, S Japan
Shklov *see* Shklow
Shklow *107 D6 Rus.* Shklov.
Mahilyowskaya Voblasts', E Belarus
Shkodër *101 C5 var.* Shkodra, *It.* Scutari,
Croatian Skadar. Shkodër, NW Albania
Shkodra *see* Shkodër
Shkodrës, Liqeni i *see* Scutari, Lake
Shkumbinit, Lumi i *101 C6 var.*
Shkumbi, Shkumbin. *river* C Albania
Shkumbi/Shkumbin *see* Shkumbinit,
Lumi i
Sholāpur *see* Solāpur
Shostka *109 F1* Sums'ka Oblast',
NE Ukraine
Show Low *48 B2* Arizona, SW USA
Show Me State *see* Missouri
Shpola *109 E3* Cherkas'ka Oblast',
N Ukraine
Shqipëria/Shqipërisë, Republika e *see*
Albania
Shreveport *42 A2* Louisiana, S USA
Shrewsbury *89 D6 hist.* Scrobesbyrig'.
W England, United Kingdom
Shu *114 C5 Kaz.* Shū. Zhambyl,
SE Kazakhstan
Shuang-liao *see* Liaoyuan
Shū, Kazakhstan *see* Shu
Shumagin Islands *36 B3* island group
Alaska, USA
Shumen *104 D2* Shumen, NE Bulgaria
Shumilina *107 E5 Rus.* Shumilino.
Vitsyebskaya Voblasts', NE Belarus
Shumilino *see* Shumilina
Shunsen *see* Chuncheon
Shuqrah *121 B7 var.* Shaqrā. SW Yemen
Shwebo *136 B3* Sagaing,
C Myanmar (Burma)
Shyichy *107 C7 Rus.* Shiichi.
Homyel'skaya Voblasts', SE Belarus
Shymkent *114 B5 prev.* Chimkent.
Türkistan/Turkestan, S Kazakhstan
Shyshchytsy *107 C6 Rus.* Shishchitsy.
Minskaya Voblasts', C Belarus
Siam *see* Thailand
Siam, Gulf of *see* Thailand, Gulf of
Sian *see* Xi'an
Siang *see* Brahmaputra
Siangtan *see* Xiangtan
Šiauliai *106 B4 Ger.* Schaulen. Šiauliai,
N Lithuania
Siazan' *see* Siyäzän
Sibay *111 D6* Respublika Bashkortostan,
W Russia
Šibenik *100 B4 It.* Sebenico.
Šibenik-Knin, S Croatia
Siberia *see* Sibir'
Siberoet *see* Siberut, Pulau
Siberut, Pulau *138 A4 prev.* Siberoet.
island Kepulauan Mentawai,
W Indonesia
Sibi *134 B2* Baluchistan, SW Pakistan
Sibir' *115 E3 var.* Siberia. *physical region*
NE Russia
Sibiti *77 B6* Lékoumou, S Congo
Sibiu *108 B4 Ger.* Hermannstadt, *Hung.*
Nagyszeben. Sibiu, C Romania
Sibolga *138 B3* Sumatera, W Indonesia
Sibu *138 D3* Sarawak, East Malaysia
Sibut *76 C4 prev.* Fort-Sibut. Kémo,
S Central African Republic
Sibuyan Sea *139 E2 sea* W Pacific Ocean
Sichon *137 C6 var.* Ban Sichon, Si Chon.
Nakhon Si Thammarat, SW Thailand
Si Chon *see* Sichon
Sichuan *129 B5 var.* Chuan, Sichuan
Sheng, Ssu-ch'uan, Szechuan,
Szechwan. *province* C China
Sichuan Pendi *129 B5 basin* C China
Sichuan Sheng *see* Sichuan
Sicilian Channel *see* Sicily, Strait of
Sicily *97 C7 Eng.* Sicily; *anc.* Trinacria.
island Italy, C Mediterranean Sea
Sicily, Strait of *97 B7 var.* Sicilian
Channel. *strait* C Mediterranean Sea
Sicuani *61 E4* Cusco, S Peru

Sidári *104 A4* Kérkyra, Iónia Nisiá,
Greece, C Mediterranean Sea
Sidas *138 C4* Borneo, C Indonesia
Siderno *97 D7* Calabria, SW Italy
Sidhirókastron *see* Sidirókastro
Sîdî Barrânî *72 A1* NW Egypt
Sidi Bel Abbès *70 D2 var.* Sidi bel Abbès,
Sidi-Bel-Abbès. NW Algeria
Sidirókastro *104 C3 prev.*
Sidhirókastron. Kentrikí Makedonía,
NE Greece
Sidley, Mount *154 B4 mountain*
Antarctica
Sidney *44 D1* Montana, NW USA
Sidney *44 D4* Nebraska, C USA
Sidney *40 C4* Ohio, N USA
Sidon *see* Saïda
Sidra *see* Surt
Sidra/Sidra, Gulf of *see* Surt, Khalīj,
N Libya
Siebenbürgen *see* Transylvania
Siedlce *98 E3 Ger.* Sedlez, *Rus.* Sesdlets.
Mazowieckie, C Poland
Siegen *94 B4* Nordrhein-Westfalen,
W Germany
Siemiatycze *98 E3* Podlaskie, NE Poland
Siena *96 B3 Fr.* Sienne; *anc.* Saena Julia.
Toscana, C Italy
Sienne *see* Siena
Sieradz *98 C4* Sieradz, C Poland
Sierpc *98 D3* Mazowieckie, C Poland
Sierra Leone *74 C4 off.* Republic of Sierra
Leone. *country* W Africa
Sierra Leone Basin *66 C4 undersea basin*
E Atlantic Ocean
Sierra Leone, Republic of *see*
Sierra Leone
Sierra Leone Ridge *see* Sierra Leone Rise
Sierra Leone Rise *66 C4 var.* Sierra
Leone Ridge, Sierra Leone Schwelle.
undersea rise E Atlantic Ocean
Sierra Leone Schwelle *see* Sierra
Leone Rise
Sierra Vista *48 B3* Arizona, SW USA
Sievierodonetsk *109 H3 prev.*
Syeverodonets'k, *Rus.* Severodonetsk.
Luhans'ka Oblast', E Ukraine
Sífnos *105 C6 anc.* Siphnos. *island*
Kykládes, Greece, Aegean Sea
Sigli *138 A3* Sumatera, W Indonesia
Siglufjörður *83 E4* Nordhurland Vestra,
N Iceland
Signal Peak *48 A2 mountain* Arizona,
SW USA
Signan *see* Xi'an
Siguatepeque *52 C2* Comayagua,
W Honduras
Siguiri *74 D4* NE Guinea
Sihanoukville *137 D6 Khmer.* Preăh
Seihânŭ; *prev.* Kompong Som.
Sihanoukville, SW Cambodia
Siilinjärvi *84 E4* Pohjois-Savo,
C Finland
Siirt *117 F4 var.* Sert; *anc.* Tigranocerta.
Siirt, SE Turkey (Türkiye)
Sikandarabad *see* Secunderābād
Sikasso *74 D4* Sikasso, S Mali
Sikeston *45 H5* Missouri, C USA
Sikhote-Alin', Khrebet *115 G4*
mountain range SE Russia
Siking *see* Xi'an
Siklós *99 C7* Baranya, SW Hungary
Sikoku *see* Shikoku
Sikoku Basin *see* Shikoku Basin
Šilalė *106 B4* Tauragė, W Lithuania
Silchar *135 G3* Assam, NE India
Silesia *98 B4 physical region* SW Poland
Silifke *116 C4 anc.* Seleucia. İçel,
S Turkey (Türkiye)
Siliguri *see* Shiliguri
Siling Co *126 C5 lake* W China
Silinhot *see* Xilinhot
Silistra *104 E1 var.* Silistria; *anc.*
Durostorum. Silistra, NE Bulgaria
Silistria *see* Silistra
Sillamäe *106 E2 Ger.* Sillamäggi.
Ida-Virumaa, NE Estonia
Sillamäggi *see* Sillamäe
Sillein *see* Žilina
Šilutė *106 B4 Ger.* Heydekrug. Klaipėda,
W Lithuania
Silvan *117 E4* Dıyarbakır,
SE Turkey (Türkiye)
Silva Porto *see* Kuito
Silver State *see* Colorado
Silver State *see* Nevada
Simanichy *107 C7 Rus.* Simonichi.
Homyel'skaya Voblasts', SE Belarus
Simav *116 B3* Kütahya,
W Turkey (Türkiye)

Simav Çayı *116 A3 river*
NW Turkey (Türkiye)
Simbirsk *see* Ul'yanovsk
Simeto *97 C7 river* Sicilia, Italy,
C Mediterranean Sea
Simeulue, Pulau *138 A3 island*
NW Indonesia
Simferopol *109 F5* Respublika Krym,
S Ukraine
Simitla *104 C3* Blagoevgrad, SW Bulgaria
Şimlāul Silvaniei/Şimleul Silvaniei *see*
Şimleu Silvaniei
Şimleu Silvaniei *108 B3 Hung.*
Szilágysomlyó; *prev.* Şimlăul Silvaniei,
Şimleul Silvaniei. Sălaj, NW Romania
Simoda *see* Shimoda
Simonoseki *see* Shimonoseki
Simpelveld *87 D6* Limburg,
SE Netherlands
Simplon Pass *95 B8 pass* S Switzerland
Simpson *see* Fort Simpson
Simpson Desert *148 B4 desert* Northern
Territory/South Australia
Sinā' *see* Sīnā', Shibh Jazirat
Sīnā', Shibh Jazirat *72 C2 var.* Sinai
Peninsula, Sinai; *Ar.* Sīnā'. *physical
region* NE Egypt
Sinai Peninsula *see* Sīnā',
Shibh Jazirat
Sinaia *108 C4* Prahova, SE Romania
Sinano Gawa *see* Shinano-gawa
Sincelejo *58 B2* Sucre, NW Colombia
Sind *see* Sindh
Sindelfingen *95 B6* Baden-Württemberg,
SW Germany
Sindh *134 B3 var.* Sind. *province*
SE Pakistan
Sindi *106 D2 Ger.* Zintenhof. Pärnumaa,
SW Estonia
Sines *92 B4* Setúbal, S Portugal
Singan *see* Xi'an
Singapore *138 B3 country capital*
(Singapore) S Singapore
Singapore *138 A1 off.* Republic of
Singapore. *country* SE Asia
Singapore, Republic of *see* Singapore
Singen *95 B6* Baden-Württemberg,
S Germany
Singida *73 C7* Singida, C Tanzania
Singkang *139 E4* Sulawesi, C Indonesia
Singkawang *138 C3* Borneo, C Indonesia
Singora *see* Songkhla
Singó *see* Shingū
Sining *see* Xining
Siniscola *97 A5* Sardegna, Italy,
C Mediterranean Sea
Sinj *100 B4* Split-Dalmacija, SE Croatia
**Sinkiang/Sinkiang Uighur
Autonomous Region** *see* Xinjiang
Uygur Zizhiqu
Sinnamarie *see* Sinnamary
Sinnamary *59 H3 var.* Sinnamarie.
N French Guiana
Sinneh *see* Sanandaj
Sînnicolau Mare *see* Sânnicolau Mare
Sinoe, Lacul *see* Sinoie, Lacul
Sinoie, Lacul *108 D5 prev.* Lacul Sinoe.
lagoon SE Romania
Sinop *116 D2 anc.* Sinope. Sinop,
N Turkey (Türkiye)
Sinope *see* Sinop
Sinsheim *95 B6* Baden-Württemberg,
SW Germany
Sint Maarten *55 G3 Eng.* Saint Martin.
*Self-governing country of the
Netherlands.* E West Indies
Sint-Michielsgestel *86 C4*
Noord-Brabant, S Netherlands
Sin-Miclăuş *see* Gheorgheni
Sint-Niklaas *87 B5 Fr.* Saint-Nicolas.
Oost-Vlaanderen, N Belgium
Sint-Pieters-Leeuw *87 B6* Vlaams
Brabant, C Belgium
Sintra *92 B3 prev.* Cintra. Lisboa,
W Portugal
Sinūjiif *73 E5* Nugaal, NE Somalia
Sinus Aelaniticus *see* Aqaba, Gulf of
Sinus Gallicus *see* Lion, Golfe du
Sinyang *see* Xinyang
Sinzyô *see* Shinjō
Sion *95 A7 Ger.* Sitten; *anc.* Sedunum.
Valais, SW Switzerland
Sioux City *45 F3* Iowa, C USA
Sioux Falls *45 F3* South Dakota, N USA
Sioux State *see* North Dakota
Siphnos *see* Sífnos
Siping *128 D3 var.* Ssu-p'ing, Szeping;
prev. Ssu-p'ing-chieh. Jilin, NE China
Siple, Mount *154 A4 mountain* Siple
Island, Antarctica

Siquirres *53 E4* Limón, E Costa Rica
Siracusa *97 D7 Eng.* Syracuse. Sicilia,
Italy, C Mediterranean Sea
Sir Edward Pellew Group *148 B2 island
group* Northern Territory, NE Australia
Siret *108 C3 var.* Siretul, *Ger.* Sereth, *Rus.*
Seret. *river* Romania/Ukraine
Siretul *see* Siret
Siria *see* Syria
Sirikit Reservoir *136 C4 lake* N Thailand
Sirjān *120 D4 prev.* Sa'īdābād. Kermān,
S Iran
Sirna *see* Sýrna
Şırnak *117 F4* Şırnak,
SE Turkey (Türkiye)
Síros *see* Sýros
Sirte *see* Surt
Sirte, Gulf of/Sirti, Gulf of *see*
Surt, Khalīj
Şurvan *117 H3 prev.* Äli-Bayramı.
SE Azerbaijan
Sisak *100 B3 var.* Siscia, *Ger.* Sissek,
Hung. Sziszek; *anc.* Segestica.
Sisak-Moslavina, C Croatia
Siscia *see* Sisak
Sisimiut *82 C3 var.* Holsteinborg,
Holsteinsborg, Holstenborg,
Holstensborg. Qeqqata, S Greenland
Sissek *see* Sisak
Sîstān, Daryācheh-ye *see* Şāberī,
Hāmūn-e
Sitaş Cristuru *see* Cristuru Secuiesc
Siteía *105 D8 var.* Sitía. Kríti, Greece,
E Mediterranean Sea
Sitges *93 G2* Cataluña, NE Spain
Sitía *see* Siteía
Sittang *see* Sittaung
Sittard *87 D5* Limburg, SE Netherlands
Sittaung *136 B4 var.* Sittoung, Sittang.
river S Myanmar (Burma)
Sitten *see* Sion
Sittoung *see* Sittaung
Sittwe *136 A3 var.* Akyab. Rakhine State,
W Myanmar (Burma)
Siuna *52 D3* Región Autónoma Atlántico
Norte, NE Nicaragua
Siut *see* Asyūţ
Sivas *116 D3 anc.* Sebastia, Sebaste.
Sivas, C Turkey (Türkiye)
Siverek *117 E4* Şanlıurfa,
S Turkey (Türkiye)
Severskiy Donets *see* Siverskyi Donets
Siverskyi Donets *109 G3 var.* Donets,
Rus. Severskiy Donets. *river* E Ukraine
Siwa *see* Sīwah
Sīwah *72 A2 var.* Siwa. NW Egypt
Six Counties, The *see* Northern Ireland
Six-Fours-les-Plages *91 D6* Var,
SE France
Siyäzän *117 H2 Rus.* Siazan'.
NE Azerbaijan
Sjar *see* Sääre
Sjenica *101 D5 Turk.* Seniça. Serbia,
SW Serbia
Skadar *see* Shkodër
Skadarsko Jezero *see* Scutari, Lake
Skagerak *see* Skagerrak
Skagerrak *85 A6 var.* Skagerak. *channel*
N Europe
Skagit River *46 B1 river* Washington,
NW USA
Skalka *84 C3 lake* N Sweden
Skarżysko-Kamienna *98 D4*
Świętokrzyskie, C Poland
Skaudvilė *106 B4* Tauragė, SW Lithuania
Skegness *89 E6* E England,
United Kingdom
Skellefteå *84 D4* Västerbotten, N Sweden
Skellefteälven *84 C4 river* N Sweden
Ski *85 B6* Akershus, S Norway
Skiáthos *105 C5* Skiáthos, Vóreies
Sporádes, Greece, Aegean Sea
Skidal' *107 B5 Rus.* Skidel'.
Hrodzyenskaya Voblasts', W Belarus
Skidel' *see* Skidal'
Skiermûntseach *see* Schiermonnikoog
Skierniewice *98 D3* Łódzkie, C Poland
Skiftet *106 C1 strait* Finland Atlantic
Ocean Baltic Sea Gulf of Bothnia/
Gulf of Finland
Skíros *see* Skýros
Skópelos *105 C5* Skópelos, Vóreies
Sporádes, Greece, Aegean Sea
Skopje *101 D6 var.* Üsküb, *Turk.* Üsküp,
prev. Skoplje; *anc.* Scupi. *country
capital* (North Macedonia)
N North Macedonia
Skoplje *see* Skopje
Skovorodino *115 F4* Amurskaya Oblast',
SE Russia

Skudnesfjorden 85 A6 *fjord* S Norway
Skuodas 106 B3 *Ger.* Schoden, *Pol.*
Szkudy. Klaipėda, NW Lithuania
Skye, Isle of 88 B3 *island* NW Scotland,
United Kingdom
Skylge *see* Terschelling
Skýros 105 C5 *var.* Skíros.
Vóreies Sporádes, Greece, Aegean Sea
Skýros 105 C5 *var.* Skíros; *anc.* Scyros.
island Vóreies Sporádes, Greece,
Aegean Sea
Slagelse 85 B7 Vestsjælland, E Denmark
Slatina 100 C3 *Hung.* Szlatina; *prev.*
Podravska Slatina. Virovitica-
Podravina, NE Croatia
Slatina 108 B5 Olt, S Romania
Slavgorod *see* Slawharad
Slavonski Brod 100 C3 *Ger.* Brod, *Hung.*
Bród; *prev.* Brod, Brod na Savi.
Brod-Posavina, NE Croatia
Slavuta 108 C2 Khmel'nyts'ka Oblast',
NW Ukraine
Slavyansk *see* Sloviansk
Slawharad 107 E7 *Rus.* Slavgorod.
Mahilyowskaya Voblasts', E Belarus
Sławno 98 C2 Zachodnio-pomorskie,
NW Poland
Slēmani *see* As Sulaymānīyah
Sliema 102 B5 N Malta
Sligo 89 A5 *Ir.* Sligeach. Sligo,
NW Ireland
Sliven 104 D2 *var.* Slivno. Sliven,
C Bulgaria
Slivnitsa 104 B2 Sofiya, W Bulgaria
Slivno *see* Sliven
Slobozia 108 C5 Ialomiţa, SE Romania
Slonim 107 B6 *Pol.* Słonim.
Hrodzyenskaya Voblasts', W Belarus
Słonim *see* Slonim
Slovakia 99 C6 *off.* Slovak Republic, *Ger.*
Slowakei, *Hung.* Szlovákia, *Slvk.*
Slovensko, Slovenská Republika.
country C Europe
Slovak Ore Mountains *see*
Slovenské rudohorie
Slovak Republic *see* Slovakia
Slovenia 95 D8 *off.* Republic of Slovenia,
Ger. Slowenien, *Slvn.* Slovenija. *country*
SE Europe
Slovenia, Republic of *see* Slovenia
Slovenija *see* Slovenia
Slovenská Republika *see* Slovakia
Slovenské rudohorie 99 D6 *Eng.* Slovak
Ore Mountains, *Ger.* Slowakisches
Erzgebirge, Ungarisches Erzgebirge.
mountain range C Slovakia
Slovensko *see* Slovakia
Sloviansk 109 G3 *prev.* Slov''yans'k, *Rus.*
Slavyansk. Donets'ka Oblast',
E Ukraine
Slov''yans'k *see* Sloviansk
Slowakei *see* Slovakia
Slowakisches Erzgebirge *see*
Slovenské rudohorie
Slowenien *see* Slovenia
Słubice 98 B3 *Ger.* Frankfurt. Lubuskie,
W Poland
Słupsk 98 C2 *Ger.* Stolp. Pomorskie,
N Poland
Slutsk 107 C6 Minskaya Voblasts',
S Belarus
Smallwood Reservoir 39 F2 *lake*
Newfoundland and Labrador,
S Canada
Smara 70 B3 *var.* Es Semara.
N Western Sahara
Smarhon' 107 C5 *Pol.* Smorgonie, *Rus.*
Smorgon'. Hrodzyenskaya Voblasts',
W Belarus
Smederevo 100 D4 *Ger.* Semendria.
Serbia, N Serbia
Smederevska Palanka 100 D4 Serbia,
C Serbia
Smela *see* Smila
Smila 109 E2 *Rus.* Smela. Cherkas'ka
Oblast', C Ukraine
Smilten *see* Smiltene
Smiltene 106 D4 *Ger.* Smilten. Valka,
N Latvia
Smøla 84 A4 *island* W Norway
Smolensk 111 A5 Smolenskaya Oblast',
W Russia
Smorgon'/Smorgonie *see* Smarhon'
Smyrna *see* İzmir
Snake 34 B4 *river* Yukon, NW Canada
Snake River 46 C3 *river* NW USA
Snake River Plain 46 D4 *plain* Idaho,
NW USA
Sneek 86 D2 *var.* Snits. Fryslân,
N Netherlands

Sneeuw-gebergte *see* Maoke,
Pegunungan
Sněžka 98 B4 *Ger.* Schneekoppe, *Pol.*
Śnieżka. *mountain* N Czechia (Czech
Republic) /Poland
Śniardwy, Jezioro 98 D2 *Ger.*
Spirdingsee. *lake* NE Poland
Snieckus *see* Visaginas
Śnieżka *see* Sněžka
Snina 99 E5 *Hung.* Szinna. Prešovský
Kraj, E Slovakia
Snits *see* Sneek
Snovsk 109 E1 *Rus.* Shchors.
Chernihivs'ka Oblast', N Ukraine
Snowdonia *see* Eryri
Snowdrift *see* Łutselk'e
Snow Mountains *see* Maoke, Pegunungan
Snyder 49 F3 Texas, SW USA
Sobradinho, Barragem de *see*
Sobradinho, Represa de
Sobradinho, Represa de 63 F2 *var.*
Barragem de Sobradinho. *reservoir*
E Brazil
Sochi 111 A7 Krasnodarskiy Kray,
SW Russia
Société, Îles de la/Society Islands *see*
Société, Archipel de la
Society Islands 145 G4 *var.* Archipel de
Tahiti, Îles de la Société, *Eng.* Society
Islands. *island group*
W French Polynesia
Soconusco, Sierra de *see* Madre, Sierra
Socorro 48 D2 New Mexico, SW USA
Socorro, Isla 50 B5 *island* W Mexico
Socotra *see* Suqutrá
Soc Trăng 137 D6 *var.* Khanh Hung. Soc
Trăng, S Vietnam
Socuéllamos 93 E3 Castilla-La Mancha,
C Spain
Sodankylä 84 D3 Lappi, N Finland
Sodari *see* Sodiri
Söderhamn 85 C5 Gävleborg, C Sweden
Södertälje 85 C6 Stockholm, C Sweden
Sodiri 72 B4 *var.* Sawdirī, Sodari.
Northern Kordofan, C Sudan
Soekaboemi *see* Sukabumi
Soemba *see* Sumba, Pulau
Soengaipenoeh *see* Sungaipenuh
Soerabaja *see* Surabaya
Soerakarta *see* Surakarta
Sofia 104 C2 *var.* Sophia, Sofiya, *Lat.*
Serdica. *country capital* (Bulgaria)
Sofiya-Grad, W Bulgaria
Sofiya *see* Sofia
Sogamoso 58 B3 Boyacá, C Colombia
Sognefjorden 85 A5 *fjord* NE North Sea
Sohag *see* Sūhāj
Sohar *see* Şuḩār
Sohm Plain 66 B3 *abyssal plain*
NW Atlantic Ocean
Sohrau *see* Żory
Sokal 108 C2 L'vivs'ka Oblast',
NW Ukraine
Söke 116 A4 Aydın,
SW Turkey (Türkiye)
Sokhumi 117 E1 *Rus.* Sukhumi.
NW Georgia
Sokodé 75 F4 C Togo
Sokol 110 C4 Vologodskaya Oblast',
NW Russia
Sokółka 98 E3 Podlaskie, NE Poland
Sokolov 99 A5 *Ger.* Falkenau an der
Eger; *prev.* Falknov nad Ohří.
Karlovarský Kraj,
W Czechia (Czech Republic)
Sokone 74 B3 W Senegal
Sokoto 75 F3 Sokoto, NW Nigeria
Sokoto 75 F4 *river* NW Nigeria
Sokotra *see* Suqutrá
Solāpur 124 B3 *var.* Sholāpur.
Mahārāshtra, W India
Solca 108 C3 *Ger.* Solka. Suceava,
N Romania
Sol, Costa del 92 D5 *coastal region*
S Spain
Soldeu 91 B7 NE Andorra Europe
Solec Kujawski 98 C3 Kujawsko-
pomorskie, C Poland
Soledad 58 B1 Atlántico, N Colombia
Isla Soledad *see* East Falkland
Soligorsk *see* Salihorsk
Solikamsk 111 D5 Permskaya Oblast',
NW Russia
Sol'-Iletsk 111 D6 Orenburgskaya
Oblast', W Russia
Solingen 94 A4 Nordrhein-Westfalen,
W Germany
Solka *see* Solca
Sollentuna 85 C6 Stockholm, C Sweden
Solo *see* Surakarta

Solok 138 B4 Sumatera, W Indonesia
Solomon Islands 144 C3 *prev.* British
Solomon Islands Protectorate. *country*
W Solomon Islands N Melanesia
W Pacific Ocean
Solomon Islands 144 C3 *island group*
Papua New Guinea/Solomon Islands
Solomon Sea 144 B3 *sea* W Pacific Ocean
Soltau 94 B3 Niedersachsen,
NW Germany
Sol'tsy 110 A4 Novgorodskaya Oblast',
W Russia
Solun *see* Thessaloníki
Solwezi 78 D2 North Western,
NW Zambia
Sōma 130 D4 Fukushima, Honshū,
C Japan
Somalia 73 D5 *off.* Federal Republic of
Somalia, *Som.* Jamhuuriyadda
Federaalka Soomaaliya, Soomaaliya;
prev. Italian Somaliland, Somaliland
Protectorate. *country* E Africa
Somali Basin 69 E5 *undersea basin*
W Indian Ocean
Somalia, Federal Republic of *see* Somalia
Somaliland 73 D5 *disputed territory*
N Somalia
Somaliland Protectorate *see* Somalia
Sombor 100 C3 *Hung.* Zombor.
Vojvodina, NW Serbia
Someren 87 D5 Noord-Brabant,
SE Netherlands
Somerset 42 A5 *var.* Somerset Village.
W Bermuda
Somerset 40 C5 Kentucky, S USA
Somerset Island 42 A5 *island*
W Bermuda
Somerset Island 37 F2 *island* Queen
Elizabeth Islands, Nunavut,
NW Canada
Somerset Village *see* Somerset
Somers Islands *see* Bermuda
Somerton 48 A2 Arizona, SW USA
Someş 108 B3 *river* Hungary/Romania
Europe
Somme 90 C2 *river* N France
Sommerfeld *see* Lubsko
Somotillo 52 C3 Chinandega,
NW Nicaragua
Somoto 52 D3 Madriz, NW Nicaragua
Songea 73 C8 Ruvuma, S Tanzania
Sŏngjin *see* Kimch'aek
Songkhla 137 C7 *var.* Songkla, *Mal.*
Singora. Songkhla, SW Thailand
Songkla *see* Songkhla
Sonoran Desert 48 A3 *var.* Desierto de
Altar. *desert* Mexico/USA
Sonsonate 52 B3 Sonsonate,
W El Salvador
Soochow *see* Suzhou
**Soomaaliya/Soomaaliya,
Jamhuuriyadda Federaalka** *see*
Somalia
Soome Laht *see* Finland, Gulf of
Sop Hao 136 D3 Houaphan, N Laos
Sophia *see* Sofia
Sopianae *see* Pécs
Sopot 98 C2 *Ger.* Zoppot. Pomorskie,
N Poland
Sopron 99 B6 *Ger.* Ödenburg.
Győr-Moson-Sopron, NW Hungary
Sorau/Sorau in der Niederlausitz *see*
Żary
Sorgues 91 D6 Vaucluse, SE France
Sorgun 116 D3 Yozgat,
C Turkey (Türkiye)
Soria 93 E2 Castilla y León, N Spain
Soroca 108 D3 *Rus.* Soroki. N Moldova
Sorochino *see* Sarochyna
Soroki *see* Soroca
Sorokyne 109 H3 *Rus.* Krasnodon.
Luhans'ka Oblast', E Ukraine
Sorong 139 F4 Papua, E Indonesia
Sørøy *see* Sørøya
Sørøya 84 C2 *var.* Sørøy, *Smi.* Sállan.
island N Norway
Sortavala 110 B3 *prev.* Serdobol'.
Respublika Kareliya, NW Russia
Sotavento, Ilhas de 74 A3 *var.* Leeward
Islands. *island group* S Cape Verde
(Cabo Verde)
Sotkamo 84 E4 Kainuu, C Finland
Souanké 77 B5 Sangha, NW Congo
Soueida *see* As Suwaydā'
Soufli 104 D3 *prev.* Souflíon. Anatolikí
Makedonía kai Thráki, NE Greece
Souflíon *see* Soufli
Soufrière 55 F2 W Saint Lucia
Soukhné *see* As Sukhnah
Soûl *see* Seoul

Soûr 119 A5 *var.* Şūr; *anc.* Tyre.
SW Lebanon
Souris River 45 E1 *var.* Mouse River.
river Canada/USA
Soúrpi 105 B5 Thessalía, C Greece
Sousse 71 F2 *var.* Sūsah. NE Tunisia
South Africa 78 C4 *off.* Republic of
South Africa, *Afr.* Suid-Afrika. *country*
S Africa
South Africa, Republic of *see*
South Africa
South America 56 *continent*
Southampton 89 D7 *hist.* Hamwih, *Lat.*
Clausentum. S England,
United Kingdom
Southampton Island 37 G3 *island*
Nunavut, NE Canada
South Andaman 133 F2 *island* Andaman
Islands, India, NE Indian Ocean
South Australia 149 A5 *state* S Australia
South Australian Basin 142 B5
undersea basin SW Indian Ocean
South Bend 40 C3 Indiana, N USA
South Beveland *see* Zuid-Beveland
South Bruny Island 149 C8 *island*
Tasmania, SE Australia
South Carolina 43 E2 *off.* State of South
Carolina, *also known as* The Palmetto
State. *state* SE USA
South Carpathians *see* Carpaţii
Meridionali
South China Basin 125 E4 *undersea
basin* SE South China Sea
South China Sea 125 E4 *sea* SE Asia
South Dakota 44 D2 *off.* State of South
Dakota, *also known as* The Coyote
State, Sunshine State. *state* N USA
Southeast Indian Ridge 141 D7
undersea ridge Indian Ocean/
Pacific Ocean
Southeast Pacific Basin 153 E5 *var.*
Belling Hausen Mulde. *undersea basin*
SE Pacific Ocean
South East Point 149 C7 *headland*
Victoria, S Australia
Southend-on-Sea 89 E6 E England,
United Kingdom
Southern Alps 151 B6 *var.* Kā Tiritiri o
Te Moana. *mountain range* South
Island, New Zealand
Southern Cook Islands 145 F4 *island
group* S Cook Islands
Southern Cross 147 B6 Western
Australia
Southern Indian Lake 37 F4 *lake*
Manitoba, C Canada
Southern Ocean 67 B7 *ocean* Atlantic
Ocean/Indian Ocean/Pacific Ocean
Southern Uplands 88 C4 *mountain
range* S Scotland, United Kingdom
South Fiji Basin 142 D4 *undersea basin*
S Pacific Ocean
South Geomagnetic Pole 154 B3 *pole*
Antarctica
South Georgia 57 D8 *island* South
Georgia and the South Sandwich
Islands, SW Atlantic Ocean
**South Georgia and the South Sandwich
Islands** 57 D8 *UK Overseas Territory*
SW Atlantic Ocean
South Goulburn Island 146 E2 *island*
Northern Territory, N Australia
South Huvadhu Atoll 132 A5 *atoll*
S Maldives
South Indian Basin 141 D7 *undersea
basin* Indian Ocean/Pacific Ocean
South Island 151 C6 *var.* Te
Waipounamu. *island* S New Zealand
South Korea 129 E4 *off.* Republic of
Korea, *Kor.* Taehan Min'guk. *country*
E Asia
South Lake Tahoe 47 C5 California,
W USA
South Orkney Islands 154 A2 *island
group* Antarctica
South Ossetia 117 F2 *former autonomous
region* SW Georgia
South Pacific Basin *see* Southwest
Pacific Basin
South Platte River 44 D4 *river*
Colorado/Nebraska, C USA
South Pole 154 B3 *pole* Antarctica
South Sandwich Islands 57 D8 *island
group* SW Atlantic Ocean
South Sandwich Trench 57 E8 *trench*
SW Atlantic Ocean
South Shetland Islands 154 A2 *island
group* Antarctica
South Shields 88 D4 NE England,
United Kingdom

South Sioux City *45 F3* Nebraska, C USA
South Sudan *72 B5 off.* Republic of South Sudan, *country* N Africa
South Taranaki Bight *150 C4 bight* SE Tasman Sea
South Tasmania Plateau *see* Tasman Plateau
South Uist *88 B3 island* NW Scotland, United Kingdom
South-West Africa/South West Africa *see* Namibia
South West Cape *151 A8 var.* Puhiwaero. *headland* Stewart Island, New Zealand
Southwest Indian Ocean Ridge *see* Southwest Indian Ridge
Southwest Indian Ridge *141 B6 var.* Southwest Indian Ocean Ridge. *undersea ridge* SW Indian Ocean
Southwest Pacific Basin *143 E4 var.* South Pacific Basin. *undersea basin* SE Pacific Ocean
Sovereign Base Area *102 C5 uk military installation* S Cyprus
Soweto *78 D4* Gauteng, NE South Africa
Sōya-kaikyō *see* La Pérouse Strait
Spain *92 D3 off.* Kingdom of Spain, *Sp.* España; *anc.* Hispania, Iberia, *Lat.* Hispana. *country* SW Europe
Spain, Kingdom of *see* Spain
Spalato *see* Split
Spanish Town *54 B5 hist.* St.Iago de la Vega. C Jamaica
Sparks *47 C5* Nevada, W USA
Sparta *see* Spárti
Spartanburg *43 E1* South Carolina, SE USA
Spárti *105 B6 Eng.* Sparta. Pelopónnisos, S Greece
Spearfish *44 D2* South Dakota, N USA
Speightstown *55 G1* NW Barbados
Spencer *45 F3* Iowa, C USA
Spencer Gulf *149 B6 gulf* South Australia
Spey *88 C3 river* NE Scotland, United Kingdom
Spice Islands *see* Maluku
Spiess Seamount *67 C7 seamount* S Atlantic Ocean
Spijkenisse *86 B4* Zuid-Holland, SW Netherlands
Spíli *105 C8* Kríti, Greece, E Mediterranean Sea
Spin Boldak *123 E5 prev.* Spīn Būldak. Kandahār, S Afghanistan
Spīn Būldak *see* Spin Boldak
Spirdingsee *see* Śniardwy, Jezioro
Spitsbergen *83 F2 island* NW Svalbard
Split *100 B4 It.* Spalato. Split-Dalmacija, S Croatia
Spōgi *106 D4* Daugvapils, SE Latvia
Spokane *46 C2* Washington, NW USA
Spratly Islands *138 B2 disputed territory* SE Asia
Spree *94 D4 river* E Germany
Springbok *78 B5* NE South Africa
Springfield *40 B4 state capital* Illinois, N USA
Springfield *41 G3* Massachusetts, NE USA
Springfield *45 G5* Missouri, C USA
Springfield *40 C4* Ohio, N USA
Springfield *46 B3* Oregon, NW USA
Spring Garden *59 F2* NE Guyana
Spring Hill *43 E4* Florida, SE USA
Springs Junction *151 C5* West Coast, South Island, New Zealand
Springsure *148 D4* Queensland, E Australia
Sprottau *see* Szprotawa
Spruce Knob *41 E4 mountain* West Virginia, NE USA
Srbija/Republika Srbija *see* Serbia
Srbinje *see* Foča
Srbobran *100 D3 var.* Bácsszenttamás, *Hung.* Szenttamás. Vojvodina, N Serbia
Srebrenica *100 C4* Republika Srpska, E Bosnia and Herzegovina
Sredets *104 D2 prev.* Syulemeshlii. Stara Zagora, C Bulgaria
Sredets *104 E2 prev.* Grudovo. Burgas, E Bulgaria
Srednerusskaya Vozvyshennost' *109 G1 Eng.* Central Russian Upland. *mountain range* W Russia
Sremska Mitrovica *100 C3 prev.* Mitrovica, *Ger.* Mitrowitz. Vojvodina, NW Serbia
Srepok, Sông *see* Srêpôk, Tônle
Srêpôk, Tônle *137 E5 var.* Sông Srepok. *river* Cambodia/Vietnam

Sri Aman *138 C3* Sarawak, East Malaysia
Sri Jayawardenapura Kotte *132 D3 administrative capital* (Sri Lanka) Western Province, W Sri Lanka
Srīkākulam *135 F5* Andhra Pradesh, E India
Sri Lanka *132 D3 off.* Democratic Socialist Republic of Sri Lanka; *prev.* Ceylon. *country* S Asia
Sri Lanka, Democratic Socialist Republic of *see* Sri Lanka
Srinagarind Reservoir *137 C5 lake* W Thailand
Srpska, Republika *100 B3 republic* Bosnia and Herzegovina
Ssu-ch'uan *see* Sichuan
Ssu-p'ing/Ssu-p'ing-chieh *see* Siping
Stabroek *87 B5* Antwerpen, N Belgium
Stade *94 B3* Niedersachsen, NW Germany
Stadskanaal *86 E2* Groningen, NE Netherlands
Stafford *89 D6* C England, United Kingdom
Staicele *106 D3* Limbaži, N Latvia
Ştaierdorf-Anina *see* Anina
Stájerlakanina *see* Anina
Stakhanov *see* Kadiïvka
Stalin *see* Varna
Stalinabad *see* Dushanbe
Stalingrad *see* Volgograd
Stalino *see* Donets'k
Stalinobod *see* Dushanbe
Stalinsk *see* Novokuznetsk
Stalinski Zaliv *see* Varnenski Zaliv
Stalin, Yazovir *see* Iskar, Yazovir
Stalowa Wola *98 E4* Podkarpackie, SE Poland
Stamford *41 F3* Connecticut, NE USA
Stampalia *see* Astypálaia
Stanislau *see* Ivano-Frankivsk
Stanislav *see* Ivano-Frankivsk
Stanisławów *see* Ivano-Frankivsk
Stanke Dimitrov *see* Dupnitsa
Stanley *65 D7 var.* Port Stanley, Puerto Argentino. *dependent territory capital* (Falkland Islands) East Falkland, Falkland Islands
Stanleyville *see* Kisangani
Stann Creek *see* Dangriga
Stanovoy Khrebet *113 E3 mountain range* SE Russia
Stanthorpe *149 D5* Queensland, E Australia
Staphorst *86 D2* Overijssel, E Netherlands
Starachowice *98 D4* Świętokrzyskie, C Poland
Stara Kanjiža *see* Kanjiža
Stara Pazova *100 D3 Ger.* Altpasua, *Hung.* Ópazova. Vojvodina, N Serbia
Stara Planina *see* Balkan Mountains
Stara Zagora *104 D2 Lat.* Augusta Trajana. Stara Zagora, C Bulgaria
Starbuck Island *145 G3 prev.* Volunteer Island. *island* E Kiribati
Stargard in Pommern *see* Stargard Szczeciński
Stargard Szczeciński *98 B3 Ger.* Stargard in Pommern. Zachodnio-pomorskie, NW Poland
Stari Bečej *see* Bečej
Starobel'sk *see* Starobilsk
Starobielsk *see* Starobilsk
Starobilsk *109 H2 Rus.* Starobel'sk. Luhans'ka Oblast', E Ukraine
Starobin *107 C7 var.* Starobyn. Minskaya Voblasts', S Belarus
Starobyn *see* Starobin
Starogard Gdański *98 C2 Ger.* Preussisch-Stargard. Pomorskie, N Poland
Starokonstantinov *see* Starokostiantyniv
Starokostiantyniv *108 D2 prev.* Starokostyantyniv, *Rus.* Starokonstantinov. Khmel'nyts'ka Oblast', NW Ukraine
Starokostyantyniv *see* Starokostiantyniv
Starominskaya *111 A7* Krasnodarskiy Kray, SW Russia
Staryya Darohi *107 C6 Rus.* Staryye Dorogi. Minskaya Voblasts', S Belarus
Staryye Dorogi *see* Staryya Darohi
Staryy Oskol *111 B6* Belgorodskaya Oblast', W Russia
State College *41 E4* Pennsylvania, NE USA
Staten Island *see* Estados, Isla de los
Statesboro *43 E2* Georgia, SE USA
States, The *see* United States of America
Station Nord *83 F1* N Greenland

Staunton *41 E5* Virginia, NE USA
Stavanger *85 A6* Rogaland, S Norway
Stavers Island *see* Vostok Island
Stavropol' *111 B7 prev.* Voroshilovsk. Stavropol'skiy Kray, SW Russia
Stavropol' *see* Tol'yatti
Steamboat Springs *44 C4* Colorado, C USA
Steenwijk *86 D2* Overijssel, N Netherlands
Steier *see* Steyr
Steierdorf/Steierdorf-Anina *see* Anina
Steinamanger *see* Szombathely
Steinkjer *84 B4* Nord-Trøndelag, C Norway
Stejarul *see* Karapelit
Stendal *94 C3* Sachsen-Anhalt, C Germany
Stepanakert *see* Xankändi
Stephenville *49 F3* Texas, SW USA
Sterling *44 D4* Colorado, C USA
Sterling *40 B3* Illinois, N USA
Sterlitamak *114 B3* Respublika Bashkortostan, W Russia
Stettin *see* Szczecin
Stettiner Haff *see* Szczeciński, Zalew
Stevenage *89 E6* E England, United Kingdom
Stevens Point *40 B2* Wisconsin, N USA
Stewart Island *151 A8 var.* Rakiura. *island* S New Zealand
Steyerlak-Anina *see* Anina
Steyr *95 D6 var.* Steier. Oberösterreich, N Austria
St. Helena Bay *78 B5 bay* SW South Africa
Stif *see* Sétif
Stillwater *49 G1* Oklahoma, C USA
Štip *101 E6* E North Macedonia
Stirling *88 C4* C Scotland, United Kingdom
Stjørdalshalsen *84 B4* Nord-Trøndelag, C Norway
St-Maur-des-Fossés *90 E2* Val-de-Marne, Île-de-France, N France
Stockach *95 B6* Baden-Württemberg, S Germany
Stockholm *85 C6 country capital* (Sweden) Stockholm, C Sweden
Stockmannshof *see* Pļaviņas
Stockton *47 B6* California, W USA
Stockton Plateau *49 E4 plain* Texas, SW USA
Stœng Trêng *see* Stung Treng
Stoke *see* Stoke-on-Trent
Stoke-on-Trent *89 D6 var.* Stoke. C England, United Kingdom
Stolbce *see* Stowbtsy
Stolbtsy *see* Stowbtsy
Stolp *see* Słupsk
Stolpmünde *see* Ustka
Stómio *104 B4* Thessalía, C Greece
Store Bælt *see* Storebælt
Storebelt *see* Storebælt
Støren *85 B5* Sør-Trøndelag, S Norway
Storfjorden *83 G2 fjord* S Norway
Storhammer *see* Hamar
Stornoway *88 B2* NW Scotland, United Kingdom
Storsjön *85 B5 lake* C Sweden
Storuman *84 C4* Västerbotten, N Sweden
Storuman *84 C4 lake* N Sweden
Stowbtsy *107 C6 Pol.* Stolbce, *Rus.* Stolbtsy. Minskaya Voblasts', C Belarus
Strabane *89 B5 Ir.* An Srath Bán. W Northern Ireland, United Kingdom
Strakonice *99 A5 Ger.* Strakonitz. Jihočeský Kraj, S Czechia (Czech Republic)
Strakonitz *see* Strakonice
Stralsund *94 D2* Mecklenburg-Vorpommern, NE Germany
Stranraer *89 C5* S Scotland, United Kingdom
Strasbourg *90 E3 Ger.* Strassburg; *anc.* Argentoratum. Bas-Rhin, NE France
Strășeni *108 D3 var.* Strasheny. C Moldova
Strasheny *see* Strășeni
Strassburg *see* Strasbourg, France
Strassburg *see* Aiud, Romania
Stratford *150 D4* Taranaki, North Island, New Zealand
Strathfield *148 E2* New South Wales, E Australia
Straubing *95 C6* Bayern, SE Germany
Strehaia *108 B5* Mehedinți, SW Romania
Strelka *114 D4* Krasnoyarskiy Kray, C Russia
Strigonium *see* Esztergom

Strofiliá *see* Strofyliá
Strofyliá *105 C5 var.* Strofilia. Évvoia, C Greece
Stromboli *97 D6 island* Isole Eolie, S Italy
Stromeferry *88 C3* N Scotland, United Kingdom
Strömstad *85 B6* Västra Götaland, S Sweden
Strömsund *84 C4* Jämtland, C Sweden
Struga *101 D6* SW North Macedonia
Struma *see* Strymónas
Strumica *101 E6* E North Macedonia
Strumyani *104 C3* Blagoevgrad, SW Bulgaria
Strymónas *104 C3 Bul.* Struma. *river* Bulgaria/Greece
Stryi *108 B2 prev.* Stryy. L'vivs'ka Oblast', NW Ukraine
Stryy *see* Stryi
Studholme *151 B6* Canterbury, South Island, New Zealand
Stuhlweissenberg *see* Székesfehérvár
Stung Treng *137 D5 Khmer.* Stœng Trêng. Stung Treng, NE Cambodia
Sturgis *44 D3* South Dakota, N USA
Stuttgart *95 B6* Baden-Württemberg, SW Germany
Stykkishólmur *83 E4* Vesturland, W Iceland
Styr *108 C1 Rus.* Styr'. *river* Belarus/Ukraine
Su *see* Jiangsu
Suakin *72 C3 var.* Sawakin. Red Sea, NE Sudan
Subačius *106 C4* Panevėžys, NE Lithuania
Subaykhān *118 E3* Dayr az Zawr, E Syria
Subotica *100 D2 Ger.* Maria-Theresiopel, *Hung.* Szabadka. Vojvodina, N Serbia
Suceava *108 C3 Ger.* Suczawa, *Hung.* Szucsava. Suceava, NE Romania
Su-chou *see* Suzhou
Suchow *see* Suzhou, Jiangsu, China
Suchow *see* Xuzhou, Jiangsu, China
Sucker State *see* Illinois
Sucre *61 F4 hist.* Chuquisaca, La Plata. *country capital* (Bolivia-legal capital) Chuquisaca, S Bolivia
Suczawa *see* Suceava
Sudak *109 F5* S Ukraine
Sudan *72 A4 off.* Republic of Sudan, *Ar.* Jumhuriyat as-Sudan; *prev.* Anglo-Egyptian Sudan. *country* N Africa
Sudanese Republic *see* Mali
Sudan, Jumhuriyat as- *see* Sudan
Sudan, Republic of *see* Sudan
Sudbury *38 C4* Ontario, S Canada
Sudd *73 B5 swamp region* N South Sudan
Sudeten *98 B4 var.* Sudetes, Sudetic Mountains, *Cz./Pol.* Sudety. *mountain range* Czechia (Czech Republic) / Poland
Sudetes/Sudetic Mountains/Sudety *see* Sudeten
Südkarpaten *see* Carpaţii Meridionali
Südliche Morava *see* Južna Morava
Sudong, Pulau *138 A2 island* SW Singapore Asia
Sudostroy *see* Severodvinsk
Sue *73 B5 river* S South Sudan
Sueca *93 F3* Comunitat Valenciana, E Spain
Sue Wood Bay *42 B5 bay* W Bermuda North America W Atlantic Ocean
Suez *see* As Suways
Suez Canal *see* Suways, Qanāt as
Suez, Gulf of *see* Khalīj as Suways
Suğla Gölü *116 C4 lake* SW Turkey (Türkiye)
Şūhāj *72 B2 var.* Sawhāj, Suliag; *Eng.* Sohag. C Egypt
Şuḩār *121 D5 var.* Sohar. N Oman
Sühbaatar *127 E1* Selenge, N Mongolia
Suhl *95 C5* Thüringen, C Germany
Suicheng *see* Suixi
Suid-Afrika *see* South Africa
Suidwes-Afrika *see* Namibia
Suixi *129 C6 var.* Suicheng. Guangdong, S China
Sujawal *134 B3* Sindh, SE Pakistan
Sukabumi *138 C5 prev.* Soekaboemi. Jawa, C Indonesia
Sukagawa *131 D5* Fukushima, Honshū, C Japan
Sukarnapura *see* Jayapura
Sukarno, Puntjak *see* Jaya, Puncak
Sukhne *see* As Sukhnah

Sukhona 110 C4 var. Tot'ma. river NW Russia

Sukhumi see Sokhumi

Sukkertoppen see Maniitsoq

Sukkur 134 B3 Sindh, SE Pakistan

Sukumo 131 B7 Kōchi, Shikoku, SW Japan

Sulaimaniya see As Sulaymānīyah

Sulaiman Range 134 C2 mountain range C Pakistan

Sula, Kepulauan 139 E4 island group C Indonesia

Sulawesi 139 E4 Eng. Celebes. island C Indonesia

Sulawesi, Laut see Celebes Sea

Sulechów 98 B3 Ger. Züllichau. Lubuskie, W Poland

Suliag see Sūhāj

Sullana 60 B2 Piura, NW Peru

Sullivan Island see Lanbi Kyun

Sulphur Springs 49 G2 Texas, SW USA

Sultānābād see Arāk

Sulu Archipelago 139 E3 island group SW Philippines

Sülüktü 123 E2 Rus. Sulyukta. Batken Oblusu/Batkenskaya Oblast', SW Kyrgyzstan

Sulu, Laut see Sulu Sea

Sulu Sea 139 E2 var. Laut Sulu. sea SW Philippines

Sulyukta see Sülüktü

Sumatera 137 B8 Eng. Sumatra. island W Indonesia

Sumatra see Sumatera

Šumava see Bohemian Forest

Sumba, Pulau 139 E5 Eng. Sandalwood Island; prev. Soemba. island Nusa Tenggara, C Indonesia

Sumba, Selat 139 E5 strait Nusa Tenggara, S Indonesia

Sumbawanga 73 B7 Rukwa, W Tanzania

Sumbe 78 B2 var. N'Gunza, Port. Novo Redondo. Cuanza Sul, W Angola

Sumeih 73 B5 Southern Darfur, S Sudan

Sumgait see Sumqayıt, Azerbaijan

Summer Lake 46 B4 lake Oregon, NW USA

Summit 93 H5 Alaska, USA

Sumqayt 117 H2 Rus. Sumgait. E Azerbaijan

Sumy 109 F2 Sums'ka Oblast', NE Ukraine

Sunbury 149 C7 Victoria, SE Australia

Sunda Islands see Greater Sunda Islands

Sunda, Selat 138 B5 strait Jawa/Sumatera, SW Indonesia

Sunda Trench see Java Trench

Sunderland 88 D4 var. Wearmouth. NE England, United Kingdom

Sundsvall 85 C5 Västernorrland, C Sweden

Sunflower State see Kansas

Sungaipenuh 138 B4 prev. Soengaipenoeh. Sumatera, W Indonesia

Sunnyvale 47 A6 California, W USA

Sunset State see Oregon

Sunshine State see Florida

Sunshine State see New Mexico

Sunshine State see South Dakota

Suntar 115 F3 Respublika Sakha (Yakutiya), NE Russia

Sunyani 75 E5 W Ghana

Suolȯčielgi see Saariselkä

Suomenlahti see Finland, Gulf of

Suomen Tasavalta/Suomi see Finland

Suomussalmi 84 E4 Kainuu, E Finland

Suong 137 D6 Tbong Khmum, C Cambodia

Suoyarvi 110 B3 Respublika Kareliya, NW Russia

Supe 60 C3 Lima, W Peru

Supérieur, Lac see Superior, Lake

Superior 40 A1 Wisconsin, N USA

Superior, Lake 40 B1 Fr. Lac Supérieur. lake Canada/USA

Suqrah see Sawqirah

Suquṭrá 121 C7 var. Sokotra, Eng. Socotra. island SE Yemen

Şūr 121 E5 NE Oman

Şūr see Soûr

Surabaja see Surabaya

Surabaya 138 D5 prev. Surabaja, Soerabaja. Jawa, C Indonesia

Surakarta 138 C5 var. Solo; prev. Soerakarta. Jawa, S Indonesia

Šurany 99 C6 Hung. Nagysurány. Nitriansky Kraj, SW Slovakia

Sūrat 134 C4 Gujarāt, W India

Suratdhani see Surat Thani

Surat Thani 137 C6 var. Suratdhani. Surat Thani, SW Thailand

Surazh 107 E5 Vitsyebskaya Voblasts', NE Belarus

Surdulica 101 E5 Serbia, SE Serbia

Sûre 87 D7 var. Sauer. river W Europe

Surendranagar 134 C4 Gujarāt, W India

Surfers Paradise 149 E5 Queensland, E Australia

Surgut 114 D3 Khanty-Mansiyskiy Avtonomnyy Okrug-Yugra, C Russia

Surin 137 D5 Surin, E Thailand

Surinam see Suriname

Suriname 59 G3 off. Republic of Suriname, var. Surinam; prev. Dutch Guiana, Netherlands Guiana. country N South America

Suriname, Republic of see Suriname

Sūriyah/Sūriyah, Al-Jumhūriyah al-'Arabīyah as- see Syria

Surkhab, Darya-i- see Kahmard, Daryā-ye

Surkhob 123 F3 river C Tajikistan

Surt 71 G2 var. Sidra, Sirte. N Libya

Surt, Khalij 71 F2 var. Gulf of Sirte, Gulf of Sidra, Gulf of Sirti, Sidra. gulf N Libya

Surtsey 83 E5 island S Iceland

Suruga-wan 131 D6 bay SE Japan

Susa 96 A2 Piemonte, NE Italy

Sūsah see Sousse

Susanville 47 B5 California, W USA

Susitna 36 C3 Alaska, USA

Susteren 87 D5 Limburg, SE Netherlands

Susuman 115 G3 Magadanskaya Oblast', E Russia

Sutlej 134 C2 river India/Pakistan

Suur Munamägi 106 D3 var. Munamägi, Ger. Eier-Berg. mountain SE Estonia

Suur Väin 106 C2 Ger. Grosser Sund. strait W Estonia

Suva 145 E4 country capital (Fiji) Viti Levu, W Fiji

Suvalkai/Suvalki see Suwałki

Suvorovo 104 E2 prev. Vetrino. Varna, E Bulgaria

Suwałki 98 E2 Lith. Suvalkai, Rus. Suvalki. Podlaskie, NE Poland

Suways, Qanāt as- 72 B1 Eng. Suez Canal. canal NE Egypt

Suzhou 129 D5 var. Soochow, Su-chou, Suchow; prev. Wuhsien. Jiangsu, E China

Svalbard 83 E1 constituent part of Norway. island group Arctic Ocean

Svartisen 84 C3 glacier C Norway

Svay Riĕng 137 D6 Khmer. Svay Riĕng. Svay Riĕng, S Cambodia

Svay Riĕng see Svay Riĕng

Sveg 85 B5 Jämtland, C Sweden

Svenstavik 85 C5 Jämtland, C Sweden

Sverdlovsk see Yekaterinburg

Sverige see Sweden

Sveti Vrach see Sandanski

Svetlogorsk see Svietlahorsk

Svetlograd 111 B7 Stavropol'skiy Kray, SW Russia

Svetlovodsk see Svitlovodsk

Svetozarevo see Jagodina

Svilengrad 104 D3 prev. Mustafa-Pasha. Haskovo, S Bulgaria

Svitlovodsk 109 F3 Rus. Svetlovodsk. Kirovohrads'ka Oblast', C Ukraine

Svizzera see Switzerland

Svobodnyy 115 G4 Amurskaya Oblast', SE Russia

Svyataya Anna Trough 155 C4 var. Saint Anna Trough. trough N Kara Sea

Svyetlahorsk 107 D7 Rus. Svetlogorsk. Homyel'skaya Voblasts', SE Belarus

Swabian Jura see Schwäbische Alb

Swakopmund 78 B3 Erongo, W Namibia

Swan Islands 53 E1 island group NE Honduras North America

Swansea 89 C7 Wel. Abertawe. S Wales, United Kingdom

Swarzędz 98 C3 Poznań, W Poland

Swatow see Shantou

Swaziland see Eswatini

Sweden 84 B4 off. Kingdom of Sweden, Swe. Sverige. country N Europe

Sweden, Kingdom of see Sweden

Sweetwater 49 F3 Texas, SW USA

Świdnica 98 B4 Ger. Schweidnitz. Wałbrzych, SW Poland

Świdwin 98 B2 Ger. Schivelbein. Zachodnio-pomorskie, NW Poland

Świebodzice 98 B4 Ger. Freiburg in Schlesien, Swiebodzice. Wałbrzych, SW Poland

Świebodzin 98 B3 Ger. Schwiebus. Lubuskie, W Poland

Świecie 98 C3 Ger. Schwertberg. Kujawsko-pomorskie, C Poland

Swindon 89 D7 S England, United Kingdom

Swinemünde see Świnoujście

Świnoujście 98 B2 Ger. Swinemünde. Zachodnio-pomorskie, NW Poland

Swiss Confederation see Switzerland

Switzerland 95 A7 off. Swiss Confederation, Fr. La Suisse, Ger. Schweiz, It. Svizzera; anc. Helvetia. country C Europe

Sycaminum see Hefa

Sydenham Island see Nonouti

Sydney 148 D1 state capital New South Wales, SE Australia

Sydney 39 G4 Cape Breton Island, Nova Scotia, SE Canada

Sydney Island see Manra

Syedpur see Saidpur

Syemyezhava 107 C6 Rus. Semechevo. Minskaya Voblasts', C Belarus

Syene see Aswan

Syeverodonets'k see Sievierodonetsk

Syktyvkar 110 D4 prev. Ust'-Sysol'sk. Respublika Komi, NW Russia

Sylhet 135 G3 Sylhet, NE Bangladesh

Synelnykove 109 G3 Dnipropetrovska Oblast, E Ukraine

Syowa 154 C2 Japanese research station Antarctica

Syracuse 41 E3 New York, NE USA

Syracuse see Siracusa

Syrdar'ya 114 B4 Sirdaryo Viloyati, E Uzbekistan

Syria 118 B3 off. Syrian Arab Republic, var. Siria, Syrie, Ar. Al-Jumhūriyah al-'Arabīyah as-Sūriyah, Sūriyah. country SW Asia

Syrian Arab Republic see Syria

Syrian Desert 119 D5 Ar. Al Hamad, Bādiyat ash Shām. desert SW Asia

Syrie see Syria

Sýrna 105 E7 var. Sirna. island Kykládes, Greece, Aegean Sea

Sýros 105 C6 var. Síros. island Kykládes, Greece, Aegean Sea

Syulemeshlii see Sredets

Syvash, Zaliv see Syvash, Zatoka

Syvash, Zatoka 109 F4 Rus. Zaliv Syvash. inlet S Ukraine

Syzran' 111 C6 Samarskaya Oblast', W Russia

Szabadka see Subotica

Szamotuły 98 B3 Poznań, W Poland

Szászrégen see Reghin

Szatmárrnémeti see Satu Mare

Száva see Sava

Szczecin 98 B3 Eng./Ger. Stettin. Zachodnio-pomorskie, NW Poland

Szczecinek 98 B2 Ger. Neustettin. Zachodnio-pomorskie, NW Poland

Szczeciński, Zalew 98 A2 var. Stettiner Haff, Ger. Oderhaff. bay Germany/Poland

Szczuczyn Nowogródzki see Shchuchyn

Szczytno 98 D3 Ger. Ortelsburg. Warmińsko-Mazurskie, NE Poland

Szechuan/Szechwan see Sichuan

Szeged 99 D7 Ger. Szegedin, Rom. Seghedin. Csongrád, SE Hungary

Szegedin see Szeged

Székelkeresztúr see Cristuru Secuiesc

Székesfehérvár 99 C6 Ger. Stuhlweissenberg; anc. Alba Regia. Fejér, W Hungary

Szeklerburg see Miercurea-Ciuc

Szekler Neumarkt see Târgu Secuiesc

Szekszárd 99 C7 Tolna, S Hungary

Szenice see Senica

Szentgotthárd 99 A7 Vas, W Hungary

Szenttamás see Srbobran

Szeping see Siping

Szilágysomlyó see Şimleu Silvaniei

Szinna see Snina

Sziszek see Sisak

Szitás-Keresztúr see Cristuru Secuiesc

Szkudy see Skuodas

Szlatina see Slatina

Szlovákia see Slovakia

Szolnok 99 D6 Jász-Nagykun-Szolnok, C Hungary

Szombathely 99 B6 Ger. Steinamanger; anc. Sabaria, Savaria. Vas, W Hungary

Szprotawa 98 B4 Ger. Sprottau. Lubuskie, W Poland

Sztálinváros see Dunaújváros

Szucsava see Suceava

T

Tabariya, Bahrat see Kinneret, Yam

Table Rock Lake 49 G1 reservoir Arkansas/Missouri, C USA

Tábor 99 B5 Jihočeský Kraj, S Czechia (Czech Republic)

Tabora 73 B7 Tabora, W Tanzania

Tabrīz 120 C2 var. Tebriz; anc. Tauris. Āžarbāyjān-e Sharqi, NW Iran

Tabuaeran 145 G2 prev. Fanning Island. atoll Line Islands, E Kiribati

Tabūk 120 A4 Tabūk, NW Saudi Arabia

Täby 85 C6 Stockholm, C Sweden

Tachau see Tachov

Tachov 99 A5 Ger. Tachau. Plveňský Kraj, W Czechia (Czech Republic)

Tacloban 139 F2 off. Tacloban City. Leyte, C Philippines

Tacloban City see Tacloban

Tacna 61 E4 Tacna, SE Peru

Tacoma 46 B2 Washington, NW USA

Tacuarembó 64 D4 prev. San Fructuoso. Tacuarembó, C Uruguay

Tademaït, Plateau du 70 D3 plateau C Algeria

Tadmor see Tadmur

Tadmur 118 C3 var. Tamar, Gk. Palmyra, Bibl. Tadmor. Ḥimṣ, C Syria

Tādpatri 132 C2 Andhra Pradesh, E India

Tadzhikistan see Tajikistan

Taegu see Daegu

Taehan-haehyŏp see Korea Strait

Taehan Min'guk see South Korea

Taejŏn see Daejeon

Tafassâsset, Ténéré du 75 G2 desert N Niger

Tafila/Ṭafilah, Muḥāfaẓat aṭ see Aṭ Ṭafīlah

Taganrog 111 A7 Rostovskaya Oblast', SW Russia

Taganrog, Gulf of 109 G4 Rus. Taganrogskiy Zaliv, Ukr. Tahanroz'ka Zatoka. gulf Russia/Ukraine

Taganrogskiy Zaliv see Taganrog, Gulf of

Taguatinga 63 F3 Tocantins, C Brazil

Tagus 92 C3 Port. Rio Tejo, Sp. Río Tajo. river Portugal/Spain

Tagus Plain 80 A4 abyssal plain E Atlantic Ocean

Tahanroz'ka Zatoka see Taganrog, Gulf of

Tahat 71 E4 mountain SE Algeria

Tahiti 145 H4 island Îles du Vent, W French Polynesia

Tahiti, Archipel de see Société, Archipel de la

Tahlequah 49 G1 Oklahoma, C USA

Tahoe, Lake 47 B5 lake California/Nevada, W USA

Tahoua 75 F3 Tahoua, W Niger

Taibei see Taipei

Taichū see Taichung

Taichung 129 D6 Jap. Taichū; var. Taizhong, Taiwan. C Taiwan

Taiden see Daejeon

Taieri 151 B7 river South Island, New Zealand

Taihape 150 D4 Manawatu-Wanganui, North Island, New Zealand

Taihoku see Taipei

Taikyū see Daegu

Tailem Bend 149 B7 South Australia

T'ainan see Tainan

Tainan 129 D6 prev. T'ainan, Dainan. S Taiwan

Taipei 129 D6 var. Taibei, T'aipei; Jap. Taihoku; prev. Daihoku. capital (Taiwan) N Taiwan

Taiping 138 B3 Perak, Peninsular Malaysia

Taitetinu 151 A7 var. Caswell Sound. sound South Island, New Zealand

Taiwan 129 D6 off. Republic of China, var. Formosa, Formo'sa. country E Asia

Taiwan see Taichung

T'aiwan Haihsia/Taiwan Haixia see Taiwan Strait

Taiwan Strait 129 D6 var. Formosa Strait, Chin. T'aiwan Haihsia, Taiwan Haixia. strait China/Taiwan

Taiyuan *128 C4 var.* T'ai-yuan, T'ai-yüan; *prev.* Yangku. *province capital* Shanxi, C China
T'ai-yuan/T'ai-yüan *see* Taiyuan
Taizhong *see* Tiachung
Ta'izz *121 B7* SW Yemen
Tajikistan *123 E3 off.* Republic of Tajikistan, *Rus.* Tadzhikistan, *Taj.* Jumhurii Tojikiston; *prev.* Tajik S.S.R. *country* C Asia
Tajikistan, Republic of *see* Tajikistan
Tajik S.S.R *see* Tajikistan
Tajo, Río *see* Tagus
Tak *136 C4 var.* Rahaeng. Tak, W Thailand
Takao *see* Kaohsiung
Takaoka *131 C5* Toyama, Honshū, SW Japan
Takapuna *150 D2* Auckland, North Island, New Zealand
Takeshima *see* Liancourt Rocks
Takhiatash *see* Taxiatosh
Takhtakupyr *see* Taxtako'pir
Takikawa *130 D2* Hokkaidō, NE Japan
Takla Makan Desert *see* Taklimakan Shamo
Taklimakan Shamo *126 B3 Eng.* Takla Makan Desert. *desert* NW China
Takow *see* Kaohsiung
Takutea *145 G4 island* S Cook Islands
Talabriga *see* Aveiro, Portugal
Talabriga *see* Talavera de la Reina, Spain
Talachyn *107 D6 Rus.* Tolochin. Vitsyebskaya Voblasts', NE Belarus
Talamanca, Cordillera de *53 E5 mountain range* S Costa Rica
Talara *60 B2* Piura, NW Peru
Talas *123 F2* Talas Oblusu/Talasskaya Oblast', SW Kyrgyzstan
Talaud, Kepulauan *139 F3 island group* E Indonesia
Talavera de la Reina *92 D3 anc.* Caesarobriga, Talabriga. Castilla-La Mancha, C Spain
Talca *64 B4* Maule, C Chile
Talcahuano *65 B5* Bío Bío, C Chile
Taldykorgan *see* Taldyqorghan
Taldyqorghan *114 C5 var.* Taldykorgan; *prev.* Taldy-Kurgan. Taldykorgan, SE Kazakhstan
Ta-lien *see* Dalian
Taliq-an *see* Tāluqān
Tal'ka *107 C6* Minskaya Voblasts', C Belarus
Talkhof *see* Puurmani
Tall Abyaḍ *118 C2 var.* At Tall al Abyaḍ, Tell Abyad, *Fr.* Tell Abiad. Ar Raqqah, N Syria
Tallahassee *42 D3 prev.* Muskogean. *state capital* Florida, SE USA
Tallin *see* Tallinn
Tallinn *106 D2 Ger.* Reval, *Rus.* Tallin; *prev.* Revel. *country capital* (Estonia) Harjumaa, NW Estonia
Tall Kalakh *118 B4 var.* Tell Kalakh. Ḥimṣ, C Syria
Tallulah *42 B2* Louisiana, S USA
Talnakh *114 D3* Taymyrskiy (Dolgano-Nenetskiy) Avtonomnyy Okrug, N Russia
Talne *109 E3 Rus.* Tal'noye. Cherkas'ka Oblast', C Ukraine
Tal'noye *see* Talne
Taloga *49 F1* Oklahoma, C USA
Tāloqān *see* Tāluqān
Tāluqān *123 E3 var.* Tāloqān, Taliq-an. Takhār, NE Afghanistan
Talsen *see* Talsi
Talsi *106 C3 Ger.* Talsen. Talsi, NW Latvia
Taltal *64 B2* Antofagasta, N Chile
Talvik *84 D2* Finnmark, N Norway
Tamabo, Banjaran *138 D3 mountain range* East Malaysia
Tamale *75 E4* C Ghana
Tamana *145 E3 prev.* Rotcher Island. *atoll* Tungaru, W Kiribati
Tamanrasset *71 E4 var.* Tamenghest. S Algeria
Tamar *89 C7 river* SW England, United Kingdom
Tamar *see* Tadmur
Tamatave *see* Toamasina
Tamazunchale *51 E4* San Luis Potosí, C Mexico
Tambacounda *74 C3* SE Senegal
Tambov *111 B6* Tambovskaya Oblast', W Russia

Tambura *73 B5* W Equatoria, SW South Sudan
Tamchaket *see* Tâmchekkeṭ
Tâmchekkeṭ *74 C3 var.* Tamchaket. Hodh el Gharbi, S Mauritania
Tamenghest *see* Tamanrasset
Tamil Nādu *132 C3 prev.* Madras. *cultural region* SE India
Tam Ky *137 E5* Quang Nam-ƒa Năng, C Vietnam
Tammerfors *see* Tampere
Tampa *43 E4* Florida, SE USA
Tampa Bay *43 E4 bay* Florida, SE USA
Tampere *85 D5 Swe.* Tammerfors. Pirkanmaa, W Finland
Tampico *51 E3* Tamaulipas, C Mexico
Tamworth *149 D6* New South Wales, SE Australia
Tanabe *131 C7* Wakayama, Honshū, SW Japan
Tana Bru *84 D2* Finnmark, N Norway
T'ana Häyk' *72 C4 var.* Lake Tana. *lake* NW Ethiopia
Tanais *see* Don
Tana, Lake *see* T'ana Häyk'
Tanami Desert *146 D3 desert* Northern Territory, N Australia
Tananarive *see* Antananarivo
Ţăndărei *108 D5* Ialomiţa, SE Romania
Tandil *65 D5* Buenos Aires, E Argentina
Tandjoengkarang *see* Bandar Lampung
Tanega-shima *131 B8 island* Nansei-shotō, SW Japan
Tanen Taunggyi *see* Tane Range
Tane Range *136 B4 Bur.* Tanen Taunggyi. *mountain range* W Thailand
Tanezrouft *70 D4 desert* Algeria/Mali
Ţanf, Jabal aṭ *118 D4 mountain* SE Syria
Tanga *73 C7* Tanga, E Tanzania
Tanganyika and Zanzibar *see* Tanzania
Tanganyika, Lake *73 B7 lake* E Africa
Tanger *70 C2 var.* Tangiers, Tangier, *Fr./Ger.* Tangerk, *Sp.* Tánger; *anc.* Tingis. NW Morocco
Tangerk *see* Tanger
Tanggula Shan *126 C4 mountain* W China
Tangier *see* Tanger
Tangiers *see* Tanger
Tangra Yumco *126 B5 var.* Tangro Tso. *lake* W China
Tangro Tso *see* Tangra Yumco
Tangshan *128 D3 var.* T'ang-shan. Hebei, E China
T'ang-shan *see* Tangshan
Tanimbar, Kepulauan *139 F5 island group* Maluku, E Indonesia
Tanintharyi *137 B6 prev.* Tenasserim. S Myanmar (Burma)
Tanjungkarang/Tanjungkarang-Telukbetung *see* Bandar Lampung
Tanna *144 D4 island* S Vanuatu
Tannenhof *see* Krynica
Tan-Tan *70 B3* SW Morocco
Tan-tung *see* Dandong
Tanzania *73 C7 off.* United Republic of Tanzania, *Swa.* Jamhuri ya Muungano wa Tanzania; *prev.* German East Africa, Tanganyika and Zanzibar. *country* E Africa
Tanzania, Jamhuri ya Muungano wa *see* Tanzania
Tanzania, United Republic of *see* Tanzania
Taoudenit *see* Taoudenni
Taoudenni *see* Tombouctou
Tapa *106 E2 Ger.* Taps. Lääne-Virumaa, NE Estonia
Tapachula *51 G5* Chiapas, SE Mexico
Tapaiu *see* Gvardeysk
Tapajós, Rio *63 E2 var.* Tapajóz. *river* NW Brazil
Tapajóz *see* Tapajós, Rio
Taps *see* Tapa
Ţarābulus *71 F2 var.* Ţarābulus al Gharb, *Eng.* Tripoli. *country capital* (Libya) NW Libya
Ţarābulus al Gharb *see* Ţarābulus
Ţarābulus/Ţarābulus ash Shām *see* Tripoli
Taraclia *108 D4 Rus.* Tarakilya. S Moldova
Tarakilya *see* Taraclia
Taranaki, Mount *150 C4 var.* Egmont. *volcano* North Island, New Zealand
Tarancón *93 E3* Castilla-La Mancha, C Spain
Taranto *97 E5 var.* Tarentum. Puglia, SE Italy

Taranto, Gulf of *97 E6 Eng.* Gulf of Taranto. *gulf* S Italy
Taranto, Gulf of *see* Taranto, Golfo di
Tarapoto *60 C2* San Martín, N Peru
Tarare *91 D5* Rhône, E France
Tarascon *91 D6* Bouches-du-Rhône, SE France
Tarawa *144 D2 atoll and capital* (Kiribati) Tungaru, W Kiribati
Taraz *114 C5 prev.* Aulie Ata, Auliye-Ata, Dzhambul, Zhambyl. Zhambyl, S Kazakhstan
Tarazona *93 E2* Aragón, NE Spain
Tarbes *91 B6 anc.* Bigorra. Hautes-Pyrénées, S France
Tarcoola *149 A6* South Australia
Taree *149 D6* New South Wales, SE Australia
Tarentum *see* Taranto
Târgovişte *108 C5 prev.* Tîrgovişte. Dâmboviţa, S Romania
Targu Jiu *108 B4 prev.* Tîrgu Jiu. Gorj, W Romania
Târgul-Neamţ *see* Târgu-Neamţ
Târgul-Săcuiesc *see* Târgu Secuiesc
Târgu Mureş *108 B4 prev.* Oşorhei, Tîrgu Mures, *Ger.* Neumarkt, *Hung.* Marosvásárhely. Mureş, C Romania
Târgu-Neamţ *108 C3 var.* Târgul-Neamţ; *prev.* Tîrgu-Neamţ. Neamţ, NE Romania
Târgu Ocna *108 C4 Hung.* Aknavásár; *prev.* Tîrgu Ocna. Bacău, E Romania
Târgu Secuiesc *108 C4 Ger.* Neumarkt, Szekler Neumarkt, *Hung.* Kezdivásárhely; *prev.* Chezdi-Oşorheiu, Târgul-Săcuiesc, Tîrgu Secuiesc. Covasna, E Romania
Tar Heel State *see* North Carolina
Tarija *61 G5* Tarija, S Bolivia
Tarim *121 C6* Y Yemen
Tarim Basin *see* Tarim Pendi
Tarim Pendi *124 C2 Eng.* Tarim Basin. *basin* NW China
Tarim He *126 B3 river* NW China
Tarma *60 D3* Junín, C Peru
Tarn *91 C6 cultural region* S France
Tarn *91 C6 river* S France
Tarnobrzeg *98 D4* Podkarpackie, SE Poland
Tarnopol *see* Ternopil
Tarnów *99 D5* Małopolskie, S Poland
Tarraco *see* Tarragona
Tarragona *93 G2 anc.* Tarraco. Cataluña, E Spain
Tarrasa *see* Terrassa
Tàrrega *93 F2 var.* Tarrega. Cataluña, NE Spain
Tarsatica *see* Rijeka
Tarsus *116 C4* İçel, S Turkey (Türkiye)
Tartous/Tartous *see* Ţarţūs
Tartu *106 D3 Ger.* Dorpat; *prev.* Rus. Yurev, Yury'ev. Tartumaa, SE Estonia
Ţarţūs *118 A3 off.* Muḥāfaẓat Ţarţūs, *var.* Tartous, Tartus. *governorate* W Syria
Ţarţūs, Muḥāfaẓat *see* Ţarţūs
Ta Ru Tao, Ko *137 B7 island* S Thailand
Tarvisio *96 D2* Friuli-Venezia Giulia, NE Italy
Tarvisium *see* Treviso
Tashauz *see* Daşoguz
Tashi Chho Dzong *see* Thimphu
Tashkent *see* Toshkent
Tash-Kömür *123 F2 Rus.* Tash-Kumyr. Jalal-Abad Oblusu/Dzhalal-Abadskaya Oblast', W Kyrgyzstan
Tash-Kumyr *see* Tash-Kömür
Tashqurghan *see* Khulm
Tasiilaq *82 D4 var.* Angmagssalik. Semersooq, S Greenland
Tasikmalaja *see* Tasikmalaya
Tasikmalaya *138 C5 prev.* Tasikmalaja. Jawa, C Indonesia
Tasman Basin *142 C5 var.* East Australian Basin. *undersea basin* S Tasman Sea
Tasman Bay *151 C5 var.* Te Tai-o-Aorere. *inlet* South Island, New Zealand
Tasmania *149 B8 prev.* Van Diemen's Land. *state* SE Australia
Tasmania *152 B4 island* SE Australia
Tasman Plateau *142 C5 var.* South Tasmania Plateau. *undersea plateau* SW Tasman Sea
Tasman Sea *142 C5 sea* SW Pacific Ocean

Tassili-n-Ajjer *71 E4 plateau* E Algeria
Tatabánya *99 C6* Komárom-Esztergom, NW Hungary
Tatar Pazardzhik *see* Pazardzhik
Tathlith *121 B5* 'Asīr, S Saudi Arabia
Tatra Mountains *99 D5 Ger.* Tatra, *Hung.* Tátra, *Pol./Slvk.* Tatry. *mountain range* Poland/Slovakia
Tatra/Tátra *see* Tatra Mountains
Tatry *see* Tatra Mountains
Ta-t'ung/Tatung *see* Datong
Tatvan *117 F3* Bitlis, SE Turkey (Türkiye)
Ta'ū *145 F4 var.* Tau. *island* Manua Islands, E American Samoa
Taukum, Peski *123 G1 desert* SE Kazakhstan
Taumarunui *150 D4* Manawatu-Wanganui, North Island, New Zealand
Taungdwingyi *136 B3* Magway, C Myanmar (Burma)
Taunggyi *136 B3* Shan State, C Myanmar (Burma)
Taungoo *136 B4* Bago, C Myanmar (Burma)
Taunton *89 C7* SW England, United Kingdom
Taupo *150 D3* Waikato, North Island, New Zealand
Taupo, Lake *150 D3 lake* North Island, New Zealand
Tauragė *106 B4 Ger.* Tauroggen. Tauragė, SW Lithuania
Tauranga *150 D3* Bay of Plenty, North Island, New Zealand
Tauris *see* Tabrīz
Tauroggen *see* Tauragė
Taurus Mountains *see* Toros Dağları
Tavas *116 B4* Denizli, SW Turkey (Türkiye)
Tavastehus *see* Hämeenlinna
Tavira *92 C5* Faro, S Portugal
Tavoy *see* Dawei
Tavoy Island *see* Mali Kyun
Ta Waewae Bay *151 A7 bay* South Island, New Zealand
Tawakoni, Lake *49 G2 reservoir* Texas, SW USA
Tawau *138 D3* Sabah, East Malaysia
Ţawkar *see* Tokar
Tawzar *see* Tozeur
Taxco *51 E4 var.* Taxco de Alarcón. Guerrero, S Mexico
Taxco de Alarcón *see* Taxco
Taxiatosh *122 C2 Rus.* Takhiatash. Qoraqalpog'iston Respublikasi, W Uzbekistan
Taxtako'pir *122 D1 Rus.* Takhtakupyr. Qoraqalpog'iston Respublikasi, NW Uzbekistan
Tay *88 C3 river* C Scotland, United Kingdom
Taylor *49 G3* Texas, SW USA
Taymā' *120 A4* Tabūk, NW Saudi Arabia
Taymyr, Ozero *115 E2 lake* N Russia
Taymyr, Poluostrov *115 E2 peninsula* N Russia
Taz *114 D3 river* N Russia
Tbilisi *117 G2 var.* T'bilisi, *Eng.* Tiflis. *country capital* (Georgia) SE Georgia
T'bilisi *see* Tbilisi
Tchad *see* Chad
Tchad, Lac *see* Chad, Lake
Tchien *see* Zwedru
Tchongking *see* Chongqing
Tczew *98 C2 Ger.* Dirschau. Pomorskie, N Poland
Te Anau *151 A7* Southland, South Island, New Zealand
Te Anau, Lake *151 A7 lake* South Island, New Zealand
Teapa *51 G4* Tabasco, SE Mexico
Teate *see* Chieti
Tebingtinggi *138 B3* Sumatera, N Indonesia
Tebriz *see* Tabrīz
Techirghiol *108 D5* Constanţa, SE Romania
Tecomán *50 D4* Colima, SW Mexico
Tecpan *51 E6 var.* Tecpan de Galeana. Guerrero, S Mexico
Tecpan de Galeana *see* Tecpan
Tecuci *108 C4* Galaţi, E Romania
Tedzhen *see* Harī Rōd
Tees *89 D5 river* N England, United Kingdom
Tefé *62 D2* Amazonas, N Brazil
Tegal *138 C4* Jawa, C Indonesia
Tegelen *87 D5* Limburg, SE Netherlands

Tegucigalpa 52 C3 *country capital* (Honduras) Francisco Morazán, SW Honduras
Te Hauturu-o-Toi 150 D2 *var.* Little Barrier Island. *island* N New Zealand
Teheran *see* Tehrān
Te Houhou 151 A7 *var.* Gerorge Sound. *sound* South Island, New Zealand
Tehrān 120 C3 *var.* Teheran. *country capital* (Iran) Tehrān, N Iran
Tehuacán 51 F4 Puebla, S Mexico
Tehuantepec 51 F5 *var.* Santo Domingo Tehuantepec. Oaxaca, SE Mexico
Tehuantepec, Golfo de 51 F5 *var.* Gulf of Tehuantepec. *gulf* S Mexico
Tehuantepec, Gulf of *see* Tehuantepec, Golfo de
Tehuantepec, Isthmus of *see* Tehuantepec, Istmo de
Tehuantepec, Istmo de 51 F5 *var.* Isthmus of Tehuantepec. *isthmus* SE Mexico
Te Ika-a-Māui *see* North Island
Tejen 122 C3 *Rus.* Tedzhen. Ahal Welaýaty, S Turkmenistan
Tejen *see* Harī Rōd
Tejo, Rio *see* Tagus
Te Kao 150 C1 Northland, North Island, New Zealand
Tekax 51 H4 *var.* Tekax de Álvaro Obregón. Yucatán, SE Mexico
Tekax de Álvaro Obregón *see* Tekax
Tekeli 114 C5 Almaty, SE Kazakhstan
Tekirdağ 116 A2 *It.* Rodosto; *anc.* Bisanthe, Raidestos, Rhaedestus. Tekirdağ, NW Turkey (Türkiye)
Te Kuiti 150 D3 Waikato, North Island, New Zealand
Tela 52 C2 Atlántida, NW Honduras
Telanaipura *see* Jambi
Telangana 134 D5 *cultural region* SE India
Tel Aviv-Jaffa *see* Tel Aviv-Yafo
Tel Aviv-Yafo 119 A6 *var.* Tel Aviv-Jaffa. Tel Aviv, C Israel
Teles Pirés *see* São Manuel, Rio
Telish 104 C2 *prev.* Azizie. Pleven, N Bulgaria
Tell Abiad/Tell Abyad *see* Tall Abyaḍ
Tell Kalakh *see* Tall Kalakh
Tell Shedadi *see* Ash Shadādah
Tel'man/Tel'mansk *see* Gubadag
Teloekbetoeng *see* Bandar Lampung
Telo Martius *see* Toulon
Telschen *see* Telšiai
Telšiai 106 B3 *Ger.* Telschen. Telšiai, NW Lithuania
Telukbetung *see* Bandar Lampung
Temerin 100 D3 Vojvodina, N Serbia
Temeschburg/Temeschwar *see* Timişoara
Temesvár/Temeswar *see* Timişoara
Temirtau 114 C4 *prev.* Samarkandski, Samarkandskoye. Karagandy, C Kazakhstan
Tempio Pausania 97 A5 Sardegna, Italy, C Mediterranean Sea
Temple 49 G3 Texas, SW USA
Temuco 65 B5 Araucanía, C Chile
Temuka 151 B6 Canterbury, South Island, New Zealand
Tenasserim *see* Tanintharyi
Ténenkou 74 D3 Mopti, C Mali
Ténéré 75 G3 *physical region* C Niger
Tenerife 70 A3 *island* Islas Canarias, Spain, NE Atlantic Ocean
Tengger Shamo 127 E3 *desert* N China
Tengréla 74 D4 *var.* Tingréla. N Ivory Coast
Tenkodogo 75 E4 S Burkina
Tennant Creek 148 A3 Northern Territory, C Australia
Tennessee 42 C1 *off.* State of Tennessee, *also known as* The Volunteer State. *state* SE USA
Tennessee River 42 C1 *river* S USA
Tenos *see* Tínos
Te-Oneroa-a-Tōhē 150 C1 *var.* Ninety Mile Beach. *beach* North Island, New Zealand
Tepelenë *see* Tepelenë
Tepelenë 101 C7 *var.* Tepelena, *It.* Tepeleni. Gjirokastër, S Albania
Tepeleni *see* Tepelenë
Tepic 50 D4 Nayarit, C Mexico
Teplice 98 A4 *Ger.* Teplitz; *prev.* Teplice-Šanov, Teplitz-Schönau. Ústecký Kraj, NW Czechia (Czech Republic)

Teplice-Šanov/Teplitz/Teplitz-Schönau *see* Teplice
Tequila 50 D4 Jalisco, SW Mexico
Teraina 145 G2 *prev.* Washington Island. *atoll* Line Islands, E Kiribati
Teramo 96 C4 *anc.* Interamna. Abruzzi, C Italy
Tercan 117 E3 Erzincan, NE Turkey (Türkiye)
Terceira 92 A5 *var.* Ilha Terceira. *island* Azores, Portugal, NE Atlantic Ocean
Terceira, Ilha *see* Terceira
Terekhovka *see* Tsyerakhowka
Te Renerga Wairua *see* Reinga, Cape
Teresina 63 F2 *var.* Therezina. *state capital* Piauí, NE Brazil
Termez *see* Termiz
Termia *see* Kýthnos
Términos, Laguna de 51 G4 *lagoon* SE Mexico
Termiz 123 E3 *Rus.* Termez. Surkhondaryo Viloyati, S Uzbekistan
Termoli 96 D4 Molise, C Italy
Terneuzen 87 B5 *var.* Neuzen. Zeeland, SW Netherlands
Terni 96 C4 *anc.* Interamna Nahars. Umbria, C Italy
Ternopil 108 C2 *Pol.* Tarnopol, *Rus.* Ternopol'. Ternopil's'ka Oblast', W Ukraine
Ternopol' *see* Ternopil
Terracina 97 C5 Lazio, C Italy
Terranova di Sicilia *see* Gela
Terranova Pausania *see* Olbia
Terrassa 93 G2 *Cast.* Tarrasa. Cataluña, E Spain
Terre Adélie 154 C4 *physical region* Antarctica
Terre Haute 40 B4 Indiana, N USA
Terre Neuve *see* Newfoundland and Labrador
Terschelling 86 C1 *Fris.* Skylge. *island* Waddeneilanden, N Netherlands
Teruel 93 F3 *anc.* Turba. Aragón, E Spain
Tervel 104 E1 *prev.* Kurtbunar, *Rom.* Curtbunar. Dobrich, NE Bulgaria
Tervueren *see* Tervuren
Tervuren 87 C6 *var.* Tervueren. Vlaams Brabant, C Belgium
Teseney 72 C4 *var.* Tessenei. W Eritrea
Tessalit 75 E2 Kidal, NE Mali
Tessaoua 75 G3 Maradi, S Niger
Tessenderlo 87 C5 Limburg, NE Belgium
Tessenei *see* Teseney
Testigos, Islas Los 59 E1 *island group* N Venezuela
Te Tai-o-Aorere *see* Tasman Bay
Tete 79 E2 Tete, NW Mozambique
Teterow 94 C3 Mecklenburg-Vorpommern, NE Germany
Tétouan 70 C2 *var.* Tetouan, Tetuán. N Morocco
Tetovo 101 D5 Razgrad, N Bulgaria
Tetschen *see* Děčín
Tetuán *see* Tétouan
Teverya *see* Tverya
Te Waewae Bay 151 A7 *bay* South Island, New Zealand
Te Waipounamu *see* South Island
Texarkana 42 A2 Arkansas, C USA
Texarkana 49 H2 Texas, SW USA
Texas 49 F3 *off.* State of Texas, *also known as* Lone Star State. *state* S USA
Texas City 49 H4 Texas, SW USA
Texel 86 C2 *island* Waddeneilanden, NW Netherlands
Texoma, Lake 49 G2 *reservoir* Oklahoma/Texas, C USA
Teziutlán 51 F4 Puebla, S Mexico
Thaa Atoll *see* Kolhumadulu
Thai, Ao *see* Thailand, Gulf of
Thai Binh 136 D3 Thai Binh, N Vietnam
Thailand 137 C5 *off.* Kingdom of Thailand, *Th.* Prathet Thai; *prev.* Siam. *country* SE Asia
Thailand, Gulf of 137 C6 *var.* Gulf of Siam, *Th.* Ao Thai, *Vtn.* Vinh Thai Lan. *gulf* SE Asia
Thailand, Kingdom of *see* Thailand
Thai Lan, Vinh *see* Thailand, Gulf of
Thai Nguyên 136 D3 Bắc Thai, N Vietnam
Thakhèk 136 D4 *var.* Muang Khammouan. Khammouan, C Laos
Thamarīd *see* Thamarīt
Thamarīt 121 D6 *var.* Thamarīd, Thumrayt. SW Oman
Thames 150 D3 Waikato, North Island, New Zealand

Thames 89 B8 *river* S England, United Kingdom
Thandwe 136 A4 *var.* Sandoway. Rakhine State, W Myanmar (Burma)
Thanh Hoa 136 D3 Thanh Hoa, N Vietnam
Thanintari Taungdan *see* Bilauktaung Range
Thanlwin *see* Salween
Thar Desert 134 C3 *var.* Great Indian Desert, Indian Desert. *desert* India/Pakistan
Tharthar, Buhayrat ath 120 B3 *lake* C Iraq
Thásos 104 C4 Thásos, E Greece
Thásos 104 C4 *island* E Greece
Thaton 136 B4 Mon State, S Myanmar (Burma)
Thayet 136 A4 *var.* Thayetmyo. Magway, C Myanmar (Burma)
Thayetmyo *see* Thayet
The Crane 55 H2 *var.* Crane. S Barbados
The Dalles 46 B3 Oregon, NW USA
The Flatts Village *see* Flatts Village
The Hague *see* 's-Gravenhage
Theodosia *see* Feodosiia
The Pas 37 F5 Manitoba, C Canada
Therezina *see* Teresina
Thérma 105 D6 Ikaría, Dodekánisa, Greece, Aegean Sea
Thermaic Gulf/Thermaicus Sinus *see* Thermaïkós Kólpos
Thermaïkós Kólpos 104 B4 *Eng.* Thermaic Gulf; *anc.* Thermaicus Sinus. *gulf* N Greece
Thermiá *see* Kýthnos
Thérmo 105 B5 Dytikí Ellás, C Greece
The Rock 93 H4 New South Wales, SE Australia
The Sooner State *see* Oklahoma
Thessaloníki 104 C3 *Eng.* Salonica, Salonika, *Croatian* Solun, *Turk.* Selânik. Kentrikí Makedonía, N Greece
The Valley 55 G3 *dependent territory capital* (Anguilla) E Anguilla
The Village 49 G1 Oklahoma, C USA
The Volunteer State *see* Tennessee
Thiamis *see* Thýamis
Thian Shan *see* Tien Shan
Thibet *see* Xizang Zizhiqu
Thief River Falls 45 F1 Minnesota, N USA
Thienen *see* Tienen
Thiers 91 C5 Puy-de-Dôme, C France
Thiès 74 B3 W Senegal
Thikombia *see* Cikobia
Thimbu *see* Thimphu
Thimphu 135 G3 *var.* Thimbu; *prev.* Tashi Chho Dzong. *country capital* (Bhutan) W Bhutan
Thionville 90 D3 *Ger.* Diedenhofen. Moselle, NE France
Thíra 105 D7 *var.* Santorini. Kykládes, Greece, Aegean Sea
Thiruvananthapuram 132 C3 *var.* Trivandrum, Tiruvantapuram. *state capital* Kerala, SW India
Thitu Island 129 C8 *island* NW Spratly Islands
Tholen 86 B4 *island* SW Netherlands
Thomasville 42 D3 Georgia, SE USA
Thompson 37 F4 Manitoba, C Canada
Thonon-les-Bains 91 D5 Haute-Savoie, E France
Thoothukudi 132 C3 *var.* Tuticorin. Tamil Nādu, SE India
Thorenburg *see* Turda
Thorlákshöfn 83 E5 Sudhurland, SW Iceland
Thorn *see* Toruń
Thornton Island *see* Millennium Island
Thorshavn *see* Tórshavn
Thospitis *see* Van Gölü
Thouars 90 B4 Deux-Sèvres, W France
Thoune *see* Thun
Thracian Sea 104 C4 *Gk.* Thrakikó Pélagos; *anc.* Thracium Mare. *sea* Greece/Turkey (Türkiye)
Thracium Mare/Thrakikó Pélagos *see* Thracian Sea
Three Gorges Reservoir 129 C5 *reservoir* C China
Three Kings Islands *see* Manawatāwhi
Thrissur 132 C3 *var.* Trichūr. Kerala, SW India
Thuin 87 B7 Hainaut, S Belgium
Thule *see* Qaanaaq
Thumrayt *see* Thamarīt
Thun 95 A7 *Fr.* Thoune. Bern, W Switzerland

Thunder Bay 38 B4 Ontario, S Canada
Thuner See 95 A7 *lake* C Switzerland
Thung Song 137 C7 *var.* Cha Mai. Nakhon Si Thammarat, SW Thailand
Thurso 88 C2 N Scotland, United Kingdom
Thýamis 104 A4 *var.* Kalamás; *prev.* Thiamis. *river* W Greece
Tianjin 128 D4 *var.* Tientsin. Tianjin Shi, E China
Tianjin *see* Tianjin Shi
Tianjin Shi 128 D4 *var.* Jin, Tianjin, T'ien-ching, Tientsin. *municipality* E China
Tian Shan *see* Tien Shan
Tianshui 128 B4 Gansu, C China
Tiba *see* Chiba
Tiber 96 C4 *Eng.* Tiber. *river* C Italy
Tiber *see* Tevere
Tiber *see* Tivoli, Italy
Tiberias *see* Tverya
Tiberias, Lake *see* Yam Kinneret
Tibesti 76 C2 *var.* Tibesti Massif, *Ar.* Tibistī. *mountain range* N Africa
Tibesti Massif *see* Tibesti
Tibet *see* Xizang Zizhiqu
Tibetan Autonomous Region *see* Xizang Zizhiqu
Tibet, Plateau of *see* Qingzang Gaoyuan
Tibistī *see* Tibesti
Tibni *see* At Tibnī
Tiburón, Isla 50 B2 *var.* Isla del Tiburón. *island* NW Mexico
Tiburón, Isla del *see* Tiburón, Isla
Tichau *see* Tychy
Tichît 74 D2 *var.* Tichitt. Tagant, C Mauritania
Tichitt *see* Tichît
Ticinum *see* Pavia
Ticul 51 H3 Yucatán, SE Mexico
Tidjikdja *see* Tidjikja
Tidjikja 74 C2 *var.* Tidjikdja; *prev.* Fort-Cappolani. Tagant, C Mauritania
T'ien-ching *see* Tianjin Shi
Tienen 87 C6 *var.* Thienen, *Fr.* Tirlemont. Vlaams Brabant, C Belgium
Tiên Giang, Sông *see* Mekong
Tien Shan 83 C3 *Chin.* Thian Shan, Tian Shan, Tien Shan, *Rus.* Tyan'-Shan'. *mountain range* C Asia
Tientsin *see* Tianjin
Tierp 85 C6 Uppsala, C Sweden
Tierra del Fuego 65 B8 *island* Argentina/Chile
Tiflis *see* T'bilisi
Tifton 42 D3 Georgia, SE USA
Tifu 139 F4 Pulau Buru, E Indonesia
Tighina *see* Bender
Tigranocerta *see* Siirt
Tigris 120 B2 *Ar.* Dijlah, *Turk.* Dicle. *river* Iraq/Syria/Turkey (Türkiye)
Tiguentourine 71 E3 E Algeria
Ti-hua/Tihwa *see* Ürümqi
Tijuana 50 A1 Baja California Norte, NW Mexico
Tikapa Moana *see* Hauraki Gulf
Tikhoretsk 111 A7 Krasnodarskiy Kray, SW Russia
Tikhvin 110 B4 Leningradskaya Oblast', NW Russia
Tikiarjuaq *see* Whale Cove
Tiki Basin 143 G3 *undersea basin* S Pacific Ocean
Tikinso 74 C4 *river* C Guinea
Tiksi 115 F2 Respublika Sakha (Yakutiya), NE Russia
Tilburg 86 C4 Noord-Brabant, S Netherlands
Tilimsen *see* Tlemcen
Tilio Martius *see* Toulon
Tillabéri 75 F3 *var.* Tillabéry. Tillabéri, W Niger
Tillabéry *see* Tillabéri
Tílos 105 E7 *island* Dodekánisa, Greece, Aegean Sea
Timan Ridge *see* Timanskiy Kryazh
Timanskiy Kryazh 110 D3 *Eng.* Timan Ridge. *ridge* NW Russia
Timaru 151 B6 Canterbury, South Island, New Zealand
Timbaki/Timbákion *see* Tympáki
Timbedgha 74 D3 *var.* Timbédra. Hodh ech Chargui, SE Mauritania
Timbédra *see* Timbedgha
Timbuktu *see* Tombouctou
Timiş 108 A4 *county* SW Romania
Timişoara 108 A4 *Ger.* Temeschwar, Temeswar, *Hung.* Temesvár; *prev.* Temeschburg. Timiş, W Romania
Timmins 38 C4 Ontario, S Canada

Timor *125 F5 island* Nusa Tenggara, C Indonesia

Timor-Leste *see* East Timor

Timor Lorosa'e *see* East Timor

Timor Sea *125 F5 sea* E Indian Ocean

Timor Timur *see* East Timor

Timor Trench *see* Timor Trough

Timor Trough *125 F5 var.* Timor Trench. *trough* NE Timor Sea

Timrå *85 C5* Västernorrland, C Sweden

Tindouf *70 C3* W Algeria

Tineo *92 C1* Asturias, N Spain

Tingis *see* Tanger

Tingo María *60 C3* Huánuco, C Peru

Tingréla *see* Tengréla

Tinhosa Grande *76 E2 island* N Sao Tome and Principe, Africa, E Atlantic Ocean

Tinhosa Pequena *76 E1 island* N Sao Tome and Principe, Africa, E Atlantic Ocean

Tinian *144 B1 island* S Northern Mariana Islands

Tínos *105 D6* Tínos, Kykládes, Greece, Aegean Sea

Tínos *105 D6 anc.* Tenos. *island* Kykládes, Greece, Aegean Sea

Tip *101 E6* Papua, E Indonesia

Tipitapa *52 D3* Managua, W Nicaragua

Tip Top Mountain *38 C4 mountain* Ontario, S Canada

Tirana *see* Tiranë

Tiranë *101 C6 var.* Tirana. *country capital* (Albania) Tiranë, C Albania

Tiraspol *108 D4 Rus.* Tiraspol'. E Moldova

Tiraspol' *see* Tiraspol

Tiree *88 B3 island* W Scotland, United Kingdom

Tîrgovişte *see* Târgovişte

Tîrgu Jiu *see* Targu Jiu

Tîrgu Mures *see* Târgu Mureş

Tîrgu-Neamţ *see* Târgu-Neamţ

Tîrgu Ocna *see* Târgu Ocna

Tîrgu Secuiesc *see* Târgu Secuiesc

Tîrlemont *see* Tienen

Tírnavos *see* Týrnavos

Tirnovo *see* Veliko Tarnovo

Tirol *95 C7 off.* Land Tirol, *var.* Tyrol, *It.* Tirolo. *state* W Austria

Tirol, Land *see* Tirol

Tirolo *see* Tirol

Tirreno, Mare *see* Tyrrhenian Sea

Tiruchchirāppalli *132 C3 prev.* Trichinopoly. Tamil Nādu, SE India

Tiruppattūr *132 C2* Tamil Nādu, SE India

Tiruvantapuram *see* Thiruvananthapuram

Tisa *see* Tisza

Tisza *103 F1 Ger.* Theiss, *Rom./Slvn./ Croatian* Tisa, *Rus.* Tissa, *Ukr.* Tysa. *river* SE Europe

Tiszakécske *99 D7* Bács-Kiskun, C Hungary

Titano, Monte *96 E1 mountain* C San Marino

Titi *151 A8 var.* Muttonbird Islands. *island group* SW New Zealand

Titicaca, Lake *61 E4 lake* Bolivia/Peru

Titograd *see* Podgorica

Titose *see* Chitose

Titova Mitrovica *see* Mitrovicë

Titovo Užice *see* Užice

Titu *108 C5* Dâmboviţa, S Romania

Titule *77 D5* Orientale, N Dem. Rep. Congo

Tiverton *89 C7* SW England, United Kingdom

Tivoli *96 C4 anc.* Tiber. Lazio, C Italy

Tizimín *51 H3* Yucatán, SE Mexico

Tizi Ouzou *71 E1 var.* Tizi-Ouzou. N Algeria

Tizi-Ouzou *see* Tizi Ouzou

Tiznit *70 B3* SW Morocco

Tjilatjap *see* Cilacap

Tjirebon *see* Cirebon

Tlaquepaque *50 D4* Jalisco, C Mexico

Tlascala *see* Tlaxcala

Tlaxcala *51 F4 var.* Tlascala, Tlaxcala de Xicohténcatl. Tlaxcala, C Mexico

Tlaxcala de Xicohténcatl *see* Tlaxcala

Tlemcen *70 D2 var.* Tilimsen, Tlemsen. NW Algeria

Tlemsen *see* Tlemcen

Toamasina *79 G3 var.* Tamatave. Toamasina, E Madagascar

Toba, Danau *138 B3 lake* Sumatera, W Indonesia

Tobago *55 H5 island* NE Trinidad and Tobago

Toba Kakar Range *134 B2 mountain range* NW Pakistan

Tobol *114 C4 Kaz.* Tobyl. *river* Kazakhstan/Russia

Tobol'sk *114 C3* Tyumenskaya Oblast', C Russia

Tobruch/Tobruk *see* Ţubruq

Tobyl *see* Tobol

Tocantins *63 E3 off.* Estado do Tocantins. *state/region* C Brazil

Tocantins, Estado do *see* Tocantins

Tocantins, Rio *63 F2 river* N Brazil

Tocoa *52 D2* Colón, N Honduras

Tocopilla *64 B2* Antofagasta, N Chile

Todi *96 C4* Umbria, C Italy

Todos os Santos, Baía de *63 G3 bay* E Brazil

Toetoes Bay *151 B8 bay* South Island, New Zealand

Tofua *145 E4 island* Ha'apai Group, C Tonga

Togo *75 E4 off.* Togolese Republic; *prev.* French Togoland. *country* W Africa

Togolese Republic *see* Togo

Tojikiston, Jumhurii *see* Tajikistan

Tokanui *151 B7* Southland, South Island, New Zealand

Tokar *72 C3 var.* Ţawkar. Red Sea, NE Sudan

Tokat *116 D3* Tokat, N Turkey (Türkiye)

Tokelau *145 E3 NZ overseas territory* W Polynesia

Tőketerebes *see* Trebišov

Tokio *see* Tōkyō

Tokmak *109 G4 var.* Velykyy Tokmak. Zaporiz'ka Oblast', SE Ukraine

Tokmok *123 G2 Rus.* Tokmak. Chüy Oblusu/Chuyskaya Oblast', N Kyrgyzstan

Tokoroa *150 D3* Waikato, North Island, New Zealand

Tokounou *74 C4* C Guinea

Tokushima *131 C6 var.* Tokusima. Tokushima, Shikoku, SW Japan

Tōkyō *130 A1 var.* Tokio. *country capital* (Japan) Tōkyō, Honshū, S Japan

Tōkyō-wan *130 A2 bay* S Japan

Tolbukhin *see* Dobrich

Toledo *92 D3 anc.* Toletum. Castilla-La Mancha, C Spain

Toledo *40 D3* Ohio, N USA

Toledo Bend Reservoir *49 G3 reservoir* Louisiana/Texas, SW USA

Toletum *see* Toledo

Toliara *79 F4 var.* Toliary; *prev.* Tuléar. Toliara, SW Madagascar

Toliary *see* Toliara

Tolmein *see* Tolmin

Tolmin *95 D7 Ger.* Tolmein, *It.* Tolmino. W Slovenia

Tolmino *see* Tolmin

Tolna *99 C7 Ger.* Tolnau. Tolna, S Hungary

Tolnau *see* Tolna

Tolochin *see* Talachyn

Tolosa *93 E1* País Vasco, N Spain

Tolosa *see* Toulouse

Toluca *51 E4 var.* Toluca de Lerdo. México, S Mexico

Toluca de Lerdo *see* Toluca

Tol'yatti *111 C6 prev.* Stavropol'. Samarskaya Oblast', W Russia

Tomah *40 B2* Wisconsin, N USA

Tomakomai *130 D2* Hokkaidō, NE Japan

Tomar *92 B3* Santarém, W Portugal

Tomaschow *see* Tomaszów Mazowiecki

Tomaschow *see* Tomaszów Mazowiecki

Tomaszów *see* Tomaszów Mazowiecki

Tomaszów Lubelski *98 E4 Ger.* Tomaschow. Lubelskie, E Poland

Tomaszów Mazowiecka *see* Tomaszów Mazowiecki

Tomaszów Mazowiecki *98 D4 var.* Tomaszów Mazowiecka; *prev.* Tomaszów, *Ger.* Tomaschow. Łódzkie, C Poland

Tombigbee River *42 C3 river* Alabama/ Mississippi, S USA

Tombouctou *75 E3 Eng.* Timbuktu. *var.* Taoudenni, N Mali

Tombua *78 A2 Port.* Porto Alexandre. Namibe, SW Angola

Tomelloso *93 E3* Castilla-La Mancha, C Spain

Tomini, Gulf of *see* Teluk Tomini

Tomini, Teluk *139 E4 Eng.* Gulf of Tomini; *prev.* Teluk Gorontalo. *bay* Sulawesi, C Indonesia

Tomsk *114 D4* Tomskaya Oblast', C Russia

Tömür Feng *126 B3 pre.* Pik Pobedy; *Kyrg.* Jengish Chokusu. *mountain* China/Kyrgyzstan

Tonezh *see* Tonyezh

Tonga *145 E4 off.* Kingdom of Tonga, *var.* Friendly Islands. *country* SW Pacific Ocean

Tonga, Kingdom of *see* Tonga

Tongatapu *145 E5 island* Tongatapu Group, S Tonga

Tongatapu Group *145 E5 island group* S Tonga

Tonga Trench *143 E3 trench* S Pacific Ocean

Tongchuan *128 C4* Shaanxi, C China

Tongeren *87 D6 Fr.* Tongres. Limburg, NE Belgium

Tongking, Gulf of *see* Tonkin, Gulf of

Tongliao *127 G2* Nei Mongol Zizhiqu, N China

Tongres *see* Tongeren

Tongshan *see* Xuzhou, Jiangsu, China

Tongtian He *126 C4 river* C China

Tonj *73 B5* Warab, C South Sudan

Tonkin, Gulf of *129 B7 var.* Tongking, Gulf of, *Chin.* Beibu Wan, *Vtn.* Vinh Băc Bô. *gulf* China/Vietnam

Tônlé Sap *137 D5 Eng.* Great Lake. *lake* W Cambodia

Tonopah *47 C6* Nevada, W USA

Tonyezh *107 C7 Rus.* Tonezh. Homyel'skaya Voblasts', SE Belarus

Tooele *44 B4* Utah, W USA

Toowoomba *149 E5* Queensland, E Australia

Topeka *45 F4 state capital* Kansas, C USA

Toplicza *see* Topliţa

Topliţa *108 C3 Ger.* Töplitz, *Hung.* Maroseviz; *prev.* Topliţa Română, *Hung.* Oláh-Topliza, Topliza. Harghita, C Romania

Topliţa Română/Töplitz *see* Topliţa

Topol'čany *99 C6 Hung.* Nagytapolcsány. Nitriansky Kraj, W Slovakia

Topolovgrad *104 D3 prev.* Kavakli. Haskovo, S Bulgaria

Topolya *see* Bačka Topola

Top Springs Roadhouse *146 E3* Northern Territory, N Australia

Tor *154 C2 Norwegian research station* Antarctica

Torda *see* Turda

Torez *see* Chystiakove

Torgau *94 D4* Sachsen, E Germany

Torhout *87 A5* West-Vlaanderen, W Belgium

Torino *96 A2 Eng.* Turin. Piemonte, NW Italy

Tornacum *see* Tournai

Torneå *see* Tornio

Torneträsk *84 C3 lake* N Sweden

Tornio *84 D4 Swe.* Torneå. Lappi, NW Finland

Tornionjoki *84 D3 river* Finland/Sweden

Toro *92 D2* Castilla y León, N Spain

Toronto *38 D5 province capital* Ontario, S Canada

Toros Dağları *116 C4 Eng.* Taurus Mountains. *mountain range* S Turkey (Türkiye)

Torquay *89 C7* SW England, United Kingdom

Torrance *46 D2* California, W USA

Torre, Alto da *92 B3 mountain* C Portugal

Torre del Greco *97 D5* Campania, S Italy

Torrejón de Ardoz *93 E3* Madrid, C Spain

Torrelavega *92 D1* Cantabria, N Spain

Torrens, Lake *149 A6 salt lake* South Australia

Torrent *93 F3 Cas.* Torrente, *var.* Torrent de l'Horta. Comunitat Valenciana, E Spain

Torrent de l'Horta/Torrente *see* Torrent

Torreón *50 D3* Coahuila, NE Mexico

Torres Strait *148 C1 strait* Australia/ Papua New Guinea

Torres Vedras *92 B3* Lisboa, C Portugal

Torrington *44 D3* Wyoming, C USA

Tórshavn *83 F5 Dan.* Thorshavn. *Dependent territory capital* Faroe Islands

To'rtkok'l *122 D1 var.* Türtkül, *Rus.* Turtkul'; *prev.* Petroaleksandrovsk. Qoraqalpog'iston Respublikasi, W Uzbekistan

Tortoise Islands *see* Galápagos Islands

Tortosa *93 F2 anc.* Dertosa. Cataluña, E Spain

Tortue, Montagne *59 H3 mountain range* C French Guiana

Tortuga, Isla *see* La Tortuga, Isla

Toruń *98 C3 Ger.* Thorn. Toruń, Kujawsko-pomorskie, C Poland

Tõrva *106 D3 Ger.* Tõrwa. Valgamaa, S Estonia

Tõrwa *see* Tõrva

Torzhok *110 B4* Tverskaya Oblast', W Russia

Tosa-wan *131 B7 bay* SW Japan

Toscana *96 B3 Eng.* Tuscany. *region* C Italy

Toscano, Archipelago *96 B4 Eng.* Tuscan Archipelago. *island group* C Italy

Toshkent *123 E2 Eng./Rus.* Tashkent. *country capital* (Uzbekistan) Toshkent Viloyati, E Uzbekistan

Totana *93 E4* Murcia, SE Spain

Tot'ma *see* Sukhona

Totness *59 G3* Coronie, N Suriname

Tottori *131 B6* Tottori, Honshū, SW Japan

Touâjîl *74 C2* Tiris Zemmour, N Mauritania

Touggourt *71 E2* NE Algeria

Toukoto *74 C3* Kayes, W Mali

Toul *90 D3* Meurthe-et-Moselle, NE France

Toulon *91 D6 anc.* Telo Martius, Tilio Martius. Var, SE France

Toulouse *91 B6 anc.* Tolosa. Haute-Garonne, S France

Toungoo *see* Taungoo

Touraine *90 B4 cultural region* C France

Tourane *see* Đà Nẵng

Tourcoing *90 C2* Nord, N France

Tournai *87 A6 var.* Tournay, *Dut.* Doornik; *anc.* Tornacum. Hainaut, SW Belgium

Tournay *see* Tournai

Tours *90 B4 anc.* Caesarodunum, Turoni. Indre-et-Loire, C France

Tovarkovskiy *111 B5* Tul'skaya Oblast', W Russia

Tower Island *see* Genovesa, Isla

Townsville *148 D3* Queensland, NE Australia

Towoeti Meer *see* Towuti, Danau

Towraghoudi *122 D4* Herāt, NW Afghanistan

Towson *41 F4* Maryland, NE USA

Towuti, Danau *139 E4 Dut.* Towoeti Meer. *lake* Sulawesi, C Indonesia

Toyama *131 C5* Toyama, Honshū, SW Japan

Toyama-wan *131 B5 bay* W Japan

Toyohara *see* Yuzhno-Sakhalinsk

Toyota *131 C6* Aichi, Honshū, SW Japan

Tozeur *71 E2 var.* Tawzar. W Tunisia

Tråblous *see* Tripoli

Trabzon *117 E2 Eng.* Trebizond; *anc.* Trapezus. Trabzon, NE Turkey (Türkiye)

Traiectum ad Mosam/Traiectum Tungorum *see* Maastricht

Traiskirchen *95 E6* Niederösterreich, NE Austria

Trajani Portus *see* Civitavecchia

Trajectum ad Rhenum *see* Utrecht

Trakai *107 C5 Ger.* Traken, *Pol.* Troki. Vilnius, SE Lithuania

Traken *see* Trakai

Tralee *89 A6 Ir.* Trá Lí. SW Ireland

Trá Lí *see* Tralee

Tralles Aydin *see* Aydın

Tran *104 B2 var.* Trŭn. Pernik, W Bulgaria

Trang *137 C7* Trang, S Thailand

Transantarctic Mountains *154 B3 mountain range* Antarctica

Transilvania *see* Transylvania

Transilvaniei, Alpi *see* Carpaţii Meridionali

Transjordan *see* Jordan

Transnistria *108 D3 disputed territory* NE Moldova

Transsylvanische Alpen/Transylvanian Alps *see* Carpaţii Meridionali

Transylvania *108 B4 Eng.* Ardeal, Transilvania, *Ger.* Siebenbürgen, *Hung.* Erdély. *cultural region* NW Romania

Trapani *97 B7 anc.* Drepanum. Sicilia, Italy, C Mediterranean Sea

Trápeăng Vêng *see* Kampong Thom

Trapezus *see* Trabzon

Traralgon *149 C7* Victoria, SE Australia

Trasimenischersee *see* Trasimeno, Lago

Trasimeno, Lago *96 C4 Eng.* Lake of
Perugia, *Ger.* Trasimenischersee. *lake*
C Italy
Traù *see* Trogir
Traverse City *40 C2* Michigan,
N USA
Tra Vinh *137 D6 var.* Phu Vinh. Tra
Vinh, S Vietnam
Travis, Lake *49 F3 reservoir* Texas,
SW USA
Travnik *100 C4* Federacija Bosna I
Hercegovina, C Bosnia and
Herzegovina
Trbovlje *95 E7 Ger.* Trifail. C Slovenia
Treasure State *see* Montana
Třebíč *99 B5 Ger.* Trebitsch. Vysočina,
C Czechia (Czech Republic)
Trebinje *101 C5* Republika Srpska,
S Bosnia and Herzegovina
Trebišov *99 D6 Hung.* Tőketerebes.
Košický Kraj, E Slovakia
Trebitsch *see* Třebíč
Trebnitz *see* Trzebnica
Tree Planters State *see* Nebraska
Trélazé *90 B4* Maine-et-Loire,
NW France
Trelew *65 C6* Chubut, SE Argentina
Tremelo *87 C5* Vlaams Brabant,
C Belgium
Trenčín *99 C5 Ger.* Trentschin, *Hung.*
Trencsén. Trenčiansky Kraj,
W Slovakia
Trencsén *see* Trenčín
Trengganu, Kuala *see* Kuala Terengganu
Trenque Lauquen *64 C4* Buenos Aires,
E Argentina
Trent *see* Trento
Trento *96 C2 Eng.* Trent, *Ger.* Trient;
anc. Tridentum. Trentino-Alto Adige,
N Italy
Trenton *41 F4 state capital* New Jersey,
NE USA
Trentschin *see* Trenčín
Tres Arroyos *65 D5* Buenos Aires,
E Argentina
Treskavica *100 C4 mountain range*
SE Bosnia and Herzegovina
Tres Tabernae *see* Saverne
Treves/Trèves *see* Trier
Treviso *96 C2 anc.* Tarvisium. Veneto,
NE Italy
Trichinopoly *see* Tiruchchirāppalli
Trichūr *see* Thrissur
Tridentum/Trient *see* Trento
Trier *95 A5 Eng.* Treves, *Fr.* Trèves; *anc.*
Augusta Treverorum. Rheinland-Pfalz,
SW Germany
Triesen *94 E2* SW Liechtenstein Europe
Triesenberg *94 E2* SW Liechtenstein
Trieste *96 D2 Slvn.* Trst. Friuli-Venezia
Giulia, NE Italy
Trifail *see* Trbovlje
Tríkala *104 B4 prev.* Trikkala. Thessalía,
C Greece
Trikkala *see* Tríkala
Trimontium *see* Plovdiv
Trinacria *see* Sicilia
Trincomalee *132 D3 var.* Trinkomali.
Eastern Province, NE Sri Lanka
Trindade, Ilha da *67 C5 island* Brazil,
W Atlantic Ocean
Trinidad *61 F3* Beni, N Bolivia
Trinidad *64 D4* Flores, S Uruguay
Trinidad *44 D5* Colorado, C USA
Trinidad *55 H5 island* Trinidad
and Tobago
Trinidad and Tobago *55 H5 off.*
Republic of Trinidad and Tobago.
country SE West Indies
Trinidad and Tobago, Republic of *see*
Trinidad and Tobago
Trinité, Montagnes de la *59 H3*
mountain range C French Guiana
Trinity River *49 G3 river* Texas,
SW USA
Trinkomali *see* Trincomalee
Trípoli *105 B6 prev.* Trípolis.
Pelopónnisos, S Greece
Tripoli *118 B4 var.* Tarābulus, Ţarābulus
ash Shām, Ţrâblous; *anc.* Tripolis.
N Lebanon
Trípoli *see* Tarabulus
Trípolis *see* Trípoli, Greece
Trípolis *see* Trípoli, Lebanon
Tripolitania *71 F3 region* NW Libya
Tristan da Cunha *see* St Helena,
Ascension and Tristan da Cunha
Triton Island *129 B7 island*
S Paracel Islands
Trivandrum *see* Thiruvananthapuram

Trnava *99 C6 Ger.* Tyrnau, *Hung.*
Nagyszombat. Trnavský Kraj,
W Slovakia
Trnovo *see* Veliko Tarnovo
Trogir *100 B4 It.* Traù. Split-Dalmacija,
S Croatia
Troglav *100 B4 mountain* Bosnia and
Herzegovina/Croatia
Trois-Rivières *39 E4* Québec, SE Canada
Troki *see* Trakai
Troll *154 C2 Norwegian research station*
Antarctica
Trollhättan *85 B6* Västra Götaland,
S Sweden
Tromsø *84 C2 Fin.* Tromssa. Troms,
N Norway
Tromssa *see* Tromsø
Trondheim *84 B4 Ger.* Drontheim; *prev.*
Nidaros, Trondhjem. Sør-Trøndelag,
S Norway
Trondheimsfjorden *84 B4 fjord*
S Norway
Trondhjem *see* Trondheim
Troódos *102 C5 var.* Troodos
Mountains. *mountain range* C Cyprus
Troodos Mountains *see* Troódos
Troppau *see* Opava
Troy *42 D3* Alabama, S USA
Troy *41 F3* New York, NE USA
Troyan *104 C2* Lovech, N Bulgaria
Troyes *90 D3 anc.* Augustobona
Tricassium. Aube, N France
Trst *see* Trieste
Trstenik *100 E4* Serbia, C Serbia
Trucial States *see* United Arab Emirates
Trujillo *52 D2* Colón, NE Honduras
Trujillo *60 B3* La Libertad, NW Peru
Trujillo *92 C3* Extremadura,
W Spain
Truk Islands *see* Chuuk Islands
Trün *see* Tran
Truro *39 F4* Nova Scotia, SE Canada
Truro *89 C7* SW England,
United Kingdom
Trzcianka *98 B3 Ger.* Schönlanke. Pila,
Wielkopolskie, C Poland
Trzebnica *98 C4 Ger.* Trebnitz.
Dolnośląskie, SW Poland
Tsalka *117 F2* S Georgia Asia
Tsamkong *see* Zhanjiang
Tsangpo *see* Brahmaputra
Tsarevo *104 E2 prev.* Michurin. Burgas,
E Bulgaria
Tsarigrad *see* İstanbul
Tsaritsyn *see* Volgograd
Tschakathurn *see* Čakovec
Tschaslau *see* Čáslav
Tschenstochau *see* Częstochowa
Tsefat *119 B5 var.* Safed, *Ar.* Safad; *prev.*
Zefat. Northern, N Israel
Tselinograd *see* Astana
Tsetsen Khan *see* Öndörhaan
Tsetserleg *126 D2* Arhangay, C Mongolia
Tshela *77 B6* Bas-Congo, W Dem.
Rep. Congo
Tshikapa *77 C7* Kasai-Occidental,
SW Dem. Rep. Congo
Tshuapa *77 D6 river* C Dem. Rep. Congo
Tsinan *see* Jinan
Tsing Hai *see* Qinghai Hu, China
Tsinghai *see* Qinghai, China
Tsingtao/Tsingtau *see* Qingdao
Tsinkiang *see* Quanzhou
Tsintao *see* Qingdao
Tsitsihar *see* Qiqihar
Tsu *131 C6 var.* Tu. Mie, Honshū,
SW Japan
Tsugaru-kaikyo *130 C3 strait* N Japan
Tsumeb *78 B3* Otjikoto, N Namibia
Tsuruga *131 C6 var.* Turuga. Fukui,
Honshū, SW Japan
Tsiurupynsk *see* Oleshky
Tsuruoka *130 D4 var.* Turuoka.
Yamagata, Honshū, C Japan
Tsushima *131 A7 var.* Tsushima-tō,
Tusima. *island group* SW Japan
Tsushima-tō *see* Tsushima
Tsyerakhowka *107 D8 Rus.* Terekhovka.
Homyel'skaya Voblasts', SE Belarus
Tu *see* Tsu
Tuamotu, Archipel des *see* Tuamotu, Îles
Tuamotu Fracture Zone *143 H3*
fracture zone E Pacific Ocean
Tuamotu, Îles *145 H4 var.* Archipel des
Tuamotu, Dangerous Archipelago,
Tuamotu Islands. *island group*
N French Polynesia
Tuamotu Islands *see* Tuamotu, Îles
Tuapi *53 E2* Región Autónoma Atlántico
Norte, NE Nicaragua

Tuapse *111 A7* Krasnodarskiy Kray,
SW Russia
Tuba City *48 B1* Arizona, SW USA
Ţūbās *119 E6* N West Bank, SW Asia
Tubbergen *86 E3* Overijssel,
E Netherlands
Tubeke *see* Tubize
Tubize *87 B6 Dut.* Tubeke. Walloon
Brabant, C Belgium
Tubmanburg *74 C5* NW Liberia
Ţubruq *71 H2 Eng.* Tobruk, *It.* Tobruch.
NE Libya
Tubuai, Îles/Tubuai Islands *see*
Australes, Îles
Tucker's Town *see* Tukums
Tuckum *see* Tukums
Tucson *48 B3* Arizona, SW USA
Tucumán *see* San Miguel de Tucumán
Tucumcari *48 E2* New Mexico, SW USA
Tucupita *59 E2* Delta Amacuro,
NE Venezuela
Tucuruí, Represa de *63 F2 reservoir*
NE Brazil
Tudela *93 E2 Basq.* Tutera; *anc.* Tutela.
Navarra, N Spain
Tuguegarao *139 E1* Luzon, N Philippines
Tuktoyaktuk *37 E3* Northwest
Territories, NW Canada
Tukums *106 C3 Ger.* Tuckum. Tukums,
W Latvia
Tula *111 B5* Tul'skaya Oblast', W Russia
Tulancingo *51 E4* Hidalgo, C Mexico
Tulare Lake Bed *47 C7 salt flat*
California, W USA
Tulcán *60 B1* Carchi, N Ecuador
Tulcea *108 D5* Tulcea, E Romania
Tul'chin *see* Tulchyn
Tulchyn *108 D3 Rus.* Tul'chin.
Vinnyts'ka Oblast', C Ukraine
Tuléar *see* Toliara
Tulia *49 E2* Texas, SW USA
Ţülkarm *119 D6* West Bank, SW Asia
Tulle *91 C5 anc.* Tutela. Corrèze,
C France
Tulln *95 E6 var.* Oberhollabrunn.
Niederösterreich, NE Austria
Tully *148 D3* Queensland, NE Australia
Tulsa *49 G1* Oklahoma, C USA
Tuluá *58 B3* Valle del Cauca,
W Colombia
Tulun *115 E4* Irkutskaya Oblast',
S Russia
Tumaco *58 A4* Nariño, SW Colombia
Tumakūru *132 C2 prev.* Tumkūr.
Karnātaka, W India
Tumba, Lac *see* Ntomba, Lac
Tumbes *60 A2* Tumbes, NW Peru
Tumkür *see* Tumakūru
Tumuc-Humac Mountains *63 E1 var.*
Serra Tumucumaque. *mountain range*
N South America
Tumucumaque, Serra *see*
Tumuc-Humac Mountains
Tunca Nehri *see* Tundzha
Tunduru *73 C8* Ruvuma, S Tanzania
Tundzha *104 D3 Turk.* Tunca Nehri.
river Bulgaria/Turkey (Türkiye)
Tungabhadra Reservoir *132 C2 lake*
S India
Tungaru *145 E2 prev.* Gilbert Islands.
island group W Kiribati
T'ung-shan *see* Xuzhou
Tung-t'ing Hu *see* Dongting Hu
Tunis *71 E1 var.* Tūnis. *country capital*
(Tunisia) NE Tunisia
Tunis, Golfe de *102 D3 Ar.* Khalīj Tūnis.
gulf NE Tunisia
Tunisia *71 F2 off.* Republic of Tunisia,
Ar. Al Jumhūrīyah at Tūnisīyah, *Fr.*
République Tunisienne. *country*
N Africa
Tunisia, Republic of *see* Tunisia
Tunisie, République *see* Tunisia
Tūnisīyah, Al Jumhūrīyah at *see* Tunisia
Tūnis, Khalīj *see* Tunis, Golfe de
Tunja *58 B3* Boyacá, C Colombia
Tuong Buong *see* Tương Đương
Tương Đương *136 D4 var.* Tuong
Buong. Nghệ An, N Vietnam
Tüp *123 G2 Kir.* Tyup. Ysyk-Köl
Oblusu/Issyk-Kul'skaya Oblast',
NE Kyrgyzstan
Tupelo *42 C2* Mississippi, S USA
Tupiza *61 G5* Potosí, S Bolivia
Turabah *121 B5* Makkah,
W Saudi Arabia

Turangi *150 D4* Waikato, North Island,
New Zealand
Turan Lowland *122 C2 var.* Turan Plain,
Kaz. Turan Oypaty, *Rus.* Turanskaya
Nizmennost', *Turk.* Turan Pesligi, *Uzb.*
Turan Pasttekisligi. *plain* C Asia
Turan Oypaty/Turan Pesligi/Turan
Plain/Turanskaya Nizmennost' *see*
Turan Lowland
Turan Pasttekisligi *see* Turan Lowland
Ţurayf *120 A3* Al Ḥudūd ash
Shamālīyah, NW Saudi Arabia
Turba *see* Teruel
Turbat *134 A3* Baluchistan, SW Pakistan
Turčiansky Svätý Martin *see* Martin
Turda *108 B4 Ger.* Thorenburg, *Hung.*
Torda. Cluj, NW Romania
Turek *98 C3* Wielkopolskie, C Poland
Turfan *see* Turpan
Turin *see* Torino
Turkana, Lake *73 C6 var.* Lake Rudolf.
lake N Kenya
Turkestan *see* Türkistan/Turkistan
Turkey *116 B3 off.* Republic of Türkiye,
var. Türkiye, *Turk.* Türkiye
Cumhuriyeti. *country* SW Asia
Turkish Republic of Northern Cyprus
102 D5 disputed territory administered
by Turkey Cyprus
Türkistan/Turkistan *114 B5 prev.*
Turkestan. Türkistan/Turkistan,
S Kazakhstan
Türkiye Turkey
Türkiye, Republic of *see* Turkey
Türkiye Cumhuriyeti *see* Turkey
Türkmenabat *122 D3 prev.* Rus.
Chardzhev, Chardzhou, Chardzhui,
Lenin-Turkmenski, *Turkm.* Chärjew.
Lebap Welaýaty, E Turkmenistan
Türkmen Aylagy *122 B2 Rus.*
Turkmenskiy Zaliv. *lake gulf*
W Turkmenistan
Türkmenbashi *see* Türkmenbaşy
Türkmenbaşy *122 B2 Rus.*
Turkmenbashi; *prev.* Krasnovodsk.
Balkan Welaýaty, W Turkmenistan
Türkmenbaşy Aylagy *122 A2 prev.* Rus.
Krasnovodskiy Zaliv, *Turkm.*
Krasnowodsk Aylagy. *lake Gulf*
W Turkmenistan
Turkmenistan *122 B2 prev.*
Turkmenskaya Soviet Socialist
Republic. *country* C Asia
Turkmenskaya Soviet Socialist
Republic *see* Turkmenistan
Turkmenskiy Zaliv *see* Türkmen Aylagy
Turks and Caicos Islands *55 E2 UK*
Overseas Territory N West Indies
Turku *85 D6 Swe.* Åbo.
Varsinais-Suomi, SW Finland
Turlock *47 B6* California, W USA
Turnagain, Cape *150 D4 headland*
North Island, New Zealand
Turnau *see* Turnov
Turnhout *87 C5* Antwerpen, N Belgium
Turnov *98 B3 Ger.* Turnau. Liberecký
Kraj, N Czechia (Czech Republic)
Türnovo *see* Veliko Tarnovo
Turnu Măgurele *108 B5 var.*
Turnu-Măgurele. Teleorman,
S Romania
Turnu Severin *see* Drobeta-
Turnu Severin
Turócszentmárton *see* Martin
Turoni *see* Tours
Turpan *126 C3 var.* Turfan. Xinjiang
Uygur Zizhiqu, NW China
Turpan Depression *see* Turpan Pendi
Turpan Pendi *126 C3 var.* Turpan
Depression. *depression* NW China
Turpentine State *see* North Carolina
Türtkül/Turtkul' *see* To'rtko'l'1
Turuga *see* Tsuruga
Turuoka *see* Tsuruoka
Tuscaloosa *42 C2* Alabama, S USA
Tuscan Archipelago *see* Toscano,
Arcipelago
Tuscany *see* Toscana
Tusima *see* Tsushima
Tutela *see* Tulle, France
Tutela *see* Tudela, Spain
Tutera *see* Tudela
Tuticorin *see* Thoothukudi
Tutrakan *104 D1* Silistra, NE Bulgaria
Tutuila *145 F4 island*
W American Samoa
Tuvalu *145 E3 prev.* Ellice Islands.
country SW Pacific Ocean
Tuwayq, Jabal *121 C5 mountain range*
C Saudi Arabia

Tuxpan 50 D4 Jalisco, C Mexico
Tuxpan 51 F4 var. Tuxpán de Rodríguez
  Cano. Veracruz-Llave, E Mexico
Tuxpán de Rodríguez Cano see Tuxpán
Tuxtepec 51 F4 var. San Juan Bautista
  Tuxtepec. Oaxaca, S Mexico
Tuxtla 51 G5 var. Tuxtla Gutiérrez.
  Chiapas, SE Mexico
Tuxtla see San Andrés Tuxtla
Tuxtla Gutiérrez see Tuxtla
Tuy Hoa 137 E5 Phu Yên, S Vietnam
Tuz, Lake 116 C3 lake
  C Turkey (Türkiye)
Tver' 110 B4 prev. Kalinin. Tverskaya
  Oblast', W Russia
Tverya 119 B5 var. Tiberias; prev.
  Teverya. Northern, N Israel
Twin Falls 46 D4 Idaho, NW USA
Tyan'-Shan' see Tien Shan
Tychy 99 D5 Ger. Tichau. Śląskie,
  S Poland
Tyler 49 G3 Texas, SW USA
Tylos see Bahrain
Tympáki 105 C8 var. Timbaki; prev.
  Timbákion. Kríti, Greece,
  E Mediterranean Sea
Tynda 115 F4 Amurskaya Oblast',
  SE Russia
Tyne 88 D4 river N England,
  United Kingdom
Tyōsi see Chōshi
Tyras see Bilhorod-Dnistrovskyi
Tyras see Dniester
Tyre see Soûr
Tyrnau see Trnava
Týrnavos 104 B4 var. Tírnavos.
  Thessalía, C Greece
Tyrol see Tirol
Tyros see Bahrain
Tyrrhenian Sea 97 B6 It. Mare Tirreno.
  sea N Mediterranean Sea
Tyumen' 114 C3 Tyumenskaya Oblast',
  C Russia
Tyup see Tüp
Tywyn 89 C6 W Wales, United Kingdom
Tzekung see Zigong
Tziá see Kéa

U

Uaco Cungo 78 B1 C Angola
UAE see United Arab Emirates
Uanle Uen see Wanlaweyn
Uaupés, Río see Vaupés, Río
Ubangi 77 C5 var. Oubangui. river N
  Dem. Rep. Congo
Ubangi-Shari see Central
  African Republic
Ube 131 B7 Yamaguchi, Honshū,
  SW Japan
Úbeda 93 E4 Andalucía, S Spain
Uberaba 63 F4 Minas Gerais, SE Brazil
Uberlândia 63 F4 Minas Gerais,
  SE Brazil
Ubol Rajadhani/Ubol Ratchathani see
  Ubon Ratchathani
Ubon Ratchathani 137 D5 var. Muang
  Ubon, Ubol Rajadhani, Ubol
  Ratchathani, Udon Ratchathani. Ubon
  Ratchathani, E Thailand
Ubrique 92 D5 Andalucía, S Spain
Ubsu-Nur, Ozero see Uvs Nuur
Ucayali, Río 60 D3 river C Peru
Uchiura-wan 130 D3 bay
  NW Pacific Ocean
Uchkuduk see Uchquduq
Uchquduq 122 D2 Rus. Uchkuduk.
  Navoiy Viloyati, N Uzbekistan
Uchtagan Gumy/Uchtagan, Peski see
  Uçtagan Gumy
Uçtagan Gumy 122 C2 var. Uchtagan
  Gumy, Rus. Peski Uchtagan. desert
  NW Turkmenistan
Udaipur 134 C3 prev. Oodeypore.
  Rājasthān, N India
Uddevalla 85 B6 Västra Götaland,
  S Sweden
Udine 96 D2 anc. Utina. Friuli-Venezia
  Giulia, NE Italy
Udintsev Fracture Zone 154 A5 tectonic
  feature S Pacific Ocean
Udipi see Udupi
Udon Ratchathani see Ubon Ratchathani
Udon Thani 136 C4 var. Ban Mak
  Khaeng, Udorndhani. Udon Thani,
  N Thailand
Udorndhani see Udon Thani
Udupi 132 B2 var. Udipi. Karnātaka,
  SW India

Uele 77 D5 var. Welle. river NE Dem.
  Rep. Congo
Uelzen 94 C3 Niedersachsen, N Germany
Ufa 111 D6 Respublika Bashkortostan,
  W Russia
Ugāle 106 C2 Ventspils, NW Latvia
Uganda 73 B6 off. Republic of Uganda.
  country E Africa
Uganda, Republic of see Uganda
Uhorshchyna see Hungary
Uhuru Peak see Kilimanjaro
Uíge 78 B1 Port. Carmona, Vila
  Marechal Carmona. Uíge, NW Angola
Uinta Mountains 44 B4 mountain range
  Utah, W USA
Uitenhage see Kariega
Uithoorn 86 C3 Noord-Holland,
  C Netherlands
Ujda see Oujda
Ujelang Atoll 144 C1 var. Wujlān. atoll
  Ralik Chain, W Marshall Islands
Újgradiska see Nova Gradiška
Újmoldova see Moldova Nouă
Ujungpandang see Makassar
Ujung Salang see Phuket
Ujvidék see Novi Sad
UK see United Kingdom
Ukhta 114 C3 Respublika Komi,
  NW Russia
Ukiah 47 B5 California, W USA
Ukmergė 106 C4 Pol. Wiłkomierz.
  Vilnius, C Lithuania
Ukraine 108 C2 off. Ukraine, Rus.
  Ukraina, Ukr. Ukrayina; prev.
  Ukrainian Soviet Socialist Republic,
  Ukrainskay S.S.R. country SE Europe
Ukraine see Ukraine
Ukrainian Soviet Socialist Republic see
  Ukraine
Ukrainskay S.S.R/Ukrayina see Ukraine
Ulaanbaatar 127 E2 Eng. Ulan Bator;
  prev. Urga. country capital (Mongolia)
  Töv, C Mongolia
Ulaangom 126 C2 Uvs, NW Mongolia
Ulan Bator see Ulaanbaatar
Ulanhad see Chifeng
Ulan-Ude 115 E4 prev. Verkhneudinsk.
  Respublika Buryatiya, S Russia
Uleåborg see Oulujoki
Uleälv see Oulujärvi
Uleträsk see Oulujärvi
Ulft 86 E4 Gelderland, E Netherlands
Ulianovka see Blahovishchenske
Ullapool 88 C3 N Scotland,
  United Kingdom
Ulm 95 B6 Baden-Württemberg,
  S Germany
Ulsan 129 E4 Jap. Urusan.
  SE South Korea
Ulster 89 B5 province Northern Ireland,
  United Kingdom/Ireland
Ulukhaktok 37 E3 var. Holman. Victoria
  Island, Northwest Territories, N Canada
Ulungur Hu 126 B2 lake NW China
Uluru 147 D5 var. Ayers Rock. monolith
  Northern Territory, C Australia
Ulyanivka see Blahovishchenske
Ul'yanovsk 111 C5 prev. Simbirsk.
  Ul'yanovskaya Oblast', W Russia
Umán 51 H3 Yucatán, SE Mexico
Uman 109 E3 Cherkas'ka Oblast',
  C Ukraine
Umanak/Umanaq see Uummannaq
'Umān, Khalīj see Oman, Gulf of
'Umān, Salṭanat see Oman
Umbrian-Michigan Mountains see
  Umbro-Marchigiano, Appennino
Umbro-Marchigiano, Appennino 96
  C3 Eng. Umbrian-Michigan
  Mountains. mountain range C Italy
Umeå 84 C4 Västerbotten, N Sweden
Umeälven 84 C4 river N Sweden
Umiat 36 D2 Alaska, USA
Umm Buru 72 A4 Western Darfur,
  W Sudan
Umm Durman see Omdurman
Umm Ruwaba 72 C4 var. Umm
  Ruwābah, Um Ruwāba. Northern
  Kordofan, C Sudan
Umm Ruwābah see Umm Ruwaba
Umnak Island 36 A3 island Aleutian
  Islands, Alaska, USA
Um Ruwāba see Umm Ruwaba
Umtali see Mutare
Umtata see Mthatha
Una 100 B3 river Bosnia and
  Herzegovina/Croatia
Unac 100 B3 river W Bosnia and
  Herzegovina
Unalaska Island 36 A3 island Aleutian
  Islands, Alaska, USA

'Unayzah 120 B4 var. Anaiza. Al Qaşīm,
  C Saudi Arabia
Unci see Almería
Uncía 61 F4 Potosí, C Bolivia
Uncompahgre Peak 44 B5 mountain
  Colorado, C USA
Undur Khan see Öndörhaan
Ungaria see Hungary
Ungarisches Erzgebirge see Slovenské
  rudohorie
Ungarn see Hungary
Ungava Bay 39 E1 bay Québec, E Canada
Ungava Peninsula see Ungava,
  Péninsule d'
Ungava, Péninsule d' 38 D1 Eng. Ungava
  Peninsula. peninsula Québec,
  SE Canada
Ungeny see Ungheni
Ungheni 108 D3 Rus. Ungeny.
  W Moldova
Unguja see Zanzibar
Üngüz Angyrsyndaky Garagum 122 C2
  Rus. Zaunguzskiye Garagumy. desert
  N Turkmenistan
Ungvár see Uzhhorod
Unimak Island 36 B3 island Aleutian
  Islands, Alaska, USA
Union 43 E1 South Carolina,
  SE USA
Union City 42 C1 Tennessee, S USA
Union of Myanmar see Myanmar
United Arab Emirates 121 C5 Ar. Al
  Imārāt al 'Arabīyah al Muttaḥidah,
  abbrev. UAE; prev. Trucial States.
  country SW Asia
United Arab Republic see Egypt
United Kingdom 89 B5 off. United
  Kingdom of Great Britain and Northern
  Ireland, abbrev. UK. country
  NW Europe
**United Kingdom of Great Britain and
  Northern Ireland** see United Kingdom
United Mexican States see Mexico
United Provinces see Uttar Pradesh
United States of America 35 B5 off.
  United States of America, var. America,
  The States, abbrev. U.S., USA. country
  North America
Unst 88 D1 island NE Scotland,
  United Kingdom
Ünye 116 D2 Ordu, W Turkey (Türkiye)
Upala 52 D4 Alajuela, NW Costa Rica
Upata 59 E2 Bolívar, E Venezuela
Upemba, Lac 77 D7 lake SE Dem.
  Rep. Congo
Upernavik 82 C2 var. Upernivik.
  Avannaata, C Greenland
Upernivik see Upernavik
Upington 78 C4 Northern Cape,
  W South Africa
'Upolu 145 F4 island SE Samoa
Upper Klamath Lake 46 A4 lake
  Oregon, NW USA
Upper Lough Erne 89 A5 lake
  SW Northern Ireland, United Kingdom
Upper Red Lake 45 F1 lake Minnesota,
  N USA
Upper Volta see Burkina
Uppsala 85 C6 Uppsala, C Sweden
Uqsuqtuuq see Gjoa Haven
Ural 112 B3 Kaz. Zhayyk. river
  Kazakhstan/Russia
Ural Mountains see Ural'skiye Gory
Ural'sk see Oral
Ural'skiye Gory 114 C3 var. Ural'skiy
  Khrebet, Eng. Ural Mountains.
  mountain range
  Kazakhstan/Russia
Ural'skiy Khrebet see Ural'skiye Gory
Uraricoera 62 D1 Roraima, N Brazil
Ura-Tyube see Istaravshan
Urbandale 45 F3 Iowa, C USA
Urdun see Jordan
Uren' 111 C5 Nizhegorodskaya Oblast',
  W Russia
Urga see Ulaanbaatar
Urganch 122 D2 Rus. Urgench; prev.
  Novo-Urgench. Xorazm Viloyati,
  W Uzbekistan
Urgench see Urganch
Urgut 123 E3 Samarqand Viloyati,
  C Uzbekistan
Urmia, Lake see Orūmīyeh, Daryācheh-ye
Uroševac see Ferizaj
Ürōteppa see Istaravshan
Uruapan 51 E4 var. Uruapan del
  Progreso. Michoacán,
  SW Mexico
Uruapan del Progreso see Uruapan
Uruguai, Rio see Uruguay

Uruguay 64 D4 off. Oriental Republic of
  Uruguay; prev. La Banda Oriental.
  country E South America
Uruguay 64 D3 var. Rio Uruguai, Río
  Uruguay. river E South America
Uruguay, Oriental Republic of see
  Uruguay
Uruguay, Río see Uruguay
Urumchi see Ürümqi
Urumi Yeh see Orūmīyeh, Daryācheh-ye
Ürümqi 126 C3 var. Tihwa, Urumchi,
  Urumqi, Urumtsi, Wu-lu-k'o-mu-shi,
  Wu-lu-mu-ch'i; prev. Ti-hua. Xinjiang
  Uygur Zizhiqu, NW China
Urumtsi see Ürümqi
Urundi see Burundi
Urup, Ostrov 115 H4 island Kuril'skiye
  Ostrova, SE Russia
Urusan see Ulsan
Urziceni 108 C5 Ialomiţa, SE Romania
Usa 114 E3 river NW Russia
Uşak 116 B3 prev. Ushak. Uşak,
  W Turkey (Türkiye)
Ushak see Uşak
Ushant see Ouessant, Île d'
Ushuaia 65 B8 Tierra del Fuego,
  S Argentina
Usinsk 110 E3 Respublika Komi,
  NW Russia
Üsküb/Üsküp see Skopje
Usmas Ezers 106 B3 lake NW Latvia
Usol'ye-Sibirskoye 115 E4 Irkutskaya
  Oblast', C Russia
Ussel 91 C5 Corrèze, C France
Ussuriysk 115 G5 prev. Nikol'sk,
  Nikol'sk-Ussuriyskiy, Voroshilov.
  Primorskiy Kray, SE Russia
Ustica 96 B6 island S Italy
Ust'-Ilimsk 115 E4 Irkutskaya Oblast',
  C Russia
Ústí nad Labem 98 A4 Ger. Aussig.
  Ústecký Kraj, NW Czechia
  (Czech Republic)
Ustinov see Izhevsk
Ustka 98 C2 Ger. Stolpmünde.
  Pomorskie, N Poland
Ust'-Kamchatsk 115 H2 Kamchatskaya
  Oblast', E Russia
Ust'-Kamenogorsk see Öskemen
Ust'-Kut 115 E4 Irkutskaya Oblast',
  C Russia
Ust'-Olenëk 115 E3 Respublika Sakha
  (Yakutiya), NE Russia
Ustrzyki Dolne 99 E5 Podkarpackie,
  SE Poland
Ust'-Sysol'sk see Syktyvkar
Ust Urt see Ustyurt Plateau
Ustyurt Plateau 122 B1 var. Ust Urt,
  Uzb. Ustyurt Platosi. plateau
  Kazakhstan/Uzbekistan
Ustyurt Platosi see Ustyurt Plateau
Usulután 52 C3 Usulután, SE El Salvador
Usumacinta, Río 52 B1 river Guatemala/
  Mexico
Usumbura see Bujumbura
U.S./USA see United States of America
Utah 44 B4 off. State of Utah, also known
  as Beehive State, Mormon State. state
  W USA
Utah Lake 44 B4 lake Utah, W USA
Utena 106 C4 Utena, E Lithuania
Utica 41 F3 New York, NE USA
Utina see Udine
Utqiaġvik 36 D2 var. Barrow.
  Alaska, USA
Utrecht 86 C4 Lat. Trajectum ad
  Rhenum. Utrecht, C Netherlands
Utsunomiya 131 D5 var. Utunomiya.
  Tochigi, Honshū, S Japan
Uttarakhand 135 E2 cultural region
  N India
Uttar Pradesh 135 E3 prev. United
  Provinces, United Provinces of Agra
  and Oudh. cultural region N India
Utunomiya see Utsunomiya
Uulu 106 D2 Pärnumaa, SW Estonia
Uummannaq 82 C3 var. Umanak,
  Umanaq. Avannaata, C Greenland
Uummannarsuaq see Nunap Isua
Uvalde 49 F4 Texas, SW USA
Uvarovichi 107 D7 Rus. Uvarovichi.
  Homyel'skaya Voblasts', SE Belarus
Uvarovichi see Uvarovichy
Uvea, Île 145 E4 island N Wallis
  and Futuna
Uviéu see Oviedo
Uvs Nuur 126 C1 var. Ozero Ubsu-Nur.
  lake Mongolia/Russia
'Uwaynāt, Jabal al 72 A3 var. Jebel
  Uweinat. mountain Libya/Sudan

Uweinat, Jebel *see* 'Uwaynāt, Jabal al
Uyo *75 G5* Akwa Ibom, S Nigeria
Uyuni *61 F5* Potosí, W Bolivia
Uzbekistan *122 D2* off. Republic of
  Uzbekistan. *country* C Asia
Uzbekistan, Republic of *see* Uzbekistan
Uzhgorod *see* Uzhhorod
Uzhhorod *108 B2 Rus.* Uzhgorod; *prev.*
  Ungvár. Zakarpats'ka Oblast',
  W Ukraine
Užice *100 D4 prev.* Titovo Užice. Serbia,
  W Serbia

V

Vaal *78 D4 river* C South Africa
Vaals *87 D6* Limburg, SE Netherlands
Vaasa *85 D5 Swe.* Vasa; *prev.*
  Nikolainkaupunki. Österbotten,
  W Finland
Vaassen *86 D3* Gelderland,
  E Netherlands
Vác *109 C6 Ger.* Waitzen. Pest,
  N Hungary
Vadodara *134 C4 prev.* Baroda. Gujarāt,
  W India
Vaduz *94 E2 country capital*
  (Liechtenstein) W Liechtenstein
Våg *see* Váh
Vágbeszterce *see* Považská Bystrica
Váh *109 C5 Ger.* Waag, *Hung.* Vág. *river*
  W Slovakia
Váhtjer *see* Gällivare
Väinameri *106 C2 prev.* Muhu Väin,
  Ger. Moon-Sund. *sea* E Baltic Sea
Vajdahunyad *see* Hunedoara
Valachia *see* Wallachia
Valday *110 B4* Novgorodskaya Oblast',
  W Russia
Valdecañas, Embalse de *92 D3 reservoir*
  W Spain
Valdepeñas *93 E4* Castilla-La Mancha,
  C Spain
Valdez *36 C3* Alaska, USA
Valdia *see* Weldiya
Valdivia *65 B5* Los Lagos, C Chile
Val-d'Or *38 D4* Québec, SE Canada
Valdosta *43 E3* Georgia, SE USA
Valence *91 D5 anc.* Valentia, Valentia
  Julia, Ventia. Drôme, E France
València *93 F3 var.* Valencia. Comunitat
  Valenciana, E Spain
Valencia *46 D1* California, USA
Valencia *58 D1* Carabobo, N Venezuela
Valencia, Gulf of *93 F3 var.* Gulf of
  Valencia. *gulf* E Spain
Valencia, Gulf of *see* Valencia, Golfo de
Valencià/València *see* Valenciana,
  Comunitat
Valenciana, Comunitat *93 F3 var.*
  Valencia, *Cat.* València; *anc.* Valentia.
  autonomous community NE Spain
Valenciennes *90 D2* Nord, N France
Valentia *see* Valence, France
Valentia *see* Valenciana, Comunitat
Valentia Julia *see* Valence
Valentine State *see* Oregon
Valera *58 C2* Trujillo, NW Venezuela
Valetta *see* Valletta
Valga *106 D3 Ger.* Walk, *Latv.* Valka.
  Valgamaa, S Estonia
Valira *91 A8 river* Andorra/Spain Europe
Valjevo *100 C4* Serbia, W Serbia
Valjok *see* Válljohka
Valka *106 D3 Ger.* Walk. Valka, N Latvia
Valka *see* Valga
Valkenswaard *87 D5* Noord-Brabant,
  S Netherlands
Valladolid *51 H3* Yucatán, SE Mexico
Valladolid *92 D2* Castilla y León,
  NW Spain
Vall D'Uxó *see* La Vall d'Uixó
Valle de La Pascua *58 D2* Guárico,
  N Venezuela
Valledupar *58 B1* Cesar, N Colombia
Vallejo *47 B6* California, W USA
Vallenar *64 B3* Atacama, N Chile
Valletta *97 C8 prev.* Valetta. *country
  capital* (Malta) E Malta
Valley City *45 E2* North Dakota,
  N USA
Válljohka *84 D2 var.* Valjok. Finnmark,
  N Norway
Valls *93 G2* Cataluña, NE Spain
Valmiera *106 D3 Est.* Volmari, *Ger.*
  Wolmar. Valmiera, N Latvia
Valona *see* Vlorë

Valozhyn *107 C5 Pol.* Wołożyn, *Rus.*
  Volozhin. Minskaya Voblasts',
  C Belarus
Valparaíso *64 B4* Valparaíso, C Chile
Valparaíso *40 C3* Indiana, N USA
Valverde del Camino *92 C4* Andalucía,
  S Spain
Van *117 F3* Van, E Turkey (Türkiye)
Vanadzor *117 F2 prev.* Kirovakan.
  N Armenia
Vancouver *36 D5* British Columbia,
  SW Canada
Vancouver *46 B3* Washington, NW USA
Vancouver Island *36 D5 island* British
  Columbia, SW Canada
Vanda *see* Vantaa
Van Diemen Gulf *146 D2 gulf* Northern
  Territory, N Australia
Van Diemen's Land *see* Tasmania
Vaner, Lake *see* Vänern
Vänern *85 B6 Eng.* Lake Vaner; *prev.*
  Lake Vener. *lake* S Sweden
Vangaindrano *79 G4* Fianarantsoa,
  SE Madagascar
Van Gölü *117 F3 Eng.* Lake Van; *anc.*
  Thospitis. *salt lake* E Turkey (Türkiye)
Van Horn *48 D3* Texas, SW USA
Van, Lake *see* Van Gölü
Vannes *90 A3 anc.* Dariorigum.
  Morbihan, NW France
Vantaa *85 D6 Swe.* Vanda. Uusimaa,
  S Finland
Vanua Levu *145 E4 island* N Fiji
Vanuatu *144 C4 off.* Republic of
  Vanuatu; *prev.* New Hebrides. *country*
  SW Pacific Ocean
Vanuatu, Republic of *see* Vanuatu
Van Wert *40 C4* Ohio, N USA
Vapincum *see* Gap
Varakļāni *106 D4* Madona, C Latvia
Vārānasi *135 E3 prev.* Banaras, Benares,
  hist. Kasi. Uttar Pradesh, N India
Varangerfjorden *84 E2 Smi.*
  Várjjatvuotna. *fjord* N Norway
Varangerhalvøya *84 D2 Smi.* Várnjárga.
  peninsula N Norway
Varannó *see* Vranov nad Topľou
Varasd *see* Varaždin
Varaždin *100 B2 Ger.* Warasdin, *Hung.*
  Varasd. Varaždin, N Croatia
Varberg *85 B7* Halland, S Sweden
Vardar *101 E6 Gk.* Axios. *river* North
  Macedonia/Greece
Varde *85 A7* Ribe, W Denmark
Vareia *see* Logroño
Varéna *107 B5 Pol.* Orany. Alytus,
  S Lithuania
Varese *96 B2* Lombardia, N Italy
Vârful Moldoveanu *108 B4 var.*
  Moldoveanul; *prev.* Vîrful Moldoveanu.
  mountain C Romania
Várjjatvuotna *see* Varangerfjorden
Varkaus *85 E5* Pohjois-Savo, C Finland
Varna *104 E2 prev.* Stalin; *anc.* Odessus.
  Varna, E Bulgaria
Varnenski Zaliv *104 E2 prev.* Stalinski
  Zaliv. *bay* E Bulgaria
Várnjárga *see* Varangerhalvøya
Varshava *see* Warszawa
Vasa *see* Vaasa
Vasilikí *105 A5* Lefkáda, Iónia Nisiá,
  Greece, C Mediterranean Sea
Vasilishki *107 B5 Pol.* Wasiliszki.
  Hrodzyenskaya Voblasts', W Belarus
Vasil'kov *see* Vasylkiv
Vaslui *108 D4* Vaslui, C Romania
Västerås *85 C6* Västmanland, C Sweden
Vasylkiv *109 E2 var.* Vasil'kov. Kyivska
  Oblast, N Ukraine
Vatican City *97 A7 off.* Vatican City
  State. *country* S Europe
Vatican City State *see* Vatican City
Vatnajökull *83 E5 glacier* SE Iceland
Vatter, Lake *see* Vättern
Vättern *85 B6 Eng.* Lake Vatter; *prev.*
  Lake Vetter. *lake* S Sweden
Vaughn *48 D2* New Mexico, SW USA
Vaupés, Río *58 C4 var.* Rio Uaupés.
  river Brazil/Colombia
Vava'u Group *145 E4 island group*
  N Tonga
Vavuniya *132 D3* Northern Province,
  N Sri Lanka
Vawkavysk *107 B6 Pol.* Wołkowysk,
  Rus. Volkovysk. Hrodzyenskaya
  Voblasts', W Belarus
Växjö *85 C7 var.* Vexiö. Kronoberg,
  S Sweden
Vaygach, Ostrov *110 E2 island*
  NW Russia

Veendam *86 E2* Groningen,
  NE Netherlands
Veenendaal *86 D4* Utrecht,
  C Netherlands
Vega *84 B4 island* C Norway
Veglia *see* Krk
Veisiejai *107 B5* Alytus, S Lithuania
Vejer de la Frontera *92 C5* Andalucía,
  S Spain
Veldhoven *87 D5* Noord-Brabant,
  S Netherlands
Velebit *100 A3 mountain range*
  C Croatia
Velenje *95 E7 Ger.* Wöllan. N Slovenia
Veles *101 E6 Turk.* Köprülü.
  C North Macedonia
Velho *see* Porto Velho
Velika Kikinda *see* Kikinda
Velika Morava *100 D4 var.* Glavn'a
  Morava, Morava, *Ger.* Grosse Morava.
  river C Serbia
Velikaya *113 G2 river* NE Russia
Veliki Bečkerek *see* Zrenjanin
Velikiye Luki *110 A4* Pskovskaya
  Oblast', W Russia
Velikiy Novgorod *110 B4 prev.*
  Novgorod. Novgorodskaya Oblast',
  W Russian Federation
Veliko Tarnovo *104 D2 prev.* Tirnovo,
  Trnovo, Tŭrnovo. Veliko Tarnovo,
  N Bulgaria
Velingrad *104 C3* Pazardzhik, C Bulgaria
Vel'ký Krtíš *99 D6* Banskobystrický
  Kraj, C Slovakia
Vellore *132 C2* Tamil Nādu, SE India
Velobriga *see* Viana do Castelo
Velsen *see* Velsen-Noord
Velsen-Noord *86 C3 var.* Velsen.
  Noord-Holland, W Netherlands
Vel'sk *110 C4 var.* Velsk.
  Arkhangel'skaya Oblast', NW Russia
Velvendós *see* Velvéntos
Velvéntos *104 B4 var.* Velvendós.
  C Greece
Velykyy Tokmak *see* Tokmak
Vendôme *90 C4* Loir-et-Cher, C France
Venedig *see* Venezia
Vener, Lake *see* Vänern
Venetia *see* Venezia
Venezia *96 C2 Eng.* Venice, *Fr.* Venise,
  Ger. Venedig; *anc.* Venetia. Veneto,
  NE Italy
Venezia, Golfo di *see* Venice, Gulf of
Venezuela *58 D2 off.* Bolivarian Republic
  of Venezuela; *prev.* Republic of
  Venezuela, Estados Unidos de
  Venezuela, United States of Venezuela.
  country N South America
Venezuela, Estados Unidos de *see*
  Venezuela
Venezuela, Gulf of *58 C1 Eng.* Gulf of
  Maracaibo, *Sp.* Golfo de Venezuela. *gulf*
  NW Venezuela
Venezuela, Gulf of *see* Venezuela,
  Golfo de
Venezuelan Basin *56 B1 undersea basin*
  E Caribbean Sea
Venezuela, Bolivarian Republic of *see*
  Venezuela
Venezuela, Republic of *see* Venezuela
Venezuela, United States of *see*
  Venezuela
Venice *42 C4* Louisiana, S USA
Venice, Gulf of *96 C2 It.* Golfo di
  Venezia, *Slvn.* Beneški Zaliv. *gulf*
  N Adriatic Sea
Venise *see* Venezia
Venlo *87 D5 prev.* Venloo. Limburg,
  SE Netherlands
Venloo *see* Venlo
Venta *106 B3 Ger.* Windau. *river* Latvia/
  Lithuania
Venta Belgarum *see* Winchester
Ventia *see* Valence
Ventimiglia *96 A3* Liguria,
  NW Italy
Ventspils *106 B2 Ger.* Windau.
  Ventspils, NW Latvia
Vera *64 D3* Santa Fe, C Argentina
Veracruz *51 F4 var.* Veracruz Llave.
  Veracruz-Llave, E Mexico
Veracruz Llave *see* Veracruz
Vercellae *see* Vercelli
Vercelli *96 A2 anc.* Vercellae. Piemonte,
  NW Italy
Verdal *see* Verdalsøra
Verdalsøra *84 B4 var.* Verdal.
  Nord-Trøndelag, C Norway
Verde, Cabo *see* Cape Verde

Verde, Costa *92 D1 coastal region*
  N Spain
Verden *94 B3* Niedersachsen,
  NW Germany
Veria *see* Véroia
Verkhnedvinsk *see* Vyerkhnyadzvinsk
Verkhneudinsk *see* Ulan-Ude
Verkhoyanskiy Khrebet *115 F3
  mountain range* NE Russia
Vermillion *45 F3* South Dakota, N USA
Vermont *41 F2 off.* State of Vermont,
  also known as Green Mountain State.
  state NE USA
Vernadsky *154 A2 Ukrainian research
  station* Antarctica
Vernal *44 B4* Utah, W USA
Vernon *49 F2* Texas, SW USA
Verőcze *see* Virovitica
Véroia *104 B4 var.* Veria, Vérroia, *Turk.*
  Karaferiye. Kentrikí Makedonía,
  N Greece
Verona *96 C2* Veneto, NE Italy
Vérroia *see* Véroia
Versailles *90 D1* Yvelines, N France
Versecz *see* Vršac
Verulamium *see* St Albans
Verviers *87 D6* Liège, E Belgium
Vesdre *87 D6 river* E Belgium
Veselinovo *104 D2* Shumen, NE Bulgaria
Vesontio *see* Besançon
Vesoul *90 D4 anc.* Vesulium, Vesulum.
  Haute-Saône, E France
Vesterålen *84 B2 island group* N Norway
Vestfjorden *84 C3 fjord* C Norway
Vestmannaeyjar *83 E5* Suðurland,
  S Iceland
Vesulium/Vesulum *see* Vesoul
Vesuna *see* Périgueux
Vesuvio *97 D5 Eng.* Vesuvius. *volcano*
  S Italy
Vesuvius *see* Vesuvio
Veszprém *99 C7 Ger.* Veszprim.
  Veszprém, W Hungary
Veszprim *see* Veszprém
Vetrino *see* Suvorovo
Vetrino *see* Vyetryna
Vetter, Lake *see* Vättern
Veurne *87 A5 var.* Furnes.
  West-Vlaanderen, W Belgium
Vexiö *see* Växjö
Viacha *61 F4* La Paz, W Bolivia
Viana del Castelo *see* Viana do Castelo
Viana do Castelo *92 B2 var.* Viana de
  Castelo; *anc.* Velobriga. Viana do
  Castelo, NW Portugal
Vianen *86 C4* Utrecht, C Netherlands
Viangchan *136 C4 Eng./Fr.* Vientiane.
  country capital (Laos) C Laos
Viangphoukha *136 C3 var.* Vieng Pou
  Kha. Louang Namtha, N Laos
Viareggio *96 B3* Toscana, C Italy
Viborg *85 A7* Viborg, NW Denmark
Vic *93 G2 var.* Vich; *anc.* Ausa, Vicus
  Ausonensis. Cataluña, NE Spain
Vicentia *see* Vicenza
Vicenza *96 C2 anc.* Vicentia. Veneto,
  NE Italy
Vich *see* Vic
Vichy *91 C5* Allier, C France
Vicksburg *42 B2* Mississippi, S USA
Victoria *79 H1 country capital*
  (Seychelles) Mahé, SW Seychelles
Victoria *36 D5 province capital*
  Vancouver Island, British Columbia,
  SW Canada
Victoria *102 A5 var.* Rabat. Gozo,
  NW Malta
Victoria *49 G4* Texas, SW USA
Victoria *149 C7 state* SE Australia
Victoria *see* Masvingo, Zimbabwe
Victoria Bank *see* Vitória Seamount
Victoria de Durango *see* Durango
Victoria de Las Tunas *see* Las Tunas
Victoria Falls *78 C3* Matabeleland
  North, W Zimbabwe
Victoria Falls *78 C2 waterfall* Zambia/
  Zimbabwe
Victoria Falls *see* Iguaçu, Salto do
Victoria Island *37 F3 island* Northwest
  Territories/Nunavut, NW Canada
Victoria, Lake *73 B6 var.* Victoria
  Nyanza. *lake* E Africa
Victoria Land *154 C4 physical region*
  Antarctica
Victoria Nyanza *see* Victoria, Lake
Victoria River *146 D3 river* Northern
  Territory, N Australia
Victorville *47 C7* California, W USA
Vicus Ausonensis *see* Vic
Vicus Elbii *see* Viterbo

Vidalia *43 E2* Georgia, SE USA
Videm-Krško *see* Krško
Vídeň *see* Wien
Vidin *104 B1 anc.* Bononia. Vidin,
  NW Bulgaria
Vidzy *107 C5* Vitsyebskaya Voblasts',
  NW Belarus
Viedma *65 C5* Río Negro, E Argentina
Vieja, Sierra *48 D3 mountain range*
  Texas, SW USA
Vieng Pou Kha *see* Viangphoukha
Vienna *see* Wien, Austria
Vienna *see* Vienne, France
Vienne *91 D5 anc.* Vienna. Isère,
  E France
Vienne *90 B4 river* W France
Vientiane *see* Viangchan
Vientos, Paso de los *see*
  Windward Passage
Vierzon *90 C4* Cher, C France
Viesíte *106 C4 Ger.* Eckengraf. Jēkabpils,
  S Latvia
Vietnam *136 D4 off.* Socialist Republic of
  Vietnam, *Vtn.* Công Hoa Xa Hôi Chu
  Nghia Viêt Nam. *country* SE Asia
Vietnam, Socialist Republic of *see*
  Vietnam
Vietri *see* Viêt Tri
Viêt Tri *136 D3 var.* Vietri. Vinh Phu,
  N Vietnam
Vieux Fort *55 F2* S Saint Lucia
Vigo *92 B2* Galicia, NW Spain
Viipuri *see* Vyborg
Vijayawāda *132 D1 prev.* Bezwada.
  Andhra Pradesh, SE India
Vila *see* Port-Vila
Vila Artur de Paiva *see* Cubango
Vila da Ponte *see* Cubango
Vila de Mocímboa da Praia *see*
  Mocímboa da Praia
Vila do Conde *92 B2* Porto,
  NW Portugal
Vila do Zumbo *78 D2 prev.* Vila do
  Zumbu, Zumbo. Tete,
  NW Mozambique
Vila do Zumbu *see* Vila do Zumbo
Vilafranca del Penedès *93 G2 var.*
  Villafranca de Panadés. Cataluña,
  NE Spain
Vila General Machado *see* Camacupa
Vila Henrique de Carvalho *see* Saurimo
Viļaka *106 D4 Ger.* Marienhausen. Balvi,
  NE Latvia
Vilalba *92 C1* Galicia, NW Spain
Vila Marechal Carmona *see* Uíge
Vila Nova de Gaia *92 B2* Porto,
  NW Portugal
Vila Nova de Portimão *see* Portimão
Vila Pereira de Eça *see* N'Giva
Vila Rial *92 C2 var.* Vila Rial. Vila Real,
  N Portugal
Vila Rial *see* Vila Real
Vila Robert Williams *see* Caála
Vila Salazar *see* N'Dalatando
Vila Serpa Pinto *see* Menongue
Vileyka *see* Vilyeyka
Vilhelmina *84 C4* Västerbotten,
  N Sweden
Vilhena *62 D3* Rondônia, W Brazil
Vília *105 C5* Attikí, C Greece
Viliya *107 C5 Lith.* Neris. *river*
  W Belarus
Viliya *see* Neris
Viljandi *106 D2 Ger.* Fellin.
  Viljandimaa, S Estonia
Vilkaviškis *106 B4 Pol.* Wyłkowyszki.
  Marijampolė, SW Lithuania
Villa Acuña *see var.* Ciudad Acuña.
  Coahuila, NE Mexico
Villa Bella *61 F2* Beni, N Bolivia
Villacarrillo *93 E4* Andalucía, S Spain
Villa Cecilia *see* Ciudad Madero
Villach *95 D7 Slvn.* Beljak. Kärnten,
  S Austria
Villacidro *97 A5* Sardegna, Italy,
  C Mediterranean Sea
Villa Concepción *see* Concepción
Villa del Pilar *see* Pilar
Villafranca de los Barros *92 C4*
  Extremadura, W Spain
Villafranca del Panadés *see* Vilafranca
  del Penedès
Villahermosa *51 G4 prev.* San Juan
  Bautista. Tabasco, SE Mexico
Villajoyosa *see* La Vila Joiosa
Villa María *64 C4* Córdoba, C Argentina
Villa Martín *61 F5* Potosí, SW Bolivia
Villa Mercedes *64 C4* San Juan,
  Argentina

Villanueva *50 D3* Zacatecas, C Mexico
Villanueva de la Serena *92 C3*
  Extremadura, W Spain
Villanueva de los Infantes *93 E4*
  Castilla-La Mancha, C Spain
Villarrica *64 D2* Guairá, SE Paraguay
Villavicencio *58 B3* Meta, C Colombia
Villaviciosa *92 D1* Asturias, N Spain
Villazón *61 G5* Potosí, S Bolivia
Villena *93 F4* Comunitat Valenciana,
  E Spain
Villeurbanne *91 D5* Rhône, E France
Villingen-Schwenningen *95 B6*
  Baden-Württemberg, S Germany
Villmanstrand *see* Lappeenranta
Vilna *see* Vilnius
Vilnius *107 C5 Pol.* Wilno, *Ger.* Wilna;
  prev. Rus. Vilna. *country capital*
  (Lithuania) Vilnius, SE Lithuania
Vilvoorde *87 C6 Fr.* Vilvorde. Vlaams
  Brabant, C Belgium
Vilvorde *see* Vilvoorde
Vilyeyka *107 C5 Pol.* Wilejka, *Rus.*
  Vileyka. Minskaya Voblasts',
  NW Belarus
Vilyuy *115 F3 river* NE Russia
Viña del Mar *64 B4* Valparaíso, C Chile
Vinarós *93 F3* Comunitat Valenciana,
  E Spain
Vincennes *40 B4* Indiana, N USA
Vindhya Mountains *see* Vindhya Range
Vindhya Range *134 D4 var.* Vindhya
  Mountains. *mountain range* N India
Vindobona *see* Wien
Vineland *41 F4* New Jersey, NE USA
Vinh *136 D4* Nghê An, N Vietnam
Vinh Loi *see* Bac Liêu
Vinishte *104 C2* Montana, NW Bulgaria
Vinita *49 G1* Oklahoma, C USA
Vinkovci *100 C3 Ger.* Winkowitz, *Hung.*
  Vinkovce. Vukovar-Srijem, E Croatia
Vinkovce *see* Vinkovci
Vinnitsa *see* Vinnytsia
Vinnytsia *108 D2 prev.* Vinnytsya, *Rus.*
  Vinnitsa. Vinnyts'ka Oblast', C Ukraine
Vinnytsya *see* Vinnytsia
Vinogradov *see* Vynohradiv
Vinson Massif *154 A3 mountain*
  Antarctica
Viranşehir *117 E4* Şanlıurfa,
  SE Turkey (Türkiye)
Virful Moldoveanu *see* Vârful
  Moldoveanu
Virginia *45 G1* Minnesota, N USA
Virginia *41 E5 off.* Commonwealth of
  Virginia, *also known as* Mother of
  Presidents, Mother of States, Old
  Dominion. *state* NE USA
Virginia Beach *41 F5* Virginia,
  NE USA
Virgin Islands *see* British Virgin Islands
Virgin Islands (US) *55 F3 var.* Virgin
  Islands of the United States; *prev.*
  Danish West Indies. *US unincorporated
  territory* E West Indies
Virgin Islands of the United States *see*
  Virgin Islands (US)
Viróchey *137 E5* Ratanakiri,
  NE Cambodia
Virovitica *100 C2 Ger.* Virovititz, *Hung.*
  Veröcze; *prev. Ger.* Werowitz.
  Virovitica-Podravina, NE Croatia
Virovititz *see* Virovitica
Virton *87 D8* Luxembourg, SE Belgium
Virtsu *106 D2 Ger.* Werder. Läänemaa,
  W Estonia
Vis *100 B4 It.* Lissa; *anc.* Issa. *island*
  S Croatia
Vis *see* Fish
Visaginas *106 C4 prev.* Sniečkus. Utena,
  E Lithuania
Visäkhapatnam *135 E5 var.*
  Vishakhapatnam. Andhra Pradesh,
  SE India
Visalia *47 C6* California, W USA
Visby *85 C7 Ger.* Wisby. Gotland,
  SE Sweden
Viscount Melville Sound *37 F2 prev.*
  Melville Sound. *sound* Northwest
  Territories, N Canada
Visé *87 D6* Liège, E Belgium
Viseu *92 C2 prev.* Vizeu. Viseu,
  N Portugal
Vishakhapatnam *see* Visäkhapatnam
Vislinski Zaliv *see* Vistula Lagoon
Visoko *100 C4* Federacija Bosna I
  Hercegovina, C Bosnia and
  Herzegovina
Visttasjohka *84 D3 river* N Sweden
Vistula *see* Wisła

Vistula Lagoon *98 C2 Ger.* Frisches Haff,
  Pol. Zalew Wiślany, *Rus.* Vislinskiy
  Zaliv. *lagoon* Poland/Russia
Vitebsk *see* Vitsyebsk
Viterbo *96 C4 anc.* Vicus Elbii. Lazio,
  C Italy
Viti *see* Fiji
Viti Levu *145 E4 island* W Fiji
Vitim *115 F4 river* C Russia
Vitória *63 F4 state capital* Espírito Santo,
  SE Brazil
Vitoria *see* Vitoria-Gasteiz
Vitória Bank *see* Vitória Seamount
Vitória da Conquista *63 F3* Bahia,
  E Brazil
Vitoria-Gasteiz *93 E1 var.* Vitoria, *Eng.*
  Vittoria. País Vasco, N Spain
Vitória Seamount *67 B5 var.* Victoria
  Bank, Vitoria Bank. *seamount*
  C Atlantic Ocean
Vitré *90 B3* Ille-et-Vilaine, NW France
Vitsyebsk *107 E5 Rus.* Vitebsk.
  Vitsyebskaya Voblasts', NE Belarus
Vittoria *97 C7* Sicilia, Italy,
  C Mediterranean Sea
Vittoria *see* Vitoria-Gasteiz
Vizcaya, Golfo de *see* Biscay, Bay of
Vizianagaram *135 E5 var.* Vizianagram.
  Andhra Pradesh, E India
Vizianagram *see* Vizianagaram
Vjosës, Lumi i *101 C7 var.* Vijosa,
  Vijosë, Gk. Aóos. *river* Albania/Greece
Vlaanderen *see* Flanders
Vlaardingen *86 B4* Zuid-Holland,
  SW Netherlands
Vladikavkaz *111 B8 prev.* Dzaudzhikau,
  Ordzhonikidze. Respublika Severnaya
  Osetiya, SW Russia
Vladimir *111 B5* Vladimirskaya Oblast',
  W Russia
Vladimirovka *see* Yuzhno-Sakhalinsk
Vladimir-Volynskiy *see*
  Volodymyr-Volynskyi
Vladivostok *115 G5* Primorskiy Kray,
  SE Russia
Vlagtwedde *86 E2* Groningen,
  NE Netherlands
Vlasotince *101 E5* Serbia, SE Serbia
Vlieland *86 C1 Fris.* Flylân. *island*
  Waddeneilanden, N Netherlands
Vlijmen *86 C4* Noord-Brabant,
  S Netherlands
Vlissingen *87 B5 Eng.* Flushing, *Fr.*
  Flessingue. Zeeland, SW Netherlands
Vlodava *see* Włodawa
Vlonë/Vlora *see* Vlorë
Vlorë *101 C7 prev.* Vlonë, *It.* Valona,
  Vlora. Vlorë, SW Albania
Vlotslavsk *see* Włocławek
Vöcklabruck *95 D6* Oberösterreich,
  NW Austria
Vogelkop *see* Doberai, Jazirah
Vohimena, Tanjona *79 F4 Fr.* Cap
  Sainte Marie. *headland* S Madagascar
Voiron *91 D5* Isère, E France
Vojvodina *100 D3 Ger.* Wojwodina.
  Vojvodina, N Serbia
Volga *111 B7 river* NW Russia
Volga Uplands *see* Privolzhskaya
  Vozvyshennost'
Volgodonsk *111 B7* Rostovskaya Oblast',
  SW Russia
Volgograd *111 B7 prev.* Stalingrad,
  Tsaritsyn. Volgogradskaya Oblast',
  SW Russia
Volkhov *110 B4* Leningradskaya Oblast',
  NW Russia
Volkoysk *see* Vawkavysk
Volmari *see* Valmiera
Volnovakha *109 G3* Donets'ka Oblast',
  SE Ukraine
Volodymyr-Volynskyi *108 C1 prev.*
  Volodymyr-Volyns'kyi, *Pol.*
  Włodzimierz, *Rus.* Vladimir-Volynskiy.
  Volyns'ka Oblast', NW Ukraine
Volodymyr-Volyns'kyy *see*
  Volodymyr-Volynskyi
Vologda *110 B4* Vologodskaya Oblast',
  W Russia
Vólos *105 B5* Thessalía, C Greece
Volozhin *see* Valozhyn
Vol'sk *111 C6* Saratovskaya Oblast',
  W Russia
Volta *75 E5 river* SE Ghana
Volta Blanche *see* White Volta
Volta, Lake *75 E5 reservoir*
  SE Ghana
Volta Noire *see* Black Volta
Volturno *97 D5 river* S Italy
Volunteer Island *see* Starbuck Island

Volzhskiy *111 B6* Volgogradskaya
  Oblast', SW Russia
Võnnu *106 E3 Ger.* Wendau. Tartumaa,
  SE Estonia
Voorst *86 D3* Gelderland, E Netherlands
Voranava *107 C5 Pol.* Werenów, *Rus.*
  Voronovo. Hrodzyenskaya Voblasts',
  W Belarus
Vorderrhein *95 B7 river* SE Switzerland
Vóreioi Sporádes *105 C5 var.* Vóreioi
  Sporádes, Vórioi Sporádhes, *Eng.*
  Northern Sporades. *island group*
  E Greece
Vóreioi Sporádes *see* Vóreies Sporádes
Vórioi Sporádhes *see* Vóreies Sporádes
Vorkuta *114 C2* Respublika Komi,
  NW Russia
Vormsi *106 C2 var.* Vormsi Saar, *Ger.*
  Worms, *Swed.* Ormsö. *island*
  W Estonia
Vormsi Saar *see* Vormsi
Voronezh *111 B6* Voronezhskaya
  Oblast', W Russia
Voronovo *see* Voranava
Voroshilov *see* Ussuriysk
Voroshilovgrad *see* Luhansk
Voroshilovsk *see* Stavropol', Russia
Võru *106 D3 Ger.* Werro. Võrumaa,
  SE Estonia
Vosges *90 E4 mountain range* NE France
Vostochno-Sibirskoye More *115 F1
  Eng.* East Siberian Sea. *sea* Arctic Ocean
Vostochnyy Sayan *115 E4 Mong.* Dzüün
  Soyonï Nuruu, *Eng.* Eastern Sayans.
  mountain range Mongolia/Russia
Vostock Island *see* Vostok Island
Vostok *154 C3 Russian research station*
  Antarctica
Vostok Island *145 G3 var.* Vostock
  Island; *prev.* Stavers Island. *island* Line
  Islands, SE Kiribati
Voznesensk *109 E3* Mykolayivs'ka
  Oblast', S Ukraine
Vranje *101 E5* Serbia, SE Serbia
Vranov *see* Vranov nad Topl'ou
Vranov nad Topl'ou *99 D5 var.* Vranov,
  Hung. Varannó. Prešovský Kraj,
  E Slovakia
Vratsa *104 C2* Vratsa, NW Bulgaria
Vrbas *100 C3* Vojvodina, NW Serbia
Vrbas *100 C3 river* N Bosnia
  and Herzegovina
Vršac *100 E3 Ger.* Werschetz, *Hung.*
  Versecz. Vojvodina, NE Serbia
Vsetín *99 C5 Ger.* Wsetin. Zlínský Kraj,
  E Czechia (Czech Republic)
Vučitrn *see* Vushtrri
Vukovar *100 C3 Hung.* Vukovár.
  Vukovar-Srijem, E Croatia
Vulcano, Isola *97 C7 island* Isole Eolie,
  S Italy
Vung Tau *137 E6 prev. Fr.* Cape Saint
  Jacques, Cap Saint-Jacques. Ba
  Ria-Vung Tau, S Vietnam
Vushtrri *101 D5 Serb.* Vučitrn.
  N Kosovo
Vyatka *111 C5 river* NW Russia
Vyatka *see* Kirov
Vyborg *110 B3 Fin.* Viipuri.
  Leningradskaya Oblast', NW Russia
Vyerkhnyadzvinsk *107 D5 Rus.*
  Verkhnedvinsk. Vitsyebskaya
  Voblasts', N Belarus
Vyetryna *107 D5 Rus.* Vetrino.
  Vitsyebskaya Voblasts', N Belarus
Vynohradiv *108 B3 Cz.* Sevluš, *Hung.*
  Nagyszőllős, *Rus.* Vinogradov; *prev.*
  Sevlyush. Zakarpats'ka Oblast',
  W Ukraine

W

Wa *75 E4* NW Ghana
Waag *see* Váh
Waagbistritz *see* Považská Bystrica
Waal *86 C4 river* S Netherlands
Wabash *40 C4* Indiana, N USA
Wabash River *40 B5 river* N USA
Waco *49 G3* Texas, SW USA
Wad Al-Hajarah *see* Guadalajara
Waddān *71 F3* NW Libya
Waddeneilanden *86 C1 Eng.* West
  Frisian Islands. *island group*
  N Netherlands
Waddenzee *86 C1 var.* Wadden Zee. *sea*
  SE North Sea
Wadden Zee *see* Waddenzee
Waddington, Mount *36 D5 mountain*
  British Columbia, SW Canada

Wādī as Sīr *119 B6 var.* Wadi es Sir.
'Ammān, NW Jordan
Wadi es Sir *see* Wādī as Sīr
Wadi Halfa *72 B3 var.* Wādī Ḥalfā'.
Northern, N Sudan
Wādī Mūsā *119 B7 var.* Petra. Ma'ān,
S Jordan
Wad Madani *see* Wad Medani
Wad Medani *72 C4 var.* Wad Madanī.
Gezira, C Sudan
Waflia *139 F4* Pulau Buru, E Indonesia
Wagadugu *see* Ouagadougou
Wagga Wagga *149 C7* New South Wales,
SE Australia
Wagin *147 B7* Western Australia
Wah *134 C1* Punjab, NE Pakistan
Wahai *139 F4* Pulau Seram, E Indonesia
Wahaybah, Ramlat Al *see* Wahībah,
Ramlat Āl
Wahiawā *47 A8 var.* Wahiawa. O'ahu,
Hawaii, USA, C Pacific Ocean
Wahībah, Ramlat Ahl *see* Wahībah,
Ramlat Āl
Wahībah Sands *121 E5 var.* Ramlat Ahl
Wahībah, Ramlat Al Wahaybah, *Eng.*
Wahībah Sands. *desert* N Oman
Wahībah Sands *see* Wahībah, Ramlat Āl
Wahpeton *45 F2* North Dakota, N USA
Wahran *see* Oran
Waiau *151 A7 river* South Island,
New Zealand
Waigeo, Pulau *139 G4 island* Maluku,
E Indonesia
Waikaremoana, Lake *150 E4 lake* North
Island, New Zealand
Wailuku *47 B8* Maui, Hawaii, USA,
C Pacific Ocean
Waimate *151 B6* Canterbury, South
Island, New Zealand
Waiouru *150 D4* Manawatu-Wanganui,
North Island, New Zealand
Waipara *151 C6* Canterbury, South
Island, New Zealand
Waipawa *150 E4* Hawke's Bay, North
Island, New Zealand
Waipukurau *150 D4* Hawke's Bay,
North Island, New Zealand
Wairau *151 C5 river* South Island,
New Zealand
Wairoa *150 E4* Hawke's Bay, North
Island, New Zealand
Wairoa *150 D2 river* North Island,
New Zealand
Waitaki *151 B6 river* South Island,
New Zealand
Waitara *150 D4* Taranaki, North Island,
New Zealand
Waitzen *see* Vác
Waiuku *150 D3* Auckland, North Island,
New Zealand
Wakasa-wan *131 C6 bay* C Japan
Wakatipu, Lake *151 A7 lake* South
Island, New Zealand
Wakayama *131 C6* Wakayama, Honshū,
SW Japan
Wake Island *152 C2* US *unincorporated
territory* NW Pacific Ocean
Wake Island *142 D1 atoll*
NW Pacific Ocean
Wakkanai *130 C1* Hokkaidō, NE Japan
Walachei/Walachia *see* Wallachia
Wałbrzych *98 B4* Ger. Waldenburg,
Waldenburg in Schlesien. Dolnośląskie,
SW Poland
Walcourt *87 C7* Namur, S Belgium
Wałcz *98 B3* Ger. Deutsch Krone.
Zachodnio-pomorskie, NW Poland
Waldenburg/Waldenburg in Schlesien
see Wałbrzych
Waldia *see* Weldiya
Wales *36 C2* Alaska, USA
Wales *89 C6 Wel.* Cymru. *cultural region*
Wales, United Kingdom
Walgett *149 D5* New South Wales,
SE Australia
Walk *see* Valga, Estonia
Walk *see* Valka, Latvia
Walker Lake *47 C5 lake* Nevada, W USA
Wallachia *108 B5 var.* Walachia, Ger.
Walachei, Rom. Valachia. *cultural
region* S Romania
Walla Walla *46 C2* Washington,
NW USA
Wallenthal *see* Hațeg
Wallis and Futuna *145 E4 Fr.* Territoire
de Wallis et Futuna. *French overseas
collectivity* C Pacific Ocean
Wallis et Futuna, Territoire de *see*
Wallis and Futuna
Walnut Ridge *42 B1* Arkansas, C USA

Waltenberg *see* Zalău
Walthamstow *89 B7* Waltham Forest,
SE England, United Kingdom
Walvisbaai *see* Walvis Bay
Walvis Bay *78 A4 Afr.* Walvisbaai.
Erongo, NW Namibia
Walvish Ridge *see* Walvis Ridge
Walvis Ridge *69 B7 var.* Walvish Ridge.
undersea ridge E Atlantic Ocean
Wan *see* Anhui
Wanaka *151 B6* Otago, South Island,
New Zealand
Wanaka, Lake *151 A6 lake* South Island,
New Zealand
Wanchuan *see* Zhangjiakou
Wandel Sea *83 E1 sea* Arctic Ocean
Wandsworth *89 A8* Wandsworth,
SE England, United Kingdom
Wanganui *150 D4* Manawatu-
Wanganui, North Island, New Zealand
Wangaratta *149 C7* Victoria,
SE Australia
Wankie *see* Hwange
Wanki, Río *see* Coco, Río
Wanlaweyn *73 D6 var.* Wanle Weyn, *It.*
Uanle Uen. Shabeellaha Hoose,
SW Somalia
Wanle Weyn *see* Wanlaweyn
Wanxian *see* Wanzhou
Wanzhou *129 B5 var.* Wanxian.
Chongqing, C China
Warangal *135 E5* Telangana, C India
Warasdin *see* Varaždin
Warburg *94 B4* Nordrhein-Westfalen,
W Germany
Ware *37 E4* British Columbia, W Canada
Waremme *87 C6* Liège, E Belgium
Waren *94 C3* Mecklenburg-
Vorpommern, NE Germany
Wargla *see* Ouargla
Warkworth *150 D2* Auckland, North
Island, New Zealand
Warnemünde *94 C2* Mecklenburg-
Vorpommern, NE Germany
Warner *49 G1* Oklahoma, C USA
Warnes *61 G4* Santa Cruz, C Bolivia
Warrego River *149 C5 seasonal river*
New South Wales/Queensland,
E Australia
Warren *40 D3* Michigan, N USA
Warren *40 D3* Ohio, N USA
Warren *41 E3* Pennsylvania, NE USA
Warri *75 F5* Delta, S Nigeria
Warrnambool *149 B7* Victoria,
SE Australia
Warsaw/Warschau *see* Warszawa
Warszawa *98 D3 Eng.* Warsaw, *Ger.*
Warschau, *Rus.* Varshava. *country
capital* (Poland) Mazowieckie, C Poland
Warta *98 B3 Ger.* Warthe. *river*
W Poland
Wartberg *see* Senec
Warthe *see* Warta
Warwick *149 E5* Queensland, E Australia
Washington *44 A2* NE England,
United Kingdom
Washington D.C. *41 E4 country capital*
(USA) District of Columbia, NE USA
Washington Island *see* Teraina
Washington, Mount *41 G2 mountain*
New Hampshire, NE USA
Wash, The *89 E6 inlet* E England,
United Kingdom
Wasiliszki *see* Vasilishki
Waspam *53 E2 var.* Waspán. Región
Autónoma Atlántico Norte,
NE Nicaragua
Waspán *see* Waspam
Watampone *139 E4 var.* Bone. Sulawesi,
C Indonesia
Watenstedt-Salzgitter *see* Salzgitter
Waterbury *41 F3* Connecticut, NE USA
Waterford *89 B6 Ir.* Port Láirge.
Waterford, S Ireland
Waterloo *40 G3* Iowa, C USA
Watertown *41 F2* New York, NE USA
Watertown *45 F2* South Dakota, N USA
Waterville *41 G2* Maine, NE USA
Watford *89 A7* E England,
United Kingdom
Watlings Island *see* San Salvador
Watsa *77 E5* Orientale, NE Dem.
Rep. Congo
Watts Bar Lake *42 D1 reservoir*
Tennessee, S USA
Wau *73 B5 var.* Wāw. Western Bahr el
Ghazal, C South Sudan
Waukegan *40 B3* Illinois, N USA
Waukesha *40 B3* Wisconsin, N USA
Wausau *40 B2* Wisconsin, N USA

Waverly *45 G3* Iowa, C USA
Wavre *87 C6* Walloon Brabant,
C Belgium
Wāw *see* Wau
Wawa *38 C4* Ontario, S Canada
Waycross *43 E3* Georgia, SE USA
Wearmouth *see* Sunderland
Webfoot State *see* Oregon
Webster City *45 F3* Iowa, C USA
Weddell Plain *154 A2 abyssal plain*
SW Atlantic Ocean
Weddell Sea *154 A2 sea*
SW Atlantic Ocean
Weener *94 A3* Niedersachsen,
NW Germany
Weert *87 D5* Limburg, SE Netherlands
Weesp *86 C3* Noord-Holland,
C Netherlands
Węgorzewo *98 D2 Ger.* Angerburg.
Warmińsko-Mazurskie, NE Poland
Weichsel *see* Wisła
Weimar *94 C4* Thüringen, C Germany
Weissenburg *see* Alba Iulia, Romania
Weissenburg in Bayern *95 C6* Bayern,
SE Germany
Weissenstein *see* Paide
Weisskirchen *see* Bela Crkva
Weiswampach *87 D7* Diekirch,
N Luxembourg
Wejherowo *98 C2* Pomorskie,
NW Poland
Welchman Hall *55 G1* C Barbados
Weldiya *72 C4 var.* Waldia, *It.* Valdia.
Āmara, N Ethiopia
Welkom *78 D4* Free State, C South Africa
Welle *see* Uele
Wellesley Islands *148 B2 island group*
Queensland, N Australia
Wellington *151 D5 country capital* (New
Zealand) Wellington, North Island,
New Zealand
Wellington *45 F5* Kansas, C USA
Wellington *see* Wellington, Isla
Wellington, Isla *65 A7 var.* Wellington.
island S Chile
Wells *36 C4* Nevada, W USA
Wellsford *150 D2* Auckland, North
Island, New Zealand
Wells, Lake *147 C5 lake*
Western Australia
Wels *95 D6 anc.* Ovilava. Oberösterreich,
N Austria
Wembley *89 A8* London, SE England
Wemmel *87 B6* Vlaams Brabant,
C Belgium
Wenatchee *46 B2* Washington, NW USA
Wenchi *75 E4* W Ghana
Wen-chou/Wenchow *see* Wenzhou
Wendau *see* Võnnu
Wenden *see* Cēsis
Wenzhou *129 D5 var.* Wen-chou,
Wenchow. Zhejiang, SE China
Werda *78 C4* Kgalagadi, S Botswana
Werder *see* Virtsu
Werenów *see* Voranava
Werkendam *86 C4* Noord-Brabant,
S Netherlands
Werowitz *see* Virovitica
Werro *see* Võru
Werschetz *see* Vršac
Wesenberg *see* Rakvere
Weser *94 B3 river* NW Germany
Wessel Islands *148 B1 island group*
Northern Territory, N Australia
West Antarctica *154 B3* physical region
Antarctica
West Australian Basin *see*
Wharton Basin
West Bank *119 A6 disputed region*
SW Asia
West Bend *40 B3* Wisconsin, N USA
West Bengal *135 F4 cultural region*
NE India
West Cape *151 A7 headland* South
Island, New Zealand
West Des Moines *45 F3* Iowa, C USA
Westerland *94 B2* Schleswig-Holstein,
N Germany
Western Australia *146 B4 state*
W Australia
Western Bug *see* Bug
Western Carpathians *99 E7 mountain
range* W Romania Europe
Western Desert *see* Şaḥrā' al Gharbīyah
Western Dvina *85 E7 Bel.* Dzvina, Ger.
Düna, Latv. Daugava, Rus. Zapadnaya
Dvina. *river* W Europe
Western Ghats *134 C5 mountain range*
SW India
Western Isles *see* Outer Hebrides

Western Punjab *see* Punjab
Western Sahara *70 B3 disputed territory*
administered by Morocco N Africa
Western Samoa *see* Samoa
Western Samoa, Independent State of
see Samoa
Western Sayans *see* Zapadnyy Sayan
Western Scheldt *see* Westerschelde
Western Sierra Madre *see* Madre
Occidental, Sierra
Westerschelde *87 B5 Eng.* Western
Scheldt; *prev.* Honte. *inlet* S North Sea
West Falkland *65 C7 var.* Gran Malvina,
Isla Gran Malvina. *island* W Falkland
Islands
West Fargo *45 F2* North Dakota, N USA
West Frisian Islands *see*
Waddeneilanden
West Irian *see* Papua
Westliche Morava *see* Zapadna Morava
West Mariana Basin *142 B1 var.* Perece
Vela Basin. *undersea feature*
W Pacific Ocean
West Memphis *42 B1* Arkansas, C USA
West New Guinea *see* Papua
Weston-super-Mare *89 D7* SW England,
United Kingdom
West Palm Beach *43 F4* Florida, SE USA
West Papua *see* Papua
Westport *151 C5* West Coast, South
Island, New Zealand
West Punjab *see* Punjab
West River *see* Xi Jiang
West Siberian Plain *see*
Zapadno-Sibirskaya Ravnina
West Virginia *40 D4 off.* State of West
Virginia, *also known as* Mountain State.
state NE USA
Wetar, Pulau *139 F5 island* Kepulauan
Damar, E Indonesia
Wetzlar *95 B5* Hessen, W Germany
Wevok *36 C2 var.* Wewuk. Alaska, USA
Wewuk *see* Wevok
Wexford *89 B6 Ir.* Loch Garman.
SE Ireland
Weyburn *37 F5* Saskatchewan, S Canada
Weymouth *89 D7* S England,
United Kingdom
Wezep *86 D3* Gelderland, E Netherlands
Whakatane *150 E3* Bay of Plenty, North
Island, New Zealand
Whale Cove *37 G3 var.* Tikiarjuaq.
Nunavut, C Canada
Whangarei *150 D2* Northland, North
Island, New Zealand
Wharton Basin *141 D5 var.* West
Australian Basin. *undersea feature*
E Indian Ocean
Whataroa *151 B6* West Coast, South
Island, New Zealand
Wheatland *44 D3* Wyoming, C USA
Wheeler Peak *48 D1 mountain* New
Mexico, SW USA
Wheeling *40 D4* West Virginia,
NE USA
Whenua Hou *see* Codfish Island
Whitby *89 D5* N England,
United Kingdom
Whitefish *44 B1* Montana, NW USA
Whitehaven *89 C5* NW England,
United Kingdom
Whitehorse *36 D4 territory capital*
Yukon, W Canada
White Nile *72 B4 Ar.* Al Baḥr al Abyaḍ,
An Nīl al Abyaḍ, Bahr el Jebel. *river*
C South Sudan
White River *44 D3 river* South Dakota,
N USA
White Sea *see* Beloye More
White Volta *75 E4 var.* Nakambé, Fr.
Volta Blanche. *river* Burkina/Ghana
Whitianga *150 D2* Waikato, North
Island, New Zealand
Whitney, Mount *47 C6 mountain*
California, W USA
Whitsunday Group *148 D3 island group*
Queensland, E Australia
Whyalla *149 B6* South Australia
Wichita *45 F5* Kansas, C USA
Wichita Falls *49 F2* Texas, SW USA
Wichita River *49 F2 river* Texas,
SW USA
Wickenburg *48 B2* Arizona, SW USA
Wicklow *89 B6 Ir.* Cill Mhantáin.
county E Ireland
Wicklow Mountains *89 B6 Ir.* Sléibhte
Chill Mhantáin. *mountain range*
E Ireland
Wieliczka *99 D5* Małopolskie, S Poland
Wieluń *98 C4* Sieradz, C Poland

Wien 95 E6 *Eng.* Vienna, *Hung.* Bécs, *Slvk.* Vídeň, *Slvn.* Dunaj; *anc.* Vindobona. *country capital* (Austria) Wien, NE Austria

Wiener Neustadt 95 E6 Niederösterreich, E Austria

Wierden 86 E3 Overijssel, E Netherlands

Wiesbaden 95 B5 Hessen, W Germany

Wieselburg and Ungarisch-Altenburg/ Wieselburg-Ungarisch-Altenburg *see* Mosonmagyaróvár

Wiesenhof *see* Ostrołęka

Wight, Isle of 89 D7 *island* United Kingdom

Wigorna Ceaster *see* Worcester

Wijchen 86 D4 Gelderland, SE Netherlands

Wijk bij Duurstede 86 D4 Utrecht, C Netherlands

Wilcannia 149 C6 New South Wales, SE Australia

Wilejka *see* Vilyeyka

Wilhelm, Mount 144 B3 *mountain* C Papua New Guinea

Wilhelm-Pieck-Stadt *see* Guben

Wilhelmshaven 94 B3 Niedersachsen, NW Germany

Wilia/Wilja *see* Neris

Wilkes Barre 41 F3 Pennsylvania, NE USA

Wilkes Land 154 C4 *physical region* Antarctica

Wiłkomierz *see* Ukmergė

Willard 48 D2 New Mexico, SW USA

Willcox 48 C3 Arizona, SW USA

Willebroek 87 B5 Antwerpen, C Belgium

Willemstad 55 E5 *dependent territory capital* (Curaçao) Lesser Antilles, S Caribbean Sea

Williston 44 D1 North Dakota, N USA

Wilmington 41 F4 Delaware, NE USA

Wilmington 43 F2 North Carolina, SE USA

Wilmington 40 C4 Ohio, N USA

Wilna/Wilno *see* Vilnius

Wilrijk 87 C5 Antwerpen, N Belgium

Winchester 89 D7 *hist.* Wintanceaster, *Lat.* Venta Belgarum. S England, United Kingdom

Winchester 41 E4 Virginia, NE USA

Windau *see* Ventspils, Latvia

Windau *see* Venta, Latvia/Lithuania

Windhoek 78 B3 *var.* Windhuk. *country capital* (Namibia) Khomas, C Namibia

Windhuk *see* Windhoek

Windorah 148 C4 Queensland, C Australia

Windsor 38 C5 Ontario, S Canada

Windsor 89 D7 S England, United Kingdom

Windsor 41 G3 Connecticut, NE USA

Windward Islands 55 H4 *island group* E West Indies

Windward Islands *see* Barlavento, Ilhas de, Cape Verde (Cabo Verde)

Windward Passage 54 D3 *Sp.* Paso de los Vientos. *channel* Cuba/Haiti

Winisk 38 C2 *river* Ontario, C Canada

Winkowitz *see* Vinkovci

Winnebago, Lake 40 B2 *lake* Wisconsin, N USA

Winnemucca 47 C5 Nevada, W USA

Winnipeg 37 G5 *province capital* Manitoba, S Canada

Winnipeg, Lake 37 G5 *lake* Manitoba, C Canada

Winnipegosis, Lake 38 A3 *lake* Manitoba, C Canada

Winona 45 G3 Minnesota, N USA

Winschoten 86 E2 Groningen, NE Netherlands

Winsen 94 B3 Niedersachsen, N Germany

Winston Salem 43 E1 North Carolina, SE USA

Winsum 86 D1 Groningen, NE Netherlands

Wintanceaster *see* Winchester

Winterswijk 86 E4 Gelderland, E Netherlands

Winterthur 95 B7 Zürich, NE Switzerland

Winton 148 C4 Queensland, E Australia

Winton 151 A7 Southland, South Island, New Zealand

Wisby *see* Visby

Wisconsin 40 A2 *off.* State of Wisconsin, *also known as* Badger State. *state* N USA

Wisconsin Rapids 40 B2 Wisconsin, N USA

Wisconsin River 40 B3 *river* Wisconsin, N USA

Wisła 98 C2 *Eng.* Vistula, *Ger.* Weichsel. *river* C Poland

Wiślany, Zalew *see* Vistula Lagoon

Wismar 94 C2 Mecklenburg-Vorpommern, N Germany

Wittenberge 94 C3 Brandenburg, N Germany

Wittlich 95 A5 Rheinland-Pfalz, SW Germany

Wittstock 94 C3 Brandenburg, NE Germany

W. J. van Blommesteinmeer 59 G3 *reservoir* E Suriname

Władysławowo 98 C2 Pomorskie, N Poland

Włocławek 98 C3 *Ger./Rus.* Vlotslavsk. Kujawsko-pomorskie, C Poland

Włodawa 98 E4 *Rus.* Vlodava. Lubelskie, SE Poland

Włodzimierz *see* Volodymyr-Volynskyi

Wlotzkasbaken 78 B3 Erongo, N Namibia

Wodonga 149 C7 Victoria, SE Australia

Wodzisław Śląski 99 C5 *Ger.* Loslau. Śląskie, S Poland

Wojerecy *see* Hoyerswerda

Wöjjä *see* Wotje Atoll

Wojwodina *see* Vojvodina

Woking 89 D7 SE England, United Kingdom

Wolf, Isla 60 A4 *island* Galápagos Islands, W Ecuador South America

Wolfsberg 95 D7 Kärnten, SE Austria

Wolfsburg 94 C3 Niedersachsen, N Germany

Wolgast 94 D2 Mecklenburg-Vorpommern, NE Germany

Wołkowysk *see* Vawkavysk

Wollan *see* Velenje

Wollaston Lake 37 F4 Saskatchewan, C Canada

Wollongong 149 D6 New South Wales, SE Australia

Wolmar *see* Valmiera

Wołożyn *see* Valozhyn

Wolvega 86 D2 Fris. Wolvegea. Fryslân, N Netherlands

Wolvegea *see* Wolvega

Wolverhampton 89 D6 C England, United Kingdom

Wolverine State *see* Michigan

Wŏnsan 129 E3 SE North Korea

Woodburn 46 B3 Oregon, NW USA

Woodland 47 B5 California, W USA

Woodruff 40 B2 Wisconsin, N USA

Woods, Lake of the 38 A3 *Fr.* Lac des Bois. *lake* Canada/USA

Woodville 150 D4 Manawatu-Wanganui, North Island, New Zealand

Woodward 49 F1 Oklahoma, C USA

Worcester 78 C5 Western Cape, SW South Africa

Worcester 89 D6 *hist.* Wigorna Ceaster. W England, United Kingdom

Worcester 41 G3 Massachusetts, NE USA

Workington 89 C5 NW England, United Kingdom

Worland 44 C3 Wyoming, C USA

Wormatia *see* Worms

Worms 95 B5 *anc.* Augusta Vangionum, Borbetomagus, Wormatia. Rheinland-Pfalz, SW Germany

Worms *see* Vormsi

Worthington 45 F3 Minnesota, N USA

Wotje Atoll 144 D1 *var.* Wōjjä. *atoll* Ratak Chain, E Marshall Islands

Woudrichem 86 C4 Noord-Brabant, S Netherlands

Wrangel Island 115 F1 *Eng.* Wrangel Island. *island* NE Russia

Wrangel Island *see* Vrangelya, Ostrov

Wrangel Plain 155 B2 *undersea feature* Arctic Ocean

Wrocław 98 C4 *Eng./Ger.* Breslau. Dolnośląskie, SW Poland

Września 98 C3 Wielkopolskie, C Poland

Wsetin *see* Vsetín

Wuchang *see* Wuhan

Wuday'ah 121 C6 *spring/well* S Saudi Arabia

Wuhai 127 C3 *var.* Haibowan. Nei Mongol Zizhiqu, N China

Wuhan 129 C5 *var.* Han-kou, Han-k'ou, Hanyang, Wuchang, Wu-han; *prev.* Hankow. *province capital* Hubei, C China

Wu-han *see* Wuhan

Wuhsien *see* Suzhou

Wuhsi/Wu-his *see* Wuxi

Wuhu 129 D5 *var.* Wu-na-mu. Anhui, E China

Wujlân *see* Ujelang Atoll

Wukari 75 G4 Taraba, E Nigeria

Wuliang Shan 129 A6 *mountain range* SW China

Wu-lu-k'o-mu-shi/Wu-lu-mu-ch'i *see* Ürümqi

Wu-na-mu *see* Wuhu

Wuppertal 94 A4 *prev.* Barmen-Elberfeld. Nordrhein-Westfalen, W Germany

Würzburg 95 B5 Bayern, SW Germany

Wusih *see* Wuxi

Wuxi 129 D5 *var.* Wuhsi, Wu-hsi, Wusih. Jiangsu, E China

Wuyi Shan 125 E3 *mountain range* SE China

Wye 89 C6 *Wel.* Gwy. *river* England/ Wales, United Kingdom

Wyłkowyszki *see* Vilkaviškis

Wyndham 146 D3 Western Australia

Wyoming 40 C3 Michigan, N USA

Wyoming 44 B3 *off.* State of Wyoming, *also known as* Equality State. *state* C USA

Wyszków 98 D3 *Ger.* Probstberg. Mazowieckie, NE Poland

Xaafuun, Raas 72 E4 *var.* Ras Hafun. *cape* NE Somalia

Xaçmaz 117 H2 *Rus.* Khachmas. N Azerbaijan

Xaignabouli 136 C4 *prev.* Muang Xaignabouri, *Fr.* Sayaboury. Xaignabouli, N Laos

Xai-Xai 79 E4 *prev.* João Belo, Vila de João Belo. Gaza, S Mozambique

Xalapa 51 F4 Veracruz-Llave, Mexico

Xam Nua 136 D3 *var.* Sam Neua. Houaphan, N Laos

Xankändi 117 G3 *Rus.* Khankendi; *prev.* Stepanakert. SW Azerbaijan

Xánthi 104 C3 Anatolikí Makedonía kai Thráki, NE Greece

Xàtiva 93 F3 *Cas.* Xátiva; *anc.* Setabis, *var.* Jativa. Comunitat Valenciana, E Spain

Xauen *see* Chefchaouen

Xäzär Dänizi *see* Caspian Sea

Xeres *see* Jeréz de la Frontera

Xiaguan *see* Dali

Xiamen 129 D6 *var.* Hsia-men; *prev.* Amoy. Fujian, SE China

Xi'an 128 C4 *var.* Changan, Sian, Signan, Siking, Singan, Xian. *province capital* Shaanxi, C China

Xiang *see* Hunan

Xiangkhoang *see* Phônsaven

Xiangtan 129 C5 *var.* Hsiang-t'an, Siangtan. Hunan, S China

Xiao Hinggan Ling 128 D2 *Eng.* Lesser Khingan Range. *mountain range* NE China

Xichang 129 B5 Sichuan, C China

Xieng Khouang *see* Phônsaven

Xieng Ngeun *see* Muong Xiang Ngeun

Xigazê 126 C5 *var.* Jih-k'a-tse, Shigatse, Xigaze. Xizang Zizhiqu, W China

Xi Jiang 124 D3 *var.* Hsi Chiang, *Eng.* West River. *river* S China

Xilinhot 127 F2 *var.* Silinhot. Nei Mongol Zizhiqu, N China

Xilokastro *see* Xylókastro

Xin *see* Xinjiang Uygur Zizhiqu

Xingkai Hu *see* Khanka, Lake

Xingu, Rio 63 E2 *river* C Brazil

Xingxingxia 126 D3 Xinjiang Uygur Zizhiqu, NW China

Xining 127 E4 *var.* Hsining, Hsi-ning, Sining. *province capital* Qinghai, C China

Xinjiang *see* Xinjiang Uygur Zizhiqu

Xinjiang Uygur Zizhiqu 126 B3 *var.* Sinkiang, Sinkiang Uighur Autonomous Region, Xin, Xinjiang. *autonomous region* NW China

Xinpu *see* Lianyungang

Xinxiang 128 C4 Henan, C China

Xinyang 129 C5 *var.* Hsin-yang, Sinyang. Henan, C China

Xinzo de Limia 92 C2 Galicia, NW Spain

Xiqing Shan 124 D2 *mountain range* C China

Xiva 122 D2 *Rus.* Khiva, Khiwa. Xorazm Viloyati, W Uzbekistan

Xixón *see* Gijón

Xizang *see* Xizang Zizhiqu

Xizang Gaoyuan *see* Qingzang Gaoyuan

Xizang Zizhiqu 126 B4 *var.* Thibet, Tibetan Autonomous Region, Xizang, *Eng.* Tibet. *autonomous region* W China

Xolotlán *see* Managua, Lago de

Xuchang *see* Xuwen

Xuddur 73 D5 *var.* Hudur, *It.* Oddur. Bakool, SW Somalia

Xuwen 129 C7 *var.* Xucheng. Guangdong, S China

Xuzhou 128 D4 *var.* Hsu-chou, Suchow, Tongshan; *prev.* T'ung-shan. Jiangsu, E China

Xylókastro 105 B5 *var.* Xilokastro. Pelopónnisos, S Greece

Ya'an 129 B5 *var.* Yaan. Sichuan, C China

Yabělo 73 C5 Oromīya, C Ethiopia

Yablis 53 E2 Región Autónoma Atlántico Norte, NE Nicaragua

Yablonovyy Khrebet 115 F4 *mountain range* S Russia

Yabrai Shan 127 E3 *mountain range* NE China

Yafran 71 F2 NW Libya

Yaghan Basin 67 B7 *undersea feature* SE Pacific Ocean

Yagotin *see* Yahotyn

Yahotyn 109 E2 *Rus.* Yagotin. Kyivska Oblast', N Ukraine

Yahualica 50 D4 Jalisco, SW Mexico

Yakima 46 B2 Washington, NW USA

Yakima River 46 B2 *river* Washington, NW USA

Yakoruda 104 C3 Blagoevgrad, SW Bulgaria

Yaku-shima 131 B8 *island* Nansei-shotō, SW Japan

Yakutat 36 D4 Alaska, USA

Yakutsk 115 F3 Respublika Sakha (Yakutiya), NE Russia

Yakymivka 109 F4 Zaporizka Oblast', S Ukraine

Yala 137 C7 Yala, SW Thailand

Yalizava 107 D6 *Rus.* Yelizovo. Mahilyowskaya Voblasts', E Belarus

Yalong Jiang 129 A5 *river* C China

Yalova 116 B3 Yalova, NW Turkey (Türkiye)

Yalpug, Ozero *see* Yalpuh, Ozero

Yalpuh, Ozero 108 D4 *Rus.* Ozero Yalpug. *lake* SW Ukraine

Yalta 109 F5 Respublika Krym, S Ukraine

Yalu 125 E2 *Chin.* Yalu Jiang, *Jap.* Oryokko, *Kor.* Amnok-kang. *river* China/North Korea

Yalu Jiang *see* Yalu

Yamaguchi 131 B7 *var.* Yamaguti. Yamaguchi, Honshū, SW Japan

Yamal, Poluostrov 114 D2 *peninsula* N Russia

Yamaniyah, Al Jumhūrīyah al *see* Yemen

Yambio 73 B5 *var.* Yambiyo. Western Equatoria, S South Sudan

Yambiyo *see* Yambio

Yambol 104 D2 *Turk.* Yanboli. Yambol, E Bulgaria

Yamdena, Pulau 139 G5 *prev.* Jamdena. *island* Kepulauan Tanimbar, E Indonesia

Yamoussoukro 74 D5 *country capital* (Ivory Coast) C Ivory Coast

Yamuna 134 D3 *prev.* Jumna. *river* N India

Yana 115 F2 *river* NE Russia

Yanboli *see* Yambol

Yanbu 'al Baḥr 121 A5 Al Madīnah, W Saudi Arabia

Yangambi 77 D5 Orientale, N Dem. Rep. Congo

Yangchow *see* Yangzhou

Yangiyo'l 123 E2 *Rus.* Yangiyul'. Toshkent Viloyati, E Uzbekistan

Yangiyul' *see* Yangiyo'l

Yangku *see* Taiyuan

Yangon 136 B4 *Eng.* Rangoon. Yangon, S Myanmar (Burma)
Yangtze 129 B5 *var.* Yangtze Kiang, *Eng.* Yangtze. *river* C China
Yangtze see Chang Jiang
Yangtze Kiang see Chang Jiang
Yangzhou 129 D5 *var.* Yangchow. Jiangsu, E China
Yankton 45 E3 South Dakota, N USA
Yany Kapu 109 F4 *Rus.* Krasnoperekopsk. Respublika Krym, S Ukraine
Yannina see Ioánnina
Yanskiy Zaliv 113 F2 *bay* N Russia
Yantai 128 D4 *var.* Yan-t'ai; *prev.* Chefoo, Chih-fu. Shandong, E China
Yaoundé 77 B5 *var.* Yaunde. *country capital* (Cameroon) Centre, S Cameroon
Yap 144 A1 *island* Caroline Islands, W Federated States of Micronesia
Yapanskoye More see Japan/Sea, Sea of
Yapen, Pulau 139 G4 *prev.* Japen. *island* E Indonesia
Yap Trench 142 B2 *var.* Yap Trough. *undersea feature* SE Philippine Sea
Yap Trough see Yap Trench
Yapurá see Caquetá, Río, Brazil/Colombia
Yapurá see Japurá, Rio, Brazil/Colombia
Yaqui, Río 50 C2 *river* NW Mexico
Yaransk 111 C5 Kirovskaya Oblast', NW Russia
Yarega 110 D4 Respublika Komi, NW Russia
Yaren 144 D2 *de facto country capital* (Nauru) Nauru, SW Pacific
Yarkant see Shache
Yarlung Zangbo Jiang see Brahmaputra
Yarmouth 39 F5 Nova Scotia, SE Canada
Yarmouth see Great Yarmouth
Yaroslav see Jarosław
Yaroslavl' 111 B4 Yaroslavskaya Oblast', W Russia
Yarumal 58 B2 Antioquia, NW Colombia
Yasyel'da 107 B7 *river* Brestskaya Voblasts', SW Belarus Europe
Yatsushiro 131 A7 *var.* Yatusiro. Kumamoto, Kyūshū, SW Japan
Yatusiro see Yatsushiro
Yaunde see Yaoundé
Yavari see Javari, Rio
Yavari, Rio 62 C2 *var.* Yavarí. *river* Brazil/Peru
Yaviza 53 H5 Darién, SE Panama
Yavoriv 108 B2 *Pol.* Jaworów, *Rus.* Yavorov. L'vivs'ka Oblast', NW Ukraine
Yavorov see Yavoriv
Yazd 120 D3 *var.* Yezd. Yazd, C Iran
Yazoo City 42 B2 Mississippi, S USA
Yding Skovhøj 85 A7 *hill* C Denmark
Ýdra 105 C6 *var.* Ídhra. *island* Ýdra, S Greece
Ye 137 B5 Mon State, S Myanmar (Burma)
Yecheng 126 A3 *var.* Kargilik. Xinjiang Uygur Zizhiqu, NW China
Yedy Kuiu 109 G5 *prev.* Lenine, *Rus.* Lenino. Respublika Krym, S Ukraine
Yefremov 111 B5 Tul'skaya Oblast', W Russia
Yekaterinburg 114 C3 *prev.* Sverdlovsk. Sverdlovskaya Oblast', C Russia
Yekaterinodar see Krasnodar
Yekaterinoslav see Dnipro
Yelets 111 B5 Lipetskaya Oblast', W Russia
Yelisavetpol see Gäncä
Yelizavetgrad see Kropyvnytskyi
Yelizovo see Yalizava
Yell 88 D1 *island* NE Scotland, United Kingdom
Yellowhammer State see Alabama
Yellowknife 37 E4 *territory capital* Northwest Territories, W Canada
Yellow River see Huang He
Yellow Sea 128 D4 *Chin.* Huang Hai, *Kor.* Hwang-Hae. *sea* E Asia
Yellowstone River 44 C2 *river* Montana/Wyoming, NW USA
Yel'sk 107 C7 Homyel'skaya Voblasts', SE Belarus
Yelwa 75 F4 Kebbi, W Nigeria
Yemen 121 C7 *off.* Republic of Yemen, *Ar.* Al Jumhūrīyah al Yamanīyah, Al Yaman. *country* SW Asia
Yemen, Republic of see Yemen
Yemva 110 D4 *prev.* Zheleznodorozhnyy. Respublika Komi, NW Russia

Yenakiieve 109 G3 *prev.* Yenakiyeve, Ordzhonikidze, Rykovo; *Rus.* Yenakiyevo. Donets'ka Oblast', E Ukraine
Yenakiyeve see Yenakiieve
Yenakiyevo see Yenakiieve
Yenangyaung 136 A3 Magway, W Myanmar (Burma)
Yendi 75 E4 NE Ghana
Yengisar 126 A3 Xinjiang Uygur Zizhiqu, NW China
Yenierenköy 102 D4 *var.* Yialousa, *Gk.* Aigialoúsa. NE Cyprus
Yenipazar see Novi Pazar
Yenisey 114 D3 *river* Mongolia/Russia
Yenping see Nanping
Yeovil 89 D7 SW England, United Kingdom
Yeppoon 148 D4 Queensland, E Australia
Yerevan 117 F3 *Eng.* Erivan. *country capital* (Armenia) C Armenia
Yeriho see Jericho
Yerushalayim see Jerusalem
Yeso see Hokkaidō
Yeu, Île d' 90 A4 *island* NW France
Yevlakh see Yevlax
Yevlax 117 G2 *Rus.* Yevlakh. C Azerbaijan
Yevpatoriia 109 F5 *prev.* Yevpatoriya. Respublika Krym, S Ukraine
Yevpatoriya see Yevpatoriia
Yeya 109 H4 *river* SW Russia
Yezerishche see Yezyaryshcha
Yezo see Hokkaidō
Yezyaryshcha 107 E5 *Rus.* Yezerishche. Vitsyebskaya Voblasts', NE Belarus
Yialousa see Yenierenköy
Yiannitsá see Giannitsá
Yichang 129 C5 Hubei, C China
Yildizeli 116 D3 Sivas, N Turkey (Türkiye)
Yinchuan 128 B4 *var.* Yinch'uan, Yin-ch'uan, Yinchwan. *province capital* Ningxia, N China
Yinchwan see Yinchuan
Yindu He see Indus
Yin-hsien see Ningbo
Yining 126 B2 *var.* I-ning, *Uygh.* Gulja, Kuldja. Xinjiang Uygur Zizhiqu, NW China
Yin-tu Ho see Indus
Yisrael/Yisra'el see Israel
Yíthion see Gýtheio
Yogyakarta 138 C5 *prev.* Djokjakarta, Jogjakarta, Jokyakarta. Jawa, C Indonesia
Yokohama 131 D5 Aomori, Honshū, C Japan
Yokohama 130 A2 Kanagawa, Honshū, S Japan
Yokote 130 D4 Akita, Honshū, C Japan
Yola 75 H4 Adamawa, E Nigeria
Yonago 131 B6 Tottori, Honshū, SW Japan
Yong'an 129 D6 *var.* Yongan. Fujian, SE China
Yongzhou 129 C6 *var.* Lengshuitan. Hunan, S China
Yonkers 41 F3 New York, NE USA
Yonne 90 C4 *river* C France
Yopal 58 C3 *var.* El Yopal. Casanare, C Colombia
York 89 D5 *anc.* Eboracum, Eburacum. N England, United Kingdom
York 45 E4 Nebraska, C USA
York, Cape 148 C1 *headland* Queensland, NE Australia
York, Kap see Innaanganeq
Yorkton 37 F5 Saskatchewan, S Canada
Yoro 52 C2 Yoro, C Honduras
Yoshkar-Ola 111 C5 Respublika Mariy El, W Russia
Yösönbulag see Altay
Youngstown 40 D4 Ohio, N USA
Youth, Isle of see Juventud, Isla de la
Ypres see Ieper
Yreka 46 B4 California, W USA
Yrendagüé see General Eugenio A. Garay
Yssel see IJssel
Ysyk-Köl 123 G2 *prev.* Ozero Issyk-Kul', *var.* Issiq Köl. *lake* E Kyrgyzstan
Ysyk-Köl see Balykchy
Yu see Henan
Yuan see Red River
Yuan Jiang see Red River
Yuba City 47 B5 California, W USA
Yucatán, Canal de see Yucatan Channel
Yucatan Channel 51 H3 *Sp.* Canal de Yucatán. *channel* Cuba/Mexico

Yucatan Peninsula see Yucatán, Península de
Yucatán, Península de 35 C7 *Eng.* Yucatan Peninsula. *peninsula* Guatemala/Mexico
Yuci see Jinzhong
Yue see Guangdong
Yue Shan, Tai see Lantau Island
Yueyang 129 C5 Hunan, S China
Yugoslavia see Serbia
Yukhavichy 107 D5 *Rus.* Yukhovichi. Vitsyebskaya Voblasts', N Belarus
Yukhovichi see Yukhavichy
Yukon 36 D3 *prev.* Yukon Territory, *Fr.* Territoire du Yukon. *territory* NW Canada
Yukon River 36 C2 *river* Canada/USA
Yukon, Territoire du see Yukon
Yukon Territory see Yukon
Yulin 129 C6 Guangxi Zhuangzu Zizhiqu, S China
Yuma 48 A2 Arizona, SW USA
Yun see Yunnan
Yungki see Jilin
Yung-ning see Nanning
Yunjinghong see Jinghong
Yunki see Jilin
Yunnan 129 A6 *var.* Yun, Yunnan Sheng, Yünnan, Yun-nan. *province* SW China
Yunnan see Kunming
Yunnan Sheng see Yunnan
Yünnan/Yun-nan see Yunnan
Yurev see Tartu
Yurihonjō see Honjō
Yuruá, Río see Juruá, Rio
Yury'ev see Tartu
Yushu 126 D4 *var.* Gyêgu. Qinghai, C China
Yuty 64 D3 Caazapá, S Paraguay
Yuzhno-Sakhalinsk 115 H4 *Jap.* Toyohara; *prev.* Vladimirovka. Ostrov Sakhalin, Sakhalinskaya Oblast', SE Russia
Yuzhnyy Bug see Pivdennyi Buh
Yuzhou see Chongqing
Ýylanly see Andalyp

Z

Zaandam see Zaanstad
Zaanstad 86 C3 *prev.* Zaandam. Noord-Holland, C Netherlands
Zabaykal'sk 115 F5 Chitinskaya Oblast', S Russia
Zabern see Saverne
Zabid 121 B7 W Yemen
Žabinka see Zhabinka
Ząbkowice see Ząbkowice Śląskie
Ząbkowice Śląskie 98 B4 *var.* Ząbkowice, *Ger.* Frankenstein, Frankenstein in Schlesien. Dolnośląskie, SW Poland
Zábřeh 99 C5 *Ger.* Hohenstadt. Olomoucký Kraj, E Czechia (Czech Republic)
Zacapa 52 B2 Zacapa, E Guatemala
Zacatecas 50 D3 Zacatecas, C Mexico
Zacatepec 51 E4 Morelos, S Mexico
Zacháro 105 B6 *var.* Zákharo, Zakháro. Dytikí Ellás, S Greece
Zadar 100 A3 *It.* Zara; *anc.* Iader. Zadar, SW Croatia
Zadetkyi Kyun 137 B6 *var.* St.Matthew's Island. *island* Myeik Archipelago, S Myanmar (Burma)
Zafra 92 C4 Extremadura, W Spain
Żagań 98 B4 *var.* Zagań, Żegań, *Ger.* Sagan. Lubuskie, W Poland
Zagazig see Az Zaqāzīq
Zágráb see Zagreb
Zagreb 100 B2 *Ger.* Agram, *Hung.* Zágráb. *country capital* (Croatia) Zagreb, N Croatia
Zagros Mountains 120 C3 *Eng.* Zagros Mountains. *mountain range* W Iran
Zagros Mountains see Zāgros, Kūhhā-ye
Zaháro see Zacháro
Zāhedān 120 E4 *var.* Zahidan; *prev.* Duzdab. Sīstān va Balūchestān, SE Iran
Zahidan see Zāhedān
Zaḥlah see Zahlé
Zahlé 118 B4 *var.* Zaḥlah. C Lebanon
Záhony 99 E6 Szabolcs-Szatmár-Bereg, NE Hungary
Zaire see Congo (river)
Zaire see Congo (Democratic Republic of)
Zaječar 100 E4 Serbia, E Serbia
Zakataly see Zaqatala

Zakháro see Zacháro
Zakhidnyi Buh/Zakhodni Buh see Bug
Zākhō 120 B2 *var.* Zākhū, Zaxo. Dahūk/Dihok, N Iraq
Zākhū see Zākhō
Zakopane 99 D5 Małopolskie, S Poland
Zákynthos 105 A6 *var.* Zákinthos, *It.* Zante. *island* Iónia Nísoi, Greece, C Mediterranean Sea
Zalaegerszeg 99 B7 Zala, W Hungary
Zaláu 108 B3 *Ger.* Waltenberg, *Hung.* Zilah; *prev. Ger.* Zillenmarkt. Sălaj, NW Romania
Zalim 121 B5 Makkah, W Saudi Arabia
Zambesi/Zambeze see Zambezi
Zambezi 78 C2 North Western, W Zambia
Zambezi 78 D2 *var.* Zambesi, *Port.* Zambeze. *river* S Africa
Zambia 78 C2 *off.* Republic of Zambia; *prev.* Northern Rhodesia. *country* S Africa
Zambia, Republic of see Zambia
Zamboanga 139 E3 *off.* Zamboanga City. Mindanao, S Philippines
Zamboanga City see Zamboanga
Zambrów 98 E3 Łomża, E Poland
Zamora 92 D2 Castilla y León, NW Spain
Zamora de Hidalgo 50 D4 Michoacán, SW Mexico
Zamość 98 E4 *Rus.* Zamoste. Lubelskie, E Poland
Zamoste see Zamość
Zancle see Messina
Zanda 126 A4 Xizang Zizhiqu, W China
Zanesville 40 D4 Ohio, N USA
Zanjān 120 C2 *var.* Zenjan, Zinjan. Zanjān, NW Iran
Zante see Zákynthos
Zanthus 147 C6 Western Australia, S Australia Oceania
Zanzibar 73 D7 Zanzibar, E Tanzania
Zanzibar 73 C7 *Swa.* Unguja. *island* E Tanzania
Zaozhuang 128 D4 Shandong, E China
Zapadna Morava 100 D4 *Ger.* Westliche Morava. *river* C Serbia
Zapadnaya Dvina 110 A4 Tverskaya Oblast', W Russia
Zapadnaya Dvina see Western Dvina
Zapadno-Sibirskaya Ravnina 114 C3 *Eng.* West Siberian Plain. *plain* C Russia
Zapadnyy Bug see Bug
Zapadnyy Sayan 114 D4 *Eng.* Western Sayans. *mountain range* S Russia
Zapala 65 B5 Neuquén, W Argentina
Zapiola Ridge 67 B6 *undersea feature* SW Atlantic Ocean
Zapolyarnyy 110 C2 Murmanskaya Oblast', NW Russia
Zaporizhzhia 109 F3 *prev.* Aleksandrovsk, Zaporizhzhya; *Rus.* Zaporozh'ye. Zaporiz'ka Oblast', SE Ukraine
Zaporizhzhya see Zaporizhzhia
Zaporozh'ye see Zaporizhzhia
Zapotiltic 50 D4 Jalisco, SW Mexico
Zaqatala 117 G2 *Rus.* Zakataly. NW Azerbaijan
Zara 116 D3 Sivas, C Turkey (Türkiye)
Zara see Zadar
Zarafshan see Zarafshon
Zarafshan 122 D2 *Rus.* Zarafshan. Navoiy Viloyati, N Uzbekistan
Zarafshon see Zeravshan
Zaragoza 93 F2 *Eng.* Saragossa; *anc.* Caesaraugusta, Salduba. Aragón, NE Spain
Zarand 120 D3 Kermān, C Iran
Zaranj 122 D5 Nīmrūz, SW Afghanistan
Zarasai 106 C4 Utena, E Lithuania
Zárate 64 D4 *prev.* General José F. Uriburu. Buenos Aires, E Argentina
Zarautz 93 E1 *var.* Zara
Zaxo see Zākhō
Zaysan Köli 114 D5 *var.* Ozero Zaysan. *lake* E Kazakhstan
Zaysan, Ozero see Zaysan Köli
Zelenogradsk 106 A4 *Ger.* Cranz, Kranz. Kaliningradskaya Oblast', W Russia
Zelle see Celle
Zel'va 107 B6 *Pol.* Zelwa. Hrodzyenskaya Voblasts', W Belarus
Zelwa see Zel'va
Zelzate 87 B5 *var.* Selzaete. Oost-Vlaanderen, NW Belgium
Žemaičių Aukštumas 106 B4 *physical region* W Lithuania

255

Zemst *87 C5* Vlaams Brabant, C Belgium
Zemun *100 D3* Serbia, N Serbia
Zengg *see* Senj
Zenica *100 C4* Federacija Bosna I Hercegovina, C Bosnia and Herzegovina
Zenta *see* Senta
Zeravshan *123 E3* *Taj./Uzb.* Zarafshon. *river* Tajikistan/Uzbekistan
Zevenaar *86 D4* Gelderland, SE Netherlands
Zevenbergen *86 C4* Noord-Brabant, S Netherlands
Zeya *113 E3* *river* SE Russia
Zgerzh *see* Zgierz
Zgierz *98 C4* *Ger.* Neuhof, *Rus.* Zgerzh. Łódź, C Poland
Zgorzelec *98 B4* *Ger.* Görlitz. Dolnośląskie, SW Poland
Zhabinka *107 A6* *Pol.* Żabinka. Brestskaya Voblasts', SW Belarus
Zhambyl *see* Taraz
Zhanaozen *see* Zhangaözen
Zhangaözen *114 A4* *var.* Zhanaozen; *prev.* Novyy Uzen'. Mangistau, W Kazakhstan
Zhangaqazaly *see* Ayteke Bi
Zhang-chia-k'ou *see* Zhangjiakou
Zhangdian *see* Zibo
Zhangjiakou *128 C3* *var.* Changkiakow, Zhang-chia-k'ou, *Eng.* Kalgan; *prev.* Wanchuan. Hebei, E China
Zhangzhou *129 D6* Fujian, SE China
Zhanjiang *129 C7* *var.* Chanchiang, Chan-chiang, *Cant.* Tsamkong, *Fr.* Fort-Bayard. Guangdong, S China
Zhaoqing *129 C6* Guangdong, S China
Zhayyk *see* Ural
Zhdanov *see* Mariupol
Zhe *see* Zhejiang
Zhejiang *129 D5* *var.* Che-chiang, Chekiang, Zhe, Zhejiang Sheng. *province* SE China
Zhejiang Sheng *see* Zhejiang
Zheleznodorozhnyy *106 A4* Kaliningradskaya Oblast', W Russia
Zheleznodorozhnyy *see* Yemva
Zheleznogorsk *111 A5* Kurskaya Oblast', W Russia
Zhëltyye Vody *see* Zhovti Vody
Zhengzhou *128 C4* *var.* Ch'eng-chou, Chengchow; *prev.* Chenghsien. *province capital* Henan, C China
Zhezkazgan *see* Zhezqazghan

Zhezqazghan *114 C4* *var.* Zhezkazgan; *prev.* Dzhezkazgan. Karagandy, C Kazakhstan
Zhidachov *see* Zhydachiv
Zhitkovichi *see* Zhytkavichy
Zhitomir *see* Zhytomyr
Zhlobin *107 D7* Homyel'skaya Voblasts', SE Belarus
Zhmerinka *see* Zhmerynka
Zhmerynka *108 D2* *Rus.* Zhmerinka. Vinnyts'ka Oblast', C Ukraine
Zhodino *see* Zhodzina
Zhodzina *107 D6* *Rus.* Zhodino. Minskaya Voblasts', C Belarus
Zholkev/Zholkva *see* Zhovkva
Zhonghua Renmin Gongheguo *see* China
Zhongshan *154 D3* *Chinese research station* Antarctica
Zhosaly *114 B4* *prev.* Dzhusaly. Kzylorda, SW Kazakhstan
Zhovkva *108 B2* *Pol.* Żółkiew, *Rus.* Zholkev, Zholkva; *prev.* Nesterov. L'vivs'ka Oblast', NW Ukraine
Zhovti Vody *109 F3* *Rus.* Zhëltyye Vody. Dnipropetrovska Oblast, E Ukraine
Zhovtnevoye *see* Zhovtneve
Zhydachiv *108 B2* *Pol.* Zydaczów, *Rus.* Zhidachov. L'vivs'ka Oblast', NW Ukraine
Zhytkavichy *107 C7* *Rus.* Zhitkovichi. Homyel'skaya Voblasts', SE Belarus
Zhytomyr *108 D2* *Rus.* Zhitomir. Zhytomyrs'ka Oblast', NW Ukraine
Zibo *128 D4* *var.* Zhangdian. Shandong, E China
Zichenau *see* Ciechanów
Zielona Góra *98 B4* *Ger.* Grünberg, Grünberg in Schlesien, Grüneberg. Lubuskie, W Poland
Zierikzee *86 B4* Zeeland, SW Netherlands
Zigong *129 B5* *var.* Tzekung. Sichuan, C China
Ziguinchor *74 B3* SW Senegal
Zilah *see* Zalău
Žilina *99 C5* *Ger.* Sillein, *Hung.* Zsolna. Žilinský Kraj, N Slovakia
Zillenmarkt *see* Zalău
Zimbabwe, Republic of *see* Zimbabwe
Zimbabwe *78 D3* *off.* Republic of Zimbabwe; *prev.* Rhodesia. *country* S Africa
Zimnicea *108 C5* Teleorman, S Romania
Zimovniki *111 B7* Rostovskaya Oblast', SW Russia

Zinder *75 G3* Zinder, S Niger
Zinov'yevsk *see* Kropyvnytskyi
Zintenhof *see* Sindi
Zipaquirá *58 B3* Cundinamarca, C Colombia
Zittau *94 D4* Sachsen, E Germany
Zlatni Pyasatsi *104 E2* *var.* Zlatni Pyasŭtsi. Dobrich, NE Bulgaria
Zlatni Pyasŭtsi *see* Zlatni Pyasatsi
Zlín *99 C5* *prev.* Gottwaldov. Zlínský Kraj, E Czechia (Czech Republic)
Złoczów *see* Zolochev
Złotów *98 C3* Wielkopolskie, C Poland
Znamenka *see* Znamianka
Znamianka *109 F3* *prev.* Znam"yanka, *Rus.* Znamenka. Kirovohrads'ka Oblast', C Ukraine
Znam"yanka *see* Znamianka
Żnin *98 C3* Kujawsko-pomorskie, C Poland
Zoetermeer *86 C4* Zuid-Holland, W Netherlands
Żółkiew *see* Zhovkva
Zolochev *108 C2* *Pol.* Złoczów, *Rus.* Zolochiv. L'vivs'ka Oblast', W Ukraine
Zolochiv *109 G2* *Rus.* Zolochev. Kharkivs'ka Oblast', E Ukraine
Zolochiv *see* Zolochev
Zolote *109 H3* *Rus.* Zolotoye. Luhans'ka Oblast', E Ukraine
Zolotonosha *109 E2* Cherkas'ka Oblast', C Ukraine
Zolotoye *see* Zolote
Zólyom *see* Zvolen
Zomba *79 E2* Southern, S Malawi
Zombor *see* Sombor
Zongo *77 C5* Equateur, N Dem. Rep. Congo
Zonguldak *116 C2* Zonguldak, NW Turkey (Türkiye)
Zonhoven *87 D6* Limburg, NE Belgium
Zoppot *see* Sopot
Żory *99 C5* *var.* Zory, *Ger.* Sohrau. Śląskie, S Poland
Zouar *76 C2* Borkou-Ennedi-Tibesti, N Chad
Zouérat *74 C2* *var.* Zouérate, Zouïrât. Tiris Zemmour, N Mauritania
Zouérate *see* Zouérat
Zouïrât *see* Zouérat
Zrenjanin *100 D3* *prev.* Petrovgrad, Veliki Bečkerek, *Ger.* Grossbetschkerek, *Hung.* Nagybecskerek. Vojvodina, N Serbia
Zsil/Zsily *see* Jiu
Zsolna *see* Žilina

Zsombolya *see* Jimbolia
Zsupanya *see* Županja
Zubov Seamount *67 D5* *undersea feature* E Atlantic Ocean
Zueila *see* Zawīlah
Zug *95 B7* *Fr.* Zoug. Zug, C Switzerland
Zugspitze *95 C7* *mountain* S Germany
Zuid-Beveland *87 B5* *var.* South Beveland. *island* SW Netherlands
Zuider Zee *see* IJsselmeer
Zuidhorn *86 E1* Groningen, NE Netherlands
Zuidlaren *86 E2* Drenthe, NE Netherlands
Züllichau *see* Sulechów
Zumbo *see* Vila do Zumbo
Zundert *87 C5* Noord-Brabant, S Netherlands
Zunyi *129 B5* Guizhou, S China
Županja *100 C3* *Hung.* Zsupanya. Vukovar-Srijem, E Croatia
Zürich *95 B7* *Eng./Fr.* Zurich, *It.* Zurigo. Zürich, N Switzerland
Zurich, Lake *see* Zürichsee
Zürichsee *95 B7* *Eng.* Lake Zurich. *lake* NE Switzerland
Zurigo *see* Zürich
Zutphen *86 D3* Gelderland, E Netherlands
Zuwärah *71 F2* NW Libya
Zuwaylah *see* Zawīlah
Zuyevka *111 D5* Kirovskaya Oblast', NW Russia
Zvenigorodka *see* Zvenyhorodka
Zvenyhorodka *109 E2* *Rus.* Zvenigorodka. Cherkas'ka Oblast', C Ukraine
Zvishavane *78 D3* *prev.* Shabani. Matabeleland South, S Zimbabwe
Zvolen *99 C6* *Ger.* Altsohl, *Hung.* Zólyom. Banskobystrický Kraj, C Slovakia
Zvornik *100 C4* E Bosnia and Herzegovina
Zwedru *74 D5* *var.* Tchien. E Liberia
Zwettl *95 E6* Wien, NE Austria
Zwevegem *87 A6* West-Vlaanderen, W Belgium
Zwickau *95 C5* Sachsen, E Germany
Zwolle *86 D3* Overijssel, E Netherlands
Żydaczów *see* Zhydachiv
Zyôetu *see* Jōetsu
Żyrardów *98 D3* Mazowieckie, C Poland
Zyryanovsk *114 D5* Vostochnyy Kazakhstan, E Kazakhstan